INTERNATIONAL HYDROLOGY SERIES

World Water Resources at the Beginning of the Twenty-First Century

Edited by

I.A. Shiklomanov

State Hydrological Institute, Russian Federation

W0077398

and

John C. Rodda

Past President, International Association of Hydrological Sciences,
Centre for Ecology and Hydrology, Wallingford, Oxon

PUBLISHED BY THE PRESS SYNDICATE OF THE UNIVERSITY OF CAMBRIDGE
The Pitt Building, Trumpington Street, Cambridge, United Kingdom

CAMBRIDGE UNIVERSITY PRESS
The Edinburgh Building, Cambridge CB2 2RU, UK
40 West 20th Street, New York NY 10011–4211, USA
477 Williamstown Road, Port Melbourne, VIC 3207, Australia
Ruiz de Alarcón 13, 28014 Madrid, Spain
Dock House, The Waterfront, Cape Town 8001, South Africa

http://www.cambridge.org

First published 2003
First paperback edition 2004

Typeface Times 9.5/13 pt   *System* LATEX2$_\varepsilon$ [TB]

A catalogue record for this book is available from the British Library

Library of Congress Cataloguing in Publication data

World water resources at the beginning of the 21st century / scientific leader and editor,
I.A. Shiklomanov, John C. Rodda.
   p.   cm. – (International hydrology series)
Includes bibliographical references and index.
ISBN 0 521 82085 5 hardback
1. Water balance (Hydrology) 2. Water use. 3. Water consumption. I. Shiklomanov, I. A.
II. Rodda, J.C. III. Series.
GB661.2 .W67 2003
553.7′09′05 –dc21   2002031201

ISBN 0 521 82085 5 hardback
ISBN 0 521 61722 7 paperback

Contents

Introduction

Water is one of the most widely distributed substances on planet Earth; in different forms and amounts it is available everywhere, interacting with the atmosphere, biosphere and lithosphere. Water and water resources occupy a special place among natural resources due to their great diversity and vital role in supporting human life and in powering many of the natural processes shaping the Earth. Water is often the most important element of the landscape for human beings, the basis for the entire organic world, and an integral part of the ecological system. However, many of the world's natural disasters and extremes are associated with water or the lack of it. The floods that occur over most of the globe cause enormous damage, kill millions of people and take away the livelihood of many more.

Of greatest significance is fresh water as this is the most important natural resource. Life is not possible without it, because it has no substitute. Human beings have always consumed fresh water and have used it for many other purposes; however, for most of historical time the human impact on water resources was insignificant or local in character. The properties of natural waters including their changing and cleansing during their movement through the hydrological cycle and their ability for self-purification allowed the fresh waters to retain their characteristic purity, quantity and quality over time. This gave birth to the illusion that water resources would always be pure and readily available: it was almost as if they were a gift from the natural environment. In these circumstances, historically, the tradition arose of a careless attitude towards the use of water and water resources; the cost of waste water treatment was kept to a minimum and little was spent on protection of water resources from pollution.

This situation has changed dramatically during recent decades. In many regions and in most countries, the results of long-term neglect and misuse of water resources have become obvious due to the increasing use of water resources and the transformation of land use in most river basins. For the first 50 years of the twentieth century the quantity of water used globally grew to 785 km^3 (156 km^3 for every 10 years), while from 1951 to 1960 it rose by 620 km^3, i.e. the growth rate increased more than fourfold. Over the next decade it increased further to 690 km^3/year.

This acceleration occurred principally because of the rapid expansion of irrigation, but also because of the growth of the volume of water used by industry and for the production of thermal power.

From 1951 to 1960 the world's irrigated areas increased by more than 40 million ha, while for the previous 50 years it rose by 54 million ha. Since the early 1950s the water used by industry has been subject to intense growth, because of the increasing production of synthetic fibres, plastic, cellulose, petro-chemical materials, etc. which require huge amounts of water. To produce 1 tonne of these materials, hundreds and even thousands of tonnes of water are needed. The waste water is often discharged into rivers and lakes without proper treatment causing the progressive pollution of the receiving waters.

Thermal and nuclear power production require huge amounts of water for cooling and a smaller amount for feeding the boilers. During the second half of the twentieth century many large reservoirs were constructed for hydropower and to regulate runoff in order to improve water supply and harness water resources. All the world's largest reservoirs with a capacity of over 50 km^3 were constructed during this period. By the 1960s the total volume of water in reservoirs had grown more than 25 times, as compared to the volume at mid-century.

During recent decades the dramatic increase in water use due to the growth of the global economy has led to serious anthropogenic changes in the characteristics of the hydrology of rivers and lakes, particularly changes in the quantity and quality of their water. These changes, which have also affected ground water, have led to alterations in the water budgets of many river basins and changes in the available water resources.

For the first time in history the availability of water resources and their distribution in space and time has begun to be determined by human activity, in addition to the natural variations in climate. Now also in many parts of the world water resources are being degraded by pollution. The consequence is that the ever-increasing demand cannot readily be met by the available water resources in many areas and the availability of water has become a prime factor limiting the growth of the population and the development of the economy. Especially severe problems arise in arid regions where the population is growing rapidly and the limited resource is already used to a high degree.

Being familiar with the accelerating growth in the use of water, many scientists predicted that during the 1960s and 1970s water use would grow exponentially. For example, for the USA this growth was predicted to be 850–1100 km^3/year by the end of the century, or 3 to 4 times the rate of the 1960s. An even greater growth rate (5 to 6 times) was planned for this period in the former Soviet Union. For the world as a whole, at that time, water use was predicted to grow to the enormous values of 9000–11000 km^3/year by the end of the century, i.e. approximately 20–25% of the global renewable water resource.

These forecasts could not remain unnoticed by the wider scientific community and the mass media. The water problem became an object of attention and anxiety in many countries, not only in those with an arid climate. Foreseeing a large population, the mass media described with ever-increasing frequency the depletion of all kinds of natural resources and, especially, fresh water. The famous French scientist Furon in the book *Water Problems on the Terrestrial Globe*, published in the early 1960s, came to the conclusion that in the future, there would remain nothing for humans to do except for "drinking sea water". At the same time, in a number of countries the construction of desalination plants started. Some scientists directed their attention to the huge resources of fresh water accumulated in the ice sheets and glaciers of the Antarctic. Ideas and plans for the transport of icebergs to regions suffering from water deficits were put forward. Other scientists and engineers have been concerned with increasing the efficiency of water use, including reducing the abstraction of fresh water and the more complete treatment of waste water, and with managing the demand for water.

In these circumstances the world water problem has become an object of consideration at the highest political and administrative level. Many international organizations, primarily the United Nations Educational, Scientific and Cultural Organization (UNESCO), the World Meteorological Organization (WMO), the International Association of Hydrological Sciences (IAHS) and the Food and Agricultural Organization (FAO), drew attention to the need for the detailed and exhaustive study of water and water resources. In particular, to stimulate the study of water resources and the water budget of the Earth and to aid the development of the scientific principles of rational use and the protection of water resources, in 1965 UNESCO established an unprecedented programme of co-operation in hydrology, the International Hydrological Decade (IHD), which ran from 1965 to 1974. At the end of the decade a decision was made to continue international co-operation in hydrology and water resources in the framework of UNESCO. This co-operation is still in progress and is known as the International Hydrological Programme (IHP).

Important achievements in this programme of international co-operation in hydrology and water resources were the detailed and comprehensive studies of the water balance and water resources of the world conducted during the IHD. The results were published in 1974 by Russian scientists under the guidance of the State Hydrological Institute (SHI), St Petersburg, as the monograph *World Water Balance and Water Resources of the Earth* (Korzun, 1974b) and in 1975 by Baumgartner and Reichel (Germany) as the monograph *World Water Balance*. The data from these monographs have been widely used by scientists as the most complete and reliable assessment of the world's water resources. All later publications concerned with water resources at the global scale give no new information as compared with the above studies. This pertains in particular to the well-known country-by-country data on water resources and water use published periodically by the World Resources Institute in Washington, DC. These figures cannot be considered as sufficiently reliable, as they are obtained from many different sources, including obsolete ones. They have also been determined by different methods for different years and for different long-term periods.

As to the global assessment of water use, beginning with the 1970s, generalized figures have been published with different levels of comprehensiveness and reliability from country to country. In some cases estimates were even given for the water use at the end of the twentieth century and into the more distant future. When assessing the reliability of these estimates, it should be recognized that since the late 1970s, in most developed countries important changes have been taking place in approaches to the use of water, especially for meeting industrial needs and for thermal power production. Increasing attention has been given to reducing the use of fresh water, such as by recycling, by using sea water for cooling, minimizing the use of water in different industrial processes, increased treatment of waste water, together with the setting of standards for discharges of effluents to protect the receiving waters and to restore their hydrochemical regimes. These measures resulted in decreased growth rates and even stabilization of the volume of water used in many developed countries and they are the reason why forecasts of future water use need to be revised. For instance, in the forecasts of long-term water use made in the 1980s for a number of developed countries (USA, Germany, Sweden), fresh water use was assumed to decrease slightly by the end of the twentieth century.

Another most important factor affecting world water use is that, since the 1970s, the growth of irrigation has not been maintained and has fallen behind the growth of population. This is due to the increasing costs of new irrigation projects, waterlogging, salinization and the need to undertake environmental protection measures and other factors of a natural and socio-economic character.

The estimates of global water use made in the 1970s and 1980s took no account of these conditions in determining the most probable trends in water use for the coming decades. For example, the forecasts of global water use in 1990 and 2000 published in 1987 by the State Hydrological Institute are significantly overestimated.

More than a quarter of a century has passed since the detailed assessments were made of world water resources based on data collected before 1965. During this time, hydrology has made considerable progress, especially in the countries of Africa and South America, where large areas were poorly studied or not examined at all. Networks of hydrological stations have grown, and time series of observations of hydrological and meteorological variables have lengthened, allowing assessment of water resources to be made with a greater reliability not only for values of water resources but also their long-term variations, which is especially important for determining water availability during dry years and droughts. However there is growing evidence during the 1990s of the decline in hydrological networks in Africa and in the former Soviet Union and elsewhere as governments have reduced spending, but in recent years new information has appeared on the use of water in developed and developing countries, together with related data on population, irrigated areas, hydropower, etc., which allows more detailed and objective assessments of water use and water availability to be made for the different regions for both the present and for the future.

To obtain new and more detailed and reliable data on water resources and water use at the global scale, the programme of international co-operation within UNESCO's IHP–IV (1991–5) included a project for preparing a fundamental and comprehensive monograph on world water resources and their use. The National Committee of the Russian Federation for the IHP took the responsibility for this project and the work of preparing the monograph was given to scientists of the State Hydrological Institute in St Petersburg.

This Monograph represents the final results of this IHP project. All the work on assessing water resources and preparing the Monograph for publication over the period 1991 to 1996 was undertaken at the SHI by the scientists of the Laboratory for Water Resources and Water Balance (Head: Professor Vladimir I. Babkin) and the Department for River Runoff and Water Management Problems (Head: Candidate of Geographical Sciences Vladimir Yu. Georgiyevsky).

The studies of global water resources were based on the following methodology:

- Assessments were made for the land surface of the globe, all the continents and natural regions, selected countries and the principal river basins.
- Use was made of annual and monthly runoff data collected by the world hydrological network as well as data on precipitation and air temperature.

- Account was taken of the water resources of areas without observational data including sites at river mouths.
- All the assessments of water resources were made for a single time period sufficiently long for the reliable determination of tendencies and trends, for sampling extreme values, and for estimating the characteristics of long-term variability. In some cases work was needed to extend the time series and fill in gaps in the observations.
- The analysis provided annual values of water resources and also their distribution during the year (by months).
- Changes in water use are given for all major fresh-water users for the twentieth century and estimates are made for several decades ahead.
- Use was made of reliable national data on water use and these data were applied to assess future water use.
- The actual population supplied with water was taken into account, with the average and minimum values of water resources as well as water consumption.
- The effects of climate change on water resources were considered.

The Monograph contains an analysis of renewable fresh-water resources as the total runoff, including the ground water represented in the upper part of aquifers drained by the river systems. These water resources provide the major source of water used globally, regionally and from country to country. Detailed data on world ground water resources are being published by UNESCO in a special monograph edited by Professor Igor S. Zektser.

It should be mentioned that the global problems of pollution and of fresh water are not treated in the present Monograph, although they are very pertinent. However, attention is given in the Monograph to water quality in regional water management caused by constructing large reservoirs (the Nile Basin) and inter-basin water transfers (Canada). At the global scale the problems of water pollution and water quality are addressed in the well-known reports published under the aegis of World Health Organization and the United Nations Environmental Programme, often within the GEMS Water Programme which is shared with UNESCO and WMO.

The structure of the present Monograph consists of 12 chapters, an Introduction and a Conclusion. The chapters can be divided into three groups. The first three chapters present general information about Earth's hydrology and the hydrosphere and a description of the methodology applied in the Monograph for assessing and forecasting water resources, water use, and water availability.

The next six chapters present material for each continent. Each of these chapters has the same structure. At the beginning of the chapter, there is a brief description of the natural conditions, the main features of economic development and the use of water, the hydrology of the continent, the data used, then the analysis of the results for water resources, water use, and the availability of water. Then follows the temporal and spatial distribution of these attributes for river basins, regions and for the different countries of the continent. In addition in certain chapters, some of the most acute water management problems are considered, in particular, the Aral Sea, the Aswan High Dam and the Nasser Reservoir on the Nile River, and large-scale transfer in Canada.

The last three chapters are devoted to generalizing and analysing the data for an appreciation of global water resources and water use. Climate change is also considered and its possible implications for water resources and the demand for water.

When preparing the Monograph, extensive use was made of observational data from the world hydrological network available in the SHI archives, the Global Runoff Data Centre (GRDC), Koblenz, Germany, and in published reports of UNESCO and WMO as well as from the reports and papers published by SHI and by many similar institutes and organizations across the world. The Monograph employs these results and the conclusions of numerous national and international studies published during the last 10–15 years by a wide range of bodies. The list of publications used is given at the end of the book.

In conclusion, I consider it necessary to express my sincere gratitude to my colleagues and assistants, and to the workers of the SHI, whose enthusiastic labour made it possible to obtain new data on the Earth's water resources and data on water use, and to prepare for publication and to translate this monograph into English. Among them are, primarily, the authors of Sections of the Monograph indicated in the List of Contents. There are also the leading scientists and engineers: Mr Alexander I. Moiseyenkov, Mrs Irena A. Nikiforova, Ms Svetlana N. Pavlova, Ms Vera N. Babkina, Ms Lyudmila V. Korotun, Ms Tatiana I. Printseva, Ms Tatiana G. Molchanova, Mrs Elena V. Golovkina and Mrs Lyubov P. Babkina, who made invaluable contributions to the preparation of the data and the manuscript, as well as the leader of the group of translators, Mrs Valentina G. Yanuta, who undertook the work of translating the Monograph into English.

On behalf of the SHI scientists I must also express my appreciation of the contribution of Dr V. Grabs, Director of GRDC, Koblenz, Germany, as well as the help of the heads of hydrological services of many nations, who were so kind as to present their data on runoff which was used to assess global water resources.

I.A. Shiklomanov

1 The Earth and its physical features

1.1 PHYSICAL GEOGRAPHY

1.1.1 The area of the Earth's surface

The total area of the surface of the Earth is 510 million km^2. Over 361 million km^2 or 71% of this area is occupied by the World Ocean and only 149 million km^2 or 29% is covered by land. Water and land are distributed unevenly over the globe. In the Northern Hemisphere land extends over 100 million km^2 (39% of its area), while there are 49 million km^2 in the Southern Hemisphere (19%). The area of water in the Northern Hemisphere is 155 million km^2 (61%), and in the Southern 206 million km^2 (81%).

1.1.2 The World Ocean

The World Ocean is divided into four separate oceans by the distribution of the land (Stepanov, 1983): namely the Pacific, Atlantic, Indian and Arctic Oceans, and into numerous seas, gulfs, bays and straits. The Southern Ocean is also identified but is less well defined than the others. Basic information on the oceans and seas (Korzun, 1974b) are presented in Tables 1.1 and 1.2 respectively. The volume of water in the World Ocean is about 1340 million km^3.

1.1.3 Continents and islands

During the present geological epoch the Earth's land consists of six continents: Eurasia, Africa, North America, South America, Australia and Antarctica. The borders between the separate continents are rather arbitrary. The border between Eurasia and Africa passes through the Strait of Gibraltar, along the Mediterranean Sea, Suez Canal, Red Sea, and the Straits of Bab el Mandeb. The boundary between North and South America passes through the Panama Canal. In this Monograph, Eurasia is subdivided into two parts which are considered as independent: namely Europe and Asia. The border between these continents extends from Matochkin Shar, in the north, along Pay Khoy, the Ural Mountains, Mugodzhary, along the River Emba, and the north and west coast of the Caspian Sea and Caucasus Mountains. Information on the continents and largest islands is given in Tables 1.3 and 1.4 (Terehov, 1981).

PRIMARY WATERSHEDS

Primary and secondary watersheds can be identified on the land surface. The primary watershed divides the land into two: the first carrying runoff to the Atlantic and Arctic Oceans (60% of the land area) and the second where runoff occurs to the Pacific and Indian Oceans (40%). The secondary watersheds are those surrounding the basins of the Pacific, Atlantic, Indian and Arctic Oceans and those delineating areas of internal runoff.

The primary watershed extends northwards from Cape Horn along the Andes and the Rocky Mountains to the Bering Strait, then across the eastern plateau of Asia in a westerly direction, and then it turns to run along the eastern edge of Africa to finish at the Cape of Good Hope.

The watersheds of ocean basins are located on individual continents in the following way. In Europe the watershed between the Arctic and Atlantic Oceans passes from the southwest coast of Norway along the Scandinavian Uplands, through the Manselkya Highland, and between Segozero and Onega. The watershed line between the Atlantic Ocean and the area of internal runoff to the Caspian Sea passes between Lakes Onega and Beloye Ozero, along the Valdai Hills, through the Central Russian and the Privolzhskaya Uplands, to Ergeny and the Caucasus Mountains.

In Asia the watershed between the Atlantic and Indian Oceans extends from the south end of the Suez Canal to the source of the Euphrates River. Then the watershed between the Indian Ocean and the area of internal runoff to the north passes along the Plateau of Serkhed, through the Hindu Kush, and from the southern part of Tibet to the Kukushili Mountains to meet the Pacific Ocean watershed. The main watershed of the basins with rivers flowing into the Pacific Ocean passes from Cape Dezhnev along the Chukot, Kolyma, Dzungur, Stanovy, Yablonovy and Hentey Ranges, along the highlands of the northern area of the Gobi and further along Great Khingan Mountains, Inshan, Nan Shan, Kukushili, Tanghla, Henduanshan, to Bilau.

Table 1.1. *Major hydrological and morphometric characteristics of the World Ocean*

Ocean	Total area (with islands), km² × 10⁶	Area of water surface, km² × 10⁶	Area of catchment, km² × 10⁶	Water volume km³ × 10⁶	Water volume %	Depth, m Average	Depth, m Maximum
Pacific	182.6	178.7	24.9	707.1	53.4	3957	11 034
Atlantic	92.7	91.7	50.7	330.1	24.6	3602	9 219
Indian	77.0	76.2	20.9	284.6	21.0	3736	7 450
Arctic	18.5	14.7	22.5	16.7	1.0	1131	5 220
World Ocean	370.8	361.3	119.0	1338.5	100	3704	11 034

Table 1.2. *Major morphometric characteristics of seas*

Sea	Area, km² × 10³	Volume, km³	Sea	Area, km² × 10³	Volume, km³
		Pacific Ocean			
Coral Sea	4791	11 470	Java Sea	480	22
South China Sea	3447	3 929	Sulawesi Sea	435	1586
Bering Sea	2344	3 796	Sulu Sea	348	553
Sea of Okhotsk	1617	1 317	Molucca Sea	291	554
Sea of Japan	1070	1 630	Seram Sea	187	227
East China Sea	752	263	Flores Sea	121	222
Yellow Sea	417	17	Bali Sea	119	49
Banda Sea	695	2 129	Savu Sea	105	178
		Atlantic Ocean			
Caribbean Sea	2754	6 860	North Sea	554	52
Mediterranean Sea	2505	3 754	Baltic Sea	448	20
Gulf of Mexico	1543	2 332	Black Sea	431	555
Hudson Bay	819	92	Sea of Azov	40	0.4
Baffin Bay	689	593	Sea of Marmara	11	4.0
		Indian Ocean			
Arabian Sea	3683	10 070	Timor Sea	615	250
Bay of Bengal	2172	5 616	Andaman Sea	602	660
Arafura Sea	1037	204	Red Sea	450	251
		Arctic Ocean			
Barents Sea	1470	268	Kara Sea	903	101
Norway Sea	1547	2 408	Laptev Sea	678	363
Greenland Sea	1205	1 740	Chukchi Sea	590	45
East Siberian Sea	926	61	Beaufort Sea	476	478
			White Sea	91	4.4

The watershed of the rivers draining to the Arctic Ocean in Asia passes from the northern end of land in the Strait of Matochkin Shar, along the Pay Khoy Range and the Ural Mountains, to the interfluvial area of Tobol, Turgay, Ishim, and to the Kazakh area of low hills, onwards to the ranges of Tarbagatay, Mongolian Altai, Tank Ola, Hangay and Hentey, and then it extends along the watershed of rivers draining to the Pacific Ocean.

In Africa the watershed between the basins of Atlantic and Indian Oceans passes from the Gulf of Suez along the peaks of mountains situated besides the Red Sea, along the eastern part of the Abyssinian Highlands, to the east of Lake Victoria between Lake Tanganyika and Lake Nyasa, along the Muchinga Mountains, between the Rivers Congo and Zambezi, Cubango and Cunene, westwards and southwards of Lake Etosha, along

Table 1.3. *Morphometric characteristics of continents*

Continent	Area with islands, km$^2 \times 10^6$	Area of islands, km$^2 \times 10^6$	Altitude above sea level, m		
			Average	Maximum	Minimum
Europe	10.5	0.7	300	5642	−28
Asia	43.5	2.7	950	8848	−392
Africa	30.1	0.6	750	5895	−150
North America	24.2	4.1	700	6193	−85
South America	17.8	0.1	580	7014	−35
Australia and Oceania	8.9	1.3	350	5029	−12
Antarctica	14.0	0.058	2040	5140	−

Damaraland, across the hills of the southwest and the southern borders of the Kalahari Desert, through the Drakensberg Mountains to Cape Agulhas.

In North America the watershed between the Arctic Ocean and the Pacific and Atlantic Oceans passes from Cape Prince of Wales along the Brooks Range, through the Richardson Mountains, Seluin, and Rocky Mountains, along the uplands between the Mississippi and Nelson Rivers, northwards of Lake Superior and Lake Huron and along the Labrador Peninsula. The watershed between the Atlantic and Pacific Oceans passes along the Rocky Mountains, around the upper parts of the Mississippi and South Saskatchewan, along the Isthmus of Tehuantepec and to the Panama Canal.

In South America the watershed separating runoff to the Atlantic and Pacific Oceans starts at the Panama Canal and passes along the Andes, through the Strait of Magellan along Tierra del Fuego to its southern tip.

In Australia the watershed between the basins of Pacific and Indian Oceans passes from Cape York along the Great Dividing Range to South East Point (Cape Otway).

Excluding the areas of internal runoff, the Arctic Ocean takes 15% of the runoff from the total land area of the globe, the Atlantic 34%, the Pacific 17% and the Indian Ocean 14%.

RIVERS

Depending on the size of the basin they drain, the length and volume of the flowing water, rivers are usually subdivided into very large, large, medium, small and very small. Table 1.5 presents information on the morphology of the principal river basins of the earth.

The largest river in the world is the Amazon with a catchment area of 6915 thousand km^2, and length 6280 km. Its total annual runoff amounts to about 15% of the total runoff of all the world's rivers. Among very large rivers are the Congo (catchment area 3680 thousand km^2 and length 4370 km) and Mississippi (2980 thousand km^2 and 4700 km). Over the world as a whole there are 20 rivers with catchment areas between 3 million

to 1 million km^2 and 89 rivers with basin areas from 1 million km^2 to 100 000 km^2. Most rivers are amongst the medium, small and very small categories. About 80% of the land surface drains to the World Ocean, while the area of internal runoff where the rivers do not reach the ocean accounts for 20% of the land surface. Most of the world's largest rivers drain to the ocean.

In Europe the area of internal runoff consists of the Caspian Sea basin, which includes the basins of Volga, Ural, and Kura Rivers. In Asia the area of internal runoff is larger and includes: the basin of the Aral Sea (Amu Darya, Syr Darya Rivers) the basin of Lake Balkhash (Ili River) and many rivers flowing into small lakes or disappearing in arid areas (Tedzhen, Murgab, Sary-Su, Turgay, Irgiz and Nura Rivers). There are also the deserts of Alashan, Gobi, and Takla-Makan in Central Asia, while parts of Asia Minor and most of the Arabian Peninsula have areas of internal runoff. There are several closed basins situated in the interfluvial area of the Indus and Ganges.

Almost one-third of Africa drains internally. These are the Sakhara, Libyan, Nubian, Kalahari, and Namib Deserts and semi-deserts, together with the basins of Lakes Chad, Rukwa and Turkana.

In North America the Great Basin (including Great Salt Lake), the deserts of the Mexican Plateau, the Colorado Plateau and the right bank of the Rio Grande have no outlets to the ocean, while in South America the internal runoff areas include the basins of the Lakes Titicaca–Poopo, the Puna de Atakama Desert, the semi-desert plateau of Patagonia and other territories.

In Australia Lakes Eyre, Amadeus, Torrens and Frome are closed, as well as the Great Sandy Desert, Gibson Desert and Great Victoria Desert. Little is known about drainage on the Antarctic continent.

The total area of internal runoff (Korzun, 1974b) amounts to 30.2 million km^2, including Europe 2.2 million km^2, Asia 12.3 million km^2, Africa 9.6 million km^2, Australia 3.9 million km^2, South America 1.4 million km^2 and North America 0.88 million km^2.

Table 1.4. *Principal islands of more than 10 000 km² in area*

Island	Area, km² × 10³	Island	Area, km² × 10³
Europe		*North America*	
Great Britain	230.0	Greenland	2176.0
Iceland	103.0	Baffin Island	519.0
Ireland	84.4	Victoria Island	213.8
Novaya Zemlya Islands	81.3	Ellesmere Island	202.7
Spitsbergen Islands	62.1	Cuba	105.0
Sicily	25.4	Newfoundland	111.0
Sardinia	23.8	Haiti	77.2
Franz Josef Land	16.1	Banks Island	69.9
		Devon Island	56.4
Asia		Southampton Island	44.1
Kalimantan	735.7	Melville Island	42.1
Sumatra	435.0	Alexander Archipelago	36.8
Honshu	223.4	Axel Heiberg Island	34.4
Sulawesi	179.4	Prince of Wales Island	33.3
Java	126.5	Vancouver Island	32.1
Luzon	105.6	Somerset Island	24.3
Mindanao	95.6	Aleutian Islands	17.7
Hokkaido	77.7	Prince Patrick Island	15.8
Sakhalin	76.4	The Bahamas	11.4
Sri Lanka	65.6	Jamaica	11.1
Kyushu	42.6	Queen Charlotte Islands	10.3
Novosibirsk Islands	38.4	Cape Breton	10.3
Severnaya Zemlya Islands	37.6		
Taiwan	35.9	*South America*	
Hainan	33.7	Tierra del Fuego	48.0
Timor	33.6	Falkland Islands	12.0
Shikoku	18.8	(Islas Malvinas)	
Seram	18.2		
Halmahera	18.0	*Australia and Oceania*	
Kuril Islands	15.6	New Guinea	829.3
Sumbawa	15.5	New Zealand	265.3
Flores	15.2	Tasmania	68.4
Palawan	11.8	Solomon Islands	40.4
Bangka	11.6	New Britain	36.6
Sumba	11.2	Fiji Islands	18.2
		Hawaii	16.7
Africa		New Caledonia	16.7
Madagascar	587.0	New Hebrides	14.8
		Bougainville Island	10.0
		Antarctica	
		Alexander I Land	43.2

LAKES

Lakes are widespread on all continents. There are about 15 million of them, and the total water surface area is about 2 million km² or 1.5% of the land area (excluding the Antarctic). Most of the lakes are small and very small. Across the world there are 88 large lakes with a water surface area exceeding 1000 km². Of these lakes 28 are located in Asia, 13 in Europe, 16 in Africa, 22 in North America, 5 in South America and 4 in Australia. The number of lakes with a surface area greater than 10 000 km² is 19; 1 in Europe (Lake Ladoga), 4 in Asia (Aral, Baikal, Balkhash, Tonle Sap), 4 in Africa (Victoria, Nyasa, Chad, Turkana), 8 in North America (Superior, Huron, Michigan, Great Bear Lake, Great Slave Lake,

Table 1.5. *Major morphometric characteristics of principal world rivers*

River	Area of catchment, km² × 10³	Length, km	River	Area of catchment, km² × 10³	Length, km
			Europe		
Volga	1380	3700	Douro	95	938
Danube	817	2850	Daugava	88	1020
Dnieper	504	2200	Garonne	56	650
Don	422	1870	Ebro	87	928
North Dvina	357	1302	Tagus	81	1010
Pechora	322	1809	Seine	79	776
Neva	281	74	Mezen	78	966
Rhine	252	1320	Po	75	652
Ural	236	2534	Dniester	72	1352
Vistula	198	1092	Guadiana	72	801
Elbe	148	1165	South Bug	64	792
Loire	120	1010	Kuban	61	907
Odra	119	907	Guadalquivir	57	680
Rhône	98	812	Onega	57	416
Neman	98	937			
			Asia		
Ob	2990	3650	Salween	325	2820
Yenisey	2580	3490	Godavari	313	1465
Lena	2490	4410	Huai He	270	1000
Amur	1855	2820	Krishna	259	1401
Yangtze	1808	6300	Helmand	250	1150
Ganges with Brahmaputra and Meghna	1746	5425	Yana	238	872
			Liao He	229	1390
			Olenek	219	2270
Amu Darya	1100	1415	Anadyr	191	1150
Indus	960	3180	Kura	188	1360
Mekong	795	4500	Pyasina	182	818
Huang He	752	5464	Chao Phraya	160	1200
Shatt al Arab (Tigris and Euphrates)	750	2900	Taz	150	1400
			Songka (Red)	145	1185
			Mahanadi	142	851
Kolyma	647	2130	Ili	140	1000
Xi Jiang	454	2214	Taimyra	124	754
Tarim	446	2000	Kerulen	120	1264
Syr Darya	440	2210	Pur	112	389
Irrawaddy	410	2300	Anabar	100	939
Khatanga	364	1634	Narmada	99	1312
Indigirka	360	1726			
			Africa		
Congo	3680	4370	Ogowe	203	850
Nile	2870	6670	Gambia	180	1200
Niger	2090	4160	Rufiji	178	1400
Zambezi	1330	2660	Cuanza	149	630
Orange	1020	1860	Ruvuma	145	800
Chari	880	1400	Qui Hon	137	830
Okowango	785	1800	Sanaga	135	860

Table 1.5. (cont.)

River	Area of catchment, km² × 10³	Length, km	River	Area of catchment, km² × 10³	Length, km
Juba	750	1600	Savi	107	680
Senegal	441	1430	Bandoma	97	780
Limpopo	440	1600	Wad Dra	95	1150
Volta	394	1600	Tana	91	720
North America					
Mississippi	2980	3780	Koksoak	133	1300
Mackenzie	1787	5472	Rio Grande de	125	960
Nelson	1132	2574	Santiago		
St. Lawrence	1026	3057	Brazos	118	2060
Yukon	850	2897	Mobile	116	1250
Columbia	668	1953	Colorado	110	1390
Colorado	637	2333	Mus	108	–
Rio Bravo del	570	2880	Hais	108	–
Norte			Goalzas	106	–
Churchill	298	1609	Severn	101	976
Fraser	233	1370	Fort George	98	–
Telon	142	–	Saguenay	90	–
Albany	134	975	Panuko	84	–
South America					
Amazon	6915	6280	Chubut	138	850
La Plata	3100	4700	Rio Negro	130	1000
Orinoco	1000	2740	Rio Dose	81	600
São Francisco	600	2800	Rio Colorado	65	1000
Parnaíba	325	1450	Paraíba	59	800
Magdalena	260	1530	Atrata	32	644
Essequibo	155	970	Bío Bío	24	380
Australia					
Murray	1072	3490	Gascoyne	79	770
Cooper Creek	285	2000	Victoria	77	570
Diamantina	115	896	Mitchell	69	520
Fitzroy	143	960	Murchison	68	700
Burdekin	131	680	Fly	64	1040
Flinders	108	930	Fortescue	55	670
Ashburton	82.0	640	Kluta	22	338
Sepik	81.0	1120			

Erie, Winnipeg, Ontario), 1 in South America (Maracaibo), and 1 in Australia (Lake Eyre).

Most lakes are situated in the Northern Hemisphere and are located in glaciated areas (there are many small lakes in the tundra). Many lakes of Europe (e.g. Ladoga and Onega) are situated in large basins, often grabens where the northern sides were eroded by ice. Tectonic depressions, glacial erosion and moraine dams form many lakes in Sweden: Vanern, Vattern, Malaren, for example. There are many lakes formed by glacial dams in the northwest of Russia, and in Finland, Poland, Germany and Canada. A large group of lakes in the south of Finland (e.g. Lakes Saimaa and Paijanne) are divided from the Gulf of Finland by a huge dam made of a double ridge of terminal moraines, known as Salpa-Uselka. The chain of large lakes in North America (Lake Winnipeg, Lake of the Woods, and the Great Lakes: Superior, Huron, Michigan, Erie and Ontario) lie behind morainic deposits left by the receding ice, which covered the whole of the north of the North America continent. A group of alpine lakes (Lake Geneva, Lake Maggiore and Lake Garda) are located in the glacially eroded basins at the foot of the Alps.

A number of lakes are located in deep tectonic depressions in mountain areas such as Baikal (1741 m), Khubsugul (267 m), Issyk Kul (702 m), Nyasa (706 m), and Titicaca (281 m). In the mountain systems of the Tien Shan, the Pamirs and the Altai there are many lakes formed from the blocking of river valleys with rock fragments during earthquakes. Among them are Lake Teletskoye in the Altai Mountains, and Lake Sarezskoye in the Pamirs in the Murghab River valley (this lake was formed in 1911 as a result of the Usoisky River being blocked).

The lakes in high mountain areas are often situated on plateaux surfaces and are mainly of a tectonic origin. Among the large lakes are Lakes Victoria (altitude 1136 m above sea level) and Tanganyika (773 m) in Africa; Titicaca (3812 m) in South America; Kara Kul (3954 m) and Chatyr Kul (3486 m) on the Pamirs, and Issyk Kul (1609 m) on the Tien Shan in Asia. One of the highest lakes is Lake Horpatso, situated in Tibet at an altitude of 5400 m.

The Caspian Sea (−27 m), and the Dead Sea (−392 m) are situated in deep depressions below sea level. The Caspian Sea and a number of other large lakes (Lakes Balkhash, Balaton etc.) are relics of former more extensive water bodies that appeared after the recession of the ice sheets.

Numerous small lakes are formed by wind action (aeolian lakes) in the hot, dry climate of the steppes such as in Western Siberia and Kazakhstan. In regions where limestone, dolomite and gypsum formations dominate the geology, there are karstic lakes, and in areas of permafrost there are thermokarstic lakes. These form when buried ice melts. Lakes of volcanic origin are frequent in Kamchatka, in the Kuril Islands, in the Armenian Highlands, in Middle and Central Asia, and in New Zealand.

Table 1.6 shows the morphological characteristics of the largest lakes. The total volume of water stored in the world's lakes is 176 400 km³; salt lakes account for 85 400 km³ and fresh lakes for 91 000 km³. The largest volume of saline waters (91% of the total volume) is found in a single water body – the Caspian Sea.

In Asia, the volume of salt lakes is only 3% of the volume of the world total; the volume of fresh waters in Asia is almost 10 times greater than the salt lakes, because of Lake Baikal which holds 27% of the total volume of the world's freshwater lakes.

In Africa all the large lakes are fresh. Lake Chad situated on the edge of the Sahara, although highly mineralized, is not related to the salt lakes. In North America among the salt lakes is the Great Salt Lake, while in South America Lake Poopo and Lake Titicaca are not salt lakes, but their water cannot be used for drinking.

RESERVOIRS

During the twentieth century the numbers of reservoirs increased markedly. They are used for public water supply, irrigation, hydropower generation and for other purposes. By the late 1980s, Avakyan *et al.* (1987) estimated there were about 30 000 reservoirs across the world with a volume of greater than 1 million m³. There were 2500 reservoirs with a capacity larger than 100 million m³, accounting for more than 90% (or 5750 km³) of both the total volume and the total surface area of all the world's reservoirs. According to the estimates available, the total volume of such reservoirs now exceeds 5750 km³, and the total surface area is about 400 000 km².

The large reservoirs constructed during the twentieth century since 1950 have substantially transformed the volume and pattern of fresh water stored on the land surface. They also allowed the development and maintenance of a large number of inter-basin transfer systems (Vugeinsky, 1991).

Of the world's reservoirs, most are valley reservoirs, which are created by damming the river channel. The biggest valley reservoir in the world in terms of volume is the Bratskoye Reservoir on the River Angara (169.3 km³), and in terms of water surface area the Volta on the Volta River (8480 km²). Since 1950, cascades of reservoirs have been constructed on many large rivers such as the Nile, Yenisei, Colorado, Euphrates, Huang He, Zambezi, Volga, Parana, Mississippi and Missouri.

Reservoirs have also been built by constructing a dam to raise the water level of an existing lake, for example, in Finland, in the northwest of the European part of Russian, and in East Africa. The largest reservoir of this type is Lake Victoria, where the dam at the Owen Falls harnesses a storage of 204.8 km³ and a surface area of 68 800 km².

Along with these two types of reservoirs there are also ones filled in natural depressions by diverting water from a river or by pumping. The largest reservoir in the world of this type is Wadi-Tartar in Iraq having a volume of 72.8 km³ and a surface area of 2000 km².

Reservoirs differ widely in their usage. Hydropower reservoirs are numerous in Africa and South America. In Asia and Latin America there are reservoirs that are used primarily for irrigation.

Besides the above usage, many reservoirs on the planet are made for public water supply. In addition there are the reservoirs constructed for navigation, flood protection, fisheries, recreation, timber rafting, and for a variety of different needs. In recent decades multi-purpose reservoirs have been constructed in many parts of the world.

The greatest proportion of the world total volume of stored water is made up from the reservoirs of the USA, Russia, Canada, India and China. Information on reservoirs with a capacity of more than 20 km³ is given in Table 1.7.

Table 1.6. *Major morphometric characteristics of principal world lakes*

Lake	Area, km^2	Maximum depth, m	Volume, km^3	Country
Europe				
Caspian Sea[a]	378 000	1025	78200	Russia, Kazakhstan, Azerbaijan, Iran, Turkmenistan
Ladoga	18 135	230	908	Russia
Onega	9 890	120	295	Russia
Vänern	5 648	106	153	Sweden
Chudsko-Pskovskoye	3 558	15.3	25.2	Russia, Estonia
Vättern	1 856	122	74	Sweden
Suur-Saimaa	1 800	58	36.0	Finland
Mälaren	1 140	61	14.3	Sweden
Päijänne	1 116	95	18.1	Finland
Inari	1 116	92	15.9	Finland
Ilmen	982	4	12	Russia
Balaton	593	12	1.9	Hungary
Geneva	584	310	88.9	Switzerland, France
Bodensee	539	252	48.5	Germany, Austria, Switzerland
Hjamaren	484	22	2.9	Sweden
Storsjon	464	74	7.38	Sweden
Asia				
Aral Sea[a,b]	64 100	68	1020	Kazakhstan, Uzbekistan
Baikal	31 500	1741	23000	Russia
Balkhash[a]	18 200	25	106	Kazakhstan
Tonle Sap	10 100[c]	12	40	Cambodia
Issyk Kul[a]	6 280	702/668	1730	Kirghizia
Dongting Hu	6 000[d]	10	–	China
Rezaieh (Urmia)[a]	5 800	16	45	Iran
Zaisan	5 510	10	53.0	Kazakhstan
Taimyr	4 560	26	13	Russia
Koko Nor[a]	4 220	38	–	China
Khanka	4 190	11	18.5	Russia, China
Van[a]	3 760	145	–	Turkey
Lop Nor	3 500	5	(5)	China
Ubsu Nur[a]	3 350	–	–	Mongolia
Khubsugul	2 770	207	381	Mongolia
Poyang Hu	2 700	20	–	China
Alakol[a]	2 650	54	58.6	Kazakhstan
Chany[a]	2 500	10	4.3	Kazakhstan
Tuz[a]	2 500	–	–	Turkey
Nam Co[a]	2 460	–	–	China
Tai Hu	2 210	–	–	China
Kara-Us-Nur	1 760	–	–	Mongolia
Tengiz[a]	1 590	8	–	Kazakhstan
Sevan	1 360	86	58.5	Armenia
Toba	1 110	529	1258	Indonesia
Marka Kul	454	27		Kazakhstan
Kara Kul	380	238	–	Kirghizia
Teletskoye	245	128	40	Russia

Table 1.6. (*cont.*)

Lake	Area, km^2	Maximum depth, m	Volume, km^3	Country
Africa				
Victoria	68 800	84	2 750	Tanzania, Kenya, Uganda
Tanganyika	32 000	1471	17 800	Tanzania, Zaire, Zambia, Rwanda, Burundi
Nyasa	30 900	706	7 725	Malawi, Mozambique, Tanzania
Chad	10 000–25 000^e	10–11	72	Chad, Niger, Nigeria
Turkana	8 660	73	–	Kenya
Albert	5 300	58	280	Uganda, Zaire
Mweru	5 100	15	32.0	Zambia, Zaire
Bangweulu	4 920^f	5	5.0	Zambia
Rukwa	4 500	–	–	Tanzania
Tana	3 150	14	28.0	Ethiopia
Kiwu	2 370	496	569	Zaire, Rwanda
Edward	2 325	112	78.2	Zaire, Uganda
Leopold II	2 325	6	–	Zaire
Katnit	1 270	60	14	Nigeria
Abaya	1 160	13	8.20	Ethiopia
Shirwa	1 040	2.6	45.0	Malawi
Tumba	765	–		Zairc
Faguibini	620	14	3.72	Mali
Gabel-Aulia	600	12	–	Sudan
Chamo	551	13	–	Ethiopia
Upemba	530	3	0.90	Zaire
Zwoi	434	7	1.10	Ethiopia
Shalla	409	266	37	Ethiopia
North America				
Superior	84 500	406	11 600	Canada, USA
Huron	63 500	229	3 580	Canada, USA
Michigan	58 000	281	4 680	USA
Great Bear Lake	31 400	137	1 010	Canada
Great Slave Lake	28 600	156	1 070	Canada
Erie	25 800	64	545	Canada, USA
Winnipeg	24 400	19	127	Canada
Ontario	19 300	236	1 710	Canada, USA
Nicaragua	8 030	70	108	Nicaragua
Athabasca	7 940	60	110	Canada
Reindeer Lake	6 640	–	–	Canada
Winnipegosis	5 360	12	16	Canada
Manitoba	4 700	28	17	Canada
Great Salt Lake^a	4 660	14	19	USA
Lake of the Woods	4 410	21	–	Canada, USA
Dubawnt	3 830	–	–	Canada
Mistassini	2 190	120	–	Canada
Managua	1 490	26	7.97	Nicaragua
Saint Clair	1 200	7	5.3	Canada
Small Slave Lake	1 190	3	–	Canada
Chapala	1 080	10	10.0	Mexico

Table 1.6. (*cont.*)

Lake	Area, km^2	Maximum depth, m	Volume, km^3	Country
South America				
Maracaibo	13 300	35	–	Venezuela
Titikaka	8 372	281	893	Peru, Bolivia
Poopo[a]	2 530	3	2	Bolivia
Buenos Aires	2 400	–	–	Argentina, Chile
Argentino	1 400	300	–	Argentina
Valencia	350	39	6.3	Venezuela
Australia				
Eyre	15 000	20		
Amadeus[a]	8 000	–	–	
Torrens	5 800	–	–	
Gairdner	4 780	–	–	
Georgi	145	3	0.3	
Taupo	611	164	60	New Zealand

[a] Salt lake.

[b] Area of the Aral Sea water surface is given before reducing its level.

[c] With low levels 3000 km^2, with high levels 30 000 km^2.

[d] With low levels 4000 km^2, with high levels 12 000 km^2.

[e] With low levels 7000–10 000, with high levels 18 000–25 000 km^2.

[f] With low levels 4000 km^2, with high levels 15 000 km^2.

1.2 THE HYDROSPHERE

1.2.1 The origins of water on the Earth

The hydrosphere[1] surrounding the Earth includes liquid, solid and gaseous forms of water. The hydrological cycle transports this water about the Earth exchanging energy and moving materials as part of the process. The hydrosphere unity is determined by not only its continuity but also the constant water exchange between all its elements. The hydrosphere includes all types of natural waters – oceans, seas, rivers, lakes and glaciers, underground, atmospheric and biologically combined waters. The lower limit of the hydrosphere is assumed to be at the level of Mokhorovichich surface, and the upper limit practically coincides with the upper atmospheric limit (Blyutgen, 1972). Sea, lake, river, glacier, underground and atmospheric waters are all interrelated and water moves from one situation to another as the hydrological cycle progresses (Glushkov, 1929; Vernadsky, 1967).

The Earth's hydrosphere is one of the oldest mantles of this planet and it appeared between 3.5 and 4 billion years ago (Klige *et al.*, 1998). It developed together with and in close relationship to the lithosphere, the atmosphere, and then with life itself. Up to the present the mechanisms of the origin of water on the Earth have not been completely explained (Kotwicki, 1991). However, the degasification theory seems to be the most likely explana-

tion (Rubey, 1951; Vinogradov, 1959; Artyushkov, 1970; Condie, 1989). According to this theory the basic mass of the hydrosphere formed as a result of the processes of melting and degassing the Earth's mantle and it was determined by geophysical processes operating at depth.

The mechanism is assumed to be that water vapour, the carbon compounds CO_2, CO and CH, ammonia, sulphur and its compounds H_2S and SO, acid halides HCl, HF, HBr, boric acid, hydrogen, argon and some other gases came to the Earth's surface during lava degassification (Monin and Shishkov, 1979; Holland, 1989). The largest part of the volcanic gases condensed and was transformed into water, forming the hydrosphere.

Acid vapours HCl, HF, HBr, ammonia, sulphur and its compounds, and a considerable part of the CO_2 dissolved in drops of condensed water and fell as acid rain to the Earth's surface. These acid flows ran to low places (oceanic depressions) on the Earth's primary surface, at the same time reacting with underlying rocks and taking out of them the equivalent amount of alkali and alkali earths. Oceanic water appeared to be saline from the very beginning, and land waters fresh as a result of the leaching occurring in

1 There are different interpretations of term "hydrosphere" and viewpoints on its origin (Hydrosphere, 1960; Belousov *et al.*, 1972; Chebotarev, 1978; Monin and Shishkov, 1979; L'vovich, 1986; Kotwicki, 1991; Hydrosphere, 1993a, b).

Table 1.7. *Principal reservoirs in the world with the capacity of more than 20 km³*

Reservoir	Continent	Country	Basin	Year of filling up	Dam backwater, m	Full volume, km³	Use[a]
Owen Falls (Lake Victoria)	Africa	Uganda, Kenya, Tanzania	Victoria–Nile	1954	31	204.8	H F I
Bratskoye	Asia	Russia	Angara	1967	106	169.3	H N T W F R
Nasser	Africa	Egypt	Nile	1970	95	169	I H A N F
Kariba	Africa	Zambia, Zimbabwe	Zambezi	1959	100	160.3	H N I F A
Volta	Africa	Ghana	Volta	1965	70	148	H N I F
Daniel Johnson	N. America	Canada	Manicouagan	1968	214	141.8	H N S
Guri	S. America	Venezuela	Caroni	1986	162	136.3	H
Krasnoyarskoye	Asia	Russia	Yenisey	1967	100	73.3	H N T W S F R
Wadi-Tartar	Asia	Iraq	Tigris	1976	–	72.8	S I
WAC Bennet	N. America	Canada	Peace	1967	183	70.3	H S N
Zeiskoye	Asia	Russia	Zeya	1974	98	68.4	S H N T F
Cabora Bassa	Africa	Mozambique	Zambezi	1977	127	62	H I N F
La Grande 2	N. America	Canada	La Grande	1978	168	61.7	H
La Grande 3	N. America	Canada	La Grande	1981	93	60.0	H
Ust-Ilimskoye	Asia	Russia	Angara	1977	88	59.4	H N T W F
Boguchanskoye	Asia	Russia	Angara	1989	70	58.2	H N T
Kuibyshevskoye	Europe	Russia	Volga	1959	29	58.0	H N W I F
Serra Da Mesa	S. America	Brazil	Tocantins	1993	144	54.4	H
Caniapiscau	N. America	Canada	Caniapiscau	1981	56	53.8	H
Upper Wainganga	Asia	India	Wainganga	1987	43	50.7	
Bukhtarminskoye	Asia	Kazakhstan	Irtysh	1967	67	49.6	H F N S R
Ataturk	Asia	Turkey	Euphrare	1995	175	48.7	H I
Serros Coloradeso	S. America	Argentina	Neuguen	1977	35	43.4	H I
Tucurui	S. America	Brazil	Tocantins	1984	65	43.0	H
Vilyuiskoye	Asia	Russia	Vilyui	1972	68	35.9	H W N
Sanmenxia	Asia	China	Huang He	1962	90	35.4	H I S N
Kouilou	Africa	Congo	Kouilou	1992	137	35	H
Hoover	N. America	USA	Colorado	1936	221	34.8	I H S A
Sobradinho	S. America	Brazil	São Francisco	1979	43	34.2	N H W
Glen Canyon	N. America	USA	Colorado	1966	216	33.3	H A S R
Kemano	N. America	Canada	Nechako	1952	104	32.7	H
Churchill Falls	N. America	Canada	Churchill	1971	32	32.3	H
Nechaco	N. America	Canada	Nechaco	1953	25	32.2	H
Jenpeg	N. America	Canada	Nelson	1975	30	31.8	N H
Volgogradskoye	Europe	Russia	Volga	1961	27	31.4	H W F I N T R
Keban	Asia	Turkey	Euphrates	1976	190	30.6	I H
Garrison	N. America	USA	Missouri	1956	62	30.1	S I W H A
Iroquois	N. America	USA, Canada	St. Lawrence	1958	20	30.0	H N A
Sayanskoye	Asia	Russia	Yenisey	1978	220	29.1	H I N T W R
Itaipu	S. America	Brazil, Paraguay	Parana	1982	165	29.0	H
Oahe	N. America	USA	Missouri	1962	75	28.8	H S N R
Kapchagaiskoye	Asia	Kazakhstan	Ili	1970[b]	41	28.1	H A
Kossou	Africa	Côte d'Ivoire	Bandama	1972	57	28	H I F
Razzaza Dyke	Asia	Iraq	Euphrates	1970	15	26.0	S I
Rybinskoye	Europe	Russia	Volga	1955	18	25.4	H N T W F R
Loma de la Lata	S. America	Argentina	Neuguen	1977	16	25.1	H
Longyangxia	Asia	China	Huan He	1990	172	24.7	H I A
Mica	N. America	Canada	Columbia	1976	175	24.7	H A
Tsimlyanskoye	Europe	Russia	Don	1952	26	23.9	H F R W N

Table 1.7. (*cont.*)

Reservoir	Continent	Country	Basin	Year of filling up	Dam backwater, m	Full volume, km^3	Use^a
Kenney	N. America	Canada	Nechako	1952	104	23.7	H
Khantaiskoye	Asia	Russia	Khantaika	1975	50	23.5	H N
Fort Peck	N. America	USA	Missouri	1937	76	23.0	S H I N
Xinanjiang	Asia	China	Xinanjiang	1960	100	21.6	H A
Ilia Solteira	S. America	Brazil	Parana	1974	85	21.2	H N
Yacyreta	S. America	Argentina, Paraguay	Parana	1991	41	21.0	H N I
Furnas	S. America	Brazil	Grande	1965	96	20.9	H A
El Chocon	S. America	Argentina	Limay	1975	65	20.2	H I A

^{a} H, hydropower; N, navigation; W, water supply; I, irrigation; F, fishery; T, timber rafting; R, recreation; A, accumulation; S, struggle with inundations.

^{b} In 1970 the filling of the reservoir was stopped. The project reservoir capacity was not attained.

the upper zone of the Earth's crust, remaining saline only in deep areas.

Some scientists (Shoemaker, 1984; Alvarez, 1987) do not agree with this theory on the origins of the hydrosphere. They consider that the Earth has experienced during its history numerous collisions with comets that were potential sources of water.

Estimates of the amount of water formed in this way at early stages of the evolution of the Earth vary from 4% to 40% (Chyba, 1987) and some suggest even higher proportions (Hoyle, 1978) of the volume.

Present-day geological studies have shown that the hydrosphere existed during most geological periods (Markov, 1960; Strakhov, 1963). According to calculations (Timofeyev *et al.*, 1988) the Earth's mantle contains 28×10^9 km^3 of water, which supports degasification as the origin of the hydrosphere.

During the early history of the Earth degasification was more intensive. The basic mass of the hydrosphere would probably have formed during the first hundreds of millions of years. Oceans appeared rapidly during this time (Kuenen, 1950).

However, Revelle (1955) was of the opinion that the oceans appeared late and quickly. According to Schopf (1980), the major volume of degasification occurred between 4.6 and 2.5 billion years ago, and according to Sorokhtin (1974), the maximum rate of growth took place during the Lower Riphean.

New studies (Staudacher and Allegre, 1982; Hydrosphere, 1993a, b) show that the Earth degasified rapidly during the 50 million years after it originated. The results of further studies in this area have been generalized by Holland (1989) and Kump (1989).

During the Archean Period the Earth's surface relief was subdued and water covered an area of about 500 million km^2 (Klige, 1992). There was a warm and humid climate without distinct latitudinal zones and with alternating periods when minor warming and cooling occurred together with glaciation (Monin and Shishkov, 1979). In the Proterozoic Era, photosynthesis became active with

the development of live matter in the hydrosphere (Alpatjyev, 1983).

The gradual increase in the land area with the growth of the thickness of the Earth's crust and the development of mountains exerted a considerable effect on the hydrological cycle. At this time conditions were more arid and ice sheets developed in distinct climatic zones, the hydrological cycle between the oceans, atmosphere and land grew more active and a river network developed (Drozdov *et al.*, 1981).

In the Palaeozoic Era the hydrological cycle became more complicated due to changes in the ratio between the area of ocean and land. During this period the ocean reached its greatest size in the Ordovician. Marine deposits show that this was the most powerful transgression in the history of the Earth. The land area was 72 million km^2 or 50% of its present size. The sea level rose by more than 250 m and 83% of our planet was covered by water (Klige, 1980). In contrast, the area of the land was greatest during the Mesozoic when the sea level was 100 m lower than at present.

Simultaneously with the decrease in the size of the ocean and an increase in the elevation of the continents, as a result of the development of mountain-forming processes, climatic conditions became more arid, runoff decreased, and a considerable part of the water became locked up in ice sheets and glaciers. At this time the character of the water cycle came close to the one that exists today. By the Mesozoic the gaseous composition of the atmosphere had changed greatly, as a result of the increase in the amount of carbon dioxide and oxygen due to the development vegetation and animal life. There is evidence of boreal, humid and subtropical climates on land in the late Triassic (Razumikhin, 1976).

The recent great oceanic transgression started at the end of the Jurassic Period and reached its maximum in the Cretaceous Period. Since then the ocean has, in general, regressed and the land area has increased (by about 35 million km^2), and this has been accompanied by powerful mountain-building of the Alpine

Table 1.8. *Water content in the hydrosphere*

Type of water	Area of distribution, km$^2 \times 10^3$	Volume, km$^3 \times 10^3$	Water layer, m	Fraction of total volume of hydrosphere, %	Fraction of fresh water, %
World Ocean	361 300	1 338 000	3700	96.5	–
Ground water (gravity and capillary)	134 800	23 400[a]	174	1.7	
Predominantly fresh ground water	134 800	10 530	78	0.76	30.1
Soil moisture	82 000	16.5	0.2	0.001	0.05
Glaciers and permanent snow cover:	16 227.5	24 064	1463	1.74	68.7
Antarctica	13 980	21 600	1546	1.56	61.7
Greenland	1 802.4	2 340	1298	0.17	6.68
Arctic Islands	226.1	83.5	369	0.006	0.24
Mountainous regions	224	40.6	181	0.003	0.12
Ground ice of permafrost zone	21 000	300	14	0.022	0.86
Water in lakes:	2 058.7	176.4	85.7	0.013	–
Fresh	1 236.4	91.0	73.6	0.007	0.26
Salt	822.3	85.4	103.8	0.006	–
Swamp water	2 682.6	11.5	4.28	0.0008	0.03
River stream water	148 800	2.12	0.014	0.0002	0.006
Biological water	510 000	1.12	0.002	0.0001	0.003
Water in the air	510 000	12.9	0.025	0.001	0.04
Total volume of the hydrosphere	510 000	1 386 000	2718	100	–
Fresh water	148 800	35 029.2	235	2.53	100

[a] With no account of underground water of the Antarctic, approximately estimated at 2 million km^3, including predominantly fresh water of about 1 million km^3.

type on most continents. This has lead to an increasing in the role for continental water in the global water cycle.

More recently, wide fluctuations in the distribution of land and water can be traced particularly in the Pleistocene, over the last million years as a result of the change in climatic conditions accompanied by alterations to the water regime on the Earth's surface. Cooling, which appeared periodically as a result of variations in the amount of solar radiation coming to the Earth's surface, led to the formation of large continental glaciations. Glaciers accumulated huge masses of water – more than 60 million km^3, which resulted in the lowering of the sea level by more than 100 m. At the same time a considerable amount of moisture (up to 1 million km^3) was accumulated in the extensive closed drainage basins on the continents. There are also changes in the volume of ground water with storage increasing considerably during the humid periods.

1.2.2 The contemporary hydrosphere

There are no large disagreements in the estimates of the volume of the present hydrosphere (Korzun, 1974a; Kotwicki, 1991; Hydrosphere, 1993a, b), since it is determined, basically, by the enormous volume of water contained in the World Ocean. The

volume of the hydrosphere is most frequently estimated to be 1370 million km^3 and this figure is practically equal to the water volume in the World Ocean. However, as more information becomes available about the relief of the ocean bottom, particularly for the Arctic Ocean, where underwater ridges have been discovered, there have been reductions to a total of 1338 million km^3, i.e. a reduction of 32 million km^3 (Frolov, 1971). The total volume of the hydrosphere, according to current data (Korzun, 1974) is 1386 million km^3 (Table 1.8). Fresh water in all its states makes up only 2.53% of the total, of which 1.74% is in the ice sheets of the Antarctic and the Arctic and in mountain glaciers.

In addition to free (gravitational) water, the lithosphere contains a large amount of physically and chemically combined water. The average content of the physically and chemically combined waters amounts to 3.5% of the rock weight, i.e. some 0.84×10^{24} g (Derpgolts, 1971). Combined water does not participate actively in the hydrological cycle, at least at recognizable time-scales, and is not taken into account in this present study.

THE WORLD OCEAN

Table 1.8 shows that the World Ocean holds by far the largest part of total volume of water on the planet. However, in recent

Table 1.9. *Present-day glaciation of continents and islands of the Earth*

Region	Area of glaciers, km²	Water volume, km³
Arctic		
Greenland	1 802 400	2 340 000
Franz Josef Land	13 735	2 530
Novaya Zemlya	24 420	9 200
Severnaya Zemlya	17 470	4 620
Arctic Islands	226 090	83 500
Canadian Archipelago	148 825	48 400
Spitzbergen (Western)	21 240	18 690
Small Islands	400	60
Total	2 028 490	2 423 500
Europe		
Iceland	11 785	3 000
Scandinavia	5 000	645
Alpes	3 200	350
Caucasus	1 430	95
Total	21 415	4 090
Asia		
Pamir-Altai	11 255	1 725
Tien Shan	7 115	735
Dzungarian Ala Tau, Sayan Mountains	1 635	140
Eastern Siberia	400	30
Kamchatka, Plateau of Koryak	1 510	80
Hindu Kush	6 200	930
Karakoram Pass	15 670	2 180
Himalayas	33 150	4 990
Tibet	32 150	4 820
Total	109 085	15 630
North America		
Alaska (Pacific Coast)	52 000	12 200
Inner Alaska	15 000	1 800
USA	510	60
Mexico	12	2
Total	67 522	14 062
South America		
Venezuela, Colombia, Andes, Tierra del Fuego	7 100	2 700
Patagonian Andes	17 900	4 050
Total	25 000	6 750
Oceania		
New Zealand	1 000	100
New Guinea	14.5	7
Total	1 014.5	107
Africa		
Kenya, Mount Kilimanjaro, Ruwenzori	22.5	3
Antarctica	13 980 000	21 600 000

years, studies have appeared (Sofer and Skirstymonskaya, 1994; Wallace, 1996) that show volumes which differ from the data here by between 0.7% and 10%. Including the water stored in the bottom silts of the oceans causes the 10% difference.

The World Ocean has accumulated 3.06×10^{25} Joules of heat (Stepanov, 1983). Every year it takes up almost twice as much solar energy as the land, and this factor determines its important role in the planetary heat exchange. The major portion of this energy is employed in evaporating over 500 000 km³ per year of water, which ensures global water exchange.

GLACIERS AND ICE SHEETS

The largest volume of fresh water is stored in the planet's glaciers and ice sheets. The total area of the present glaciation exceeds 16.2 million km² (Kotlyakov, 1997). The mean ice thickness on this area is 1700 m, and the maximum is more than 4000 m (in Antarctica). The distribution of ice sheets and glaciers and the water stored in them is given in Table 1.9. The data on glacier thickness and water storage are approximate. To estimate the mean thickness of ice data were used from the few measurements from ice drilling and seismic sounding (Korzun, 1974b). These data were applied by analogy to other glaciers taking into account their morphological features. The accuracy of the assessment of that water storage in the Antarctica, for example, is about ±3.0 million km³. The total water volume in the ice across the globe is estimated to exceed 24 million km³ (Korzun, 1974b). Most of the water stored in the ice cover is concentrated in Antarctica (almost 90%), while the remainder is found in Greenland (almost 10%) and in mountain glaciers.

Glaciers are giant "water reservoirs" and "coolers" greatly influencing the climate and water regime of the Earth (Kotlyakov, 1979). Their state and the changes from this state over time are an important indicators of global climatic and hydrological changes – past, present and future. Cooling and warming and the advance and recession of glaciers result in the change of all the elements of the hydrological cycle: precipitation, runoff and evaporation, and the volume of water stored on land and in the ocean. During glaciation a large amount of water becomes locked up as snow and ice on the land. As a result the volume of runoff decreases, the World Ocean level falls by tens of metres, uncovering extensive areas of the continental shelves. With the decline of glaciation, river flow increases, the volume of water in the ocean becomes larger, the level rises and the land area diminishes.

UNDERGROUND ICE

Areas of permafrost extend over northeast Europe and the north and northeastern parts of Asia, including the Arctic islands; they cover northern Canada and the fringes of Greenland and

Antarctica, as well as higher parts of South America. The total area of permafrost is about 21 million km², some 14% of the land area. In the Southern Hemisphere (Antarctica, South America) permafrost covers about 1 million km². The depth of permafrost ranges from 400 to 650 m. Underground ice within this range is found as vein formations and strata. The water stored as underground ice can be estimated only approximately due to lack of data and few studies (Grave, 1968) but the most likely figure is 300 thousand km³ (Korzun, 1974b). In the permafrost areas 150–200 km³ of water occurs in the form of river ice.

The annual snowfall over the Earth is about 1.7×10^{13} tonnes, and this snow covers an area of between 100 and 126 million km². The distribution of snow varies considerably from year to year depending on climatic conditions.

UNDERGROUND WATER

The volume of gravitational water contained in the pores, fissures and fractures of the water-saturated strata of the Earth's crust represents the natural storage of water underground. The geographical distribution of ground water is closely related to the geological structure of the Earth's crust. It also depends considerably on the climatic factors: precipitation, condensation and evaporation, and particularly on the infiltration. Since runoff also depends on these factors, there is a strong relationship between ground water and runoff: ground water draining to rivers are included in the volume of runoff, being its most stable contribution to the hydrograph, especially during dry periods and drought.

The reliable estimation of ground water storage is very difficult (Garmonov et al., 1974). The water content of water-bearing strata can be obtained approximately by multiplying the volume of water-bearing table by a water loss factor and effective porosity. The natural storage of ground water is determined down to the absolute depth of 2000 m – the depth of the isobath which indicates approximately the distribution of the Earth's continental crust.

Three zones of ground water movement can be distinguished vertically:

1. A zone of active water exchange is located above the local base level and is highly dynamic. Movement of water in this zone increases with height above the base level. Here the character of the water is most closely related to the nature of the overlying soil and to the rock strata containing them and also to climatic factors. The effective porosity of this zone is about 15%.

2. A zone of less active water exchange is located below this first zone down to sea level. This zone is situated below local base levels and the water here is only affected by large rivers which may have deep channels. Drainage of ground water in this zone is also related to basins and depressions.

Table 1.10. *Natural ground water resources in the upper layer of the Earth's crust by hydrodynamic zones*

Continent	Zone[a]	Ground water resources, km³ × 10⁶	Total resources of ground water, km³ × 10⁶
Europe	1	0.2	1.6
	2	0.3	
	3	1.1	
Asia	1	1.3	7.8
	2	2.1	
	3	4.4	
Africa	1	1.0	5.5
	2	1.5	
	3	3.0	
North America	1	0.7	4.3
	2	1.2	
	3	2.4	
South America	1	0.3	3.0
	2	0.9	
	3	1.8	
Australia and Oceania	1	0.1	1.2
	2	0.2	
	3	0.9	

[a] For explanation of zones see text.

Where these lie under the sea the discharge of water from this zone occurs into the sea. Less movement provides for higher mineralization of ground waters, however here they are basically fresh or weakly mineralized. The nature of the waters in this zone is determined by the occurrence of aquifers and aquicludes and their juxtaposition in the form of depressions, troughs, synclines and monoclines forming artesian basins. The effective porosity of this zone is 12%.

3. A third zone lies in the crust from sea level to the absolute depth – 2000 m. The waters of the upper part of this zone are only influenced by the biggest rivers at depth, and by large-scale features such as depressions in the relief of land and the ocean. In the upper part of this zone water is fresh or weakly mineralized, with saline water and brines below. The effective porosity is 5%.

The mean altitude of each continent was used for calculating the total volume of ground water stored in the Earth's crust (Korzun, 1974b). The total storage of ground water to the 2000 m level in the Earth's crust was estimated to be 23.4 million km³ (Table 1.10). With 3.6 million km³ in the first zone, 6.2 million km³ in the second and 13.6 million km³ in the third, rivers are fed mainly from water stored in the first zone.

Table 1.11. *Water resources in the principal lakes of the Earth*

Continent	Number of lakes	Total area, km^2 × 10^3	Water resources, km^3 Fresh	Water resources, km^3 Salt
Europe	34	430.4	2 027	78 000
Asia	43	209.9	27 782	3 165
Africa	21	196.8	30 000	–
North America	30	392.9	25 623	19
South America	6	27.8	913	2
Australia and Oceania	11	41.7	154	174
Total	145	1300	86 500	81 360

Table 1.12. *Area of bog over the Earth*

Continent	Bog area, km^2 × 10^3
Eurasia	925
Africa	341
North America	180
South America	1232
Australia and Oceania	4

Table 1.13. *Water volume in river channels of the Earth*

Continent	Water volume in river channels, km^3
Europe	80
Asia	565
Africa	195
North America	250
South America	1000
Australia and Oceania	25

LAKES AND RESERVOIRS

There are 145 large lakes across the globe with an area of 100 km^2 and holding 168 thousand km^3 of water (Korzun, 1974b) (Table 1.11). This is 95% of the total volume of all the world's lakes, giving a total volume of lake water of 176.4 thousand km^3. Of this total 91 thousand km^3 is fresh water, and 85.4 thousand km^3 is salt. The hydrology of about 40% of the world's large lakes has not been studied and their volumes are estimated approximately.

Some studies (L'vovich, 1986; Wallace, 1996) exaggerated estimates of the water stored in lakes and these vary from 200 000 km^3 to 278 000 km^3. The hydrosphere includes water held in reservoirs. Their total volume exceeds 6000 km^3 and they regulate about 15% of the Earth's total runoff.

WATER STORED IN SWAMPS, CHANNEL NETWORKS, SOIL, LIVING ORGANISMS, PLANTS AND THE ATMOSPHERE

Swamps and bogs are widespread across the Earth with a total area of approximately 2.7 million km^2 or about 2% of the land area. The most swampy continent is South America (Table 1.12). The total volume of water in the world's swamps and bogs is estimated to be about 11 470 km^3 (Korzun, 1974b). This value has been obtained on the assumption that the mean thickness of the peat bogs is 4.5 m, their volume is 12 070 km^3, and that they are 95% water.

The hydrosphere also includes the water stored in the river channel network. The total volume of this water – 2120 km^3 – was estimated by the State Hydrological Institute (Korzun, 1974b) taking into account the volume of runoff and the lengths of the main rivers and their tributaries (Table 1.13). According to L'vovich (1986) this volume is 1200 km^3. In spite of the very small volume of water in the river channels, it is this water which is continuously renewed and which is most important for human use.

The soil moisture is an integral part of the hydrosphere. This water occurs mainly in the top 2 metres of the soil. The total volume of soil moisture is estimated to be approximately 16 500 km^3 (Korzun, 1974b). This figure assumes that soil moisture is 10% of the 2-m layer, and that the area of soil containing moisture covers 55% of the land area or 82 million km^2. L'vovich (1986) estimated the total volume of soil moisture to be 83 thousand km^3; however he did not state the method used to estimate it.

Biological water (the water included in living organisms such as plants and animals) is an active link in the hydrologic cycle. Part of the water that evaporates from the land and enters into the atmosphere is due to transpiration of soil moisture by vegetation. Alpatjyev (1969) gives the volume of living matter in the biosphere as 1.4×10^{12} tonnes. The water content of living matter is about 80% (Derpgolts, 1971), i.e. 1.12×10^{12} tonnes or approximately 1120 km^3.

The water contained in the atmosphere, as water vapour, water drops and ice crystals, is an important part in the hydrosphere, possibly the most active part. The total volume of moisture in the atmosphere, according to the different estimates (Korzun, 1974b; L'vovich, 1986; Wallace, 1996), varies from 12 900 km^3 to 14 000 km^3.

1.2.3 Global water exchange in the hydrosphere – the hydrological cycle

The waters of the hydrosphere are in constant, usually cyclic, motion under the effects of solar radiation, the energy released from the Earth's interior and gravitational forces. Due to geological processes about 1 km^3 of water a year is released from the mantle through degasification and this rises gradually to the Earth's surface. As a result of convection in the mantle, part of this matter can emerge through breaks in ocean rift zones related to oceanic ridges (Monin, 1977). The global process of water exchange provides some stability in the distribution of waters between the land, the oceans and the atmosphere. This equilibrium is relative and can change in time, and these changes can lead to corresponding changes in hydrological and climatic conditions.

Water evaporating from the surface of reservoirs, soil and vegetation enters into the atmosphere as water vapour where it is dissipated upwards by turbulent diffusion and is transported by air currents from one place to another. With a temperature decrease, water vapour is condensed, transforming it to a liquid or solid. During rainfall from clouds, part of the water returns to the Earth's surface (inland cycle), and part of it returns to reservoirs in the form of runoff. Some precipitation can fall into the ocean.

Water evaporated from the surface of the oceans and seas mostly (90%) falls back into the sea, short-circuiting the cycle. A smaller part of it (10%) participates in the major cycle, being transported by atmospheric circulation to the land where, as rainfall, it can be involved in a number of smaller versions of the complete hydrological cycle when surface and ground water and ice drainage reaches the World Ocean, closing the complete cycle. Part of the water is combined and decomposed by plants.

Part of the water contained by the Earth is in chemical compounds, such as crystal hydrate, sorbate and many other forms which are found in porous deposits in the Earth's crust. This chemically combined water can be removed from the total water exchange for thousands of years. The crustal rocks lose water during the process of metamorphization and subduction under the effects of high pressure and high temperature. This water rises through rock pores and appears on the Earth's surface (Vinogradov, 1973).

The global hydrological cycle is not a closed system. Solar energy and energy from space, together with cosmic dust, meteorites and meteors, arrive from space. The Earth in its turn gives back part of its energy to space and dissipates hydrogen and helium to it (Alpatjyev, 1983; Kulp, 1951). This exchange of matter and energy brings about 0.01 km^3 of water per year (Derpgolts, 1971; Alpatjyev, 1969) from space to the Earth. At the same time part of the hydrosphere is lost due to the dissipation of light gases, and their escape beyond the limits of the Earth's gravitational field,

Table 1.14. *Periods of renewal of water resources on the Earth*

Water of hydrosphere	Period of renewal
World Ocean	2 500 years
Ground water	1 400 years
Polar ice	9 700 years
Mountain glaciers	1 600 years
Ground ice of the permafrost zone	10 000 years
Lakes	17 years
Bogs	5 years
Soil moisture	1 year
Channel networks	16 days
Atmospheric moisture	8 days
Biological water	several hours

amounting to about 0.1 km^3 per year (from 0.03 to 0.27 km^3: Yuri, 1959; Pavlov, 1977; Alpatjyev, 1983).

Every year human influences grow and cause more and more changes to natural processes, including the hydrological cycle. These changes bring about alterations to the water balance and to water resources and their availability. The rapid growth of population, the development of industrial production and the rise of agriculture have resulted in the increased use of water, reaching a global total of about 4 thousand km^3 per year (Shiklomanov, 1997) by 1990. Some 80% of this water is used for agriculture, primarily for irrigation, and this causes more evaporation and an intensification of the hydrological cycle.

Human activities have also changed the character of ground water. Although there are some examples of artificial recharge of aquifers, more often the water table has been lowered to provide water for drinking. Every year up to 20 thousand km^3 of ground water is abstracted (Plotnikov, 1976), which results generally in the reduction of aquifer storage and the lowering of ground water levels, and in some cases in land subsidence.

The construction of reservoirs has led to the slowing down of the movement of river waters (Kalinin, 1974). Slowing the movement of water can influence its quality particularly by the accumulation of pollutants. Because the World Ocean water is contaminated by oil products, this leads to the reduction of evaporation from the water surface by about 10% (Duvanin, 1981) and this contributes to the reduction in the rate of exchange of water between the ocean and the land surface.

Of course water in the hydrosphere is connected by the hydrological cycle; however the rates of movement and residence times are very different for water in its different states (Table 1.14). Table 1.14 shows that biological waters included in plants and living organisms are renewed most rapidly – perhaps over a period of a few hours. Plants transpire this water. Atmospheric water, which forms due to evaporation from any water surface, is renewed on average over 8 days. Water stored in the channel network is also renewed on average over a period of 16 days. The soil water is renewed over a period of a year and is spent mainly for evaporation and partly on runoff. Water stored in swamps has a 5-year residence time.

Most lake water is renewed on average over a period of 17 years. However different lakes have different renewal times. For example, for Lake Baikal this time is 380 years. All other types of natural waters (glaciers, ground waters, ocean waters etc.) are renewed more slowly, possibly over periods of thousands and even tens of thousands of years. The largest period is for the ice in the tundra and in Antarctica, which may be renewed only over several hundreds of thousands of years (Kotlyakov, 1984).

The time data presented here for the exchange of natural water in the global hydrological cycle (Korzun, 1974b; L'vovich, 1986) are very approximate and typical of the lower limits of the exchange process (Kalinin, 1972).

2 Water resources assessment

2.1 ASSESSMENT METHODS

World economic development and population growth are closely linked to the use of fresh water resources. Consequently there is a pressing and continuing need to assess water resources and to develop and maintain the hydrological network which is the basis of this assessment.

However network development has been rather slow and uneven from country to country. As a result the adequate and reliable assessment of water resources has always been difficult and it has become even more so during recent decades, while the growth in demand for water has been accelerating.

At the present time, the hydrological network, in its most advanced state, includes precipitation and evaporation stations as well as those where runoff, soil water, ground water, sediment and water quality are measured across the river basin. For some basins and regions the hydrological network is well developed and it is possible to estimate the annually renewed water resources directly from data captured by the network, particularly runoff data. For others where the hydrological network is rudimentary water resources have to be assessed from meteorological data.

In recent years a number of approaches have been developed in hydrology for assessing water resources. Most of these approaches are based on the law of conservation of mass as presented most simply in the water balance equation for a river basin:

$$Y = P - E, \qquad (2.1)$$

where Y is the runoff; P is the precipitation; E is the evaporation.

The water balance equation for assessing water resources has been used by Baumgartner and Reichel (1975) and by L'vovich (1974) and partly in the monograph *World Water Balance and Water Resources of the Earth* (Korzun, 1974b).

When using the water balance equation to assess water resources, it is assumed that the appropriate maps of precipitation and evaporation are available. There are, however, large errors in the measurement of precipitation, while there are also large errors in evaporation estimation by the Turc, Thornthwaite, Penman,

Budyko and other methods (Konstantinov, 1968; Budyko, 1971; Babkin, 1979).

Therefore, assessments of water resources using equation 2.1, which are based on precipitation and evaporation maps even for long-term means, are only approximate, especially for arid regions. In this case the errors of determining precipitation and evaporation exceed runoff values. But of course, these runoff estimates should be checked against estimates determined by other methods where this is possible. The equations relating runoff, precipitation and evaporation are usually used to check these maps and in estimating the figures for long-term water resources (Budyko, 1971; Babkin, 1979). The long-term mean and annual water resources estimates often use regression and models along with information on climatic and catchment characteristics for the basins concerned (Babkin *et al.*, 1972; Rozhdestvensky and Chebotaryev, 1974; Kouzin and Babkin, 1979; Babkin, 1998; Shiklomanov *et al.*, 2000).

In the simplest regression model it is assumed that observed values of runoff and the factors determining the runoff are available for a particular set of river basins. The problem then is reduced to the minimization of the sum of squares of deviations of the computed runoff values Y, from the observed ones Y_l, i.e.

$$\sum_{i=1}^{n} (Y_i - Y_{li})^2 = \min \qquad (2.2)$$

This minimization is attained with the use of the method of least squares allowing assessment of the parameters $a_j (j = 0, 1, 2, \ldots, k)$, the coefficients of the regression equation:

$$Y = a_0 + a_1 x_1 + a_2 x_2 + a_k x_k \qquad (2.3)$$

with a certain reliability

$$S = \sigma_0 (1 - r^2) \qquad (2.4)$$

In equations 2.3 and 2.4, S is the root-mean-square error of the equation; σ_0 is the root-mean-square deviation of runoff; r is the correlation coefficient; $x_1, x_2, \ldots, x_k$ are the variables.

The drainage area, mean basin elevation, mean basin slope, swamp, lake and forest areas and measurements of other features serve as independent variables for assessing the long-term mean runoff.

To estimate annual runoff, precipitation, relative humidity, air temperature and other climatic factors are used. When the regression equations are being calculated, the following factors can be taken into account: the number of river basins (for long-term means) or the number of years of observations (for annual runoff values) should be more than twice the number of variables. For the equations calculated the errors of the coefficients are important. Equation 2.3 is assumed to be reliable if its error, S, is at a minimum and the errors of its coefficients are 1.5 to 3 times less than the values of determined parameters.

Regression equations are widely applied in many countries to assess the runoff of individual river basins. A multiple regression model is an extension of this method. This entails the successive inclusion of certain variables with the successive calculation of the regression equation depending on these factors. The resulting equation or equations express runoff from the combination of individual factors (e.g. of underlying surface only: lake, swamp and forest areas, mean basin slope etc., or climatic elements: precipitation, air humidity deficit, air temperature etc.) and on the latitude and longitude.

In factor analysis, the models are subdivided into two classes: the factor and the component. Unlike regression analysis where the variables are used as runoff factors, the factors are unknown in the factor model, but many features on which these factors are dependent are included. The problem is reduced to searching for these factors from these features and partly to interpreting these factors. Basin and climatic characteristics are usually applied as these features (Babkin *et al.*, 1972). From the whole set of these features, the factors F_2 are determined by optimization from the expression:

$$X_i = \sum_{r=1}^{k} l_{ir} F_r + e_i \qquad (2.5)$$

where l_{ir} is the load of ith variable on rth factor.

This expression is true for the factor model and for the component model (the major component method) but the random error e_i in this expansion is absent. The difference between the two types of this one model is that in the component model, the variance of all features is fully settled by the component variances, while the factor model assumes the presence of e_i.

The optimization procedure is such that usually the first two or three components (or factors) contain up to 90% of the total generalized information (the variances of variables) allowing the relationships to be built easily between the runoff characteristics and the generalized factors. The entire procedure of using the different types of factor analysis can reveal the global factors that are important. In practice estimates of water resources for individual

years by the factor and regression models do not differ from each other (Babkin *et al.*, 1972).

Using these models (factor and regression) it is difficult to estimate world and continental water resources because large amounts of data are required. To assess the water resources of individual river basins, or to solve some specific problems related, for instance, to flood forecasting, or to assessing the efficiency of operating water management systems, runoff models with lumped or distributed parameters are often used (Denisov and Denisov, 1961; Kouchment, 1980; Lindsley, 1985; O'Donnell, 1986; Vinogradov, 1988; Eagleson, 1991). To use these models successfully however, it is necessary to have a large amount of hydrological data and information on basin characteristics. The consequence is that it is usually very difficult or impossible to use these models, for reliably assessing water resources for basins and regions. These difficulties focus attention on the importance of the measurements of the hydrological variables and particularly on observations of runoff.

However, many national hydrological networks have deficiencies, especially those in Africa, South America and Asia. The lack of observations does not allow water resources assessment to be based solely on observational data. Therefore, methods have to be employed which are based not only the law of conservation of mass (the water balance method) but on other methods which may be available. The application of these methods is justified for regions without any data, or where hydrological observations are sparse.

In this study, the assessments of the world's water resources have been determined by these methods using hydrological data for each year for the period 1921 to 1985, i.e. a period of 65 years' duration. The long-term mean and annual assessments of water resources were made, so that changes could be studied over the 65 years.

2.2 RIVER BASINS AND CONTINENTS

In order to assess the water resources of river basins and continents, ideally runoff data are needed for all the rivers within these areas. However, often there are many rivers without gauging stations and the volume of runoff has to be estimated for these rivers. Then for some rivers the flow is not measured near their mouths and the runoff has also to be estimated for these sites. There is also the problem of areas which drain to the sea where in the lower courses of the rivers concerned, the runoff can divide into many channels crossing extensive flat areas and where the runoff may suffer losses by evaporation and infiltration. For such areas the runoff can be estimated from the flows of the upstream tributaries. Particular difficulties arise in the case of islands, again because few observations are made and this has required the development of methods for the annual assessment of water resources of islands. Studies addressing this assessment of the water resources of large river basins, nations, islands and continents has to be preceded by work on filling the gaps in records of runoff and reducing them to a single time period.

The time series of runoff records for monthly and annual time intervals were analysed to reveal observational errors, misprints, gaps in the observations and other problems. Such analyses have been carried out manually, such as by scanning the records by eye, calculating monthly means and comparing the runoff and precipitation records to establish that the runoff was less than the precipitation. Gaps in monthly runoff values were filled by data calculated by regression analysis; such regression equations were characterized by correlation coefficients with values of $r \geq 0.80$. If necessary, these estimates were checked against the data obtained by comparing them with observed monthly values for years with an equal volume of runoff. Annual runoff volumes for the years with gaps in observations were determined by summing the monthly volumes. Extending the time series and fitting them into a single period (1921 to 1985) was carried out by linear regression again where the correlation coefficient had values of $r \geq 0.80$. Use was made of river analogues which were selected because of similar physiographic conditions and river regimes. In these studies, use was made of simple and multiple linear correlation. The results obtained by regression analysis were checked frequently against the estimates calculated by means of graphical correlation.

Assessment of the water resources of basins with rivers flowing into the sea or into the World Ocean and where the hydrological network was well developed, were determined from the sum of the measured volumes of flow made at stations nearest the mouths of those rivers. For basins where hydrological data were unavailable, as a first approximation long-term mean water resources were determined from the appropriate runoff map given in the *Atlas of World Water Balance* (Korzun, 1974a). Then river analogues were selected, and by the method of proportions of long-term mean runoff a factor was determined for calculating the runoff from that part of the basin without measurements for the years concerned. The water resources of the entire basin were determined as the sum of the runoff from its measured and unmeasured parts. This method of analogues was applied to basins without data with corrections for the corresponding factors, while basins where the course of the river divides into a number of channels, the flow reaching the ocean may be estimated from the flow of the tributaries in the higher parts of the basins. Where there is no observational data, because there are no river gauging stations, the long-term mean can be estimated from the runoff maps in the *Atlas of World Water Balance* (Korzun, 1974a) and annual values of water resources can be estimated by the analogue method.

To assess runoff Q for certain rivers such as the Zambezi and Brahmaputra, the following expression (Babkin, 1979) was used:

$$Q = P - \Delta S - \beta \cdot E_{\max}\theta(P - \Delta S / E_{\max}) \qquad (2.6)$$

where P is the precipitation, ΔS is the change of water storage in the river basin over a year; $E_{\max}$ is the maximum evaporation; β is the coefficient of evaporation; θ is the symbol of hyperbolic tangency.

The maximum evaporation $E_{\max}$ was estimated by the formula:

$$E_{\max} = (R \cdot T / 6\pi)^{1/2} \cdot e^{-(2+L/RT)} \qquad (2.7)$$

where R is the specific gas constant; L is the specific heat of evaporation; T is the temperature of the basin surface, in degrees Kelvin.

Runoff is used in many human activities: irrigation, industrial water supply, municipal water supply and so on. The volumes of water used in irrigation for example can reach several tens of cubic kilometres per year with water taken from rivers such as the Syr Darya, Amu Darya, Kura, Nile, and Indus. Due to the abstraction of these large volumes of water and the likelihood that they will increase in the future, it is very important to assess water resources reliably and regularly (Shiklomanov and Markova, 1987). Runoff at a river site below an abstraction point, whether this is for irrigation, water supply, or another purpose, is often considerably less than upstream and records of flow at the downstream sites should be naturalized, i.e. the volume abstracted should be added to the measured discharge (Shiklomanov, 1979). Naturalization of flow records is usually carried out using one or more methods. In the present study use was made of the regression method and the method for stream water balance (Babkin and Voskresensky, 1977; Shiklomanov and Markova, 1987; Shiklomanov, 1997, 1998a; Babkin, 1998; Shiklomanov *et al.*, 2000).

In the regression method a relationship is established between the meteorological variables over the basin and runoff (such a relationship usually has high correlation coefficient $r \geq 0.80$) for the period before abstraction started. This relationship can then be employed with meteorological data for the period after abstraction commenced to establish the naturalized runoff. The water balance method estimates runoff for individual years as the residual term in the water balance equation for the basin in question. Using these methods, naturalized flows were calculated for river basins in the Ukraine, Moldova, Tajikistan, Uzbekistan, Kazakhstan, Georgia, Azerbaijan, Russia, India and China and in other countries.

Along with the mean values of flow for the time interval under consideration, to comprehensively characterize the river flow time series in the study, two parameters more were used: the coefficient of variation C_V and the coefficient of skewness C_s (Sokolovsky, 1972; Richey *et al.*, 1989a; Finlayson and McMahon, 1991). The coefficient of variation C_V is determined as the ratio of the root-mean-square deviation to the mean runoff. The coefficient of skewness characterizes the asymmetry of the distribution curve of flows, but long time series of observations are required to determine it with accuracy. These three parameters were determined along with exceedance probability for hydrological stations with long periods of observations. This information as well as the maximum and minimum flows for the period in question are cited for different rivers in Chapters 4 to 9.

The total flow to particular oceans was determined as the sum of the flows from all basins including any areas existing between them. Runoff from these areas was calculated by the analogue and geographical interpolation methods. These same methods were applied to islands and where this was impossible, because of absence of data and mean runoff had to be determined from the runoff map, and for individual years by expressing the runoff as a coefficient of precipitation. The analogue method was used for Arctic Ocean islands and for certain continental basins.

The total river flow into the World Ocean was determined from the records for gauging stations nearest to the ocean; where these stations were sites located at a distance from the ocean, estimates were made for points near to the mouths of these rivers. For most rivers the runoff increases towards the mouth and these have correlation coefficients greater than 1 but for the others where the runoff decreases towards the mouth, the correlation coefficients are less than 1. In this study, the inflow to the World Ocean was determined on all the continents for each 10° of latitude. The analogue method of estimating runoff was used for areas without data.

2.3 NATURAL REGIONS, COUNTRIES AND CONTINENTS

Water resources of natural regions and individual countries are made up of the following components: the water derived from precipitation on the area concerned which produces the local runoff, the volume of inflow from outside the area, and water flowing out of the area, region or country.

To assess these components for each area, use was made of the method of linear runoff equations (Babkin and Voskresensky, 1977). The parameters of these equations as applied to particular records from hydrological stations were determined for every river along its length to the border of the region or country. When deriving these equations, the hydrological analogue and geographical interpolation methods were used (Sokolovsky, 1972; SHI, 1987) as well as the runoff maps given in the *Atlas of the World Water Balance* (Korzun, 1974a). The use of these equations allowed estimates to be made of the local water resources for all regions and countries, together with the runoff to the oceans and runoff across the borders for each year from 1921 to 1985.

In assessing the water resources of the northern and southern regions of the European part of Russia and adjacent countries within the basins of the Rivers Belaya, Pechora, Kama and Don, as well as within the Ural River basin (the region of Kazakhstan and Central Asia), use was made of the method for integral averaging of runoff surface equations (Babkin *et al.*, 1974). This method derives surface runoff equations for an area with their subsequent integration within that area. Spatial runoff equations Q (linear and non-linear) are determined from the conventional rectangular coordinates X and Y of the centres of weight of river basins and the average height of catchments Z. The use of this method is effective for areas with homogeneous physiographic conditions. Where there are considerable physiographic differences between catchments within the region in question, a preliminary zoning is carried out and the parameters of the equations are determined for every zone individually. For a small number of river basins the surface runoff equations are linear.

In cases when the observational data are sufficient a non-linear expression is used ($N > n$ by three and more, where N is the number of runoff data, and n is the number of parameters in the equation). The parameters of the linear and non-linear equations were determined by the least-squares method as in regression analysis. The resulting linear or non-linear spatial runoff equations derived were integrated within the area or region in question. For this purpose the region was divided into geometrical figures (triangles, rectangles, parallelograms, trapeziums) within which the integration was carried out. The total area of these figures was equal to the area of the region in question.

The calculation of the runoff layer was reduced to assessing the double integral from the equation of its surface for the average height of the catchment. This means that instead of the coordinate Z, its numerical value for the basin was substituted:

$$h = \frac{1}{S} \iint_S f(x, y) dx dy \qquad (2.8)$$

For the remaining part of the regions the water resources were determined by using the linear runoff equations. The water resources of the continents were determined as the sum of the local runoff for all the regions. Data are given for each continent on what material was used and which method was applied to assess water resources.

2.4 THE DISTRIBUTION OF WATER RESOURCES WITHIN THE YEAR

Each of the world's rivers has its own regime with its particular characteristics determined by a number of factors – especially the manner in which runoff is caused, e.g. rain snowmelt, glacier melt and groundwater discharge. To estimate the distribution of runoff during the year for river basins with different characteristics, use was made of time series of observed runoff. For the different basins the time series were arbitrarily divided into three groups (years with average water content, dry and wet years). From each group one year was chosen which was the most typical and that could comprehensively characterize the monthly distribution of runoff. Monthly values were expressed as a percentage of the annual runoff for the years selected.

The monthly distribution of runoff was determined for a number of the largest rivers draining into each of the different oceans and one typical year was chosen for each large river. For this year the sums of the monthly and the annual runoff volumes were calculated for the rivers in question. Monthly sums were expressed as a percentage of the annual sum. By analogy the distribution of runoff during the year was determined for the different natural regions and for individual countries.

Because the continents occupy vast areas, there can be enormous contrasts in the regimes of different basins during the same year. To assess the distribution of runoff during the year for a continent, one year was selected for the corresponding group of basins to apply to the entire continent. For this year the month-by-month distribution of runoff volumes was found for the natural regions, then it was summed for every month for the continent as a whole and expressed as a percentage of the volume of runoff for the selected year. By this means the monthly distribution of runoff was determined for the continent for the three types of year. The results are rather approximate and smoothed but in general they reflect the conditions pertaining to each continent for the period concerned.

2.5 METHODS FOR STUDYING LONG-TERM VARIATIONS IN WATER RESOURCES

Mathematical and statistical methods are widely used to study the long-term variations in water resources in river basins and in certain regions, as well as the inflow to the World Ocean. Runoff is considered as a stochastic process. However the mathematical description of runoff has to take account of its specific features. Long time series of runoff are characterized by cyclicity (irregular periodicity) causing the presence of series (groups) of dry or wet periods often differing in duration, whose occurrence depends on combinations of meteorological factors, often on the global scale and the characteristics of the different river basins (e.g. the presence of lakes, swamps, glaciers and aquifers). Ideas about the variations in long-term runoff are based on the assumption that they comprise a continuous stochastic process, although in practice this runoff is usually considered as a succession of independent random values with a time-step of one year.

Methods for analysing these series are based on the assumption of stationarity within these long-term variations, i.e. the statistical properties of the initial time series, such as the mean and

variance, are unchanging with time. On this assumption the inference is drawn that design parameters determined from these observations would not change in the future. Therefore, an assessment of stationarity of runoff time series is one of the primary tasks in investigating their long-term variation.

In the present study, the long-term variations in water resources were investigated with regard to the following:

1. Assessing the stationarity (stability) of the main parameters.
2. Revealing any linear trends in the runoff series.
3. Estimation and analysis of the parameters in the distribution.
4. Analysis of cyclicity including revealing the periods with increased, decreased and average values.
5. Revealing synchronism and asynchronism of long-term variations in runoff series.
6. Revealing the features of runoff variations.

2.5.1 Methods for assessing stationarity and trends

Statistical methods for assessing the stationarity of the time series used in this monograph can be divided into two groups: parametric and non-parametric (Rozhdestvensky, 1988). Parametric methods are based on knowledge of the distribution function and its parameters, including the classic methods for checking hypotheses such as the Student and Fisher tests. A large number of methods are included in the non-parametric (ordinal) methods. They include the Kolmogorov–Smirnov, Kendall, Spearman and a number of others tests (Panovsky and Brier, 1972; Rozhdestvensky and Chebotaryev, 1974; Kendall and Student, 1976; David, 1979). The non-parametric methods are based on analysing the ranked samples and they do not depend on the type of distribution of the initial sample.

The conditions for applying these methods are different and depend, primarily, on the statistical structure of the series in question. The main differences between the parametric methods and the non-parametric is a requirement for a priori assumptions about the family of the type of distribution function for the samples compared and about equal variance, when assessing the stationarity of mean values, and about equal means, when assessing the stationarity variance. This condition provides a considerable advantage for the non-parametric methods as compared to the parametric. However for non-parametric methods the maximum efficiency of estimation is reached in the case of random and normally distributed samples. Thus, each group of the methods has its advantages and disadvantages, and it is unreasonable to give preference to one or other of these groups. Therefore, for more reliable analyses, the entire suite of known methods has been used taking into account the individual features of each of them.

In the initial stage of stationarity analysis of some of the runoff time series, the stability of the variance was checked against

Fisher's criterion. For this purpose the runoff series $(y_1, y_2, \ldots y_n)$ was divided into two groups of observations, depending on the degree of alteration suspected, if this was known, or into two equal parts if there appeared to be no changes. The dates of any changes that might be recognized were: the year of start of reservoir filling, the date that and abstraction commenced and so on.

For the two groups of observations of more than 10 years in length estimates were made of the means and variances:

$$\bar{y}_1 = \frac{1}{m_1} \sum_{i=1}^{m_1} y_i; \quad \bar{y}_2 = \frac{1}{m_2} \sum_{i=m_1+1}^{n} y_i; \tag{2.9}$$

$$\sigma_1^2 = \frac{1}{m_1 - 1} \sum_{i=1}^{m_1} (y_i - \bar{y}_1)^2; \quad \sigma_2^2 = \frac{1}{m_2 - 1} \sum_{i=m_1+1}^{n} (y_i - \bar{y}_2)^2; \tag{2.10}$$

where m_1 and m_2 are the number of observations; $n = m_1 + m_2$ is the length of initial series. Fisher's statistic was calculated as the ratio of the variances σ_1^2/σ_2^2 (the greater variance is in the nominator) (Rozhdestvensky, 1988):

$$F_1 = \sigma_1^2/\sigma_2^2 \tag{2.11}$$

The stationarity of the variance of the initial series was checked against the relationship between the calculated F_1 and critical F_α values of these statistics, taking into account the accepted significance level α. The critical value F_α was determined from tables constructed from the Fisher distribution, depending on m_1 and m_2. Where $F_1 > F_\alpha$ and with the probability $P = 1 - \alpha$ the conclusion was that the variance of the initial series was non-stationary. Similarly an estimate was made of the stationarity of the mean initial series by means of the Student's t statistic. This criterion was determined from:

$$t = \left[\frac{(y_1 - y_2)}{\sqrt{m_1 \sigma_1^2 + m_2 \sigma_2^2}} \right] \cdot \sqrt{\frac{m_1 \cdot m_2(n-2)}{m_1 + m_2}} \tag{2.12}$$

The calculated values of t were compared with their critical values t_α that were taken from tables. Where $t > t_\alpha$ and with the probability $P = 1 - \alpha$, the conclusion was that the mean value of the initial series was non-stationary.

To assess stationarity and trends in runoff series, use was made of criteria developed by Kolmogorov–Smirnov, Kendall, Spearman, the technique devised by the World Meteorological Organization (WMO), and the test proposed by Polyak (estimating the significance of linear trend parameters).

The Kolmogorov–Smirnov statistic was estimated by:

$$d = max/P_1 - P_2/, \tag{2.13}$$

where P_1 and P_2 are the empirical probabilities of the two sets of observations that were calculated as

$$P_i = m_i/(n + 1) \tag{2.14}$$

where P_i is the empirical probability of ith element ranked by increase; m_i is its ordinal number.

As earlier, the calculated values of statistics d were compared to their critical values d_α taken from the table depending on the level of significance and the length of the series n.

Kendall and Spearman's (Kendall and Student, 1976) tests are very useful non-parametric methods for analysing the stationarity of mean values. The main advantage of these tests is their efficiency amongst the different non-parametric tests.

The Kendall and Spearman tests are based on calculating estimates of rank correlation coefficients. The Kendall rank correlation coefficient was determined by:

$$r_k = 1 - 4Q/n(n - 1) \tag{2.15}$$

where rank $Q = \sum_{i<j}^{n} H_{ij}$ and indicators H_{ij} were determined as

$$H_{ij} = 1, \text{ if } y_i > y_j$$
$$H_{ij} = 0, \text{ if } y_i < y_j \tag{2.16}$$

It is known that $Mr_k = 0$, and the variance can be determined as

$$Dr_k = 2(2n + 5)/9n(n - 1) \tag{2.17}$$

Alternatively if rank is to be determined in another way then:

$$u = \sum_{i<j}^{n} (i - j)H_{ij} \tag{2.18}$$

and then we obtain Spearman's rank correlation coefficient (Kendall and Student, 1976)

$$r_s = 1 - 12u/n(n - 1) \tag{2.19}$$

For $Mr_s = 0$, and the variance $Dr_s = 1/n - 1$.

The significance of the coefficients r_k and r_s can be determined from the table of Student's t distribution. For time series with $n > 10$, the critical region for the significance level was prescribed by:

$$|r| \cdot \sqrt{\frac{n-2}{1-r}} > t_\alpha \cdot (n - 2) \tag{2.20}$$

where $t_\alpha(n - 2)$ is the quantile of the t distribution with $(n - 2)$ degrees of freedom.

The stationarity of estimates for mean values depends on the availability or unavailability of trend in the structure of the series, i.e., a regular shift through the series. In the analyses of long-term variations in runoff series, the appearance of a linear trend would mean a significant tendency to increase or decrease the values in the series within a given time span and this would be as a linear dependence. To test series for the presence of a linear trend, it is better to use non-parametric tests based on the analysis of the significance of trend parameters. To check for the presence of a linear trend, use was made of the WMO method and the Polyak's technique (estimating the significance of linear trend parameters).

The WMO technique is based on calculating the following statistics (Semyenov, 1986):

$$y = 3\sqrt{\frac{mQ}{n+2.5}} * P \qquad (2.21)$$

where $mQ = 1/4n(n-1)$, and probability

$$P = \frac{\sum_{i=1}^{n-1} m_i - m_Q}{mQ} \qquad (2.22)$$

Here m_i is the number of terms in the series, and n is the length of the series. The existence of a trend was accepted if $y > 95\%$. Polyak's technique (Polyak, 1975a, b) was applied to quantitatively estimate the stationarity of a mean value and to test for the presence of a linear trend. By this technique, linear regressions were calculated of the following form:

$$y(t) = a_0 + a_1 t \qquad (2.23)$$

where $y(t)$ and t are the magnitudes of the observed value, and its ordinal number, and a_0 and a_1 are the regression coefficients.

Equation 2.23 shows the average intensity of the runoff variations with time. The significance of the regression coefficients shows what significance can be attached to the calculated linear trend.

To accept the hypothesis of a linear trend in a time series (Polyak's technique), it is necessary to fulfil the following conditions:

(1) $\bar{\sigma}^2 < \sigma^2$;

(2) $/a_1/ > 2\sigma_{a1}^2$ for the 5% significance level.

Here σ^2 is the variance of the observed values from an average value; $\bar{\sigma}^2$ is the variance of the departure of observed values from the trend line. It was determined as:

$$\bar{\sigma}^2 = \frac{1}{n-2} \sum_{i=1}^{n} \left[y_i - a_0 - a_1 \left(i - \frac{n-1}{2} \right) \right]^2 \qquad (2.24)$$

and σ_{a1}^2 is the variance of the regression coefficient a_1 that can be determined as follows:

$$\sigma_{a_1}^2 = \frac{12\sigma^2}{n(n-1)} \qquad (2.25)$$

In the case where these conditions were not fulfilled, the linear trend is insignificant. The above non-parametric methods have a maximum efficiency in the case of normally distributed samples (white noise). However runoff series as a rule do not meet this condition. Therefore, in estimating a linear trend, and consequently reaching a conclusion about the stationarity of the initial series is determined by the extent that the initial sampling differs from white noise. This assumption is very difficult to make, therefore, the final decision has to be made on the basis of the analysis of the results obtained by all the different methods used.

2.5.2 An assessment of cyclic variations

To study the statistical structure of long-term variations in natural processes, correlation and spectral analyses are often used. The analysis and recognition of spectra reveals the periods of cyclicity (in a different way, an irregular periodicity) in the processes being studied. There are a few ways of calculating spectral functions from observational data. The most convenient and widely applied is the method for estimating the spectral density by transforming the correlation function by the Fourier method (Panovsky and Brier, 1972). Not considering the correlation and spectral theories in detail, we can only mention that the time series spectrum shows the contribution of variations with different frequencies to the variance of the time series. From the spectrum characteristics it becomes clear where the greater part of variance is concentrated in spite of individual sharply expressed extremes, and it is possible to qualitatively specify the nature of the process under examination.

A runoff series cannot be considered as a succession of independent random values, one reason being that solitary wet years or dry years are observed less frequently than can be expected in a purely random independent sample. Therefore, smoothing or filtering of the initial series allows the character of the variations in long-term runoff to be revealed and tendencies towards the occurrence of long series of dry and wet years can be observed.

Many studies have recently appeared that investigate the occurrence of wet and dry years and the series they form, as well as outlining the description of the methods applied to them (Kaisl, 1972; Rozhdestvensky and Chebotaryev, 1974; Ratkovitch, 1976). To smooth the series studied and to reveal in them the periods with differences, use is made of:

- the difference integral curve
- harmonic analysis
- the flow duration curve.

Use of the difference integral curve is widespread in hydrology (Voronchuk, 1970). Harmonic analysis, in spite of its simplicity and obviousness, is used infrequently in hydrological studies, and in particular to study the cyclic variations and reveal the periods with different wet and dry periods (Kaisl, 1972). The flow duration curve of long-term runoff is based on averaging runoff values at time-spans of variable length for each phase of the record (Afanasjyev, 1967). As a result of this averaging procedure it is possible to build the curve obtained by summing those elements that fall within the current phase of a local increase or decrease. Building this curve is a rather simple and efficient way of smoothing long-term variations in runoff time series. This way of smoothing overcomes many of the disadvantages inherent in other methods. In particular, unlike the method for constructing the dynamic average on the basis of binomial coefficients, it is free of the choice of averaging period and does not depend on the subjective choice of the order of the harmonic in harmonic analysis.

In the present studies, wide use was made of difference integral runoff curves, although in a number of cases the analysis of variations was based on harmonic analysis and genetic curves.

The analysis of synchronism and asynchronism of long-term runoff variations is based on the time coincidence of runoff series and on estimation of their quantitative characteristics by using the correlation and spectral analyses.

An assessment of the probability of occurrence of groups of wet or dry years was made by using a specially developed technique (Babkin and Serkov, 1974). The essence of this technique is to reveal the three groups of years (runoff below normal, normal, above normal) in estimating the number of years of observations referring to these groups, and in estimating the probabilities of the appearance of year groups with different levels of runoff of τ years in duration, $P_{i\tau}$, by the formula:

$$P_{i\tau} = 100\tau P_{\varphi i}(1 - P_{ii})P_{ii}^{(\tau-1)} \qquad (2.26)$$

where P_{ii} is the diagonal elements of the probability matrix of changing years with the corresponding level of runoff (low, average, high); and $P_{\varphi i}$ is the final probabilities of the groups of years of different wetness.

The average duration of these different groups of years was estimated by the expression:

$$\bar{\tau} = 1/(1 - p_{ii}) \qquad (2.27)$$

The structure of the time series describing the long-term runoff variations for an area depends on the nature of that area and the degree of generalization. Thus, the runoff time series characterising a river basin can have a distinct cyclic structure with long dry and wet periods, an inter-series correlation, and it can be strongly skewed. For a time series describing long-term runoff variations for an area including a few river basins, independent random variations are more typical, as a rule, without intra-series correlation within the series. The larger the region, the greater is the probability that the time series will have a random independent structure.

3 Methods for assessing and forecasting global water use and water availability

3.1 THE HUMAN IMPACT ON WATER RESOURCES

In order to obtain a reliable and detailed assessment of the present water resources and the water availability for a basin or region and to determine future needs, it is necessary to assess the changes caused by human activities in the basins concerned.

Ever since *Homo sapiens* appeared on this planet, rivers have had great importance in developing and moulding human society itself. This is due to the fact that river systems are so widespread a feature, and they are so easy to use. They provide humans with potable water and at the same time they carry away waste. At the present time in the most populated regions of the Earth, there is no large river system where the hydrology has not been disturbed by human activities.

Within large river basins, many factors can simultaneously affect runoff, such as abstractions for irrigation and other agricultural purposes, as well as for industrial and municipal water supply. There can be soil drainage, deforestation and afforestation, agrosilviculture, urbanization, opencast mining, and mine water pumping, stream bank straightening, and excavation of sand and gravel from river channels and other activities. There may be large-scale diversions of flow from one basin to another and river flow control by reservoir operation.

These factors can be placed in four groups:

1. Factors mainly affecting runoff because of abstractions from water sources (rivers, lakes, reservoirs, aquifers); the use of these waters and their discharge back into the river network or areas of internal drainage. These may include intakes for irrigation and water supply for the needs of population and for industry, and diversions beyond the limits of the basin.
2. Factors influencing the hydrological regime and water resources because of channel changes (construction of reservoirs and ponds, embankments and straightening of channels).
3. Factors changing the conditions of runoff and other water balance components on the surface of basins (agricultural practices, drainage of waterlogged soils, deforestation, afforestation and urbanization).
4. Factors influencing the hydrological regime through climate change at the global and regional scales (changes in atmospheric gas composition, altering the hydrological cycle because of large-scale water management measures).

The way the effects of the first group of factors are determined is by the volume of water used (volume abstracted, water consumption, discharge and diversions) relative to the natural runoff or lake volume. Depending on these relationships, there can be pronounced effects on small and medium-size basins and also on large river system, or the effects may be small. It can be noted that the volume of discharge and the diversions from it are the most important characteristics for estimating the threat of pollution and changes in water quality.

The second group factors include reservoirs whose effects on runoff are especially large. Constructing enormous single reservoirs and locating several in cascades can radically change a river's hydrology, including its water quality and the total water resources of the basin especially in regions with a hot dry climate. Their effects increase with increasing reservoir volume in relation to the total runoff and to the total additional water surface area. Large reservoirs have considerably transformed the hydrology of such river systems as the Mississippi and Missouri, Yenisey and Angara, Volga, Nile and Dnieper. Construction of smaller reservoirs and ponds in large numbers, particularly in arid and semi-arid regions usually exerts a noticeable effect on runoff in small and medium rivers.

The influence of such factors as embankments, channel straightening and excavation is, as a rule, of local character and produces the greatest effect on the hydrology of individual river reaches and on water quality. However, in arid regions, these measures can produce a profound effect on the water balance of river basins and on the annual runoff in downstream reaches of medium-sized rivers (Shiklomanov, 1989).

The third group factors is concerned with changes in land use, including deforestation and urbanization, as these changes strongly affect river runoff. For example, the results of studies by L'vovich (1986) show that if an urban area increases from 0% to 60% in the European territory of Russia (ETR), the surface runoff can rise by 250% and the total (surface and ground) runoff by up to 120%. The same conclusions have been reached in studies of urbanization in humid equatorial climates (Rahman, 1989).

The urban effect is mainly of a local character however, because it is restricted by the city boundary or similar physical limits (e.g. roads) and the total urbanized area of the Earth is not very large. In certain countries, urban areas cover 6–8% of the territory. According to some estimates, by the year 2000 urban areas in the USA, Germany, The Netherlands and in the former Soviet Union will comprise 2.4%, 6.5%, 12–15% and 0.75%, respectively, of the total area of these countries (Koupriyanov, 1977).

Rather large changes in runoff can be observed in mining regions. For example, in western Upper Silesia in Poland, one of the largest European industrial regions, industrial and municipal waste and mine water discharges increase runoff coefficients from 0.30 to 0.70 (Holda et al., 1989). The results of SHI studies show that in the central part of the ETR (Dobroumov and Ustyuzhanin, 1980), due to the diversion of used ground waters, annual runoff on individual rivers can increase by 10–20%. However, the influence on river water content diminishes as the distance from the place of discharge increases while at the same time, water resources can be redistributed over the basin. In regions, where water-table levels are much lower, the total runoff can fall by up to 15–25%, and no noticeable changes are observed at downstream gauging stations.

Drainage measures can also produce different effects on renewable water resources. Results from the extensive literature on the hydrological role of wet lands and their effects show that on small catchments the impact of drainage works vary depending on the hydrogeological conditions of the catchments, the type and area of swamp and the method of drainage, and its duration. On large and medium hydrologically closed catchments that drain all the categories of ground waters, during the first years after draining, as a rule, annual and seasonal runoff slightly increase due to a decreasing total evaporation and depletion of the ground water storage. Subsequently, with the intensive cultivation of the drained areas the runoff regime becomes more uniform. The annual runoff approaches its initial value and can even slightly decrease (Shiklomanov, 1989).

Studies of the effects of deforestation and subsequent afforestation are based on long-time series, usually covering tens of years, as forest grows. As SHI studies show, in the forest zone of ETR, the total water use by forests is closely related to the biological productivity, primarily to the annual growth of trunk wood and the tree-top phytomass volume (Krestovsky, 1986). Similar inferences have been derived from the results obtained in England and North Germany from studies of planted forest, which have been widespread in Europe in recent years (Robinson, 1989).

For a hundred years after felling the trees and during the subsequent forest regeneration annual runoff totals can be 10–15% less than the normal (Shiklomanov, 1989). However, over large areas trees are usually felled gradually with an intensity rarely reaching 5% and mostly comprising 0.7–2% of the area of a river basin. In addition, the species and age structures of forests are very diverse, and this smoothes the dynamics of runoff change. Open sites in catchments are far from similar, therefore different combinations of forests and fields can give very diverse combinations of evaporation and runoff from a forest and a field (Ivanov and Penkova, 1987). From SHI estimates, for large river basins of the forest zone in the former USSR, the annual runoff change, due to the intensive forest management measured over a period of 100 to 140 years, can be both positive and negative and not above 4–6% (Krestovsky, 1986; Shiklomanov, 1989).

The largest changes over river basins can be observed as the land use intensifies: ploughing, field management measures, increasing crop yield etc. In the world's crop belts, the fraction of the area being cultivated can reach 70–80% of the total area. Detailed assessments of the possible influence of field management measures on the hydrology of large river basins have been made in Russia (Vodogretsky, 1979; Shiklomanov, 1989). They show that ploughing and cultivation and field management of 60–70% of the total catchment area can lead to a decrease in the annual runoff of 3–5% (e.g. Dnieper, Don, Ural, Irtysh with catchment areas of 200 000 to 500 000 km^2). These values refer to field management practices in the continental climates of the forest–steppe and steppe zones of Eurasia where the runoff maximum occurs in the spring due to snow melt.

Field management can influence the hydrological characteristics of river catchments differently depending on the physiographic conditions. For instance, from American research in Iowa, where the annual precipitation is 800 mm with a summer maximum, the change from pasture with a deep-root system to ploughed fields causes an increase in annual runoff of up to 30% due to smaller losses in evaporation (Shiklomanov, 1976). Slightly different data are available for other regions of the USA. For example Jones (1966) obtained data for two small catchments in Pennsylvania, where the intensive field management and anti-erosion measures have not led to noticeable changes in the annual volume of runoff compared to the control catchments. According to studies on four experimental catchments in the northern Appalachians from 1939 to 1967 (Ricca et al., 1970) due to the development and the increase in crop yield, the annual and seasonal runoffs decreased slightly with time, however subsequently this decrease stopped.

In many regions of the world, there is a very real problem of the effects of increased crop yield on total moisture and river water content due to the observed and potential growth in productivity caused by applying chemicals, crop selection of and the introduction of advanced technologies. A large number of studies from different countries treat the yield effects on the processes and volumes of water used by plants (and thus on water resources). A review of these studies is given by Rakhmanov (1973) and in Shiklomanov (1989). The water use/biomass yield relationship is considered in detail by Alpatjyev (1974).

Results from numerous investigations show that progress in the application of land use technologies and growth in yield is accompanied by a reduction in specific water use and less plant/water use as well as less evaporation from the soil (Samarakoon and Gifford, 1996; Khaydarova et al., 1998; Meinzer and Zhu, 1998; Singh et al., 1998; Horie et al., 2000; Penkova, 2000). As a result, the total water used by this type of agriculture remains the same or has increased only slightly. These conclusions are confirmed by the results of analysing runoff changes. For example, Rakhmanov's (1973) studies on changes in runoff in the Don, because of in-field agricultural practices and increasing crop yield (crops cover more than 60% of the area), showed a more than a three-fold growth in yield between 1880 and 1970 and that this had exerted no noticeable effect on the basin's water resources.

Many researchers consider that assessing the changes in evaporation and runoff that result from field management practices and increasing crop yield is a very complicated task. The inferences drawn are to be thoroughly and objectively analysed. To some extent, so far these problems still remain unsolved, especially as regard to specific physiographic conditions and certain field management methods. The analysis of the results obtained in different countries allows only approximate conclusions to be drawn about the effects of increasing yield on the catchment water balance. Studies show that, other things being equal, the growth of biological productivity with low and average initial crop yields leads to a considerable augmentation of water expenditure due to evapotranspiration. With a high initial yield, a further increase in productivity exerts no practical effect on it. However, with reduced moisture, the water consumption by plants is much less than is needed, and evapotranspiration is determined by soil moisture storage. In areas where the ground water table is deep the direct effect of growth in yield on runoff can be practically neglected (Khaydarova et al., 1998; Singh et al., 1998; Penkova, 2000).

In general, in regions with intensive crop production without irrigation, even a few-fold increase in the crop yield would entail no considerable rise in the total evaporation and no reduction of river water content. This refers to large river systems and regions, where intensively used land occupies a small fraction of the total area.

The fourth group of factors is related to anthropogenic effects on the atmosphere. Various processes cause these effects, with considerable differences in intensity and area of coverage. They can result in changes to local and global climate characteristics and alterations in the hydrological regime.

Changes in local climate and meteorological conditions can occur as a result of human impacts on vegetation, urbanization, the construction of large reservoirs and the expansion of irrigated and drained areas. Where these measures are undertaken, there are changes in the Earth's surface reflectivity (albedo), in evaporation and soil moisture, aerodynamic roughness and in other conditions relevant to the meteorological regime, which can lead to changes in the water balance and hydrology of small and sometimes medium-size rivers. However, there are no noticeable changes in the water resources of large river basins.

Both the global water cycle and the climate can change as a result of large-scale fresh water use. Adding moisture to the atmosphere by evaporation from large irrigated areas and from reservoirs stimulates more precipitation, which in turn leads to the formation of extra water resources. Of course, this effect could take place within the water balance for continents with a water cycle coefficient[1] of more than one. As some studies, show (Kalinin, 1969; Korzun, 1974b; Shiklomanov, 1989), these are, as a rule, regions above 2 to 3 million km^2 in area.

The most important aspect of contemporary human impacts on climate is the context of economic activity and, primarily, the ever-increasing consumption of fossil fuel in changing the composition of atmospheric gases, namely in increasing the concentration of carbon dioxide and the so-called trace gases (freons, nitrogen oxides, methane etc.). The last necessarily leads to an air temperature rise, changes in global circulation patterns, alterations in precipitation and evaporation regimes, leading subsequently, to alterations in water resources and ultimately to their use. The scale of these phenomena depends on the expected air temperature rise and the likely changes in precipitation. According to the recent assessments of the Intergovernmental Panel on Climate Change (IPCC, 1995, 2001, 2001a), before the end of the twenty-first century the concentration of carbon dioxide in the Earth's atmosphere is expected to double and the global temperature to increase by between 1 °C and 44.5 °C. This scale of warming can result in a very serious changes in the water balance, with severe impacts on water resources in many countries. However, for the next 20 to 30 years when assessing water resources and their use at the global scale, one can assume that the climate is stationary. Moreover, this is justified by the fact that, so far, there are no reliable data on the predicted changes in the climate (especially precipitation) for the different physiographic zones and regions of the Earth, such that could be used to assess future water resources.

Analysis of the different anthropogenic factors influencing the hydrological regime leads to conclusions about the need to consider the role of all factors related to water abstractions from water

---

1 The water cycle coefficient for any area is determined by the ratio of total precipitation falling on that area to the precipitation caused by vapour from outside.

bodies, including the control of runoff. It is necessary to do so in order to estimate the human impact on water resources at the global scale. The factors causing a decrease in surface runoff and runoff from ground water are widely distributed. They are capable of exerting an especially pronounced effect on the state of water resources over large regions. In this connection the present Monograph treats the changes in global use of fresh water for public water supply, industrial production and for agriculture, as well as water losses due to evaporation from reservoirs.

All estimates for the future are made on the basis of a stationary climate. However, in Chapter 12 data are given on the possible human impacts on the hydrological regime and on water resources due to global warming.

3.2 WATER USE

A common feature of changes in water use during recent decades is the considerable growth in use in most regions of the world. However, in the last few years, this process has needed greater capital investment to be directed towards economizing in the use of water and in environmental protection. These factors have to be taken into account in forecasts. Trends in future water use not only depend on the expected increases in demand from the different sectors such as industry and the domestic sector, but also on the effectiveness of the different means of reducing water demand (Voropayev *et al.*, 1989; Penkova and Shiklomanov, 1998; Penkova, 2000a).

Analyses of recent trends show, that under conditions of an increasing deficit in water resources, considerable changes have been observed in world water management practices. These are associated, first of all, with the development of the price of water resources and the need to preventing environmental degradation. These trends are observed in changes in water use across the main sectors.

3.2.1 Municipal water use

Municipal water use includes abstraction of water and its treatment and distribution mostly for domestic purposes to cities and towns and to public and private enterprises. The public supply also includes water for industry, which consumes high-quality fresh water from the city water supply system. Much of the domestic consumption in certain countries is for watering lawns and gardens.

The volume of municipal water use depends on the number of people served and the degree to which they are equipped with services and utilities, i.e. the availability or unavailability of pipelines and conduits and a centralized hot-water supply. It also depends

considerably on the climatic conditions. Usually 150–250 l/day per head of water (including 2.5 to 3 litres for drinking water and water for cooking) are considered to be sufficient to meet all personal demands. For operation of municipal enterprises and maintenance of cleanliness and hygiene in cities, additionally 150 to 200 l/day per head are required. Water use above these values is usually due to demand by industry and for garden-watering. In small towns and villages without effective distribution systems, water use is in the region of 75 to 100 l/day per head.

Different countries have developed different design norms for municipal water supply. These usually depend on the extent of the services provided and the climate conditions. For northern countries water supply norms are less, while for southern countries with hot dry climates they are considerably more. In some countries the norms of municipal water supply depend on the population served and on the sphere of activities. For example, in Japan, the water supply standards for small cities with a population below 10 000 are 150–300 l/day per head and these increase to 400–500 l for large cities with a population above 1 million. In addition there may be some variation for water in volume (between 5% and 10%) if there are large demands for water for a particular industry. These norms correspond, in general, to actual water use in many modern cities, where they are 300–600 l/day per head (Zaroubayev, 1976; Shiklomanov and Markova, 1987).

Due to increasing urbanization, and a rising standard of living, together with higher cultural levels, more and more water is abstracted in most countries to meet municipal needs. In Russia, at the beginning of the twentieth century, in the populated areas, piped water consumption was 15 to 30 l/day per head. At the present time, most of the population of Russia uses 300 l/day per head. In the USA, over the period from 1900 to 1970 the urban water supply increased from 100–150 to 400–500 l/day per head (Murray and Reeves, 1972; Water Resources Council, 1968), i.e. by three to four times; while in Western European countries, it grew by more than twice. By the end of the twentieth century, in the industrially developed countries of Europe and North America, the abstraction per capita was expected to increase to 400–1000 l/day. A volume of 1000 l/day per head is assumed to be the maximum urban water supply in a hot climate where a complete range of domestic devices are available and where no restrictions are imposed on water use for swimming pools, garden-watering and car-washing. A figure of 600 l/day would apply in a moderately warm climate.

At the present time, in most cities of the developed world, the water used reaches the prescribed quality. The quality is attained by primary treatment, secondary and even tertiary treatment and because of the costs involved, there is a trend towards economizing in water use and towards reducing abstractions (Flemming and Daniell, 1994; Seckler *et al.*, 1998; Rijsberman, 2000). In developed countries there have been sharp rises in water costs. For example, in the USA, from 1980 to 1990, both for public and private water use, water prices increased, in some cases by two to three times because of inflation, the higher costs of capital building and stricter requirements for protection of the natural environment (Russell and Woodcock, 1993). However, as discussed by Gleick (1993), only the countries with a relatively high per capita income, of at least $US2000 per year, are in a position to take effective measures for reducing abstractions for public services.

In the developing countries of Asia, Africa, and Latin America, the public water service supplies 50 to 100 l/day. Where water resources are under stress, it is usually not more than 10 to 40 l/day per head.

The volume abstracted and the size of the population usually determines the total volume of the water used in public services. Annual values of municipal water supply are given in many national and international publications including Shiklomanov and Markova (1987); WRI (1990, 1996, 2000); Kulshreshtha (1992); Gleick (1993, 1998); and World Bank (1993, 1995). The data published for certain countries, however, are not always comparable because some sources give the total water used by both the urban and rural population (e.g. in the USA, Australia and Brazil), while for others, only figures for the urban water use is provided (e.g. Russia, East European and African countries).

When calculating the water balance in order to determine the volume of waste water and the water used, of great importance are the values of water consumption for public services and water diversion volumes. With an effective sewage system the greater part of the water put into the supply system is returned as waste water (treated or not) to the rivers. A large part of the water consumed consists of water losses due to evaporation, leaks in the water supply and sewerage systems, and that water used for watering gardens, cleaning streets, for recreation areas, and allotments. In hot dry regions, these losses are certainly larger than in cold and humid ones. The water consumed for human use is often insignificant as compared with water losses due to evaporation in these areas.

The loss of water expressed as a percentage of water abstracted depends to a considerable extent on the volume abstracted. In modern cities equipped with public, well-managed and relatively new systems for water supply and sewage disposal, losses can be between 5% and 10% of the total intake water. For small cities with a large number of individual buildings not fully provided with such systems and where abstractions are in the region of 100–150 l/day per head, then losses can reach 40–60% of the water intake. Again, these losses are less in colder areas and higher in warmer ones.

Values for losses vary across a wide range for individual cities, regions and countries. For Russia, for example, losses for urban public services are presently estimated to be between 15% and 20%, but over a large river basin they may be from 10% to 30%. For the USA, on the average, these values are considerably above 20–35% because they include the water used in rural areas. One investigation for 46 settlements in the USA showed that losses in pipe lines vary from 2% to 50% (Khadam et al., 1991). In the industrial countries of Western Europe, water losses in public services is estimated to be between 5% and 30% of the intake water, the latter figure applying in particular to areas where the water supply is distributed in systems more than 100 years old.

In the developed world most towns and cities and many rural areas are now provided with water supply and sewage systems maintained by municipal, public and private bodies. In the future the demand for water is expected to increase, and the losses to decrease. For example, in the former USSR in 1980 these values amounted to 15–20% of the water abstracted; by the end of this century they were expected to decrease to 10–15% (Shiklomanov and Markova, 1987; Shiklomanov, 1988). These conditions have to be taken into account in the forecasts of water use.

3.2.2 Water use by industry and for power production

Water in industry is used for cooling the equipment, mechanisms and instrumentation heated in the production process; for transportation and washing; as a solvent; and in some industries it is part of the composition of the finished product. Some water is used for maintaining the necessary sanitation and for meeting the standards of hygiene in the workshops and in other parts of the different industrial enterprises and for meeting the demands of the working personnel. The largest industrial use of water is in the generation of electricity in thermal and nuclear power stations. Here vast amounts of water are needed for cooling and smaller volumes to feed the boilers.

The volumes of industrial water used differ not only for the type of industry, but they also depend on the technology of the manufacturing process and also on climatic conditions. Next to power production, the principal water users are the chemical and petrochemical industries followed by ferrous and non-ferrous metal, the wood pulp and the paper industry, and in machine-building. For example, in the former USSR in 1980, of the 107 km^3 of water used by industry 66% was for thermal power, and the five sectors of industry mentioned above took as much as 89% of the total industrial water.

The data presented in *The Global 2000 Report* (Barney, 1980) showed that in 1977 in the USA, thermal power generation took 76% of the total water used, in Japan 72%, Australia 60%, Brazil 14%, and in India 11%. The water used for thermal power production also depends on the type of fuel employed.

In studies of water used by industry, use is often made of certain indices of water consumption (e.g. for 1 tonne of finished production, for 1 kW hour or for 1 million roubles, etc.). For example, for mining and enriching 1 tonne of ore between 2 and 4 m^3 of water is used; to produce 1 tonne of cast-iron 40–50 m^3, rolled metal 10–15 m^3, copper 500 m^3, and nickel 4000 m^3. Very much fresh water is used for wood pulp and paper, and for the petroleum industries. For example, to produce 1 tonne of cellulose some 400–500 m^3 of water is required, viscose silk 1000–1100 m^3, synthetic rubber up to 2800 m^3, synthetic fibre and plastics 2500–5000 m^3, capacitor paper up to 6000 m^3. For a thermal power plant with a capacity of 1 million kW between 1.0 and 1.6 km^3 of water is required per year. Even more water is required for nuclear power plants of the same capacity – some need 1.5 times as much, others twice and some 3 to 4 times as much. Vast quantities of water are needed for the thermal and nuclear plants of 3–5 million kW capacity and for the even larger ones which are being designed. For a wood pulp and paper plant producing 500 000 tonnes per year, some 435 million m^3 of fresh water are required, and for an average metallurgical plant about 250 million m^3 are needed (Levin, 1973).

During the past two to three decades, industrial water use has risen sharply, largely because electric power production has grown. The production of synthetic fibres, artificial rubber and plastics has also increased, with a concomitant increase in demand for water.

The nature of industrial water use depends to a very large extent on the type of water supply scheme being used. There are two basic schemes: inflow and circulating. In the first, water is abstracted from the source and after use it is discharged either treated or untreated into the receiving waters. With the circulating system the used water is cooled, treated, and returned to the water supply system for reuse. The fresh water intake for a circulating water supply is small and only sufficient to make up for the water lost in the production process or to periodically replenish water in circulation.

Technological progress in industrial water use, in terms of the rational use of water resources, relies not only on the wider application of circulating water but also on the introduction of dry technology to the production processes. Additionally there has been progress to reduce water use: for example where water is used for cooling, to substitute air cooling. By these and similar means water use can be reduced by 50–70% in the different types of industrial processes (Shiklomanov and Markova, 1987).

The quantity of water actually consumed by industry is usually a small fraction of the intake water. However, it varies considerably depending on the type of industry, the nature of the water supply, the technology involved in the process, and the climatic conditions. In thermal power production, this quantity is about 0.5–3.0% of intake water. In most of the sectors of industry, it is 5–20% and it can reach 30–40% in certain instances. It is obvious that with the inflow water supply system, water consumption,

expressed as a percentage of the intake water, is considerably less than with the circulating system.

The rapid rise in water use by industry is one of the main causes of the growth of water pollution. This growth is explained, first, by the pace of industrial development generally and then by the growth of production of synthetic fibres and petrochemical products, wood pulp and paper; next, by the rapid expansion of thermal power production and the construction of power plants; lastly, by the increasing volume discharged as waste water. In most cases, these discharges are not treated or are only part treated, often resulting in the serious pollution of the receiving waters.

The water used by thermal and nuclear power plants is discharged into rivers and lakes at a temperature some 8–12 °C above the ambient temperature. This disturbs the natural thermal regime of these water bodies significantly, changing many natural processes and raising the profile of the so-called "heat contamination" problem.

The power industry is the second highest user of fresh water after irrigation amongst the different sectors. The *Global 2000 Report* (Barney, 1980) contains data showing that in 1977, the world total industrial water use was 805 km^3, including 502 km^3 for thermal power production. Shiklomanov and Markova (1987) estimated that in 1980, industry and thermal power production used 710 km^3 per year, including water consumption of 62 km^3 or 8.7% of the intake water. About 75% of the industrial water was used in Europe and North America. Analysis of changes in industrial water use over the last 30–40 years points to a considerable growth for the globe as a whole. However, in many developed countries beginning with the 1980s, the demand for water by industry has not increased and has even decreased in some. This trend has been taken into account in estimates for the coming decades. An example of this trend is shown in Fig. 3.1, where in Sweden a more than three-fold decrease in industrial water use (and for other needs) is obvious (Falkenmark, 1977). This change was promoted by legislation that obliged industry to install recirculating systems in order to decrease the cost of water treatment.

Estimates of the future volume of industrial water use should take into account these differing trends. On the one hand, this volume should increase due to the growth of industry and power production. On the other hand, this increase is not directly proportional to industrial growth, because in most developed countries there is a tendency to use recirculating systems, and dry technologies.

In some countries, there is the increasing use of sea water for industrial purposes and power production (e.g. in the USA, Japan and Germany). In the USA as long ago as 1970 almost 20% of industrial water was sea water, and by 2000 one-third of the total volume of industrial water was expected to be sea water. In spite of these tendencies the total world water use by industry seems likely to increase, at least over the next 15 to 20 years. However,

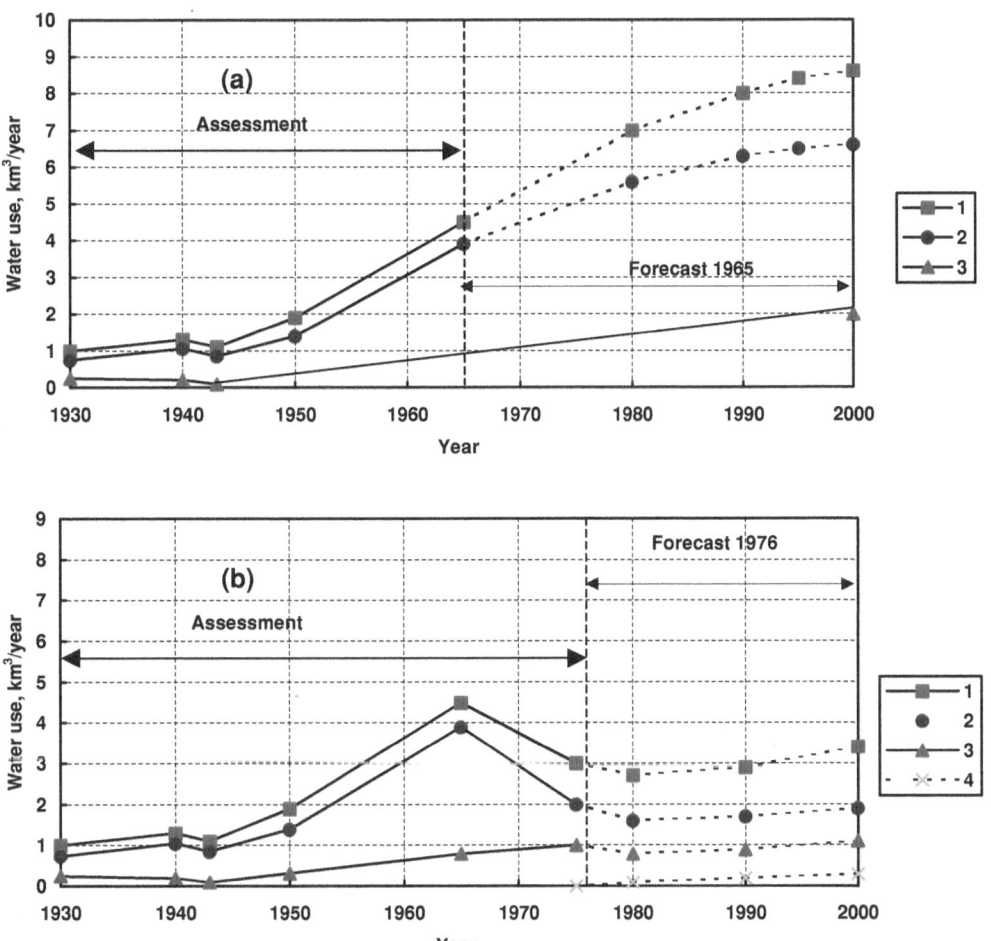

Fig. 3.1 Development of use of water resources in Sweden: (a) 1965; (b) 1975. 1, total water use (km³/year); 2, water use in industry (km³/year); 3, public and domestic water consumption (km³/year); 4, water use for dry land irrigation (km³/year). Solid line, an actual condition, dotted line, a forecast. (From Falkenmark, 1977.)

the rates of increase could be 1.5–3 times less than the increase in the volume of industrial production.

The water losses incurred by industry and during power production can be divided into:

- losses due to the additional evaporation occurring when water is moved from its source to the enterprise concerned along with those taking place in cooling and during the discharge of the warm cooling waters into the river channel
- losses by evaporation inside the enterprise during the manufacturing process
- losses due to its inclusion of water in the finished product.

The second and third groups do not depend on climatic conditions, unlike the first which is likely to be more significant for drier climates nearer the equator rather than for the more humid areas farther away.

3.2.3 Water use by agriculture

Far more water is used by agriculture than by any of the other sectors and most of this water is employed for irrigation. The demands of irrigation can place water resources under stress, particularly in dry years.

Irrigation has been practised for millennia. However, today's extensive irrigation systems were mainly introduced during the twentieth century especially from 1930 onwards. During the twentieth century the world irrigated area has increased more than five times. More than 50% of the modern irrigated lands are concentrated in four countries: China, India, the USA and Pakistan. At present approximately 15% of all the world's cultivated areas are irrigated; production from them amounts to about 36% of the entire agricultural production. This is explained both by higher yields and more intensive crop rotation.

The efficiency of land use, expressed as the cost of production per hectare, depends on the intensity of use and primarily on the quantity of fertilizers applied. Land use efficiency is found to increase with an increasing irrigated area. FAO estimated that this characteristic would apply at least until 2000 (Alexandratos,

1988, 1995). By this time the proportion of the total agricultural production from irrigated lands was expected to exceed that from dry land agriculture.

Before the 1980s almost all developed and developing countries tended to pursue a policy for the intensive development of irrigation. This irrigation caused the growth of the irrigated areas and a guarantee of increased crop production. A feature of irrigation development was its advance into the humid regions. There irrigation was considered as an integral part of the system of agrotechnical methods that provide high and stable yields of all crops independent of the meteorological conditions. At the present time in Europe, there is no country without irrigation. Irrigation is widely practised in Poland, Great Britain, Germany and The Netherlands. Irrigation is developing in Canada. In the northern regions of Europe, the effective control of the soil water regime is aimed at allowing for the combining of irrigation and drainage.

In the 1980s, the rate of increase in irrigated areas dropped sharply. This drop occurred in both developed and developing countries. Gleick (1993, 1998) showed that from 1960 to 1979 the increase of the global irrigated area corresponded to or exceeded the growth in population numbers. Subsequently the area of irrigation per capita began to decrease, i.e. the rate of growth of irrigated land decreased. This was, primarily, because of the very high cost of the construction of irrigation systems and because of soil salinization, the depletion of sources of water, and the problems of environmental protection. In a number of developed countries at the present time the irrigated area has been constant or has even tended to reduce. In countries with a distinct shortage of water and land resources (Japan, Hong Kong), irrigated areas are shrinking. It seems that there is no need for these countries to develop irrigation further. They can meet their needs at present levels of agricultural production or they find it easier to buy produce from other countries.

Considering the problem at the global scale, the development of irrigation is driven by the need for food. With population growth increasing and serious food shortages affecting almost two-thirds of the globe's population, irrigation plays an important role. Therefore, more irrigation can be expected in the future, especially in these countries with a rapid growth of population and sufficient water and land resources. The world total irrigated area is likely to increase although the rate will be less than in the 1970s. Therefore, the volumes of water used for irrigation are also expected to increase.

Irrigation is the main user of water taking about 65% of the total water use at present. Consequently the accuracy of forecasts of the increase in irrigation also determines the accuracy of future global water use, especially for such continents as Asia, Africa and South America, where irrigation comprises about 70–90% of the total water use.

Making a reliable estimate of the long-term increase in the global irrigated area is a difficult task. These data are lacking in many countries and for others it is only very approximate and even contradictory. They are cited in a variety of sources, observations are often for different years and different indicators are used. Fairly reliable information is published by FAO, beginning with the period 1961–4. However, these data are presented annually by individual countries and are far from homogeneous. With different calculations and estimates of changes in water use, it is expedient to use the data on irrigated areas in the FAO Production Yearbook, 1965–94 (FAO, 1995). This yearbook indicates that for 1993, the total world irrigated area was 248 million hectares. This is considerably less than estimates given in the past (Shiklomanov and Markova, 1987).

There are a number of predictions for the development of irrigation over the next 20 or 30 years. For example, during the last 20 years the following values for the world irrigated area (in millions hectares) have been made up to the year 2000: L'vovich (1974) – 500, Batisse (1976) – 420, Kalinin and Shiklomanov (1978) – 420, Ambroggi (1980) – 302, Zonn and Nosenko (1981) – 538, ICID (1982) – 400, and Shiklomanov and Markova (1987) – 317. All these forecasts were based on the high rates of increase that were predicted in the 1970s, and the ambitious national forecasts during that decade. Obviously, they are not reliable in the light of today's trends in the development of irrigation and the accompanying complex of socio-economic and physiographic factors.

The water used for irrigation is determined by the extent of the irrigated area and in addition by the application rate in m^3/ha and the volume of the water returned to the source. Also important are the physiographic conditions, the condition of irrigation system, the application technique and the type and nature of the crop. Data on the water abstracted and diverted for irrigation are available only for certain countries. Their accuracy is generally low especially in terms of the volumes of the return water and these usually have to be estimated.

Information about the volume of water abstracted and the size of the irrigated areas are available for a number of countries and these allow the calculation of the volume of water applied under the different physiographic conditions. After analysis and the generalization of the results, estimates can be made of the volume of water for countries, regions and continents. It is natural that the smallest values are observed in countries farthest from the equator. For instance, in the north of Europe – in Sweden, Germany, the UK, Finland, Belgium and Switzerland – they are from 300 to 4000 m^3/ha, in the south of France, Italy and Spain they vary from 5000 to 6000 and up to 8000–10 000 m^3/ha. Approximately the same values are typical on the average of East European countries (without the former USSR). In this case, the return waters are approximately 20–30% of the intake water. In the USA, water for irrigation is estimated by different authors at 8000–10 000 m³/ha, and return waters as 40–50% of the intake water. Rather high values occurred in the former USSR, about 10 000 m³/ha on average, due to diversions of large amounts of water over distances of hundreds of kilometres from the water source. For the former USSR the values of the return water after irrigation are estimated, on average, at 20–30% of the intake water.

In the countries of Asia, due to the diverse climatic conditions and the type and nature of the crop, the volumes of water are quite different. For example, for Iran and Iraq they amount, on the average, to 11–12 000 m³/ha, India 9–12 000 m³/ha, Indonesia 10 000 m³/ha, Israel and Jordan 5000–6000 m³/ha. An even larger range of values pertains to the countries of Central and South America (from 8000–9000 to 15 000–17 000 m³/ha) and more especially Africa (from 7000 to 20 000–25 000 m³/ha).

When estimating the likely future volumes of water used for irrigation these large variations have to be taken into account. Changes may also be expected because more advanced irrigation systems will be employed with improved watering regimes and application techniques. From this point of view much importance is attached to the measures for changing small and medium-size open canals to pipelines and to lining the beds of the large main canals with concrete. These measures allow an increase in the efficiency of the irrigation network from 0.4–0.6 to 0.8–0.9 (Botzak, 1988). Using more efficient modern methods for applying the water, such as drip irrigation, which help to increase productivity and decrease water use, can make savings.

During recent decades sprinkling systems have been developed which allow watering to be fully mechanized and automated, leading to reductions in water losses. Reducing losses is especially important in terms of conserving water resources, while a more even distribution of water is obtained. In addition, losses are reduced in the channels and pipes of the supply system.

At the present time the world total area irrigated by sprinkling is about 20 million hectares. The largest areas are in the USA, Italy, France and Bulgaria and in other developed countries of Europe. In the USA, for example, the sprinkler-irrigated area amounts to one-third of the total irrigated area – approximately the same fraction as in Russia. In Italy, sprinkler irrigation is used on 25% of all irrigated lands. This technique is being used more and more widely.

In some countries, radically new sprinkling technology has been developed recently and applied, e.g. mist sprinkling, in which water is sprinkled in tiny droplets. As a result, the microclimate of the surface layer is regulated: moisture content increases, air and plant temperatures decrease thus creating favourable conditions for plant development, reducing total evaporation and specific water losses in irrigation by approximately half, with a simultaneous increase in crop yield.

A new technique already used and promising considerable future reduction in water use is drip irrigation, when the water sup-plied to a plant is almost fully used for transpiration; water losses by infiltration and evaporation are reduced to a minimum; the raising of the ground water table and the salinization of irrigated areas are eliminated. An important advantage of this method is the large amount of water saved (25–90%) as compared to surface irrigation techniques, as well as providing an increase in crop yield by 50–100%. Drip irrigation is widely applied in a number of countries (Australia, USA, Israel, Italy, Mexico and Tunisia) to water grapes and vegetables, and in horticulture generally, especially where rows are widely spaced. However, across the world, drip irrigation occupies only 0.7% of all irrigated areas (Postel, 1992). The systems of drip irrigation and mist irrigation are still very expensive (about 3–5 times more expensive than the more usual sprinkler systems). However, they both reduce water use by half and increase the crop production. For the future they are both expected to become more widely used than at present.

The mean values of specific water abstraction and use of water for irrigation obtained at present for the different countries demonstrate how advanced are the irrigation systems being employed. In the future, with irrigation being introduced into new areas, the technological improvements and new technologies designed to increase productivity and to economize in the use of water will undoubtedly be applied more widely. This same target may also be achieved by improving the existing irrigation systems, increasing their efficiency and general effectiveness. These circumstances need to be taken into account when estimating the future volumes of world use of water for irrigation.

In agriculture, in addition to irrigation, water is needed for a number of services and for livestock and to supply rural populations. Water use for agriculture depends mainly on the climatic conditions and on the availability of effective distribution systems and efficient sanitation systems, and it varies from 20–30 to 200–250 l/day per head. The problem of supplying the rural population and livestock with high-quality fresh water is of great importance to many developing countries especially those in arid regions. However, the total volume of water used for these purposes is insignificant as compared to irrigation. In the main data on the water used for agriculture include the water used for irrigation. For example, for the former USSR in 1980 the water supplied to the rural population and for livestock comprised about 7% of the nation's total water use and 5% of the water consumption for irrigation. In future, rural water supply is expected to become a smaller percentage of the total water used for agriculture.

Estimates of the total water abstracted for agriculture are based on the value of the specific volume of water abstracted (in litres per day per head) and the population number. The water consumption determined as a percentage of the water abstracted depends, like municipal water use, primarily, on volumes and climatic conditions. With an abstraction rate of 100–200 l/day per head, water consumption is usually not above 15–30% of the intake water, whereas with smaller volumes, such as 25–50 l/day, it can amount to 70–100%. Therefore assessments of future volumes need to take into account that values of specific water abstraction for agricultural use will increase, with a simultaneous decrease in the relative values of water consumption.

3.2.4 The role of reservoirs as users of fresh water

The construction of large reservoirs can lead to a radical transformation in the space/time- distribution of runoff and to increasing water resources, especially during low flow periods and in dry years. However, reservoirs can also make a considerable contribution to evaporation particularly in arid and semi-arid regions. This leads to a decrease in the total water resources of the regions concerned. Thus reservoirs can be considered one of the large users of fresh water. Necessarily, this role has to be taken into account in estimating total water consumption of these regions but many authors do not do so.

Reservoirs were constructed by the ancient Egyptians and by the Romans. However, as objects on a global scale, they appeared only during the second half of the twentieth century. Over the last 40 years reservoirs have been built with a volume of more than 50 km^3. At present the total capacity of the world's reservoirs is about 6000 km^3, and their total surface area some 500 000 km^2 (see Table 1.7).

In the former USSR, more than 1000 reservoirs have been built with a total effective capacity of more than 500 km^3, a total volume of 1000 km^3, and a surface area above 70 000 km^2 (without including the area of dammed lakes). Reservoirs form cascades on most of the large rivers. The largest cascades are on the Volga–Kama, the Dnieper and the Angara–Yenisey. For instance, 11 large reservoirs have been constructed in the Volga basin. Their total volume is 180 km^3, effective capacity 94 km^3 and surface area 25 000 km^2. In the Dnieper basin, there are six large reservoirs with a total volume of 40 km^3, effective capacity 18 km^3 and surface area 7000 km^2. These cascades have completely changed the hydrology of the river systems. The total effective capacity of US reservoirs, like those of the former USSR, is about 500 km^3, against a total volume of almost 800 km^3. However, the total annual runoff in the US is about half that of the former USSR, so that the extent of control of US rivers is rather high. In Canada, there is approximately the same reservoir volume. These three countries possess about 40% of the total reservoir volume of the entire world (Shiklomanov, 1989).

Large reservoirs with a volume of more than 20 km^3 and an area greater than 1000 km^2 built mainly after 1970 are located in: China, Iran, Iraq, India, Pakistan, Norway, Brazil and Uruguay. In European countries, the largest reservoir volumes are in Spain, Bulgaria, Portugal, and in the Czech and Slovak Republics. In a number of European countries, the total reservoir volume is 15–50% of the total annual runoff. The world's largest reservoirs in volume are the Victoria on the Nile (205 km^3) and the Bratskoye on the Angara (169 km^3); the largest in area are the Volta in Ghana (8500 km^2) and the Kuibyshevskoye in Russia (6500 km^2).

Many reservoirs were constructed in developed countries reservoirs from 1950 to 1970. By that time, runoff was almost fully regulated in many regions. Subsequently the rates of reservoir construction decreased considerably. However, in the countries with copious runoff, they are still high.

Four out of the world's seven largest reservoirs (over 130 km^3) were built during the 1970s and 1980s in developing countries. They are the Victoria and Nasser on the Nile, the Volta on the Volta River and the Kariba on the Zambezi.

In the twenty-first century the world total reservoir volume is estimated to increase to 6800–7000 km^3 with a high rate of reservoir construction in certain regions. This rise will be caused by the increasing role of hydropower, which is particularly important for covering peak loads of power consumption and which cannot be achieved by either thermal or nuclear power plants. In addition, reservoirs can provide water for industrial use, and for cooling thermal and nuclear power plants, as well as to supply water for agriculture. They are the basis for large-scale water management systems providing regulation, as well as protecting populated areas from floods. However, the future types of reservoirs likely to be constructed may change as well as their locations.

Reservoirs will be constructed in regions which are mountainous, in piedmont areas, and in areas where large areas of fertile land suitable for agricultural use do not have to be flooded. Predominantly small and medium-size reservoirs will be built in developed countries.

The construction of reservoirs will result in additional losses by evaporation and thus to a decrease in fresh water resources. The volume of additional losses due to reservoir evaporation can be calculated from the difference between the value of evaporation from an open water surface and from the evaporation from the same area before flooding. As a rule, reservoirs built in arid and semi-arid zones decrease the water resources in the lower part of the basin because of the greater evaporation from the water surface.

For small valley-type reservoirs this decrease is usually insignificant because the additional water surface area is not large and evaporation from the flood plain is close to the evaporation from the water surface. But when vast areas of flood plain covered with intensively evaporating moisture-loving vegetation are flooded, the construction of reservoirs cannot be associated with decreasing water resources, even in regions with an arid climate.

Estimates have been made of the volume of additional losses by evaporation from the world's reservoirs (Shiklomanov and Markova, 1987; Shiklomanov, 1988). In 1980, this figure was about 120 km^3. In the twenty-first century it is expected to increase by a factor of 1.8.

The influence of large reservoirs on evaporation manifests itself not only within the flooded areas but also on the rivers below the reservoirs, due to the changing regime. According to data cited by Shiklomanov (1976), the construction of the cascade of reservoirs in the basin of the Volga led to a decrease in the flood duration and the area flooded in the Volga–Akhtyubinsk flood plain and in the delta. Due to these changes, losses by evaporation in these regions were lowered by 1.4 km^3/year, on average. Similar effects can take place in other basins and are more significant nearer the equator but they are usually small in humid areas.

In contrast, raising ground water levels in areas adjacent to reservoirs usually results in increasing evaporation from the basin. Taking into account these two additional factors, which act in opposite directions and compensate for each other to some extent, one can suppose that neglecting them would not have a pronounced effect on the estimates of the changes in average values of water resources due to reservoir construction. This suggests that only the increased evaporation needs to be considered.

3.3 METHODS FOR ASSESSING AND FORECASTING WATER USE AND AVAILABILITY

To estimate possible long-term changes in water resources and the future availability of water at the global scale, it is necessary to study current water use (abstraction and consumption) and the pattern of use during the twentieth century, in order to forecast water use for the period from 2010 to 2025. In this Monograph, the analysis of temporal and spatial changes in water use together with the assessments of the volumes of the renewable water resources have been made for the globe as a whole, for all continents and large regions, as well as for selected countries, developed and developing.

Natural regions are identified within each continent starting with the physiographic conditions and the level of socio-economic development, based on commonly accepted principles of geographical and economic zoning. As a rule, the boundaries of these regions coincide with international boundaries for the countries in the region. This is because all the statistics necessary for estimating and forecasting water use (population, irrigated area and indices of industrial development) are cited by country rather than by major river basin, or by the basins draining to a particular ocean. Exceptions are the largest countries: the former USSR, the USA and China. The territory of the former USSR is divided into five regions, while the USA and China are composed of two regions. The world as a whole is divided into 26 regions (Fig. 3.2).

For each continent, region or selected country, estimates were made of water abstracted and the consumption in urban public services (municipal water use), industry (including power

production), agriculture (irrigation and agricultural water supply), as well as the water losses by evaporation from reservoirs. These estimates apply to 1900, 1940, 1950, 1970, 1980, 1990; the time when this Monograph was compiled, namely 1995, and (looking to the future) for 2000, 2010 and 2025. This approach demonstrates changes in water use in both space and time for the twentieth century and for the beginning of the twenty-first century.

First, the use of water was assessed for approximately 160 countries, and then the values obtained were generalized for the regions, continents and for the globe as a whole.

These estimates relied on the availability of national data, either actual or calculated, and water use data from individual countries or group of countries. Some of the most complete and reliable data have been published by international organizations and by individual authors, for instance, for the USA, the former USSR, for all the countries of Europe, and for Canada, China, India, Pakistan, Mexico, Cuba, Brazil, Argentina, Egypt, Kenya, South Africa, Japan, Turkey and Australia. For many of these countries, there are also predictions of the use of water, usually up to the year 2000.

The principal studies containing these data are referred to in the references in Chapters 4 to 9. Recently publications have appeared which have included summary data on abstractions for most of the countries for the 1970 to 1980s and for 1990, such as WRI (1996, 2000), Kulshreshtha (1992), Margat 1994, Gleick (1993, 1998) as well as UN (1993), Seckler et al. (1998) and Rijsberman (2000). Detailed analysis of recent and future water supplies have been produced for the Mediterranean and Arab and Asian countries (Margat, 1995; ROSTAS, 1995; UN/DPCSD, 1995a; Pearce, 1996; Shahin, 1996; FAO, 1997; ESCAP, 1998; Ragab, 2001).

When summarizing the published national data, serious difficulties often arise, because the results cited by the different authors are frequently contradictory and are not comparable with each other. Information is given for water that covers only certain sectors and the period to which the data pertain is not stated. Data on water consumption is usually very scarce being unavailable for most countries; consequently these data are not quoted in most publications. There are almost no data on water use before the 1960s and1970s and on forecasts for beyond 2000. Economic data for water and data on the additional evaporation from reservoirs are very rare.

For this Monograph estimates of water use for the current period, the past and the future were made based on specially developed methods which take into account the major factors determining the volumes of water used and changes in these volumes (abstractions and consumption) (Shiklomanov, 1997, 1998, 2000a, b; Penkova and Shiklomanov, 1998; Shiklomanov et al., 2000). The characteristics of water use nationally and regionally were found from knowledge of the following factors: the level of economic and social development, population number,

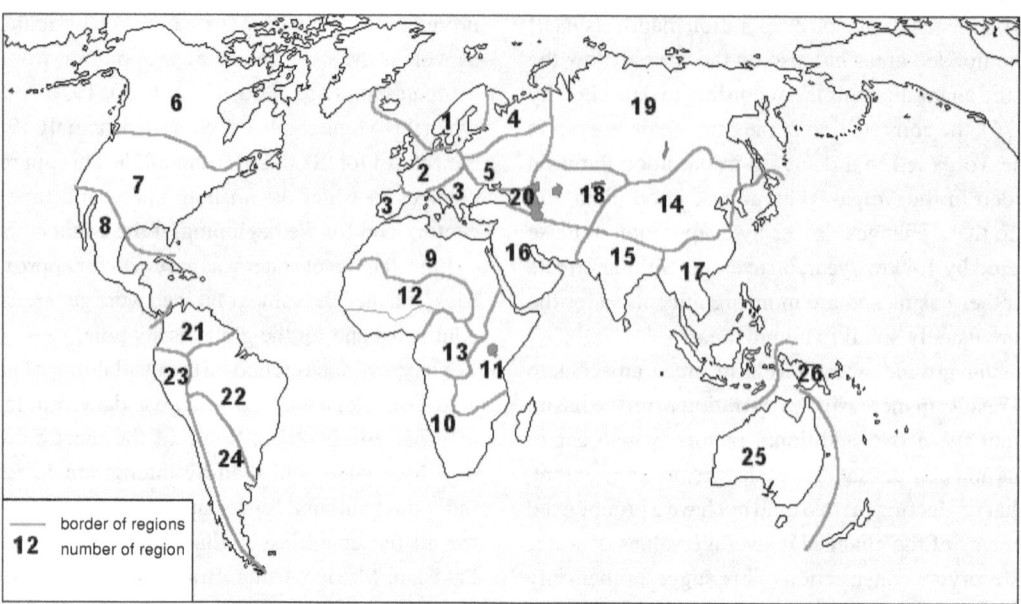

Fig. 3.2 The natural–economic regions of the world. *Europe*: 1, Northern; 2, Central; 3, Southern; 4, North of the European part of the former USSR; 5, South of European part of the former USSR. *North America*: 6, Canada and Alaska (Northern); 7, USA (Central); 8, Central America and Caribbean (Southern). *Africa*: 9, Northern; 10, Southern; 11, East; 12, West; 13, Central. *Asia*: 14, North China and Mongolia; 15, Southern; 16, Western; 17, South East; 18, Central Asia and Kazakhstan; 19, Siberia and Far East of Russia; 20, Transcaucasia. *South America*: 21, Northern; 22, Eastern; 23, Western; 24, Central. *Australia and Oceania*: 25, Australia; 26, Oceania.

physiographic (including climatic) features and the area of the region or country. The combination of these factors was considered in order to determine the volume and structure of water use, its changes and likely future trends. Where no national data existed, estimates were made from the data for countries with reliable data on water use with similar physiographic, social and economic conditions.

Abstractions for urban and rural areas were determined from population totals and their changes (urban and rural) and from per capita use of water. The population data were taken from the statistical yearbooks, for the present period and the future using UN forecasts (UN, 1995) to 2050. Per capita specific abstractions and the different national figures for the consumption of water were taken from the published national data and the data published by international organisations; or where they were absent these figures were determined from country-analogues. Likely changes in the future use of water (Section 3.2) by the urban and rural population were also taken into account.

Assessments of the water used for irrigation were made by analysing the changes in the following characteristics from 1960 to 1994: the population (using UN data), area irrigated by years (using FAO data) including specific values (i.e. hectares per 1000

inhabitants) and the annual gross national product (GNP) expressed in US dollars per capita. The volumes abstracted and consumed were taken from national estimates or determined from country analogues. Water consumption for irrigation (as a percentage of the water intake) varies by country and region in the range of 50% to 95% depending on the technique employed and the physiographic conditions. In a number of cases use was made of estimates of potential evaporation taken from the national sources and from Korzun (1974b) to check the values obtained.

The estimates of water needed for irrigation for 2000, 2010 and 2025 were based mainly on forecasts of the likely expansion of irrigated areas (F_{ir}). These forecasts were based on an approach which made allowance for the major factors determining, on the one hand, irrigation needs, and on the other, the potential for its development in each of the different countries. The first group of factors included: the climate and the features of the history of the development of the countries concerned, population growth as a whole and separately for the urban and rural populations, features of the internal and external markets and the functioning of countries in the world economic system, together with national strategies for water use. The second group included: the level of economic development of each country (expressed as the GNP), the available water resources, the value of irrigated lands and the extent of their cultivation, and any limitations associated with environmental pollution and land degradation.

A forecast was obtained for each country by analysing the long-term (1965–95) trends in the annual gross national product (GNP in US dollars) (IMF, 1994), the irrigated land use (F_{ir}, ha $\times 10^3$), the specific area of irrigated lands (F_{ir} specific, ha/1000 persons) with their extrapolation to 2025. This extrapolation took into account the long-term forecasts of trends in population N and the GNP, and of national forecasts of the development of irrigation as well as the tendencies in irrigation development, the population and the GNP for a 30-year period for groups of countries.

It was found that in most countries, independent of the GNP, there is a relationship between its trend and the rate of change and the character of changes in F_{ir}. Developed industrial countries such as Japan, Germany and Hong Kong, with very limited land resources, were excluded from this relationship; these are countries where the F_{ir} value is quite stable and in practice does not depend on the GNP. Figure 3.3a presents as an example a diagram for Japan with the extrapolation of the area irrigated to the year 2025.

For developed countries with a large GNP, traditionally developed irrigation and a large stock of virgin land, the rate of increased F_{ir} is considerably reduced during periods of economic crisis, because of the lack of sufficient resources to continue the planned development of irrigation. In these countries, during recent decades, there is a distinct reduction in the rate of F_{ir} increase, due either to the need for expensive measures to protect the natural environment (e.g. in Australia, the USA and New Zealand) or to the lack of population growth (Fig. 3.3b, c).

The development of irrigation in industrial–agricultural countries with an average income is determined to a large extent by the level of utilization of the new land available for irrigation. Where such areas are few, there are low rates in the increase of F_{ir} depending on changes in income. Obviously, in this case considerable funds have to be spent on the development of less convenient areas and on the improvement of existing irrigation systems. Figure 3.3d presents the diagram for Morocco.

In developing countries with a high rate of growth of the population and with significant areas with a potential for irrigation, the construction of new irrigation schemes is frequently carried out in accordance with the state plans and these plans will be a major factor in changes in F_{ir} in the future (China, India, Pakistan). In this group of countries, especially high rates of increase in F_{ir} are typical of agricultural countries, where the production obtained in the agricultural sector is one of the principal sources of the national income (e.g. Bolivia, Tanzania). The growth of irrigated areas is characterized by a high sustainability and slows down only during periods of a severe decrease in the GNP (see Fig. 3.3e, f and g for China, Bolivia and Tanzania). For most countries the time series showing the increase in the area of irrigated land is usually too short for forecasting and their use for this purpose can produce large errors. More reliable forecasts can be obtained from information on likely population changes, because population data exist for every country, and there are even forecasts of future population numbers. The more useful information is the value of the specific area irrigated ($F_{ir.spec.}$, in ha/1000 persons). This value is more sustainable and more sensitive to changing GNP-values than the irrigated area in absolute terms (see diagrams in Fig. 3.3). Therefore, for most countries, the $F_{ir.spec.}$ values

were extrapolated to determine the absolute value of the irrigated area.

When forecasts were made of the size of the world's irrigated areas, both the area of land suitable for irrigation and the volume of water resources accessible for use were taken into account. In a number of regions of the world, in particular the "rich" countries, the most important factor that limits the development of irrigation was considered to be the national strategy in the field of water resources. As a rule, such strategies are primarily aimed at meeting the demands of the priority water user – the public water services, with reductions in the use of water for agriculture.

To assess the future volume of water abstracted for irrigation, it is necessary to know the likely values of the application rate per hectare as these may change considerably with time. These rates are assumed to reduce slightly due to the measures undertaken to economize in the use of water. These reductions were assumed to be different for the different regions and countries depending on their level of economic development and physiographic conditions, but they were thought to be in the range of 10% to 25%. The greatest values were expected to be found in the relatively "rich" and fast-developing countries with restricted water resources.

Industrial water use was calculated from the trends in industrial production in the different regions. Use was made of data on industrial water use where it was available, including data for countries with different levels of economic development and physiographic conditions. The current and past use of water for thermal power production and other industries was determined and summed for every region. Water consumption for thermal power was assumed to be 1–4% of the total and 10–40% for other industries, depending on the level of industrial development, the availability of water supply and the various climatic conditions.

Estimates were made to 2025 for every country, taking into account the study carried out by UNIDO (Strzerek and Bowling, 1995). In this study, using forecasts of GNP, estimates are given for the increase in abstractions of industrial water from 1990 to 2025 for the principal countries. Figures are given for high, average and low levels of the global production of electricity, including four different variants of global development (Global Shift, European Renaissance, Global Balance, Global Crisis). For this Monograph, the most optimistic variant of development (Global Balance) was selected, however, for an average level of growth in electricity. For this variant UNIDO considered that abstractions in developed countries will increase by 1.4 and 2.9 times and by 3 to 10 times in developing countries. Similar results are also obtained for the other optimistic scenario – the European Renaissance. When analysing the UNIDO values for the growth of abstraction of industrial water (by all the variants), we came to the conclusion that they were considerably overestimated and not entirely realistic reflections of trends in water use. Consequently we reduced the UNIDO data

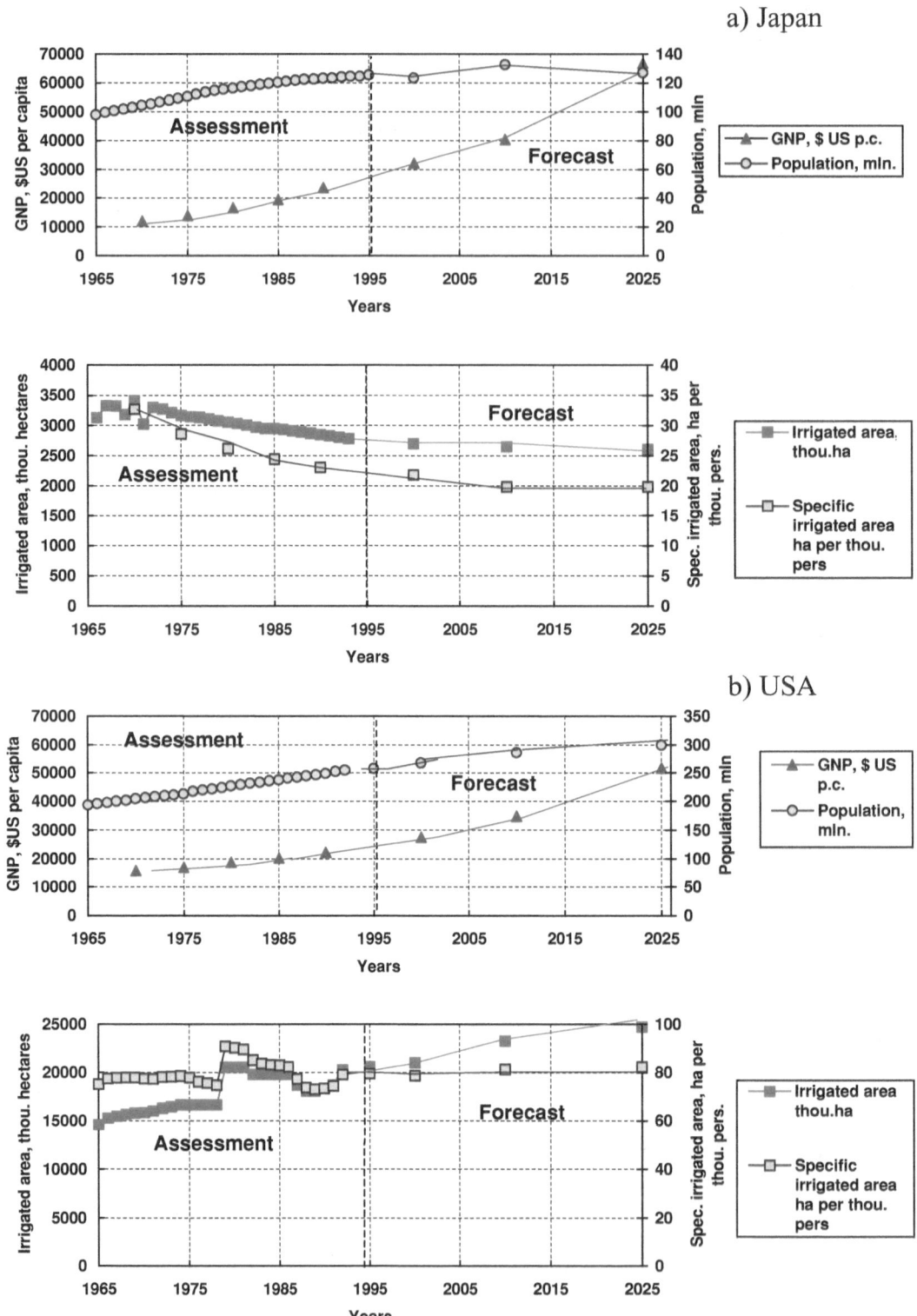

Fig. 3.3 Dynamics of main indicators of irrigation development in individual countries of the world. (a) Japan; (b) USA; (c) New Zealand; (d) Morocco; (e) China; (f) Bolivia; (g) Tanzania.

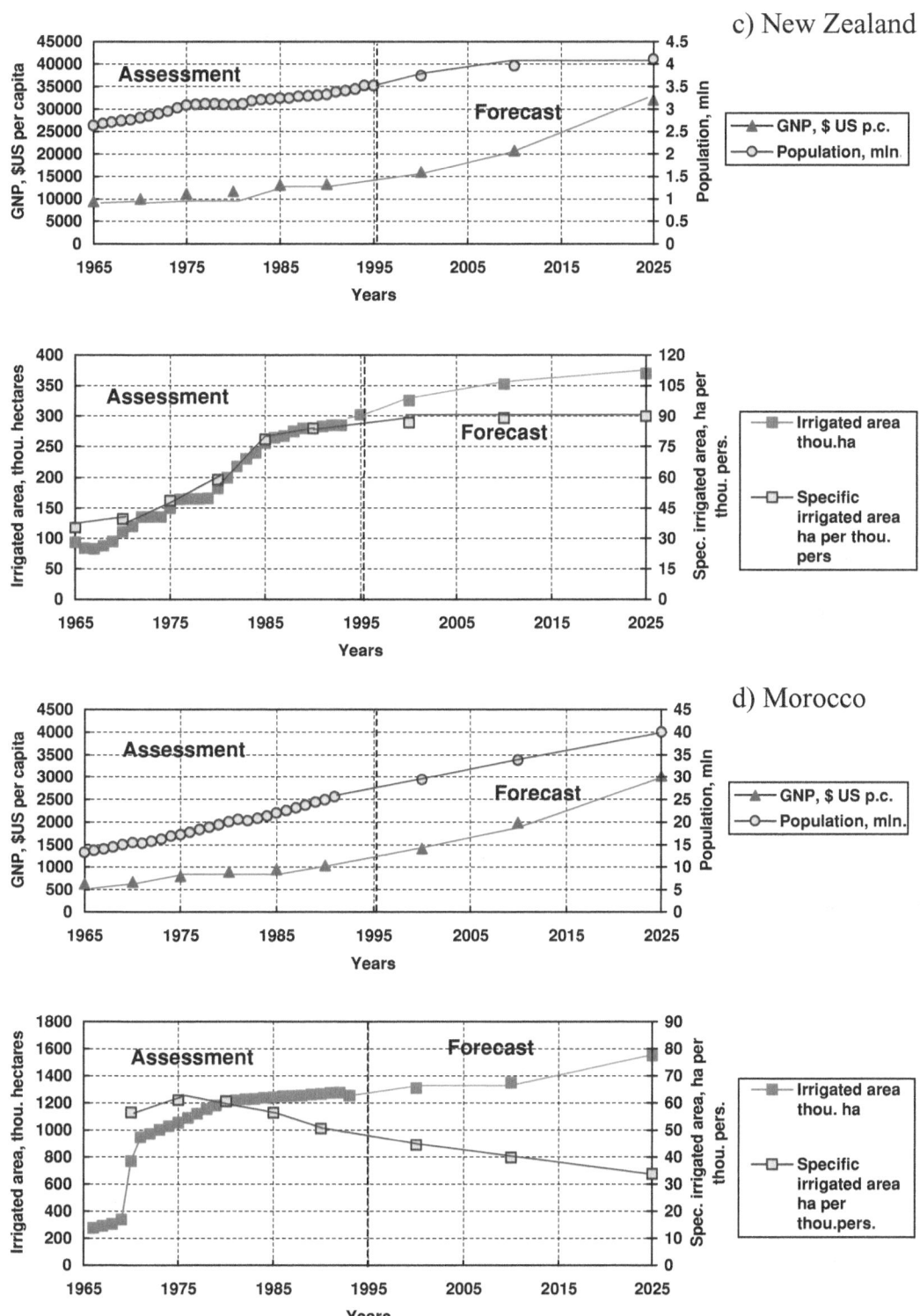

c) New Zealand

d) Morocco

Fig. 3.3 (*cont.*).

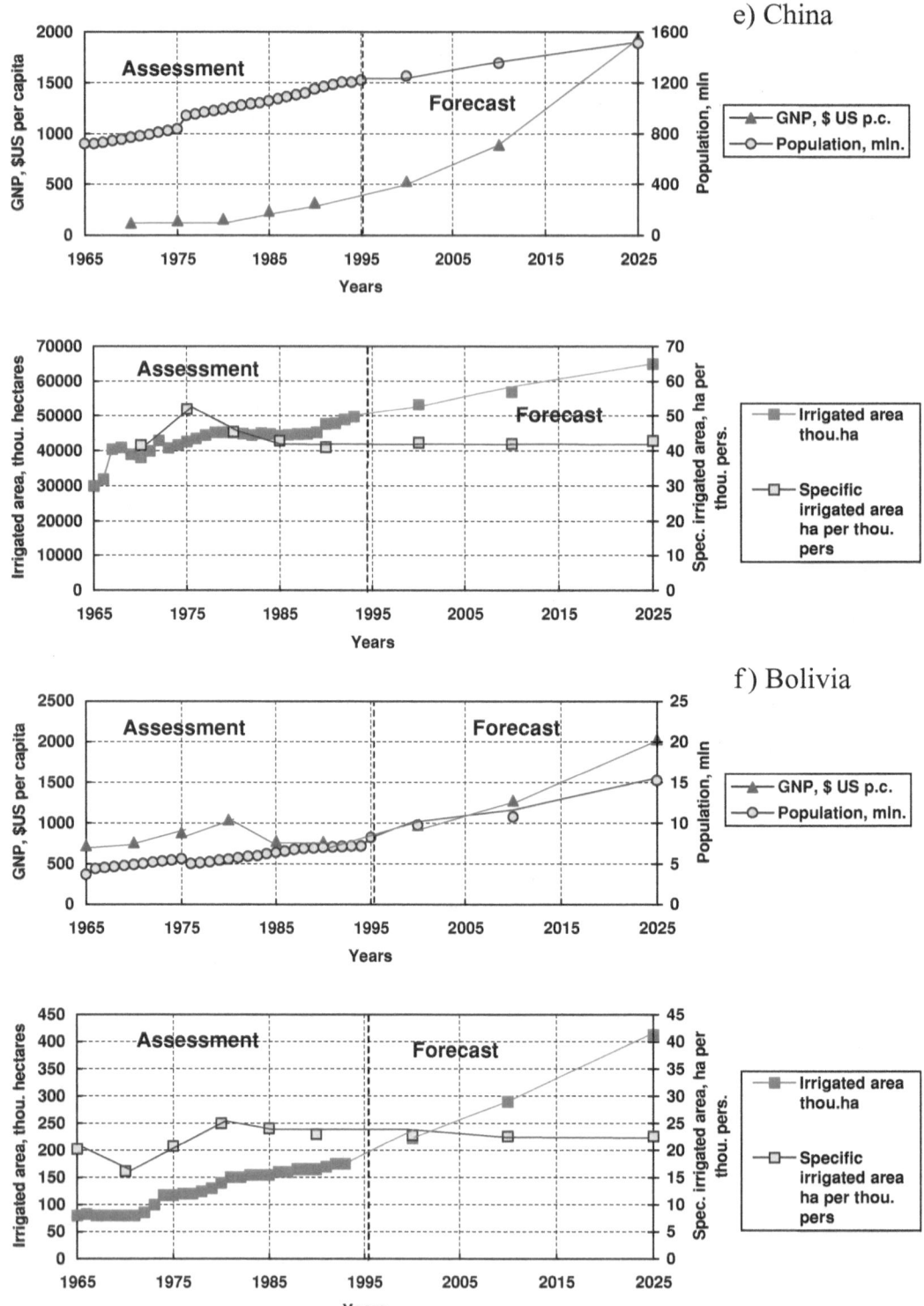

Fig. 3.3 *(cont.).*

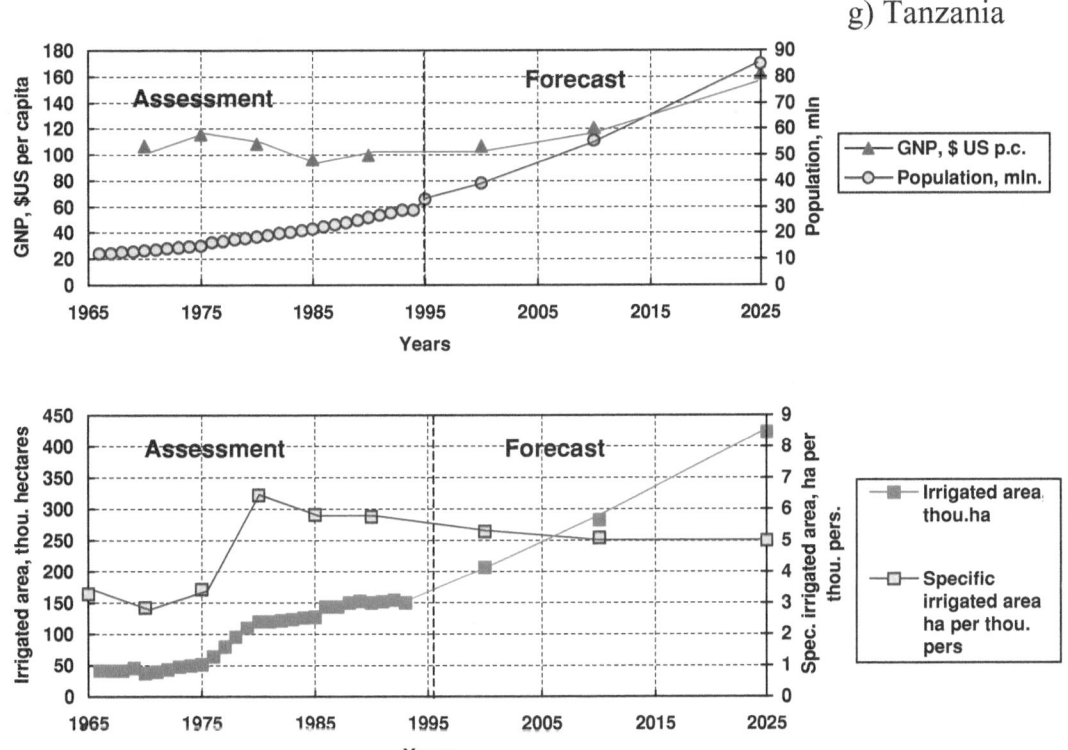

g) Tanzania

Fig. 3.3 (*cont.*).

for 2025 by 25–30% for developing countries and by 40–60% for developed countries.

The values of abstractions of water for industrial use for 2000 and 2010 were interpolated, taking account of national forecasts of GNPs for the Global Balance scenario cited by UNIDO.

The additional water losses due to evaporation from reservoirs were calculated for all the principal reservoirs of the world of more than 5 km³ in volume using the difference between the average evaporation from a water surface and from land. The coefficient K_r, showing the ratio of the additional surface area of the reservoir to the total water surface area, was taken into account in these calculations with $K_r = 0.90$–0.95 for lake reservoirs, and $K_r = 0.60$–0.80 for stream reservoirs. Research on the basins of the Volga and Dnieper, where there are various sizes of lakes and reservoirs, showed that, on average, the value of this coefficient was 0.80 (Shiklomanov, 1979). This value of the coefficient was assumed to be constant for all regions (excluding the reservoirs constructed on lakes). The initial data on reservoirs (area, volume, location, date of construction and other characteristics) were taken from the literature, for example from Korzun (1974b), Voropayev and Vendrov (1979), Gleick (1993, 1998), World Bank (1993, 1995) as well as from other publications.

The average evaporation from the water surface and from the land was determined from the *Atlas* supplemented by data in

Korzun (1974b). The total volume of additional losses by evaporation was calculated for every region by summing the data from every large reservoir (>5 km³) and by increasing this figure by 20%, because the results obtained from the large reservoirs only include about 80% of the total volume and area of water of the world's reservoirs.

The future losses by evaporation from reservoirs were estimated from the trends in reservoir construction and from plans for constructing large reservoirs. In particular, the trends in the number of reservoirs in developed and developing countries that are presented in Section 3.2 were taken into account.

WATER AVAILABILITY

Usually the water availability of a region, country or basin is defined as the quantity of annually renewable water resources per head or per unit area. In the present Monograph, the figures for several kinds of water availability are cited.

POTENTIAL WATER AVAILABILITY

Potential water availability is assessed by dividing the assessed value of the water resources by the population (water availability per head, usually in 10^3 m³ per year). In this case water resources are estimated from the annual mean for a year (season, month) or from the minimum values for these periods.

ACTUAL WATER SUPPLY

The value of the potential water availability does not take into account the water used for public water supplies and when the water use is rather high, it does not reflect the actual situation with water supply. Therefore, for a more objective view this Monograph cites the values of actual water supply. The actual water supply (per head) is estimated by dividing the water resources minus the values of consumption by the number of the population.

Thus the term actual water supply means the residual amount of fresh water that after being used is available to one person, i.e. if all the fresh water resources are consumed for public water supply, then the actual water availability is close to zero. The actual water availability can also be assessed as the average or minimum value of the water resources.

When calculating the water availability of a certain area or region (potential and actual), the problem arises as to what water resources are to be classed in this way: only the resources originating in the area itself or the total taking account the inflow from adjacent areas. In the first case, water availability is underestimated, while in the second water availability is obviously overestimated.

This problem has no easy solution, and obviously has to be solved by special agreements among the riverine countries. As such agreements, treaties or protocols have not been established for most of the countries of the world, water availability here is taken as the values of the local water resources added to half the fresh water inflow from adjacent areas. This is a conventional value, of course; however, it allows the objective study of water availability for the different regions and countries of the world.

4   Water resources, water use and water availability in Europe

4.1 INTRODUCTION

Europe is part of Eurasia, the largest of the continents. Despite a significant length of shared border, Europe and Asia are regarded as separate due to the notable differences between them. Europe is rather diverse and it has several features which are characteristic only of this particular continent. The total area of Europe is 10 460 000 km^2; the continuous land area is 93% of this figure, while the area of the islands is 7%.

Europe is one of the world's most densely populated and developed regions. There are 45 nation–states in Europe, among them the new countries formed after the disintegration of the USSR, Czechoslovakia and Yugoslavia in the mid-1990s. Russia, Kazakhstan and Turkey are in Europe, and also in Asia. The largest countries are Russia, its European part covering about 40% of the continent: France 5% and Sweden 4%. The population of Europe (as of 1995) was about 685 million. The average population density in the region is 65/km^2; the highest density is recorded in The Netherlands (376/km^2); and the lowest in Iceland (2/km^2). Fifteen percent of the population lives in the European part of Russia; 11% in Germany; 8.5% in the UK, while France and Italy have 8.4%; and there are 7.5% in the Ukraine. All the other countries together contain less than 50% of the total population of Europe. Most of the western European countries, such as the Nordic states, Germany and France, have an advanced economy and a high living standard. The former socialist states (Russia, Hungary, Poland, etc.), which started implementing large-scale reforms in the mid-1990s, are countries with transitional economies. Most of the southern European countries (Spain, Italy, etc.) are industrial–agricultural states with a developed raw material base and a rather high industrial potential.

Europe has the densest network of hydrometric stations as compared to the other continents. There are over 6000 hydrological stations in the Europe, on average one station per 1750 km^2. The different levels of economic development across the continent result in differences between the monitoring networks. For example, the countries of Western Europe where the network is most modern and best developed, compared to some of the southern states and northern Russia where network densities are least.

Runoff from Europe is divided between the Atlantic and Arctic Oceans and to areas not connected to the World Ocean such as the Caspian Sea. Drainage to the Atlantic is made up of that from the Baltic Sea basin, the Atlantic coast proper including the North Sea basin; as well as the Black Sea and Azov Sea basins and the Mediterranean Sea basin. On the other hand, due to the diversity of both the natural conditions and socio-economic factors, five regions can be identified in Europe: northern, central and southern Europe, and the northern and southern parts of the European territory of the former Soviet Union (ETSU in its abbreviated form). The boundaries of the first three regions coincide with the borders of states, whereas the boundary between the northern and southern parts of the ETSU is the main watershed in this area.

The total annual runoff from Europe, for the period of 1921 to 1985, is estimated as 2900 km^3/year. Out of this total the runoff to the Arctic Ocean is 622 km^3 (19%), 1638 km^3 (59%) to the Atlantic Ocean (excluding the islands), the runoff from the islands is 329 km^3 (11%), while inland drainage, including the Caspian Basin, is 311 km^3 (11% of the total runoff). The largest rivers are the Volga (250 km^3), Danube (225 km^3), Pechora (137 km^3), Northern Dvina (105 km^3), Rhine (86 km^3) and Rhône (65 km^3), their total runoff accounting for almost 30% of the water resources.

The water resources are evenly distributed between the natural regions of Europe: northern Europe is richest (705 km^3/year), and the southern part of ETSU is the poorest (566 km^3/year).

Extensive human activity has drastically changed the natural pattern of runoff from the continent, particularly in southern and central Europe. In these regions, the human influence on water resources is most pronounced, due to the high density of population, the high industrial output and the large amounts of energy produced.

Since the beginning of the twentieth century, both abstraction and consumption in Europe have been increasing. Presently (1995), abstractions per year amount to 511 km^3 (17%) with the consumption of water amounting to 187 km^3 (6%). By 2025, these volumes are expected to be 619 km^3 (21%), and 217 km^3 (8%

of water resources). Abstractions of water across Europe are expected to grow consistently to 2025. But for consumption, the picture is slightly different: rapid growth is likely in the southern ETSU, whereas in other regions, consumption should increase more slowly following the growth of abstractions.

The pattern of water use, in terms of the human activity or sector, has changed across the continent. At the beginning of the century, agriculture took the largest proportion of water; then, with economic development, the share of industrial and domestic water use increased markedly. Presently (1995), abstractions for industrial purposes average 45%; about 39% is used for agriculture, and not more than 14% of the volume abstracted goes to the domestic sector. A minor share (about 2–3%) is accounted for by the construction industry and by evaporation from reservoirs. By 2025 the proportion of abstractions for industry will increase to 50%, that for agriculture will drop to 34%, whereas the domestic sector is expected to continue to take 14%.

Water use in Europe is unevenly distributed. The greatest share goes to southern and central Europe, and the southern ETSU. These areas account for 36%, 32% and 27%, respectively; the smallest proportions are for northern Europe and the northern ETSU (about 2–3%). This tendency is likely to persist in the future.

Despite the fact that most of Europe lies in the humid zone, the mean specific water availability is low and presently amounts to about 4000 m^3 per capita per year. This water availability figure is one of the lowest in the world.

Specific water availability is unevenly distributed over Europe. The Nordic states, Iceland, and northern ETSU are noted for the highest availability of water resources (20–30 000 m^3 per capita per year); the lowest water availability is recorded in central and southern Europe (2–3000 m^3). By 2025, these figures will slightly reduce for all regions of Europe, except the northern ETSU.

4.2 PHYSICAL CONDITIONS

All the information in this section is taken from the literature (Grigoryev, 1961; Eramov, 1973; Korzun, 1974b; Nemec and Schaake, 1982; Losev, 1989; ILEC/UNEP, 1987–9, 1991, 1993). The data presented here have been carefully checked and compared with new data, published recently.

4.2.1 Relief

The relief of Europe is quite diverse, due to differences in the tectonic structure and in the geological history of the continent. Generally, the relief of Europe represents frequent and random alternations of mountains, hills, low plains and depressions. Mountains in Europe cover about 17% of the continent, including 2% that are mountains over 2000 m (Dobrynin, 1943).

In northwestern Europe, the largest area of mountains is the Scandinavian Uplands, extending along the Norwegian coast and covering most of the Scandinavian Peninsula (the highest point is Jotunheimen, 2481 m).

The German–Polish Plain is located south of the Baltic coast, its relief being characterized by terminal moraines, alternating with low areas of fluvioglacial sands. South and west of the German–Polish Plain, a broad belt of small low plains and extensive depressions occurs, covering much of Germany, Belgium, Denmark, The Netherlands, the northern part of France and southeastern England.

Farther south, there is a belt of medium-height uplands and mountain ridges, separated by lowlands and depressions. The most significant among them are the Massif Central in France with a highest elevation of 1886 m, and the Czech or Bohemian Massif where elevations reach 1600 m. Mountain areas in the centre of Europe alternate with fault-bounded lowlands, filled with Mesozoic and Cenozoic deposits. Such alternations of mountains and lowlands are, for instance, the Paris Basin with the North French Lowlands, and more obviously the Swabian–Franconian continuation of the Rhine rift valley. Further south, there are folded mountains, with the Alps in a central position. The Carpathians are the continuation of the Alps, separated from them by the Vienna Depression.

There are lowlands adjoining the Alps and the Carpathians, namely those of Venice–Padua, the Middle Danube and the Lower Danube, which are confined to tectonic depressions and mainly filled with marine deposits.

Three large peninsulas make up the south of Europe, i.e. the Iberian, Italian and Balkan Peninsulas, which are dominated by mountains. Southern Europe is noted for ancient, block-folded and folded mountains characterized by thick limestone layers, where karst is extensively developed.

The central part of the Iberian Peninsula is made up of high mountainous land with high ridges and uplands. In the south of the peninsula, there are the Andalusian Mountains reaching 3481 m; and in the north, the Pyrenees (highest elevation 3404 m), separating the peninsula from the rest of Europe.

The Italian Peninsula is separated from the continent by the sweep of the Alps. The Apennine Ridge extends along the peninsula reaching 2921 m in elevation.

The Balkan Peninsula is only partly separated from the rest of Europe in the northeast by the Balkan Mountains. The western part of Balkan Peninsula is taken up by the Dinarides with extensive karst phenomena, causing the marked reduction of surface drainage and formation of subterranean river courses.

Southern Europe is a young area with continuing intense tectonic movements, frequent earthquakes and recent volcanism (the active volcanoes are Etna, Vesuvius and Stromboli).

The relief of eastern Europe is noted for less diversity than the west with alternating elevated areas and lowlands, divided by river systems. The Russian Plain is extensive with an area of about 4 million km^2, which extends south from the Arctic Ocean to the Black Sea and the Caspian Sea; and eastwards, from the western border of the Baltic states to the Ural Mountains. The plain is noted for the alternation of hilly and flat uplands and lowlands. The main features of its relief are accounted for by the structure of the Russian Platform, the occurrence of horizontally bedded deposits, overlying the folded crystalline basement of the Platform over extensive areas. The Platform proper is made up of different elements: shields, anticlines, synclines, and other smaller structures. In many cases, the uplands correspond to the projection of the Platform in the relief; and lowlands, to depressions.

The most extensive surface exposures of the Platform are the Ukrainian and the Baltic Shields. The Ukrainian Shield extends from the coast of the Azov Sea to the Dnieper Upland and to southern Polesje. It is covered with the Tertiary deposits, while the granites and gneisses from which it is composed outcrop mainly near the river valleys. The crystalline Baltic Shield, within which the Karelia and Kola Peninsulas are located, is closely linked with folded structures of the Scandinavian Highland. Crystalline rocks outcrop frequently, except where there is the thin discontinuous cover of Quaternary sediments.

Gently sloping uplands are located within the Russian Plain: at Valdai, Middle Russia, Podolian, and the Volga, with the elevations up to 200–400 m. Many rivers of eastern Europe have their sources in these uplands. The most extensive lowlands on the Russian Plain, i.e. the Pechora, Moscow, Black Sea, Dnieper, Oka–Don and Caspian lowlands, lie below 100–200 m, while the southern half of the Caspian Lowland is even below sea level.

The Russian Plain is joined in the east by the Ural Mountains, extending longitudinally from the Kara Sea to the Kazakhstan steppes; the highest elevation is 1894 m (Narodnaya Mountain) in the north. In the southwest, the Russian Plain is bounded by the Eastern (Ukrainian) Carpathians, running from northwest to southeast from the sources of the San River (a tributary of the Vistula) to the sources of the Siret River (a tributary of the Danube) for 250 km. Unlike the Urals, the Carpathians are a young folded range of Alpine age.

In the southeast of Europe, there is an extensive highland, the Caucasus Mountains (the highest point is Mount Elbrus, 5642 m). The dividing ridge of the Great Caucasus is the southern boundary of the European continent.

4.2.2 Soils and vegetation

The soils and vegetation of Europe occur in distinct zones controlled by latitude and altitude. They are: the Arctic desert zone in the Arctic belt; the tundra and forest–tundra zone in the Subarctic zone; the forest zone, including taiga and mixed broadleaved forest subzones; the forest–steppe and steppe zones; the temperate semi-desert zone; and in the subtropical belt, the Mediterranean zone of evergreen xerophytic forests and shrubs.

The Arctic desert zone occurs on islands of the Arctic Ocean. There are no trees and shrub vegetation; only lichens, mosses and some other flowering plant species prevail, but without forming a closed cover.

The tundra zone lies in the extreme north of the Russian Plain, and in the Kola and Scandinavian Peninsulas, as well as in coastal Iceland. There is an extensively developed moss, lichen and shrub vegetation. Poor stunted woodland zones with thin tundra and gley soils, as well as bogs, dominate the northern river valleys. In the northwest of the continent (Iceland and coastal Norway), Subarctic meadows are widespread. In the south of the Subarctic zone, in the Russian Plain, a forest–tundra subzone can be distinguished, where the tundra combines with stunted or open woodland.

The taiga subzone covers most of the forest zone and extends southwards to approximately 60° N. The taiga forests are predominantly composed of spruce and pine in the west, with fir, spruce and larch in the east. Spruce forests most often grow on clayey and loamy varieties of podsolic and dern podsolic soils. On sandy and outwash varieties of these soil types, as well as in swamp areas, pine forests are widespread. In the mountains, the taiga forests grade into a belt of stunted birch woodland, passing into the mountain tundra belt at elevations of 1100 m in the south and 450 m in the north.

A subzone of mixed broadleaved and coniferous forests covers the north of the UK, the south of the Scandinavian Peninsula and the north of the German-Polish and Russian Plains, where it extends southwards to approximately 53° N. Towards the Urals, it is characterized by broad-leaved and coniferous forests. There are podsolic and dern podsolic soils in the west of the subzone, and grey wood soils in the east.

Southwards, a subzone of broadleaved forests is to be found, being most pronounced in the west of central Europe with its humid marine climate and ending abruptly in the Russian Plain. There beech and oak forests dominate the cover, and pine forests often grow on sandy soils. Zonal types of soils are brown wood soils in the west and grey wood soils in the east; in limestone areas, dern carbonate soils occur. Broadleaved forests are common in the mountains of this subzone, particularly the Alps. There are distinct altitude zones: oak forests of the piedmont area grade into beech forests in the lower parts of slopes, then (approximately at an elevation of 2000 m), they are followed by the belt of broadleaved and coniferous forests; still higher are found the shrub grasslands of the Subalpine and Alpine meadow belt.

The forest–steppe zone is located mainly on leached and thick chernozem soils. It is situated in continental areas of Europe, i.e. in the southern Russian Plain and Middle Danubian Lowland. In the south, the forest–steppe grades into the steppe zone dominated by forb–grass and grass steppes on common and southern chernozems, and on the chestnut soils of the southern Ukraine and Volga.

Over most of the plains, which are in the subzones of mixed and broadleaved forests, forest–steppe and steppe, the natural vegetation is replaced by crops.

The semi-desert zone in Europe covers a small area in the southeast of the Lower Volga Region. It is dominated by xerophytic shrubs and low shrubs, forming a sparse cover. Brown soils of the semi-deserts frequently alternate with solonetz, white alkali soils and sand masses.

The types of vegetation found in the Mediterranean zone of the subtropical belt are light evergreen sclerophyllous xerophyte forests and shrubs. The evergreen shrub formations and deciduous xerophyte shrubs, which occur mainly in the Balkan Peninsula and on the southern coast of the Crimea, are most often secondary, appearing in place of the cleared forests. Xerophyte scleroid herbs and shrubs grow on stony sunny slopes, without forming a continuous cover. Moderately fertile, slightly leached brown soils develop under forests and shrubs. Red soils are found in areas of limestone. The forests and shrubs, which previously occupied the lower parts of mountain slopes to a height of 700 m in the south, to 400 m in the north, and on the small plains in the Mediterranean, have been cleared and have been replaced by citrus plantations, vineyards, orchards, maize and wheat fields.

4.2.3 Climate

The main features of the climate in Europe are determined by atmospheric circulation patterns and the distribution of solar radiation over the continent. Distance from the Atlantic and altitude are also important factors.

The circulation of the atmosphere over Europe represents a balance between the Azores high and the Icelandic low pressure systems. During the winter, an offshoot of the Siberian anticyclone intrudes westwards into Europe. However, air masses of an Asian origin have a minor significance, and the weather and climate of the continent, including the transport of moisture, are dominated by an easterly movement (Chernogayeva, 1969). Air masses in summer come from the Atlantic and from the Arctic. The main process of the summer is the transformation (heating and humidification) of air masses: Atlantic air in the central zone and Arctic air in the northern zone. In winter, the transport of humid air masses induces precipitation in the temperate zone. In the eastern regions of Europe, dry and cold weather is common at this time.

The moisture content of air masses and the transport of moisture in the atmosphere above Europe are dependent on many factors, the main ones among them being the nature of the relief, the position of the Azores high and the Icelandic low, and the time of year. The highest initial moisture content (about 18 mm) is recorded in air masses coming from the south Atlantic and Mediterranean during the summer (from April to October). The transport of this moisture averages at 8 m/s. The lowest moisture content (about 5–6 mm) is typical of air masses of the winter (from November to March) which originate over the Arctic Ocean and move at 10 m/s. During winter, there is a strong flow of warm Atlantic air from the southwest, mainly covering the western and northwestern parts of Europe. The initial moisture content of air masses arriving during the winter is about 10 mm, and they move at 9–10 m/s (Chernogayeva, 1969). In all the northern regions of Europe during the winter, the moisture originates from the ocean and not from local evaporation. In the southern regions, there is an increase in local evaporation during the summer.

There are notable differences in solar radiation between northern and southern Europe (Korzun, 1974a). In the winter months, the potential total radiation ranges from zero in the north to 350 cal/cm^2 per day in the south. In the middle of the summer, the potential solar radiation over the entire continent exceeds 750 cal/cm^2 per day. Due to cyclonic activity causing cloudiness, the actual solar radiation during certain periods of the year differs markedly from the potential, reaching 60–65% of it. Because of the predominantly maritime climate, typical of many European regions, the actual solar radiation for the whole of the year is about 60–75% of the potential. In mountainous areas (e.g. the Alps), actual radiation in the winter can be up to 80% of the potential, while in summer, it decreases to 60%. In the south, in the semi-arid climates of the Iberian, Balkan, and Italian Peninsulas, actual and potential radiation are similar.

The indented character of the coastline and the location of the main mountain ranges favour the easy penetration of the humid air masses over the continent. Being located in the middle latitudes, Europe is dominated by westerly airflows from the Atlantic. This results in the dominance of a humid maritime climate over most of Europe, particularly western Europe, which is the closest to the ocean. The Gulf Stream enhances the effect of the mild, moist Atlantic air masses on the climate as it approaches the northwestern coast of Europe. Here, the influence of Gulf Stream results in outstanding positive temperature anomalies during the winter. These factors cause much of Europe to experience an oceanic type of climate, except for the peninsulas of the Mediterranean and Black Seas, as well as the southeastern Caspian Sea coast.

The major part of western Europe has an annual range of temperature of 10–20 °C, whereas in Eastern Europe, temperature ranges of 25–35 °C prevail. These ranges increase from west to east. Over most of western Europe, mean annual precipitation varies from 600 to 1000 mm, whereas in the east, totals are from 400 to 600 mm, decreasing to the northeast and southeast (UNESCO, 1970). Precipitation maps of Europe display a complex pattern in which latitude, distance from the Atlantic and altitude are important factors. There are some areas in southeast Spain where annual totals are less than 400 mm and others in Ireland, Wales, Scotland, Iceland and Norway where more than 4000 mm is recorded.

4.2.4 Hydrology

RIVERS

The diversity of relief, vegetation and soils, as well as climate, affects Europe's river systems, giving different regimes and different patterns of flow across the continent. The system is marked by a high density of rivers fed by abundant moisture, freely transported over the entire continent. The melting of ice and snow in the high mountains plays an important role in regulating the rivers and particularly the high flows.

The rivers of northern Europe, mainly in the Scandinavian Peninsula (Glomma, Torne, etc.), have shallow valleys with rapids and lakes; they are fed by the meltwater of glaciers and by rain, and display a spring and early summer peak flow. The rivers of the western fringe of Europe are usually short with high flows in winter and low flows in the summer.

The rivers of central Europe mainly flow in three directions: to the northwest, to the south, and to the southeast. The largest rivers of the Baltic Basin are the Vistula and Oder; the largest rivers flowing to the Atlantic coast are the Elbe, Rhine, Seine and Loire. The rivers flowing south are the Ebro and Rhône; and to the southeast, the Po and Danube.

Rivers draining the Alps are fed by meltwater from glaciers with a summer maximum and a winter minimum discharge. They have steep gradients and waterfalls. Rivers flowing in upland areas are fed by snowmelt in mountains and rain. They have a maximum discharge in spring and the first half of summer, and low flows in winter.

The rivers in southern Europe (Douro, Tagus, Guadiana, etc.) have a winter maximum, and a distinct summer minimum discharge. Shallow rivers become dry in summer.

The large east European rivers are concentrated on the Russian Plain and flow in two main directions, i.e. to the north and to the south. Some rivers flow north towards the Barents and White Seas; others south towards the Baltic, Black, Azov and Caspian Seas. The regimes of eastern European rivers are largely controlled by snowmelt in the spring. Despite this, north-flowing rivers differ markedly from those flowing south, because precipitation exceeds evaporation. The largest northern rivers are the Pechora, Northern Dvina and Onega; while those flowing south are the Volga, Ural, Don, Dnieper, Southern Bug and Dniester. Within the Baltic Sea Basin, the largest rivers are the Neva, Western Dvina and Neman.

Large rivers of the Russian Plain (e.g. the Volga and Northern Dvina) have a steady uniform flow where gradients are small and the river channels meander. These rivers are characterized by a spring maximum and a summer–autumn minimum. They are

Table 4.1. *Major morphometric characteristics of the principal rivers of Europe*

Basin	River	Area of catchment, km^2	Length, km
Baltic Sea	Neva	281 000	74
	Vistula	198 500	1092
	Odra	118 900	907
	Neman	98 200	937
	Daugava	87 900	1020
	Vuoksi	61 300	165
	Narva	56 200	78
	Kemijoki	52 000	550
	Gota–Alv	50 100	720
	Tornealven	40 000	408
	Kymijoki	37 200	600
	Lulealven	25 200	450
North Sea	Rhine	252 000	1320
	Elbe	148 000	1165
	Weser	46 000	432
	Glomma	41 400	587
	Thames	15 300	332
Atlantic Ocean	Loire	120 500	1010
	Douro	94 500	938
	Tagus	81 000	1010
	Seine	79 000	776
	Guadiana	72 000	801
	Guadalquivir	57 000	680
	Garonne	56 000	650
Mediterranean Sea	Rhône	98 000	812
	Ebro	87 000	928
	Po	75 000	652
	Jucar	22 400	506
Black Sea	Danube	817 000	2850
	Dnieper	504 500	2200
	Don	422 000	1870
	Dniester	72 100	1352
	South Bug	63 700	792
	Kuban	61 000	907
Arctic Ocean	North Dvina	357 300	1302
	Pechora	322 000	1809
	Onega	56 900	416
Caspian Sea	Volga	1 380 000	3700
	Ural	236 000	2534
	Terek	43 700	600

frozen for seven months in the north and two months in the south and west. A list of the main European rivers is given in Table 4.1.

LAKES

Europe has a large number lakes, and their combined area is about 5% of the total area of the continent. The number of lakes in a

Table 4.2. *Major morphometric information of the principal lakes of Europe*

Basin	Lake	Area of surface, km^2	Area of catchment and lake, km$^2 \times 10^3$	Maximum depth, m	Volume of water mass, km^3
Baltic Sea	Ladoga	18 135	70.1	230	908.0
	Onega	9 890	51.5	120	295.0
	Vanern	5 648	41.2	106	153.0
	Chudskoye-Pskovskoye	3 558	44.2	15	25.2
	Vattern	1 856	4.5	122	74.0
	Suur-Saimaa	1 800	59.0	58	36.0
	Malaren	1 140	21.5	61	14.3
	Inari	1 116	13.4	92	15.9
	Pajanne	1 116	25.4	95	18.1
	Ilmen	982	66.4	4.4	12.0
	Izo-Kalla	898	–	90	8.0
	Oulu-Yarvi	893	–	35	6.8
	Pilinen	867	12.8	60	8.5
	Keitele	500	–	65	36.0
	Hjaimaren	484	3.6	22	2.9
	Storsjon	464	12.1	74	7.4
	Sniurdwy	331	–	47	2.8
	Tornetrask	330	3.3	168	17.1
	Vortsjarv	271	3.1	6	1.0
Atlantic Ocean	Geneva	584	8.0	310	88.9
	Boden See	539	10.9	252	48.5
	Lough Neagh	396	–	31	–
Mediterranean Sea	Scadar	372	5.5	10	1.9
	Garda	370	–	348	50.0
	Ohrid	350	–	256	61.0
Black Sea	Balaton	593	5.2	12.2	1.9
	Atter-Zee	456	4.6	171	3.9
	Noizidler-Zee	323	1.3	2	0.2
Arctic Ocean	Vygozero	1 140	–	18	7.2
	Imandra	900	11.2	67	11.2
	Segozero	815	6.6	103	4.1
Caspian Sea	Beloye	1 200	14.0	20	5.2
	Seliger	222	2.3	24	1.3
	Caspian	378 000	3600	1025	78 200

particular area is dependent on a range of factors, the most important being the relationship between the water balance elements (precipitation, runoff, evaporation) and the relief of the area. Most lakes occur in the more humid areas.

The largest lakes of Europe are given in Table 4.2. The characteristics of these lakes depend on their natural state. In recent years, the sizes of lakes may have changed because some have been dammed. Most of the largest lakes in Europe, except the Caspian Sea, are found in the Baltic Basin. Lakes are most prevalent as part of the area of the country in Finland (12%) and Sweden (9%). Lakes in Europe are irregularly distributed. In the countries, where lakes are numerous, they form an essential and characteristic element of the landscape, being closely associated with the Quaternary glaciation. These are mainly the Nordic states, as well as the Alpine countries. However, the origin of lake basins in these regions is not only glacial, but also tectonic. Most large lakes in northern Europe lie in basins generated by faults, which have subsequently been shaped by glaciation and erosion.

In the Alps, lakes are mainly of glacial origin and belong to two main groups – large lakes in piedmont areas and minor karst lakes of the mountainous zone. Areas with small glacial lakes are found in many of the mountain regions of central and southern Europe, which were covered by the Quaternary glaciation, for instance, in the mountains of Scotland, Wales and England, as well as in the

Carpathians, and the mountains of the Iberian, Italian and Balkan Peninsulas.

The largest lakes of the Balkan Mountains and Italy (Ohrid, Garda and Prespa) have a partly tectonic and partly relic origin: they are remains of larger water bodies that existed at the end of the Tertiary. Italy is characterized by a great number of crater lakes in the calderas of extinct volcanoes.

There are a number of lakes in the European part of the former USSR, particularly the northwest, and in the Azov–Black Sea and Caspian Sea regions.

The northwestern lake district (in the European part of Russia), known as the Lake Land, covers Karelia, the Kola Peninsula, Leningrad, Pskov and Novgorod Oblasts. Within this area, along with a great number of small and medium-size lakes, such lakes as Ladoga, Onega, Beloe, Ilmen, Chudskoe–Pskov, Vygozero, Segozero and Imandra are to be found. These lake basins are of glacial origin and the boundary of this region almost coincides with the edge of the last glaciation. Along with lakes of a glacial origin, there are also tectonic lakes, among them many lakes in Karelia and the Kola Peninsula. Numerous lakes of a secondary origin occur in bogs and marshes and karst lakes are to be found where readily soluble rocks (limestones) are located, among them the numerous small lakes between Lake Onega and Lake Ladoga.

The Azov–Black Sea lake district contains a great number of peculiar lakes known as limans, which are located along the Black Sea and Azov Sea coasts. Their origin is connected with the invasion of the land by the sea and the flooding of river mouths. Limans characteristically extend along the flooded river valleys.

The Caspian lake district comprises a group of lakes in the Caspian Lowland. Most of lakes in this district are recharged from the flooding of steppe rivers during the spring flood period. There are also abundant temporary water bodies (limans), formed in places where meltwater accumulates in lows and sink holes.

4.3 SOCIO-ECONOMIC CONDITIONS AND THE USE OF WATER RESOURCES

4.3.1 Population

Europe is one of the most densely populated and highly developed regions of the world; general data on the continent are given in Table 4.3. In 1994 the population of Europe was estimated to be 684.63 million. The average density of population is 65/km^2. In western Europe, the highest density is in The Netherlands (376/km^2); the lowest in Iceland (2/km^2); and among the countries in ETSU, the highest density is in Moldavia (129/km^2), and the lowest in Russia (24/km^2), if no account is taken of a small part of the territory of Kazakhstan, located on the steppes beside the Urals.

Rates of population growth are shown in Fig. 4.1 for some European states from 1930 up to 1990s. For these countries, namely Albania, Sweden, Spain, Italy and France, there was a regular increase up to the 1980s, and then stabilization of the size of population, with the exception of Albania, where there was a small growth over the entire period. There is evidence (Anonymous, 1984; UN, 1995) that in most western European countries during the last two decades, population growth has ceased. In some countries (for instance, Germany and Hungary) there has even been a small reduction, which is probably due to the economic and political situation. Changes in the population of Europe as a whole from 1950 are shown in Fig. 4.1b. This diagram demonstrates that the population grew to the beginning of the 1990s, when it stabilized at 685 million.

Europe is one of the most urban continents; about 80% of the total population lives in towns and cities. The northern and central parts of western Europe are the most urbanized (90–95% urban population) while in southern Europe the comparable figure is about 80%. The urban population in countries within the ETSU is approximately 70–75% of the total. The most urbanized countries are the UK (98%), Sweden, Switzerland and The Netherlands (96%); the least urbanized are Albania (51%), Russia (63%), Poland and Romania (82%) and Ireland (87%). For the most developed countries, these percentages were reached in the 1940s and the 1950s and have remained unchanged. Modern urbanization in these countries is usually in the form of extensive suburban zones around large cities.

4.3.2 Economic development

The economy of Europe is one of the most advanced in the world with a high output of manufactured goods and energy. Special features to note at the millennium are advances in the growth of production and serious structural shifts in the economy, pronounced differences in the development of the main European regions, growing contrasts between the revenues of the different countries and increasing differences between the rich and poor (Alayev and Kolosov, 1989).

Scientific and technological progress promotes processes that use smaller quantities of raw materials. Some of the producing and processing industries are in difficulties in many of the more developed countries because there is a gradual displacement of the centres of production and the initial processing of raw materials from the developed countries to the developing ones. This displacement is due not only to a higher cost of production and processing generally in the more developed countries, but also because to the increasingly stringent nature of the regulations for environmental protection.

It should be noted that scientific and technological progress in recent years has resulted in the emergence of new branches of

Table 4.3. *Major information about the countries of Europe*[a]

Country	Area, km² × 10³	Population (1994), people × 10³	Density of population, number of people per km²
Sweden	450	8 340	19
Norway	387	4 320	11
Finland	337	5 080	15
Iceland	103	230	2
Denmark	45	5 170	115
France (+ Monaco)	547	56 800	106
Germany	356	79 200	222
Poland	313	39 200	126
United Kingdom	244	58 000	238
Austria	84	7 900	94
Czech Republic	79	10 300	130
Ireland	70	3 600	51
Slovakia	49	5 500	112
Switzerland	41	7 200	176
Netherlands	41	15 400	376
Belgium (+ Luxembourg)	33	10 300	312
Spain	505	39 600	78
Italy	301	57 700	190
Yugoslavia, former	256	24 470	93
Romania	238	22 900	96
Greece	132	10 400	79
Bulgaria	111	8 800	79
Hungary	93	10 200	109
Portugal	92	9 930	107
Turkey (European part)	34	–	–
Albania	29	3 600	118
Russia (European part)	4 240	(103 700)	(24)
Ukraine	604	51 900	86
Kazakhstan (European part)	230	(2 200)	(9)
Belarus	208	10 300	50
Lithuania	65	3 700	57
Latvia	64	2 700	42
Estonia	45	1 600	36
Moldavia	34	4 400	129
Europe as a whole	10 460	684 640	65

[a] Values in parentheses are approximate data.

the national economy in the most developed countries, of a type aimed at the extraction of natural resources. This trend increases the demand for new materials, and for science-intensive equipment, precise mechanical engineering devices and for special constructions. The resulting problems have led to the establishment of the so-called science parks or technological centres, where applied scientific institutions of this type are located (Panov, 1990). Such scientific centres have been set up in the UK, Germany,

France, Italy, Belgium, The Netherlands, and in some other countries.

The production potential of the agricultural sector in the developed countries of Europe is not used to its full extent. By applying modern approaches and new methods, agricultural production has reached a high degree of efficiency, including the reduction of production costs, increasing the yield of crops, and improving the productivity of animal breeding. In practically all west European

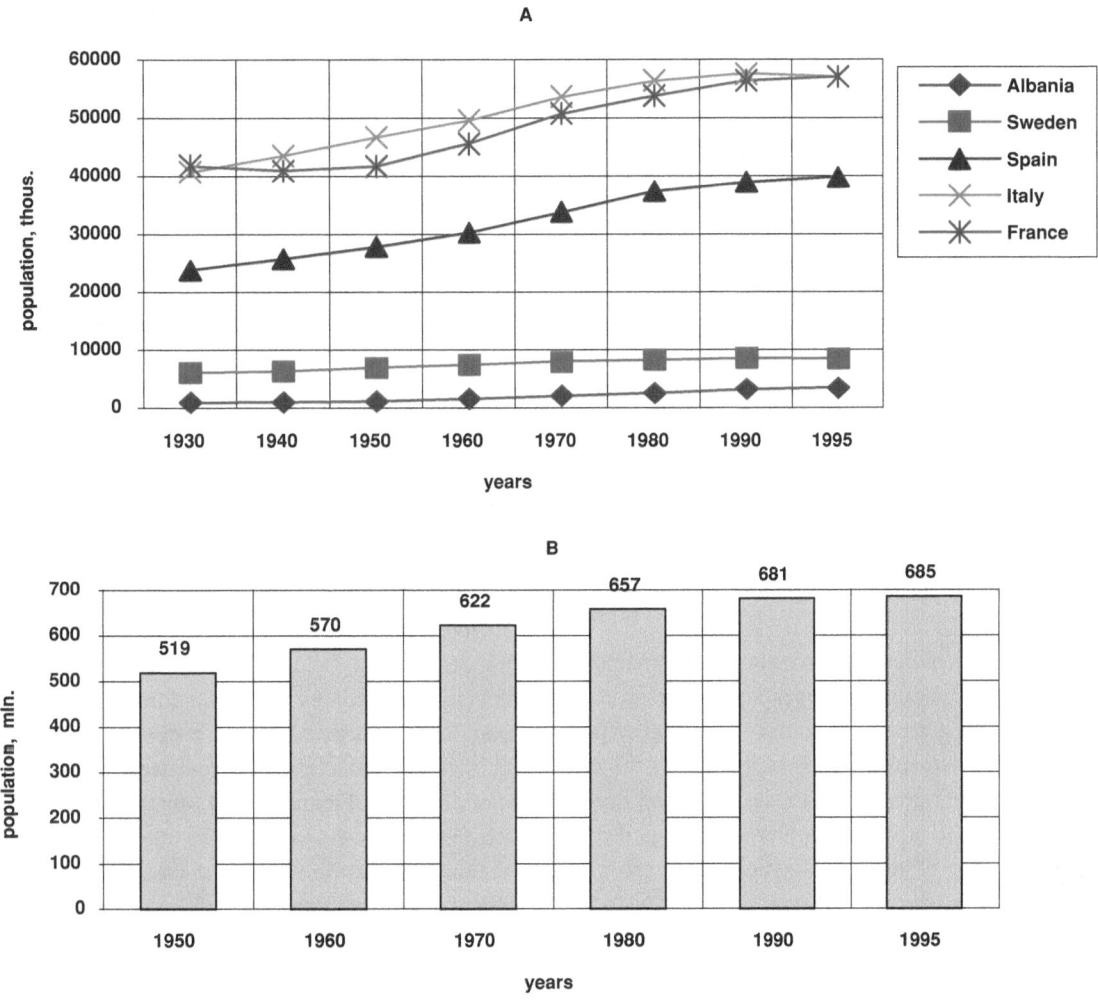

Fig. 4.1 Dynamics of the population of Europe. A, individual countries; B, Europe as a whole.

countries, national government and European Union support is given to agriculture, such as by maintaining stable prices for products, direct payments to farmers, the subsidizing of exports, etc. Trying to avoid overproduction in the favourable years with a high yield and a reduction in prices for agricultural products, the governments of many countries take various measures to restrict agricultural production. As a result, in the 1990s, the changes in the production of food generally conform to the rate of population growth. A negative factor is the worsening state of the environment, since the erosion of soils, the reduction of the areas of forest and the increasing pollution may result in a decrease in agricultural productivity; and there is also the threat of climate change, which may also reduce productivity (Budyko, 1980; Stanners and Bourdeau, 1995; Kundzewicz and Parry, 2001).

Measures for the protection of the environment taken by national and international organizations are becoming increasingly important in the economic development of west European states.

Under their influence, stricter requirements are imposed on enterprises (primarily chemical and metallurgical) causing damage to the environment and natural resources; limitations are imposed on the consumption of mineral fuel and use of ecologically "dirty" types of products; and implementation of waste-free technology and easily degradable packaging materials is stimulated. Considerable emphasis (advertising, continuous information on the state of environment, etc.) is placed on these issues in Germany, Denmark, The Netherlands and other developed countries (Alayev and Kolosov, 1989).

In the mid-1990s the European states of the former socialist camp started implementing large-scale economic reforms, including the extensive use of market tools for managing the economy (change in the level of prices, influence of taxes, cost of loans), the modernization of production, the accelerated development of science-based industries, and the extension of participation in the international division of labour.

The most commonly used indicator of the level of a nation's economic development is its gross national product (GNP), which is

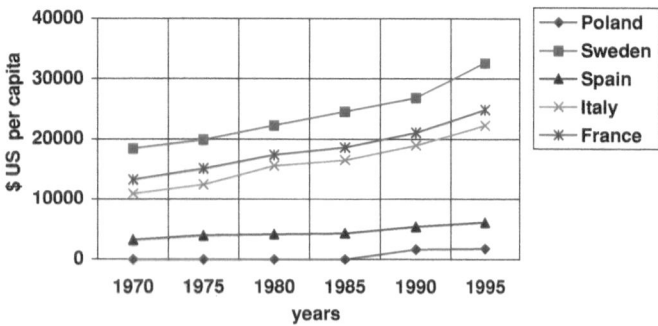

Fig. 4.2 Dynamics of gross national product ($US per capita) for some countries of Europe.

usually given in US dollars per head. GNP values for each country are based on the World Bank, UN and International Monetary Fund data (UN, 1993; World Bank, 1993, 1995; IMF, 1994). For comparison, the trends in GNPs over the period 1970 to 1995 are given in Fig. 4.2 for some countries; reliable figures are not available for all nations.

As generally recognized, national attitudes to the environment, including water resources, are usually dependent on the level of economic development of a particular country. The higher that level, the more emphasis is placed on the protection of the environment including natural resources. In the advanced countries of western Europe, the attitude to water has changed radically since the beginning of the 1970s. The energy crisis at that time, the later crisis in the economy, and shortages of foodstuffs and other factors have resulted in a new attitude to water resources.

4.3.3 Use of water resources and water use assessment

Of all continents, probably the water resources situation in Europe is most striking, due to the high density of population, the size of the industrial production and the large output of electrical energy. From the beginning of the last century to the present time, the water resources situation has changed substantially. The rapid development of industry, the intensification of agriculture, and the growing domestic need for a wholesome and reliable water supply for an increasing population have led to the increasing use of water. This is demonstrated by the 15-fold rise in the demand for water in Europe from the beginning to the end of the twentieth century. According to Shiklomanov (1988), total water use on the continent in 1900 averaged 38 km³ a year, while by 1990 it had reached 555 km³.

The rapid rise in water use, due to intense industrial and agricultural development and the growth of population, has had an ever-increasing effect on the runoff and its fluctuations and thus on the state of the water resources of the continent. In the most developed regions of Europe, at present, no large river systems

are left in a natural state, i.e. one where the hydrological regime is not disturbed to a certain extent (Shiklomanov, 1988).

The stresses in the water supplies in some European countries have been accompanied by a marked reduction in the available water resources, caused by pollution of surface and ground water from discharges of waste water from urban areas, industry and agriculture. Until the beginning of the 1970s, such pollution had largely been on a small scale and mainly concentrated around large industrial centres or urban areas. However, in recent decades, there has been a rapid increase in the total volume and extent of water pollution in certain areas. Some rivers, lakes and aquifers have already reached and even passed the "limits of reversibility", which implies they have lost their ability for self-purification so that they have become dead water bodies. This growing volume of pollution has resulted in an additional demand for purer water for dilution of the polluted waters.

In many European countries, lack of water resources is becoming a serious obstacle to economic development (for example in Greece, Spain and The Netherlands). In Belgium, the UK, Italy, France, Germany and some other states, the growth of abstractions in some regions has resulted in shortages during prolonged drought and this poses serious problems and provides arguments for regional water transfers and large capital investments (De Mare, 1977; Shiklomanov and Markova, 1987).

In addition to reasons of cost, the state of the environment is receiving increasing attention in most regions, because of the impact of human activities on ecological systems. The combination of ecological and economic problems should result in a more rational evaluation of the efficiency and effectiveness of planned projects. In many European countries environmental impact studies are necessary prior to any development, while national and European Union requirements for environmental protection have to be met. This includes meeting water quality standards and objectives, which often demand the expansion of existing treatment plants and the construction of new ones, the development of waste-free technologies, and other initiatives (Alayev and Kolosov, 1989; EEA, 1999; Kundzewicz and Parry, 2001).

Of the total volume of water abstracted, more is used by industry and agriculture and less for domestic purposes. The volume used by the individual sectors varies with location, climate, the level of development of the country and the availability of water resources. For most southern European countries, irrigation is the main user, while for northern and central states it is industry, including electricity generation.

INDUSTRY

Industry is one of the most important users of water. The amount of water abstracted to meet industrial needs has grown steadily since the beginning of the century: according to Shiklomanov and Markova (1987), this volume increased 33 times between 1900 and

1990. About 45% of the total abstracted was used for industrial purposes in the 1980s and 1990s, with the consumption of water at about 11% (Shiklomanov and Markova, 1987); but according to Gleick (1993, 1998), this value is about 54%. The most water-intensive industries are thermal power production, processing and the mining industry.

Thermal power production needs enormous volumes of water for cooling the generators. A modern thermal plant with a capacity of 1 million kW requires 1.2–1.6 km^3 of water per year, the consumption amounting to 10% of this volume (ICOLD, 1987, 1992). Nuclear power plants with the same capacity take even more – up to 2.5–3 km^3 of water per year. The cooling systems of these plants account for 50–60% of the total volume of water used by industry. With the improvement of technology in these plants the volume of water needed will be reduced (Holing, 1981).

Countries such as France and the UK have developed nuclear power to produce electricity but economic problems and the search for alternative sources of energy reinforce those who argue against the further development of this source. Despite these recent moves to stop the construction of nuclear power plants, nuclear power engineering was the most dynamic part of the power industry (Pretro and Fedorov, 1993). In Europe, the number of reactors is as follows: France 55, the UK 40, Germany 30, and Sweden 12 (Davis, 1985). However, despite the fact that nuclear energy is more environmentally friendly, the unsolved problem of what to do with the spent fuel and other difficulties raise considerable public pressure against the expansion of nuclear power.

The proportion of water abstracted for thermal power varies extensively across the continent. Among the OECD states in 1980, the largest intake of water for thermal power production, including nuclear power, was recorded in Germany and The Netherlands (64–68% of industrial water); about 50% of the water is used by industry in France and Austria; and less in other countries: the UK 19%, Italy 13%, Finland 3%, and Sweden 0.3%. In the former USSR, approximately 17% of the total abstraction for industry was used for thermal power production.

Branches of industry vary extensively in water use, even within a single country. Among processing industries, a significant amount of water is used by the food, chemical and petro-chemical, metallurgical, and pulp and paper industries. Mining normally uses a small proportion of industrial water. Unlike in the processing industries, salt water can be used in the mining industry, and it sometimes accounts for up to half of the total water used.

Because many industries have developed techniques for using fewer raw materials, including water, during the last decade, the rate of growth of water needed by industry has decreased in many of the developed countries. For instance, in Sweden, as far back as the 1970s, the growth rate was lowered and amounted to 5–10% against 15% in previous decades (Anonymous, 1992b). In Finland, in the 1990s, a 10–15% reduction occurred as compared to the 1980s. A similar situation has been experienced in the Netherlands, the UK, Belgium, Austria, and other countries (WRI/IIED, 1986, 1988; IIED/WRI, 1987). However, for many of the developing countries, industrial water needs continued to increase within the total amount of water abstracted, especially in Spain and Portugal, and in the former socialist states.

HYDROPOWER ENGINEERING AND RESERVOIRS

Water is used as a source of power over much of Europe but there are several regions where there are many hydropower plants, such as Scandinavia, the Alps, the Balkans and the Pyrenees. The published data on the theoretical, technical and economic hydropower potential of rivers for certain countries are often contradictory. Data on the energy potential of most of the world rivers (ICOLD, 1992) are among the latest and appear to be the most objective. Table 4.4 was compiled from these data and gives the hydropower reserves of countries in Europe. This table shows that the highest degree of utilization of the hydropower potential of rivers exceeds 70% in France, Switzerland and Germany and 55% in Finland and Sweden. However, the average figure for Europe is about 15% (ICOLD, 1992). If small rivers are included, then for Russia the theoretical potential output is increased from 2395 to 2898 TW hour/year, which is about 21% (Pretro and Fedorov, 1993).

Because many of the sites with a large power potential have been utilized, small-scale hydroelectric power schemes (HPP) are being developed in a number of countries. The proportion of small HPP in the energy balance can be estimated from Swedish data. In 1990, 1100 HPP produced 65.5 TW hour/year. Therefore, the percentage of small HPP is about 6% (Anonymous, 1992b).

In some countries with large rivers, such as the Danube and Rhône, cascades of low-head power units are being constructed. Some nations are noted for output of hydropower plants, such as Norway (25 GWh/year); France (27 GWh/year), together with Sweden, Spain and Italy (about 16 GWh/year).

Generally, despite a growing interest in the development of sites on small rivers which have not been utilized, the increase in capacity has continued due to the construction of large and medium-size plants, as well as by increasing the peak capacity by the construction of pumped storage schemes. The construction of pumped storage, which started in the 1950s, is due to the fact that a number of developed countries have virtually exhausted their hydropower resources under the continuing growth of peak loads. There is also the view that hydropower is more environmentally friendly than thermal power, but during construction considerable changes can be made to the aquatic environment and these may continue when the plant is operating.

Development of hydropower depends largely on an increase in the number of reservoirs, which was particularly rapid after the 1950s. In 1995 the total number of reservoirs in Europe was more than 4500; their total volume approached 650 km^3, and the

Table 4.4. *Potential reserves of hydroelectric power of European countries*[a]

Country	Hydropotential, TW hour per year			Extent of use of economic potential, %
	Theoretical (full)	Technically possible	Economically possible	
Austria	150	75	54	43.3
Finland	46	20	20	54.5
France	266	72	–	(79.7)
Germany	120	27	20	72.2
Greece	84	25	16	–
Italy	340	150	65	23.4
Norway	550	(270)	200	(44.8)
Romania	70	40	17	27.5
Russia[b]	(2896)	(1670)	(852)	(10.0)
Spain	150	70	66	37.4
Sweden	200	(130)	95	(56.2)
Switzerland	144	41	37	75.6
Ukraine	45	21	17	–
Yugoslavia	84	50	–	40.2

[a] Values in parentheses are approximate data.
[b] Data given are for the entire territory of the country.

area they covered was about 85 000 km^2 (Avakyan *et al.*, 1987). Generally, there is about 900 m^3 per head in the reservoirs of Europe.

Reservoirs are utilized in all European countries and they are noted for great diversity in terms of size, type and purpose. There are three main types: the relatively small and medium-size stream reservoirs with a total volume of 100 million m^3 which are located in mountains and on piedmont rivers; medium-size and large river reservoirs; and medium-size and large reservoirs based on lakes.

Channel reservoirs of the first type are used for hydropower production and irrigation in the mountains and on the piedmont rivers in Spain, Portugal, and in other southern European countries. In Switzerland, Austria and Germany, such reservoirs are used for storing floodwater and for navigation. In the UK, Ireland and Germany, relatively small reservoirs (with a volume less than 100 million m^3) are widely used for industrial and community water supply.

Reservoirs of the second type are, as a rule, multi-purpose (power production, navigation and water supply) and are to be found in Russia and the Ukraine. On the Volga River, for example, there is a unique cascade of 11 reservoirs with a total volume of over 185 km^3 (over one-quarter of the total volume of Europe's reservoirs); and in the Dnieper Basin, there is a cascade of six reservoirs operating, which has a total volume of about 44 km^3.

The lake reservoirs (the third type) are predominantly located in the northwest of Russia, Finland and Sweden and were mainly designed for electric power generation, and for navigation and timber rafting. These reservoirs have a total volume exceeding

35 km^3. Data on the main reservoirs are given in Table 4.5. In addition to the large reservoirs, there is also a great number of small reservoirs with a combined volume which is approximately one-quarter of that of the large reservoirs. Presently, most new reservoirs are being constructed for water supply for industry, or for domestic purposes supply. However, the trend in the design and construction of reservoirs is towards conjunctive use of surface and ground water storage.

Reservoirs regulate approximately 10% of surface water resources in Europe. Bulgaria ranks first in this respect (38%), while there is significant regulation of discharge in Spain (30%), Portugal (21%) and Slovakia (15%). Data on the number of reservoirs in the different countries and their main characteristics are given in Table 4.6 (Korzun, 1974b; Avakyan *et al.*, 1987).

More of the water consumed in Europe today comes from reservoirs than it did in 1900. According to Shiklomanov and Markova (1987) and Shiklomanov (1988), there has been, approximately, a 150-fold growth: from 0.1 km^3 in 1900 to 15 km^3 in 1990, which accounts for approximately 3% of the total water abstracted on the continent.

Water is also used for river and canal transport. There is an extensive network of waterways in the Danube Basin and in the river basins connected to the Atlantic, Mediterranean and Baltic Seas. The total length of this network is about 100 000 km. For instance, Poland has 4600 km of waterways, Hungary has 1600 km, and Germany 4200 km (Vintse, 1979; Yaskovyak, 1979). The longest route in the European territory of Russia is 14 500 km (CMEA, 1977). Transport by water developed from the 1950s to the 1970s.

Table 4.5. *Principal reservoirs of Europe*

Country	Basin	Reservoir	Year of filling up	Head at dam, m	Full/efficient volume, km^3	Type of use[a]
Russia	Volga	Kuibyshevskoye	1959	29	58.0/34.6	H N W I F
		Volgogradskoye	1961	27	31.4/8.25	H N W I F T R
		Rybinskoye	1955	18	25.4/16.7	H N W F T R
		Cheboksarsckoye	1986	15	12.6/5.7	H N W I
		Gorkovskoye	1961	17	8.8/2.8	H W I F R
	Kama	Nizhne-Kamskoye	1981	15	12.9/4.4	H N W I F
		Saratovskoye	1971	15	12.9/1.8	H N W I F
		Kamskoye	1964	21	12.2/9.2	H N W I
		Votkinskoye	1966	23	9.4/3.7	H N W R
	Don	Tsimlyanskoye	1952	26	23.9/11.5	H N W F R
	Svir, Oneg.oz.	Onezhskoye	1960	17	–/13.3	H N F T R
	Kuma-Kovda	Kumskoye	1965	19	9.8/8.7	H
	Tuloma	Verkhne-Tulomskoye	1964	46	11.5/3.9	H W F T R
	Imandra	Imandrovskoye	1936	13	11.2/2.3	H
	Sheksna	Sheksninskoye	1964	15	6.5/1.9	H W
	Niz.Vyg	Vygozerskoye	1933	6	6.4/0.7	H
Ukraine	Dnieper	Kakhovskoye	1958	16	18.2/6.8	H N W F R
		Kremenchugskoye	1963	17	13.5/9.1	H N W F R
Finland	Vuoksi, Saimaa	Saimaa	1925	–	7.4/–	H T
Sweden	Gota–Alv, Vanern	Vanern	1915	–	7.2/–	H N W R
Netherlands	Rhine, Zuiderzee	Afsluitdijk (Isselmer)	1932	–	6.0/–	W

[a]H, hydropower; N, navigation; W, water supply; I, irrigation; F, fishery; T, timber rafting; R, recreation.

Table 4.6. *Principal information[a] about reservoirs by countries of Europe*

Country	Number of reservoirs with $V_o > 1 \times 10^6$ m^3		Total V_o, km^3		Total V_e, km^3		Total area of water surface, km^2	
	1970	1990	1970	1990	1970	1990	1970	1990
Austria	28	47	–	–	–	–	–	–
Belgium	3	4	0.14	0.25	0.10	0.16	9	–
Bulgaria	44	226	2.7	10.7	2.5	9.5	–	–
Czech Republic + Slovakia	119	215	3.3	6.8	2.3	4.8	378	670
Finland	–	–	220	–	60	–	32 000	–
France	–	–	2.0	–	–	–	–	–
Germany (former DDR)	131	–	0.9	2.3	0.64	1.70	–	–
Germany (former FRG)	133	–	2.26	–	–	–	–	–
Greece	32	45	8.7	15.5	4.5	9.15	800	1020
Hungary	44	150	–	–	0.23	2.0	170	800
Poland	74	–	26	59	2.3	6.3	320	1010
Portugal	46	70	5.3	11.6	4.7	9.0	258	665
Romania	–	–	2.6	22.0	1.7	14.5	–	–
Sweden	26	26	27.1	27.1	–	–	14	14
Switzerland	81	–	3.68	–	3.17	–	113	–
Turkey (European part)	46	–	13.9	–	–	–	–	–
United Kingdom	189	203	1.5	2.2	–	–	145	190
Yugoslavia	71	–	12	–	–	–	–	–

[a]V_o, full volume of reservoir; V_e, efficient volume of reservoir.

Table 4.7. *Specific norms of water supply by countries of Europe*

Country	Fraction of urban population, %	Specific water expenditure, l/day per head
Albania	51	12
Austria	95	220
Finland	92	250
France	95	300
Greece	78	160
Ireland	87	50
Italy	94	380
Netherlands	96	140
Portugal	83	440
Romania	82	250
Russia	73	280
Spain	90	360
Sweden	96	470
Switzerland	97	320
United Kingdom	94	170

However, the proportion of the total volume transported by water is gradually decreasing due to the development of the railways, road haulage, aviation and pipelines.

DOMESTIC WATER SUPPLY

An enormous amount of water is used for domestic purposes. Abstraction in the 1980s and 1990s averaged 13% of the total with about 12% consumed (Shiklomanov and Markova, 1987; Gleick, 1993, 1998).

Domestic use of water ranges from 10 to 800 l/day per head, with great differences between towns and rural areas, and between the developed and developing countries. There is considerable variety in the published water supply data available for individual countries and for Europe in general and some figures can differ by an order of magnitude (CMEA, 1981; Mikhura, 1982). Table 4.7 shows published national data for domestic water use and population for 1980: the greater the urban population the higher is the use of water. However, the application of new technology for reducing demand can limit and even reduce the abstraction and use of water. Over Europe as a whole more than 68% of the total population is served by piped water supply. In the more advanced European countries, such as Germany and Switzerland, this figure approaches 100%; while in Portugal, Spain and Romania it is around 30% (Kosicheva and Odesser, 1979). During the 1990s, water supply systems have been constructed in many European countries where such services were lacking.

In many advanced states, potable water supply is based on ground water resources, particularly in countries where surface runoff is small. For instance, in Hungary, the annually renewable ground water resources are estimated to be 10.3 million m^3, which is 1.7 times higher than the reserves of surface water (Kirgizov, 1982a). In other countries, even where ground water reserves are much smaller, the importance of ground water is increasing. Ground water reserves in Belgium and Czechoslovakia are about 10% of surface water resources; in Finland, less than 2%; in the United Kingdom, about 8%; in Bulgaria, 16%; and in Sweden, 3% (Plekhach, 1976; Margat, 1990). The largest ground water reserves are located in Russia (over 200 km^3) (SHI, 1987). While most of the ground water is used for the domestic water supply, it is also used for industry, and in some countries for irrigation.

The proportion of ground water and surface water in the domestic water supply are given as an example in Fig. 4.3 for 1988 (Gleick, 1993). For many western European countries, more ground water is used to meet domestic requirements than surface water, except in the United Kingdom, Norway, Spain, Bulgaria, Greece and Sweden. However, even in these countries, reserves are being developed and their use is gradually increasing (Margat, 1990). In a number of European countries urban water supplies are based on artesian wells, since this water has a good quality in terms of its physical, chemical and biological properties. Due to this high quality, in many countries measures are taken to protect ground water from pollution and salt water intrusion. In Austria, Denmark and Portugal, studies are being conducted on the ground water in unused aquifers. In Belgium, Germany, France, Switzerland and Sweden, extensive government programmes are being implemented on ground water resources (Shvartsev, 1995).

Obviously, in future, the demand for water in this sector will grow in proportion to the increase in population, as well as with rising standards of living. However, for Europe in general, there has been little growth in abstractions during recent decades. In those countries with a high level of economic development, the demand for water has ceased and in some there is even a reduction of the domestic demand because of the low growth of population. There are also changes in national strategies for water, in particular moves towards saving water in all sectors of the economy. Applying new techniques for the re-use of water, by reducing leakage from distribution systems and by demand management, significant savings can be attained.

AGRICULTURE

The largest demand for water comes from agriculture; indeed water use in this sector has been steadily increasing from the beginning of the twentieth century to the present day. According to Shiklomanov and Markova (1987), use of water by agriculture increased approximately 9 times from 1900 to 1990. In the 1980s and 1990s, about 38% of the total water abstracted was for

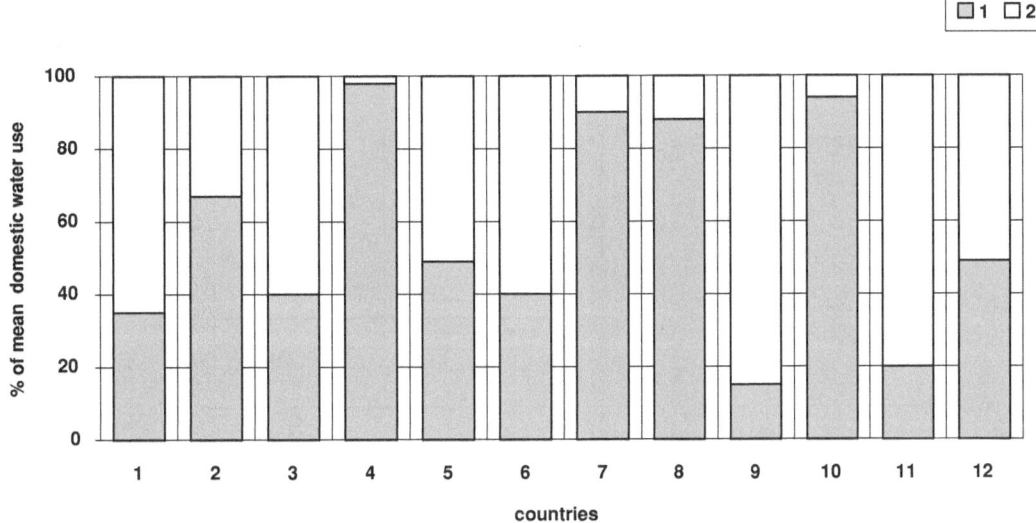

Fig. 4.3 The fraction of ground (1) and surface (2) components in the total domestic water use in 1998 for the European countries. 1, United Kingdom; 2, Belgium; 3, Bulgaria; 4, Denmark; 5, Finland; 6, Greece; 7, Hungary; 8, Italy; 9, Norway; 10, Portugal; 11, Spain; 12, Sweden.

agriculture, with about 70% of this being consumed (Shiklomanov and Markova, 1987). Other sources show that the water used for agriculture is about 33% of the total (Gleick, 1993, 1998; Margat and Vallée, 2000). Across the different regions, from 1% to 50% of the total water abstracted is used for agriculture, and, depending on the climate, more than half of this amount is consumed.

Particularly large amounts of water are used for irrigation. Irrigated areas continue to increase in size so that the production of foodstuffs equals or exceeds the growth of population. For instance, between 1900 and 1980, the water used for irrigation in the USSR increased 4 to 5 times (Shiklomanov, 1988) so that in 1980 it amounted to one-half of the total amount of water abstracted. In Hungary and Bulgaria, where irrigated agriculture is well developed, this proportion is much greater and can reach 80% (Stamenov, 1977; CMEA, 1981; Laslo, 1984).

Assessments of the water abstracted for irrigation are often based on information on the extent of irrigated areas, as well as the actual figures for the specific water requirements for irrigation (ICID, 1981, 1982). The main irrigated areas are in the south of the continent and have been developed during the second half of the twentieth century.

Recently human activities have had an increasing impact on both land and water resources, due to the growth of population and technological progress. The land area per person has reduced annually by about 2%; and the area of productive land by 6–7% (De Mare, 1977). The relative availability of land resources in Europe is quite variable for the different countries. The largest

land resources exist in the United Kingdom, Germany, Italy and in the European territory of Russia; while the smallest ones are to be found in Iceland and Ireland. Presently, Europe has sufficient land resources, and reserves of productive land account for about 80% of all land areas. However, the increase in the area farmed is curtailed by land degradation due to erosion, waterlogging and salinization, to urbanization and the construction of roads and reservoirs.

Reliable evaluation of the long-term changes in the irrigated area is a difficult task, since there are no data for many countries; while the data that exist may be misleading. The most reliable country information is published by FAO; however, this information can also be contradictory. The growth of the total area irrigated, and the prediction of future trends, has been studied by researchers from different countries (Holy, 1971; Zonn and Nosenko, 1981; Alayev and Kolosov, 1989). The results of some of these studies agree reasonably well and Fig. 4.4. presents the changes that have taken place for individual countries for the period from 1970 to 1995, and for the entire continent (1940–95), using data mainly obtained from FAO (FAO, 1995a). It is also obvious that the growth of irrigation slowed or stopped in the 1990s.

Irrigation expanded north from the semi-arid areas as part of the push towards higher yields. This is confirmed (Zonn and Nosenko, 1981) by the presence in 1980 of irrigation in Sweden (86 000 ha), Norway (50 000 ha) and Finland (100 000 ha).

Presently, in western European countries, there is an arc of approximately 15–16 million ha of irrigated lands, the largest areas being found in Italy (3.5 million ha), Spain (2.8 million ha), Romania (2 million ha) and France (1 million ha). In the ETSU, there is about 6.7 million ha of irrigated land, most of which (99%) is in the southern region. In evaluating the volumes of water needed for irrigation, account should be taken of the improvements in

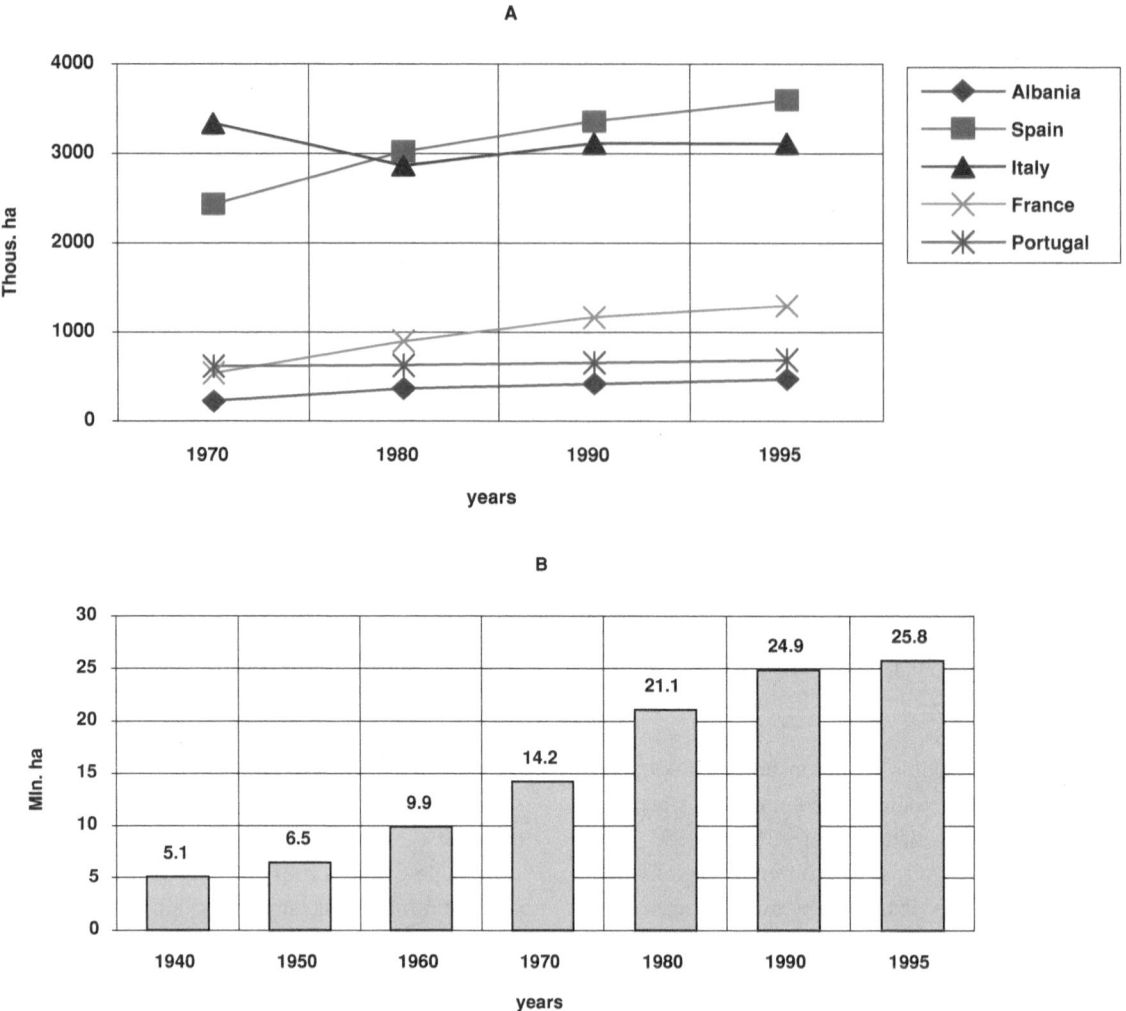

Fig. 4.4 Dynamics of irrigated areas in Europe. A, selected individual countries; B, Europe as a whole.

technology, technical facilities, and irrigation techniques, which have been aimed at saving water.

While there is greater pressure on European water resources because of increasing demand for water for domestic purposes and for industry, there is also a greater recognition that water is needed to maintain the environment. Future estimates of water use need more precise information on water resources, and on better assessments of abstractions as well as on the reliable and comprehensive listing of the relevant socio-economic factors. The actual demand for water is primarily dependent on the size of the population and on its future growth, as well as on the likely domestic needs, and on the level of demand by industry and agriculture, together with the improvements of technology and the procedures for managing the economy. In fact a comprehensive approach is vital to the solution of current water management problems and for the prediction of the future demand (Stanners and Bourdeau, 1995).

4.4 HYDROLOGICAL DATA

The main source of information on the state of the water bodies and on the hydrological cycle is the Europe-wide hydrometeorological network. The importance of this observational network is growing steadily with the increasing role played by water in the economy of all European states.

The network of stations and gauges is generally the most dense compared to other continents. There are about 6000 hydrological stations in Europe, roughly one for each 1750 km². The first data to be collected on river discharge was for the Geta–Elv at Vaenersborg, in Sweden, in 1807. However, most of the systematic observations of hydrological variables were started in the second half of the nineteenth century. Hydrological observations were needed first in many European countries because of the use of rivers for navigation and the construction and use of canals. Floods were also a stimulus; floods that caused the inundation of large cities such as Paris, Amsterdam and Saint

Petersburg led to the establishment of services to provide warnings based on observations from the hydrological network. For instance, after several major floods in the Seine Basin in the 1840s, a warning service was organized which was based on 11 local offices, 120 water level gauges and 423 rain gauges (Spengler, 1964).

The most complete observations of runoff, including measurements of associated phenomena (the nature of the channel network, the characteristics of the ice regime, low flows, geology and hydrochemistry), started after the Second World War in most countries. Since then the form of the hydrological network has remained fairly static but knowledge of hydrology generally is very variable. By the end of the twentieth century, most countries in western Europe, central Europe and the European part of the former USSR appeared to be adequately covered by networks of hydrological instruments and well-established systems for collecting and processing the data. These networks were generally poorest in the southern countries and in the northeast of Russia. A dense and well-organized network exists in Switzerland, Germany and some other countries, with each hydrological station sampling an area of between 100 and 300 km^2. In countries of southern Europe such as Albania and Greece, the network is, as a rule, much sparser.

However in some countries systematic observations of certain hydrological variables started relatively recently. For example, in the United Kingdom, many river gauging stations in the present well-organized and extensive network were installed less than 40 years ago.

The longest of the long-term systematic observations of runoff have been made at: Geta–Elv at Vaenersbor (Sweden), from 1807; Rhine at Basel (Switzerland), from 1808; Neman at Smalininkay (Lithuania), from 1812; Vltava at Prague (Czech Republic), from 1825; Elbe at Decin (Czech Republic), from 1847; Danube at Orsova (Rumania), from 1838; Vuoksa at Imatra (Finland), from 1851; Neva at Petrokrepost (Russia), from 1859. There are also long-term observations on the following rivers: Maas, Weser, Seine, Garonne, Loire, Inn, Rhône, Kemi, Glomma, Vistula, Zapadnaya, Dvina, Narva and Volga. Generally, there are about 30 hydrological sites where the observations commenced in the nineteenth century.

Table 4.8 gives an idea of the state of hydrological knowledge for Europe, using 1986 data for a number of meteorological and hydrological stations operated by the hydrological services (FRIEND, 1993). It should be also noted, that the Rhine (515 stations), the Danube (248 stations) and the Elbe (87 stations) are the rivers with the densest station network.

The pattern of organization of national hydrometeorological services in Europe states has features reflecting the way in which the governments and the economy of the particular country has developed, as well as its history and geography. For instance, in

Table 4.8. *Number of hydrometeorological stations by the countries of Europe (for 1986)^a*

Country	Meteorological stations	Hydrological stations		
		Runoff	Ground water level	Water balance
Austria	1025	711	2060	–
Finland	–	950	50	60
France	–	2000	–	–
Germany (former FRG)	2454	1800	–	–
Hungary	1320	830	2370	–
Ireland	–	1000	–	–
Sweden	–	250	–	–
Switzerland	100	750	–	–
United Kingdom	–	1200	220	–
European territory of the former Soviet Union	2780	3030	16000	12

^{a}A dash indicates lack of information for 1986.

Sweden, and several other countries, the hydrological network is controlled by a centralized hydrometeorological service. In other countries (France and the United Kingdom) similar functions are shared by state organizations accountable to different Ministries and Departments, and by public and private agencies.

The organization of the hydrometeorological service of Sweden, France and the Russian Federation are now considered in more detail. In Sweden, there are 118 main river basins, with over 250 flow gauging stations. The hydrometeorological network of Sweden is co-ordinated by the Swedish Institute of Meteorology and Hydrology (SMHI, Norrkoping), under the Ministry of Transport and Communications. The national hydrometeorological archive is held by SMHI with observations of water levels and discharges on rivers and lakes starting at the beginning of the nineteenth century. The Swedish Geological Survey has responsibility for ground water while the quality of surface water is dealt with by the National Board on Environmental Protection. The management of large regulated rivers, mainly those flowing in the north of Sweden, is co-ordinated by private companies. The National Board of Environmental Protection is engaged in evaluating changes in water resources and in the registration of water discharges (Falkenmark, 1977; Gottschalk, 1985). For this purpose, the country has been subdivided into 300 water management regions; within each region, 2900 river or lake areas are distinguished, called water bodies. For each water body, a programme has been developed for the rational use of water, discharge and intake control, hydropower engineering construction, navigation, fisheries, irrigation and other measures.

France has a hydrological network of some 2000 stations. These stations are operated by numerous agencies, mainly belonging to the Ministry of Agriculture, or the Ministry of Environmental Protection. A network of 1100 stations is controlled by 22 regional Water Management Agencies (SPRAE), located on small and minor river catchments in 22 water management regions, and they belong to the Ministry of Agriculture. These agencies co-ordinate water policy in each water management region and are responsible for the rational use of water resources, including the collection of a fee for abstractions, for discharges of different pollutants, and for the violation of water legislation. These agencies also prepare the programmes for the development of water resources (Shvartsev, 1995). The Ministry of Agriculture operates the Regional Archive of Hydrological Information (ARHMA), where the runoff series, as a rule, are short because the period of observation starts in the 1960s and 1970s. Hydrological stations, accountable to the Ministry of Environmental Protection, are located on large rivers and have the longest time series of observations of runoff, beginning in 1863. From 1986, these series have been included in the ARHMA archive. There are other hydrological stations which are the responsibility of six basin agencies that cover the whole of France.

The hydrological network of the Russian Federation inherited the reference hydrological network of the former Soviet Union. The Hydrometeorological Service of the Soviet Union was organized at a rather high level and had been functioning since 1859. In 1985, the reference network consisted of 7152 stations, about one-half being located in the European part of the country. Table 4.9 presents the changes in the network since 1940 (Shiklomanov, 1989). Up to 1985, the network had been developing at a high rate but since 1985 the number of stations has reduced markedly.

In addition to the hydrological network for surface water observations, of great importance is the network of hydrogeological stations, which is located in the different geographical zones and economic regions of the country. Perhaps even more important is the network of stations that register water use.

The study of hydrological processes is conducted through the specialized observational network, set up at the beginning of the 1950s and consisting of water balance, bog, mudflow runoff, and snow avalanche stations, plus stations for determining evaporation from water, land and snow surfaces, and sites for studying the fluviomorphological processes in rivers. Water balance stations, constructed in the 1940s to sample the different geographical zones, are employed to study of the processes of runoff formation.

In the USSR, as in other countries, the data produced by the networks gave rise to various studies and to the control of surface water quality, which have been widely applied, giving the most extensive, objective and systematic information on the state of the nation's water bodies. One result was the establishment of the State Service of Observations and Control over the Pollution of Natural Environment.

Table 4.9. *Dynamics of the number of hydrological network in the European territory of the former Soviet Union*

	Number of stations	
Year	Rivers	Lakes and reservoirs
1940	2410	41
1960	3260	235
1975	2660	346
1985	3030	414

However, from 1986, during the transition to the new form of economic management, the number of stations in the hydrological network has decreased due to the economic difficulties experienced by Russia. A similar situation has existed in all the republics of the former Soviet Union up to the present time.

An important contribution to the study of water resources globally and particularly to those of Europe has been made through UNESCO and the International Hydrological Programme (IHP). Improvements in the collection, management and assessment of data on water resources, the publication of the results of such studies, and the provision of information on the use of water resources and environmental protection on a global scale are amongst the topics addressed. Numerous UNESCO publications offer detailed information on current hydrological studies in Europe. The majority of European countries are presently involved in different international projects within the IHP, such as the FRIEND Project, which have the improved the availability, study and publication of hydrometeorological information. International archives and data banks, accessible to all users, are set up within the framework of these international programmes and in the different projects that make them up. The best-known hydrological archives include those set up in the institutes at Koblenz (Germany) and Wallingford (UK).

Presently, the hydrological network in western European countries is a well-established system for the collection, management and assessment of most if not all the necessary hydrological information. In these countries there have been a series of improvements to the instruments, including better means for transmitting data, and to the hardware and software for information management and data transfer. However the republics of the former Soviet Union, as well as the nations of Eastern Europe, have been suffering from a lack of necessary funds for the maintenance and development of their hydrological networks. Here the priority is to reach and maintain an adequate level of funding.

In most countries, the results of hydrological research is published in periodical reports, yearbooks, and in other material, issued by national research institutes as well as in scientific journals,

Table 4.10. *Comparison of water resources (km³) assessment by the individual countries of Europe*

Country	L'vovich (1974), total water resources	Engelman and LeRoy (1993), total water resources	WRI (1992) Local	WRI (1992) Foreign inflow
Albany	22.2	21.0	10.0	11.3
Austria	57.1	90.0	56.3	34.0
Belgium	9.4	16.9	8.4	4.1
Bulgaria	17.8	205.0	18.0	187.0
Czechoslovakia	27.5	90.9	28.0	62.6
Denmark	15.1	13.0	11.0	2.0
Germany (former DDR)	19.4	–	17.0	17.0
Germany (former FRG)	85.6	200.0[a]	79.0	82.0
Finland	106.0	113.0	110.0	3.0
France	232.0	185.0	170.0	15.0
Greece	66.0	58.9	49.4	13.5
Hungary	8.4	115.0	6.0	109.0
Iceland	66.4	170.0	170.0	0
Ireland	40.2	50.0	50.0	0
Italy	160.0	187.0	185.0	2.0
Netherlands	10.2	90.0	10.0	80.0
Norway	376.0	413.0	405.0	8.0
Poland	56.3	56.0	49.5	6.8
Portugal	26.7	66.0	34.0	31.6
Romania	35.6	208.0	37.0	171.0
Spain	129.0	111.0	110.0	0.1
Sweden	194.0	180.0	176.0	4.0
Switzerland	47.5	50.0	42.5	7.5
United Kingdom	152.0	120.0	120.0	0
Yugoslavia	123.0	265.0	129.0	115.0

[a]Data for United Germany.

such as the *Journal of Hydrological Sciences*. In these publications, the necessary data on the nation's water bodies are set out together with supporting information, while reports are also published on assessments of the water resources of river basins and certain regions sometimes in the context of national studies on the natural environment. There are a great number of such reports, prepared by individual authors, national institutes and international groups, dealing with the analysis of water resources of river basins and the different regions. These studies are being extended and updated.

Of great scientific value for the promotion of hydrological knowledge are the assessments of water resources for certain river basins and for Europe as a whole. Since these studies are so numerous only some of the results are presented here. According to the earliest studies, the water resources of Europe were

Table 4.11. *Distribution of areas in Europe by watersheds*

Runoff region	Area km² × 10³	Area %
Continental slope of the Arctic Ocean	1 400	12
Arctic Ocean islands	131	2
Continental slope of the Arctic Ocean with islands	1 531	14
North Sea basin and Atlantic slope	1 434	14
Baltic Sea basin	1 729	17
Black Sea basin	2 151	21
Mediterranean Sea basin	866	8
Continental slope of the Atlantic Ocean	6 180	60
Atlantic Ocean islands	589	5
Continental slope of Atlantic with islands	6 769	65
Caspian Sea basin	2 160	21
Continental slope of Europe	9 740	93
European continent islands	720	7
Continental slope of Europe with islands	10 460	100

found to be: 2830 km³ (Chernogayeva, 1969), 3210 km³ (Korzun, 1974b), 3154 km³ (L'vovich, 1974), 2800 km³ (Baumgartner and Reichel, 1975); and in the later publications: 3200 km³ (Shiklomanov and Markova, 1987), 2321 km³ (WRI, 1990, 1992, 1996, 2000; Gleick, 1993, 1998). The differences between these assessments are mainly due to the amount of hydrological information used. Assessments of the water resources of the main river basins, lakes and seas obtained by different authors, as a rule, have smaller differences, as they are based on measurements of runoff.

Table 4.10 presents some data on water resources of certain European countries, which are obtained from different sources (L'vovich, 1974; WRI, 1992; Engelman and LeRoy, 1993). Again these assessments differ markedly, due to the way these assessments are calculated and the amount of information used.

In this Monograph the water resources of Europe are assessed to the limits of the continent, taking account of the positions of the main watersheds separating the runoff to the different oceans and seas. In addition areas of inland drainage are identified, such as the basin of the Caspian Sea. Four large basins can be distinguished draining to the northwest (Baltic Sea), to the west (Atlantic coast, including the North Sea), to the southeast (Black Sea and Azov Sea), and to the south (Mediterranean Sea). The boundaries are shown in Fig. 4.5 and more information is given in Table 4.11.

In contrast to these divisions into river basins, Europe can also be divided into economic regions. Shiklomanov and Markova (1987) give five such regions namely: northern (1), central (2) and

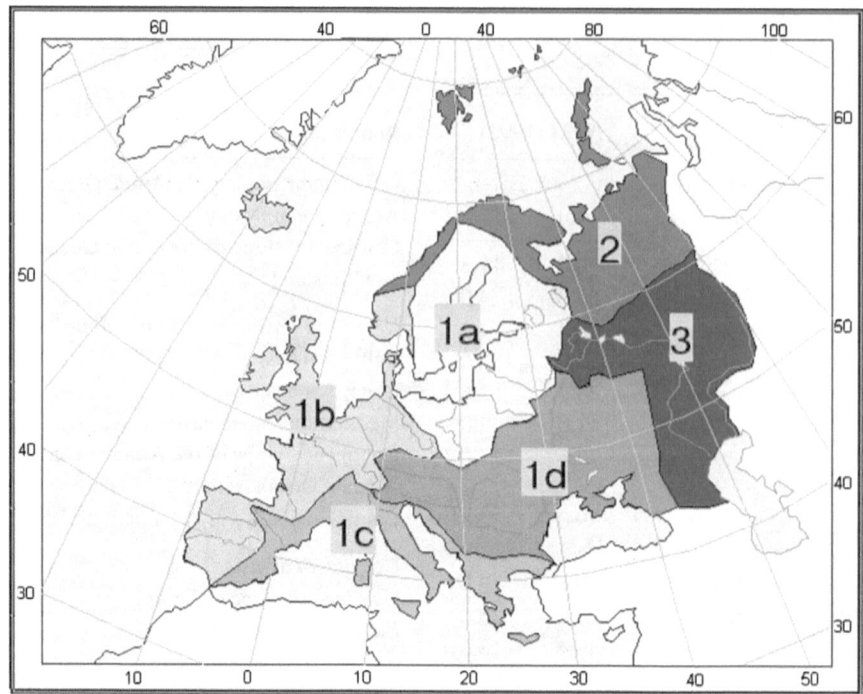

Fig. 4.5 Map of catchment basins of oceans and large seas in Europe. 1, Atlantic Ocean, including the Baltic Sea basin (1a), the North Sea basin and Atlantic coast (1b), the Mediterranean Sea basin (1c) and the Black and Azov Seas basins (1d); 2, the Arctic Ocean slope; 3, the interior runoff region (the Caspian Sea basin).

southern (3) Europe, as well as the northern (4) and southern (5) parts of the European territory of the former Soviet Union, referred to as the northern ETSU and the southern ETSU, respectively.

The boundaries of these regions are shown in Fig. 4.6. For the first three regions, the boundaries coincide with the national frontiers, while the boundary between the northern and southern parts of the ETSU is the main watershed. Table 4.12 presents a classification of all European countries in terms of these regions but only the European parts of Russia, Kazakhstan and Turkey are considered.

The areas of these natural economic regions are broadly similar, and they are employed as the basis for assessing water resources, water use and other aspects of water availability.

4.5 DISTRIBUTION OF WATER RESOURCES IN TIME AND SPACE

4.5.1 Initial data and methodology

In this Monograph water resources are taken as the annually renewable surface fresh waters contained in rivers and lakes, as well as in reservoirs, canals and other water courses. Water re-

sources will be considered in the following categories: river water resources, resources of the basins draining to the seas and oceans, and water resources of the different administrative units (natural and economic). The largest value of runoff at the mouth of the basin is interpreted as the river's water resources; this is, most often, the average annual flow at the outlet from the basin.

Basins that are noted for their industrial activity often require their flows to be naturalized. These naturalized runoffs are calibrated from knowledge of abstractions from and discharges to the river and from data on diversions from this basin to others, and on changes to storage in reservoirs due to evaporation. Obviously, for these calculations a large amount of information on the use of water in the basin is required. In practice, the necessary data on water use are either lacking or insufficient; therefore the naturalization of runoff has to be undertaken by indirect methods. As many basins are large and used intensively, the most acceptable procedure for naturalization of runoff is the water balance method (Shiklomanov, 1979), which gives the normalized runoff by taking the difference from the main water balance elements. This information can often be obtained from the relevant archives or from maps (Russel and Miller, 1990; Russel and Richard, 1990).

For basins with few data or no data at all, the average water resources can be determined from maps showing annual flow in the form of isopleths and flow variability, or by applying the procedure of hydrological analogy (Rozhdestvensky and Chebotaryev, 1974). The maps in the *Atlas of the World Water Balance* (Korzun, 1974a) showing annual runoff and other hydrometeorological

Table 4.12. *Distribution of European countries by natural–economic region*

Country, region	Area	
	km² × 10³	%
North Europe	1 322	12.6
Denmark	45	0.4
Finland	337	3.2
Iceland	103	1.0
Norway	387	3.7
Sweden	450	4.3
Central Europe	1 857	17.8
Austria	84	0.8
Belgium (+ Luxembourg)	33	0.3
Czech Republic	79	0.8
France (+ Monaco)	547	5.2
Germany	356	3.4
Ireland	70	0.7
Netherlands	41	0.4
Poland	313	3.0
Slovakia	49	0.5
Switzerland	41	0.4
United Kingdom	244	2.3
South Europe	1 791	17.1
Albania	29	0.3
Bulgaria	111	1.1
Greece	132	1.3
Hungary	93	0.9
Italy	301	2.9
Portugal	92	0.9
Romania	238	2.2
Spain	505	4.8
Turkey (European part)	34	0.3
Yugoslavia	256	2.4
Northern part of European territory of the former Soviet Union (ETSU)	2 710	25.9
Belarus (north)	76	0.8
Estonia	45	0.4
Latvia	64	0.6
Lithuania	65	0.6
Russia (north)	2 460	23.5
Southern part of European territory of the former Soviet Union (ETSU)	2 780	26.6
Belarus (south)	132	1.3
Kazakhstan (European part)	230	2.2
Moldavia	34	0.3
Russia (south)	1 780	17.0
Ukraine	604	5.8
Europe as a whole	10 460	100

characteristics are most useful for this purpose. Assessments of water resources of large basins draining to the sea employ the discharges measured at the lowest gauging stations in these basins. Runoff from inter-basin areas without gauging stations can either be evaluated by analogy or from maps as described above. For rivers much affected by human activity it is necessary to estimate the naturalized flow.

For rivers in the humid north and west of Europe, the naturalized flow is the same or only slightly more than the measured flow at the lowest gauging station. For rivers flowing into the Mediterranean, Black, Azov and Caspian Seas, as well as some reaching the Atlantic Ocean, there are large natural losses due to abstraction for agriculture and industry. For the assessment of water resources for basins or other units, local runoff, the inflow from contiguous basins and the total runoff have to be considered.

Local runoff is made up of the total runoff in rivers and water courses arising within the given area or basin. Inflow to the given basin from contiguous areas is the total runoff of inflowing rivers and water courses at the boundary of these territories. If the river serves as a boundary between two administrative units, its water resources are divided equally between them. The local runoff, together with the inflows from contiguous areas, make up the total annually renewable water resources of a basin or area.

Water resources assessment is based on the analysis of time series of annual and monthly runoffs, preferably those which are long and reliable, together with time series of the other hydrological variables and important information (Babkin and Voskresensky, 1977; Shiklomanov, 1997, 1998a, 2000a, 2000b; Babkin, 1998; Zaretskaya, 1998; Shiklomanov *et al.*, 2000).

For this Monograph a Specialized Archive of the monthly runoffs of European rivers was compiled using the published information from national sources (Romanova, 1968; Sarukhanyan and Smirnov, 1971; L'vovich, 1974, etc.), as well as the relevant material from the Global Runoff Data Centre (Koblenz, Germany). The Specialized Archive consisted of over 600 time series of monthly discharges, half of them for stations in western Europe, and the other half for stations in the European territory of the former Soviet Union.

The main criteria for selecting the stations were: the duration of the period of observation, the uniformity of the distribution of sites and their proximity to river mouths, the trustworthiness and reliability of the observations, and stations with a minimum number of gaps in the time series. For the stations which were chosen, the average duration of the series was 32 years, with a range from 4 to 69 years, while the period of observation can vary between 5 and 178 years. About 35% of all series represented in the Specialized Archive have a duration of over 40 years. A small number (within 10%) were noted for omissions in the observations.

As might be expected, the longest series are for the rivers of northern and central Europe (Sweden, Norway, Finland, Germany

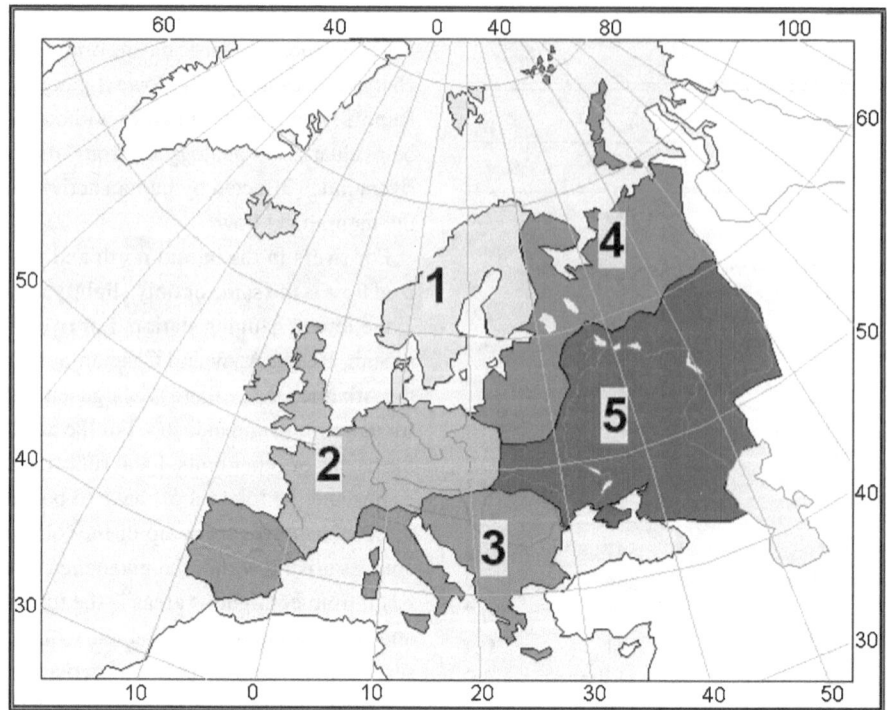

Fig. 4.6 Natural–economic regions of the European continent. North (1), central (2) and south (3) Europe; northern (4) and southern (5) parts of the European territory of the former Soviet Union (ETSU).

and Poland); and the shortest series for rivers in southern Europe (Greece, Albania and Turkey). Runoff series selected for the Specialized Archive were carefully checked for their accuracy and reliability. Figure 4.7 shows their distribution. Some 40 reference sites were selected which are located on the largest rivers: 25 in western Europe and 15 in the European territory of the former Soviet Union. Flow statistics for all reference sites are given in Table 4.13. The different initial series were reduced to a single observational period from 1921 to 1985, i.e. 65 years inclusive.

Stations with time series for a shorter period than 65 years had their records extended by regression against a series for analogue catchments with similar characteristics (Rozhdestvensky and Chebotaryev, 1974) where the correlation was not less than 0.8. If no analogue catchment was found with this correlation then the missing records were estimated from precipitation, accepting that the correlation coefficient with the remainder of the series was at least 0.7.

Gaps in the observations for the annual and monthly series over 65 years were filled by simple or multiple linear regression. Gaps in the monthly series were most often eliminated by regression against the annual series. Checks of the stationarity of each series and tests for trends and cycles and for other statistical features were carried out as described earlier (see Section 2.2). The importance of carrying out such tests cannot be stressed too much when

analysing long-term runoff fluctuations, of particular significance is the stationarity of the statistical parameters. The stability of the other parameters (coefficients of variation, asymmetry, etc.) does not have a major effect on the runoff variability (Ratkovitch, 1976).

As is shown in Section 2.2, checking was undertaken of the initial runoff series for stationarity using Student's and Fischer's tests, as well as those of Kendall and Spearman, and using the procedure of the World Meteorological Organization (WMO), by Polack's method (evaluating the significance of linear trend parameters) and by the Kolmogorov–Smirnov test. The conclusion, which was drawn by combining the results of the four methods allowed the assessment of increases or decreases within the assigned time interval.

To distinguish periods with different river regimes, three techniques were employed: double mass curve, harmonic analysis, and the averaged genetic curve. This led to periods being identified when there was a series of wet and dry years and when the regime was average. The analysis confirms an intra-series cohesiveness in runoff, a feature which was investigated in a number of papers (Kaisl, 1972; Rozhdestvensky, 1988).

An analysis of the statistics of the fluctuations of runoff was performed on the 40 reference sites for the period 1921 to 1985. All the series were found to be homogeneous and stationary and to contain periods of high, low and average flow. Further analyses showed to provide a good basis for constructing the series for characterizing the water resources of the continent.

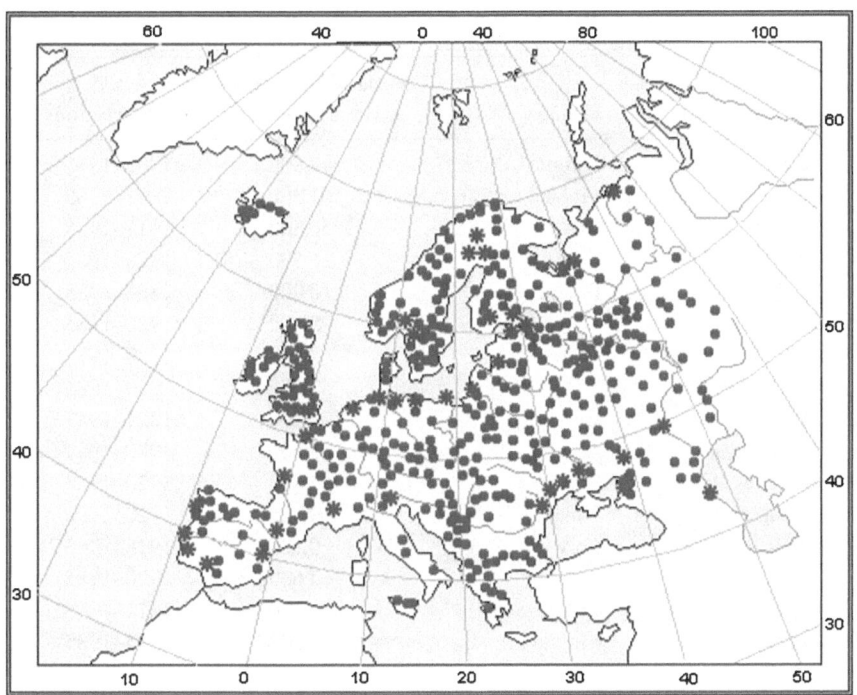

Fig. 4.7 Monitoring sites on European rivers, *, the sites included in the Specialized Archive of the present study; •, reference sites on main rivers.

4.5.2 Water resources of rivers and basins draining to the different seas and oceans

WATER RESOURCES OF RIVERS

Table 4.14 shows the statistical parameters that were calculated – average runoff, coefficients of variation, skewness, and the quantiles of 50%, 75% and 90% availability, taken from a three-parameter gamma-distribution (Rozhdestvensky and Chebotaryev, 1974). If the records were longer than the 65 years, the calculations were performed for the longer period (for instance, for the Thames to 1990 and for the Volga to 1992). The resulting estimates were compared with the estimates published in other papers (for instance, SHI, 1987; WRI, 1990, 1992, 1996, 2000; UNESCO, 1993, 1996). Generally, the results are similar, any differences being within 5–10%.

The patterns of runoff for certain rivers are shown in Figs. 4.8 and 4.9. As an example, Fig. 4.9a illustrates the trend of long-term variability of the Volga runoff by two curves: curve 1 reflects the observed fluctuations; and curve 2 smoothed or averaged runoff fluctuations. Curve 2 shows the trend of Volga flows – with high levels from 1922 to 1931 and also from 1977 to 1990, low flows from 1932 to 1944, and average flows between 1945 and 1954 and also from 1958 to 1971.

The fact that the territory of Europe is very heterogeneous in terms of its geography and climate markedly affects the regimes of its rivers. The regime of a particular river can undergo certain changes as it flows from one geographic zone to another. All the main types of regime are found, each of them being characterized by its own specific features of diurnal, intra-year and inter-year variability. Clearly, it would be desirable to have the probabilistic characteristics of the hydrological regime for all time intervals; however, it is particularly important to assess them for the annual runoff.

Most rivers draining to the Atlantic Ocean have the highest precipitation in the autumn and summer periods, and the minimum is in spring (see Subsection 4.2.3). Winter precipitation falling as rain often causes successive floods especially when the soil is saturated or frozen. Further eastwards, the autumn–winter precipitation maximum gradually changes into a summer maximum and a winter minimum which falls as snow, especially in the rivers draining into the Arctic Ocean. During the spring, snowmelt occurs and spring floods take place, their duration being dependent on the volume of snow stored and the intensity of the melting. Summer precipitation partly infiltrates into the soil, but mostly is lost by evaporation. Significant rainfall floods are recorded in northern regions and in mountain areas when precipitation rates are high.

The relief of the area, particularly the location of the mountain masses, strongly affects the intra-year distribution of runoff. With altitude, a redistribution of precipitation and temperature occurs, resulting in the changing character of the distribution of runoff in time as compared to lower areas.

Karst landscapes affect the intra-year distribution of runoff, decreasing the surface runoff and increasing the percolation to

Table 4.13. *Reference sites of the principal rivers of Europe*

River	Site	Country	Area of catchment at site, km^2	Initial period of observations	Volume of runoff at site, km^3
Narva	Vasknarva	Estonia	47 800	1903–1985	10.2
Vuoksi	Imatra	Finland	61 300	1847–1985	18.6
Kemijoki	Taivalkoski		50 800	1911–1985	16.9
Kymijoki	Pernoo		36 500	1900–1985	9.55
Loire	Montjean	France	110 000	1863–1985	26.5
Rhône	Beaucaire		95 600	1920–1985	53.6
Seine	Paris		44 300	1928–1985	8.48
Garonne	Mas-D.Agenais		52 000	1920–1985	18.9
Rhine	Kaub	Germany	103 700	1821–1987	50.6
Elbe	Wittenberg		123 500	1920–1989	22.8
Weser	Intschede		37 800	1920–1986	9.97
Thames	Teddington/Kingston	United Kingdom	9 900	1920–1990	2.50
Po	Pontelagoscuro	Italy	70 100	1918–1985	47.6
Daugava	Daugavpils	Latvia	64 600	1920–1985	13.2
Neman	Smalininkai	Lithuania	81 200	1812–1985	18.3
Dniester	Bendery	Moldavia	72 100	1895–1985	10.6
Glomma	Langnes	Norway	40 200	1901–1985	21.2
Vistula	Tczev	Poland	193 900	1900–1987	33.1
Odra	Gozdowice		109 400	1900–1987	16.9
Douro	Regua	Portugal	91 500	1933–1985	17.5
Tagus	Vila Velha de Rodao		59 200	1913–1985	10.5
Guadiana	Pulo do Lobo		60 900	1947–1985	5.90
Danube	Orsova/T.Severin	Romania	578 300	1840–1988	176.0
Volga	Volgograd	Russia	1 360 000	1914–1992	252.0
Don	Razdorskaya		378 000	1891–1985	27.5
North Dvina	Ust-Pinega		348 000	1882–1987	97.2
Pechora	Oksino		317 000	1928–1987	137.0
Neva	Novosaratovka		281 000	1859–1987	76.9
Ural	Kushum		190 000	1930–1985	8.9
Kuban	Krasnodar		45 900	1949–1985	13.8
Onega	Porog		55 700	1943–1987	15.4
Terek	Mozdok		20 700	1930–1985	8.0
Ebro	Tortosa	Spain	84 200	1913–1985	48.9
Guadalquivir	Alcada del Rio		47 000	1913–1989	17.4
Jugar	Masia del Mompo		17 900	1913–1987	4.9
Gota–Alv	Vaenersborg	Sweden	46 800	1807–1988	17.2
Torneallven	Pello		27 700	1918–1985	10.0
Lulealven	Waterworks		24 500	1900–1985	15.8
Dnieper	Kakhovsky	Ukraine	482 000	1930–1985	52.4
South Bug	Alexandrovka		46 200	1914–1985	3.2

Table 4.14. *Water resources of principal rivers of Europe (for study period of 1921–85)*

River	Area of catchment, km²	Water resources, km³/year	Coefficient of variation, C_v	Coefficient of asymmetry, C_s	Quantiles of different water availability, km³		
					50%	75%	90%
Narva	56 200	11.3	0.29	0.61	11.3	9.22	7.89
Vuoksi	61 300	18.6	0.22	0.03	18.6	16.3	14.3
Kemijoki	52 000	17.7	0.19	0.32	16.9	14.7	12.9
Kymijoki	37 200	10.1	0.27	0	10.1	8.19	6.55
Loire	120 500	33.6	0.32	0.05	33.9	26.0	19.5
Rhône	98 000	64.9	0.21	0	64.9	55.4	47.2
Seine	79 000	16.5	0.35	0.13	16.6	12.4	8.94
Garonne	56 000	21.4	0.28	0	21.6	17.1	13.5
Rhine	252 000	85.9	0.22	0	86.1	72.6	61.2
Elbe	148 000	34.6	0.16	0	34.6	30.7	27.3
Weser	46 000	13.9	0.29	0.54	13.8	11.0	8.89
Thames	15 300	3.05	0.30	0.04	3.08	2.41	1.86
Po	75 000	52.0	0.26	1.01	50.3	42.5	36.3
Daugava	87 900	20.1	0.24	0.56	19.7	16.6	14.3
Neman	98 200	18.5	0.17	0.37	18.5	16.2	14.2
Dniester	72 100	10.7	0.28	0.61	10.2	8.37	6.89
Glomma	41 400	21.7	0.16	0.29	21.6	19.3	17.3
Vistula	198 500	33.3	0.22	0.63	32.8	28.0	24.1
Odra	118 900	18.1	0.27	0.59	17.7	14.5	12.2
Douro	94 500	29.9	0.46	1.69	27.2	20.2	15.2
Tagus	81 000	12.5	0.64	1.24	10.8	6.65	3.98
Guadiana	72 000	5.90	0.75	1.07	5.07	2.16	0.77
Danube	81 700	225	0.18	0.51	223	197	175
Volga	1 380 000	250	0.20	0.32	250	219	194
Don	422 000	27.3	0.36	0.69	26.1	19.9	15.3
North Dvina	357 300	105	0.17	0.03	105	92.4	81.8
Pechora	322 000	137	0.12	0.04	136	124	115
Neva	281 000	77.9	0.16	0.17	75.7	66.5	58.6
Ural	236 000	9.73	0.66	1.23	8.26	5.05	3.02
Kuban	61 000	13.9	0.14	0.02	13.6	12.0	10.6
Onega	56 900	15.4	0.25	0.33	15.0	13.0	11.4
Terek	43 700	10.0	0.15	0.26	10.0	8.94	8.09
Ebro	87 000	38.5	0.49	0.01	38.9	23.5	12.6
Guadalquivir	57 000	20.0	0.79	0.94	16.7	6.38	1.94
Jugar	22 400	4.98	0.59	0.95	4.64	2.67	1.33
Gota–Alv	50 100	20.3	0.18	0	20.3	17.8	15.6
Tornealven	40 000	12.4	0.13	0.38	12.4	11.2	10.3
Lulealven	25 200	15.8	0.12	0.24	15.7	14.3	13.2
Dnieper	504 500	52.4	0.26	0.65	51.8	43.7	37.5
South Bug	63 700	3.38	0.39	0.79	3.15	2.37	1.80

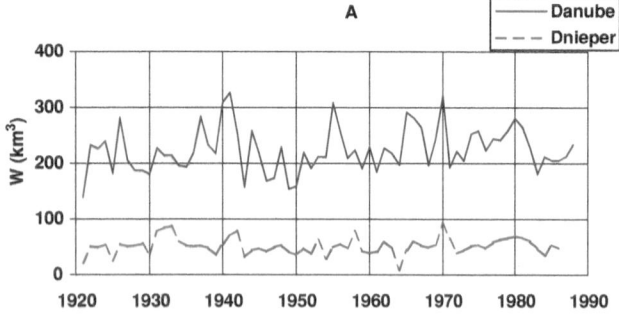

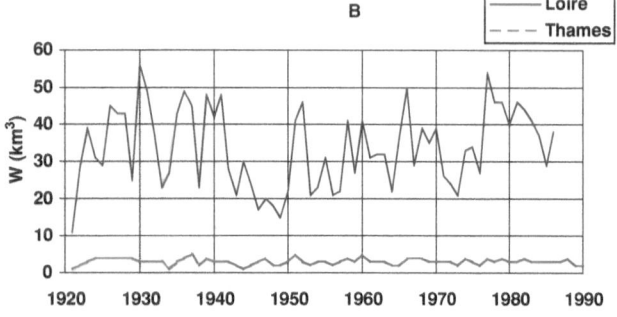

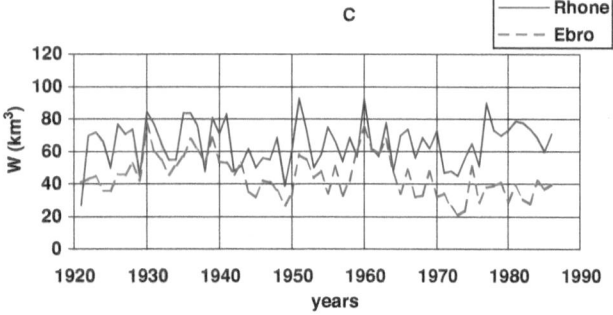

Fig. 4.8 Dynamics of annual runoff for European rivers. A, Black Sea basin; B, Atlantic Ocean basin; C, Mediterranean Sea basin.

ground water. The effect of karst can be seen in the regimes of rivers in the Balkan Peninsula, on the slopes of Dinarides, in the Pyrenees and in the Crimea.

Data from the Specialized Archive were used to classify the regimes of European rivers using the most modern classifications, such as that developed within the FRIEND Project (FRIEND, 1993). The FRIEND classification was created with the aim of mapping the different types of the river regimes in western Europe in time and space, using all the available hydrometeorological information. Figure 4.10 shows the most characteristic regimes, divided in accordance with the FRIEND classification as follows:

1. The northern Scandinavian type (with a first and second maxima of monthly runoff in April to August; and the first and the second minima between January and April). This type

is characteristic of rivers flowing to the Arctic Ocean. The example shown is the Vossa (Bulken).

2. The northern Scandinavian continental type (the first and the second maxima of monthly runoff occur from April to August; the first minimum takes place in January and the second minimum between May and August). This type is found in the northern rivers of the Nordic states in northwestern Russia. The example here is the Kjminginjoki (Haukipudas).

3. The Baltic continental type (the first and the second maxima occur between April and August; the first and the second minima from July to September). This type is characteristic of most rivers in the Baltic Basin and many rivers of central Europe. The Elbe (Wittenberge) is shown as an example.

4. The mountainous or mountainous–transitional type where the river basin is at an elevation over 500 m (the first and the second maxima occur between April and August; the first minimum in January to April, and the second from May to August). This type of regime is recorded on rivers in the piedmont areas of the Scandinavian Plateau, the Alps and the Carpathians. The example shown here is the Ebro (Tortosa).

5. The northern continental type (here the first maximum occurs in March to August, and the second or the third from September to November; while the first and the second minima are between January and April). Such a type is rarely found in central Europe, but is characteristic of many northern rivers in Scandinavia and Russia. The example is the Dalalven (Faggeby).

6. The southern continental type (the first period of maxima is between March and August, the second or the third between September and November; the first minimum falls in January and lasts to April, and the second minimum occurs between May and August). This type occurs in the Pyrenees, the Apennines, in the Balkan and Crimean Peninsulas, and along the Black Sea coast of Russia. The Po (Pontelagoscuro) is given as an example.

7. The Baltic type (the highest flows occur in March and April, and others from November to December; the lowest take place between July and September). The Baltic type is noted for a smoother distribution of monthly runoff than the Baltic continental type, and is characteristic of southern rivers in the Baltic Basin. The Odra (Gozdowice) is an example.

8. The Atlantic type (here, the highest discharges take place between January and February; and the lowest from July to September). This type is found in most rivers draining to the Atlantic and on contiguous islands, and is generally the most characteristic regime of European continent. The Thames (Teddington) is used as an example.

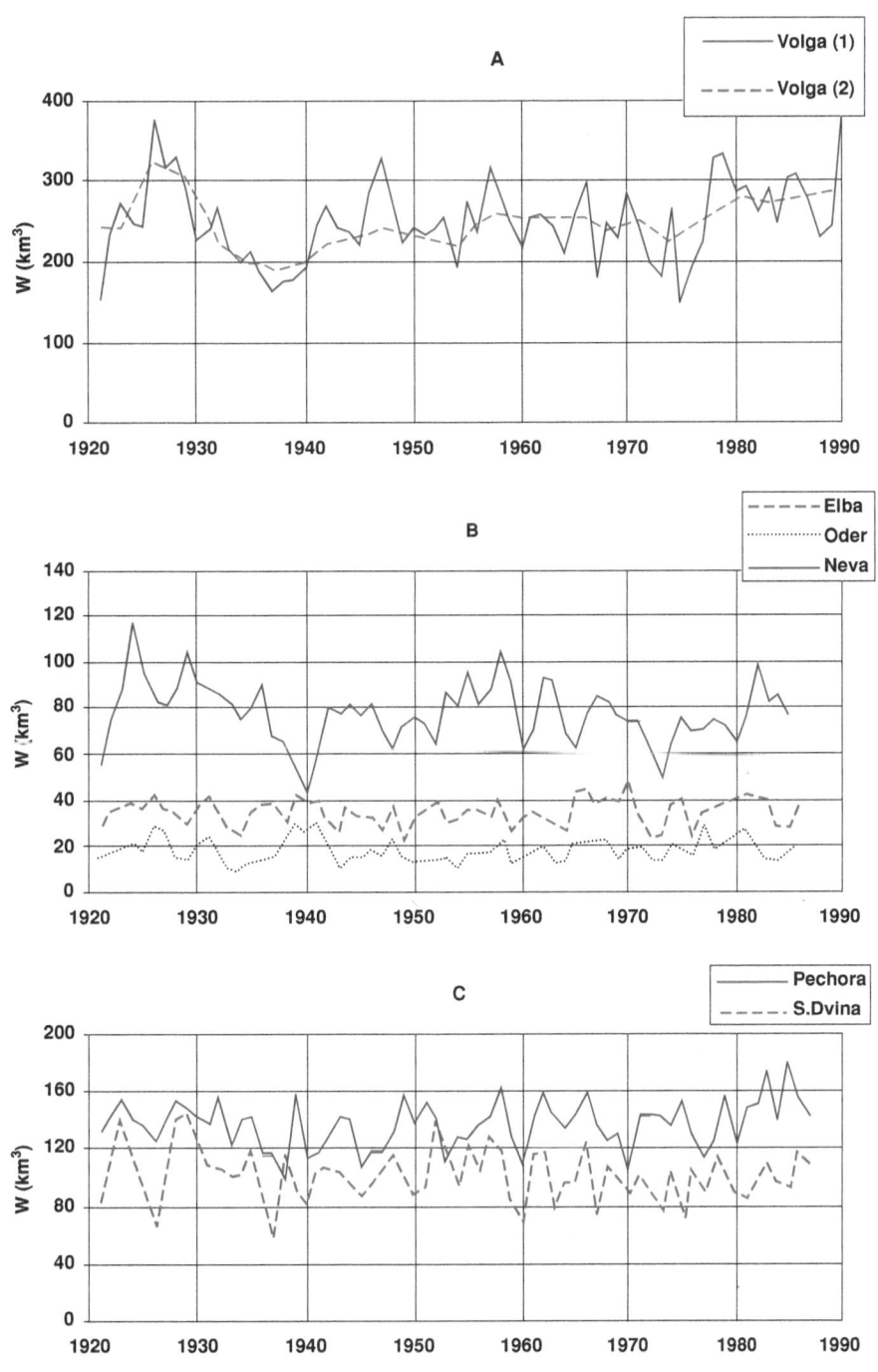

Fig. 4.9 Dynamics of annual runoff for European rivers. A, Caspian Sea basin: (1) initial series, (2) smoothed series; B, Baltic Sea basin; C, Arctic Ocean basin.

It should be emphasized that for the largest rivers, such as the Danube, Rhine, Rhône and others, traversing several climatic and geographical zones, it is rather difficult to characterize the type of regime because it changes from the source to the mouth. Such rivers have a complex regime.

THE WATER RESOURCES OF MAJOR DRAINAGE BASINS

As is shown in Tables 4.11 and 4.5, Europe is divided between basins flowing to the Arctic and Atlantic Oceans and to areas with no outlets, such as the Caspian Basin. The Atlantic-flowing basins can be subdivided into the basins of the Baltic, Black and Mediterranean Seas, as well as those of the Atlantic coast, including the North Sea basin.

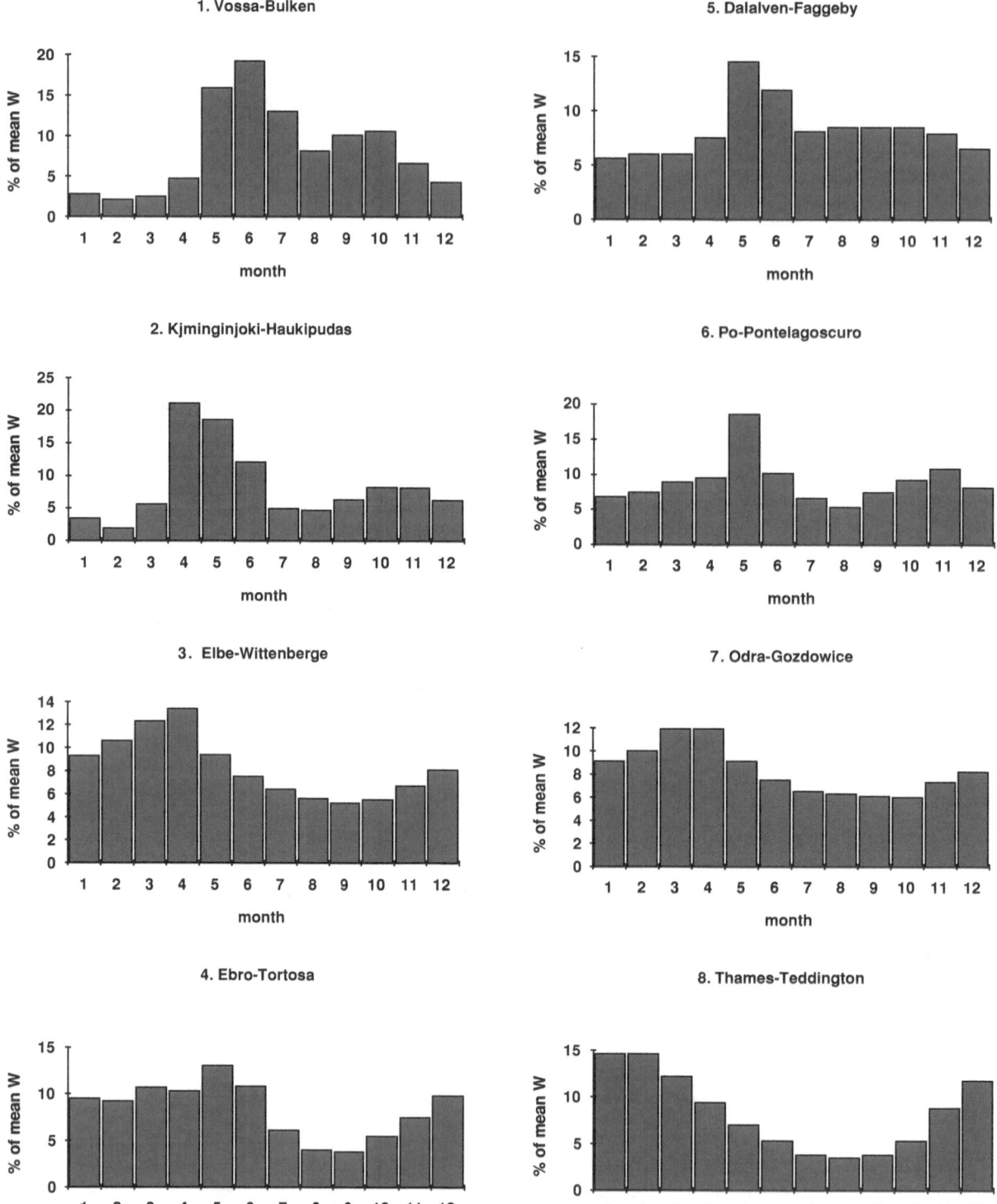

Fig. 4.10 Average monthly runoff distribution of some European rivers.

Table 4.15. *An assessment of water resources of the Baltic Sea basin (for study period of 1921–85)*

| Hydrological region | Area of basin, $km^2 \times 10^3$ | Principal rivers of the Baltic regions | | | Runoff volume (km^3), by estimates of different authors | | |
		River	Area of catchment, $km^2 \times 10^3$	Mean annual runoff, km^3	Present Monograph	Mikulski (1988)	Bergstrom and Carlsson (1994)
Bothnian Bay	261	Kemijoki	52.0	17.7	87	104	98
		Tornealven	40.0	12.4			
		Lulealven	25.2	15.8			
		Other rivers	143.8	41.0			
Bothnian Sea	230	Angerman	31.0	12.3	80	104	91
		Dalalven	33.0	12.9			
		Other rivers	166.0	55.0			
Gulf of Finland	421	Neva	281.0	77.9	105	116	112
		Narva	56.2	11.3			
		Kymijoki	37.2	10.1			
		Other rivers	46.6	6.0			
Gulf of Riga	132	Daugava	87.9	20.1	24	31	32
		Other rivers	44.1	4.0			
Baltic Proper	584	Neman	98.2	18.5	101	117	114
		Vistula	198.5	33.3			
		Odra	118.9	18.1			
		Other rivers	168.4	31.0			
Denmark Straits and Kattegat	101	Gota–Alv	50.1	20.3	28	43	37
		Other rivers	50.9	8.0			
Baltic Sea as a whole	1729				425	515–520	483

Assessments of the water resources of these basins is based on the analysis of long-term fluctuations in the runoff of rivers draining to into these seas. These assessments are made from data in the Specialized Archive (see Section 4.5.1). Runoff from interfluves is determined by the methods described earlier.

For the Baltic Sea basin, assessments made for this Monograph differ from earlier ones. This is of special interest and requires a detailed presentation of the results. The Baltic Sea is a semi-closed water body with a total water area of 377 400 km^2 and volume of 21 200 km^3. It is connected to the North Sea by the narrow and shallow Denmark Straits, which allows it to be part of the Atlantic Ocean basin.

The northern part of the Baltic region is characterized by long cold winters, which result in a low winter runoff and smoothed long-term runoff fluctuations; the southern part is noted for milder winters and a greater variability of runoff. There are a great number of lakes, particularly in Sweden, Finland and Russia, which affect the runoff, regulating and smoothing the fluctuations and increasing the evaporation. The Geta–Elv, Kemi and Neva rivers are typical of this lake-regulated type. The Baltic region has a high population density and a considerable extent of urbanization, a high level of economic and agricultural development, and prolonged anthropogenic loads to the environment. The Baltic Sea coastline is divided among nine states with a total population of over 70 million (HELCOM, 1986; UN, 1995a). Due to these factors, the study of water resources in the Baltic Sea basin is of great practical and scientific significance and has been the subject of a large number of papers, for example HELCOM (1986), Mikulski (1988), Bergstrom and Carlsson (1994).

In this Monograph the Baltic Sea basin is considered as the sum of six hydrological regions (Mikulski, 1988). Each hydrological region plays a hydrologically independent role. The major part of the fresh water inflow into the Baltic Sea is from 13 rivers, each with an annual discharge exceeding 10 km^3. Table 4.15 shows the average runoff data from the Specialized Archive. Runoff from the interfluve areas was calculated and was found to be about 35%, whereas Bergstrom and Carlsson (1993) estimated it to be 37% of the total.

The three assessments show differences between the individual hydrological regions and thus those for the entire basin. The differences are due largely to the estimates of runoff from the interfluvial areas, one reason being the different lengths of the records employed. Mikulski (1988) used the period 1921 to 1975, while Bergstrom and Carlsson (1993) employed 1950 to 1990, as

Table 4.16. *Renewable water resources of the European continent (for study period of 1921–85)*

	Area of catchment, $km^2 \times 10^3$	Runoff volume, km^3/year	Coefficient of variation, C_v	Coefficient of asymmetry, C_s
Arctic Ocean continental slope	1 400	622	0.10	0.02
Arctic Ocean islands	131	72	0.10	0.24
Arctic slope with islands	1 531	694	0.10	0.08
Baltic Sea basin	1 729	425	0.10	0.19
Basin of North Sea and Atlantic Ocean	1 434	493	0.14	0.17
Black Sea basin	2 151	403	–	–
Mediterranean Sea basin	866	316	–	–
Basins of Black and Mediterranean Seas	3 017	719	0.10	0.18
Atlantic Ocean continental slope	6 180	1637	0.10	0.04
Atlantic Ocean islands	503	236	0.10	0.07
Mediterranean Sea islands	86	22	0.10	0.84
Atlantic slope with islands	6 769	1895	0.10	0.04
Caspian Sea basin	2 160	311	0.17	0.01
Continental slope of Europe	9 740	2570	0.10	0.02
Continental slope of Europe with islands	10 460	2900	0.10	0

opposed to this Monograph where the period 1921 to 1985 has been used. Mikulski (1988) adopted the water balance method as the basis for calculating the inflow to the sea, where the normal inflow is accepted as being equal to the difference between the total precipitation and evaporation within the hydrological region being considered. All the climatic characteristics were taken from mapped data, but practice shows that such a procedure overestimates the water resources and this requires the correction and comparison of the results with the assessments obtained from the actual runoff data.

In this Monograph the average runoff to the Baltic Sea is estimated to be 425 km^3; according to Mikulski (1988) it is approximately 520 km^3, while Bergstrom and Carlsson (1993) found it to be 483 km^3.

RUNOFF TO THE OCEANS AND TO AREAS OF INTERNAL DRAINAGE

Table 4.16 presents assessments of the total water resources for areas draining to the Arctic and Atlantic oceans, including the islands, and for the Caspian Sea basin, in order to provide an assessment for Europe as a whole. This Table gives estimates of the average runoff during the period 1921 to 1985, as well as other statistics. The annually renewable runoff from Europe is 2900 km^3; the runoff to the Arctic Ocean (excluding the runoff from islands) is 622 km^3 (19%) of the total, runoff to the Atlantic Ocean (excluding the islands) is 1638 km^3 (59%), while the total runoff from offshore islands is 329 km^3 (11%). The flow to the Caspian Sea is 311 km^3 (11%).

It is interesting to compare these assessments with other published results. There are, for example, values of: 2830 km^3 (Chernogayeva, 1969), 3154 km^3 (L'vovich, 1974), 3210 km^3 (Korzun, 1974b) and 2800 km^3 (Baumgartner and Reichel, 1975); and in the later publications, 3200 km^3 (Shiklomanov and Markova, 1987) and 2321 km^3 (WRI, 1992, 1996, 2000; Gleick, 1993, 1998). The main cause is the differences between the periods of the records used and in the maps employed showing the hydrological characteristics. Probably the calculations made for this Monograph are more reliable, since they were performed on real data and for a longer period. Figure 4.11 shows the variations of total annual runoff to the oceans and to the Caspian Sea; and Fig. 4.12 presents continent-wide results for the period 1921 to 1985. The same data are also shown for Korzun (1974b). It appears that the estimated total annual runoff to the oceans calculated for this Monograph is different from the values in the other publications, but it is practically the same for the areas of inland drainage.

It is also interesting to consider the latitudinal distribution of flows to the oceans from the continent. Inflow to the Arctic Ocean gradually increases from north to south; at latitudes of 70° to 80° N, runoff occurs from the islands (10% of the total runoff); from 60° to 70° N the inflow increases due to runoff from the northwestern side of the Scandinavian Peninsula, facing the Norwegian Sea (about 28% of the total runoff), together with the discharges of rivers on the Russian Plain flowing into the Barents Sea, i.e. Pechora, Northern Dvina, etc. (62% of the total). The distribution of the inflow to the Atlantic Ocean mainly follows latitude. Thus,

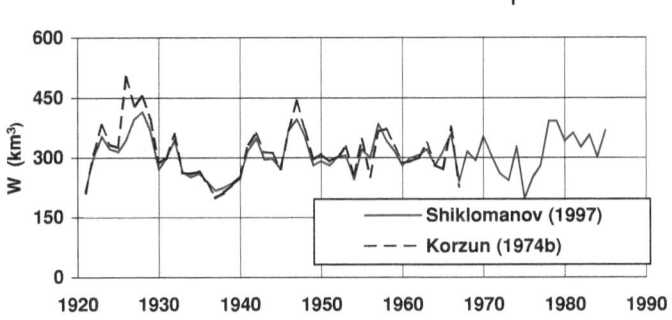

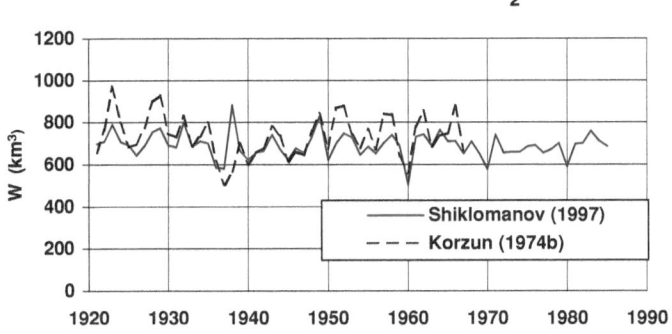

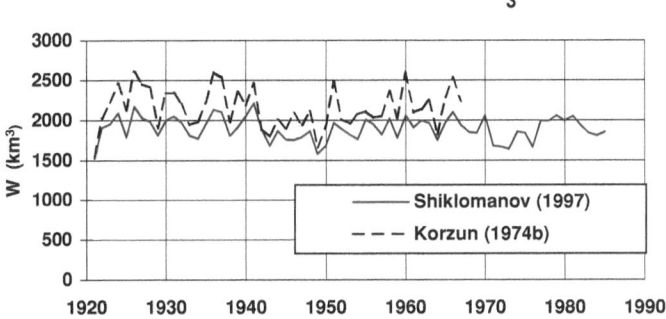

Fig. 4.11 Dynamics of water resources for Europe. 1, interior runoff region (Caspian Sea basin); 2, runoff into the Arctic Ocean; 3, runoff into the Atlantic Ocean.

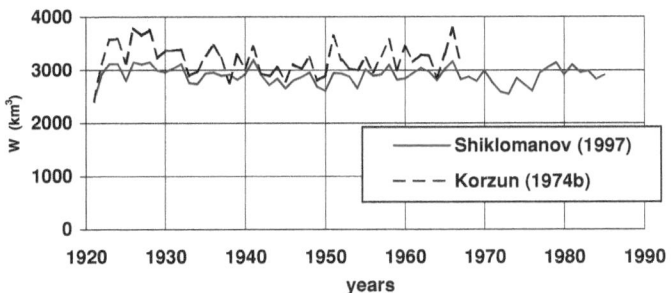

Fig. 4.12 Dynamics of water resources for the European continent as a whole.

at latitudes of 60° to 70° N the inflow is only 10% of the total runoff and is mainly the discharges to the Gulf of Bothnia. At latitudes of 50° to 60° N, a significant proportion of the inflow is contributed by the Rhine, Weser, Elbe, Oder, Vistula, Zapadnaya Dvina and Neva (about 26% of the total). The largest inflow to the Atlantic Ocean occurs from 40° to 50° N, about 61% of the total. This is due to the rivers Danube, Dnieper and Don (Black and Azov Seas); Po, Rhône and Ebro (Mediterranean Sea); Seine, Loire, Garonne and Douro (Atlantic coast). At latitudes of 30° to 40° N, the inflow comes from the Balkan, Italian and Iberian Peninsulas (Guadalquivir and Tagus) and amounts to about 3% of the total (Fig. 4.13).

The seasonal distribution of runoff is interesting and is shown in Figs. 4.14 and 4.15. Obviously, the monthly distribution has a more marked pattern from west to east and from north to south. In the north of the continent, the highest runoff is recorded in the spring and summer months; and the lowest in the winter and at the end of the summer. Runoff to the Atlantic is characterized by a quite even distribution throughout the year, which reflects strongly the even distribution of precipitation and the small range of temperature on the Atlantic coast and across contiguous areas. Here, the maximum runoff is recorded in March and April; and the minimum in July and August. The flow of meltwater is characterized by an irregular runoff distribution during the year: the highest values occur in the spring and summer months, and the lowest in autumn and winter. For the European continent as a whole, the distribution of runoff is generally regular with a slight increase in May and June (Figs. 4.14 and 4.15). Of course runoff from some of the large basins may not exhibit these tendencies because they often sample areas with contrasting hydrological regimes with different patterns of runoff, as shown in Fig. 4.10.

4.5.3 The water resources of certain countries and regions

Table 4.17 gives assessments of the water resources of specified countries and estimates of local water resources, inflows across borders and the total water resources. For various reasons the water resources of all European states cannot be calculated. In a number of publications (e.g. L'vovich, 1974; Engelman and LeRoy, 1993; Gleick, 1993, 1998; Seckler *et al.*, 1998; Rijsberman, 2000), data can be found on the water resources of most European countries. However, these data, as a rule, differ markedly. These dissimilarities are associated both with different calculation procedures and with the amount of information used; but they are due largely to the source of the runoff data. Table 4.10 (Section 4.4) presents data obtained from three different sources (L'vovich, 1974; WRI, 1992; Engelman and LeRoy, 1993) to illustrate this point. This emphasizes that information for individual countries or regions, obtained from different publications, requires very careful checking.

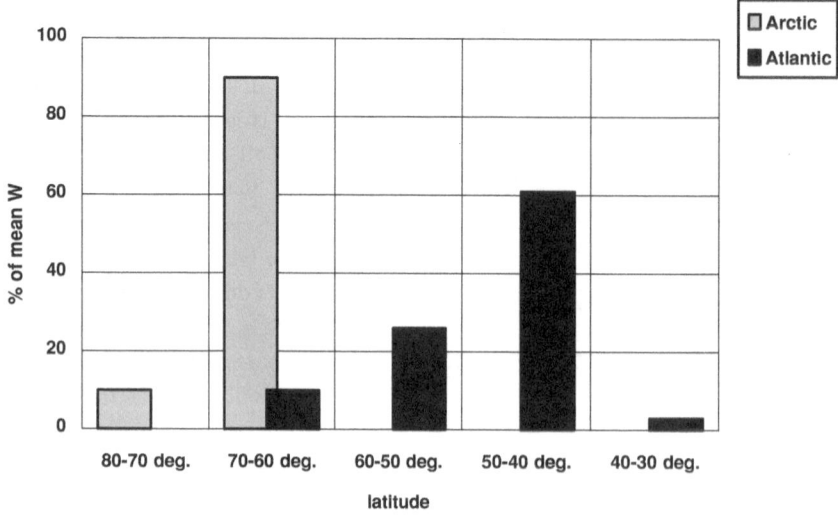

Fig. 4.13 Ocean basins inflow (as percentage of mean total runoff W) by zones of latitude (10° intervals) on the European continent.

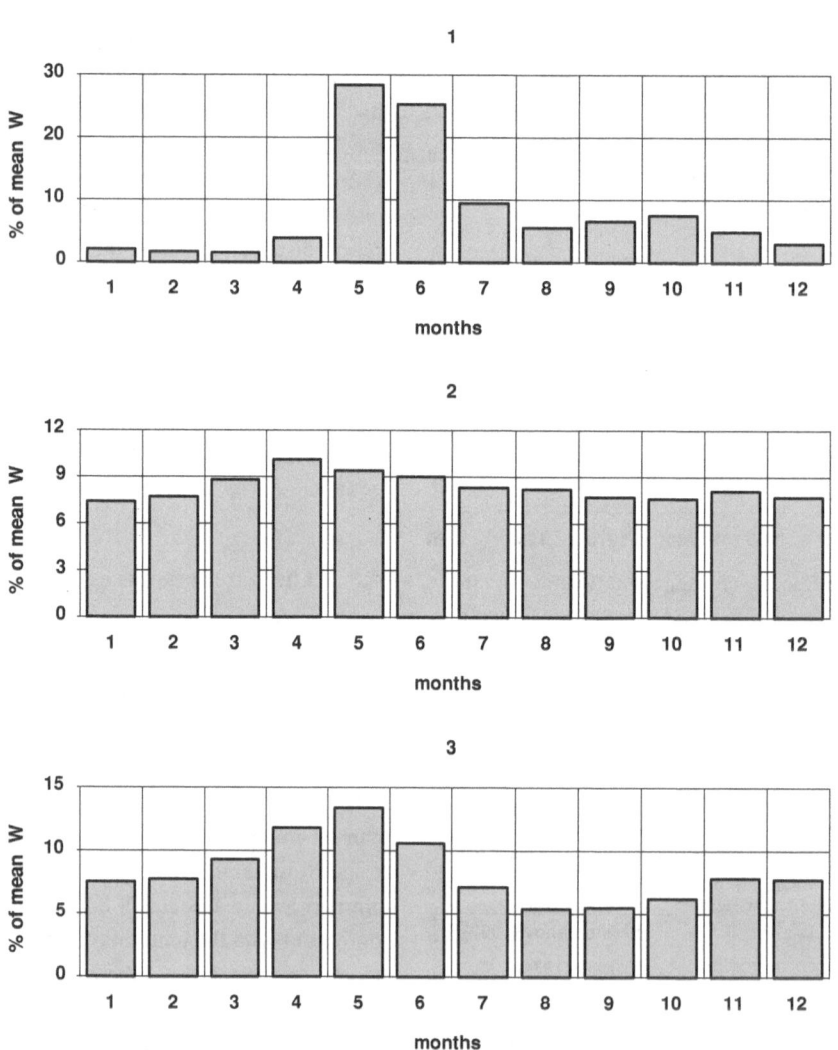

Fig. 4.14 Average monthly runoff distribution in the European continent. 1, Barents Sea basin; 2, Baltic Sea basin; 3, Mediterranean Sea basin.

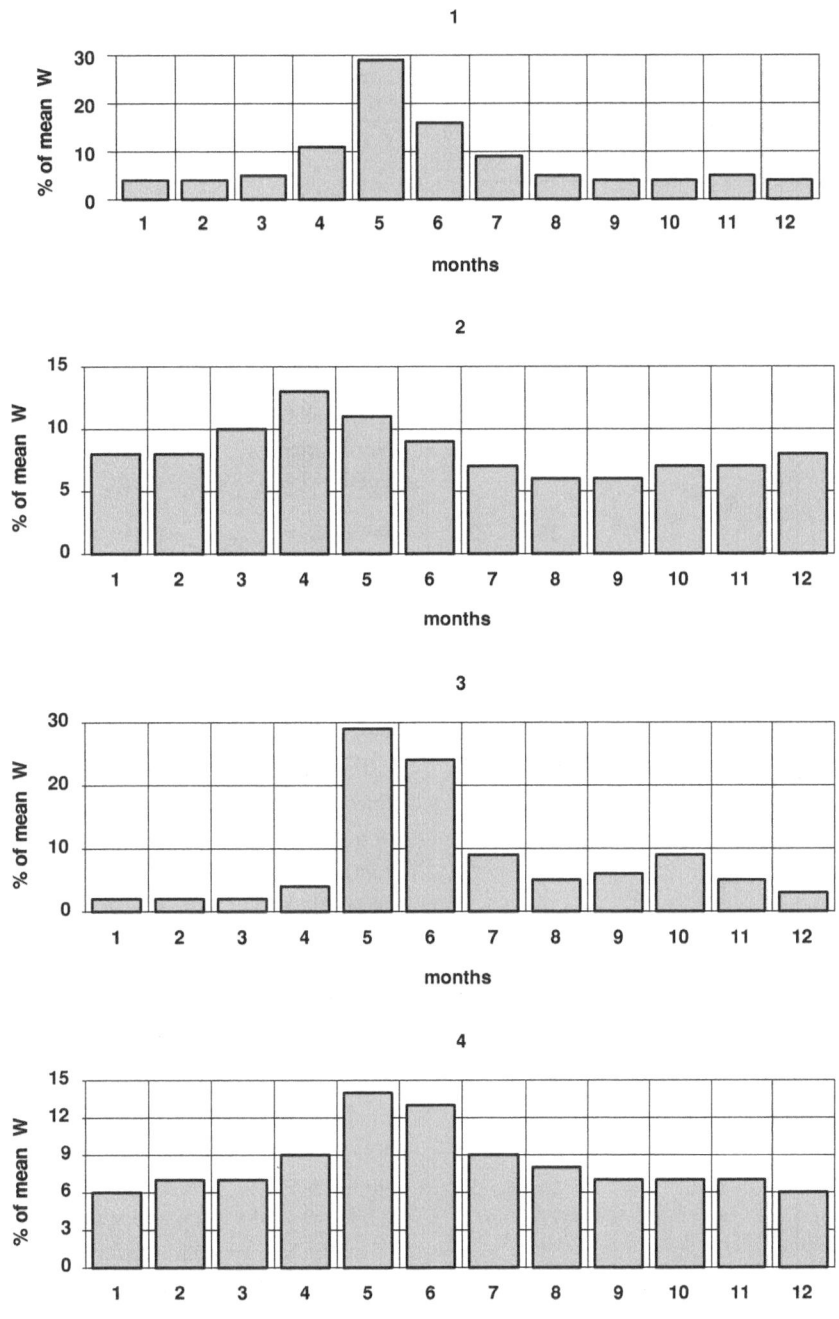

Fig. 4.15 Average monthly runoff distribution in the European continent. 1, Caspian Sea basin (interior runoff region); 2, Atlantic Ocean basin; 3, Arctic Ocean basin; 4, Europe as a whole.

Using these different publications, it is possible to obtain an entirely different assessment for each European state. However, national assessments of water resources, which are published in the different national reports and monographs, are undoubtedly the most reliable. Taking this into account, most of the assessments of water resources given in Table 4.17 are derived from different national publications (Jones, 1975; Falkenmark, 1977; Stamenov, 1977; Kaczmarek, 1980; Stancik and Jovanovich, 1988; Kwadijk and van Deursen, 1994; and many other papers). These assessments were carefully analysed and compared with other published evidence.

Using the annual and monthly runoff series from the Specialized Archive (Subsection 4.5.1) and other necessary hydrometeorological information, the water resources were calculated

Table 4.17. *Renewable water resources of European countries*

Country	Area, km² × 10³	Water resources, km³/year		
		Total	Local	Inflow
Sweden	450	176	164	12
Norway	387	340	340	–
Finland	337	106	80	26
Iceland	103	110	110	–
Denmark	45	11	11	–
France (+ Monaco)	547	195	168	27
Germany	356	209	96	113
Poland	313	56	50	6
United Kingdom	244	120	120	–
Czech Republic (+ Slovakia)	128	59	28	31
Austria	84	77	56	21
Ireland	70	40	40	–
Netherlands	41	89	10	79
Switzerland	41	50	43	7
Belgium (+ Luxembourg)	33	11	6	5
Spain	505	108	108	–
Italy	301	185	185	–
Yugoslavia, former Republic of	256	227	104	123
Romania	238	207	36	171
Greece	132	58	49	9
Bulgaria	111	112	18	94
Hungary	93	115	6	109
Portugal	92	53	19	34
Turkey (European part)	34	2	2	–
Albania	29	24	19	5
Russia (European part)	4240	931	902	29
Ukraine	604	211	52	159
Kazakhstan (European part)	230	10	2	8
Belarus	208	58	35	23
Lithuania	65	23	13	10
Latvia	64	32	15	17
Estonia	45	17	12	5
Moldavia	34	13	1	12

Table 4.18. *Renewable water resources of natural-economic regions of Europe (for study period of 1921–85)*

Regions	Area, km² × 10³	Water resources, km³/year		
		Total	Local	Inflow
Western Europe	4970	1877	1868	9
North Europe	1322	705	705	–
Central Europe	1857	623	617	6
South Europe	1791	655	546	109
European territory of the former Soviet Union (ETSU)	5490	1223	1032	191
North slope	2710	616	589	27
South slope	2780	566	443	123
Europe	10460	2900	2900	–

State Hydrological Institute (SHI), compiled between 1985 and 1990. These papers give assessments of surface water resources of all the republics within the former Soviet Union for 1980 and 1985.

From these estimates the water resources were calculated for all European states, as well as the total water resources of the regions of northern, central and southern Europe, and for the northern and southern parts of the European territory of the former Soviet Union as of 1980 and 1985. The subdivisions used are discussed above (Table 4.12 and Fig. 4.6). The water resources for the regions were taken as the sum of local runoff for all the countries within the region, with the inflow from the contiguous regions. The results are given in Table 4.18, with the water resources for the whole of Europe.

The variations in the regional water resources from 1921 to 1985 are shown in Figs. 4.17 and 4.18. The northern regions display the smallest fluctuations (1.1 to 1.2 relative to the average runoff); while those of central and southern Europe, and the southern part of the ETSU, are characterized by larger changes (1.5 to 1.8 relative to the average runoff).

The monthly distribution of runoff for the regions is shown in Fig. 4.19 as a percentage of the annual runoff and for France, Spain, Portugal, Albania, Sweden, Poland and Italy in Fig. 4.20. These Figures show that the most uniform distribution of runoff is characteristic of Central Europe. The ratio of runoff in the wet period to that in the dry period ranges from 1.4 to 1.8. Eastwards from the Atlantic coast and in the countries of southern Europe, depending on the precipitation regime, the distribution of runoff becomes less regular. Here, the ratio of runoff of the wet to the dry periods reaches 2.1 to 2.9. For some countries in southern Europe, this ratio can increase to 3.3. The most irregular distribution occurs in eastern European regions. For the northern part of the ETSU,

for six countries (France, Portugal, Spain, Albania and Sweden). Figure 4.16 shows as an example the variations of the water resources of Portugal and France. The results were compared with the published national assessments. The differences were below 5% and are mainly associated with the different periods of observation.

The estimates of the water resources of Estonia, Latvia, Lithuania, Moldavia, Belarus and Ukraine, which were parts of the Soviet Union until 1986, as well as parts of European Russia, together with Kazakhstan, were obtained mainly from published works (SHI, 1967, 1987), as well as from the research reports of the

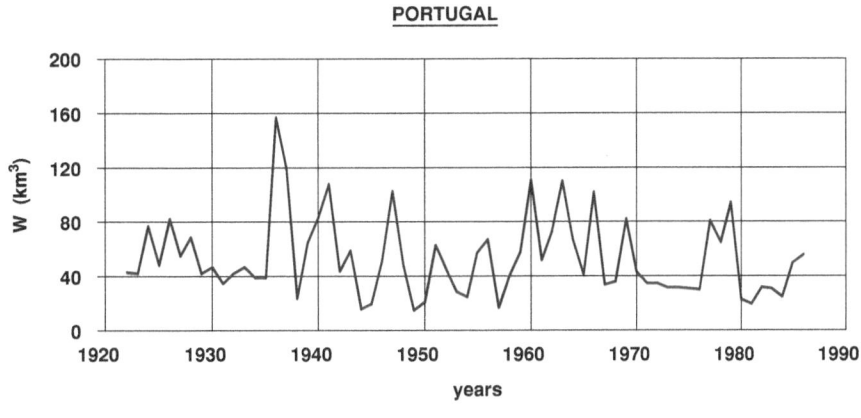

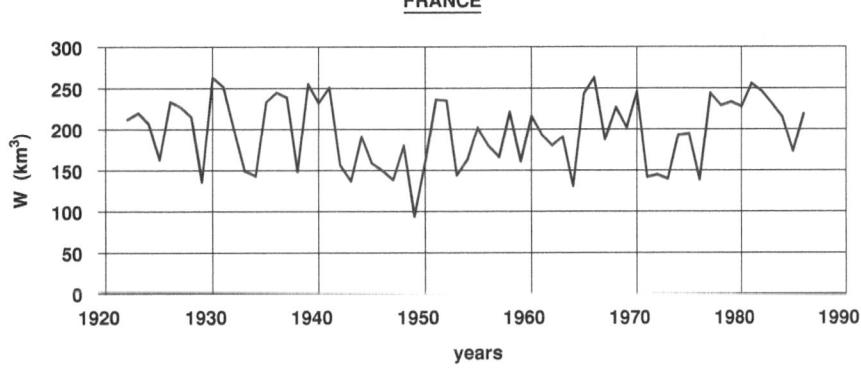

Fig. 4.16 Dynamics of water resources in Portugal and France.

the wet/dry ratio increases to 3.5 to 4.0; and for the southern part it ranges from 2.5 to 3.5. In the north of Russia, the highest runoff is recorded in the spring and summer months; and the lowest in autumn and winter. For the southern ETSU, the maximum runoff occurs in the spring; and the minimum in winter months.

It should be noted that the distribution of runoff within the year in a large natural region represents an average which does not reflect the wide diversity of the regimes that might be recorded in the different locations in these regions.

4.6 CHANGES IN WATER USE AND WATER AVAILABILITY

4.6.1 Initial data and preconditions

This Monograph gives an assessment of contemporary and potential water use and water availability for most European countries and regions, and for Europe as a whole. The main data on water use for the period from 1900 to 1980 for the regions and the continent were taken from Shiklomanov and Markova (1987). Their figures agree well with the data presented in other publications (CMEA, 1977; Yermolina and Klige, 1979; Klige, 1982; Mikhura, 1982; Smerdon, 1982).

Information on water use in the different sectors was collected starting from 1980, together with information published in the different national reports and in the publications of international organizations such as UN, UNESCO, UNIDO, ECE, WHO, IWMI and FAO. Some of these reports are: Karev and Shtyka (1973); Falkenmark (1977); Jorgulesku (1979); Vanchura (1979); Vintse (1979); Borodavchenko and Mikhura (1981); Kirgizov (1982b); Zenkov (1982); Brenning and Platon (1983); Turnock (1986); IIED/WRI (1987); SHI (1987); Soyuzvodproekt (1988); Vasilyev and Pretro (1989); Margat (1990); WRI (1990, 1992, 1996, 2000); *Water International* (1991); Strzepek and Bowling (1995); Seckler *et al.*, (1998); Gleick (2000a); Margat and Vallée (2000); Rijsberman (2000). These data were carefully analysed and checked, then assessments of water use were made for the regions and the continent for 1980, 1990 and 1995. These assessments differ slightly from those published by Shiklomanov and Markova (1987), and were taken as the basis for forecasts of water use.

These forecasts were made of water use for 2000, 2010 and 2025 for the main sectors, namely industry, agriculture and domestic purposes, as well as for additional losses from reservoirs. The approaches to the forecasts were described in detail in Chapter 3 (Shiklomanov and Markova, 1987; Shiklomanov, 1988, 2000a, 2000b; Penkova and Shiklomanov, 1998; Zaretskaya, 1998; Shiklomanov *et al.*, 2000). Data on water use were collected

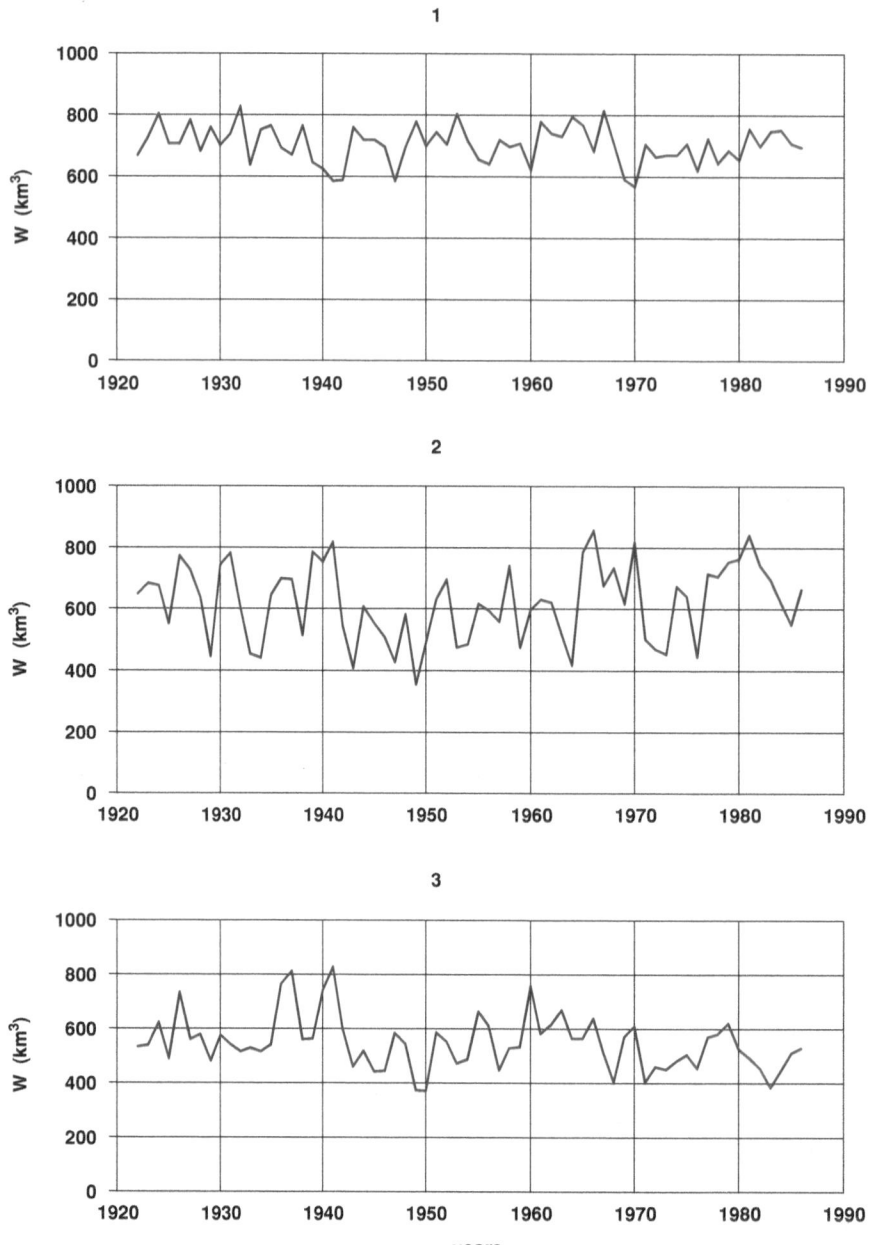

Fig. 4.17 Dynamics of water resources by natural–economic regions of
Europe. 1, North; 2, Central; 3, South Europe.

from the different sources and were carefully analysed. Likely
changes due to growth in industrial production, re-use of water,
reduction of the amount of water per unit product, increases in the
size of the irrigated area and changes in the volume of irrigation
water, as well as population growth and increase in domestic water
use were noted.

The following data were used to assist the forecasting:

•   forecasts of population in the years 2000, 2010 and 2025
using data published by the UN (1995)

•   trends in GNP (in US dollars per capita) for the years 2000,
2010 and 2025, using figures taken from the UN (1993a)
and International Monetary Fund (IMF, 1994) (unfortunately,
these data are not available for all European countries)

•   UNIDO coefficients, i.e. coefficients of growth in industrial
water use for 2025 relative to 1990 (Strzepek and Bowling,
1995)

•   data on irrigated areas for 1994, taken from FAO sources and
other publications (Zonn and Nosenko, 1981; FAO, 1995a).
The likely extent of irrigated areas for the years 2000, 2010
and 2025 were calculated for all European countries taking into account economic and population growth factors by

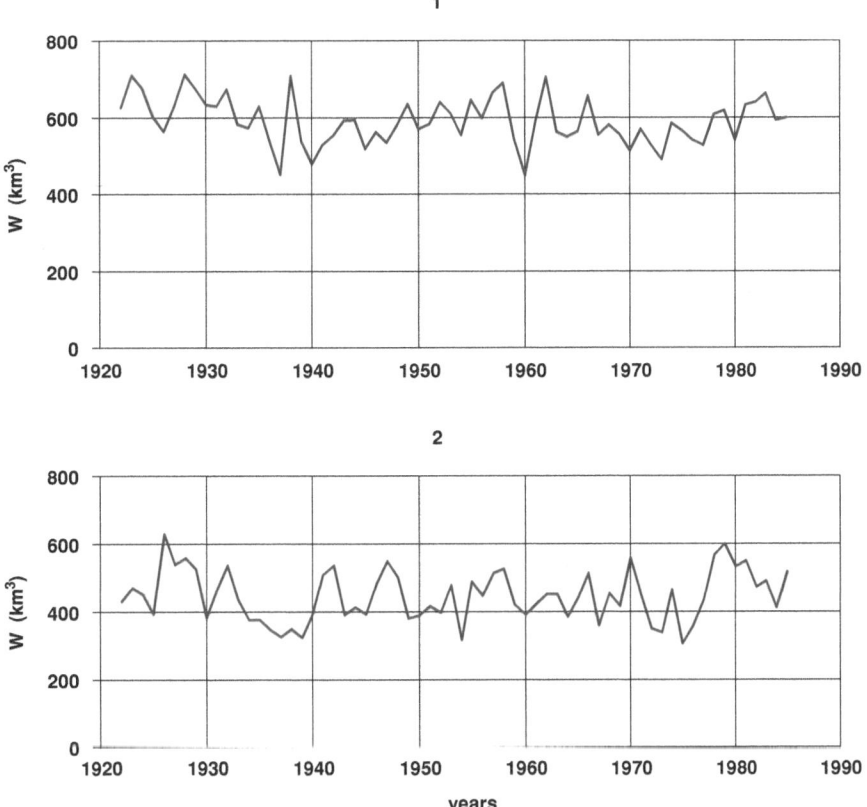

Fig. 4.18 Dynamics of water resources by natural–economic regions of the European territory of the former Soviet Union. 1, Northern slope; 2, Southern slope.

extrapolating data for the future, pursuant to recommendations presented in Section 3.3. Figure 4.21 shows the changes in population and in the extent of irrigated areas for 2000, 2010 and 2025 in Poland, Sweden, Spain, Italy and France.

Initially, water use forecasts were made separately for each country, and then summed for the regions and the continent. For northern and central Europe, there were data on abstractions and on the detailed characteristics of water use. The analysis of this data showed that for many western European states, the period of highest intake has passed, and that abstractions have now levelled out or that reductions in demand for water are occurring. For southern European countries and the southern part of the ETSU, there were insufficient data for drawing conclusions and assumptions had to be made from other factors.

Industrial water use for the regions was determined for the whole of the industrial sector including thermal power production. Forecasts of water use take into account the UNIDO coefficients of growth in industrial water use, calculated for 2025 relative to 1990 (Strzepek and Bowling, 1995). The analysis of changes and trends in recent years and the possibilities of the rationalization of industrial water use demonstrated that, in most European countries,

growth in industrial production will be accompanied by saving more water. Consequently the UNIDO coefficients were lowered for some states. Thus, for the countries of northern Europe, the UNIDO coefficients of industrial water use growth are 2.6 to 2.9 (Strzepek and Bowling, 1995). However, the UNIDO coefficients would more correctly reflect modern tendencies for saving water if they were to be between 1.6 to 1.7. For most developed countries of central and southern Europe, these coefficients range from 1.2 to 2.0, against 1.9 to 2.3 in Strzepek and Bowling (1995). For former socialist states, the UNIDO coefficients are also reduced from 1.7 to 1.9, to values ranging from 1.2 to 1.6. For the years 2000 and 2010, industrial water use was taken to be proportional to GNP. The following regional values of industrial water consumption were adopted for the base level of 1980: 9% in northern and central Europe, and 7% in southern Europe. For 2025, these values were estimated to be 12%, 11% and 6%, respectively. In the ETSU regions for 1980, these values were 12% (northern) and 18% (southern), with a likely growth to 20% and 34% by 2025.

Forecasts of agricultural water use were primarily based on changes in the area irrigated for the different horizons. The likely changes in irrigated areas were estimated with regard to population growth and growth in GNP, also taking account of the funding that might be available, as well as the rationalization and improvement of irrigation systems. Contemporary abstractions, calculated for

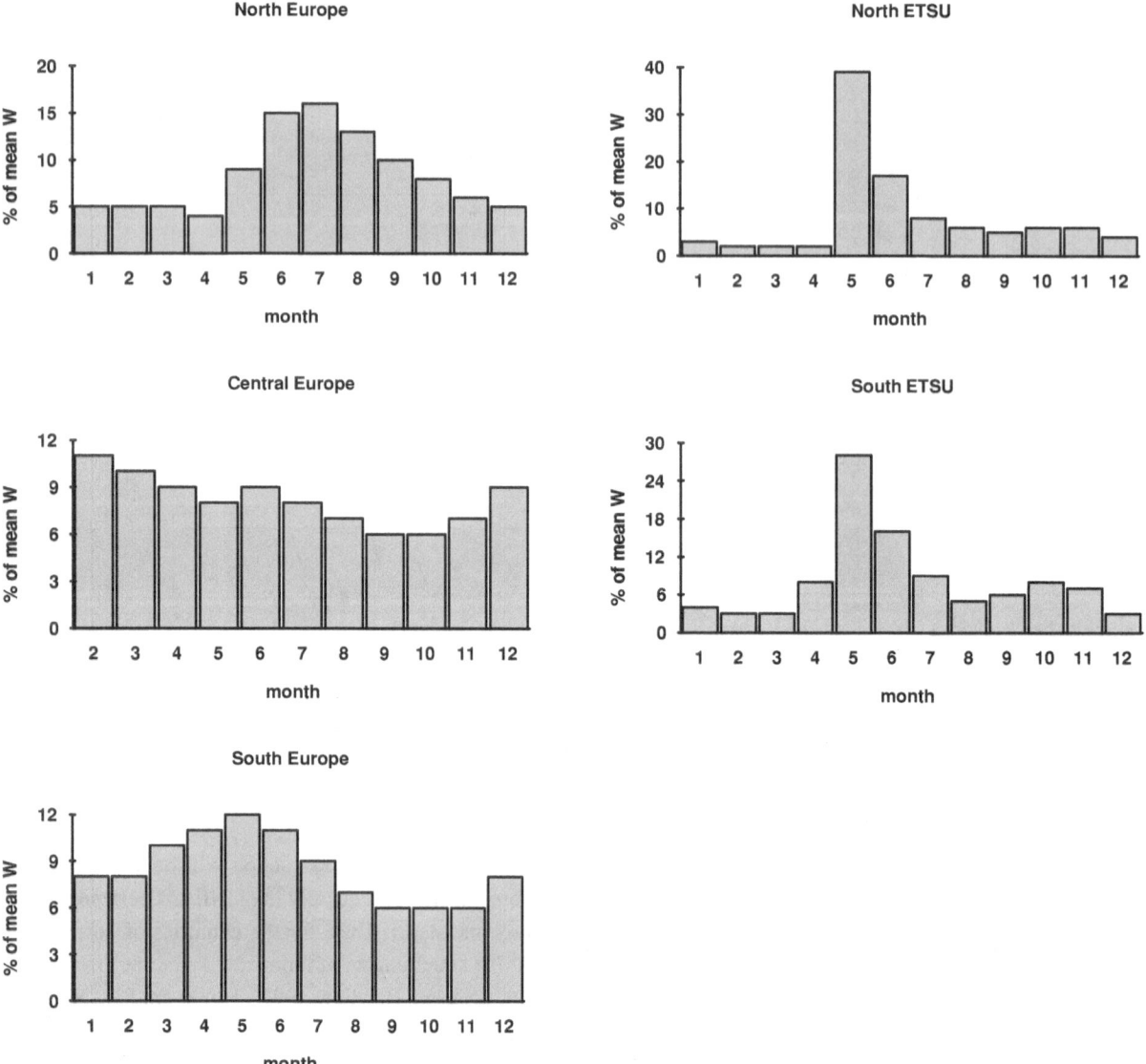

Fig. 4.19 Average monthly runoff distribution by European regions.

European countries range from 11 600 m³/ha in Italy to 670 m³/ha in Albania.

The following values of consumption were adopted for the agricultural sector for 1980: 62% in northern, 70% in central and 68% in southern Europe; while for 2025, taking population growth into account, the values are expected to be slightly lower amounting to 60%, 66% and 63%, respectively. For the ETSU regions the following values were taken for 1980: 47% (northern) and 65% (southern) and 92% and 75% for 2025.

Forecasts of water use for domestic purposes were made using the UN predictions of population growth with regard to changes in consumption per head. In 1995 average daily water use per head ranged from 12 l/day in Albania and 50 l/day in Ireland, to 470 l/day in Sweden. For the future, these averages were raised to

take into account population growth, as well as the increase in the population served by public water supply systems. Average water usage figures for the former socialist states were calculated from the evidence of 1980; and from national plans for providing drinking water supply systems and sewerage systems (CMEA, 1981).

In 1980 domestic water supply accounted for 8% of abstractions for northern and central Europe and 9% for southern Europe, while for 2025 these values were estimated to be 6%, 8% and 8%, respectively. The values for the ETSU regions are: 30% (northern) and 24% (southern), and 20% and 22%, respectively, for 2025. The additional water losses caused by evaporation from reservoirs were analysed and taken into account.

The average consumption per head was calculated for countries, regions and the continent following the method set out in

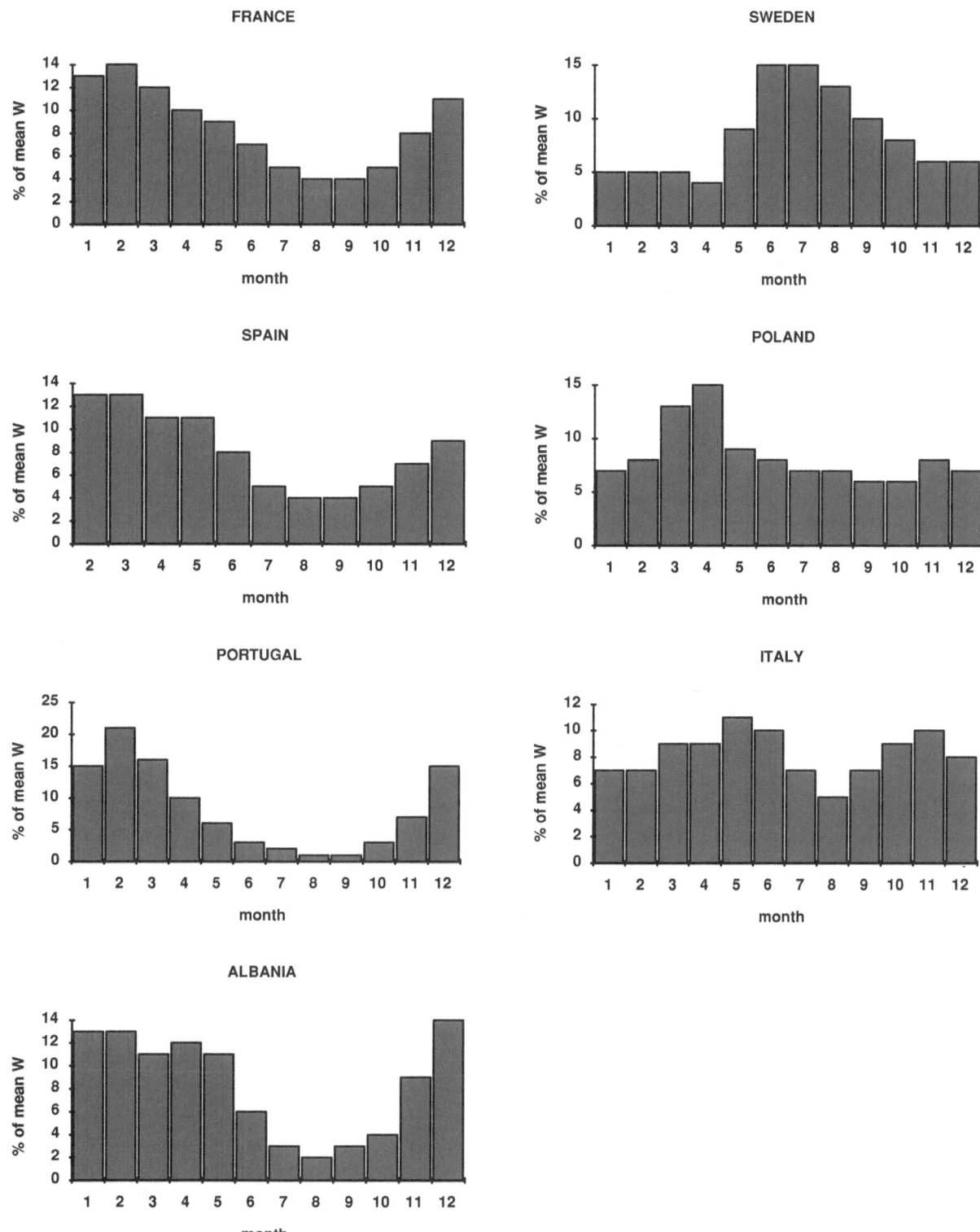

Fig. 4.20 Average monthly runoff distribution by selected European countries.

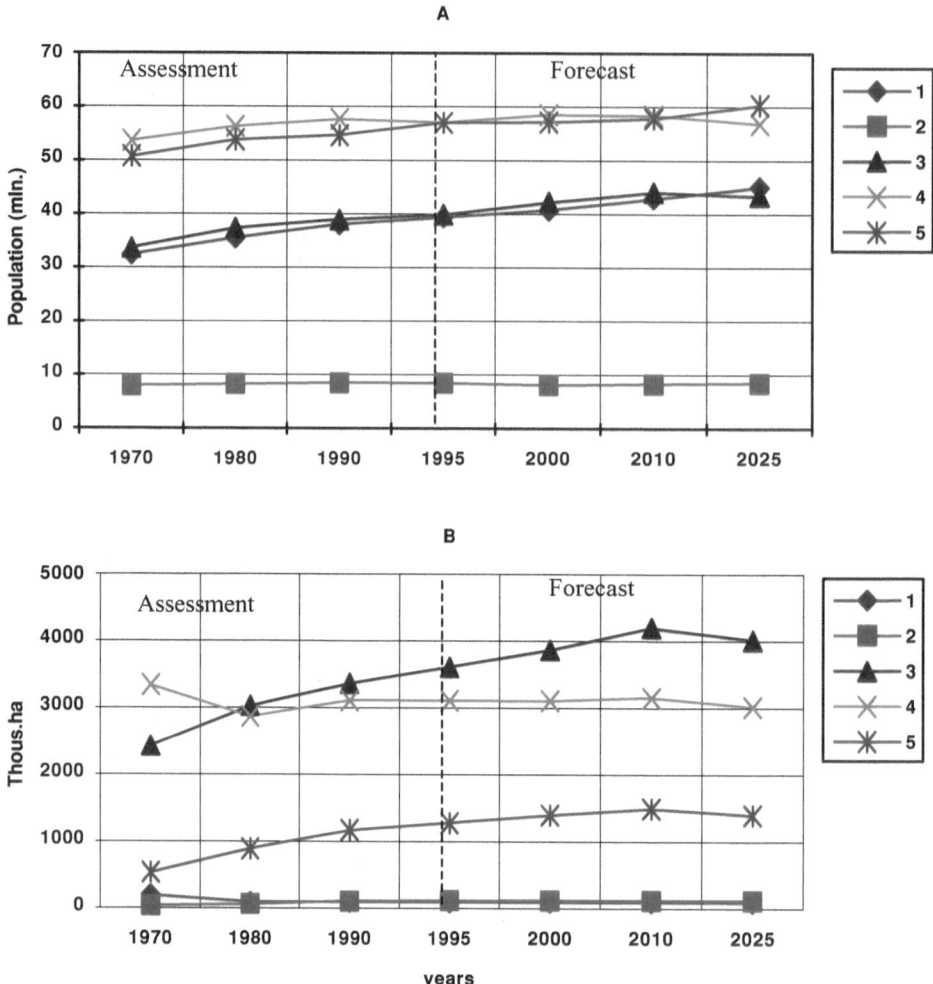

Fig. 4.21 The forecast of population dynamics (A) and irrigated areas (B) for selected European countries. 1, Poland; 2, Sweden; 3, Spain; 4, Italy; 5, France.

Section 3.3. Growth rates of population and water consumption were adopted as the average values for the country concerned, without taking into account whether the population was urban or rural.

Assessment of the available water is a basis for approaching the different water management problems in a region. Specific water availability is assessed by dividing water resources minus water consumption by the population of the area or the region (Shiklomanov and Markova, 1987). This value characterizes the residual amount of water per unit area or per person in the region. Specific water availability correlates with the selected time horizon. Comparison of the different characteristics of specific water availability at different times enables the variability in the distribution of water resources to be determined in space and time, trends to be determined and future resource problems to be identified.

4.6.2 Changes in water use

Water use was estimated for the period of 1990 to 2025 for certain types of economic activity, as well as for the different countries, regions, and for the continent. Contemporary and potential water use for most states was based on published national assessments (Jorgulesku, 1979; Vanchura, 1979; Vintse, 1979; Brenning and Platon, 1983; SHI, 1987; Anonymous, 1992a; Margat and Vallée, 2000). Table 4.19 shows changes in water use by sector from 1900 to 2025 and Table 4.20 the same changes by area, using data from Shiklomanov and Markova (1987), Zaretskaya (1998), Shiklomanov (1997, 1998a, 2000a, b) and Shiklomanov et al. (2000). Figures 4.22 and 4.23 illustrate these Tables.

Table 4.21 shows the present and forecast use of water for the period of 1980 to 2025 for the different sectors and each region. The total water use is given in Table 4.22 for the period 1980 to 2025 for certain countries. Since the beginning of the twentieth century, with population growth, economic development and a

Table 4.19. *Dynamics of water use by type of economic activity in Europe (km³/year)ᵃ*

	Assessment								Forecast		
	1900	1940	1950	1960	1970	1980	1990	1995	2000	2010	2025
Population, ×10⁶	–	–	519	570	622	657	681	685	690	694	685
Irrigation area, ha ×10⁶	2.9	5.1	6.5	9.9	14.2	21.1	24.9	25.8	26.6	28.7	30.6
Agriculture	19.6	34.5	40.9	53.9	82.2	169	195	198	203	209	212
	14.6	25.0	31.5	38.4	55.6	117	133	135	136	139	142
Public Supply	8.5	12.7	15.6	21.0	33.7	58.5	67.1	69.9	72.6	78.7	84.5
	1.8	2.3	2.7	3.0	4.2	7.2	8.4	8.6	8.8	9.2	9.5
Industrial	9.3	23.4	36.3	104	168	206	214	228	242	273	305
	1.1	2.2	3.2	7.0	11.6	22.3	26.9	28.5	30.0	37.5	47.4
Reservoirs	0.1	0.3	1.0	5.5	10.5	11.4	14.9	15.4	15.9	16.8	17.6
Total	37.5	71.0	93.8	185	294	445	491	511	534	578	619
	17.6	29.8	38.4	53.9	81.8	158	183	187	191	202	217

ᵃNominator, total water withdrawal; denominator, water consumption.

Table 4.20. *Dynamics of water use by natural–economic regions in Europe (km³/year)ᵃ*

	Assessment								Forecast		
Regions	1900	1940	1950	1960	1970	1980	1990	1995	2000	2010	2025
North Europe	1.4	2.8	3.9	7.5	9.8	11.0	11.4	12.3	13.2	14.8	16.4
	0.2	0.3	0.4	0.7	1.2	2.0	2.4	2.6	2.7	2.9	3.0
Central Europe	12.8	21.5	31.5	87.2	120	142	150	161	173	192	208
	2.7	4.2	6.0	9.5	15.1	28.2	32.5	34.8	37.2	40.1	40.4
South Europe	16.0	27.1	37.4	53.9	88.6	155	174	184	194	208	212
	11.0	18.4	25.2	29.5	39.1	74.2	82.1	83.9	85.7	85.1	82.0
North slope of European territory of former Soviet Union	0.3	0.8	0.9	1.8	3.1	13.9	16.3	15.4	14.9	16.8	20.3
	0.2	0.2	0.2	0.4	0.6	2.2	2.8	2.9	2.9	3.4	5.0
South slope of European territory of former Soviet Union	6.9	18.8	20.2	34.4	72.0	123	139	139	139	146	162
	3.5	6.7	6.6	13.8	25.7	51.2	63.2	62.8	62.4	70.8	86.1
Europe as a whole	37.5	71.0	93.8	185	294	445	491	511	534	578	619
	17.6	29.8	38.4	53.9	81.8	158	183	187	191	202	217

ᵃNominator, total water withdrawal; denominator, water consumption.

rising standard of living, both abstraction and consumption have been increasing (Tables 4.19 to 4.22).

From the analysis of Table 4.19 it is obvious that there have been considerable changes in the use of water. At the beginning of the twentieth century, most water was used by agriculture, until with growth of population and increasing industrial activity, the proportion of industrial and domestic water increased sharply. Presently (1995), industrial needs average at 45%; agriculture uses about 39%; and the domestic sector accounts for 14% of the volume abstracted. A small part (about 3%) is accounted for by evaporation from reservoirs. By 2025, the percentage of industrial water use

is expected to increase to 50%, while use by agriculture will drop to 34% and domestic water is likely to remain at 14% of the total abstracted.

Table 4.23 presents the proportion of the total available water resource which is used, both for regions and the continent at the present time (1995) and for 2025. Forecasts show (Tables 4.21 to 4.23), that water use will grow both for the regions, and the continent. Thus in 1995 abstractions amounted to 511 km³ (not more than 17% of the total water resource) with consumption averaging some 187 km³ (about 6% of water resources). By 2025, these figures are likely to be 619 km³ for abstractions, slightly more than

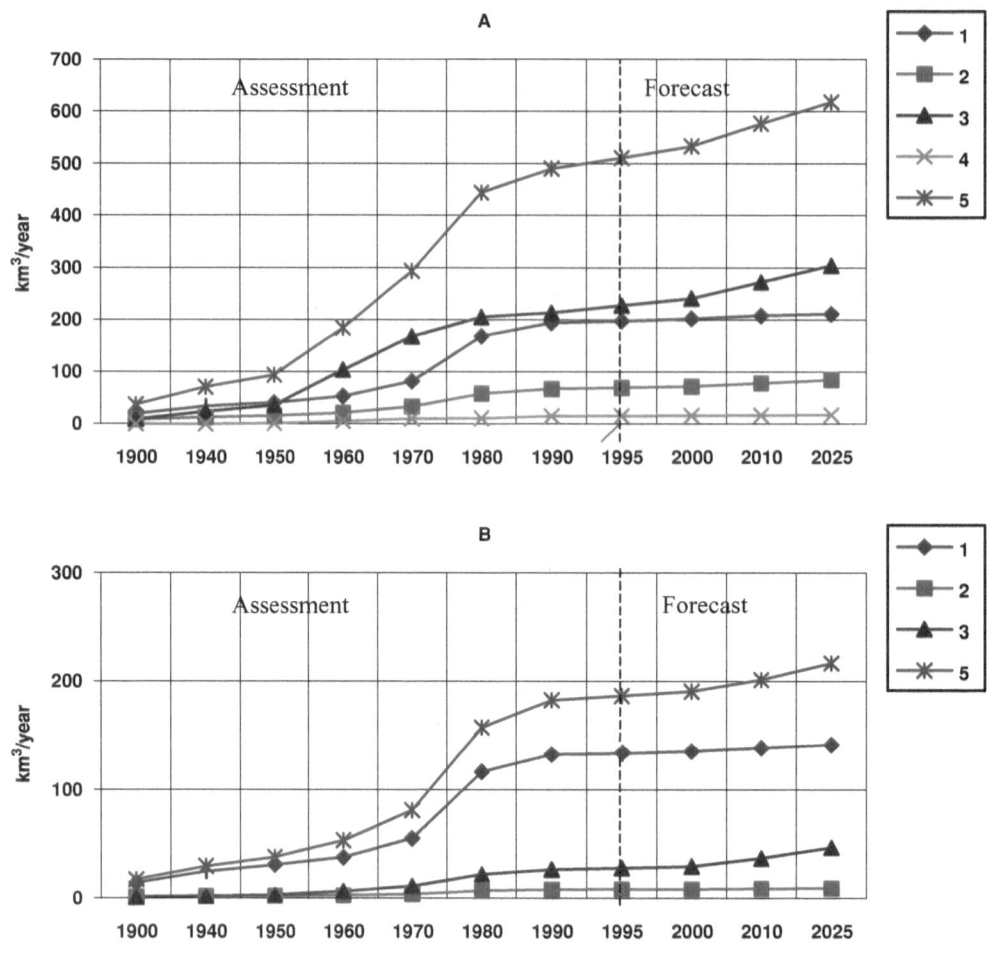

Fig. 4.22 Dynamics of water use: (A) withdrawal, (B) consumption in Europe by type of economic activity. 1, Agricultural; 2, public supply; 3, industrial; 4, reservoirs; 5, total water use.

21% of the resource, with water consumption reaching 217 km³ (about 8% of resources). Calculations show that abstractions in the regions, similar to the entire continent, are growing uniformly. For consumption, the picture is slightly different: rapid growth is taking place only in the southern part of the ETSU, and in other regions it will increase slightly by 2025.

As might be expected, the use of water resources is uneven across the continent. The greatest percentage of the resource is utilized in southern and central Europe, as well as in the southern regions of the ETSU, where in 1995 it was 36%, 32% and 27%, respectively. The smallest percentages (within the range 2–3%) occur in northern Europe and in the northern ETSU (Table 4.23). This pattern is common to the entire period. Table 4.23 shows that future water use should accelerate in the south of European Russia, particularly in the Republics of Ukraine and Moldavia. The increase in demand is due in part to the lack of application of new technologies in industry and agriculture.

The pattern of abstraction across Europe differs markedly for the different sectors, being primarily dependent on the level of development and on its potential from country to country within these regions (see Table 4.21).

During the 1990s in northern and central Europe, most water was abstracted for use by industry (50–70%); the domestic sector took about 30%; while agriculture accounted for 20–30%. For the countries in these regions, this pattern of water use is likely to continue to 2025. The opposite picture pertains in the countries of southern Europe. There industry takes about 24% of total water abstracted, agriculture about 60% and the domestic sector from 10% to 15%. For the ETSU regions, the greatest percentage is used by industry (80% in the north, 45% in the south); agriculture takes from 4% in the north to 44% in the south; while the domestic sector requires about 15% of the total water use. This pattern is expected to be maintained in the future.

Analysis of the above tables shows that the total water intake by the industrial sector including hydropower production presently averages 228 km³/year with irrecoverable losses at 28.5 km³/year. By 2025 this figure will probably increase to 305 km³/year while

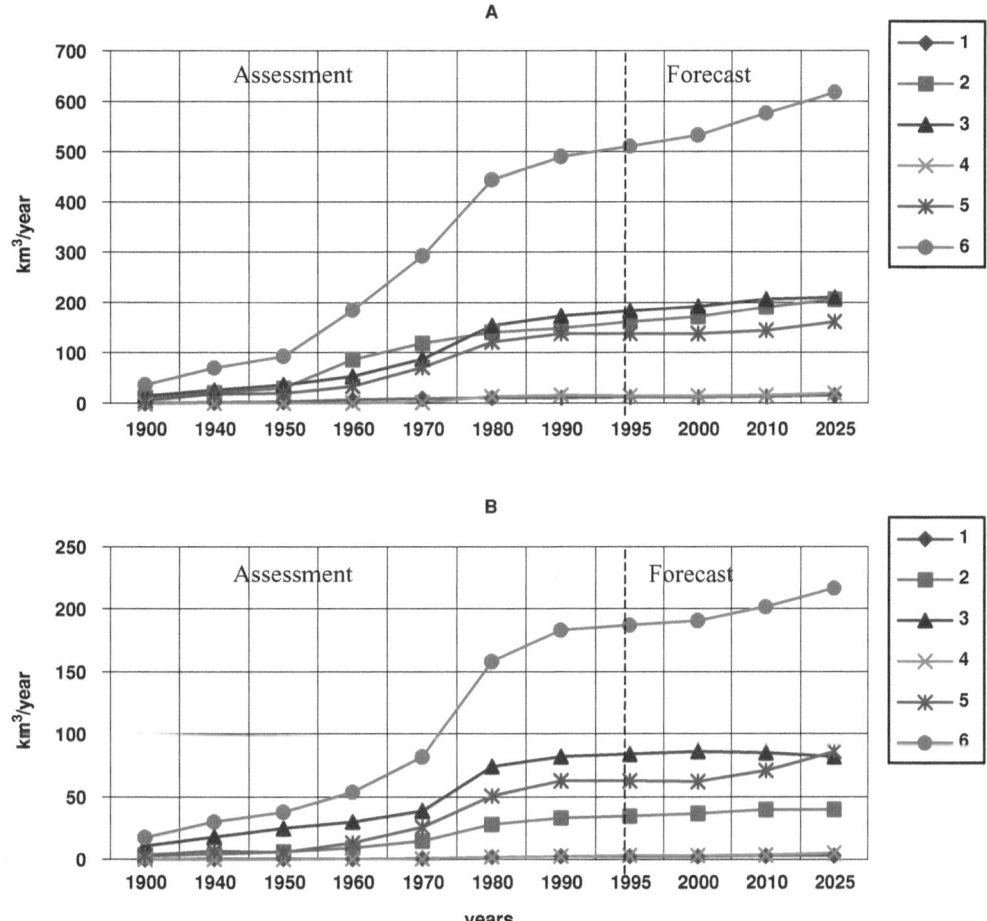

Fig. 4.23 Dynamics of water use: (A) withdrawal, (B) consumption in Europe by natural–economic regions. 1, North; 2, Central; 3, South Europe; 4, northern slope and 5, southern slope of the European territory of the former Soviet Union; 6, water use for the continent as a whole.

consumption is expected to reach about 15% (Table 4.19). These needs are mainly connected with production processes. From Table 4.20 it follows that, for the countries of northern and central Europe, industrial water needs will increase by 1.5 times to 2025, and for southern Europe by 1.7 times. In Belgium and Finland, up to 85% of abstractions meet industrial needs; in France and some other western European countries, this figure stands at about 70%; however, in Greece, Romania and Italy, it is not more than 30%. Industrial water use is predicted to grow by 1.4 times by 2025 in the northern part of the ETSU, and by 1.8 times in the south.

Contemporary abstractions for the agricultural sector average at 198 km³/year with actual consumption standing at 135 km³/year (68%). Unfortunately agriculture is inefficient in its use of water and there are also the technical difficulties in treating waste water. It is expected that by 2025 abstractions will increase to 212 km³/year, and losses to 142 km³/year (Table 4.19).

The percentage of water used by the agricultural sector varies across Europe. In northern and central Europe it fails to reach 20%, in contrast to southern Europe, where it amounts to 60%. There are also variations from country to country. For instance, in the Nordic states, agriculture takes less than 10%; in France and Germany, 20% on average; in Spain, Portugal and Italy, about 70%; in Albania 80%. In northern Russia, agricultural water use is only 4%; in the southern region of the ETSU, it is about 40% of the total water abstracted (Table 4.21). Forecasts show that in northern and central Europe, this type of water use will remain practically unchanged to 2025 (for some countries, it will even reduce); in the ETSU regions, it should increase, in the north by 1.8 times, and in the south by 1.2 times.

Current total abstractions for domestic needs average 70 km³/year with consumption amounting to 8.6 km³/year. By 2025 these volumes are expected to be 85 km³/year and 9.5 km³/year, which is less than 11% of the total. In the northern European countries, use of water for domestic purposes will increase by 1.1 times by 2025 and in the central and southern European countries by some 1.3 times, as compared to the 1990s.

Table 4.21. *Dynamics of water use by type of economic activity for the regions of Europe*[a]

	Assessment			Forecast		
	1980	1990	1995	2000	2010	2025
North Europe						
Population $\times 10^6$	22.5	23.2	23.2	22.8	22.3	23.4
Water use, km³/ year						
Agricultural	1.23	1.57	1.63	1.69	1.71	1.70
	0.70	0.97	1.01	1.05	1.03	1.02
Public supply	2.72	2.98	3.01	3.04	3.20	3.40
	0.22	0.24	0.22	0.21	0.19	0.20
Industrial	6.64	6.29	7.01	7.73	9.21	10.70
	0.60	0.57	0.67	0.77	1.01	1.08
Reservoirs	0.40	0.60	0.65	0.70	0.70	0.70
Total	11.0	11.4	12.3	13.2	14.8	16.4
	2.0	2.4	2.6	2.7	2.9	3.0
Central Europe						
Population $\times 10^6$	281	290	293	293	294	298
Water use, km³/ year						
Agricultural	26.3	29.7	30.7	31.7	32.4	30.9
	18.4	20.8	21.6	22.2	22.0	20.4
Public supply	21.9	25.1	26.5	27.9	30.9	33.7
	1.7	2.0	2.1	2.2	2.5	2.7
Industrial	94.3	93.8	102.0	111.0	126.0	142.0
	7.5	8.4	9.8	11.1	13.9	15.6
Reservoirs	0.5	0.9	1.3	1.6	1.7	1.7
Total	142.0	150.0	161.0	173.0	192.0	208.0
	28.2	32.5	34.8	37.2	40.1	40.4
South Europe						
Population $\times 10^6$	180	186	188	198	203	193
Water use, km³/ year						
Agricultural	93.8	108.0	112.0	116.0	117.0	114.0
	66.6	73.6	75.2	76.7	75.0	71.9
Public supply	19.8	21.7	23.2	24.6	26.7	25.9
	1.8	1.7	1.8	1.9	2.1	2.1
Industrial	38.3	40.5	45.1	49.6	59.7	67.7
	3.1	2.8	2.9	3.0	3.6	4.1
Reservoirs	2.8	3.9	4.0	4.1	4.4	4.0
Total	155.0	174.0	184.0	194.0	208.0	212.0
	74.2	82.1	83.9	85.7	85.1	82.0
Northern slope of ETSU[b]						
Population $\times 10^6$	26.6	28.5	28.5	27.6	26.8	25.6
Water use, km³/ year						
Agricultural	0.68	0.72	0.71	0.70	0.85	1.30
	0.32	0.65	0.65	0.65	0.80	1.20
Public supply	2.10	2.60	2.55	2.50	2.70	3.00
	0.60	0.60	0.60	0.60	0.60	0.60
Industrial	11.20	13.00	12.20	11.70	13.20	16.00
	1.30	1.60	1.60	1.60	2.00	3.20
Reservoirs	0.00	0.00	0.00	0.00	0.00	0.00
Total	13.9	16.3	15.4	14.9	16.8	20.3
	2.2	2.8	2.8	2.8	3.4	5.0

Table 4.21. *(cont.)*

	Assessment			Forecast		
	1980	1990	1995	2000	2010	2025
Southern slope of ETSU						
Population $\times 10^6$	145.4	151.6	150.5	149.4	147.2	144.0
Water use, km³/ year						
Agricultural	47.7	54.9	53.7	52.5	57.4	64.0
	30.9	36.5	36.1	35.6	40.0	48.0
Public supply	11.8	14.7	14.7	14.7	15.2	17.5
	2.8	3.8	3.8	3.8	3.8	3.9
Industrial	55.9	60.0	60.5	61.0	65.0	69.0
	9.8	13.4	13.4	13.4	17.0	23.2
Reservoirs	7.7	9.5	9.5	9.6	10.0	11.0
Total	123.0	139.0	139.0	139.0	146.0	162.0
	51.2	63.2	62.8	62.4	70.8	86.1

[a] Nominator, total water withdrawal; denominator, water consumption.
[b] ETSU, European territory of former Soviet Union.

Table 4.22. *Dynamics of water use by selected European countries (km³/year)*[a]

	Assessment			Forecast		
Countries	1980	1990	1995	2000	2010	2025
Albania	0.20	0.34	0.42	0.50	0.60	0.70
	0.11	0.20	0.22	0.24	0.26	0.29
France	33.60	35.60	36.70	37.70	40.00	42.20
	6.16	7.53	7.71	7.90	8.20	8.44
Italy	50.60	54.60	55.60	56.60	58.00	58.90
	25.10	25.30	24.70	24.10	22.00	20.90
Poland	15.10	15.50	16.50	17.40	19.30	21.20
	3.71	3.82	3.88	3.93	4.27	4.39
Portugal	9.45	9.60	10.20	10.70	11.20	11.30
	3.95	3.76	3.92	4.07	4.04	3.99
Spain	36.70	40.20	43.60	47.00	51.70	53.2
	19.00	20.00	20.90	21.80	22.30	21.70
Sweden	3.79	4.00	4.10	4.21	4.45	4.68
	0.500	0.48	0.55	0.59	0.67	0.74

[a] Nominator, total water withdrawal; denominator, water consumption.

For the ETSU regions, by 2025, use is expected to be 1.2 times the present figure.

Generally in western European countries abstractions for domestic purposes range from 15% to 35%, but in southern Europe about 13% of total abstractions are used for domestic purposes, obviously a percentage strongly dependent on climatic conditions and on the level of development. For instance, in Sweden and

Table 4.23. *Water use as a percentage of water resources by natural–economic regions of Europe*

Region	Water resources, km³/year		Water use as a percentage of water resources			
			1995		2025	
	Local runoff	Inflow	Withdrawal	Consumption	Withdrawal	Consumption
North Europe	705	–	1.7	0.37	2.3	0.4
Central Europe	617	6.0	26.0	5.6	33.5	6.5
South Europe	546	109.0	30.6	14.0	35.3	13.6
North slope of ETSU[a]	589	26.7	2.6	0.5	3.4	0.8
South slope of ETSU	443	123.0	27.5	12.5	32.1	17.0
Europe as a whole	2900	–	17.6	6.4	21.3	7.5

[a] ETSU, European territory of the former Soviet Union.

Table 4.24. *Water availability per population with surface water resources by natural–economic regions of Europe*

Regions	Population (1994), ×10⁶	Water resources, km³/year		Water use (1995), km³/year		Water availability, m³ × 10³/year per head								
		Local runoff	Inflow	Total abstraction	Consumption	Assessment						Forecast		
						1950	1960	1970	1980	1990	1995	2000	2010	2025
North Europe	23	705	–	12.3	2.6	37.70	34.90	32.40	31.20	30.30	30.50	30.80	31.50	30.00
Central Europe	293	617	6	161.0	34.8	2.69	2.47	2.24	2.11	2.02	2.01	2.00	1.98	1.94
South Europe	188	546	109	184.0	83.9	4.08	3.73	3.38	2.92	2.79	2.76	2.61	2.54	2.69
North slope of ETSU[a]	28	589	27	15.4	2.9	31.40	27.3	24.60	22.90	21.10	21.40	21.70	22.30	23.30
South slope of ETSU	152	443	123	139.0	62.8	4.49	3.87	3.46	3.12	2.92	2.93	2.96	2.94	2.91
Europe as a whole	685	2900	–	511	187	5.51	4.99	4.53	4.17	3.99	3.96	3.93	3.89	3.92

[a] ETSU, European territory of the former Soviet Union.

Denmark, up to 30–35% of the water abstracted is used to meet domestic needs; in France about 20%; however, in Greece and Albania, it is not more than 8%. In the ETSU regions it is about 11% of the total water intake.

Current and future water use has been estimated in a number of studies for the different countries, regions, and continents and for the several sectors. Comparison of these results is not always possible, however, due to the different sources and amounts of information used, the methodology and other factors. For instance, in Korzun (1974), abstractions for 1995 were forecast to be 380 km³; but in a later publication (Shiklomanov and Markova, 1987), they were estimated to be 358 km³ for the year quoted and 435 km³ for 1980. Obviously, with the appearance of new data, such assessments, particularly forecasts, have to be corrected.

One of the latest publications containing forecasts of water use is the monograph by Margat (1994). Margat presents water use assessments for the end of the 1980s and gives forecasts for 2025 for southern, western, eastern and northern Europe, as well as for the USSR and Europe in general. However, Margat's assessments of industrial water use do not account for thermal power production, but it is often impossible to distinguish the volumes used for thermal power because of lack of data. Such matters complicate comparisons of current water use data and the forecasts made from them.

4.6.3 Water availability for the regions and for individual countries

Specific water availability (m³/year per head) is given for all regions and for Europe in Table 4.24, and for selected countries in Table 4.25. In these tables, values of water availability were calculated from population data (present and future), assessments of

Table 4.25. *Water availability per population with surface water resources by selected countries in Europe*

Countries	Population (1994), ×10⁶	Water resources, km³/year		Water use (1995), km³/year		Water availability, m³ ×10³/year per head								
						Assessment						Forecast		
		Local runoff	Inflow	Withdrawal	Consumption	1950	1960	1970	1980	1990	1995	2000	2010	2025
Albania	3.60	18.6	5.2	0.42	0.22	–	–	–	7.90	6.46	5.84	5.15	4.41	4.21
France	56.80	168.0	27.0	36.70	7.71	4.13	3.77	3.37	3.25	3.07	3.06	3.03	2.99	2.86
Italy	57.70	185.0	–	55.60	24.70	–	–	–	2.84	2.77	2.77	2.73	2.74	3.02
Poland	39.20	49.50	6.4	16.50	3.88				1.38	1.29	1.24	1.20	1.13	1.09
Portugal	9.93	18.5	34.5	10.20	3.92	–	–	3.67	3.22	3.15	3.14	2.84	2.71	2.91
Spain	39.60	108.0	–	43.60	20.90	–	–	2.74	2.38	2.26	2.19	2.06	1.97	2.06
Sweden	8.34	164.0	12.2	4.10	0.55	24.10	22.70	21.10	20.40	19.80	20.20	20.70	20.20	19.80

water resources, and assessments of abstraction and consumption (present and future). Specific water availability was calculated as the sum of local water resources and a half of the inflow from outside the area concerned for the particular time period.

Despite the fact that most of Europe has a humid climate and large water resources, the mean specific water availability is low, amounting to about 4000 m³/year per head in 1995. This value is one of the lowest in the world, due mainly to a high population density as well the uneven distribution of water resources (Tables 4.24 and 4.25) The largest values are recorded in the Scandinavian Peninsula, Iceland and in the northern part of Russia. Where population density is high, such as in Germany and The Netherlands, despite the fact that these countries have large rivers flowing through them, water availability per capita is rather low.

Most of the countries in central and southern Europe have average water availability values ranging between 1000 and 7000 m³/year per head. In central Europe, despite large rivers with a higher population density, the mean water availability per capita is less than in the southern region.

In the northern part of the ETSU, specific water availability per head averages 21 000 m³/year; and in the south, about 3000 m³/year (in 1995), but the distribution of water resources here is uneven. In the north and northwest, river flows are large and the population density is low, while over most of the south there are numerous arid areas with little surface runoff, and the water resources are insufficient to meet the growing demands of the population.

It is valuable to look at the probable changes in specific water availability across the continent for the coming years. Population forecasts were employed for this purpose and the results are illustrated in Fig. 4.24. This clearly shows that the specific water availability drops by 7–9% during the 1990s, then from 2000 to 2025 it remains relatively stable for all regions and for the continent as a whole except for minor fluctuations within the 3–5% range. This agrees well with trends in population growth.

In many European countries, due to the uneven distribution of runoff by month and over longer periods, deficiencies are recorded during periods of low water and during droughts. The longest periods of low water are most often recorded in central and southern Europe, as well as in the southern regions of the ETSU. Water resources deficiencies in southern European countries, for instance, Greece and Spain, are now becoming a significant hindrance to development. Therefore it is important to have reliable assessments of water availability during low-water periods because this enables the lower limit of the available water to be determined.

Tables 4.26 and 4.27 show these lower limits across the continent for shorter periods and for the whole year with the lowest flows. Water availability is calculated for 1995 and 2025. The forecasts for 2025 demonstrate a reduction in the available water for central and southern Europe, and for the southern ETSU, as well as for the entire continent. While reductions will become more severe for some southern countries, the picture of water availability over the period shows the fluctuations and variations characteristic of most European countries. However, in the northern regions there will be slight increases by 2025 as compared to present values of available water.

4.7 CONCLUSIONS

The total annually renewable water resources in Europe, assessed for the period 1921 to 1985, amount to 2900 km³/year. This volume is divided between the Arctic and Atlantic Oceans, and the Caspian Sea. The runoff to the Atlantic amounts to 1895 km³ (65% of total water resources) and to the Arctic 694 km³ (24%), while the Caspian receives 311 km³ (12%). The largest annual runoff from the continent during the period was 3210 km³ (1941), and the smallest was 2442 km³ (1921).

The distribution of water resources among the regions is quite uniform: in northern Europe, 705 km³/year (24% of total water

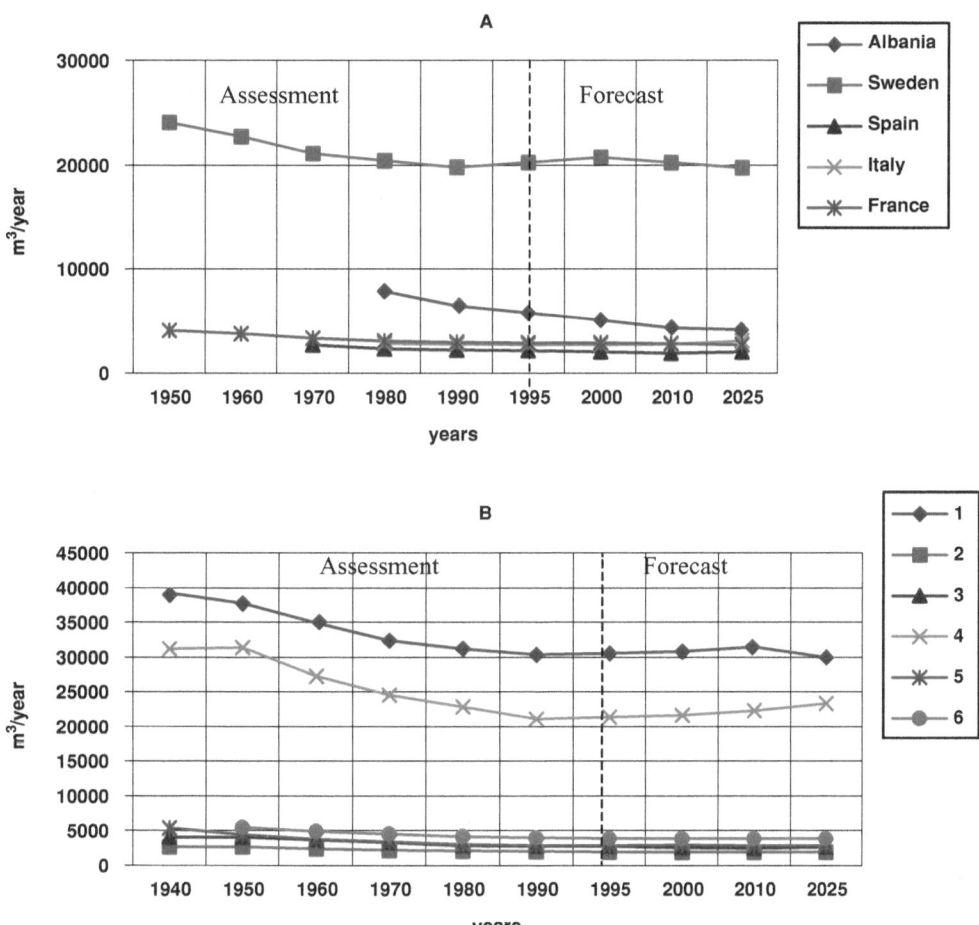

Fig. 4.24 Dynamics of per capita water supply (A) in selected European countries and (B) by natural–economic regions. 1, North; 2, Central; 3, South Europe; 4, northern and 5, southern slopes of the European territory of the former Soviet Union; 6, the continent of Europe as a whole.

resources); in central Europe, 617 km³ (21%); in southern Europe, 546 km³ (19%); in the northern part of the ETSU, 589 km³ (20%); and in the southern part of the ETSU, 443 km³ (16%).

Annual and monthly runoff data were used to assess water resources. The reliability of the assessments was based on the use of time series which were long and representative. Unlike previous research (Chernogayeva, 1969; Baumgartner and Reichel, 1972, 1975; Korzun, 1974b; L'vovich, 1974; Shiklomanov and Markova, 1987), water resources data were used for the period 1980 to 1990, acquired in the main from different national sources.

At the end of the twentieth century, the countries of western Europe appeared to be most adequately studied in terms of hydrology, while southern countries and the northeast of Russia appeared to be the least understood. As a result of these different levels of understanding the different assessments water resources across the continent will have variable reliability.

The study of the 40 main rivers found that the main statistical parameters were stationary, with a lack of linear trend over the period from 1921 to 1985. Consequently it must be assumed that long-term runoff is stationary and that there is little or no evidence of the impact of climate change on water resources at the continental scale. However, due to the high density of population and the large amount of industrial activity, the impact of human activities on water resources has changed the natural pattern of long-term runoff and its variations in many basins and regions. These changes show up most vividly in the south and centre of the continent. Analyses of water use showed that both abstractions and consumption are increasing with population growth, the rising standard of living and economic development. Water use was estimated for the period 1900 to 2025 for the sectors, as well as for most countries and regions, and for the continent as a whole.

For 1995, abstractions were estimated to be 511 km³/year, about 17% of total water resources, with consumption standing at 187 km³ (6%). Forecasts showed that by 2025 abstractions should amount to 619 km³ (21%), and consumption to 217 km³ (8%). Growth of abstractions and of consumption display slightly different patterns, with rapid growth in the south of the ETSU

Table 4.26. *Water availability by natural–economic regions of Europe throughout long-term dry period and driest year*

Region	Population, ×10⁶		Water resources, km³/year				Consumption, km³/year		Water availability, m³ × 10³/year per head			
			Driest period		Driest year				Period		Year	
				Mean annual								
	1995	2025	Dates	runoff	Year	Runoff	1995	2025	1995	2025	1995	2025
North Europe	23.2	23.4	1938–44	590	1970	567	2.6	3.0	25.50	25.10	24.50	24.10
Central Europe	293.0	298.0	1971–75	460	1949	356	34.8	40.4	1.46	1.41	1.10	1.06
South Europe	188.0	193.0	1971–76	405	1950	372	83.9	82.0	1.72	1.67	1.54	1.50
North slope of ETSU[a]	28.5	25.6	1938–45	485	1960	458	2.9	5.0	17.20	18.80	16.30	17.90
South slope of ETSU	152.0	144.0	1935–40	325	1975	306	62.8	86.1	1.74	1.66	1.62	1.53
Europe as a whole	685	685	1971–75	2550	1921	2440	187	217	3.44	3.41	3.29	3.24

[a]ETSU, European territory of the former Soviet Union.

Table 4.27. *Water availability in selected European countries in the driest year*

Country	Population, ×10⁶		Water resources in driest year		Consumption, km³/year		Water availability, m³ × 10³/year per head	
	1995	2025	Year	Runoff, km³	1995	2025	1995	2025
Albania	3.6	5.0	1975	13.1	0.22	0.29	3.58	2.56
France	56.8	60.4	1921	90.3	7.71	8.44	1.45	1.36
Portugal	9.9	10.9	1948	15.2	3.92	3.99	1.14	1.03
Spain	39.6	42.3	1981	27.2	20.90	21.70	0.16	0.13

while for other regions consumption increases with the growth in abstraction.

During the period in question the pattern of water use in the different sectors has changed markedly: at the beginning of the twentieth century, agriculture was dominant; then, with population increase and economic development, demand in the industrial and domestic sectors started to increase. Presently (1995), water for industry averages 45%, agriculture makes up 39%, while domestic purposes require less than 14%, and evaporation from reservoirs is about 2–3% of the total quantity abstracted. Forecasts indicate that by 2025, industry is likely to take 50% and agriculture 34%, while the domestic sector will stay at 14% of the total.

These forecasts for the period 1995 to 2025 are, of course, tentative due to a number of uncertain factors. Obviously, in forecasts, it is impossible to take account of extreme situations, such as wars, natural disasters etc. However natural climatic variability and global warming over this time interval could also result in alterations and changes.

A careful analysis of the continent's water resources and water use enabled water availability to be assessed in specific countries, regions, and for the continent in general. This showed that the mean value of specific water availability in Europe is low and presently (1995) amounts to about 4000 km³/year per head. Despite the fact that most of the continent lies in the humid zone and has large water resources, availability of water is one of the lowest in the world. Water availability is unevenly distributed across Europe: the largest water resources are recorded in Nordic states, Iceland and northern Russia (20 000 to 30 000 km³/year per head), and the smallest in many countries of central and southern Europe (for instance, Poland and The Netherlands), where the specific water availability is in the range 3000–5000 km³/year per head. By 2025, a reduction of specific water availability can be expected in central and southern Europe, and in the southern part of the ETSU; for northern Europe and the northern part of the ETSU, a slight increase in this value is likely.

The analysis of water availability at the present time and in the future allows four typical situations to be identified.

1. Water reserves are large and demands are limited. This is the most favourable situation, which occurs in areas with a humid climate and low population density, a relatively small volume of industrial production, or where water saving measures are

used by industry. Among such regions are the Nordic states, Iceland and northern Russia.

2. Water resources are limited, and the demands are also low. Such a situation is characteristic of sparsely populated areas with an arid climate and with little development. This situation is found in the southern and southeastern regions of Russia and Kazakhstan. Presently water availability here is adequate; however, forecasts show that in the near future, the growth of the population and of the economy, particularly irrigation and petroleum processing, will make the finding of additional freshwater resources most urgent.

3. Water resources are large, but the demands are also high. The most developed and densely populated countries come into this category. Problems of water availability are quite crucial here and require extensive measures for the planning and management of water use, particularly in connection with the increasing pollution of water bodies. Thus, according to forecasts, by 2025 the demand for water in such countries will amount to a significant proportion of runoff.

4. Water resources are limited, and the demands are extremely high. Such a situation of stress is currently prevalent in many southern European countries with developed farming (Spain, Portugal, France and Italy), as well as in southern Russia and the Republics of Moldavia and the Ukraine. In these countries, abstractions for industry can amount to more than half of the water resources of the country concerned. The extremely high demand for water means that demand is outstripping the resource. It is constantly growing and threatens to accelerate in the near future. Forecasts show that, because of increasing consumption, the water required for dilution of industrial and agricultural discharges and the increasing volume of water pollution, by 2025, the water resources available to some European countries might not be sufficient to satisfy the total demands for a clean and reliable supply of water. Particularly serious difficulties are likely to arise when low flows occur and there are lengthy periods of drought. Under these conditions, water is the major factor restricting opportunities for growth and development, restraining industrial production and limiting the spread of irrigation. Local conditions of water availability presuppose the matching of the quantity and quality of existing reserves with the expected demand for water.

5 Water resources, water use and water availability in Asia

5.1 INTRODUCTION

Asia is the world's largest continent occupying one-third of the land surface or 43.5 million km². In this vast territory extending over 10 000 km from north to south and over 11 000 from west to east, almost all the different types of climate, vegetation and soils that exist on Earth, from the Arctic to the Equator, are to be found. It is characterized particularly by the unusually widespread continental climate. In the western part of the continent the predominantly continental tropical air masses have resulted in a predominance of arid landscapes. In the east and south, bordering the Pacific and Indian Oceans, there is the largest area of monsoon climate that exists in any part of the world. Climatic differences are intensified by sharp contrasts in relief. The structure of the continent is very complicated. Asia is characterized by hills, mountains and plateaux (75% of the area) and many areas of inland drainage (28% of the land surface). Some of the largest rivers in the world (the Ganges, the Brahmaputra, Yangtze, Yenisey, Lena, Ob, Amur and Mekong) flow across the continent to the north, east and south.

The population of Asia amounts to 3445 million and continues to grow rapidly, with a mean density approaching 80 inhabitants/km² and the most densely populated areas in the regions with a monsoon climate. While there are 51 states within Asia, about 30% of the area is occupied by Russia. The other large states include China (with an area of 9.6 million km² and a population of 1209 million) and India (3.27 million km² and 919 million). The highest level of economic development is to be found in Japan, Hong Kong, Singapore, Kuwait, Israel and the United Arab Emirates. Seven large regions can be defined within Asia on the basis of the nature of the water resources and their use. These are: (1) Siberia and the Far East; (2) Northern China (without the Yangtze River Basin), Korea and Mongolia; (3) South East Asia; (4) Southern Asia; (5) Middle Asia and Kazakhstan; (6) Transcaucasia; and (7) Western Asia (Table 5.1).

The average renewable water resources of the continent over the period 1921 to 1985 are estimated to be 13 500 km³/year. The largest value (15 000 km³/year) was observed in 1937 and the least (11 800 km³/year) in 1979. Almost half the volume, some 6650 km³/year, occurs within South East Asia (6.95 million km²). Considerable water resources of 3110 km³/year exist within the largest region, namely Siberia and the Far East (12.8 million km²). The water resources of Southern Asia (4.49 million km²) are estimated to be 1990 km³/year. Transcaucasia (68 km³/year) and Middle Asia and Kazakhstan (181 km³/year) are characterized by scarce resources, but they are even more scarce in the Arabian Peninsula (less than 7 km³/year). For the continent as a whole, the potential available water resource is, on average, 311 000 m³/km² per year. South East and Southern Asia have the largest figures, namely 956 000 and 443 000 m³/km² per year, respectively. Middle Asia and Kazakhstan and especially the Arabian Peninsula have the least, some 45 400 m³/km² per year and less than 5000 m³/km² per year, respectively. About 6560 km³/year drains to the Pacific Ocean and 3920 km³/year to the Indian Ocean. Runoff to the Arctic Ocean is 2420 km³/year, to the Atlantic Ocean 200 km³/year and to the areas of inland drainage 406 km³/year.

The water resources of Asian rivers vary appreciably in space and time. The coefficient of variation of total runoff (C_v) lies between 0.1 and 0.3 in the humid regions and in arid areas it is >1. While the water resources in the regions draining to the different oceans and the continent on the whole show insignificant variations in time ($C_v = 0.06$–0.30). The variations of water resources have a cyclic character. The most frequent duration of the periods of high or low runoff does not exceed 10 years. Time periods of a different duration, namely of 1–5 years, have been identified with a sufficiently high probability ($P \geq 70$–90%). Most often the periods of dry or wet years are followed by years with average conditions. The water resources of the basins with the largest rivers poorly correlate with each other ($r < 0.60$). The distribution of runoff within a year is most uniform in the rivers of South East Asia and Transcaucasia. A non-uniform distribution of runoff is observed in the rivers of Siberia and the Far East, Middle Asia and Kazakhstan, and in Southern and Western Asia. In these regions more than half the annual runoff is discharged during the three summer months.

Table 5.1. *Distribution of Asian countries by natural–economic regions*

Region	Countries (or parts of countries) included
Siberia and Far East	Asian part of Russia
Northern China and Mongolia	China (north of Yangtze basin), Mongolia and Korea
South East Asia	Vietnam, Eastern Timor, Brunei-Darussalam, Hong Kong, Indonesia, Cambodia, China (southeastern part and Yangtze basin), Laos, Macao, Malaysia, Myanmar, Singapore, Thailand, Taiwan, Philippines, Japan
Southern Asia	Bangladesh, Bhutan, India, Maldives, Nepal, Pakistan, Sri Lanka
Middle Asia and Kazakhstan	Kazakhstan, Kyrgyzstan, Tajikistan, Turkmenistan, Uzbekistan
Transcaucasia	Azerbaijan, Armenia, Georgia
Western Asia	Afghanistan, Bahrain, Yemen, Iraq, Iraq–Saudi Arabia neutral zone, Iran, Israel, Jordan, Qatar, Cyprus, Kuwait, Lebanon, United Arab Emirates, Oman, Saudi Arabia, Syria, Turkey

Around 175 million ha of Asia is occupied by irrigated land. The largest areas are centred in the basins of the Indus, Huang Ho, Yangtze, Amu-Darya, Syr-Darya and Mekong. By 2025 irrigated areas are expected to extend to 230 million ha. Some countries use non-traditional water sources for irrigation, including ground water. The use of water across the continent is currently some 2160 km^3/year. Agriculture is the main user (81%), especially for irrigation. Water consumption losses are about 1560 km^3/year (12% of water resources). By 2025 the volume of water abstracted is expected to reach 3100 km^3/year and water consumption about 1970 km^3/year (15% of water resources). Water use in the industrial and domestic sectors is likely to double by 2025 relative to the 1995 level to about 410 and 340 km^3/year, respectively.

For the continent as a whole the water available is currently around 3500 m^3/year per head. Three regions (Northern China and Mongolia, Southern Asia and Western Asia) are characterized by very low water availability (<2000 m^3/year per head). The Arabian Peninsula has extremely low water availability (300 m^3/year per head). In the future very low availability of water is likely to occur in Middle Asia and Kazakhstan, whereas in Southern and Western Asia it is expected to be even lower. For some countries in the Near East and the Arabian Peninsula the trends in the future availability of water are likely to be altered by changes in demographic policy, economic growth and development and by exploiting non-traditional water sources. The prob-

lem of the region surrounding the Aral Sea is the most acute in the entire continent.

5.2 PHYSICAL CONDITIONS

5.2.1 Relief, climate and vegetation

Mountains and highlands occupy about 75% of Asia. The belt of the highest mountains is in Central Asia (the Hindu Kush, Karakorum, the Himalayas, etc.) and intercepts moisture from the Indian Ocean and also from the Atlantic and the Pacific Oceans. The system of ranges (Tien Shan, the Altai, Sajhan, etc.), the Greater Khingan Range located in the east and the associated plateaux and highlands of Central Asia enclose large areas occupied by deserts and semi-deserts (Takla Makan, Alashan, Gobi, etc.). Many of the mountain tops of the ranges in Central Asia exceed 7000 m and some even 8000 m, while in the other ranges most summits are below 4000 m. The mountainous regions of Asia are mainly centred on two massifs. One of them extends latitudinally from Asia Minor to Eastern China and the Malay archipelago, while the other is aligned from southwest to northeast from the northern margins of the highlands of Central Asia to the Chukchi Peninsula. A third mountain belt passes through the East Asian islands.

Lowlands occupy about 25% of the continent area. One of the most extensive lowlands in the world, the West Siberian Plain, extends over the northwestern part of the continent and includes the Turan Lowland east of the Caspian Sea. The other lowlands are located on the maritime margins of the continent, in the belt of piedmont depressions (Mesopotamia, the Indus–Ganges lowlands), in a number of isolated depressions in Eastern Asia (Penzhina and Anadyr Plains, along the lowlands of the Amur River, North China and Manchurian plains) and on the shores of the Arctic coast (Yana-Indigirka and North Siberian Lowlands). Within the Central Asian highlands, there are elevated plains confined to the intermontane depressions (Dzungaria, Kashgar, Tsaidam). Some vast plateaux surfaces (the Gobi, Middle Iran, Anatolia) are to be found at heights of about 1000 m, but in Tibet in the Tien Shan and the Pamirs the high land exceeds 4000 m.

Since Asia has a large extent and a complicated orographic structure, together with a wide range of radiation and circulation conditions governing the heat and moisture regimes, these factors cause the climate and water balance to be extremely variable. Almost the entire continent is characterized by sharp contrasts between humid and arid conditions controlled by the interception of oceanic moisture by mountain ranges. The close proximity to the sea of a number of mountain ranges with heavy rainfalls on their windward slopes contrasts with the arid and semi-arid

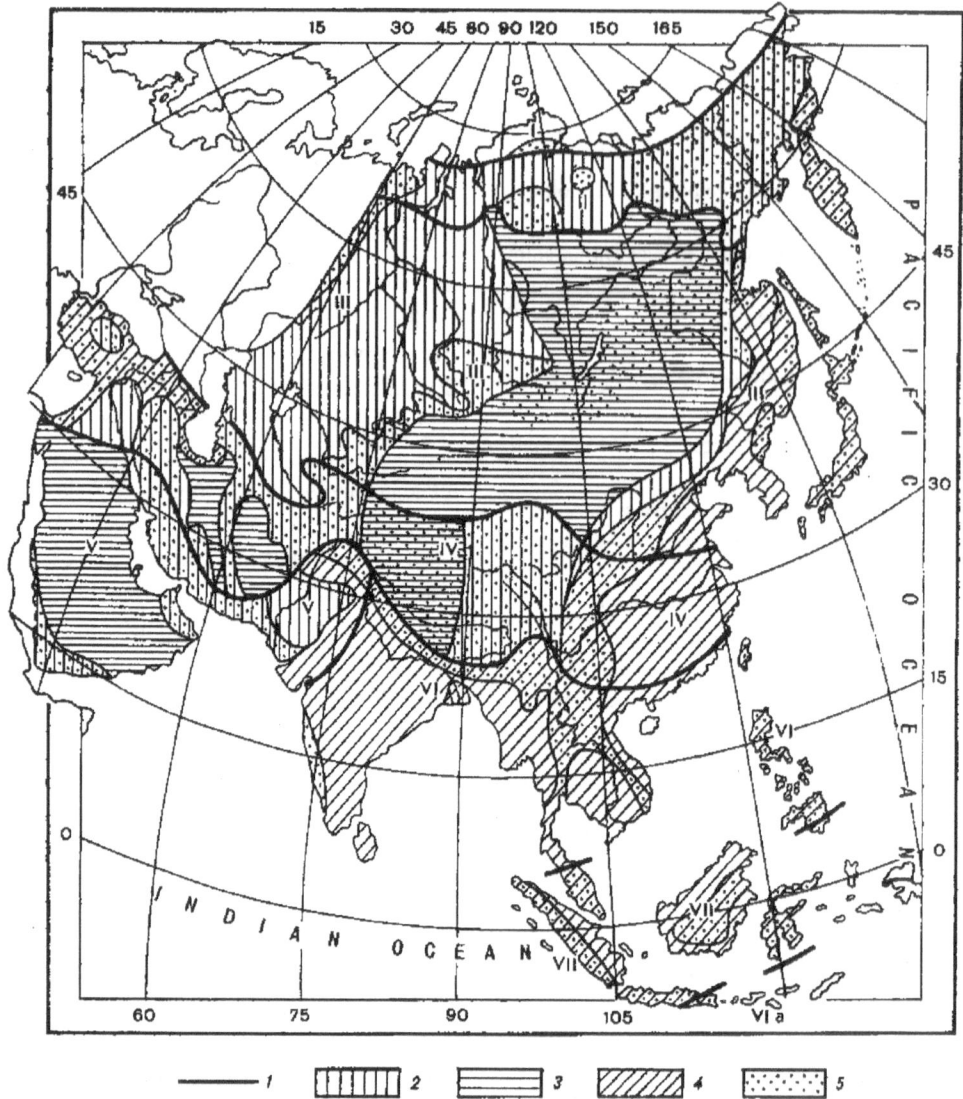

Fig. 5.1 Geographical belts in Asia (from GUGK, 1964, pp. 75, 116). I,
arctic; II, subarctic; III, temperate; IV, subtropical; V, tropical;
VI, subequatorial (VIa, of the Southern Hemisphere); VII, equatorial.
1, boundaries of the belts; 2, areas with a continental or moderately
continental climate; 3, areas with a markedly continental climate;
4, near-oceanic areas or areas subject to marked (permanent or seasonal)
influence of the ocean on the climate; 5, mountainous territory,
characterized by vertical climatic and natural vegetation zones.

conditions in the inner or even in nearby coastal regions, which
are in the "rain shadow".

Asia extends from the Arctic to equatorial latitudes (Fig. 5.1).
Hence the annual values of total solar radiation and the radiation
balance vary over a wide range. Total solar radiation changes from
290 kJ/cm^2 per year in the far north up to 670 to 840 kJ/cm^2 per
year in the equatorial regions of South East Asia (Korzun, 1974).
The radiation balance changes from almost zero values up to 336
to 420 kJ/cm^2 per year, respectively (Ryabchikov, 1988; Sedunov,

1991). During the year, arctic, temperate and tropical air masses
move across the continent. Over most of the continent, especially
in the inner regions, continental air masses are formed in the winter and summer. They are cold in the winter and hot in the summer.
The annual amplitudes of air temperatures reach quite large values. Tropical air dominates for the whole year over southwestern
and Western Asia (Arabian Peninsula, Thar Desert, southern Iran
Uplands). Maritime tropical air penetrates in summer into Southern and South East Asia.

A typical feature of the climate is its pronounced seasonal character. In arctic, subarctic and temperate latitudes, it is determined
by the thermal regime, whereas at equatorial and tropical latitudes it is governed by the moisture regime. During the winter,
the Asian High (Siberian, Mongolian) has a dominant influence
on climate. Within Asia it occupies a large area – Siberia, part of
Kazakhstan and farther south, the area of northern Mongolia as

far as the Himalayas and southern China. A strong flow of cold air moves southward in the winter as the northwestern winter monsoon. It forms unusually cold and dry winters in the east of Asia in the temperate and subtropical belts. In the inner regions of Asia the winter is especially severe. The absence of north–south barriers in the north increases the width of the zone of cold arctic air, which often penetrates the intensely cool southern Siberia and the Gobi Plateau. A high pressure area is also formed above southwest Asia which covers the Arabian Peninsula and the southern part of Iran – a branch of the Azores High. Here very dry continental tropical air masses prevail. Active cyclones develop in the winter at the polar (temperate) front. The pathways of winter cyclones moving from the Mediterranean Sea are mainly across Asia Minor. Cyclonic activity also develops in the Iranian sector of the polar front. However the Mediterranean air penetrates here only in rare cases; in this region the continental air masses interact, the tropical air of southern Iran and the Arabian Peninsula and the polar air coming at this time of the year from Middle Asia or Iran (Ryabchikov, 1988). Southern Asia, principally the Indian subcontinent, which is protected from the penetration of cool continental air from Central Asia by high mountain systems, is characterized by a warm winter (Singh, 1980). Above Indostan and Indochina, southern China and over much of Iran, tropical air and winter tradewinds prevail. The weather in this season has a stable anticyclonic character. Mean air temperatures in January change from −32 to −36 °C in the north of the continent and in the coastal regions of the Arctic Seas to +25 °C in the subequatorial regions. During winter a temperature inversion is widespread in the mountains of Central Asia. South of the Northern Tropics below-zero temperatures are observed only at heights exceeding 1500 to 2000 m. In the equatorial regions a typical minimum is +20 to 22 °C during January, when average temperatures are +24–25 °C.

During the summer atmospheric conditions are very different. The winter Asian High breaks down and over the hot continent an area of decreased pressure is established with its centre in the southwest (the eastern part of the Arabian Peninsula, southern Iran Uplands, southern Pakistan and the northwest of India). The Azores High intensifies considerably during the summer. Dry westerly winds bring tropical continental air to the Arabian Peninsula from Africa. As a result, the tropical desert climate of Arabia and western India differs sharply from the neighbouring tropical monsoon climate of India. The trajectory of cyclones is displaced to northern Eurasia. Over large parts of Middle and Central Asia in the summer the polar front is not clearly pronounced. It is more pronounced in the mountains of Southern Siberia, Middle Asia and northern Central Asia causing a summer precipitation maximum in these areas. Over the remainder of these parts the cyclonic activity is weak: hot cloudless weather is predominant. The prevailing desert landscapes in Asia are determined by these

climatic features. The summer monsoon is established over southern, southeastern and eastern Asia. In eastern Asia it is mainly related to the northward seasonal motion of the Pacific polar front and the motion of the high-pressure field over the Pacific Ocean (Ryabchikov, 1988). The Pacific Polar front, where the masses of polar maritime and tropical maritime air are in contact, passes across southern Korea, the north of Japan and south of the Kuril Islands. Weak cyclones move slowly along it during the summer. The southerly and southeasterly winds bring warm moist tropical air from more southerly Pacific Ocean latitudes. This flow is the summer Pacific monsoon. When it is in contact with the cold currents (Oyashio, Primorskiy and Sakhalin) that wash the coast of eastern Asia, this air is intensely cooled. The cool temperate latitude summers which characterize the shores of the Sea of Okhotsk and also, in part, the Sea of Japan, with much fog and drizzle, are in response to this cooling. Only south of 38 °N, where the warm Kuroshio current approaches the shores of Japan, has the summer monsoon a warming effect, bringing humid and suffocating heat with continuous rain. Quite frequently, especially in autumn, tropical cyclones with very high wind velocities pass along the Pacific polar front.

In Indostan, Indochina, over the Malay Archipelago and in southern China, the circulation of the summer monsoon is related to the extratropical convergence zone and the formation of tropical depressions. This zone moves large distances seasonally along the meridian between 25 and 30° in the northern Indian Ocean and in southern Asia. The humid equatorial monsoon brings cloud and abundant rain. Along the tropical front in the Philippines and in the South China Sea there are typhoons. Over Sri Lanka, southern Malacca and much of Indonesia equatorial maritime air masses always dominate. Here the typical equatorial climate with abundant rain and high temperatures occurs at all seasons of the year with little variation. During summer over the northern half of Asia air masses move in from the west. Because of the low relief of this part of the continent, the Atlantic air penetrates far to the east, as far as the Verkhoyansk and Zabaikalsk Ranges, and is gradually transformed into continental air. The inflow of Atlantic air masses is accompanied by cooling and precipitation. Also, in the northern part of the continent the arctic air masses move southward towards the heated land reaching as far as western Siberia and the Turan Lowland. In the temperate zone the arctic air is heated and dried and is gradually transformed to continental air. Over the plains of Middle Asia this air becomes very hot and creates the deserts to be found there.

During the summer the temperature contrasts between the northern and southern parts of the continent are small. Over much of Asia mean July temperatures range from 16 to 24 °C. Over the Arabian Peninsula and in Western and Southern Asia they reach 24–32 °C and higher (Lebedev and Kopanev, 1975). In the mountains of Central Asia mean monthly temperatures of the warmest month vary from 8 to 12 °C. Extremely low temperatures are observed in the mountains of northern Mongolia, in the Tibetan Uplands and in the adjoining ranges. Over the islands of Japan mean monthly July temperatures vary from 20 to 26 °C. In the high mountains they fall to 18–20 °C, but reach 27 °C in the Philippines. Over the Malay archipelago, the Sri Lanka islands and in Taiwan the air temperature varies from 24 to 28 °C. The distribution of precipitation and the different temperature regimes as driven by the global circulation processes are quite diverse across Asia (Korzun, 1974a). The largest amounts of precipitation (2000–4000 mm and more per year) fall in the mountain regions of South and South East Asia where summer monsoons dominate. The world's highest total annual precipitation has been recorded (10 800 mm) (Gopal Singh, 1980) on the Shillong Plateau at Cherapungi.

In contrast there are large areas in west, southwest and Central Asia that are arid and semi-arid. The driest parts are in the deserts of the Arabian Peninsula where the annual precipitation is less than 100 mm, and similar amounts are recorded in the Takla Makan Desert and in Inner Mongolia in the Gobi Desert, where less than 50–70 mm of precipitation occurs at some locations. The coastal strip along the Persian Gulf is also very dry and temperatures are very high (maximums reach 50–55 °C). Less than 100 mm of precipitation falls in the desert around the mouth of the Indus (Main Air Force Headquarters, 1992). Small amounts of precipitation (150–200 mm) are also observed in the plains of Middle Asia. The Kara Kum, Kyzyl Kum and Bet Pak Dala areas are especially dry, as are the western coast of Lake Balkhash and the Sarakamysh Depression where the annual precipitation is less than 100 mm.

North of 65 °N the precipitation decreases eastwards with the increasing continental character of the climate. The only exception is the northeastern region of the continent near the Pacific. In the Arctic the amount of precipitation decreases from 250 mm at the coast to 100 mm inland (at the interfluve of the Yana and Indigirka Rivers and Verkhoyansk). Between 200 and 250 mm of the precipitation that falls at Yakutiya is due to the anticyclonic regime of the winter and the insulating effect of mountain ranges. On the windward mountain slopes totals reach 400–500 mm per year (Kopanev and Shvert, 1991). Over the rest of Asia the annual precipitation varies from 400 to 1000 mm increasing on the windward slopes of the mountains to more than 1200–2000 mm. The islands of South East Asia experience much more precipitation: there totals reach 4000–5000 mm (Korzun, 1974a).

All the bioclimatic zones of the Northern Hemisphere are represented within the continent. Arctic deserts and tundra bordered in the south by a narrow strip of forest extend to the Arctic islands and along the coast of the Arctic Ocean. Southward there is taiga (mainly taiga forest in the west and light coniferous forest in the east) which is replaced southwards by mixed and broadleaved forests, forest–steppe and steppe. Semi-desert and desert land-

scapes are pronounced features of the Arabian Peninsula, the inner regions of the Iran Uplands and in Middle, Central and Southern Asia. In the subtropics of Western Asia Mediterranean vegetation is widespread, with monsoon mixed and broadleaved forests in eastern Asia. In the tropical latitudes of eastern and Southern Asia there are monsoon deciduous forests and savannah. On the windward mountain slopes there are evergreen forests and in equatorial latitudes (mainly in Indonesia) there are swamps and tropical forests.

In many regions of Asia the natural landscape has been considerably changed by human activities. For example, in Southern Asia 80% of woodlands have been cleared completely or replaced by less productive sparse deciduous shrub complexes. The landscape consists of different types of agri-landscapes with areas where erosion, soil degradation, secondary salinization and desertification are prevalent (Ryabchikov, 1988). The Indus–Ganges plain is one of the most densely settled areas and is almost entirely cultivated. The forests which covered the plain in the past have not been conserved. The landscapes of the plains and foothills of Indochina have also undergone significant changes. The alluvial plains are cultivated mainly for rice instead of the natural woodland. Only the Malay Archipelago, in spite of the large population density, remains one of the most afforested regions of the globe.

5.2.2 Hydrology

Asian rivers are generally large, and are amongst the largest rivers of the world in terms of the discharge, basin area and length: for example the Ganges with the Brahmaputra, the Yangtze, Yenisey, Lena and Mekong. However there are large parts of the continent occupied by deserts and almost completely devoid of the surface runoff. The river systems drain to the four oceans: the Arctic, Pacific, Indian and Atlantic (Fig. 5.2, Table 5.2), with a vast area of inland drainage located at the centre. Part of it, in Central Asia, is circled by the world's highest mountain ranges and uplands. In the western part of Central Asia are the basins of the Aral Sea and Lake Balkhash. The largest rivers of Siberia carry their waters to the Arctic Ocean: these are the Ob, Yenisey, Lena and Kolyma, whose runoff makes up the main part of the water resources of Russia.

The Pacific Ocean basin includes the Amur, Yangtze, Zhujiang, Red and Mekong Rivers. The islands, including Japan, the Philippines and the northern part of the Malay Archipelago, have a well-developed river network with short rivers carrying discharges often used for hydropower. The Indian Ocean basin includes the Ganges with the Brahmaputra, Irrawaddy, Indus, Tigris and Euphrates. Most the mouths of these rivers are deltas with numerous arms. For example, the area of the Indus delta is 8000 km^2, the Ganges with the Brahmaputra – 80 000 km^2, the Irrawaddy – 48 000 km^2, the Mekong – 93 000 km^2, the Red River – 5000 km^2

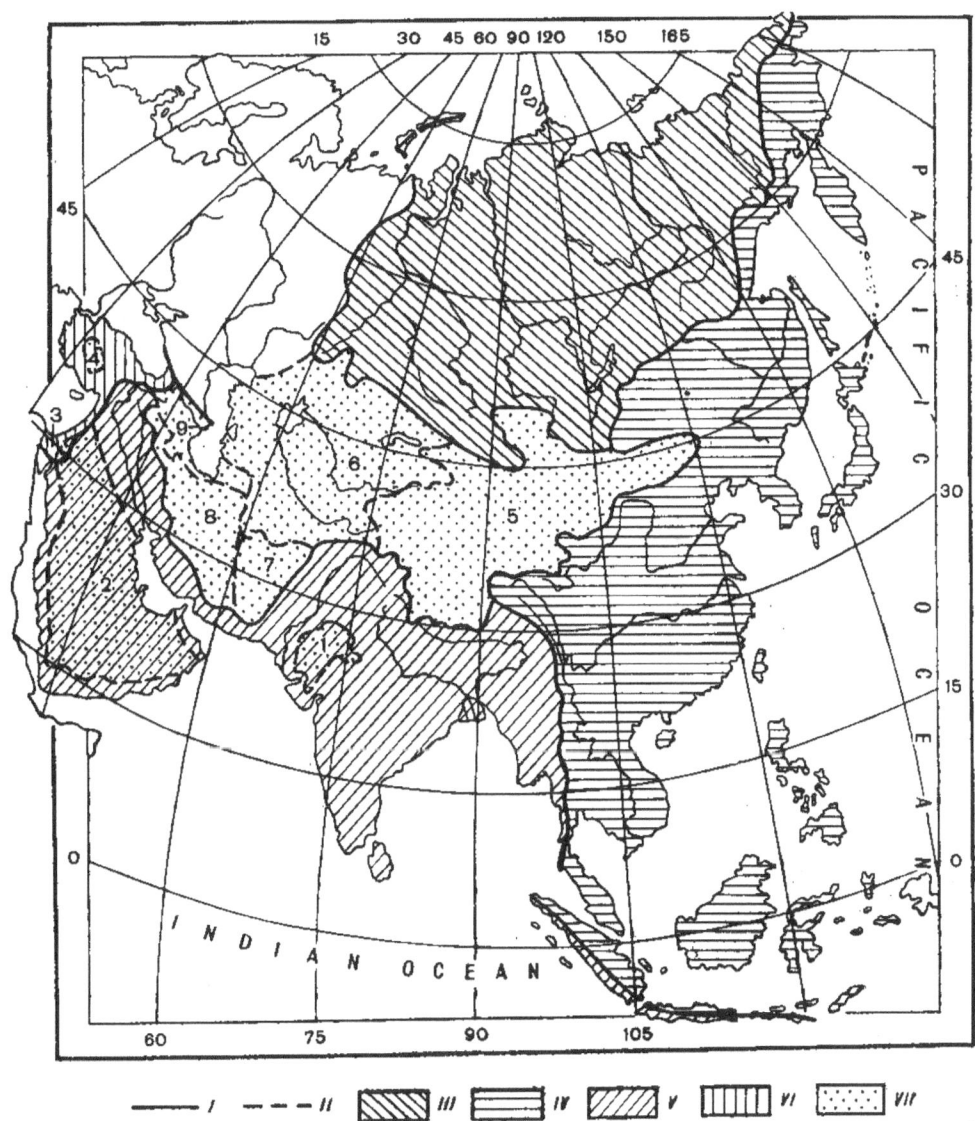

Fig. 5.2 Ocean drainage basins of Asia. I, boundaries of drainage basin; II, boundaries of inland runoff regions; III, Arctic ocean; IV, Pacific ocean; V, Indian ocean; VI, Atlantic ocean; VII, internal runoff regions: 1, Thar Desert; 2, Arabian Peninsula; 3, Dead Sea basin; 4, Inland Anatolia; 5, Central Asia; 6, Kazakhstan and Middle Asia; 7, Seistan Depression and adjacent regions; 8, Iran Uplands; 9, Pre-Caspian area.

(Gavrilov, 1950; Nguyen Van Ky, 1990; Main Air Force Headquarters, 1992).

The character of the rivers and their regimes has been radically changed by the network of irrigation canals in those countries where irrigation is practised (Pakistan, India, Bangladesh, Myanmar, Vietnam, China and the Middle Asia states). For example, in the Indus basin the length of irrigation canals is more than 60 000 km. It includes 43 main channels transporting water from the Indus over distances of up to 300 km. The network of canals in China is constantly growing. The Grand Canal, built as

Table 5.2. *Distribution of the area of Asia (including islands) by oceanic slopes*

Runoff region	Area, km² × 10³	Percentage of total area
Arctic Ocean	11 672	26.9
Pacific Ocean	11 907	27.4
Indian Ocean	9 656	22.2
Atlantic Ocean	742	1.7
Central region of endorheic runoff	9 500	21.8
Asia as a whole	43 475	100

early as the seventh and eighth centuries is the longest main canal connecting the downstream parts of the Yangtze, Huang He, Huai He and Hai He.

Information on the rivers of Asia (Table 5.3) is taken from many sources, including Sokolov, Korzun, *Encyclopedia Britannica*, Czaya, Szestay, Milliman and Meade, Jang Jicheng, Marcinek, Ministry of Water Resources of China, Probst and Tardy, SHI, Subramanian, Zezhen and Shangshi, Meybeck, Yevseyeva *et al.*, He Shangde, Kuznetsov, Showers, Vatkar, Abu Zeid and Biswas, Jing Zhang *et al.*, Kammerer, Milliman, Nguyen Van Ky, ESCAP, Harden and Sundborg, Zhang *et al.*, Vaithiyanathan *et al.*, Chaturvedi, Gleick, Shengquan *et al.*, Central Water Commission, Amiotte Suchet, Meybeck and Ragu, Milliman *et al.*, Takeuchi *et al.*, Zhang Qishum and Zhang Xiao, Grabs and De Couet, Shiklomanov, Gordeev and Tsirkunov.

A brief description of the two largest river systems follows. The Ganges–Brahmaputra–Meghna (Fig. 5.3) is one of the largest river systems in the world. By volume of runoff it is second after the Amazon. This large international basin occupies an area of 1.75 million km^2 in five states: China, Bhutan, Nepal, India and Bangladesh. The total length of the main Ganges is 5425 km. The Ganges proper originates in the Himalayas at a height of 7000 m. Its length is 2525 km (Chaturvedi, 1993; Central Water Commission, 1994) and its area is 1.078 million km^2 (Chaturvedi, 1993). In its upper reaches the Ganges flows through the Himalayas at the bottom of deep canyons and it has the character of a mountain river. Here the river falls from 2 to 20 m/km.

In its middle and downstream sections the Ganges flows across the vast alluvial Indus–Ganges Plain crossing a region strongly dissected by the channels of other rivers flowing down from the Himalayas. Then it flows across a flat plain in a valley 8 to 12 km wide. The river gradients are very small here; there are many oxbow lakes, branches, swamps and irrigation canals in the flood plain. The channel width is 400 to 600 m. The largest tributary of the Ganges is the Jamuna and the downstream section of the Ganges begins from its confluence with this river. Its height in this area gradually decreases from 200 m to 25 m at beginning of the delta. In some areas mountain spurs approach the right bank and the shore becomes high and rocky. In the other places the channel is wide with a pronounced valley. Several large tributaries enter the Ganges, draining basins in the Himalayas with areas from 10 000 to over 100 000 km^2 with lengths of about 1000 km. With the water from the left bank tributaries the Ganges becomes a large river whose width in some places exceeds 2 km. The Jamuna, Son and Damodar are right-bank tributaries of the Ganges. Below the Rajmahal Rise the Ganges flows along the Bengal Lowlands forming with the Brahmaputra one of the largest deltas in the world.

The water level in the river begins to rise at the end of April or the beginning of May and continues to rise up to the end of August

Table 5.3. *Major morphometric characteristics of principal rivers of Asia*

River	Drainage area, km$^2 \times 10^3$	Length, km
Ob	2990	3650
Yenisey	2580	3490
Lena	2490	4410
Amur	1855	2820
Yangtze	1808	6300
Greater Ganges	1746	5425
Ganges	1078	2525
Brahmaputra	577	2900
Meghna	91	900
Indus	960	3180
Mekong	795	4500
Huang He	752	5464
Shatt al Arab	750	2900
Kolyma	647	2130
Xi Jiang	454	2214
Tarim (Yarkand)	446	2000
Irrawaddy	410	2300
Khatanga	364	1634
Indigirka	360	1726
Salween	325	2820
Godavari	313	1465
Amu-Darya	309[a]	1415
Huai He	270	1000
Krishna	259	1401
Helmand	250	1150
Yana	238	872
Liao	229	1390
Olenek	219	2270
Syr-Darya	219[b]	2210
Anadyr	191	1150
Kura	188	1360
Pyasina	182	818
Chao Phraya	160	1200
Taz	150	1400
Songka (Red)	145	1185
Mahanadi	142	851
Ili	140	1000
Anabar	100	939
Narmada	99	1312
Cavery	81	800
Kyzyl-Irmak	76	1151
Penjina	73	713
Tapi	65	724
Nadym	64	545
Kamchatka	56	704
Fuchumtzjen	54	494
Beijang	47	468

[a] Area in Kerki.

[b] Area in Tyumen-Aryk.

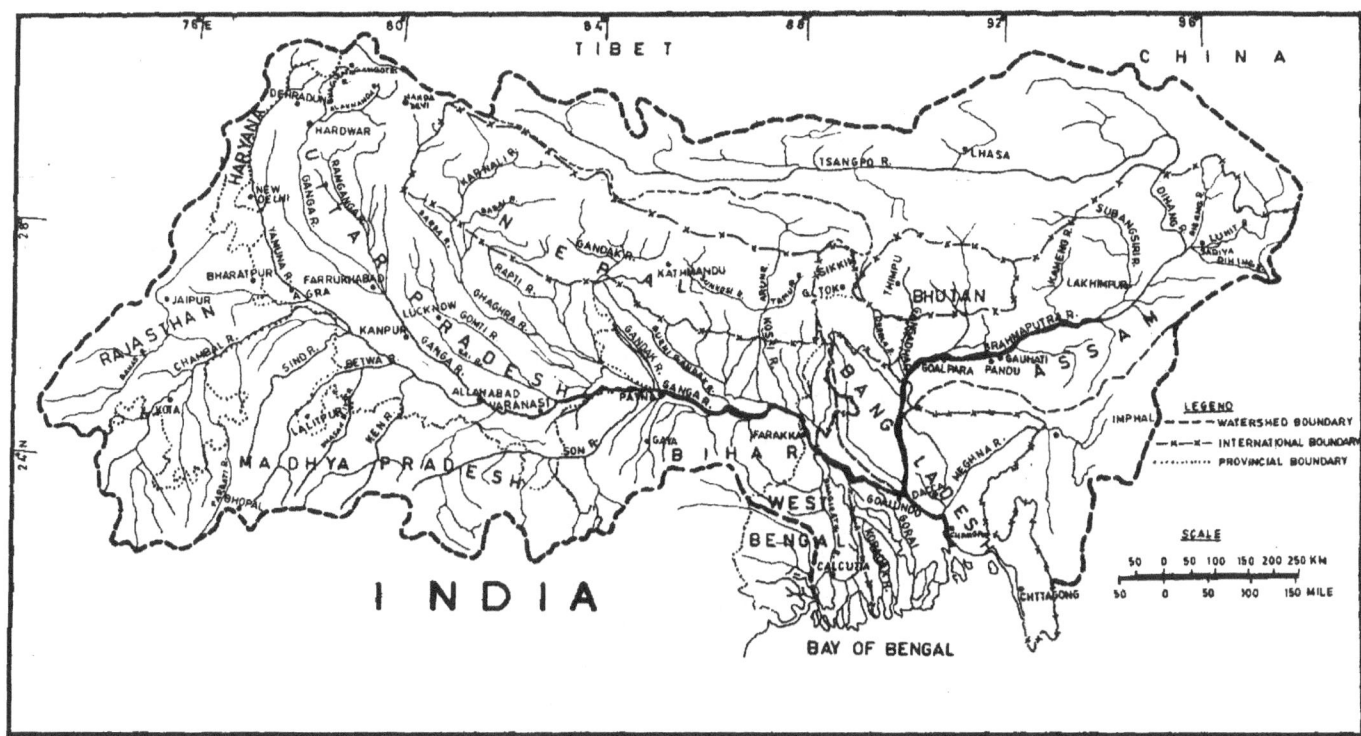

Fig. 5.3 The Ganges–Brahmaputra–Meghna river system (from Chaturvedi, 1993).

or the first half of September when maximum levels occur. The mean rise is 10 to 12 m but it can reach 15 m and more during especially large floods when the Ganges overflows its banks. The northern bank is most subject to floods. The tributaries also top the banks and change channels. As a result, up to 1 million ha is flooded annually but during catastrophic floods the flooded area is four times as large.

The Brahmaputra River is 2900 km long (Vatkar, 1989) with an area of 577 000 km^2 (Chaturvedi, 1993). The sources of Brahmaputra are several rivers flowing down the northern slopes of the Himalayas and from the southern slopes of the Kailas Range which connect as one channel at a height of 4860 m. The river flows eastward for 1700 km across southern Tibet within the Himalayas and exits the mountains in the Assam Plateau at a height of 155 m. After its junction with the Dibang and Luhit, the river is called the Brahmaputra and it has numerous tributaries in its Indian section. Then the river enters the alluvial Bangladesh Plain where it joins the Ganges. Below the confluence the river has the name of Padma and it flows southeast for 105 km to the confluence with the Meghna. In its lower reaches the river forms an extensive delta with an area of 59 570 km^2 (Chaturvedi, 1993), which is on average less than 7 m above sea level. The Brahmaputra valley is 80–90 km wide but the channel width varies from 6.5 to 10 km and in some places is up to 16 km.

The river is fed by a mixture of precipitation and the meltwater from snow and glaciers in the Himalayas. An enormous mass of water passes through an unstable geomorphological zone and causes extensive floods, migration of channels and erosion. The water levels are subject to significant variations of tidal character. The height of the tide at the mouth reaches 5 m, and tides can reach a distance of up to 300 km upstream. The river has a significant potential for hydropower (150 000 MW) shared between the Ganges (63%) and the Brahmaputra (36%). The largest potential exists on the rivers in Nepal (83 000 MW) but at the present time only 5% of the potential has been developed in India and 0.3% in Nepal (Chaturvedi, 1993).

The Yangtze River is the largest river of China and one of the largest rivers on Earth. The basin of the Yangtze extends over 1.81 million km^2 and the river has a length of 6300 km (Ministry of Water Resources of China, 1987; Meybeck and Ragu, 1995). The Yangtze originates in the central Tibetan Uplands at a height of about 5000 m and after flowing eastwards for several thousand kilometres, it reaches the East China Sea. In its upper reaches the river flows in narrow, deep canyons and falls 90% of its total height difference. In the middle section it exits the mountains and flows gently along the margin of the Sichuan Depression with a width of 300 to 500 m.

Upstream of the town of Ichang the Yangtze crosses a mountainous section of the depression and flows through three gorges (Sanxia) with a total length of 100 km. In these gorges the channel narrows to a width of 150 to 200 m and the depth reaches

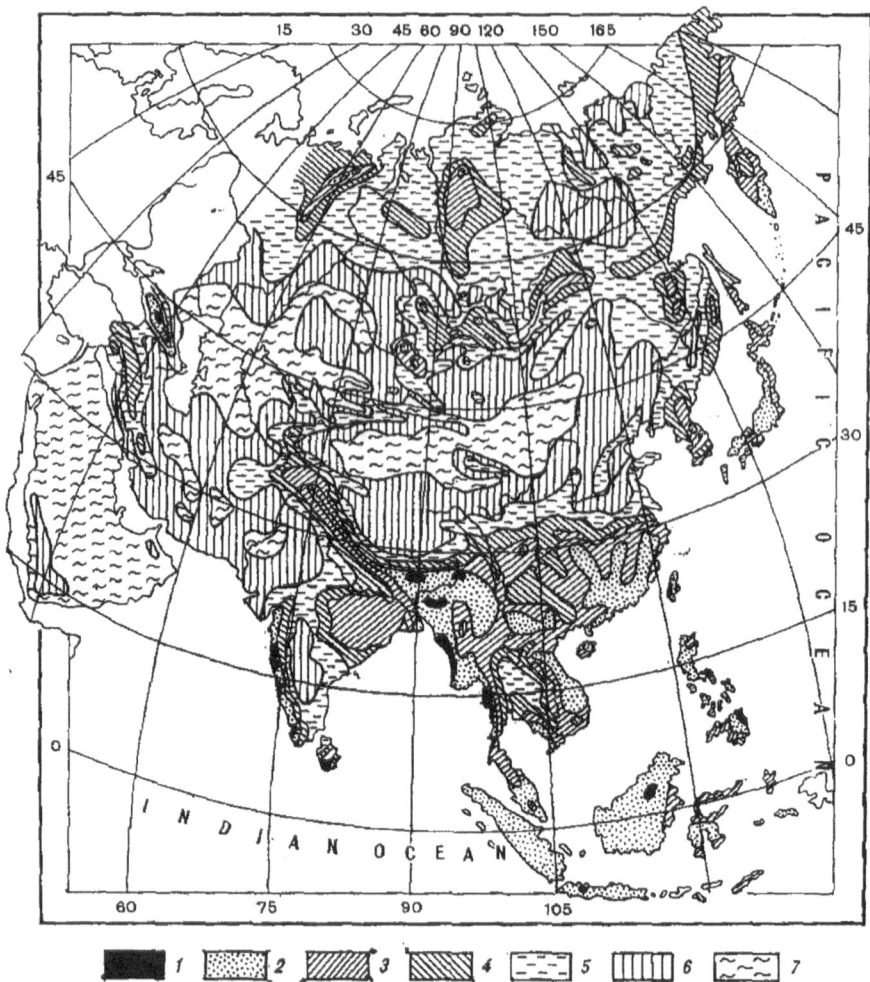

Fig. 5.4 Pattern of distribution of mean annual runoff (mm) in Asia.
1, More than 3000; 2, 3000–1000; 3, 1000–500; 4, 500–300;
5, 300–100; 6, 100–5; 7, less than 5.

100 m, but because of this narrow width the level in the river can rise very rapidly reaching from 30 to 80 m above normal. The downstream section of the Yangtze from Ichang to the mouth is 1735 km long and the river flows in a well-developed valley with numerous branches and arms. The channel width is 1–2 km and the depth is 20–30 m. In this section of the river there are large lakes, Poyanghu and Dongtinghu, which are connected in many places to the Yangtze. Both lakes provide natural regulation of the Yangtze. In the city of Wuhan the Yangtze reaches a width of 1.6 km, and in Nanching 18 km. At its mouth the river forms a delta 200 km long and 80 km wide. Many canals that have a triple purpose – drainage, irrigation and transport – dissect the delta. The tidal influence reaches 470 km inland. The Yangtze and its tributaries provide a considerable amount of hydropower (40% of the resources of China).

The changes in the mean annual runoff over the continent vary enormously, from 4000 mm or more down to almost zero

(Fig. 5.4). The areas with the largest runoff are in the Malay Archipelago, the Philippines, the southern part of Japan, and in the mountain regions of Southern and South East Asia. Because of the differences in the distribution of precipitation across Southern and South East Asia, there are large differences in the distribution of runoff. On the windward slopes of the Western Ghats totals reach 3000–3500 mm but only 400 to 500 km to the east they decrease to 50–100 mm (part of the Deccan Plateau adjoining the mountains from the east). A dramatic decrease in runoff is also observed between the western Burma Ranges and the central lowlands to the east.

In those parts of Asia furthest from the sea the runoff is between 100 and 300 mm per year, but even in these regions values of more than 1000 mm can be observed in some parts of the mountain systems, for example on the Putoran Plateau, on the windward slopes of the Tien Shan and in the mountainous parts of Siberia. Altai, Sayan, and the plateau and highlands of Siberia are the main source regions of the major rivers of Siberia. Across Asia runoff in general decreases from east to west but values of 1000–1500 mm are typical of the highlands bordering the Mediterranean

Table 5.4. *Major morphometric characteristics of the principal lakes of Asia*

Lake	Country	Volume, km³	Surface area, km²	Depth, maximum, m
Baikal	Russia	23 000	31 500	1741
Issyk-Kul	Kyrgyzstan	1 730	6 280	702
Toba	Indonesia	1 258	1 100	529
Aral Sea[a]	Kazakhstan, Uzbekistan	1 064[b]	66 400[b]	68
Khubsugul	Mongolia	381	2 770	267
Van[a]	Turkey	–	3 760	145
Dead Sea[a]	Israel, Jordan	188	940	400
Khantaiskoye	Russia	–	822	420
Balkhash[a]	Kazakhstan	106	18 200	25.6
Kukunor[a]	China	–	4 220	38.0
Kirgiz-Nor[a]	Mongolia	66	1 480	80
Alakol[a]	Kazakhstan	59	2 650	54
Sevan	Armenia	56	1 360	86
Zaisan	Kazakhstan	53	5 510	10
Karakul[a]	Tajikistan	–	380	238
Rezaijeh (Urmia)[a]	Iran	45	5 800	16
Teletskoye	Russia	40	223	325
Tonle Sap	Cambodia	40	3 000–30 000	12
Fumibkhal	Thailand	30	300	123
Poyang Hu	China	–	2 700	20

[a] Salt lake.
[b] For the level of 1960.

Sea, the southeast coast of the Black Sea fringing the Caucasus (2000–2500 mm), the slopes of the Elbrus Mountains draining to the Caspian Sea (500–700 mm), as well as for the Zagros and Taurus Mountains where the Tigris and the Euphrates rise. A distinguishing feature of Asia is the presence of extensive areas of extremely low annual runoff (1–5 mm) in Central Asia, over the plains of Middle Asia and Kazakhstan, in the deserts and semi-deserts of the Near and Middle East and over almost the entire Arabian Peninsula, where the runoff is represented by temporary water courses or wadis (UN/DPCSD, 1995b). Over Western Asia there is a deficit in the water balance which reaches 2000 mm in the Arabian Peninsula (Korzun, 1974a).

Asia is the continent with the largest number of lakes and these differ both in size and origin. At the present time there is detailed information available on more than 60 lakes with an area greater than 100 km². The volume of water in these lakes is more than 32 000 km³. Table 5.4 contains data on the largest lakes of Asia (Chang, 1987; ILEC/UNEP, 1987–9, 1991, 1993). In the tundra, thermokarstic and glacial lakes are widespread. The thermokarstic

lakes are small, shallow water bodies. Their origin is related to the processes of melting of the permafrost and they are fed by precipitation and the runoff of surface water during snowmelt. Most lakes, especially the small ones, freeze to the bottom in winter. The largest lake of glacial origin is Khantaisk, while Lake Taimyr, beyond the Polar Circle, is of tectonic origin. In the forest zone there is one of the largest tectonic lakes in the world, the unique freshwater Lake Baikal. One of the largest Asian lakes is Issyk-Kul Lake situated in the Tien Shan Mountains. It is also of tectonic origin but contains brackish water.

The Aral Sea occupies the lowest part of the Turan Lowland. It is in the arid zone and is fed by the flows of the Amu-Darya and Syr-Darya and by precipitation on the lake surface. Due to the changed climatic conditions and intense agricultural development, the area of the Aral Sea continues to decrease (see Section 5.7). The Dead Sea occupies the world's deepest depression in the land surface and is located between Jordan and Israel. Its level is 392 m below the level of the World Ocean and it contains strongly mineralized, bitter saline water.

Reservoirs are an integral part of the hydrographic network of the continent, and Asia has more than any other continent. According to rough estimates there are about 16 000 reservoirs, and more than 600 of these have a capacity greater than 100 million m³. The total storage is 1700 km³ (Table 5.5) and their total area is estimated as 45–50 000 km² (Goldsmith and Hildyard, 1985; Avakyan *et al.*, 1987; ICOLD, 1987, 1992; Van der Leeden *et al.*, 1990). Reservoirs are not uniformly distributed over the continent. They are predominantly located in the Indochina and Indian Peninsulas, in China and on the periphery of the Arabian Peninsula, with more in China and India than elsewhere. The Bratsk reservoir on the Angara River in the Asian part of Russia has the largest volume of any artificial water body in the world.

Most of the reservoirs are in the river valleys, usually for the seasonal regulation of runoff and more rarely for regulation over longer periods. Some reservoirs are used for flood control, for example the Vadi-Tartar reservoir in Iraq which stores the flood waters of the Tigris. The first reservoirs were constructed in arid central, southern and southwestern Asia many centuries ago. Until the middle of the twentieth century they were used for irrigation and flood control. Subsequently reservoirs have been built for the dual purposes of irrigation and hydropower. Many of these reservoirs have been built over the last 20 to 30 years in China, Vietnam, Laos, Indonesia, Malaysia, India, Pakistan, Bangladesh, Iran, Afghanistan and Turkey. This is also the period when a number of large multipurpose reservoirs were constructed in Russia, many with a capacity exceeding 20 km³. There are also some reservoirs constructed for fisheries, navigation and recreation. In many countries the reservoirs form cascades, for example, a cascade of four reservoirs on the Krishna River (India) and a cascade of nine reservoirs on the Sakarya River (Turkey), while the largest series

Table 5.5. *Principal reservoirs in Asia*

Reservoir	Country	Basin	Year of filling up	Dam backwater, m	Full volume, km^3	Type of use[a]
Bratskoye	Russia	Angara	1967	106	169.3	E N T S F R
Krasnoyarskoye	Russia	Yenisey	1967	100	73.3	E N T S H F R
Vadi-Tartar	Iraq	Tigris	1976	–	72.8	H I
Zeiskoye	Russia	Zeya	1974	98	68.4	H E N T F
Ust-Ilimskoye	Russia	Angara	1977	88	59.4	E N T F S
Boguchanskoye	Russia	Angara	1989	70	58.2	E N T
Upper Wainganga	India	Wainganga	1987	43	50.7	
Bukhtarminskoye	Kazakhstan	Irtysh	1967	67	49.6	E F N H R
Ataturk	Turkey	Euphrates	1995	175	48.7	E I
Vilyuiskoye	Russia	Vilyui	1972	68	35.9	E S N
Sanmenxia	China	Huang He	1962	90	35.4	E I H N
Keban	Turkey	Euphrates	1976	190	30.6	I E
Sayanskoye	Russia	Yenisey	1978	220	29.1	E I N T S R
Kapchagaiskoye	Kazakhstan	Ili	1970[b]	41	28.1	E I N F R
Razzaza Dyke	Iraq	Euphrates	1970	15	26.0	H I
Longyangxia	China	Huan He	1990	172	24.7	E I W
Khantaiskoye	Russia	Khantaika	1975	50	23.5	E F
Xinanjiang	China	Xinanjiang	1960	100	21.6	E W
Toktogulskoye	Kyrgyzstan	Naryn	1974	180	19.5	E I N W
Srinagarind	Thailand	Quae Yai	1981	140	17.8	E I H

[a] E, hydropower; N, navigation; S, water supply; I, irrigation; F, fishery; T, timber rafting; R, recreation; W, water storage for different purposes; H, flood control.
[b] In 1970 the filling of the reservoir was stopped. The project reservoir capacity was not attained.

of cascades by volume are the reservoirs in the Angara–Yenisey cascade (Russia).

5.3 SOCIO-ECONOMIC CONDITIONS AND THE USE OF WATER RESOURCES

Due to adverse relief and climate, particularly aridity (Romanova *et al.*, 1994), Asia has the largest area of any continent which is not suitable for agriculture, some 75% of the total land area against 45% for Europe. Agriculture is however the mainstay of the economy of most Asian countries. Farming in Asia has been reasonably efficient for several millennia. Old farming regions formed the cores of eastern civilizations, in areas with high population densities: the valleys of the Yangtze and the Huang Ho Rivers in China, the Ganges Valley in India and Bangladesh, and these differ in soil fertility, especially where irrigation is undertaken. At the present time, of the 20 countries with the most irrigation 11 are located in Asia. Some countries use the most modern technologies in their irrigation systems (e.g. Israel, Jordan, Japan and Kuwait).

The type of farming determines water use to a considerable extent. Future trends and patterns in the use of water will be governed to a great extent by changes in the agricultural sector or stimulated by increasing demands for food from a rising population (Rijsberman, 2000, 2001; Brown, 2001). Intense demographic growth was observed on the continent during the second half of the twentieth century, predominantly as a result of the decreased mortality level but also because of the continued high birth rate. According to UN data, between 1960 and 1970, the population of Asia (except for Russia, China and some socialist countries) increased by 324 million (Fig. 5.5). A 70% increase occurred due to the rise in three countries, namely, India, Pakistan and Indonesia and but almost half was due to the increase due to in India alone (Gaphurov, 1977).

From 1970 to 1980 and from 1980 to 1992 the greatest annual population growth was observed in the countries of Western Asia (from 3% to 15.5%). During these decades in China the growth rate was 1.8% and 1.4%, respectively, and in India 2.1% to 2.3%. Similar growth rates were observed in most other countries except for the republics of the former USSR where from 1980 to 1992 they were the least on the continent, from 0.6% to 1.8% (World Bank, 1995). At the present time 60% of the world's population lives in Asia (UN, 1995a). The two most populous countries are China (1209 million) and India (919 million). According to estimates, the populations of China and India are likely to reach 1890 and 1680 million respectively by 2025. In recent years growth

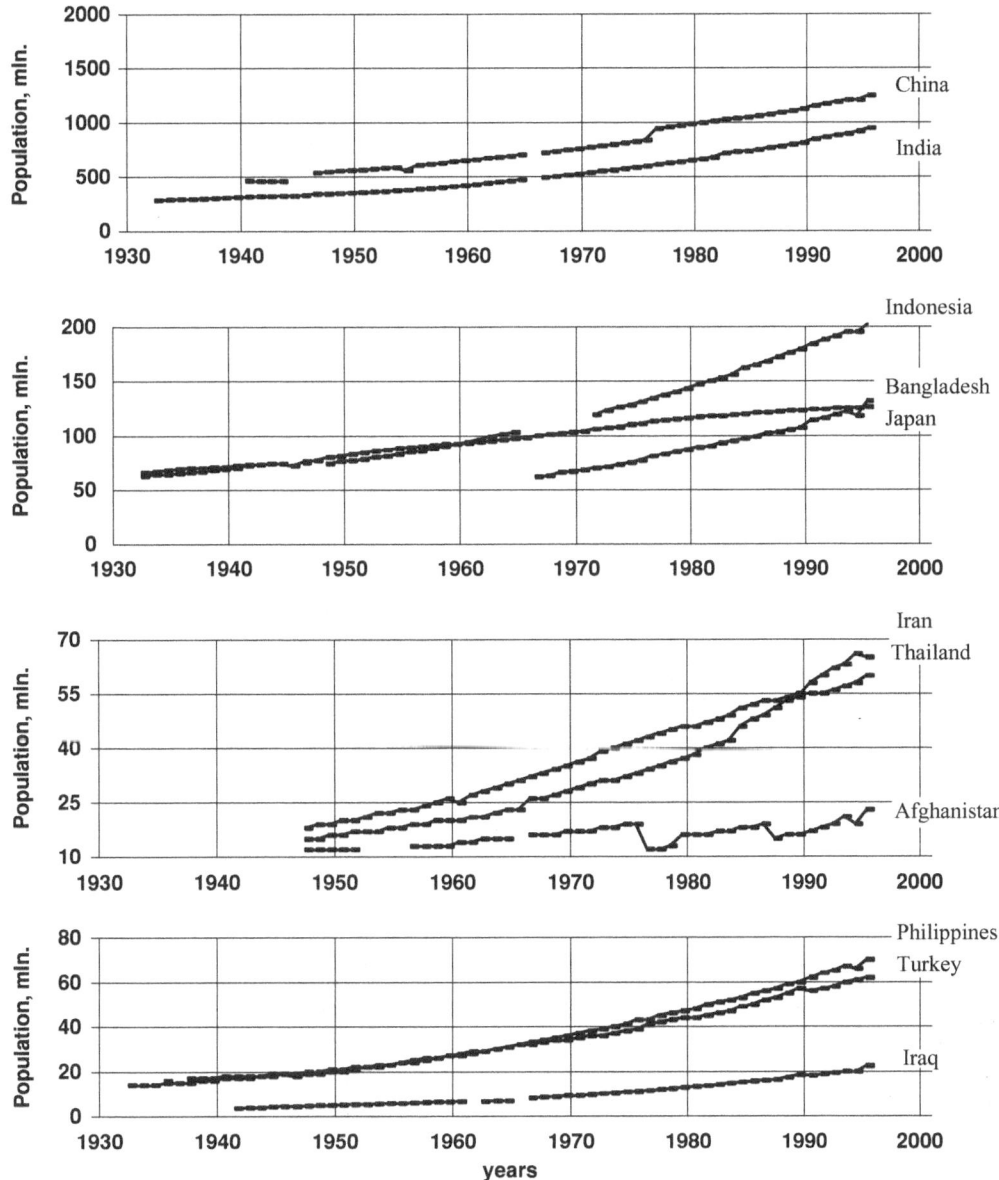

Fig. 5.5 Dynamics of population in selected countries of Asia.

rates have slowed as a result of "family planning" programmes in some countries such as China, Indonesia, Korea, Singapore, Sri Lanka and Hong Kong. For the period 1990 to 2000 growth rates are expected to be 0% to 0.7% in some countries. For Russia the population is expected to decrease, whereas in the Near East growth rates are likely to remain quite high (2% to 4% per year).

About 80% of the population of developing countries is rural. This percentage has decreased relatively slowly in spite of migration to towns and a sharp increase in the urban population. From 1951 to 1970 the urban population increased from 15.5% to 20.8% in Southern Asia, from 13.7 to 19.7 in South East Asia and from 23.3 to 38.7% in Western Asia. In Japan it has increased over this period from 29.5% to 55.5%. Special geographical and

historical conditions, the presence of unique resources, large size and overpopulation have determined the development of "non-standard" urbanized countries with urban enclaves: Bangladesh, Indonesia, India and Pakistan (with a population greater than 100 million people); newly urbanized countries (Iran, Saudi Arabia, United Arab Emirates, Brunei, Bahrain, Qatar and Kuwait); and completely urbanized countries (Singapore, Hong Kong and Macao). The former are characterized by small urban populations (20% to 30%), but with very large numbers living on the coast (e.g. Karachi, Mumbai, Chennai, Calcutta, Dacca, Jakarta and Surabaya.). In the countries of the second type the large urban percentage (>60%) already existed in the mid twentieth century and the rates of its growth from the 1950s exceed all other groups of

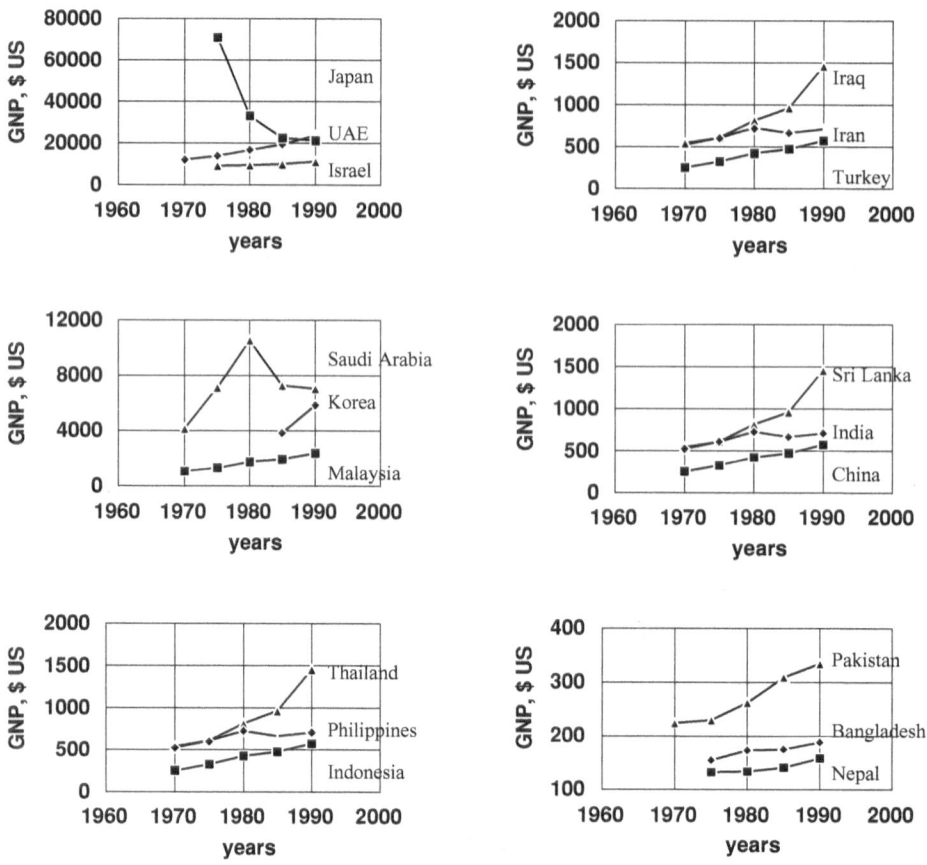

Fig. 5.6 Dynamics of GNP (in $US per capita) for selected Asian countries (from IMF, 1994).

countries (Gaphurov, 1977; Komar, 1990; Simoniya, 1990; World Bank, 1993; Parnikel', 1995). Between 1980 and 1990 there was a decrease in the growth of the percentage urban population in most countries as compared to the period 1965 to 1980. The exceptions were the increased growth rates in Oman, Malaysia, Indonesia, Pakistan, Nepal, Bhutan and Laos (World Bank, 1993).

A range of factors including climate, relief, level of development and the degree to which the available water resources are employed determines water use in Asian countries (Shiklomanov, 1997, 1998, 2000a, 2000b; Penkova, 2000b; Shiklomanov et al., 2000). During recent decades certain countries have had a considerable measure of economic growth (Fig. 5.6). According to the World Bank, in Hong Kong, Indonesia, Japan, Malaysia, Singapore, Korea, Taiwan and Thailand from 1965 the annual growth rates were twice as large as in other Asian countries and three times as large as those in Latin America. The Chinese economy also grew from 1980 to 1990. The growth in the GNP was 9.6%, whereas from 1952 to 1972 it was 2% to 4%. The oil-producing states of Arabia (Saudi Arabia, United Arab Emirates, Qatar, Oman and the Sultanate of Brunei) have a special economic position with the per capita income exceeding the indicators for industrially developed countries (Zevelev, 1985; Chudodeyev, 1993;

Ganshin and Remyga, 1993; Markova, 1995; Rastyannikov and Shirokov, 1995; Russian Academy of Science, 1995; Yakovlev, 1995).

A country's level of economic development governs to a great extent the level of water use in the domestic sector. Between 96% and 100% of the population is served by public water supplies in the richest countries (Israel, Hong Kong, Singapore, United Arab Emirates and Japan). In countries with an average income, for example, Jordan, Korea, Saudi Arabia and Iran, it is within the range 46% to 84%. In the countries with a low income (up to $US700) only in India, China, Myanmar and Bangladesh does the population served exceed 70%, and it is close to 30% in Bhutan, Laos and Yemen. The highest domestic water use occurs in countries that have a higher than average income (Fig. 5.7), such as Malaysia, Singapore, the Philippines, Hong Kong and Japan (300 to 480 l/day per head). In the most developed countries of Western Asia (except for Bahrain) it is 150 to 200 l/day per head. In Southern Asia it is the least, some 15 to 60 l/day per head due to the rural character of much of this area.

An important factor in economic growth is the development of the electricity supply. For the period 1960 to 1980 electricity production in China and India increased five-fold and six-fold respectively. A similar increase was observed in Indonesia and the Philippines. In some countries there was an unprecedented growth

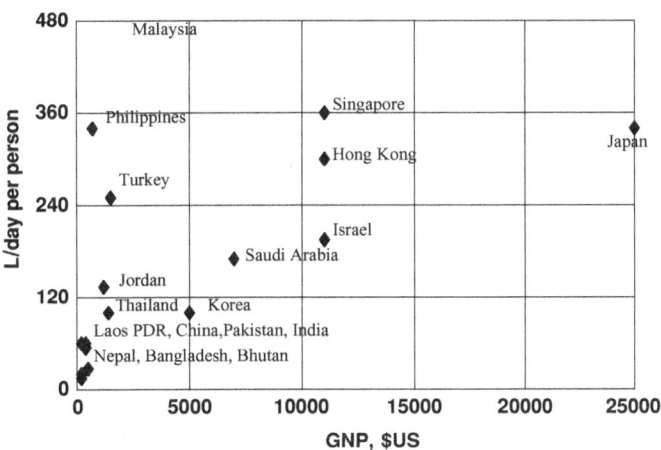

Fig. 5.7 The relationship between specific public and domestic water use (l/day per head) and GNP in countries of Asia (1990).

in electric power production: Thailand, Korea, Nepal (more than 20 times), Turkey, Syria, Iraq, Hong Kong and Singapore (8 to 10 times). The development of electricity continued throughout the 1980s, in spite of the economic difficulties experienced during this period; in most countries the production of electricity has increased 2 to 3 times and in some countries (Indonesia and Oman) by 5 to 6 times. In 1993 the total output of electricity in Asia was 763 090 GW (more than 25% of the power production of the entire world). It is largest in Japan, China, Russia, India, Korea, Iran and Turkey (175–200 GW). Thermal power produces 71% of all electricity and hydropower some 21% (Gleick, 1993, 1998; Muranova, 1993; ESCAP, 1994; World Bank, 1995).

The development of hydropower needs reservoirs. Most reservoirs were constructed after the Second World War often for flow regulation, the rates of construction increasing more than 10 times. During the period 1921 to 1950 13 large dams (>100 million m^3) were constructed every decade. During the 1951–60 and 1961–70 periods, 51 and 79 reservoirs were constructed, respectively. In India, China and Japan the increase in reservoir storage over these decades was about 240 km^3. From the World Register of Dams as of 1970, Japan and India had the largest number of dams, 1900 and 1000, respectively. From 1949 up to the end of the 1980s more than 86 000 reservoirs of different sizes were constructed in China, with a total volume of about 430 billion m^3 (World Dams Today, 1970; Ryabchikov, 1976; Voropayev and Vendrov, 1979; Chen and Wu, 1987; ICOLD, 1987, 1992; Soyuzvodproekt, 1988). Table 5.4 presents data on the largest reservoirs of Asia.

Construction of large reservoirs has produced quite a noticeable and sometimes a significant influence on the economy of large areas, especially if they form cascades and systems. Many hydrosystems have created or improved the conditions for water supply, industrial development and agricultural exploitation of the countries concerned, for example the Angara–Yenisey cascade, irrigation–

energy systems in India, the reservoirs of the Asad hydropower stations in Syria, Keban and Karakay in Turkey (Voropayev and Vendrov, 1979; Malik, 1990; Urazov, 1991). The reservoirs constructed for irrigation of arid lands have a special importance for they are the only source of water in many areas. Thousands of dams have been built in the Near East, Middle, Central and South East Asia. Small reservoirs are numerous in India, especially around Chennai and in the countries of the Arabian Peninsula. According to ROSTAS (1995), in the last 10 years about 260 reservoirs with 369 million m^3 capacity were built in Saudi Arabia, Yemen, the United Arab Emirates and Oman for regulating surface runoff, flood protection and replenishment of ground water. Most of these reservoirs are built in mountainous regions.

The industrial development of Asian countries has been reflected in the increase in the total industrial water demand. Up to 1960 it increased by about 10–20 km^3 a decade, but in the 1960s and 1970s the increase reached about 50 km^3 a decade (Shiklomanov and Markova, 1987). The largest use was in the production of thermal energy. In Japan this made up 57% of the total. In India, from the second half of the 1970s along with the development of primary and secondary industry there was increased development of the fuel energy area and a change in the type of energy used (in the national energy budget up to that time the non-industrial fuel types comprised about 50%, including wood at 40%). The development of industries using much water (the oil and chemical industries, metallurgy and machine-building) resulted in an increase in the proportion of water used by the industrial sector. Large industrial cities such as Mumbai, Calcutta, Jangipur and Baroda, where industry used 40% to 60% of the water, began to establish quotas restricting the growth of water use. In towns where the requirements exceeded the capacity of the water supply networks, the need for water-saving technologies became obvious (Levitanus 1986a, b; Kutsobin, 1990).

A reduction in the increase in water use by industry has occurred in Asia since the 1980s. This is due no doubt to the slower development of industry, decreased prices of raw materials, problems with debt payments, limited local markets in Asian countries, the cycle of trade and crises in the world economy. However water and energy saving, the development of environmental protection measures and water pollution control are also becoming important. In Japan for example, since the early 1970s, when major environmental protection laws were adopted, the volume of water abstracted by industry has decreased 2.5 times. The world's lowest use of energy per unit of GNP has been reached, the use of raw materials has been halved and water is being recycled (Suzuki, 1973; Ryabchikov, 1976; Environment Agency, 1987; Alpatov, 1992; Russian Academy of Science, 1994; World Bank, 1994).

In arid and semi-arid countries where local surface and ground water sources are insufficient, the re-use of waste water is important; and in many areas desalination plants are in operation especially around the Persian Gulf. Saudi Arabia is the world's leader in large-scale desalination (with a production of more than 20 000 m³/day). Other methods for obtaining non-traditional water resources are also used. For example, in northern Israel cloud seeding has been used to increase precipitation (up to 13%) and replenish of ground water over the last 30 years (UN, 1992, UN/DPCSD, 1995a; Gleick, 1993, 1998; Harpaz, 1994; UNESCO, 1995a; Shahin, 1996; FAO, 2000). However such costly measures are available only in relatively rich countries. In the 1980s most Asian countries experienced decreased rates of economic growth. In some countries (Jordan, United Arab Emirates, Saudi Arabia, Iran and the Philippines) between 1980 and 1992 economic growth rates were negative. In particular, the income of oil-exporting countries decreased by a factor of three due to their decreased share of the market and the general decrease in the demand for energy (World Bank, 1995).

In Western Asia the situation is complicated by the lack of regulation of the allocation and use of water resources. Some authorities believed that by 2000 freshwater resources could be a more important factor in development than oil. About 88% of the water resources of the Euphrates and Tigris originate in Turkey. At the present time Turkey is implementing a programme for the construction of hydropower and irrigation works in the headwaters of these basins. Upon completion of this programme the inflow to Iraq may be significantly reduced and Syria may be experiencing reductions even now. An important aspect for Iraq is the possible deterioration of river water quality consequent on the increased irrigation in the upper part of the basin, since farming in Iraq has already been suffering from soil salinization.

These water resources and water supply problems stimulate international co-operation over the water resources of the Near East. Under consideration is the construction of the "world waterline" which would be more than 2000 km long from the Rivers Ceyhan and Jeyhan in Turkey to supply water to Syria, Jordan and Saudi Arabia, with a second line to Kuwait, Qatar, Bahrain, the United Arab Emirates and Oman. The capacity of both waterlines would be about 2 km³ of water per year (Davydov, 1985; Anonymous, 1988; Anonymous, 1989; Creton, 1991; Kliot, 1994).

Large volumes of alluvial and deep ground water are used in the Arabian Peninsula. According to Abdul Razzak (1995), the total resources of alluvial aquifers are assessed at 115.5 km³, with the largest volume (84 km³) located in Saudi Arabia. The water in these aquifers is usually of good quality and is the main source of drinking water in the towns and villages of central Saudi Arabia, Yemen and Oman. In Kuwait, Qatar and the United Arab Emirates it is used in addition to desalinated water. Of special concern is the over-exploitation of fossil water from deep horizons where the total water resources is assessed to be 2300 km³. For the 1980 to 1995 period the total volume abstracted from them was about 90 km³, resulting in the decrease of water levels and piezomet-

ric surfaces in some areas by up to 70 m (Abdul Razzak, 1995; ROSTAS, 1995; UN/DPCSD, 1995a; UNESCO, 1995a).

Ground water has been used for a source of irrigation and water supply for a long time in the countries of the Levant and the Iran Upland. In Syria, Jordan and Iraq a great deal of attention is paid to the joint use and protection of surface and ground water. However, in some regions salinization and contamination occur in some cases as a result of sea water intrusion. The development of efficient ground water protection programmes is handicapped because of insufficient knowledge (Noor Rahman Rahmani, 1989; Usmanov and Ignatikov, 1990; World Bank, 1995; UNESCO, 1995a). In monsoon Asia there is considerable small-scale irrigation with over 30% of the water in India coming from wells, where there are more than 5 million open wells and several tens of thousands of drilled wells. There are tens of thousands of wells in Pakistan, the Philippines and other countries (Ryabchikov, 1976; ESCAP, 1989b). In 27 large cities in northern China, ground water sources provide 88% of the water used; in the south of the country many towns use ground water as the main source due to pollution of surface waters. It was forecast that by 2000, at least one-third of the water used in China would be ground water. Around the capital, as well as in some other towns (Tianjin, Zhengzhou) water supplied by canals, from rivers and reservoirs (Mengxiong and Zunhuang, 1994) reduces the demand for ground water.

During the 1980s and 1990s many Asian countries faced unprecedented demand for water due to the expansion of industrial and agricultural production, improvement of the economy and increasing integration into the international market. The impact was greatest on the countries of Southern and South East Asia. China achieved great success as a result of economic reforms and from the end of the 1970s because of the open regulated market economy (especially in the southern provinces). Radical changes occurred to the economy of some other countries (Russia, Vietnam, Laos and Mongolia). In some countries considerable success has already been achieved. In most of them the construction of many hydraulic structures took place during the last 30 to 35 years and the area irrigated increased. In some countries (India, Bangladesh, Pakistan, Turkey, etc.) along with the development of irrigation, highly productive varieties of cereals and cotton were introduced and there was a greater application of fertilizers ("green revolution") from the late 1960s (Anonymous, 1984b; Kadyrov and Shalomayev, 1984; Lukichev, 1990; Chou Vichitkh, 1994; FAO, 1995; Markova, 1995; World Bank, 1995; Chaturvedi, 2000).

In the early 1990s, many countries became the largest producers of some of the most important foods and raw materials; natural rubber, copra, tea, tobacco, jute, vegetable oil and rice. One of the main factors for this increased production was the extension of arable lands and the land under multi-year crops. Now in some regions only limited extensions are possible, while in others there are no further areas to utilize. For example, in Bangladesh in the early 1970s arable land and fallow occupied almost 97% of the area suitable for cultivation. In India the proportion of arable land reached the extremely high level of 57%, which is five times higher than the general world level (Kutsobin, 1990; Muranova, 1993). In China with a large population and a shortage of arable land there is competition between arable and livestock farming.

The need for more arable land is one of the reasons for disappearance of large tracts of forest on the continent. Only in a few countries (Japan, Singapore and Turkey) was there no decrease in the area of forest (Muranova, 1993; World Bank, 1995). However, large areas remain unused due to desertification, soil degradation and other reasons. Such areas are often characterized by the low efficiency of the irrigation systems (0.6 to 0.8) and the large losses of irrigation water (Turkey and Syria up to 70%, Iran up to 80% of the annual abstraction) (ESCAP, 1978, 1987, 1989a, b; ESCWA, 1978; Government of India, 1981; Gerasimov, 1986; Maslennikova, 1987; UNEP, 1988a, b, c; Milovsky, 1990; World Bank, 1995; Chaturvedi, 2000).

Along with improved land use, improvement of water use systems is considered to be one of the main measures for control of land degradation. In many Asian countries the governments devote much attention to the optimization of water and land resources. In India consideration has been given to medium-scale and large-scale transfers to arid and semi-arid regions such as Rajasthan and the Kutch and Kathiawar Peninsulas. There are also possibilities for integrated river basin development in areas with copious water resources such as Madhya Pradesh, Andhra Pradesh, Maharashtra and Karnataka. India also faces the task of extending the "green revolution" to the zone of rain-fed farming, the development of agriculture and the increased productivity of crops by growing them on an industrial basis. Currently the main volume of agricultural production in the majority of Asian countries is obtained from small semi-natural farms with a low productivity. Any increased production appears small and it may be that agricultural development is barely in advance of population growth. There is a decrease in labour availabilty and a decreased number of ploughs per inhabitant as well as financing difficulties for new agricultural areas (Bolotin and Sheinis, 1983; Lukichev, 1990; Karagodin and Elyanov, 1992; Russian Academy of Science, 1995).

In the deltas and valleys of large rivers – the Tigris and Euphrates, Indus, Ganges, Yangtze, Irrawaddy and Mekong – the irrigated lands are continuously cultivated, especially in such countries as India and Indonesia (Java); in Vietnam, China, North Korea and Japan, they occupy 95% of the area and they are harvested two to three times a year. However at higher elevations irrigated areas form a mosaic of terraces and usually they are in use only during the wet season. Primarily rice (in the alluvial plains of India, Indochina and southern China, and in Indonesia and the Philippines), citrus and garden cultures (Mediterranean), cereals and cotton (Mesopotamia, the Syrian plateau, and the el-

evated plains of India and Indochina) are irrigated by traditional methods. However some developed countries use new methods to simulate plant growth (Russian Academy of Science, 1994; Isayev, 1990). Typically the use of river water for irrigation in the monsoon tropics is inefficient; large rivers are under-used while small rivers are over-used. Further development of irrigation has to be based on large-scale water transfers (the largest project is the construction of the Great Irrigation Canal from the Ganges). China and India are the leading countries in irrigation development (Figs. 3.3, 5.8). Each of them has approaching 50 million ha under irrigation (FAO, 1995a).

In China, from the time the People's Republic was founded, the construction costs for reclamation have comprised 60% to 70% of state investments in agriculture. From 1965 to 1990 the irrigated areas increased annually by 1.2% reaching 11% of all agricultural lands and about half of all arable lands (World Bank, 1993). The water use in agriculture reached 90% of the total for all uses. Irrigation had developed in all regions, including Tibet. The largest areas are situated south of 32°N in the main rice region of the country (Sichuan and Gansu). However large rivers are used less than small ones. One of the main obstacles to using river water in irrigation is its high turbidity (Ryabchikov, 1976). The growth of irrigation in China has experienced considerable variations; the peak in 1978 reached 48 million ha but by 1985 there were 40 million ha. The causes for this reduction include the decrease in the ground water levels, competition with other water users and destruction of public irrigation systems due to the agricultural reforms in 1979 to 1989, whose main feature was a bias to family farms. During the last decade the earlier extensive system of large and medium-size reservoirs and irrigation canals has declined and the land under cereals has reduced due to increased construction activities in villages and the growth of the area in other crops. Over the period 1978–86 the area occupied by cereals has decreased by 9.1 million ha (7.4%). Hence in 1989 it was proposed to increase the capital investment in agriculture by 68%, allocating a significant portion of this money for repair of hydraulic structures and drainage–irrigation systems. By 1994 more than 950 000 ha of new irrigated lands had been introduced (Frazer, 1986; Brown, 1990; Korkunov, 1992; Ganshin and Remyga, 1993; Gleick, 1993, 1998; Russian Academy of Science, 1995).

In India in the early 1990s the irrigated area was 24% of all agricultural land and covered about half the land suitable for irrigation. From 1965 the annual growth was 2.3% (World Bank, 1993) with agricultural water use exceeding 90% of the total water use. The country has rich traditions of water use; however the most active development of water resources dates from the 1950s with the construction of the 600-km Rajasthan Canal, the Farakk dam on the main channel of the Ganges in Calcutta, and the desalination plant on the Gulf of Cambay. Irrigation is especially important in the Indus–Ganges plains and in lowlands of the southeastern coast.

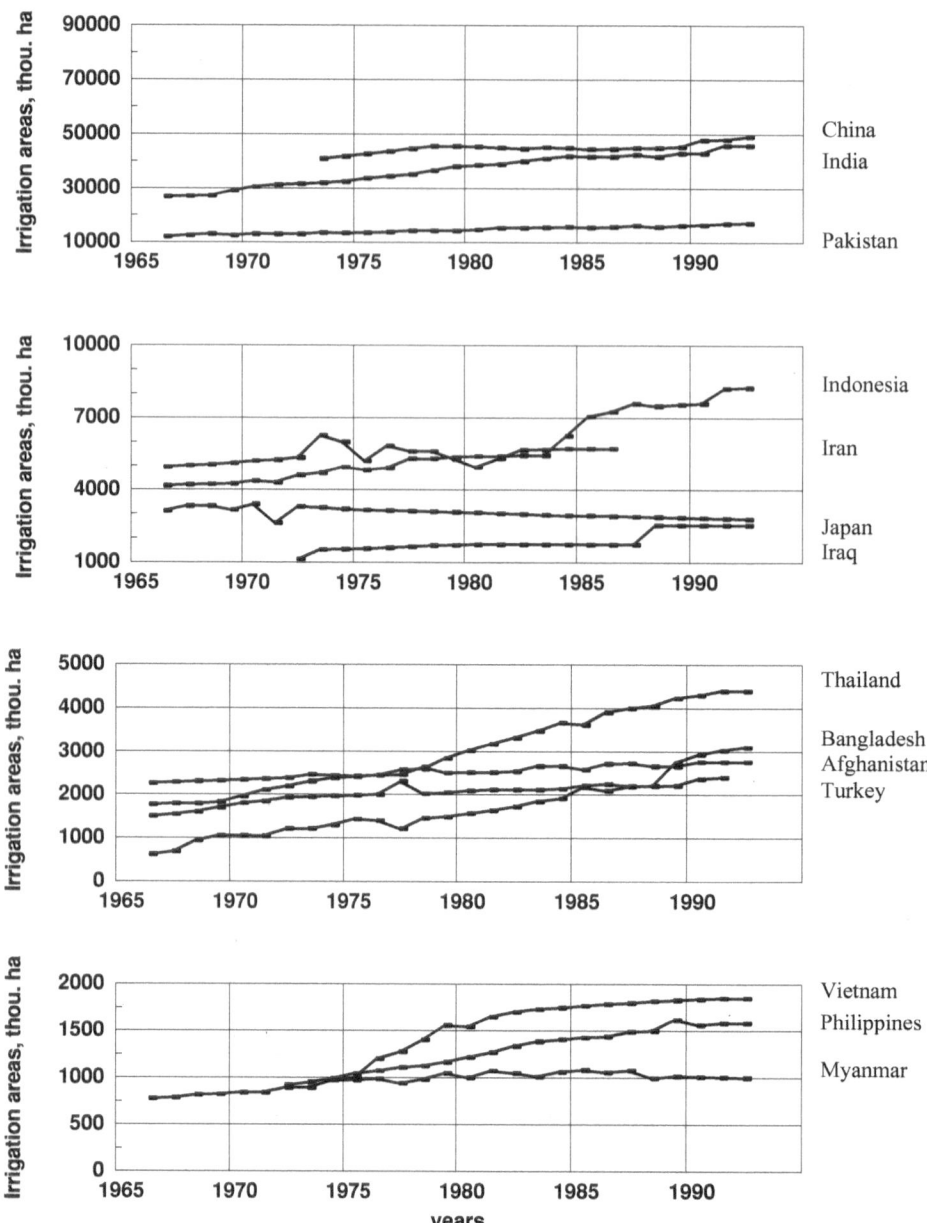

Fig. 5.8 Dynamics of irrigation areas for selected Asian countries.

In recent decades the largest growth of irrigation has occurred in the northwestern region of the country (Levitanus, 1986a, b).

In the early 1980s more than half the available surface water resources were being exploited in India. All the water resources in the northwestern regions (the upper reaches of the Ganges, Bias, Sutlej, Ravi, Jamuna, Chambal, Sabarmati, Mahi, Tapi, etc.), in the southeastern region (Kaveri, Penner, Krishna, etc.) and in the eastern region (Damodar, Son) were being used. The potential of the Rivers Narmada, Godavari, Mahanadi and Brahmani was nearly all in use. According to data from the national Irrigation Commission, the exploration of the potential of the Brahmaputra River is planned for the twenty-first century (Gerasimov, 1986).

Up to the early 1980s an increase occurred in India not only in the absolute but also of the specific irrigation area. It was close to 60 ha/1000 inhabitants. From 1980 to 1990 the rate of construction of irrigation systems slightly decreased, and the specific area has decreased to 50 ha/1000 inhabitants. In many regions the depletion of traditional water supply sources is considered the main cause. This leads to the need for water transfers, increased investment in hydraulic construction, increased centralized management of water use and the solution of international problems in the use of border rivers. To work out a national water policy, India set up the National Council on Water Resources. In 1982 the national agency for development of water resources was established with the responsibility for assessing water resources, actual water use and preparing forecasts for up to 2025. In addition to traditional hydraulic engineering projects a great deal of attention is given to investigating the possibilities for increasing the potential of dry farming and to the study of non-traditional water supply sources, for example rainwater harvesting, irrigation by sea and waste water, artificial recharge, cloud seeding and other measures. In 1954 the country adopted the National Programme for flood control with most attention devoted to the valley of the Brahmaputra and its tributaries (Wunderlich and Prins, 1987; Golubev and Leontyev, 1988).

Of the other countries in the eastern and southern Asia, Pakistan has the largest irrigated area (about 17 million ha), Indonesia has more than 8 million ha, Thailand about 4.5 million ha, Bangladesh more than 3 million ha and Japan about 2.8 million ha (FAO, 1995a). In Pakistan, irrigation provides 90% of all agricultural production which, in turn, provides about 70% of the hard currency income. It is predominantly based on the water resources of the Indus River basin which the British began to develop at the end of the nineteenth century. At the present time the world's largest irrigation system in the world is situated in Pakistan supplying water to more than 3 million farms (Scogerboe, 1983; Badruddin, 1987; McQuenn, 1992). Pakistan shares the resources of the Indus with India under the 1960 Agreement in the proportion of 80:90 (Tolokonnikova, 1972). The 5-year plans for the development of agriculture and the drainage work programme are evolved at the federal level. Financing of the Programme for Monitoring and Assessment of Water Resources is partly undertaken by the World and Asian Banks.

The main water management problems in the Indus River basin are salinization, rising ground water levels and floods. The main source of salinization is irrigation by ground water. From estimates for 1981, more than 5 million ha of irrigated land were strongly salinized and unsuitable for farming, and the total waterlogged area was about 7 million ha. According to the 1974 Programme, by the mid-1980s the soils over an area of about 5.3 million ha were to be restored. However, even with completion of this programme, about 5.3 million ha were still subject to salinization and waterlogging. For the complete restoration of these soils a large investment will be needed over a period of 5 to 20 years (Morozova, 1983). Construction of the necessary drainage began in 1975 and by 1982 50 km of main drains had been completed and studies for optimization of the irrigation regime had been made. Studies by British specialists have shown that the levels of ground water around some dams (Sukkur in 1932, Guddu in 1962) had been stabilized and the accumulation of salts had not resulted in losses of arable areas. This questions the need for investment in drainage but the need for water saving was confirmed while the rice yield increased by 10% to 15% (Badruddin, 1987; McCruady, 1988; Bhath and Kijne, 1991; ESCAP, 1992; Clark and Anid, 1994).

In Indonesia Java is the centre of irrigation and rice production in an area which occupies only 7% of the country. However the largest plantations are on Sumatra. Rice occupies 60% of the total area under crops and more than four-fifths of land under rice is irrigated (Simoniya, 1983). From the early 1980s there was a rapid increase in irrigation, but from 1985 the specific area irrigated decreased because of the intense population growth. The problem of developing backward regions is quite severe.

In Thailand the most rapid development of agriculture began in the 1950s and continued into the early 1970s, predominantly as a result of the mass migration of the agricultural population to new lands, and the growth of irrigation. In the 1970s in some provinces the irrigated land reached 90% of all land under cultivation. In the 1950s the construction of year-round irrigation systems with reservoirs began and by 1970 such systems (including systems of gravity irrigation) covered 723 000 ha; however, up to the mid-1970s they were not completely used (Simoniya, 1976). In the 1980s the high rates of growth of irrigation occurred in the northern (50% of the country's growth) and the northeastern regions. However most water is used in the central region where the main large-scale irrigation systems are located (the largest water user is the Chao Phraya system). By the early 1990s the total volume of water used for irrigation exceeded 30 km^3/year, mainly drawn from rivers. Ground water provides about 0.2 km^3, about 22% of total water used (ESCAP, 1991). During the period 1980–90 the specific irrigation area in Thailand increased from 65 to 80 ha/1000 inhabitants, although more water is being used by the domestic and industrial sectors and there are growing regional problems of water shortages. These shortages are stimulating the programmes for saving water including the introduction of higher tariffs. Plans for transferring water from the Mekong to the Chao Phraya basin are also being considered.

The rapid growth in irrigation is also typical of Bangladesh, Nepal, Malaysia, the Philippines, Mongolia, Saudi Arabia, Yemen, Qatar, Jordan, Iraq, Turkey, Syria, Afghanistan and Cyprus. Between 1965 and 1990 the area under irrigation in North Korea increased by more than 900 000 ha, in Syria by 190 000 ha and in Mongolia from 0 to 80 000 ha. During this period Vietnam became the world's third largest exporter of rice (after the USA and Thailand). Most of this rice is produced in numerous small irrigation systems covering about 2 million ha, predominantly in the valley of the Hongha River and in the northern part of the central region. Their development continues within the framework of the plans for the comprehensive use of water resources. The countries of the Iranian Uplands contain the largest specific irrigation area on the continent (about 100 to 150 ha/1000 inhabitants); however, in Iran itself in the 1980s the area irrigated remained the same and even showed a decline. However, the efficiency of irrigation in Asia remains unsatisfactory. The low efficiency of most canals means that a significant volume of water (40–60%) is lost due to seepage and evaporation. However these problems require substantial investment to provide a solution.

5.4 HYDROLOGICAL DATA

According to data published by international organizations such as WMO, data from National Committees for the International Hydrological Programme, reports of governmental bodies and separate studies, there are 12 000 stations for measuring discharge in Asia. The largest number of stations, some 5130 (Sokolov, 1986) is in northern Asia in the former USSR where station density is, on average, one station per 3600 km^2. The distribution of stations in the hydrological network is very heterogeneous. The largest density (one station per 300 km^2) is in Japan, Cyprus and Israel. In the Asian part of the former USSR there is one station per 5300 km^2. In Sri Lanka and Hong Kong the average density is one station per 500 km^2, on average; in Thailand 890 km^2, in Bangladesh 1200 km^2, in South Korea 1600 km^2, in Iran 2000 km^2, in Iraq 2650 km^2, in North Vietnam 3700 km^2, in India 2930 km^2, in Laos 5900 km^2 and in Myanmar 10 000 km^2. The Mongolian People's Republic has a sparse hydrological network, with one station on average per 37 000 km^2. A low density is typical of the countries of the Arabian Peninsula. In Saudi Arabia there is one station per 8050 km^2; in Syria 1600 km^2, in Jordan 2970 km^2, in the United Arab Emirates 4400 km^2, in Oman 4080 km^2 and in Qatar 260 km^2.

Limited observations of discharge were made during the nineteenth century on some rivers: from 1869 to 1979 on the Irrawaddy River, from 1874 to 1875 on the Amur River and from 1886 to 1887 on the Angara River. Continuous observations are available on the following rivers: Yangtze from 1865, Irtysh from 1900, Yenisey from 1903, Godavari, Indus and Red from 1900, Krishna from 1902, Mekong from 1905, Sungari from 1898 and Xijiang from 1901. In Turkey, Syria, Israel, Iraq, India, Sri Lanka, Malaysia and Japan the runoff measurements were only started in the 1930s and 1940s, and not until the 1950s in Afghanistan, Pakistan, Burma and Mongolia. From the 1960s measurements of discharge were being carried out in Indochina, as well as in Nepal, on Cyprus and in Jordan, Saudi Arabia and Kuwait. Observations commenced in Qatar in 1966, in Oman from 1970 and in the United Arab Emirates only in 1980 (UN, 1992). For the majority of rivers, especially in southern Asia, the observations of runoff are short and sometimes discontinuous. For the assessment of water resources 800 representative gauging sites were selected (UNESCO, 1993, 1996b), and the discharge data from these was placed in a database. Figure 5.9 shows the location of the stations while Table 5.6 provides more information. Asia was divided into seven large regions

(Shiklomanov and Markova, 1987) each differing in its hydrology but grouping countries with similar natural and climatic conditions and social and economic characteristics.

The Northern China and Mongolia region occupies an area of 8.29 million km^2 including China north of the Yangtze River, Mongolia, South Korea and North Korea. Within this region there are the Central Asian plains with the great deserts (Takla Makan, Gobi, Alashan, Ordos), and much of the highest parts of the world such as the Tibetan Uplands and a large part of eastern Asia. The entire region is arid or semi-arid and constantly faces the problems of seasonal and regional water deficits. The population of this region is approaching 480 million people. It is a region with high rates of growth contrasting with ancient traditions including high rates of water use. The total irrigated area exceeds 23 million ha and continues to increase.

The Southern Asian region with an area of 4.5 million km^2 includes the following countries: Bangladesh, Bhutan, India, the Maldives, Nepal, Pakistan and Sri Lanka. It is crossed by the fertile valleys of the largest rivers of the continent – the Indus, Ganges and Brahmaputra. Much of the region (except for the areas containing the Himalayas and rivers draining from them) has a moisture deficit reaching 1500 mm. Runoff of even large rivers is over-regulated and is used for communal needs. The population of the region is almost 1.2 billion people. These are mainly (70–80%) engaged in agriculture, the major product of the economy. The countries differ in their ancient traditions of irrigation with the largest irrigation system situated in the basin of the Indus. India and Pakistan are among the five countries with the largest areas of irrigation, the total exceeding 70 million ha.

Western Asia extends over 6.82 million km^2 and includes the countries of the Near and Middle East with a total population of about 230 million, the largest being Saudi Arabia, Iran, Turkey and Afghanistan. Most of these nations have grown rapidly and since the 1970s their population growth has exceeded the average growth rate for Asia by 10–13% and this trend is predicted to persist. The rapid growth of industry, urban development and the intensification of agriculture increase the need for water. However in many countries this need is difficult or impossible to meet because of the lack of water resources. Due to this situation it is vital to find ways of overcoming shortages. Abstraction of ground water, including water from deep aquifers, is widespread, runoff is captured and advanced technologies (micro-irrigation, hot houses, hydroponics, etc.) have been introduced. The total irrigated area is approaching 7 million ha.

The South East Asian region has an area of 6.95 million km^2 and its population exceeds 1.2 billion. It includes southeastern China (including the Yangtze River basin), Myanmar, the countries of the Indochina Peninsula, Japan, the Philippines and Indonesia and several small island states. The region is mainly humid with a

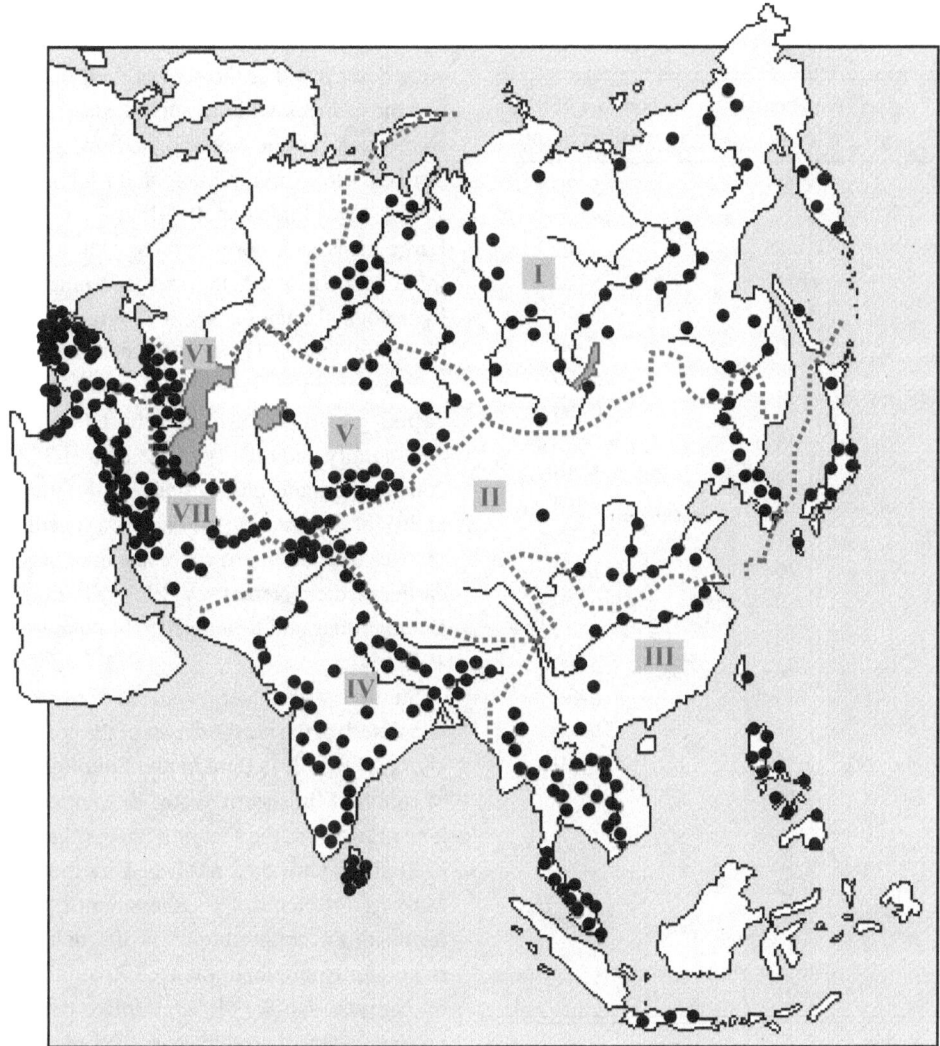

Fig. 5.9 Distribution of Asian territory over natural–economic regions and location of main runoff gauge stations. I, Siberia and Far East; II, North China and Mongolia; III, South East Asia; IV, Southern Asia; V, Middle Asia; VI, Transcaucasus; VII, Western Asia. •, Gauge station.

moisture excess of 1500–2000 mm in Malaysia and Japan. The total runoff is half the entire runoff of Asia. Industry is either well established or advancing rapidly and agriculture is highly developed. The gross national income is already high and increasing while the same applies to the standard of living.

Middle Asia and Kazakhstan includes Kazakhstan, Kyrgyzstan, Tajikistan, Turkmenistan and Uzbekistan. Its area is 3.99 million km^2 while its population is 54 million. This is an arid and semi-arid region with a moisture deficit of 1000–1500 mm. It is also an area of internal drainage. Runoff varies from values close to zero to 500–1000 mm in the mountains. At certain times, a large number of rivers and temporary water courses flow to closed depressions and lakes where the water infiltrates or evaporates. The main component of the water resources of this region is the rivers of Middle Asia – the Amu-Darya and Syr-Darya. However, this runoff is intensively used, predominantly for irrigation at the expense of the Aral Sea.

Siberia and the Far East is the largest region in area with 12.8 million km^2, but its population is only 42.5 million. The region has a moisture excess while its drainage is to the Arctic and Pacific Oceans through the Ob, Yenisey, Lena and Amur Rivers.

Transcaucasia includes three republics of the former Soviet Union – Azerbaijan, Armenia and Georgia. The region is small, only 0.19 million km^2, and the population is 16.5 million. Across the plains and lowlands there is a lack of water resources, while in the high mountains there is a surplus. The runoff within Transcaucasia ranges from 2000 mm in the Caucasus to 5–10 mm in the Apsheron Peninsula. Irrigation is the main water use.

Table 5.6. *Number of river runoff observation stations used to estimate water resources in some Asian countries*

Country	Number of stations	Number of years of observations
Afghanistan	32	5–20
Bangladesh	3	6–10
Cambodia	8	5–30
China	74	5–120
Cyprus	6	5–30
India	50	5–90
Iran	58	5–70
Iraq	14	6–50
Israel	34	5–20
Japan	65	5–50
Korea	15	5–30
Laos	11	5–50
Malaysia	26	5–50
Myanmar	6	6–30
Mongolia	10	6–30
Nepal	33	5–10
Pakistan	5	6–90
Philippines	11	5–20
Sri Lanka	14	5–50
Syria	10	5–50
Thailand	12	6–50
Turkey	86	5–50
Vietnam	12	6–70
Countries of the former Soviet Union	206	50–100
Asia as a whole	801	5–120

5.5 DISTRIBUTION OF WATER RESOURCES IN TIME AND SPACE

5.5.1 Data and methodology

Data from the network of the 800 selected river gauging stations were used to assess the water resources of the continent. However these data suffer because the length of the period of observation is not the same in for all stations and because there are many gaps in the series. These gaps were filled using the techniques described earlier such as regression equations and the selection of analogue rivers. The gaps were eliminated in both the annual and monthly runoff series. Where necessary, data on precipitation and temperature were used for assessing the annual runoff of some rivers and for the extension of certain time series (the Ganges, Brahmaputra, Indus and Yangtze, for example). These data change along the course of the river depending on climate, the use of water and other factors. For the rivers originating in the mountains and where the use of water is mainly in the plains (the Amu-Darya, Syr-Darya, Ili, Kura, Liao He, Hai He and Indus), their resources were determined as the sum of their discharges in the mountains. For those rivers without such abstractions, their water resources were taken as the volume of flow at their mouths (the Lena, Yenisey, Ob, Amur, Xijiang and Mekong).

The hydrological network does not cover Asia adequately. Large areas lack observations, such as many of the islands. For these problem areas the methods employed previously, namely hydrological analogues and geographical interpolation, were applied (Chapter 2). Using these methods long-term water resources were assessed as well as the resources for individual years. The accuracy of runoff estimates based on maps depends largely on the quantity and quality of the data and information used in their construction and on the scale of the map. The increasing availability of information on runoff in recent years has promoted the increased reliability of water resources assessment as compared to earlier studies (Nemaltsev, 1969; Korzun, 1974b; L'vovich, 1974; Baumgartner and Reichel, 1975; Nikolayeva and Chernogayeva, 1977).

Studies of the water resources of the former Soviet Union have been used for the northern part of the continent (SHI, 1987; Babkin *et al.*, 1990, 1991; Babkin and Sokolov, 1992). The Chinese map of runoff (Ministry of Water Resources of China, 1987) and the map of runoff of the Mekong Basin (Nguyen Din Tien, 1980) were used along with data published by the UNESCO Committee on Mekong Problems. The assessment of water resources has also taken into account studies in the deltas of the most important rivers, for example, in the Red River and the Mekong carried out by Nguyen Van Ky (1990). Unlike deltas in temperate and arid regions where there are losses, in the deltas of Vietnam and in general in the humid tropics runoff increases. These studies were based on the 1 : 2 000 000 map of runoff of Vietnam (Chan Than Suan, 1978) using data from 125 basins and 15 basins from the border regions.

Certain national assessments were also based on maps. For Sri Lanka the assessment employed the 1 : 500 000 maps of the water balance using 58 representative river basins (Pelenda Appukhamilage Piyasiri Karunatilaka, 1990), for the Mongolian People's Republic the runoff map from the national Atlas was employed (Tuvdendorzh and Myagmarzhav, 1985). Runoff was estimated from precipitation for Myanmar and for some of the islands, and from numerous national and international publications for Java, Sumatra, Borneo and Sulawesi, as well as islands in the Philippines Archipelago. For the Rivers Hai He, Luan He, Huang He and Huai He the runoff losses were determined (Ministry of Water Resources of China, 1987) and the mean runoff calculated taking the losses into account. The runoff from India, Bangladesh, Myanmar, Pakistan, Iraq and some other countries was estimated from meteorological data.

Table 5.7. *Renewable water resources and water availability by natural–economic regions of Asia*

Region	Area, km² × 10⁶	Population (1994), ×10⁶	Water resources, km³/yr				Coefficient of variation, C_v	Potential water availability, m³ × 10³/year	
			Inflow	Local				Per km²	Per head
				Average	Minimum	Maximum			
Siberia and Far East	12.76	42	218	3 107	2 628	3 500	0.06	243	76.60
North China and Mongolia	8.29	482		1 029	590	1 735	0.23	124	2.13
South East Asia	6.95	1404	120	6 646	5 342	7 607	0.09	956	4.77
Southern Asia	4.49	1214	300	1 988	1 535	2 458	0.10	443	1.77
Middle Asia and Kazakhstan	3.99	54	46	181	121	265	0.17	45	3.78
Transcaucasia	0.19	16	12.1	68	51	89	0.12	358	4.63
Western Asia	6.82	232		490	227	931	0.35	72	2.11
Asia as a whole	43.50	3445		13 510	11 800	15 000	0.06	311	3.92

Table 5.8. *Renewable water resources and water availability of individual countries of Asia*

Region	Area, km² × 10⁶	Population (1994), ×10⁶	Water resources, km³/yr				Coefficient of variation, C_v	Potential water availability, m³ × 10³/year	
			Inflow	Local				Per km²	Per head
				Average	Minimum	Maximum			
Armenia	0.03	3.55	2.1	6.19	4.80	8.53	0.12	207	2.04
Azerbaijan	0.09	7.47	20.2	7.78	5.08	12.8	0.15	86.4	2.39
China	9.60	1209		2700	1970	3930	0.12	281	2.23
Georgia	0.07	5.45	6.9	53.3	40.8	67.7	0.13	761	10.4
India	3.27	919	581	1456	1065	1794	0.11	445	1.90
Kazakhstan	2.72	16.7	56	70.2	39.3	111	0.24	25.8	5.88
Kyrgyzstan	0.20	4.67		48.7	37.3	70.1	0.19	244	10.4
Pakistan	0.81	137	186	40	22.5	63.9	0.21	49.4	0.61
Tajikistan	0.15	5.93	47.9	47.4	36.4	65.8	0.13	316	12.0
Turkmenistan	0.49	4.01	69.8	1.13	0.92	1.14	0.22	2.3	8.99
Uzbekistan	0.45	20.3	98.1	9.52	4.98	19.7	0.27	21.2	2.89

5.5.2 The water resources of natural regions and individual countries

The renewable water resources of Asia with its adjacent islands were estimated to be 13 510 km³/year – the largest of any continent (Table 5.7). However, the potential water availability, some 3920 m³/year per head, is the least. Its large size, the presence of the high mountain ranges and massifs, and the wide range of climatic conditions cause the water resources to be very unevenly distributed over the continent. Waterless deserts contrast with many areas where precipitation is abundant, especially in South East Asia where almost half the water resources (6650 km³/year) occur. Significant water resources of 3110 km³/year also originate within Siberia and the Far East, in contrast to Transcaucasia, Middle Asia and Kazakhstan where the resources are very small.

Table 5.8 presents data on the water resources of some of the large countries. The two largest, namely India and China, whose combined area is 30% of the total for Asia and population 62%, contain the headwaters of rivers that produce 30% of all the runoff. The variability of their water resources as well as the availability of water per head are similar. By comparison the Arabian Peninsula has the smallest water resources. Countries here use ground water, as well as desalinized sea water and recycled water (Abdul Razzak, 1995; ROSTAS, 1995) and the relevant data are presented in Table 5.9.

Basins draining to the Pacific and the offshore islands have the largest amounts of available water. They account for much of the size of the water balance, since they make up 87% of the area of the continent, but there are also differences between them

Table 5.9. *Renewable water resources of surface and ground water and water availability of the countries of Arabian Peninsula (1992)*

Country	Renewable water resources, km³/year			Population, ×10⁶	Potential water availability, m³ × 10³/year per head	Desalinated water, km³	Used water, km³	Total water resources, km³	Water supply with total water resources, m³ × 10³/year per head
	Surface	Ground	Total						
Saudi Arabia	2.2300	3.85	6.080	11.78	0.516	0.795	0.217	7.092	0.602
Yemen	3.5000	1.55	5.050	13.12	0.385	0.009	0.006	5.065	0.386
United Arab Emirates	0.1500	0.12	0.275	1.81	0.152	0.342	0.062	0.679	0.375
Oman	0.9180	0.55	1.468	1.51	0.972	0.032	0.010	1.510	1.000
Kuwait	0.0001	0.16	0.160	1.52	0.105	0.240	0.083	0.483	0.318
Bahrain	0.0002	0.10	0.100	0.52	0.192	0.056	0.032	0.188	0.362
Qatar	0.0014	0.05	0.052	0.50	0.102	0.083	0.023	0.157	0.314
Arabian Peninsula as a whole	6.8000	6.38	13.180	30.76	0.428	1.557	0.433	15.170	0.493

Source: Based on data from Abdul Razzak (1995); ROSTAS (1995).

Table 5.10. *Renewable water resources and water availability of large islands of Asia*

Island	Area, km² × 10³	Population, ×10⁶	Water resources, km³/year	Potential water availability, m³ × 10³/year	
				Per km²	Per head
Kalimantan	539.4	6.3	734	1360	116.0
Sumatra	473.6	25.5	696	1470	27.3
Irian Jaya	421.9	1.1	469	1110	426.0
Sulawesi	199.1	10.5	176	884	16.8
Java and Madura	132.2	89.0	198	1500	2.2
Maluku	74.5	1.3	76	1020	58.5
Bali and Tengara	73.6	8.1	31	421	3.8

(Table 5.10). The largest amounts of available water are on Irian Jaya Island (eastern Indonesia). These figures are more than 100 times greater than the mean value for the whole continent and amongst the highest in the world. For most islands, the available water per unit area is twice as large as the mean for the onshore basins and three to five times that for the continent. The central area of inland drainage has the least water availability: on average, 42 700 m³/km² per year and on dry plains less than 5000 m³/km² per year. The least water is available (6 m³/km² per year) in Kuwait and in the other countries of the Arabian Peninsula.

The water resources of the continent and many Asian countries have been assessed a number of times (Table 5.11). The most complete data on water resources were presented by the Institute of World Resources (WRI/IIED, 1986, 1988; IIED/WRI, 1987; WRI, 1990, 1992, 1996, 2000); by IWMI (Seckler *et al.*, 1998); in Gleick (1993, 1998); and in Engelman and LeRoy (1993). In fact they are identical data for the 37 countries where the values were taken from Forcasiewicz and Margat (1980). Data for Cyprus, Israel, Lebanon, Syria and Turkey were those given by Margat (personal communication), FAO (1997) and Margat and Vallée (2000). Data on Bhutan, Vietnam, and North Korea were communicated by Belyayev (personal communication). As indicated by Gleick (1993), these data should be considered with some scepticism. The Institute of World Resources notes that data on water resources collected by Margat were obtained from published documents, including national reports, UN publications and scientific publications. Data for small countries and countries in the arid and semi-arid zones seem to be less reliable, compared with data from humid countries.

Of course, data on water resources has to be used rather carefully, since the assessments were often made with different data for different periods, usually by different methods. Frequently the sources do not indicate the methods used and the period of observations. Shahin (1989) presented data in *Water in Crisis* (Gleick, 1993) on the renewable water resources of 12 countries in the Arab region. The mean annual runoff includes data for ephemeral streams. The assessment of ground water presented by the author is much less than the total supply of ground water and is doubtful according to Gleick (1993, 1998). Data for surface and ground water runoff are given as long-term means without an indication of the period of observations and without taking into account the water abstracted. Countries such as Kuwait, Saudi Arabia, Bahrain and Qatar also take account of desalinated and re-used water and

Table 5.11. *Water resources of the countries of Asia (km^3/yr) by data of different authors*

Country	Shahin (1989) In: Gleik (1993)		Forkasiewicz and Margat (1980); Margat (1988); Margat and Vallec (2000); Belyaev (pers. comm.) In: WRI/IIED (1986); IIED/WRI (1987); WRI (1990, 1992, 1996, 2000); Gleick (1993)			Engelman and Le Roy (1993)	Nikolayeva and Chernogayeva (1977); L'vovich, (1974)			Jalal (1987) In: ESCAP (1978)	Van der Leeden (1975)
	Surface	Ground	Local	Inflow	Outflow	Local + Inflow	Total	Surface	Ground		
Afghanistan			50			50.0	76	48	28	50[a]	50
Bahrain		0.09	0			0.09					
Bangladesh			1357	1000		2357	129[b]	95	34	123[c]	
Bhutan			95[d]			95.0					
Cambodia			88	410		498	118	88	30	88.1[e]	80
China			2800	0.0		2800	2857[f]	1927	930	2680	
Cyprus			0.9	0.0	0.0	0.9	0.4	0.3	0.1		
Hong Kong										1.32[g]	
India			1850	235		2085	1590	1218	372	1780[e,g]	1508
Indonesia			2530			2530	1510[h]	1279	231	2530	1940
Iran			117			118	173	109	64	117.5[g]	100
Iraq	43.20		34	66.0		109	36	21	15		
Japan			547	0.0	0.0	547	397	212	185	547[g]	
Jordan	0.90	0.41	0.7	0.4		1.3					
Korea (North)			67[d]			67.0	67	46	21		70
Korea (South)			63			63.0	60	42	18	63[g]	
Kuwait		0.16	0	0.0		0.16					
Laos			270			270	228	178	50	273	270
Lebanon	4.40	0.60	4.8	0.0	0.86	4.98					
Malaysia			456			456	486	415	71	456[e]	650
Mongolia			24.6			25.0	47	24	23	24.6	25
Myanmar			1082			1082	680	524	156	1082	680
Nepal			170			170	119	86	33	170	170
Oman	1.37	0.56	2.0	0.0		1.93					
Pakistan			298	170		468	188[i]	137	51	183[g]	298
Philippines			323	0.0	0.0	323	391	210	181	323[e,g]	390
Qatar		0.05	0.02	0.0		0.05					
Saudi Arabia	2.20	2.35	2.2	0.0		4.55					
Singapore			0.6	0.0	0.0	0.60				0.6[g]	
Sri Lanka			43	0.0	0.0	43.0	59	42	17	43.2	43.0
Syria	22.10	3.67	7.6	27.9	30.0	25.8	17	14	5		
Thailand			110	69.0		179	171	127	44	111[e,g]	220
Turkey			196	7.0	69.0	203	170	110	60		
Turkmenistan											
United Arab Emirates	0.10	0.39	0.3	0.0		0.49					
Vietnam			376[d]			376	376	292	84		
Yemen (North)	2.10	1.00	1.0	0.0		5.2					
Yemen (South)	1.70	0.40	1.5	0.0							

[a] The Oxus River is not included.

[b] 138 (an estimate of Nikolayeva and Chernogayeva, 1977).

[c] Reports of the country presented at the International Conf. "Water for Peace" (Washington DC, 23–31 May 1967); 1234 km^3, inflow from other countries.

[d] Data of Belyaev (personal communication).

[e] Reports of countries to the World Water Conference.

[f] 2880, Lvovich's estimate islands included.

[g] Data presented for UN Water Conference "Water Development and Management" (Mar-del-Plata, Argentina, March 1977).

[h] Without western Irian Jaya.

[i] 73, Lvovich's estimate.

other non-conventional sources in their assessments of water resources, but these were not considered by Shahin.

The data on water resources of these countries presented in ROSTAS (1995) and the data of Abdul Razzak (1995) can be considered the most reliable (see Tables 5.9 and 5.11). In the Proceedings of the Regional Symposium on Water Resources, Jalal (1987) refers to data of the Report prepared before 1973 and presented by ESCAP for the UN Water Conference (ESCAP, 1978). Van der Leeden (1975) published data on water resources for Asian countries, excluding the inflows from other states. L'vovich (1974) and Nikolayeva and Chernogayeva (1977) give data on the water resources of Asian countries based on the maps of the water balance. These data were later presented in Barney (1980).

Using these and other available sources the following conclusions can be drawn:

- Current estimates of water resources presented by the Institute of World Resources (WRI, 1992, 1996, 2000) by Gleick (1993, 1998) and by Engelman and LeRoy (1993), and presented in Table 5.11, are based on data more than 20 years old. This concerns primarily such countries as Afghanistan, Indonesia, Iran, Iraq, Japan, Cambodia, South Korea, Laos, Malaysia, Mongolia, Nepal, the Philippines, Sri Lanka and Thailand.
- The SHI data on the water resources of China and India from 1920 to 1985 (this Monograph), do not differ significantly from the current national estimates (Table 5.8).
- The water resources of Pakistan (WRI, 1992, 1996, 2000) were estimated to be 298 km³/year. A similar assessment of the runoff with the indication that these resources exclude the inflows from adjacent countries is given in Van der Leeden (1975) although at the World Water Conference (ESCAP, 1978) they were assessed as 183 km³/year. According to SHI estimates, the average water resources of Pakistan for the period 1920–85 are 226 km³/year. Of these figures, the inflows are 186 km³/year.
- The estimates of water resources of Bangladesh presented by the Institute of World Resources (WRI, 1992, 1996, 2000) are dubious: the local runoff is estimated as 1357 km³, the inflow 1000 km³ and the total as 2357 km³. According to Jalal (1987), the local runoff in Bangladesh is 123 km³, the inflow from other countries is 1234 km³, so the total runoff is 1357 km³. Nikolayeva and Chernogayeva (1979) estimated the local runoff in Bangladesh as 138 km³, and L'vovich (1974) as 129 km³. The estimate of local runoff of 1357 km³/year (from a depth of 9420 mm and the specific runoff of 299 l/s per km²) is not realistic. It probably includes the inflow from other countries. In addition the SHI estimate of the water resources of the Ganges, Brahmaputra and Meghna together is 1390 km³/year, and this figure compares well with national estimates.

Unfortunately the estimates of water resources for a number of Asian countries are in doubt, due to the different data used and the different methods of making the estimates. For these and several other reasons, the estimates of the average water resources of the continent differ: 13 190 km³/year (L'vovich, 1974; Nikolayeva and Chernogayeva, 1979); 14 410 km³/year (Korzun, 1974b); and 12 205 km³/year (Baumgartner and Reichel, 1975). Asian water resources also vary in time ($C_v = 0.06$) but those of the regions except for Siberia and the Far East have much greater variability (Table 5.7). Figures 5.10 and 5.11 show the changes in the water resources of the different regions, certain countries and for the

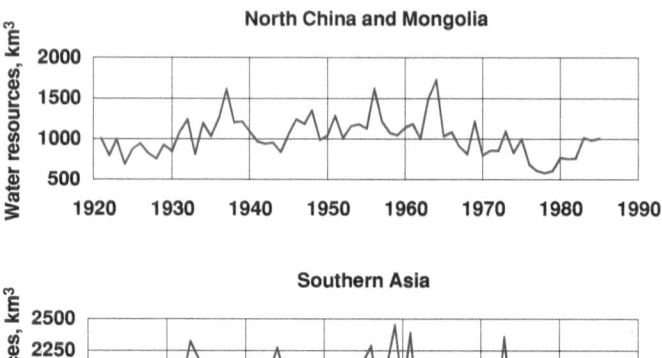

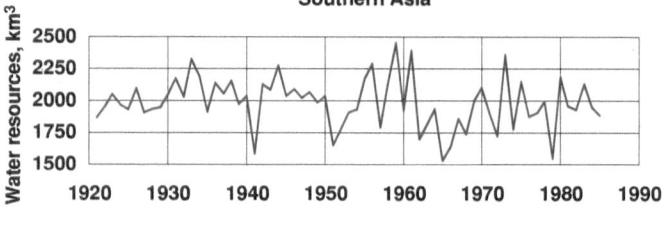

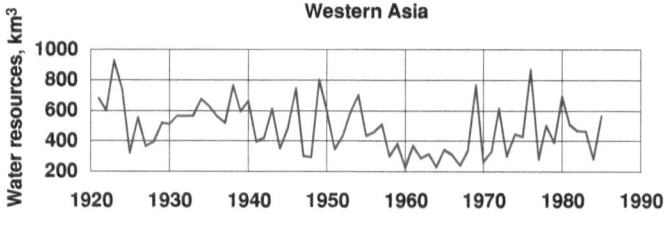

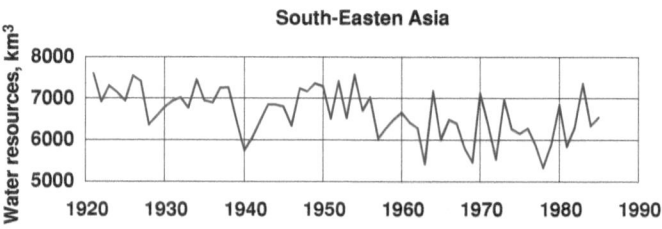

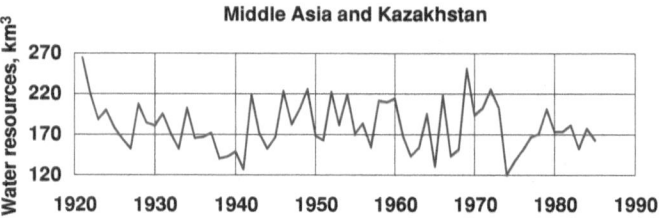

Fig. 5.10 Dynamics of water resources by regions of Asia.

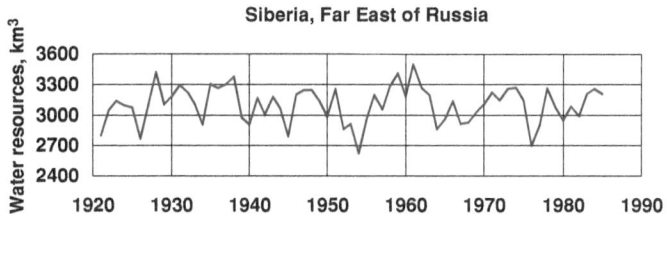

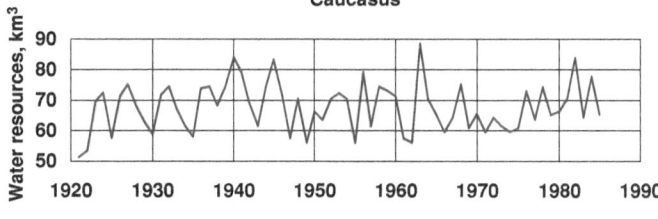

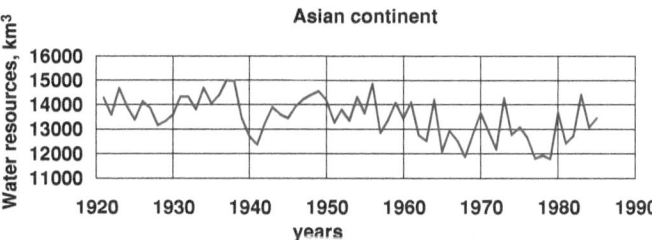

Fig. 5.10 (*cont.*).

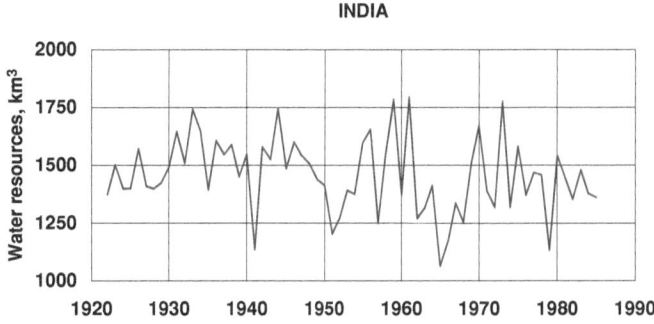

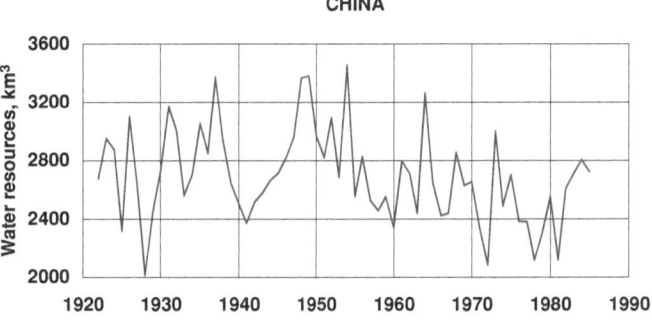

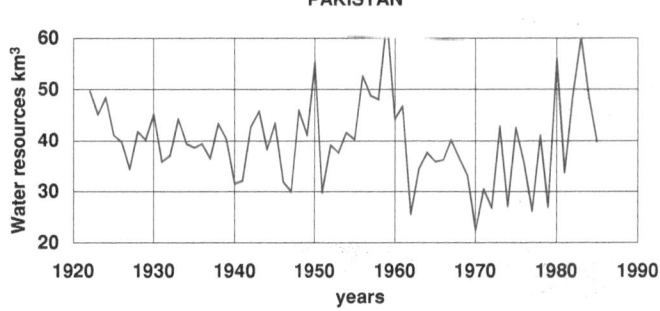

Fig. 5.11 Dynamics of water resources for selected Asian countries.

whole of Asia for the period 1921 to 1985. There seems to be a tendency for runoff to decrease in Southeast Asia, Northern China and Mongolia over the last two decades and for this decrease to be reflected in the figures for the continent as a whole.

Checks for stationarity on these values were performed using the different tests mentioned in Chapter 2 (Shiklomanov, 1997, 1998, 2000a, 2000b; Babkin, 1998; Shiklomanov *et al.*, 2000). These tests showed the non-stationary state of annual water resources time series for the continent and for Western and South East Asia, as well as for China and India. Other tests showed the series for northern China and Mongolia, Southern Asia and Pakistan to be non-stationary with a probability ($P \geq 95\%$), indicative of the presence of possible trends. While these trends may indicate the direct effects of climatic change over the period for the regions under consideration, the nature of the statistics of these series such as the mean, the variation coefficient the skewness are only approximate.

Figure 5.12 and Table 5.12 shows cycles of different durations for the different regions, the continent and for several countries with phases of increased and decreased runoff. From the early 1960s there was a dry phase in Siberia and the Far East, South East Asia, Southern Asia, Western Asia, and Northern China and Mongolia. This drier period is shown in the records for China, India, Pakistan and other countries. It may well be why different estimates have been obtained for Asian water resources during the

last 20 to 25 years. Over some parts of Asia, the duration of dry and wet cycles varies between 10 and 18 years (Siberia and the Far East for example), but in the south of these regions the duration of cycles increases to between 20 and 26 years (such as in Northern China and Mongolia, Middle Asia and Kazakhstan, and Transcaucasia). The longest cycles are those for South East Asia (39 years). The duration of dry phases varies from 4 to 29 years with individual years being 3% to 23% below the long-term mean. The duration of wet phases varies from 3 to 34 years with individual years being 2% to 19% above the long-term mean. Table 5.12 contains detailed data on these and other trends and cycles. These studies revealed the features of the hydrological regimes and the available water resources against assessment of their probability of occurrence. The probability of occurrence of groups of years which are wet or dry with a duration from 1 to 5 years for the continent as a whole is 90%. Some 25% of the groups are

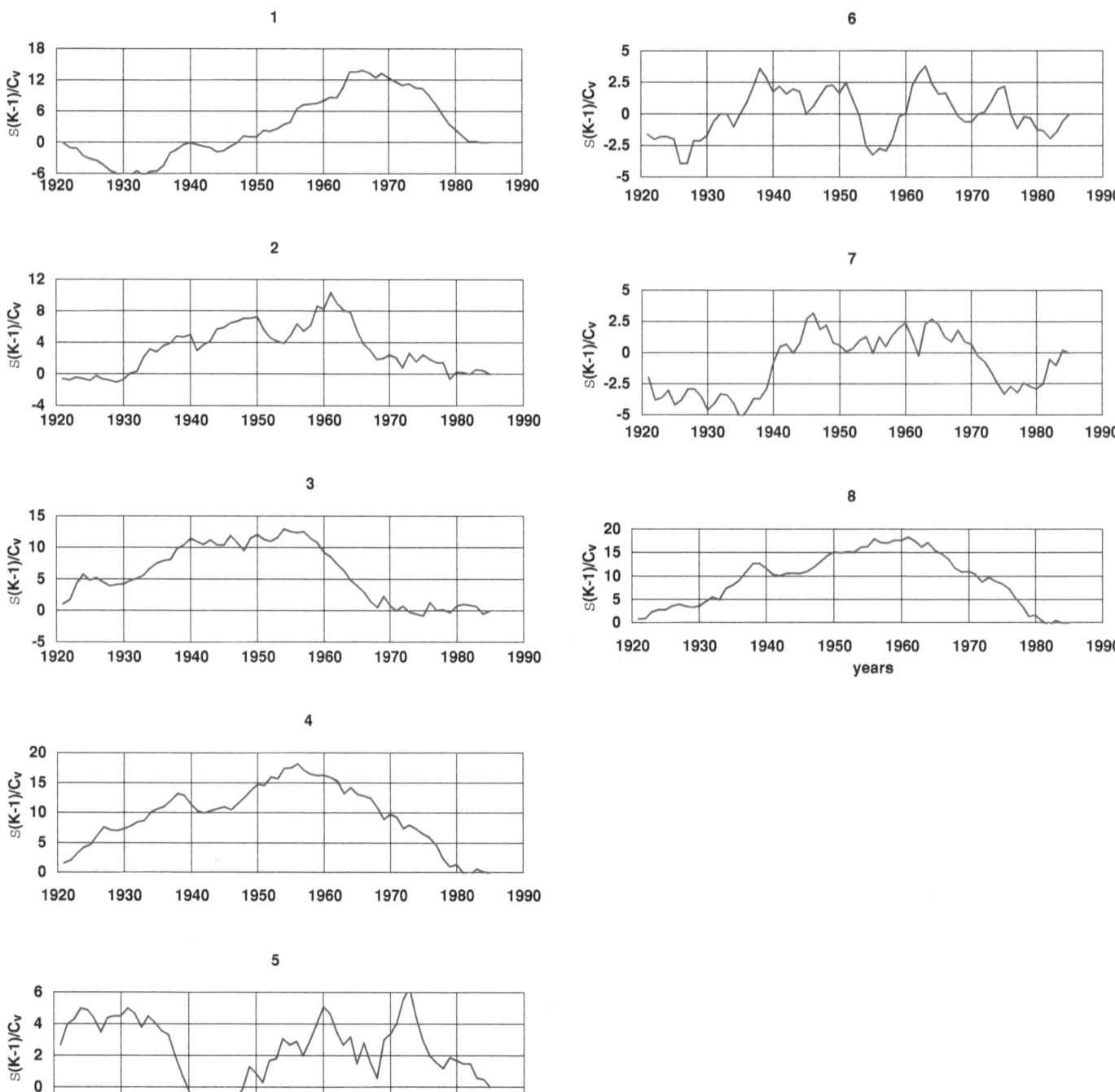

Fig. 5.12 Normalized difference integral curves of runoff for Asian regions. 1, North China and Mongolia; 2, Southern Asia; 3, Western Asia; 4, South East Asia; 5, Middle Asia and Kazakhstan; 6, Siberia, Far East of Russia; 7, Transcaucasus; 8, Asian continent.

Table 5.12. *Periods of different water content by natural–economic regions, selected countries and the continent of Asia*

Territory	Periods of high water content			Periods of low water content			Periods of average water content		
	Years	Ratio of mean for period to long-term mean, K_{av}	Number of years	Years	Ratio of mean for period to long-term mean, K_{av}	Number of years	Years	Ratio of mean for period to long-term mean, K_{av}	Number of years
Regions									
Siberia and	1928–38	1.04	11	1921–27	0.97	7			
Far East	1946–51	1.02	6	1939–45	0.97	7			
	1956–63	1.05	8	1952–55	0.92	4			
	1970–75	1.03	6	1964–69	0.96	6			
	1983–85	1.04	3	1976–82	0.97	7			
Northern China	1931–40	1.15	10	1921–30	0.85	10			
and Mongolia	1945–66	1.17	22	1941–44	0.90	4			
				1967–85	0.83	19			
South East Asia	1921–38	1.06	18	1939–46	0.97	8			
	1947–56	1.06	10	1957–85	0.95	29			
Southern Asia	1931–61	1.04	31	1962–85	0.96	24	1921–30	1.00	10
Middle Asia	1921–31	1.08	11	1932–45	0.91	14			
and Kazakhstan	1946–60	1.08	15	1961–68	0.90	8			
	1969–73	1.19	5	1974–85	0.91	12			
Transcaucasia	1936–46	1.09	11	1921–35	0.96	15			
	1952–64	1.02	13	1947–51	0.93	5			
	1976–85	1.04	10	1965–75	0.92	11			
Western Asia	1921–54	1.13	34	1955–75	0.77	21			
	1976–85	1.03	10						
Countries									
China	1930–38	1.09	9	1921–29	0.99	9			
	1946–54	1.13	9	1939–45	0.95	7			
				1955–85	0.94	31			
India	1930–61	1.04	32	1921–29	0.98	9			
				1962–85	0.96	24			
Pakistan	1921–24	1.13	4	1925–55	0.98	31			
	1956–61	1.27	6	1962–79	0.84	18			
	1980–85	1.20	6						
Asia as a whole	1921–39	1.04	19	1940–45	0.98	6			
	1946–61	1.03	16	1962–85	0.95	24			

below average, 41% are average and 24% above average. There is a high probability of occurrence of groups of years of different states of wetness or dryness for the regions of Northern China and Mongolia (92%), Middle Asia and Kazakhstan (96%), Western Asia (87%) and for Pakistan (97%), as presented in Table 5.13.

The variations in the water resources of the continent are shown in Table 5.14. The probability that dry years will be followed by dry years is 50%, and by average years some 31%. The probability of wet years following dry years is only 19%. An average year being followed by another average year has a probability of 58%, while an average year being followed by a dry and a wet

year has a probability of 19% and 23%, respectively. There is a 53% probability that a wet year will be followed by an average year, and 35% by a wet year. The probability is only 12% for a wet year to follow a dry year. Table 5.14 shows the variations in water resources across Asia and it reveals considerable differences:

- The probability of occurrence of years with average resources is much higher (54–89%) for humid regions (Siberia and the Far East, Southern Asia, South East Asia, Transcaucasia) than for drier regions (Western Asia, Middle Asia and

Table 5.13. *Probability (as percentage) of appearance of groups of years with different water content of 1 to 5 years in duration*

Territory	Probability of appearance of year groups of 1 to 5 years in duration			
	Dry years	Years with average water content	Wet years	Total
Regions				
Siberia and Far East	18	39	15	72
Northern China and Mongolia	31	41	20	92
South East Asia	14	41	12	67
Southern Asia	16	29	16	61
Middle Asia and Kazakhstan	29	40	27	96
Transcaucasia	21	47	13	81
Western Asia	39	14	34	87
Countries				
China	20	43	17	80
India	36	16	28	80
Pakistan	31	39	27	97
Asia as a whole	25	41	24	90

Kazakhstan, Northern China and Mongolia) where it ranges from 8% to 44%.

- Changes from dry years or average years to dry years have a lower probability for humid regions (13–33%) than for dry regions (30–52%).
- Where there is a change from average to dry years, this probability is higher for humid regions (0–20%) than for dry regions.
- The mean duration of groups of years with average conditions ranges from 1 to 3 years.

Due to the large diversity in conditions across Asia, the runoff from different basins and regions is very different from their neighbours and frequently they do not correlate. High correlations are observed between the water resources of the continent and the water resources of South East Asia ($r = 0.86$) and Southern Asia ($r = 0.62$), China ($r = 0.71$), India ($r = 0.57$) and the region of Northern China and Mongolia ($r = 0.52$).

The water resources of some regions have a close correlation to the runoff of the largest rivers, as might be expected. For example, the water resources of Siberia and the Far East have significant correlations with the runoff of the Yenisey ($r = 0.53$) and the Amur ($r = 0.54$). The water resources of Northern China and Mongolia correlate well with the runoff of the Amur ($r = 0.57$),

Table 5.14. *Water content transition from one grouping to another by natural–economic regions and selected individual countries of Asia*

Territory	Water content characteristic[a]	Probabilities of characteristics transition from one grouping to another			Average duration of groups, years
		→1	→2	→3	
Regions					
Siberia and Far East	1	31	54	15	1.44
	2	20	67	13	3.00
	3	0	58	42	1.71
Northern China and Mongolia	1	52	38	10	2.10
	2	35	35	30	1.53
	3	10	35	55	2.22
South East Asia	1	33	67	0	1.50
	2	13	72	15	3.54
	3	0	89	11	1.12
Southern Asia	1	20	60	20	1.25
	2	13	76	11	4.18
	3	25	62	13	1.14
Middle Asia and Kazakhstan	1	41	24	35	1.70
	2	30	44	26	1.80
	3	10	60	30	1.43
Transcaucasia	1	22	64	14	1.27
	2	24	61	15	2.56
	3	0	89	11	1.12
Western Asia	1	52	17	31	2.07
	2	36	37	27	1.57
	3	42	8	50	2.00
Countries					
China	1	33	50	17	1.50
	2	20	62	18	2.67
	3	0	83	17	1.20
India	1	50	11	39	2.00
	2	40	30	30	1.42
	3	38	15	46	1.86
Pakistan	1	37	47	16	1.58
	2	30	48	22	1.93
	3	22	28	50	2.00
Asia as a whole	1	50	31	19	2.00
	2	19	58	23	2.40
	3	12	53	35	1.54

[a] 1, low; 2, average; 3, high.

Table 5.15. *Runoff of individual rivers of Asia and its variability*

River	Station	Area of watershed, $km^2 \times 10^3$	Study period	Runoff, km^3/yr Average	Minimum	Maximum	Coefficient of variation, C_v
Amur	Khabarovsk	1630	1897–85	267	135	422	0.22
Yangtze	Hanko	1488	1865–86	735	451	1009	0.14
Yenisey	Khabarovsk	1400	1903–85	238	197	322	0.14
Irtysh	Tobolsk	969	1900–88	66.7	38.8	123	0.27
Ganges	Farakka	935	1902–85	380	244	543	0.15
Ind	Sukkur	900	1902–85	133	72.6	207	0.20
Lena	above Aldan mouth	897	1930–85	225	159	331	0.16
Huang He	Litzin	752	1919–85	43.7	9.1	76.8	0.36
Si(Kiang)	Wuchzhou	592	1901–85	227	102	347	0.19
Ob	Kolpashevo	486	1900–87	125	77.3	178	0.18
Brahmaputra	Pandu	405	1902–85	574	396	881	0.12
Mekong	Mukdakhan	391	1905–87	260	175	357	0.14
Sungari	Harbin	390	1898–85	37.9	12.1	84.2	0.39
Godovari	Polavaram	299	1902–85	98.3	27.5	210	0.31
Krishna	Vidjavada	251	1901–87	51.6	5.4	96.2	0.31
Krasnaya	Sontyay	136	1902–70	120	73.8	154	0.16
Menam	Chainat	118	1905–87	26.7	12.9	57.9	0.27
Rioni	Sakochikidze	13.3	1920–85	12.7	9.0	17.2	0.15

the Huang He ($r = 0.50$) and China ($r = 0.52$). The water resources of South East Asia in addition to the correlations indicated above also correlate quite closely with the runoff of the Yangtze ($r = 0.63$) and China ($r = 0.69$). The water resources of the Southern Asia region are significantly correlated with the runoff of the Ganges ($r = 0.67$), the Godavari ($r = 0.72$), the Indus ($r = 0.53$), and with Pakistan ($r = 0.53$) and they are closely correlated with the water resources of India ($r = 0.97$). The water resources of Middle Asia and Kazakhstan and Western Asia do not have significant correlations with those of other regions. The water resources of Transcaucasia correlate only with the water resources of the Rioni River. The water resources of China have a significant correlation with the runoff of the Yangtze ($r = 0.84$), while the water resources of India are related to the runoff of the Godavari ($r = 0.69$). The water resources of Pakistan are completely correlated with the flow of the Indus ($r = 1.0$).

5.5.3 The water resources of major rivers and the inflow to the ocean

Studies of the water resources of separate river basins are important for an understanding of the general features of the hydrological regime of the continent and its regions. In Asia there are 15 rivers that discharge more than 100 km^3/year. These contribute almost 50% of the total runoff volume and 15% of the Earth's runoff. The largest Asian rivers are the Ganges, Yangtze, Yenisey,

Lena, Mekong and Ob. Data on these rivers and others are listed in Table 5.15, while Table 5.16 presents data on the water resources of the largest river systems of Asia.

By volume the Ganges–Brahmaputra–Meghna is the largest river system of Asia and second in the world after the Amazon. These systems resources are governed by the climatic and orographic conditions of the basin. In the west the annual precipitation is 350 mm, increasing to 1000 to 1200 mm in the east. The world's record precipitation is recorded in the Meghna River (10 800 mm). In the Brahmaputra Basin the precipitation is also very large although its distribution is quite uniform: from 2500 mm in the north to 2000 mm in the south. The flow of the Ganges proper averages 550 km^3/year, the Brahmaputra 645 km^3/year and the Meghna 195 km^3/year. The potential water availability per km^2 for the entire system is the largest in Asia. However there is less water per head than on average over the continent, due to the very large population of the basin (Table 5.16). Of the water originating in the basin, about 1250 km^3/year is discharged to the Bay of Bengal. About 140 km^3 of water per year does not reach the mouth. The losses are due to irrigation and seepage of river water during floods into the sand–pebble deposits (up to 500–600 m thick) which are widespread in the valleys of the Ganges and Brahmaputra and their tributaries.

The Mekong Basin is a region with a pronounced monsoon climate. The mean annual precipitation ranges from 1000 to 1300 mm in the upper part of the basin and 1000 to 2500 mm

Table 5.16. *Renewable water resources of principal rivers of Asia*

River	Area, km² × 10⁶	Population (1994), ×10⁶	Water resources, km³/yr			Coefficient of variation, C_v	Potential water availability, m³ × 10³/year	
			Average	Minimum	Maximum		Per km²	Per head
Ob	2.99	22.5	404	270	586	0.16	135	18.0
Yenisey	2.58	4.8	642	466	749	0.08	249	134.0
Lena	2.49	1.9	539	424	670	0.11	216	284.0
Amur	1.86	4.5	355	225	538	0.21	191	78.9
Yangtze	1.81	346	1003	610	1410	0.15	554	2.9
Ganges (with Brahmaputra and Meghna)	1.75	439	1389	1220	1690	0.04	794	3.2
Indus	0.96	150	220	126	359	0.19	229	1.5
Mekong	0.79	100	505	376	610	0.16	639	5.0
Huang He	0.75	82.0	66	22.1	97	0.38	88	0.8
Aldan	0.73		168	117	212	0.16	230	
Kolyma	0.65		128	74	203	0.23	197	
Khatanga	0.36		90	51	172	0.25	251	
Indigirka	0.36		55	39	89	0.21	152	
Amu-Darya	0.31[a]	15.5	69	57	118	0.14	224	4.5
Yana	0.24		29	19	74	0.20	122	
Syr-Darya	0.22[b]	13.4	37	26	75	0.21	170	2.8

[a] Area at Kerki.
[b] Area at Tyumen-Aryk.

Table 5.17. *Water resources of the Mekong Basin*

River	Station	Area of catchment, km² × 10⁶	Water resources, km³/year			Coefficient of variation, C_v
			Average	Minimum	Maximum	
Mekong	Chiang Saen	189	86.5	65.9	126.0	0.15
Mekong	Vientiane	299	143.0	84.0	194.0	0.15
Mekong	Thakhek	373	235.0	158.0	289.0	0.14
Mekong	Mukdahan	391	260.0	175.0	357.0	0.14
Mekong	Pakse	545	318.0	2.6	382.0	0.12
Mekong	Kratieh	646	459.0	357.0	564.0	0.12
Mun	Ubon	104	19.7	7.9	37.2	0.31
Chi	Yasthon	431	8.0	3.7	14.6	0.31

in the centre, while in northeast Thailand it reaches 4000 mm in the mountains between Laos and Vietnam. The annual precipitation in the Mekong Delta varies from 750 to 3200 mm. During the wet season – the 6 months from mid May to mid November or sometimes mid December – some 80–90% of the discharge occurs. Table 5.17 shows that the coefficients of variation are small ($C_v = 0.12$–0.15) for the Mekong, and a little larger ($C_v =$

0.30) for its tributaries. Specific runoff is 460 000 m³/km² per year in the upstream part of the basin and 700 000 m³/km² per year downstream.

The average flow of the Yangtze at 1003 km³/year places it second to the Ganges (Table 5.18). The climate of the Yangtze Basin is subtropical monsoon. The mean annual precipitation increases from west to east, varying from 300–600 mm in the upper part of

Table 5.18. *Water resources of the Yangtze Basin*

River	Station	Area of catchment, km² × 10³	Water resources, km³/year			Coefficient of variation, C_v
			Average	Minimum	Maximum	
Jingtagjiang	Chinshan	458	147.0	107.0	195.0	0.12
Yangtze	Chongqing	875	369.0	254.0	470.0	0.11
Yangtze	Yichang	1005	448.0	315.0	587.0	0.10
Yangtze	Hankou	1488	735.0	451.0	1009.0	0.14
Yangtze	Datong	1705	898.0	518.0	1356.0	0.15
Yangtze	mouth	1808	1003.0	610.0	1410.0	0.15
Minjiang	Gaochang	135	98.7	73.8	133.0	0.11
Jialingjiang	Beipei	156	68.5	49.9	99.8	0.21
Hanjiang	Hanshui	142	51.1	19.2	107.0	0.37
Ganjiang	Waizhu	81	68.2	26.5	109.0	0.31
Xiangjiang	Xiangtang	82	68.4	32.0	100.0	0.31

Table 5.19. *Water resources of the Ob Basin*

River	Station	Area of catchment, km² × 10³	Water resources, km³/year			Coefficient of variation, C_v
			Average	Minimum	Maximum	
Ob	Barnaul	169	46.6	32.2	72.5	0.20
Ob	Novosibirsk	252	57.0	38.2	80.7	0.20
Ob	Kolpashevo	486	125	77.3	178	0.18
Ob	Belogorjye	2690	325	236	454	0.17
Ob	Salekhard	2950	404	270	586	0.16
Tom	Tomsk	62	33.4	20.3	47.9	0.19
Irtysh	Tobolsk	969	66.7	38.8	123	0.27

the basin, to 800–1200 mm in the middle and to more than 1200–1600 mm in the lower part. Mean precipitation over the whole basin is 1100 mm and the mean runoff 526 mm. Some 20–30% of the runoff occurs between November and May, and 70–80% between June and October. The coefficient of variation for the annual discharge of the Yangtze is 0.10 to 0.15, but it is higher in the tributaries and ranges from 0.21 to 0.37 (Table 5.18). The specific runoff in the upper part of the basin is 300 000 m³/km² per year, in the middle 400–500 000 m³/km² per year and in the lower part more than 500 000 m³/km² per year.

The mean runoff of the Ob Basin is only 130 mm per year while the precipitation is 543 mm (Table 5.19). Consequently the runoff coefficient of the Ob is 0.24, which is about half that of the Yenisey (0.42), the Lena (0.46) and the Kolyma (0.50). The Ob basin is the largest in Asia. The coefficient of variation (C_v) of the annual discharge is 0.16–0.20, while the value for the Irtysh Basin is 0.27. The water resources of the Ob River are presented in Table 5.19.

The runoff from Asia flows to the Arctic, Indian and Atlantic Oceans (Fig. 5.2, Table 5.2) but a significant part of its area, some 12.3 million km² or 28%, is in internal drainage. The most extensive area (regions 5–9 in Fig. 5.2) includes Central Asia (4.32 million km²), Kazakhstan and Middle Asia with the basins of the Amu-Darya, Syr-Darya, Ili, Gilmend and Murgab Rivers (3.41 million km²), together with the areas in the Iran Uplands (0.88 million km²), the basin of the Sistan Depression with the adjoining region (0.55 million km²) and the coastal Caspian area (0.34 million km²) in Iran. An area of significant inland drainage (2.7 million km²) is situated within the Arabian Peninsula and the Thar Desert. Table 5.20 illustrates just how heterogeneous is the distribution of Asia's water resources. The area draining to the Pacific with the adjacent islands has the largest water resource while the areas with inland drainage are the driest parts of the continent.

Some 54% of the continent drains to the Pacific and Indian Oceans, with 78% of the water resources. The total runoff to the Arctic Ocean is 18% and to the Atlantic Ocean only 1.5%.

Table 5.20. *Renewable water resources and water availability of oceanic slopes and endorheic regions of Asia*

Region	Area, km² × 10⁶	Water resources, km³/year	Potential water availability, m³ × 10³/year per km²
Arctic Ocean slope	11.7	2 418	207.0
Pacific Ocean slope	11.9	6 565	552.0
Indian Ocean slope	9.7	3 918	404.0
Atlantic Ocean slope	0.7	201	279.0
Central region of endorheic runoff	9.5	406	42.7
Asian continent with islands	43.5	13 509	310.0

Table 5.21. *Distribution of ocean basin inflow (km³/year) by latitude in Asia*

Latitude	Pacific Ocean	Indian Ocean	Arctic Ocean	Atlantic Ocean	World Ocean
70–80° N			1647		1 647
60–70° N	326		771		1 097
50–60° N	691				691
40–50° N	248			200	448
30–40° N	1 681			1	1 682
20–30° N	749	1 574			2 323
10–20° N	410	1 166			1 576
0–10° N	1 544	440			1 984
0–10° S	866	426			1 292
10–20° S		4			4
Total	6 515	3 610			12 744

The water resources of the central part of Asia where the rivers do not reach the ocean comprise 3% of the runoff for all of Asia. The difference between the volume of runoff draining from the land surface and the inflow to the oceans is quite large. This difference is especially pronounced for drainage to the Indian Ocean where losses due to evaporation, seepage and irrigation are estimated to be 308 km³/year. For the rivers draining to the Pacific, the total losses are about are about 50 km³, less than 1% of the total. There are also the losses on tropical islands due to increased evaporation and transpiration, but these losses have not been estimated. There are also differences within the different parts of the continents draining each of the oceans. For example, for the Arctic Ocean the distribution of runoff indicates that most originates in the basins of the Kara (57%) and Laptev (32%) Seas while that for the East Siberian Sea makes up 10% of the total inflow and the Chukot Sea basin only 1%.

The variability of runoff is small ($C_v = 0.07$–0.22). Some 29% of the water resources of Asia originate in that part of the continent draining to the Indian Ocean, but the differences across this area are quite large. In particular, the western part includes the desert and semi-deserts of the Arabian Peninsula and the Middle East where less than 10% of the resources occur. Compare the Ganges–Brahmaputra–Meghna system, which makes up 35% of the total and where the specific runoff is 795 000 m³/km² per year, with Saudi Arabia where it is several hundred times less.

The total inflow to the Arctic Ocean (with the islands) differs by only 2% from the value calculated earlier (Korzun, 1974b) while that for the Pacific Ocean including the islands is 12% less than in Korzun (1974b). The difference in the estimates is due to the fact that the runoff from the islands was overestimated. The availabilty of national estimates published in the 1980s and 1990s provided improved estimates of the water resources of these islands. The water resources of the Chinese basins have also been estimated

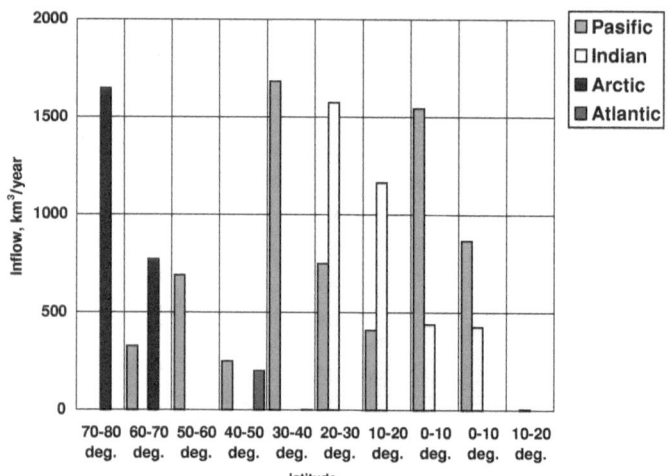

Fig. 5.13 Inflow of ocean basins by latitudinal zones (every 10 degrees) in Asia.

with greater reliability and in more detail. The total water resources of the area of interior drainage are 8% below the previous estimates in Korzun (1974b). The inflow to the oceans from Asia differs significantly from the estimates of the water resources of the different basins. The differences are due to: (1) losses due to evaporation; (2) losses due to abstractions for irrigation and for use by industry.

Table 5.21 and Fig. 5.13 present data on the inflows to the oceans surrounding Asia, for every 10° of latitude. The main areas of inflow are from 0 to 40° N; from 70 to 80° N and from 0 to 10° S. The variability in time of runoff across Asia is quite large and reflects the diverse natural conditions. The coefficients of variation of annual runoff (C_v) range from 0.10–0.30 in the humid regions to 1.0–1.5 in the dry areas. The lowest values (0.10–0.20) are typical

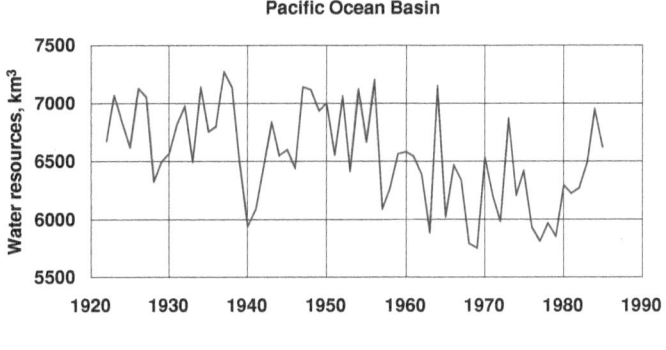

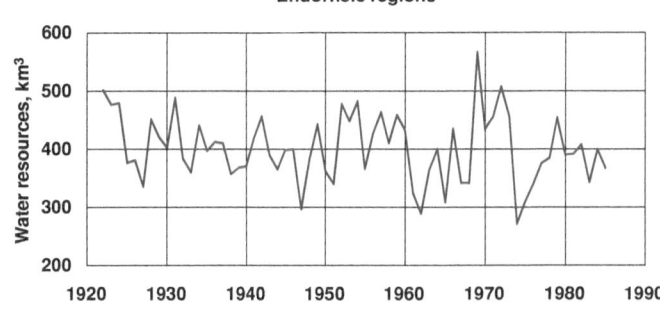

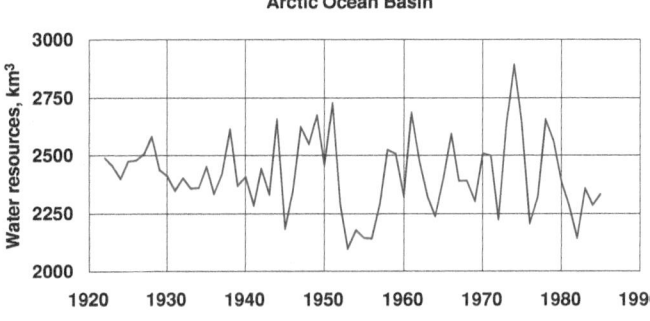

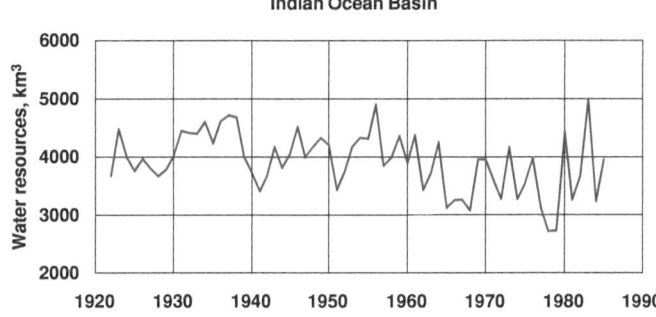

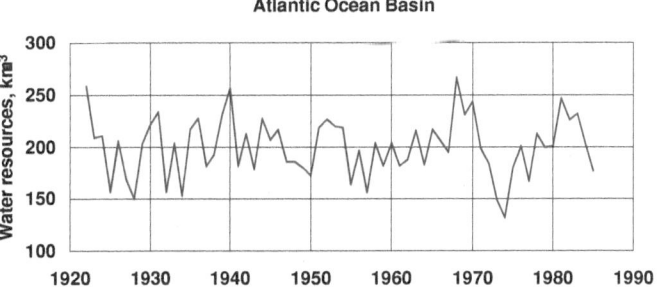

Fig. 5.14 Dynamics of water resources of Asia by ocean slopes.

of the rivers of Northern Siberia and the Far East, while the annual variability of runoff in the Himalayas and other wet mountain regions is also small. As regions get drier the coefficients of variation increase to 0.40–0.50. The largest C_v values are observed for the rivers of Kazakhstan (Ishim 0.82) and the basins of Hai He and Luan He (up to 0.89) in Northern China. Sharp variations in runoff characterize the temporary water courses in the semi-desert regions of Central Asia, the Arabian Peninsula and Iran. As the area of a basin increases it includes more and different tributaries, often with different patterns of runoff. There is a tendency for the coefficients of variation for the complete area draining to an ocean to be small (0.06–0.14), while those for the whole continent are even smaller ($C_v = 0.06$).

The uneven distribution of water resources in time for basins draining to the Pacific and Indian Oceans is revealed by the pattern of mean values and the distribution of individual values. For the rivers Yangtze–Hankou, Mekong–Mukdahan, Ganges–Farakka,

Brahmaputra–Pandu, Indus–Sukkur and Amur–Khabarovsk linear trends were determined in the non-stationary series with a probability of 95%. Figure 5.14 presents the changes in the runoff from Asia to the surrounding oceans. Due to the diverse conditions, the water resources of most Asian river basins have individual regimes. There is a poor correlation or no correlation at all between basins. The only exception is the significant correlation ($r = 0.59$) between the areas draining to the Pacific and to the Indian Oceans. A high correlation ($r = 0.88$) exists between the water resources of the continent and the Pacific and Indian drainage areas. The correlation coefficient between the water resources of Asia and the runoff of the Yangtze is 0.54.

The runoff to the Arctic Ocean has a significant correlation with the runoff in the Ob ($r = 0.59$), the Yenisey ($r = 0.61$) and the Lena ($r = 0.54$). The runoff to the Indian Ocean correlates with the runoff in the Ganges ($r = 0.55$), while the runoff to the Pacific Ocean correlates with the Yangtze runoff ($r = 0.56$). The runoff to the Atlantic Ocean does not have a significant correlation with

Table 5.22. *Average monthly distribution of water resources for selected rivers of Asia*

River	Station	Monthly runoff distribution, % of mean annual value												Wet period		Dry period	
		1	2	3	4	5	6	7	8	9	10	11	12	Months	%	Months	%
Amu-Darya	Kerki	3.3	2.8	3.0	5.6	13.7	19.0	20.0	12.4	7.4	5.3	4.0	3.5	5–7	52.7	1–3	9.1
Amur	Khabarovsk	1.3	.6	.6	1.4	14.0	14.7	17.4	19.5	14.2	10.8	4.0	1.5	6–8	51.6	1–3	2.5
Brahmaputra	Pandu	1.9	1.6	1.8	2.9	9.5	10.0	17.8	22.0	15.2	10.7	4.2	2.4	7–9	55.0	1–3	5.3
Ganges	Farakka	1.6	1.2	1.1	0.8	0.9	2.0	17.0	34.0	24.4	10.4	4.0	2.6	7–9	75.4	3–5	2.8
Godavari	Polavaram	1.0	0.7	0.5	0.3	0.4	0.5	27.1	25.1	35.4	6.2	1.9	0.9	7–9	87.6	3–5	1.2
Huang He	Lianjiang	2.9	3.6	5.2	4.1	4.0	4.1	10.5	23.9	14.0	14.3	8.9	4.5	8–10	52.2	1–3	11.7
Indigirka	Vorontsovo	0.2	0.1	0.1	0.0	1.7	30.0	29.6	22.2	12.3	2.7	0.7	0.4	6–8	81.2	2–4	0.2
Indus	Darband	1.4	1.3	2.0	2.7	7.6	18.4	28.9	21.1	10.0	3.6	2.2	1.8	6–8	67.4	1–3	4.7
Kolyma	Srednekolymsk	0.4	0.3	0.2	0.2	7.2	39.2	19.7	15.0	12.2	3.7	1.2	0.7	6–8	73.9	2–4	0.7
Kura	Tbilisi	6.4	5.8	9.5	17.4	16.0	10.6	7.0	4.7	5.1	5.5	5.2	6.8	3–5	44.0	8–10	15.3
Lena	Kyusyur	1.3	1.0	0.7	0.6	2.6	37.3	20.0	13.8	12.5	7.1	1.7	1.4	6–8	71.1	2–4	2.3
Mekong	Mukdahan	2.9	2.1	1.7	1.7	2.5	7.5	18.0	23.3	23.8	9.3	4.2	3.0	7–9	65.1	2–4	5.5
Ob	Salekhard	3.1	2.5	2.2	2.2	9.8	21.6	19.5	15.2	9.5	6.9	4.0	3.5	6–8	56.3	2–4	6.9
Red	Yenbai	2.9	2.6	2.3	2.0	4.2	9.1	22.0	21.3	12.3	11.5	6.1	3.7	7–9	55.6	2–4	6.9
Xijiang	Uchzhou	1.9	2.5	1.8	3.9	7.1	13.6	33.8	12.4	11.6	6.6	2.8	2.0	6–8	59.8	1–3	6.2
Yana	Dzhangky	0.0	0.0	0.0	0.0	5.6	33.4	27.2	21.0	10.9	1.5	0.3	0.1	6–8	81.6	2–4	0.0
Yangtze	Hankou	3.4	2.2	3.2	5.0	9.2	12.1	17.0	14.8	13.1	9.5	6.4	4.1	7–9	44.9	1–3	8.8
Yenisey	Igarka	3.3	3.5	3.5	3.4	9.5	38.9	11.7	7.4	7.4	5.9	2.9	2.6	5–7	60.1	1–3	10.3

the runoff to the other Oceans. The runoff in one river does not normally correlate with that in others. One exception is the Ob and its tributary the Irtysh ($r = 0.83$). The runoff in the Godavari correlates quite closely with the runoff in the Ganges ($r = 0.73$) and the runoff in the Brahmaputra.

5.5.4 The distribution of runoff within the year

In the southern part of the continent (excluding the mountainous regions) where the rivers drain areas of high rainfall, the distribution of runoff within the year is governed the precipitation regime. In the northern part of the continent and in the mountain regions in addition to precipitation, temperature influences runoff and in particular the period of snow melt. Ground water plays a small part in determining the water resources of Asia. Exceptions are in the karst regions of the Levant and Turkey, the Shansk Upland and southeastern China where limestones contribute ground water flow to the runoff and decrease the surface runoff in medium and small rivers.

Along the southwestern slopes of the Himalayas ground water runoff comprises 30–40% of the total. In areas of the Philippines with permeable volcanic formations, the ground water component reaches 30%, and these formations regulate the river regimes. Changes in the monthly runoffs for different basins are presented in Table 5.22 and in Fig. 5.15, and for the drainage areas in Table 5.23. and in Fig. 5.16. In most southern Asian rivers 60–80% of the annual runoff takes place during the summer and early autumn months while the monsoon controls the runoff in rivers in the southern, southeastern and eastern regions. In some regions (eastern Indochina) the highest runoff occurs in the autumn and winter months (September to November) due to the northeasterly monsoon. Runoff is low for six to seven months over the winter in the Ganges, Narmada, Mahandi, Krishna and Godavari. Some 60–70% of the annual runoff of the Irrawaddy, Menam, Mekong, Yangtze, Xijiang and Red Rivers occurs during the summer. Floods take place from July to September or from August to October in the Huang He, Liao He and Hai He Rivers and in the basin of the Sungari River due to the monsoon. Here snowmelt contributes 10–15% of the annual runoff. However snowmelt governs the regimes of many rivers in the northern part of the continent and in the mountains.

In the middle of the Ob Basin floods take place between April and June. Farther east and downstream in the Ob, the floods occur from May to July and in the northernmost parts between June and August. In the northeast of the continent and in the mountains in temperate latitudes, the flood season begins in June and lasts to August. Summer runoff is a very large part of the flow in the northern rivers (more than 80% of the annual total). In the mountains

Table 5.23. *Average monthly distribution of water resources by continental slopes and endorheic runoff region of Asia*

Ocean slope	Monthly runoff distribution, % of mean annual value												Wet period		Dry period	
	1	2	3	4	5	6	7	8	9	10	11	12	Months	%	Months	%
Arctic Ocean	2.0	1.7	1.9	1.8	13.0	31.0	18.8	11.4	8.6	5.8	2.0	2.0	6–8	61.2	2–4	5.4
Pacific Ocean	6.4	5.5	5.7	6.5	8.2	9.0	13.3	12.3	10.7	7.8	8.8	5.8	7–9	36.3	1–3	16.6
Indian Ocean	1.8	1.6	1.9	2.4	5.8	7.0	19.0	25.5	19.8	9.3	3.6	2.3	7–9	64.3	1–3	5.3
Atlantic Ocean	11.9	10.6	11.9	11.9	7.2	5.9	5.0	5.1	5.6	7.2	8.7	9.0	1–3	34.4	7–9	15.7
Central region of endorheic runoff	3.9	3.5	4.7	8.6	14.6	18.1	16.0	10.0	6.8	5.2	4.4	4.2	5–7	48.7	12–2	11.6

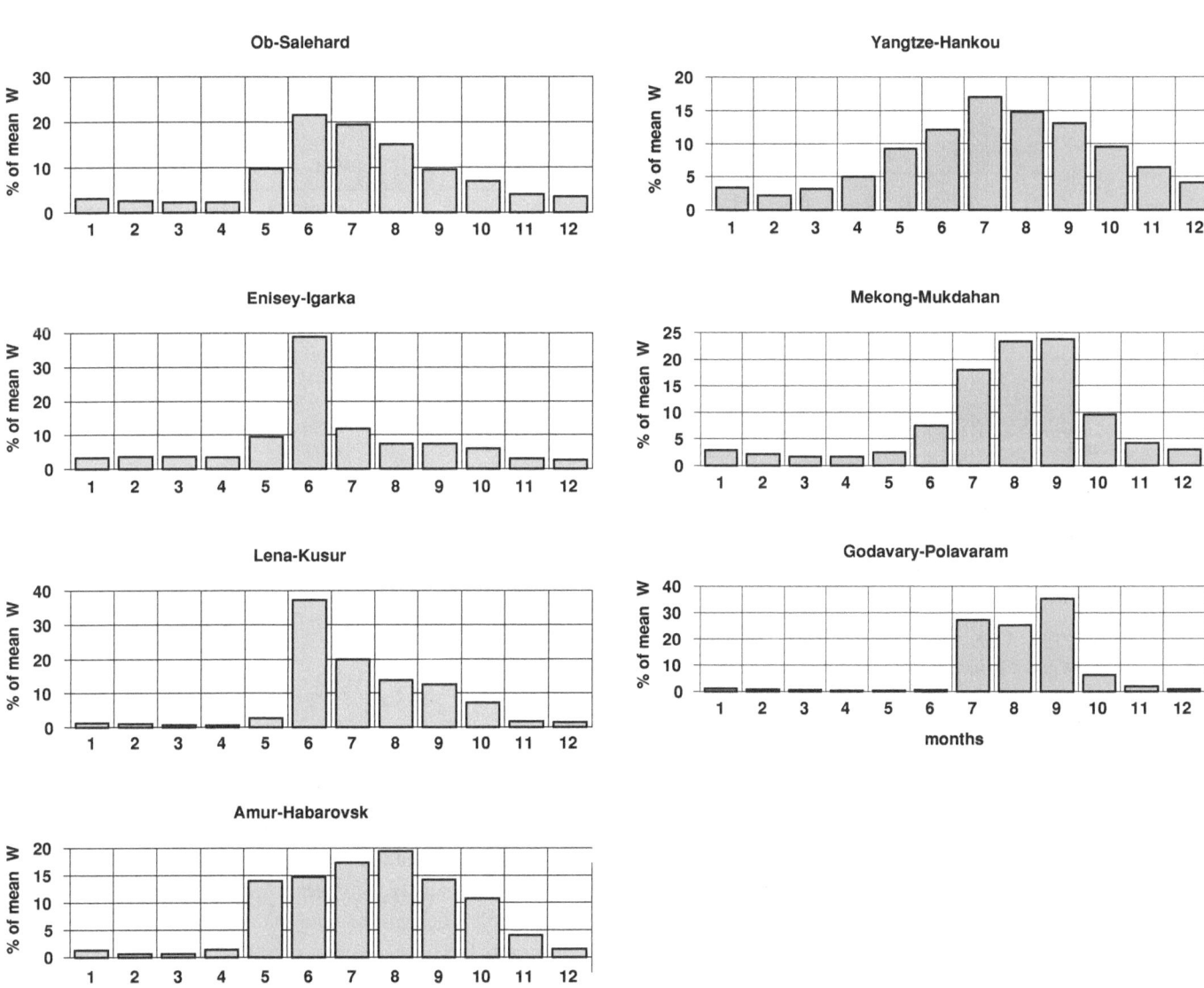

Fig. 5.15 Average monthly runoff distribution of selected Asian rivers.

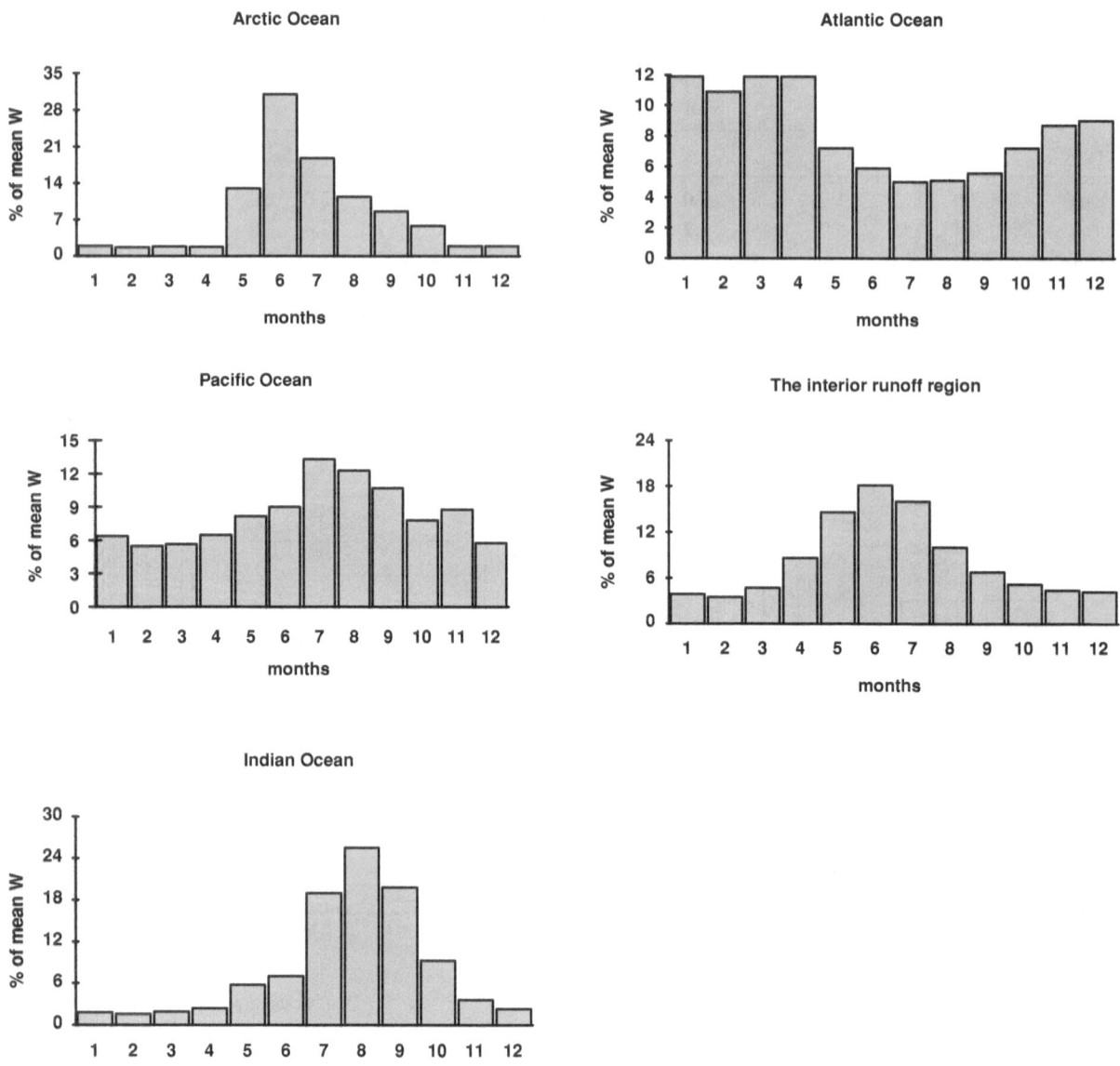

Fig. 5.16 Average monthly water resources distribution by oceanic basins of Asia.

situated more to the south it is from 50% to 60% due to earlier snow melting. The wet period gradually moves westwards and southwestwards from an area with maximum runoff in summer to a spring maximum to spring to summer months and in the extreme southwest to a winter maximum. In the rivers of the plains and low mountains, the winter runoff reaches 80–90% of the annual value due to the increased cyclonic activity at this time. In the mountain regions of the Near and Middle East it decreases to 50%–60% due to winter precipitation falling partly in the form of snow. Some 40–50% of the annual runoff occurs in the rivers of Japan, Taiwan, the Philippines and Sri Lanka in the three wettest months.

The driest season over much of the continent is the winter. As a result of the severe conditions in the far north many rivers

freeze over for five to six months. In Southern Asia low flows take place between February and April before the onset of the summer monsoon. Low flows during the summer and autumn months is typical of the rivers of the southwestern part of the continent. Some 0–10% of the annual runoff occurs in the three driest months in the islands to the southeast of Asia and about 10–20% in the islands to the east. The distribution of runoff during the year means that for most rivers there are a number of months when 80% of the annual runoff occurs. The duration of this period varies across Asia over a range from between one to three months up to eight or nine months. The most irregular distribution is in the rivers of the northern part of the continent, in the southern regions of western Siberia, Kazakhstan, much of the Indian Peninsula, and

Table 5.24. *Average monthly distribution of water resources during the years with average water content by natural–economic regions of Asia*

Ocean slope	Monthly runoff distribution, % of mean annual value												Wet period		Dry period	
	1	2	3	4	5	6	7	8	9	10	11	12	Months	%	Months	%
Siberia and Far East	2.0	1.7	1.9	1.8	13.0	31.0	18.8	11.4	8.6	5.8	2.0	2.0	6–8	61.2	2–4	5.4
Northern China and Mongolia	2.3	2.2	3.0	4.7	6.2	9.2	14.7	20.6	14.3	11.3	7.5	4.0	7–9	49.6	1–3	7.5
South East Asia	8.0	7.0	7.2	7.4	7.7	7.9	11.3	10.4	9.2	6.9	10.2	6.8	7–9	30.9	12–2	21.8
Continental part	3.7	2.8	3.2	4.9	7.7	10.6	18.2	15.1	14.1	9.1	6.2	4.4	7–9	47.4	1–3	9.7
Islands	10.7	9.6	9.7	9.4	7.7	6.2	6.7	7.5	6.1	5.5	12.8	8.3	1–3	29.8	8–10	19.0
South Asia	1.6	1.3	1.4	1.8	5.2	6.9	19.5	26.2	20.8	9.5	3.6	2.2	7–9	66.5	1–3	4.3
Middle Asia and Kazakhstan	3.4	2.9	3.5	6.6	13.1	19.4	18.6	11.7	7.4	5.4	4.2	3.7	6–8	49.7	1–3	9.8
Transcaucasia	6.4	5.8	9.5	17.4	16.0	10.6	7.0	4.7	5.1	5.5	5.2	6.8	4–6	44.0	7–9	16.8
Western Asia	7.5	7.8	13.1	16.4	20.5	10.1	5.3	3.7	3.2	3.6	4.0	5.0	3–5	50.0	8–10	10.5
Asia as a whole	5.1	4.1	4.7	5.1	8.8	13.7	14.9	13.8	11.2	7.2	6.8	4.6	6–8	42.4	12–2	13.8

in the plains and low mountains of the Near and Middle East. Here between 80% of the runoff and almost the entire total in the arid regions is observed during a period between one and three months long. During the rest of the year the runoff is extremely small. In some years there is no runoff due to rivers freezing over or drying up.

A more uniform distribution of runoff is observed in the mountain regions of temperate and southern latitudes where 80% of the runoff occurs within a space of six to seven months. A relatively uniform distribution of runoff is characteristic of the islands of the Pacific Ocean and Malaysia (80% of the annual runoff in eight to nine months). Of course large variations of flow are typical, not only in medium sized rivers, but in large rivers as well. The serious consequences of floods on the large rivers (the Yangtze, Ganges, Brahmaputra, Mekong and Amur) are widely known; thousands of people may die and an immense amount of damage can be caused. In contrast during dry periods there is not enough flow in the Indus and in other rivers to meet the demand for water. Table 5.24 and Fig. 5.17 show the within-year distribution of runoff for the regions of Asia. Some 80% of the total runoff occurs over a period of eight months (from April to November). About 40% of the runoff occurs over the three months (June to August) of the wet period and 14% during the dry period (December to February).

5.6 CHANGES IN WATER USE AND WATER AVAILABILITY

5.6.1 Data and preconditions

The reliable assessment of changes in the use of water and in water supply is difficult because of lack of data. Some countries publish their national statistics on water, but information on water

use is available only for some countries and only for the last 20–25 years. These countries (except for the former Soviet Union) include Japan, Turkey, Iraq (from the early 1970s), Iran, India, Israel, Jordan, Saudi Arabia (from the early 1970s), China (from the early 1980s) (UN, 1992; Central Water Commission, 1994; Margat, 1994). Hence estimates of the use of water and the changes from year to year had to be made from numerous alternative sources and by indirect methods.

Asian trends in water use were presented in the mid-1980s studies by SHI (Shiklomanov and Markova, 1987; Shiklomanov, 1988). They contain figures for abstraction and consumption for the continent and for seven regions for the period from 1900 to the 1980s, with forecasts up to 2000. These estimates used data from a number of countries on the different aspects of water management and water use (Van der Leeden, 1975; Yermolina and Kalinin, 1975; Nikolayeva and Chernogayeva, 1979; Yermolina and Klige, 1979; Zonn and Nosenko, 1981; Makarov and Marchenko, 1982; Zevelev, 1985) as well as estimates for irrigation and for domestic purposes: China (Ross, 1983), India (Kayastha, 1980; Levitanus, 1986a, b), Pakistan (Morozova, 1983), Bangladesh (Sarker and Sarker, 1979), Iraq (Badry *et al.*, 1980), Iran (Shtepa, 1973; Shakhbazyan, 1979), Syria (Makarov, 1976), Yemen (Kadzayev, 1984), Korea (Makarov and Marchenko, 1982) and Vietnam (Hoang Tien, 1978). The areal extent of irrigation for the period 1960–81 was taken from FAO data (FAO, 1995b), while values for earlier years were estimated from other sources and forecasts were made using data from national publications. The specific water use for irrigation was assessed for individual countries and averaged by region (except for the countries of the former Soviet Union). The value varies from $10\,000$–$12\,000\ m^3$/ha to 9000–$11\,000\ m^3$/ha, assuming about 20% of the water intake is reused.

Since Shiklomanov and Markova (1987) and Shiklomanov (1988) appeared a large number of publications have been

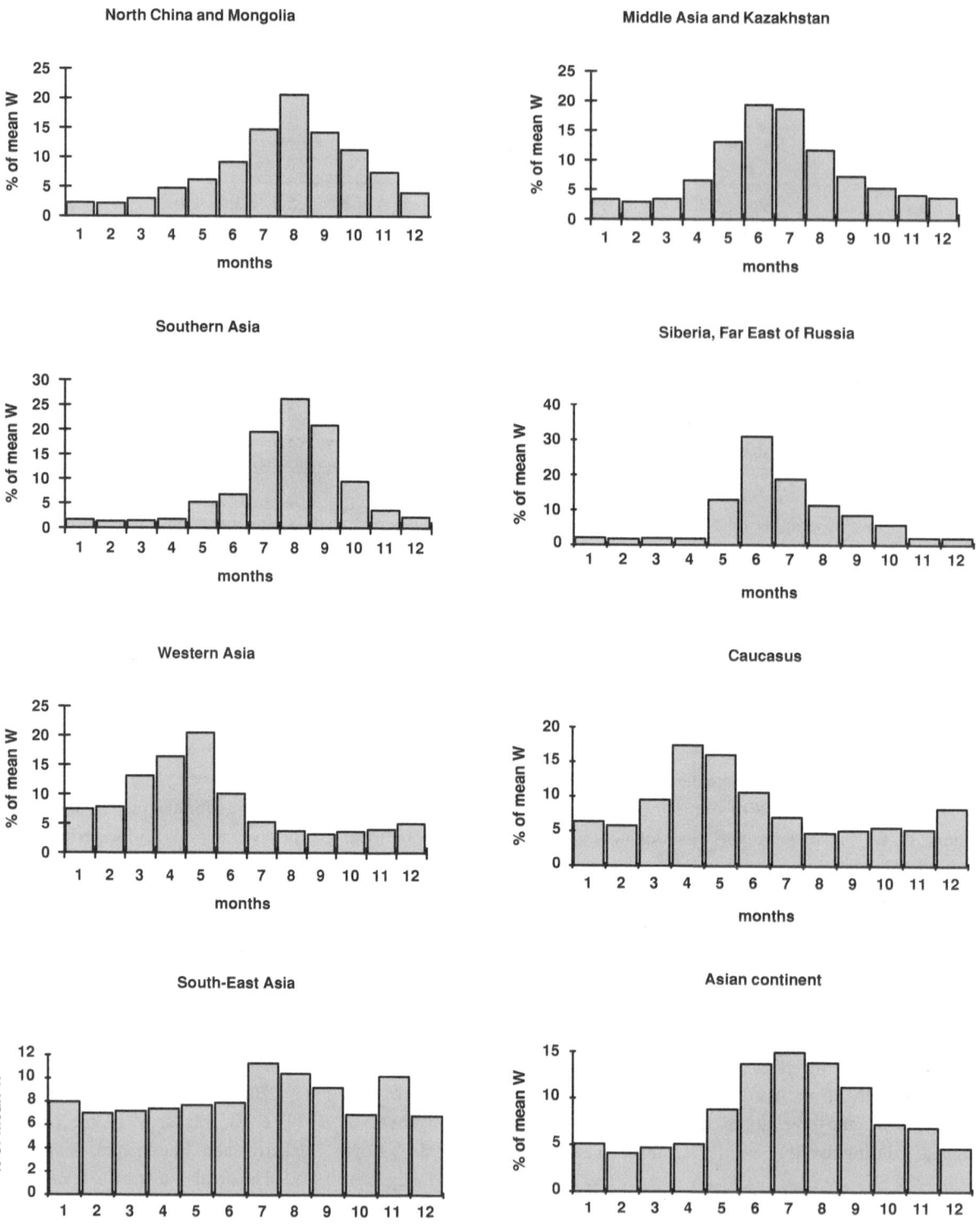

Fig. 5.17 Average monthly water resources by regions of Asia.

concerned with problems of water in Asia. They include studies and handbooks that were undertaken by international organizations (World Bank, 1986, 1993, 1994; WRI 1990, 1992, 1996, 2000; Kulshreshtha, 1992; Andressian, 1993; UN, 1993; Margat, 1994; Strzepek and Bowling, 1995; Seckler *et al.*, 1998; Rijsberman, 2000; FAO, 2000). They also include studies and reports by groups of countries belonging to different regional bodies, some under the aegis of the UN (ESCAP, 1987, 1989a, b, 1998; ESCAP/UN, 1989; ROSTAS, 1994, 1995; UN/DPCSD, 1995b; UNESCO, 1995), on the countries of the Mediterranean (Margat, 1990, 1992; UN, 1992; UN/DPCSD, 1995b; Pearce, 1996; Shahin, 1996; Margat and Vallée, 2000; Ragab, 2001). There are also the detailed national and international studies on trends in water use for India (Central Water Commission, 1994), China (FAO, 1993), Thailand (ESCAP, 1991) and some other countries, as well as a large number of papers that cover different aspects of the economy and water management problems (Gleick, 2000).

During preparation of this Monograph a comparison was made of the data on water use contained in Shiklomanov and Markova (1987), Shiklomanov (1988) and in later publications. This comparison confirmed the SHI estimates of water use in for the different sectors, for the regions and for the whole continent up to 1980. Fresh studies were devoted to confirming the forecasts of water use and water availability values made in the earlier study by Shiklomanov for the years 1990–5 and 2000 and forecasting the values for 2010 and 2025. These estimates were made using the methodology presented in Section 3.3 (Shiklomanov, 1997, 1998a, 2000a,b; Penkova and Shiklomanov, 1998; Shiklomanov, *et al.*, 2000). Trends in recent years differ from those which were taken into account in the previous SHI forecasts. In particular for the regions of Southern and Western Asia the predicted volumes for 2000 turned out to be overestimated by 24% and 5.6%. The previous forecasts for 2000 were also overestimated: for Middle Asia and Kazakhstan, Siberia and the Far East, and Transcaucasia by 15%, 63% and 43%, respectively. The causes of the discrepancies obviously differ for the different regions. In Western Asia and Southern Asia they are mainly connected with the economic problems of the last 15 years and with the depletion of the available water resources. In the regions of the former Soviet Union it was predominantly due to perestroyka and the reorganization of water management.

For the continent as a whole, the main factor in determining the use of water remains the development of irrigation. The largest land reserve suitable for rain fed farming is in Indonesia, Malaysia, Myanmar, Laos and Cambodia. However it is predicted that this resource will be depleted by 2000 and further expansion of agriculture will only be possible with irrigation. Large areas of suitable land are available in the most arid regions of some countries of monsoon Asia, but they will require much greater water resources than are available. However they may require water from some of the large international rivers, and this will need international agreements. Nevertheless, the high rate of population growth, which is predicted, suggests that a significant increase in the area of irrigation will be required. This is indicated by numerous water projects that are being developed in many countries. For example, in India the irrigated area according to data from the Central Water Commission (1994) is estimated to be 115.6 million ha. The greatest possibilities for the expansion of irrigation exist in the Ganges Basin which has 40% of the total irrigation potential of the country (Ryabchikov, 1976; Chaturvedi, 1991, 2000). The Ministry of Irrigation has developed a project for creating a complex of reservoirs and canals in the lower reaches of the Ganges and Brahmaputra, connecting the basins of these rivers and facilitating the transfer of water to the western regions of India during the monsoon. It was planned to start this project in the 1990s with the participation of Nepal and Bangladesh. The project when implemented would redistribute sufficient water for any future requirements in northern and northwestern India including those in the Thar Desert (Levitanus, 1986a, b; Chaturvedi, 1993, 2000).

It is planned to implement a large number of other projects for developing the resources of large river basins in Southern Asia and South East Asia. One ambitious project is on the Lower Mekong whose potential is sufficient to provide a reliable source of water and to raise the living standards of the population of the basin and of the countries adjoining it (Chomchai, 1993; Anonymous, 1995b). The World Bank has been funding a large number of programmes concerned with water management, for example, in India the exploration of mountain watersheds and the development of small hydro power sources and the use of the water resources in Orissa State; in Pakistan, the control of water logging and land salinization in the Punjab; and in Indonesia, the exploration of the watersheds of western Java. The funding also includes a number of large water management programmes in China such as the construction of a dam on the Yalung River, the development of irrigation in the basin of the Sulo River, the development of the water resources in Yunnan Province in the Yangtze Basin (World Bank, 1996).

The water management projects developed in the Yangtze Basin are of special importance from the perspective of assessment of the changes in water use in China. The economic significance of this basin for the country as a whole is very noticeable. About 30% of China's population lives in the basin. It has almost 30% of all irrigation and more than 70% of rice crops – the staple food of the country (FAO, 1993; Kurbatov, 1993). The Yangtze River and its tributaries provide the water for irrigation and there are some transfers to adjoining basins such as the Hwan He and Huai He Rivers. About 40% of hydropower resources of the country come from the Yangtze. In 1992 the Three Gorges Project on the upper Yangtze River was approved and included in a 10-year development plan from 1991 to 2000 (Jiasheng and Jiasheng, 1994). The main purpose of the project is flood regulation and electric power generation. A very large hydropower station (Sansi) is being constructed, fed by a reservoir with a capacity of not less than 323 km^3. The annual power production is expected to be 110 109 kWh using a 25 000–30 000 MW power plant. This will be one of the largest power stations of its type anywhere in the world, providing electricity to a wide area and contributing to the economic development a vast region from Beijing in the north to Kuangchou in the south and as far as Shanghai to the east. In the Yangtze Basin the production of metals and chemicals consume the most water. Chinese industry plans further development of the production of raw materials and energy and this should contribute to increased water use in the industrial sector (Ganshin and Remyga, 1993).

Across South East Asia the increased demand and increased prices for oil and gas in the world market can stimulate the demand for water. For the countries that import oil, the real way out of the energy dilemma would be the expansion of alternative energy sources, primarily hydropower. In this situation more reservoirs would be needed, especially in countries where there has been little development of water and hydropower. In the countries where there has been this development, even more dams would be required. In particular, the rise in the construction of hydropower projects is evident in Pakistan (Jaensch, 1995), for example, the large project which is under way in the Indus below the Tarbella Dam.

There is an obvious relationship between increased reservoir construction and the expansion of irrigation. For example in Turkey in 1990 the Atatyurk Dam was completed and filling of the reservoir commenced, the dam of which is the largest in Western Asia. Water from this reservoir will irrigate about 900 000 ha of arable land (Urazov, 1991). During the coming decade about 60 reservoirs are planned for construction in countries in the Arabian Peninsula (ROSTAS, 1995). Dams are being constructed in Iraq, Afghanistan and Syria (Gleick, 1993). In many Asian countries there will be an expansion of irrigation and increased agricultural water use; the exceptions are the highly developed urbanized states and those with a shortage of land (Japan, Hong Kong).

In this Monograph the forecasts of the likely water use for irrigation have been based on the analysis of water use data for the agricultural sector and analysis of national trends in the areas irrigated, both in absolute (thousands of ha) and specific values (ha/1000 inhabitants). Such factors as changes in GNP, the funds available for investment, the available water resources and the national water strategy were taken into account (see Section 3.3). The different scenarios for the future development of the agricultural sector produced by different national and international organizations and perspectives on the implementation of large projects and programmes were widely used. Mean weighted irrigation averages

were determined as the ratio of the actual abstraction for irrigation to the total area of land, prepared in the country concerned. The values obtained were checked against national estimates and figures for the distribution of irrigation averages from numerous sources. The specific volume of water used for irrigation in Asia varies from country to country over a very large range: from 2000–2500 m^3/ha to 21 000–23 000 m^3/ha. These values are typical of Malaysia, the Philippines, Sri Lanka, Pakistan and India where the largest areas are occupied by rice or sugar-cane and year-round irrigation is practised. In lower-income countries where irrigation is less developed, such as Laos, Indonesia, Burma, Cambodia and Bangladesh, the irrigation averages decrease to 5000–8000 m^3/ha. The lowest specific use is in Nepal where it is about 3000 m^3/ha.

In the republics of Middle Asia the irrigation averages range from 9000–10 000 m^3/ha in Kyrgyzstan to 20 000–22 000 m^3/ha in Turkmenistan and Tajikistan. High irrigation values (10 000–16 000 m^3/ha) occur in Iraq, Iran and Afghanistan, whereas in Israel, Jordan, Turkey and Syria they are within 7500–8000 m^3/ha. In Saudi Arabia the figure is 5500 m^3/ha, and 12 000–13 000 m^3/ha in Oman and in South Yemen. In Northern China, Mongolia, South Korea and North Korea the figures range from 7500 to 8500 m^3/ha.

The consumption of water is highest in arid climates, and this is where irrigation technology is usually the most advanced. The techniques include sprinkling over large areas, drip irrigation and recycling with a consequent increase in efficiency of up to 90% and more. Because of increasing shortages of water and rising pollution many countries have the problem of how to save water. This Monograph takes into account a decrease in the specific water use values (by 5–10%) with a simultaneous increase in the percentage of water consumed. The forecasts of water use for community and domestic needs were made by analysing trends in specific water use per head and in the likely growth of population (UN, 1992; UN/DPCSD, 1995b). Some studies also suggest reconsidering the concepts of irrigation in large rivers basins, such as the Ganges, where greater attention could be given to using ground water for drinking water (Chaturvedi, 1991, 2000). In most countries the use of water in this sector makes up to 15–25% of the total, a percentage that is likely to decrease in future.

Data on the volumes of industrial water used up to 1980 were taken from Shiklomanov and Markova (1987) and Shiklomanov (1988), while the levels in 1990 and in 1995 were obtained from relevant national and international publications. The forecasts for 2000, 2010 and 2025 were based on the studies by UNIDO (Strzepek and Bowling, 1995). Most scenarios predict an unprecedented increase in industrial water use in Asia. Even with the slowdown in growth of the use of water in the east and southeast, by 2025 the demand for water is expected to increase as compared to 1990. These increases are thought to be by a factor of 10.1 in Korea, 9.8 in Singapore, 6.5 in Indonesia, 5.7 in Thailand, more

Table 5.25. *Dynamics of fresh water use in Asia by type of economic activity (km³/year)[a]*

Water use	Assessment								Forecast		
	1900	1940	1950	1960	1970	1980	1990	1995	2000	2010	2025
Population, $\times 10^6$			1464	1763	2103	2646	3229	3498	3762	4291	4906
Irrigation area, ha $\times 10^6$	36.1	58.6	72.5	102	118	131	169	175	182	199	231
Agriculture	408	665	816	1144	1331	1526	1688	1743	1794	1925	2245
	320	521	643	907	1066	1247	1411	1434	1457	1553	1762
Industry	4	18	33	51	107	153	176	184	193	248	409
	1	4	6	9	13	19	29	30	32	40	58
Domestic	2	6	11	20	38	65	143	160	177	218	343
	1	3	5	9	14	18	29	31	33	36	44
Reservoirs	0	0.1	0.23	7	23	40	60	70	81	92	107
Total	414	689	860	1222	1499	1784	2067	2157	2245	2483	3104
	322	528	654	932	1116	1324	1529	1565	1603	1721	1971

[a] Nominator, total water withdrawal; denominator, water consumption.

than 5 in Bangladesh and Malaysia, and 4.3 in Nepal. Somewhat slower growth rates (3.3 to 3.7) are likely to occur in China, India, Pakistan, Thailand and Sri Lanka; and the smallest growth (2.3 to 2.9) will take place in the Near and Middle East, and Japan.

According to UNIDO, the geographical distribution of the growth of abstractions for industry reflects the development of industry itself, including the mining industry (Muranova, 1994). However, in general, these values of growth appear to be over-estimates. Taking into account the observed trends during the 1990s, country values for 2025 were reduced by 15–35%. For 2000 and 2010 the water used by industry was assumed to be proportional to the forecasts of GNP growth, which were also based on the scenarios in the UNIDO Global Balance. Based on these assumptions, for the coming period the GNP growth rates in the Asian "tiger economies" namely Hong Kong, Indonesia, Malaysia, Philippines, Singapore, South Korea and Thailand are expected to be 4.6% a year, 3.2% in the Middle East, 3.1% in Japan, 2.7% in Russia, and 6.1% in the other countries of Asia (IMF, 1994; Strzepek and Bowling, 1995). Water consumption by industry in the relatively developed countries was assumed to be more or less constant and to be about 5–10% of the water abstracted with some possible reductions with the introduction of recycling. For less developed countries these figures are likely to be somewhat larger (15–20%) with no savings likely of the limited water resources.

5.6.2 Changes and trends in the use of water

Assessments of the volumes of water used at the present time and in the future were made for each country and region. The results of these assessments by sector are given in Table 5.25

and in Fig. 5.18. Table 5.26 and Fig 5.19 show changes in the total volume of water abstracted and consumed. These tables and figures indicate that the agricultural sector, primarily irrigation in semi-arid and arid countries, is and will remain the main water user. The increase in abstraction for agriculture in 2000 and 2025, as compared to 1995, is forecast to be 3.1% and 29%, respectively, while for consumption it is 2.8% and 23%. In absolute terms these increases will reach extremely high levels comparable with agricultural water use in other continents despite slower rates of growth.

Growth rates for total water use for 2000 and 2010 are expected to slacken, but by 2025, provided that forecasts of the population growth are correct, water use should increase. The stress on water resources, improved water use in agriculture and the requirements in the domestic and industrial sectors should result in a changed pattern of water use. By 2025 agricultural water use could decrease by about 9–10% (it should be noted that in 1900 agriculture used 98% of the total, whereas in 1995 it was 81% and in 2010 and 2025 it could be 77% and 72%, respectively, of the total water use). Increases in the percentage water used by the industrial and domestic sectors occurred at different times. The biggest increase in the abstractions for industry took place in the 1960s (from 4% to 7%) but subsequently it has changed little. The largest increase in the domestic sector was in the 1990s (from 4% to 7%) and it may also increase in the future as the population increases and living standards rise. However water consumption in these sectors is relatively small and by 2025 it may be a little more than 5% of the total. Evaporation from reservoirs is also relatively small.

The geographical distribution of the volumes abstracted and consumed reflects the distribution of population and irrigation.

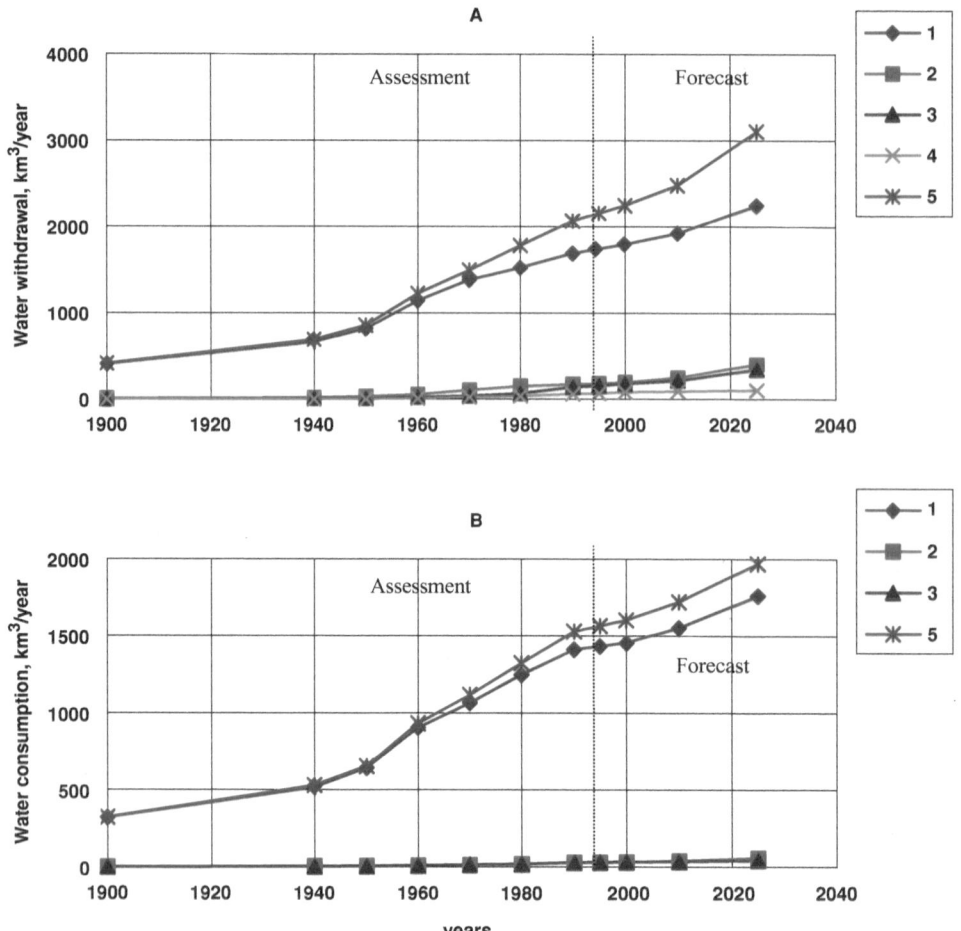

Fig. 5.18 Dynamics of (A) water withdrawal, (B) water consumption in Asia over the kinds of economic activities. 1, agricultural; 2, domestic; 3, industrial; 4, reservoirs; 5, total.

Of the two most populated regions of South Asia and South East Asia (1.2 and 1.4 billion inhabitants), the total volume abstracted in the former is almost twice as large as in the latter, whereas the increase in irrigation is now almost 20 million ha less. In total, these two regions account for 68% of the water abstracted and more than 69% of water consumed in Asia. By 2025 both percentages could be 69%. The percentage for the Southern Asia region should increase much more, due to the increase in irrigation in countries with a low income (e.g. Nepal, Bangladesh). The lowest amount of water is used in Middle Asia and Kazakhstan, Siberia and in the Far East, and in Transcaucasia. These three regions account for 9.6% of the abstractions and 8.5% of the water consumed. By 2025 these figures are expected to decrease to 7.5% and 8.2% respectively. From 1990 in all three regions water use declined and this decline was thought likely to persist at least into the twenty-first century because of the continued structural changes in the economy.

The transformation of economic mechanisms and the restructuring of economics are reflected in the changes in water use in other regions of Asia. There was a 7 km³/year decrease in abstractions in Northern China and Mongolia from 1980 to 1990 as a result of the Chinese agrarian reforms. The same reforms contributed to the decrease in the growth of the use of water in Southeast Asia. However, this decrease is also due to diversification of production, the industrial growth in most southern and eastern states and the decrease in the growth of irrigation, such as in Japan and Korea. On the whole, for the period from 1950 the largest relative growth in water use was in Siberia and the Far East where it increased by 5.5 times. However in absolute terms this growth was very small, being only 25 km³/year. In Transcaucasia water use has increased by 2.1 times and in the Southeast Asia by 2.3 times. In the other regions the growth was close to the mean for the continent (2.5 times) or a little more, varying from 2.54 to 2.69 times. By 2025 it is expected that in most regions the increase in the use of water will be close to the Asian mean (an increase of 1.4 times as compared to the current level). The only exception will be the regions of the former Soviet

Table 5.26. *Dynamics of fresh water use by natural–economic regions of Asia (km³/year)*[a]

Region	Assessment								Forecast		
	1900	1940	1950	1960	1970	1980	1990	1995	2000	2010	2025
Northern China and Mongolia	37 / 30	67 / 53	98 / 75	165 / 129	217 / 163	241 / 172	234 / 179	254 / 182	273 / 185	305 / 194	373 / 210
South Asia	201 / 160	312 / 249	367 / 293	429 / 341	524 / 412	668 / 518	895 / 664	932 / 687	969 / 710	1060 / 767	1370 / 944
Western Asia	43 / 34	69 / 54	91 / 71	136 / 107	158 / 123	192 / 147	227 / 167	238 / 174	248 / 181	283 / 201	346 / 229
South East Asia	99 / 77	170 / 129	230 / 170	399 / 299	469 / 337	484 / 365	499 / 383	525 / 388	551 / 393	617 / 413	781 / 425
Middle Asia and Kazakhstan	29 / 19	55 / 35	57 / 37	67 / 43	94 / 61	151 / 97	156 / 103	154 / 102	151 / 102	160 / 110	169 / 122
Siberia and Far East	0.7 / 0.4	4.9 / 1.1	5.6 / 1.3	10.4 / 2.8	16.3 / 7.9	25.4 / 10.9	31.3 / 15	30.6 / 15	30 / 15	32 / 17	38 / 21
Transcaucasia	4.2 / 2.1	11.3 / 7	11.4 / 7.1	15.8 / 10	20.7 / 11.9	23 / 14.1	24.4 / 18.1	23.7 / 17.5	23 / 17	26 / 19	27 / 20
Total	414 / 322	689 / 528	860 / 654	1222 / 932	1499 / 1116	1784 / 1324	2067 / 1529	2157 / 1565	2245 / 1603	2483 / 1721	3104 / 1971

[a] Nominator, total water withdrawal; denominator, water consumption.

Union where it will be lower (an increase of 1.1 to 1.2 times is expected).

The mean rates of increase in abstractions per year up to 2025 are expected to continue in all regions (except those of the former Soviet Union) and are likely to be 10–15% greater than those for the 1950 to 1995 period (up to 30% in South East Asia). It should be noted that the forecasts of water use in Tables 5.26 and 5.27 reflect, to a great extent, the national plans and forecasts of the largest countries (China, India and Thailand) that are based mainly on forecasts of water demands. Experience shows that the planned levels may only be reached at a later date. The trends and changes in the development of water vary considerably from country to country (Table 5.27). The data indicate that from the 1980s the growth in the use of water lagged in China compared to India and there are indications that this trend will continue. The largest volume used is in Pakistan where the 1995 total exceeds 240 km³/year and by 2025 this figure could reach almost 300 km³/year.

In the leading countries of the east and the southeast of Asia (Table 5.27) abstractions range from 17 to 90 km³/year and by 2025 these figures may increase to 35 to 110 km³/year. In Japan the increased water use by the industrial and domestic sectors is compensated for by the decreased use in agriculture, due to the reduced area irrigated and the introduction techniques for saving water. In Japan a rapid rise in water use should be expected to start in 2010. Countries of the Near and Middle East abstract large volumes of water. By 2025 Iran will be close to using 100 km³/year, mainly due to increases in the domestic and industrial sectors.

For agriculture abstractions may rise by 5–6 km³/year compared to 1995 levels. A steady rise in the volume of water used has occurred in Turkey, Syria and especially in Saudi Arabia where between 1980 and 1990 it increased almost seven-fold from 2.4 to 16.3 km³/year, and this trend is expected to continue. For Iran, Iraq, Turkey, Afghanistan, Saudi Arabia and Syria the total volume of water used is about 300 km³/year.

In Kazakhstan and Uzbekistan the most significant increase in water use took place between 1960 and 1980 when there was a two-fold increase. Since 1980 rates of growth have declined and during the 1990s because of the economic reforms there was a decrease in water use. Countries in Western Asia show a constant rise in the volume of water used, especially from ground water and non-traditional water resources. From 1980 to 1990 it doubled in Qatar, Kuwait and Yemen, and forecasts indicate that this growth will continue. By 2025 in Kuwait, the United Arab Emirates and Oman the volume used may be 3.4, 2.3 and 2.1 times the 1995 figure, but considerable capital investment will be needed not only for water supply but also for environmental protection. Israel has the lowest rates of growth.

It should be noted that the data presented here are only approximate, since the available estimates are few and often contradictory. The most recent estimates were produced by the Department for Energy and Natural Resources of the UN Commission for Sustainable Development and the Regional Office for Science and Technology (ROSTAS). There are also the data presented at the regional symposia on water resources. An analysis of these data

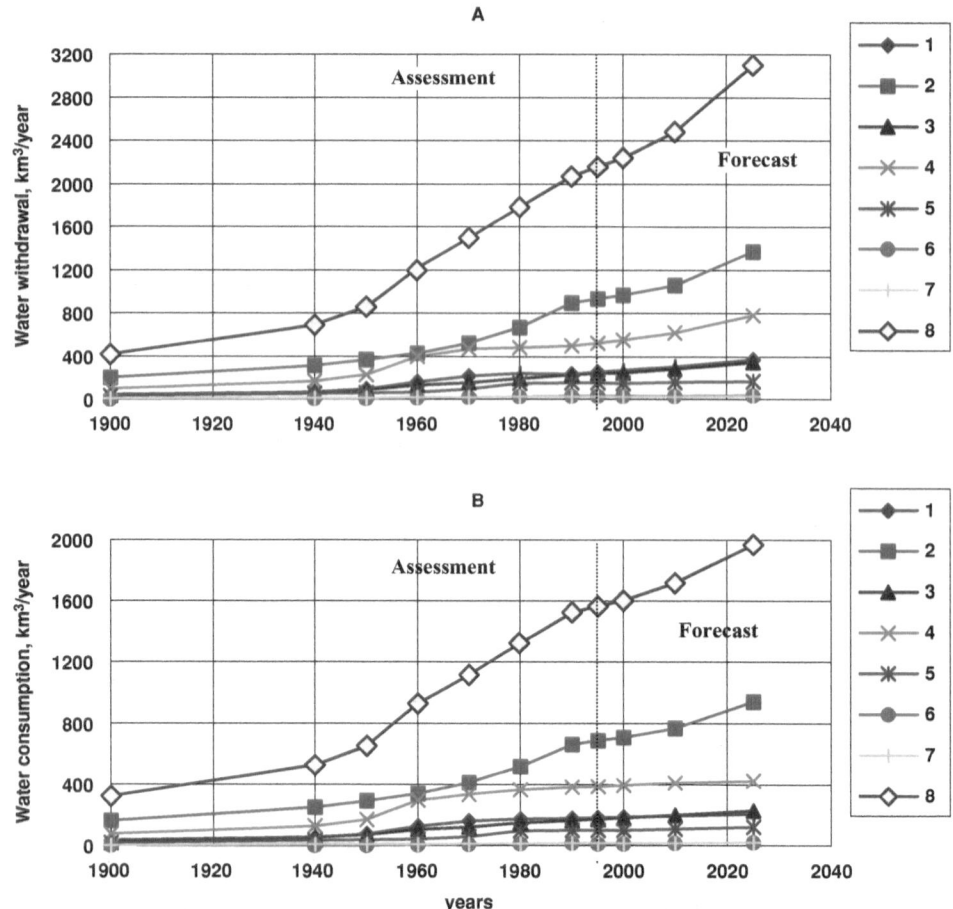

Fig. 5.19 Dynamics of (A) water withdrawal, (B) water consumption in Asia by natural–economic regions. 1, Northern China and Mongolia; 2, Southern Asia; 3, Western Asia; 4, South East Asia; 5, Middle Asia and Kazakhstan; 6, Siberia and Far East of Russia; 7, Transcaucasia; 8, Asia as a whole.

showed that for some countries the estimates for 1990 differed several times. Hence the present study took into account the material contained in a large number of other publications. Forecasting the water use in the Arabian countries took account of national strategies and the possibilities for investment, manpower and the development of institutes within the framework of recommendations of Agenda 21 (UN/DPCSD, 1995b).

Figure 5.20 presents data on changes in water use in Saudi Arabia and the United Arab Emirates obtained in SHI studies (Shiklomanov, 1997), and given in several publications of the UN Commission for Sustainable Development (UN/DPCSD, 1995b) and of ROSTAS (ESCAP, 1989a; Abdul Razzak, 1995; UNESCO, 1995). These forecasts differ by between 3 and 7 times. For these countries (especially Saudi Arabia) the UN Commission gives a high priority to the developments in the industrial and domestic sectors, but substantially underestimates the current use of water in agriculture. In view of the problems facing the governments

of these countries in developing backward agricultural regions, it should be recognised that such rapid increase in use of water in the industrial and domestic sectors is unlikely. Comparisons of the forecasts for other countries is also of interest.

Chinese data were published by Margat (1994) giving forecasts made in 1992 of abstractions for 2000. These showed totals reaching 650 to 690 km^3/year. This Monograph uses later Chinese data obtained by FAO from the Global Water Information System or GWIS (FAO, 1993). These show that totals are unlikely to exceed 540 km^3. Similar forecasts for 2000 (Margat, 1994) for Japan and Turkey are 104 and 46.5 km^3/year, respectively (Kulda and Adanali assessment of 1990). Based on later FAO and UNIDO studies, this Monograph gives totals of 90 and 33 km^3/year and these reflect more correctly the trends in the 1990s. Margat (1994) also presents the forecasts for Syria for 2010 and 2030 (using data from Wakil assessment of 1993) of 20.8 and 26.5 km^3/year respectively, but this study assumes values of 10.0 and 13.7 km^3/year for 2010 and 2025. These amounts are more realistic since the rise in irrigation from 1991 is not likely to continue into the future.

Forecasts for India made by the Central Water Commission for 2025 (1988 assessment) gave a figure of 1050 km^3/year (Margat,

Table 5.27. *Dynamics of fresh water use by selected countries of Asia (km³/year)[a]*

Country	Assessment				Forecast		
	1970	1980	1990	1995	2000	2010	2025
Afghanistan			28/21	29/21	29/22	30/23	32/24
Bangladesh			24/18	27/20	30/22	38/27	50/35
China		460/329	477/355	507/360	537/365	586/369	718/401
India	293/220	455/339	518/384	542/400	565/416	628/457	829/578
Indonesia			18/11	19/12	20/12	24/13	34/18
Iran		53/32	79/60	77/58	79/59	83/61	98/72
Iraq		43/27	46/34	51/39	55/43	65/50	79/53
Israel			2/1.3	2/1.3	2/1.3	2.1/1.3	2.3/1.4
Japan		90/45	91/45	89/43	90/43	99/46	112/55
Jordan		0.5/0.3	0.7/0.4	1/0.5	1.2/0.7	1.7/1	2.5/1.4
Kazakhstan	24/15	39/25	41/26	40/26	40/26	42/28	45/31
Lebanon		0.8/0.5	1/0.7	1.4/0.9	1.8/1	2.2/1.2	3/1.5
United Arab Emirates		0.6/0.5	1/0.8	1.3/1.1	1.6/1.3	2.1/1.5	3/1.9
Pakistan			242/175	244/177	246/178	254/181	298/203
Philippines			43/22	45/22	47/23	52/24	75/34
Saudi Arabia		2.36/1.65	16.3/12.8	19.7/15.2	23/18	27/19	32/21
Syria		4/2.6	6.4/4.3	7.5/4.9	8.5/5.6	10/6.5	14/9
Thailand		20/13	33/22	39/25	45/29	51/33	67/44
Turkey	13/8	18/10	29/15	31/16	33/16	37/16	46/18
Uzbekistan	50/30	80/48	82/52	81/52	79/51	84/55	88/58
Yemen		1.7/1.4	2.9/2.2	3.2/2.2	3.5/2.1	4.4/2.6	5.8/3.6

[a] Nominator, total water use; denominator, water consumption.

1994) but according to SHI, this should be about 830 km³/year, i.e. 21% less. A similar reduction (about 19%) was made for the forecasts for 2025 for Israel (2.3 km³/year) compared to the value in the Plan Blue/Tahal of 1987 which was 2.84 km³/year. For the Lebanon, the value of 2.91 km³/year reported to the conference of the Mediterranean countries in Rome in 1987, appears realistic. In this study however, a slightly larger value of 3.0 km³/year was obtained for 2025. It appears that most Asian forecasts of the use of water made by extrapolating the growth of earlier decades will produce slight overestimates, including those of Shiklomanov and Markova (1987) and Shiklomanov (1988). They estimated the total water use in Asia would be close to 3140 km³ by 2000, but now it seems likely this total will be reached only by 2025. In view of the extreme paucity of data, it is worth noting the agreement between this forecast and that of Margat (1994). As to the water use in the domestic sector, Margat's (1994) value of 233 km³/year appears to be underestimated if account is taken of the substantial improvements in drinking water in many Asian countries particularly within the framework of the International Drinking Water Supply and Sanitation Decade (UN, 1991).

Table 5.28 presents a comparison of current and forecast water use by region with estimates of the renewable water resources. For Asia as a whole, 15% of the water resources have been developed with 12% being consumed. By 2025 these values can be expected to be 23% and about 15%, respectively, with the smaller developments taking place in regions rich in water resources such as South East Asia (about 8%), Siberia and the Far East of Russia (1%). However in regions short of water resources there is the highest level of utilization. A high level also exists in Western Asia where almost half the resources are utilized, and in Middle Asia where the figure is more than 75%. These percentages are expected to reach 71% and 83% during the next 25 to 30 years.

5.6.3 Regional and national patterns of the availability of water

The burgeoning population of Asia is the primary cause of the continents water problems. According to the UN this population is expected to increase by half as much again during the next 30 years (Fig. 5.21). The largest growth is expected to be in Western Asia, Southern Asia, and Middle Asia and Kazakhstan, where it will rise by 1.9, 1.7 and 1.6 times, respectively. These are unfortunately, the regions with the largest stress on water resources at present and their situation is likely to worsen in the future. The need for food for the increasing population stimulates the use of water, primarily the expansion of irrigation. The area irrigated is predicted to increase by 30% or more during the next three decades. With the development of industry and improved living standards

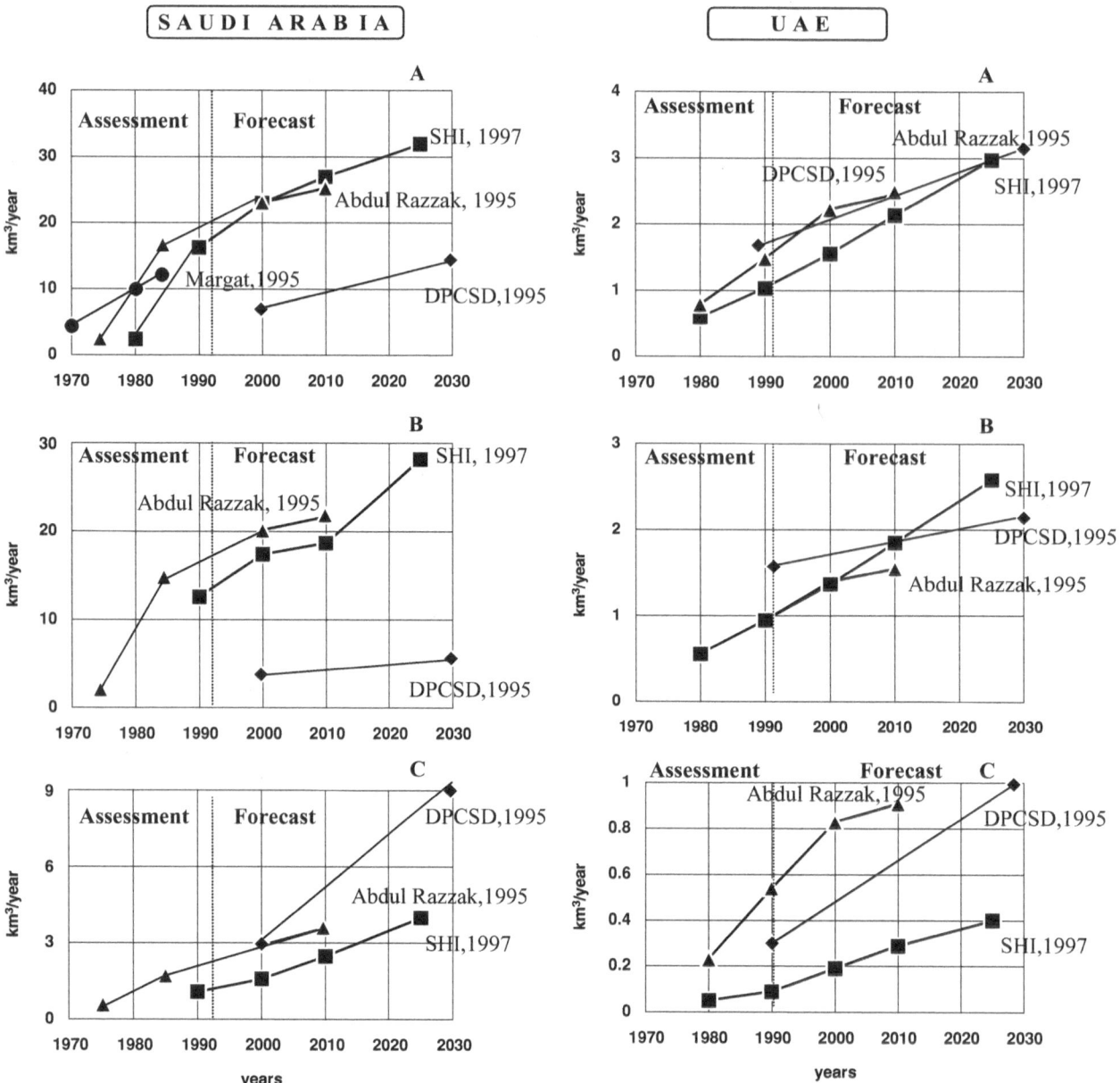

Fig. 5.20 Dynamics of (A) total, (B) agricultural, (C) public–industrial water use (km³/year) for Saudi Arabia and the United Arab Emirates according to different sources.

new problems will arise as the availability of the resource diminishes, highlighting the need for water-saving technologies and for environmental protection measures.

Table 5.29 and Fig. 5.22 present changes in the specific water supply by renewable surface water resources in the Asian regions. In 1950 in only one region (Siberia and the Far East of Russia) was the availability of water very high (>20 000 m³/year per head) while in two regions (South East Asia, Middle Asia and Kazakhstan) it was high. In the other regions it was average and

low. By 1995 the availability of water was low for the whole continent but there were four regions with very low water availability (less than 2000 m³/year per head). By 2025 all of Asia will be in the very low category. An analysis of the trends and changes in water availability by region and country provides a basis for some general conclusions. Table 5.29 indicates that by 1995 the available surface water resource per head of population had decreased 2.5 times since 1950 while by 2025 it will have declined by almost by 3.7 times. This reduction is most evident in Western Asia and Southern Asia where there is 9.6 and 6.1 times less water available, relative to 1950.

Table 5.28. *Water use as a percentage of water resources by natural–economic regions of Asia*

Region	Water resources, km³/year		Water use, km³/year				Water use as percentage of water resources			
			1995		2025		1995		2025	
	Inflow	Local runoff	Withdrawal	Consumption	Withdrawal	Consumption	Withdrawal	Consumption	Withdrawal	Consumption
Northern China and Mongolia		1 029	254	182	373	210	24.7	17.7	36.2	20.4
South Asia	300	1988	932	687	1370	944	43.6	32.1	64.1	44.2
Western Asia		490	238	174	346	229	48.5	35.5	70.6	46.7
South East Asia	120	6646	525	388	781	425	7.8	5.8	11.6	6.3
Middle Asia and Kazakhstan	46	181	154	102	169	122	75.5	50.0	82.8	59.8
Siberia and Far East of Russia	218	3 107	31	15	38	21	1.0	0.5	1.2	0.6
Transcaucasia	12	68	24	17	27	20	32.0	23.7	36.5	27.0
Asia as a whole		13 510	2157	1565	3104	1971	16.0	11.6	23.0	14.6

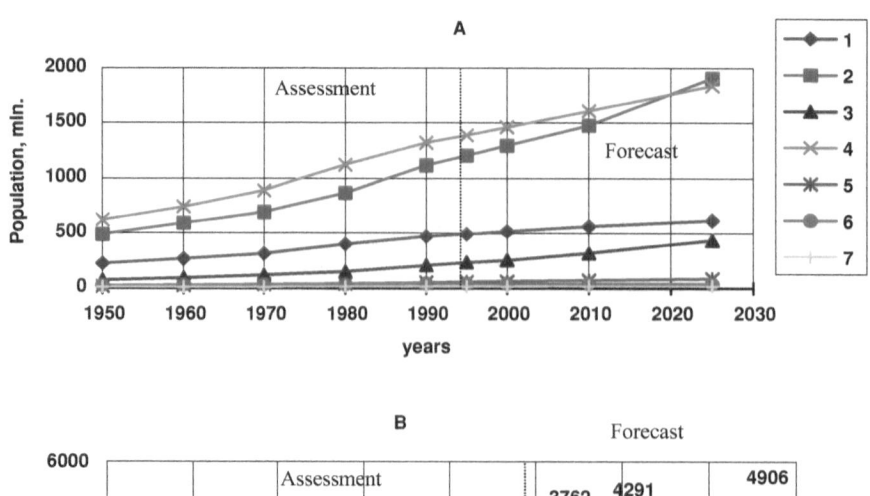

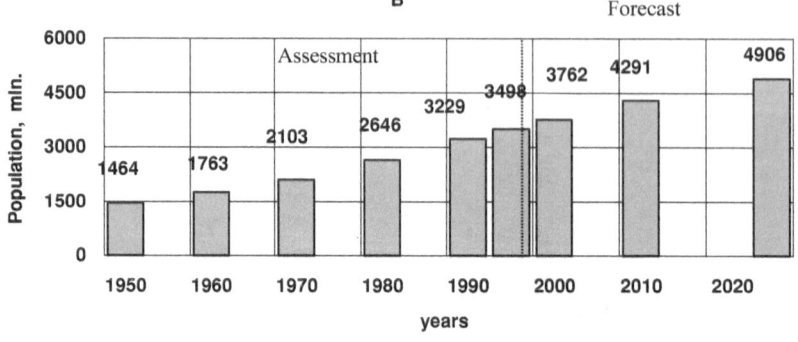

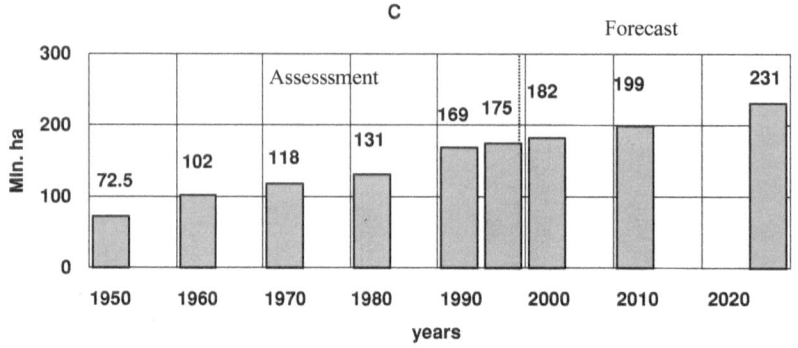

Fig. 5.21 (A) The dynamics of population for Asian regions.
1, Northern China and Mongolia; 2, Southern Asia; 3, Western Asia;
4, South East Asia; 5, Middle Asia and Kazakhstan; 6, Siberia and Far
East of Russia; 7, Transcaucasia. (B) Dynamics of population and
(C) irrigated areas for Asia as a whole.

For Asia as a whole, the decrease to 2025 is 90% governed by the growth of population and 10% by the increase in consumption for domestic needs. These percentages vary considerably, for example, for Western Asia and Southern Asia they are about 60% and 40% and for Transcaucasia, Northern China and Mongolia, and South East Asia the decrease is 80–96% due to the population growth and only 4–20% to increased consumption. The smallest decrease in the availability of water due to increased consumption is in Siberia and the Far East of Russia where it is less than 1%, while the decline in availability expected by 2025 is only 1.56

times and this is also the smallest. In the 1990s an increase in water availability occurred here which may be extended into the future due to a slower growth of population and better water management. The differences in the rate of change in the availability of water in the regions results from a number of factors. In 1950 the least water was available for Southern Asia, Northern China and Mongolia, and Western Asia, where the percentages were 43%, 49% and 64%, respectively, of the mean. In 1995 the availability has decreased severely in Western Asia and Southern Asia, as well as in Middle Asia and Kazakhstan where it was 39%, 34% and 54%, respectively. Forecasts for 2025 indicate that the current trends in the availability of water will continue. In Western Asia, Southern Asia, and Middle Asia and Kazakhstan, water availability will continue to decrease whereas in the remaining four regions it will not decline as rapidly.

Table 5.29. *Dynamics of population and water availability by natural–economic regions of Asia*

Region	Area, $km^2 \times 10^6$	Year	Population, $\times 10^6$	Water availability, $m^3 \times 10^3$/year per head
Northern China and Mongolia	8.29	1950	228	4.18
		1960	268	3.35
		1970	317	2.73
		1980	406	2.11
		1990	471	1.80
		1995	487	1.74
		2000	509	1.66
		2010	554	1.51
		2025	597	1.37
South Asia	4.49	1950	488	3.78
		1960	594	3.02
		1970	692	2.49
		1980	865	1.87
		1990	1117	1.32
		1995	1239	1.17
		2000	1363	1.05
		2010	1611	0.85
		2025	1910	0.62
Western Asia	6.82	1950	74	5.66
		1960	93	4.11
		1970	122	3.00
		1980	153	2.24
		1990	209	1.54
		1995	239	1.32
		2000	275	1.12
		2010	346	0.84
		2025	443	0.59
South East Asia	6.95	1950	662	9.87
		1960	740	8.65
		1970	889	7.16
		1980	1124	5.64
		1990	1322	4.78
		1995	1419	4.45
		2000	1496	4.22
		2010	1650	3.81
		2025	1812	3.47
Middle Asia and Kazakhstan	3.99	1950	18.2	9.18
		1960	25.5	6.31
		1970	34.5	4.14
		1980	42.8	2.50
		1990	51.5	1.96
		1995	55.0	1.85
		2000	60.2	1.69
		2010	70.5	1.33
		2025	83.4	0.98

Table 5.29. (*cont.*)

Region	Area, $km^2 \times 10^6$	Year	Population, $\times 10^6$	Water availability, $m^3 \times 10^3$/year per head
Siberia and Far East of Russia	12.76	1950	25.7	125
		1960	32.1	100
		1970	34.9	91.9
		1980	38.3	83.8
		1990	42.4	75.5
		1995	42.0	76.2
		2000	41.7	76.8
		2010	41.0	78.0
		2025	40.0	79.9
Transcaucasia	0.19	1950	8.0	8.36
		1960	10.3	6.21
		1970	12.7	4.88
		1980	17.2	3.48
		1990	16.2	3.45
		1995	16.6	3.40
		2000	17.4	3.27
		2010	18.9	2.91
		2025	20.6	2.62
Asia as a whole	43.5	1950	1464	8.78
		1960	1763	7.13
		1970	2103	5.89
		1980	2646	4.60
		1990	3229	3.71
		1995	3498	3.41
		2000	3762	3.16
		2010	4291	2.75
		2025	4906	2.35

There are changes in the patterns of water availability from country to country (Table 5.30, Fig. 5.23). The situation in India, China and Kazakhstan is similar in many respects to that in their particular regions. Kazakhstan had a higher water availability 1995 and this is expected to remain so for the future. Thailand is the opposite, with a lower water availability than the region. Uzbekistan and Pakistan have especially low values, even lower than for nations where water resources are most deficient. Despite the high growth of population in Asia the decrease in the availability of water will probably be slower in the future. Reduced rates of abstraction and consumption in Southern Asia and South East Asia are probably not the most important factors, because water and land resources are already depleted and techniques are being introduced to ensure the more rational use of water. However the pattern of the availability of water is very heterogeneous

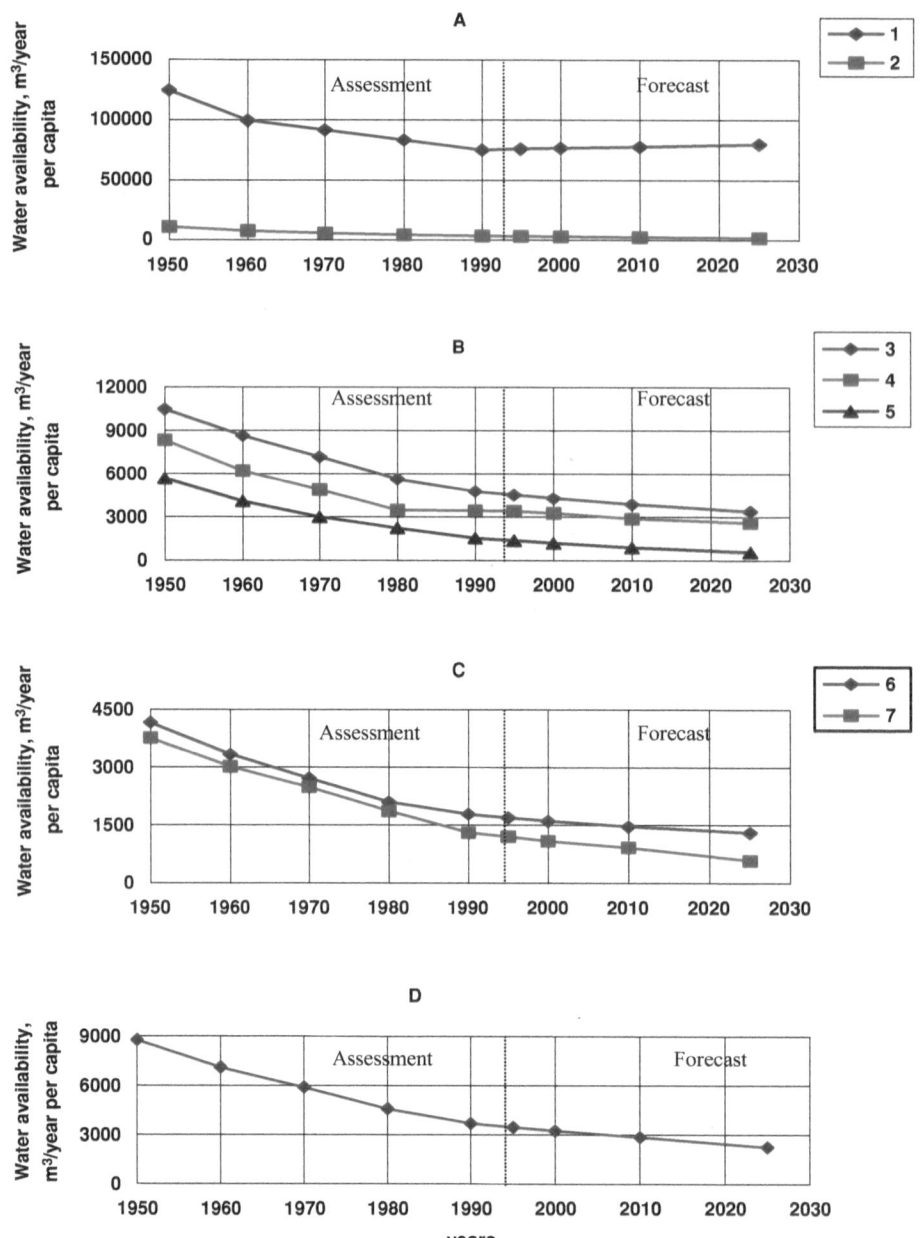

Fig. 5.22 Dynamics of water availability by natural–economic regions of Asia. (A) Siberia and Far East of Russia (1), Middle Asia and Kazakhstan (2); (B) South East Asia (3), Transcaucasia (4), Western Asia (5); (C) Northern China and Mongolia (6), Southern Asia (7); (D) Asia as a whole.

across the continent with the largest decreases occurring in the most arid regions. Even greater differences are observed during long droughts especially those extending over several years and this also gives rise to the increased variability of runoff in dry areas (Table 5.31). In Western Asia the available water may be 2.6 times less during a drought, as compared with the level for 1995 and by 2025 this figure could increase 9.4 times. In some very

dry years this decrease could be 5.5 times less than the present, while in the future the availability of water could approach zero hundreds of times. In Middle Asia and Kazakhstan there could be 1.4 and 3 times less water during a dry period some 4 to 5 years long and more than 3.1 and 12 times for an individual year.

It is obvious that most countries require additional storage to regulate runoff and that they need to obtain additional water resources to meet the demand. Table 5.32 presents the changes and trends in water availability in 10 countries from 1980, taking into account the use of ground water, desalinated water and recycled

Table 5.30. *Water availability of renewable water resources by population in individual countries of Asia*

Country	Population (1994), $\times 10^6$	Water resources, km³/year		Water use (1995), km³/year		Water availability, m³ × 10³/year per head								
		Local runoff	Inflow	Withdrawal	Consumption	Assessment						Forecast		
						1950	1960	1970	1980	1990	1995	2000	2010	2025
China	1220	2700		507	360				2.38	2.02	1.92	1.83	1.68	1.54
India	935	1456	581	542	400			2.83	2.12	1.61	1.44	1.31	1.09	0.85
Pakistan	141	40	186	170[a]	123[a]					0.093	0.071	0.054	0.033	0.01
Thailand	59	199	70	39	25				4.78	3.88	3.56	3.34	3.01	2.60
Kazakhstan	17.2	70	56	40	26	13.6	9.10	6.40	4.95	4.32	4.18	4.02	3.61	3.13
Uzbekistan	22.8	9.5	98	81	52	6.34	4.34	2.42	0.68	0.33	0.28	0.30	0.11	0.01

[a] With surface water withdrawal only (UN/DPCSD, 1995).

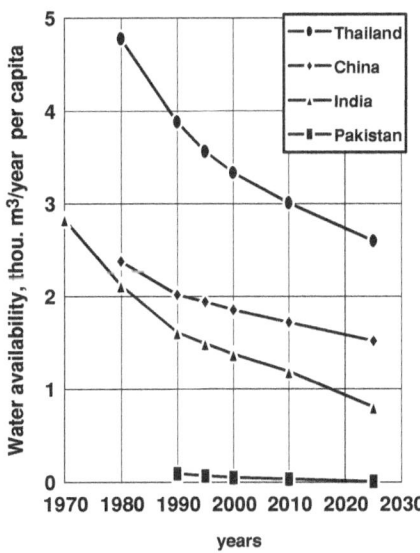 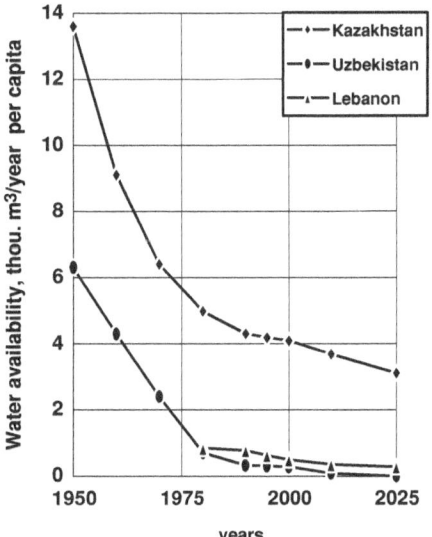

Fig. 5.23 Dynamics of water availability for selected Asian countries.

waste water. The estimates for 1980 and 1990 were obtained by analysing a large amount of data on abstractions, while the forecasts take account of the UN and Agenda 21 (1992). Table 5.32 shows that availability of water in these countries remains low, for example in Saudi Arabia there was an unprecedented increase in not only water use but also in water availability from 1980 to 1990. A similar trend can be expected to 2000 and beyond. Then, provided that the forecasts of the population growth are correct, water availability may decrease. In Oman and the United Arab Emirates the trend towards a constant decrease in availability is most probable, while in Yemen the changes in availability are largely connected with variations in the population. In the countries of the Persian Gulf that do not have surface water resources, the trend towards a small increase in the availability of water will probably persist. However the high rate of develop-

ment in these countries is quite important because it often includes costly programmes for the development of non-traditional sources of water, including the construction of desalination plants.

5.7 THE PROBLEMS OF THE ARAL SEA BASIN

The problems of the Aral Sea are the most acute among the large-scale water management problems facing Asia and the world today. The development of irrigation in the basin over the last 30 years has resulted in a significant increase in the consumption of water. In turn, this has led to a dramatic drop in sea level, fundamental changes in the water and salt balance, and the degradation of the hydrological and ecological conditions in the cotton growing regions adjoining the sea. This catastrophe is the direct result

Table 5.31. *Water availability throughout long-term dry periods and the driest year by natural–economic regions and selected countries of Asia*

Territory	Population, ×10⁶		Water resources during dry period			Water resources during dry years			Water consumption, km³		Water availability, m³ × 10³/year per head			
											Period		Years	
	1994	2025	Period	Inflow	Local runoff	Year	Inflow	Local runoff	1995	2025	1995	2025	1995	2025
Regions														
North China and Mongolia	482	597	1974–82	–	738	1978	–	587	182	210	1.15	0.88	0.84	0.63
South Asia	1214	1910	1962–68	308	1750	1965	275	1535	687	944	1.00	0.50	0.81	0.38
Western Asia	232	443	1962–67	–	290	1960	–	230	174	229	0.50	0.14	0.24	0.002
South-Eastern Asia	1404	1812	1974–79	107	5975	1978	118	5342	388	425	4.02	3.09	3.57	2.75
Middle Asia and Kazakhstan	54	83.4	1935–41	42.3	152	1974	27.2	121	102	122	1.32	0.61	0.60	0.15
Siberia and Far East	42.5	40.0	1952–55	170	2847	1954	160	2628	15	21	68.6	72.8	63.4	67.2
Transcaucasia	16.5	20.6	1965–75	11.4	63.4	1971	11.3	51.5	18	20	3.10	2.38	2.38	1.80
Countries														
China	1209	1488	1976–82	–	2354	1928	–	2015	360	401	1.65	1.31	1.37	1.08
India	919	1371	1962–68	573	1261	1965	573	1065	400	578	1.25	0.71	1.04	0.56
Asia as a whole	3445	4906	1974–79	–	12354	1979	–	11800	1565	1971	3.13	2.12	2.97	2.00

Table 5.32. *Dynamics of water withdrawal, water consumption and water availability in countries of the Arabian Peninsula and Arab countries of the Levant*

Country[a]	Parameter[b]	1980	1990	1995	2000	2010	2025
Arabian Peninsula	1	9.88	26.0	30.3	34.6	40.3	48.7
as a whole	2	5.79	21.9	26.2	30.4	36.2	44.9
	3	4.32	16.8	19.6	22.3	24.4	28.1
	4	0.288	0.281	0.300	0.318	0.308	0.239
Saudi Arabia	1	3.69	17.6	21.0	24.4	27.9	32.8
	2	2.36	16.3	19.7	23.1	26.6	31.5
$Q = 2.23$ km^3/year	3	1.65	12.8	15.2	17.7	19.1	21.0
	4	0.218	0.328	0.335	0.342	0.331	0.264
Yemen	1	3.75	4.94	5.30	5.67	6.40	7.88
$Q = 3.50$ km^3/year	2	1.70	2.89	3.18	3.47	4.35	5.83
	3	1.39	2.19	2.16	2.12	2.57	3.62
	4	0.394	0.235	0.280	0.326	0.260	0.151
United Arab Emirates	1	0.685	1.119	1.376	1.634	2.210	2.833
$Q = 0.15$ km^3/year	2	0.610	1.044	1.302	1.559	2.135	2.980
	3	0.486	0.826	1.072	1.317	1.452	1.894
	4	0.714	0.443	0.439	0.435	0.424	0.403
Oman	1	1.308	1.441	1.570	1.698	1.973	2.593
$Q = 0.918$ km^3/year	2	0.665	0.798	0.926	1.055	1.330	1.950
	3	0.561	0.621	0.666	0.712	0.840	1.031
	4	0.759	0.538	0.518	0.499	0.439	0.329
Kuwait	1	0.186	0.383	0.512	0.640	1.071	1.721
	2	0.186	0.383	0.512	0.640	1.071	1.721
	3	0.049	0.099	0.124	0.148	0.199	0.264
	4	0.100	0.132	0.148	0.164	0.231	0.385
Bahrain	1	0.138	0.225	0.266	0.306	0.365	0.425
	2	0.138	0.225	0.266	0.306	0.365	0.425
	3	0.084	0.118	0.122	0.126	0.132	0.138
	4	0.156	0.213	0.236	0.260	0.273	0.287
Qatar	1	0.128	0.250	0.270	0.290	0.349	0.475
	2	0.128	0.250	0.270	0.290	0.349	0.475
	3	0.102	0.102	0.122	0.143	0.160	0.196
	4	0.116	0.347	0.302	0.258	0.262	0.324
Countries of the Levant	1	110	111	111	112	112	113
as a whole	2	48.2	54.1	60.4	66.5	79.3	89.2
	3	30.4	39.9	45.1	50.5	58.6	64.1
	4	2.891	1.951	1.554	1.156	0.767	0.495
Lebanon	1	2.80	2.80	2.80	2.81	2.82	2.82
$Q = 2.8$ km^3/year	2	0.800	1.00	1.38	1.75	2.25	3.00
	3	0.527	0.665	0.858	1.05	1.24	1.49
	4	0.852	0.779	0.632	0.486	0.364	0.283
Jordan	1	0.859	0.964	1.20	1.44	1.96	2.75
$Q = 0.839$ km^3/year	2	0.520	0.712	0.951	1.19	1.71	2.50
	3	0.281	0.399	0.528	0.657	0.965	1.44
	4	0.198	0.172	0.146	0.121	0.109	0.133
Syria	1	25.1	25.3	25.3	25.4	25.5	25.6
$Q = 22.1$ km^3/year	2	4.00	6.40	7.45	8.50	9.91	13.7
	3	2.58	4.28	4.94	5.60	6.46	8.62
	4	2.587	1.701	1.41	1.112	0.818	0.498

[a] Q, Surface water resources.

[b] 1, Total water resources available (km^3/year); 2, water withdrawal (km^3/year); 3, water consumption (km^3/year); 4, water availability (thou m^3/year per head).

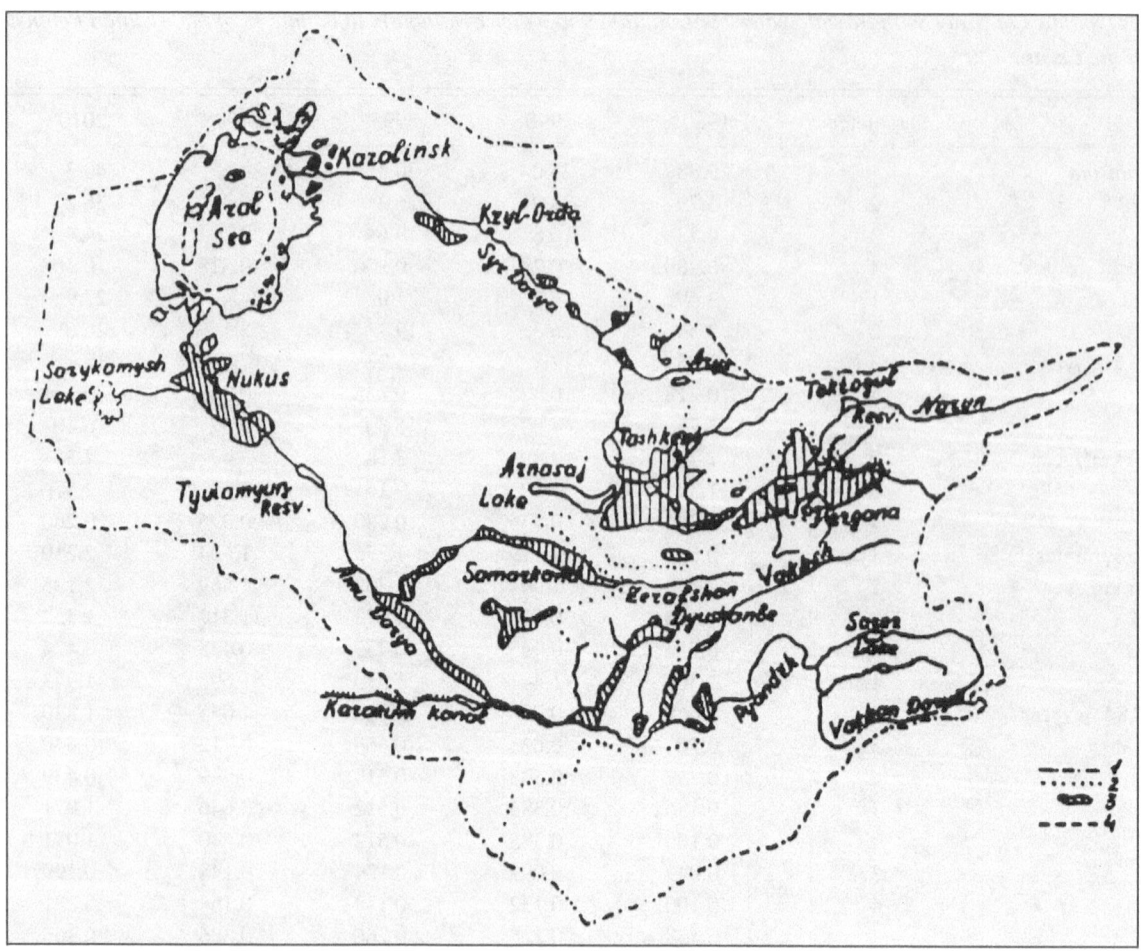

Fig. 5.24 Map of the Aral Sea basin. 1, boundary of the basin; 2, lower boundary of flow formation zone; 3, irrigated lands; 4, present-day boundary of the Aral Sea.

of the extreme changes to the regimes and flows of the Amu-Darya and Syr-Darya Rivers which provided the main inputs to the water balance of the sea. There is some inflow of ground water to the Sea, but this has been estimated at not more than 1–2% of the surface runoff (Bortnik and Chistyayeva, 1990). Conflicts have arisen between the agricultural interests, which use large volumes of water, and those interested in maintaining sufficient volumes of water for the natural environment of the basin and for the Aral Sea itself. The shortage of water resources has also been the cause of disputes over how water is divided between the independent countries which were formerly part of the Soviet Union and between the main users of water – irrigation, hydropower, fisheries and so on. The attention of many nations and organizations has been captured by this problem and effort is being focussed on finding a solution to the problems of the Aral Sea. Some scientists consider that the Aral Sea has ceased to exist as a geographical body (Bortnik et al., 1991; Voropayev, 1992). For a number of years it has been impossible to provide a sufficient inflow of river water to

balance the evaporation from the sea, with the consequence that the problem of regeneration may be an intractable one. The scientific literature on the Aral Sea is extensive; but many prominent scientists are of the opinion that the sea will gradually disappear (Berg, 1908; Geller, 1969; Voropayev and Gerasimov, 1982).

5.7.1 The water resources of the Aral Sea Basin

The Aral Sea is situated in the centre of the vast European–Asian continent, remote from the seas and oceans. Its basin (Fig. 5.24) is characterized by a continental climate, abundance of heat, high levels of solar radiation and extremely small amounts of precipitation (100–200 mm/year in the lowlands around the sea). Much of the area is desert, namely the Turan Lowlands, which are bordered in the south by the Tien Shan and Pamir Alai mountains. Firn fields, glaciers and increased amounts of precipitation in the mountainous area are the main sources of the major rivers in the region: the Amu-Darya with a drainage area of 1.1 million km^2 and the Syr-Darya with an area of 0.44 million km^2. In contrast to the mountains where runoff is generated, the plains are the zone of runoff dispersion (Shultz, 1965), where the water draining from the mountains is lost by natural evaporation and by evaporation from the irrigated areas. The waters of the Amu-Darya and Syr-Darya reached the Aral Sea, until quite recently, to form the largest water body of Middle Asia, where they were lost by evaporation. Up to 1961 the mean sea level was at about 53.0 m (BS), the water surface area was 66 000 km², the volume was 1064 km³, and the salinity was 10–11%. The inflow to the sea was estimated to be 50–60 km³/year.

The water resources of the Aral Sea Basin in terms of surface water, was determined as the sum of the mean discharges as the rivers left the mountains. These water resources were estimated to be 114 km³/year consisting of the Amu-Darya Basin with 76 km³/year and the Syr-Darya Basin with 38 km³/year (Table 5.33). Over the period 1932–60, i.e. before the onset of the steady drop in the level of the sea, the flows of the Amu-Darya and Syr-Darya was assessed to be 116 km³/year, and during the period 1961–90 112 km³/year. After 1960 there was a slight decrease in river flows compared to the preceding period. The largest volumes were observed in 1969 (193 km³/year) and the least in 1974 (83 km³/year). Four more river systems entered the Aral Sea Basin in the distant past and their water probably reached the channels of the Amu-Darya (the Tedzhen and Murgab) and the Syr-Darya (the Chu and Talas). The total volume of their flows is estimated to be 2.8 km³/year and 5.7 km³/year, respectively. However under present conditions these resources cannot be taken into account in addressing the problems of Aral Sea, since they are almost completely used for irrigation and the remains of these systems are lost in the sands of adjoining deserts.

5.7.2 Changes and trends in the use of water

There were 2.4 million ha under irrigation in the Aral Sea Basin at the beginning of the twentieth century, but the area reduced to 1.2 million ha in the early 1920s due to the Civil War (Table 5.34). However, from the late 1920s the area irrigated started to increase and this trend continued for more than 50 years. The largest increase in irrigation was due to the large-scale state programme for the reclamation of arid lands based on the construction of dams and reservoirs and the expansion of the network of irrigation canals from the 1960s onwards. By 1990 about 35% of all irrigation in the former Soviet Union was centred on the Aral Sea Basin, comprising 6.7 million ha. After 1990 there were no noticeable changes in the area irrigated (Anonymous, 1993). With this expansion there were changes in the crop types. Cereal crops prevailed (60–80%) in the 1920s and 1930s, but cotton (50–60%) and forbs (27%) dominated by the 1990s. The increase in the area irrigated and the changes in the type of crops required a sharp increase in abstractions (Table 5.35). The total water used exceeded the renewable resource after 1971, the difference being made up by some recycling. From 1961 to 1990 abstractions for irrigation

Table 5.33. *Water resources of the Aral Sea basin (km³/year)*

River and lake basins	Water resources		
	Long-term means	1932–60	1961–90
Syr-Darya	37.9	38.2	37.7
Amu-Darya	75.9	78.0	73.9
Aral Sea	114	116	112

Table 5.34. *Area of irrigated lands in the Aral Sea basin (ha × 10⁶)*

	1932–40	1941–50	1951–60	1961–70	1971–80	1981–90
Area	2.58	2.90	3.34	3.78	4.75	6.16

Table 5.35. *Water resources use in the Aral Sea basin (km³/year)*

Parameter	Mean values for the period		
	1961–70	1971–80	1981–90
Surface water resources	117	108	115
Total water withdrawal	82.0	111	127
Water withdrawal for irrigation	74.5	101	109

(including water for the Kara Kum Canal) were 85–90% of the total volume of water used. The volume employed for other purposes increased from 7.5 to 18 km³/year over the last 30 years. The ground water abstracted was only 1–2% of the total.

Any study of the changes in the area irrigated and the use of water in the Aral Sea Basin should take the following fact into consideration. At the beginning of the twentieth century irrigation in the Aral Sea Basin was based mainly on abstractions from the tributaries of the Amu-Darya and Syr-Darya. However many tributaries have long since lost their direct connection with the main river. In the Syr-Darya Basin river water was used for not more than 5% of all irrigation. About 40% was abstracted from the Amu-Darya (Dingelshtedt, 1893; Tsinzerling, 1927). At present, as a result of large-scale hydraulic works, the main areas of irrigation are supplied with water directly from the channels of the Amu-Darya and Syr-Darya in their middle and downstream sections. The location of the irrigated areas relative to the sources of water has a direct influence on the changes in runoff and on the extent of the network of channels and water courses.

In addition to the assessment of water resources as a basis for countering the crisis in the Aral Sea, it is extremely important to determine the water losses due to irrigation and particularly the losses through evaporation and crop growth. The total losses for irrigation include evaporation from fallow areas, from small

Table 5.36. *Water withdrawal and the structure of runoff expenditure for irrigation in the Aral Sea basin*[a]

	Mean values for the period[b]		
Parameter	1961–70	1971–80	1981–90
Water withdrawal	68.7	90.8	97.9
Water consumption, net	31.9 / 46	41.4 / 46	54.2 / 55
Water consumption, gross	46.6 / 68	61.8 / 68	72.9 / 74
Non-productive runoff expenditure	14.7 / 22	20.4 / 22	18.7 / 19
Returnable water	22.1 / 32	29.0 / 32	25.0 / 26

[a] Information is given with no account of water inflow to the Kara Kum Canal because of lack of data on the structure of water consumption in its zone.
[b] Nominator is the absolute value (km³/year), denominator, % of water intake.

lakes and ponds, from waterlogged areas, and from depressions with no outlets. The difference between the gross and net losses represents the non-productive losses and these might be considered as a source for saving water (Volftsun *et al.*, 1988a, b). In irrigation some water returns to the rivers and some infiltrates (Soyuzvodproekt/SHI, 1984). The volume of this water depends on the volume abstracted, the type of crops and their stage of development, the irrigation regime and the efficiency of the irrigation systems. The returned water from the irrigated areas to the rivers contributes a certain amount to flow.

Studies of the irrigation system allowed the assessment of water consumption and returned water, based on water balance calculations taking into account the weather, the soil characteristics, the depth of ground water and the volume abstracted for irrigation (Soyuzvodproekt/SHI, 1984). Such assessments were made monthly and annually, and they were averaged for the areas concerned, taking a number of other factors into account. The volume of water returned water was determined from the difference between the amount abstracted and the water consumption (Sumarokova and Degtyarev, 1985; Tsytsenko and Vonsovskaya 1985; Levchenko *et al.*, 1990). Over the last 30 years the net water consumption was 30–55 km³/year or about 50% of the water abstracted (Table 5.36), taking into account the non-productive losses of up to 70%. These losses approach 18–20 km³/year, around 75% of their volume being lost by evaporation from the so-called irrigation-discharge lakes located on the periphery of the irrigated areas. The largest of these is the Sarykamysh Lake, which is situated in the lower reaches of the Amu-Darya. In late 1980s its area was more than 3000 km². The Arnasay Lakes are located in the middle of the course of the Syr-Darya (Volftsun

Table 5.37. *Averaged values of water resources and anthropogenic runoff expenses in the Aral Sea basin*

	Anthropogenic runoff expenses[a]				Runoff expenses, as % of water resources
Period	Irrigation	Other economic branches	Water resources, km³/year	Total	
1932–40	25.3 / 97	0.8 / 3	26.1 / 100	109.7	24
1941–50	30.4 / 95	1.7 / 5	32.1 / 100	114.6	28
1951–60	38.7 / 93	2.8 / 7	41.5 / 100	121.0	34
1961–70	52.4 / 88	6.9 / 12	59.3 / 100	117.0	51
1971–80	71.7 / 92	6.5 / 8	78.2 / 100	107.8	72
1981–90	83.5 / 89	10.8 / 11	94.3 / 100	110.0	86
1932–60	31.7 / 95	1.8 / 5	33.5 / 100	116.0	29
1961–90	70.9 / 90	8.0 / 10	78.9 / 100	112.0	70

[a] Nominator, Km³/year; denominator, percentage.

et al., 1988a). The volume of the returned water increased up to the 1980s. In 1981 there was some decrease both in its absolute value and relative to the water supplied, from 32% to 26% (Table 5.36) which is probably connected with some improvements in the efficiency of water use. However this large volume of returned water downgraded the quality of the river water, especially in the lower reaches of the Amu-Darya and Syr-Darya. Taking into account the water abstracted from the Amu-Darya for the Kara Kum Canal, the total losses caused by human activities for the period 1961 to 1990 were close to 80 km³/year (Table 5.37).

5.7.3 Changes in the level of the Aral Sea and in the river flow regime

Over the last 60 years irrigation accounted for 90–95% of the total volume of water lost from the Aral Sea. From 1981 to 1990 about 90% of all water resources were used for a variety of purposes and this caused river flows to diminish and the Aral Sea to dry up. There is no need to consider other possible causes for the decrease in the level of the Aral Sea, for example, the hypotheses about the existence of a significant ground water flow from the Aral Sea to the Caspian Sea and to other regions. The problems of evaluating the effect of irrigation on the rivers of the Aral Sea Basin was highlighted as early as the 1920s, when plans were made for the reconstruction and development of the irrigation systems destroyed during the war. At that time it was believed that the planned expansion of the irrigation system could induce a significant decrease of downstream flows in the Amu-Darya and Syr-Darya (Dubovikov, 1922). The absence of this decrease, in spite of the considerable increase in the irrigated area, caused heated discussions among scientists on how irrigation influences runoff and whether this factor should be taken into account in planning further works. This discussion also continued in the 1960s and 1970s (Dunin-Barkovsky, 1967; Yunusov, 1974; Kharchenko, 1975; Sumarokova and Tsytsenko, 1978; Shiklomanov, 1979) and it gave impetus to extensive theoretical and field studies on the quantitative assessment of human-induced loads and the compensating capabilities of river basins.

Estimates of the changes in the water balance of the Aral Sea Basin were made relative to the water balance during the so-called "natural period" when, in spite of irrigation, its impact on runoff was insignificant. For this purpose the method of runoff analysis in the zone where the water was used and in the zone where it originated was applied over similar time intervals (Sumarokova and Tsytsenko, 1978; Georgiyevsky and Vladimirova, 1991). These studies have shown that the end of the "natural period" occurred roughly in 1960. Up to that date the influence of irrigation on the runoff of the Amu-Darya and Syr-Darya Rivers was small, if undetected, and this is attributed to the action of some compensating factors, including:

- replacement of natural runoff losses (evaporation from wetlands) by the evaporation from irrigated areas
- construction of a collector–drainage network providing inputs to the rivers which was responsible for significant volumes of return water and groundwater over many years
- the location of the irrigated areas at the different stages of reclamation relative to sources that simultaneously receive return waters.

Along with these compensating factors due to human influences, the natural period from 1932 to 1960 was also wetter in the mountainous parts of the basin where the runoff originates (see Table 5.33). The absence of changes in runoff was the cause of a relatively stable level in the Aral Sea whose oscillations were also influenced by a reduction in the annual evaporation from its surface (Golubev and Zmeikova, 1991). According to Shiklomanov (1989) these compensating resources of the Aral Sea Basin totalled 36 km³/year, 26 km³/year for the Amu-Darya and 10 km³/year for the Syr-Darya. The unstable balance between the increasing water consumption and the compensating factors was upset in the Syr-Darya when the Kairakum reservoir began operating, and in the Amu-Darya by the high intake of water for the Kara Kum Canal (Sumarokova and Tsytsenko, 1978).

Estimates of the human-induced changes in the annual flow to the deltas were determined in absolute and relative terms

Table 5.38. *Anthropogenic change in river water content[a] for the "natural period"*

River	Mean long-term inflow to apices of delta	Anthropogenic runoff reduction, average		
		1961–70	1971–80	1981–90
Amu-Darya – Samanbai (Chatly)	$\frac{47.0}{100}$	$\frac{10.9}{23}$	$\frac{25.7}{55}$	$\frac{38.1}{81}$
Syr-Darya – Kazalinsk	$\frac{15.0}{100}$	$\frac{6.24}{40}$	$\frac{11.1}{74}$	$\frac{12.9}{86}$

[a] Nominator, km³/year; denominator, percentage.

(Table 5.38). At the end of the 1980s the decrease in the flow of both rivers exceeded 80% of the flow for the "natural period" reaching a total of 51 km³/year. The sharp fall in flow was especially noticeable in the changes in the hydroecological state of the distributaries. Open water hydrophytic vegetation and swamps characterized the deltas during the natural regime and were a source of loss due to evapotranspiration. It was thought that the planned irrigation and the subsequent decrease in flow to the deltas would lead to drying and reduced evaporative losses. In reality, this did not take place, since these effects were not uniform across the delta, as shown by satellite imagery and aerial photography.

From 1961 to 1980, as expected, the areas of open water, reed communities and tugai bushes began to reduce. Simultaneously there was a rapid development of irrigation in the deltas accompanied by the construction of a network of irrigation canals. The increased area of irrigation has resulted in inflow to the delta not only along the river channels, but also from the drainage network. During dry periods however there was only flow in the canals: up to 90% in the Amu-Darya delta. Contrary to what might be expected in the early 1980s the deltas were affected by flooding. The flooding occurred because of the increased supply to the irrigated areas and because of the construction of dykes, dams and drains to collect water to flood reed beds which were producing forage for cattle. Due to this, the area of open water and hydrophytes in the Amu-Darya delta has increased by 1.5 to 2 times as compared to 1982 (Sumarokova et al., 1991). In the Syr-Darya delta, only an insignificant increase was observed. Hence the total losses by evaporation for the period 1981–1990 in the Amu-Darya delta were 9 km³/year, and in the Syr-Darya delta 2 km³/year. These conditions were also noted in Kuksa (1994) and Ratkovitch (1993).

The development of the basins and the deltas required quite significant proportions of river water, resulting in a dramatic decrease in the flow to the sea. This was especially pronounced during the 1974–6 dry period when the flow of 20–40 km³/year reduced to 7–10 km³/year (Fig. 5.25). In the 1980s more water was needed

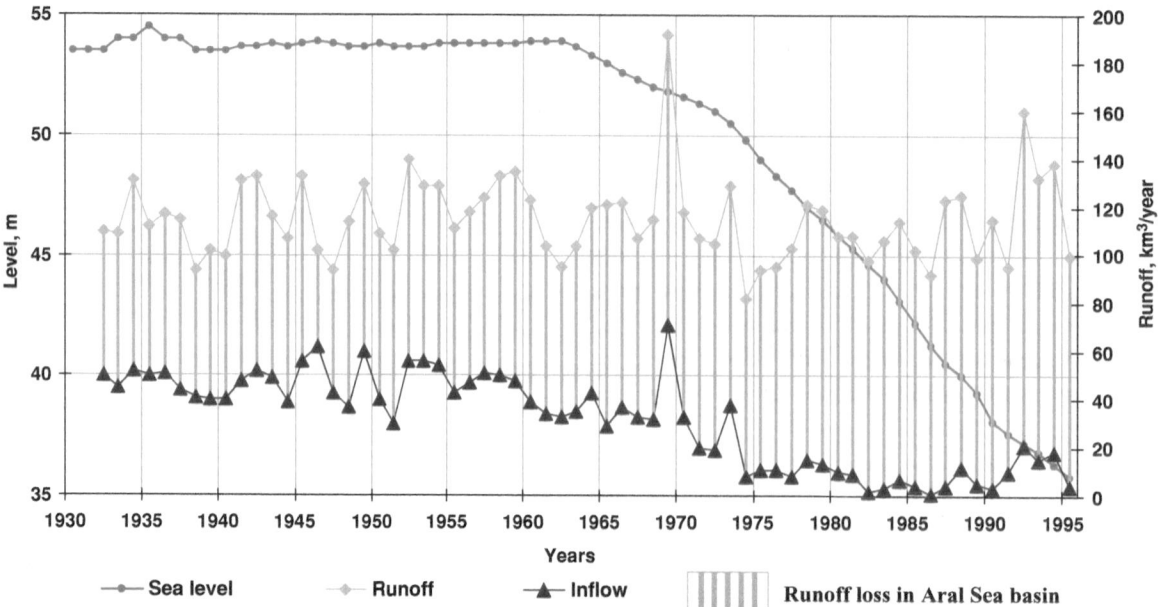

Fig. 5.25 Long-term variability of the inflow to the Aral Sea and its level.

Table 5.39. *Mean long-term water balance components of the Aral Sea (km³/year)*

| | Income | | | |
Period	Inflow to the sea	Precipitation on to sea surface	Discharge	Closure (deficit)
1911–60	56.0	9.1	66.1	− 1.0
1960–70	43.3	8.0	65.4	−14.1
1971–80	16.7	6.3	55.2	−32.5
1981–90	4.8	5.9	43.0	−32.3
1991–95	14.1	(4.3)	(30.5)	(−12.1)

[a] Values in parentheses are approximate data.

because of development and the runoff to the sea actually stopped or remained very low (0–2 km³/year). From 1992 to 1994 the flow has increased slightly as a result of increased precipitation over the mountains and more runoff, rather than due to successes in saving water (Fig. 5.25).

Table 5.39 shows the changes in the components of the water balance of the water body using the State Oceanographical Institute (GOIN, Russia) data (Bortnik *et al.*, 1991). The discrepancy in the balance over the quasi-stationary period (1911 to 1960) is attributed to errors in calculating its components. During the years that follow, the discrepancy explicitly characterizes the lack of flow to the Aral Sea. The water balance components for the last 5-year period were determined only approximately due to re-

duced observations in the basin. In the years with the sharpest fall in flow to the Aral Sea and the steepest drop in level, the Government began to make the first enquiries to scientific and research organizations about the causes of this phenomenon and about forecasts of the future state of the water body. The results of the assessment by the State Hydrological Institute were summarized in a monograph (Shiklomanov, 1989). The 1975 calculations were based on the assumption that more efficient methods of using water resources and water-saving measures would be introduced to water management practice. Unfortunately this has not taken place and the decrease in the flow to the Aral Sea has been worse than expected (Fig. 5.26). According to Shiklomanov (1989), by extrapolating the present water management situation to 2010, under average climatic conditions not more than 4 km³/year of surface water (Fig. 5.26) will flow to the Aral Sea. Similar results were obtained by Raskin *et al.* (1992) – about 3 km³/year by 2008.

In the absence of measures to promote the overall updating of the entire water management system in the Aral Basin, there has been a further drop in sea level, which by 1990 had fallen almost 15 m more. At the end of 1987 the Aral Sea became divided into two water bodies: the Large and Small Seas. The total area of these water bodies was 36 500 km² and the volume was 330 km³, while the salinity approached 30 to 40‰. After 1990 the sea level continued to decline. From 1992 there are no more observations of sea level as they were terminated. Using indirect methods, the level of the Large Sea in the mid-1990s was estimated to be roughly at the 36 m mark (BS) and the Small Sea at 38.5 to 39.0 (BS).

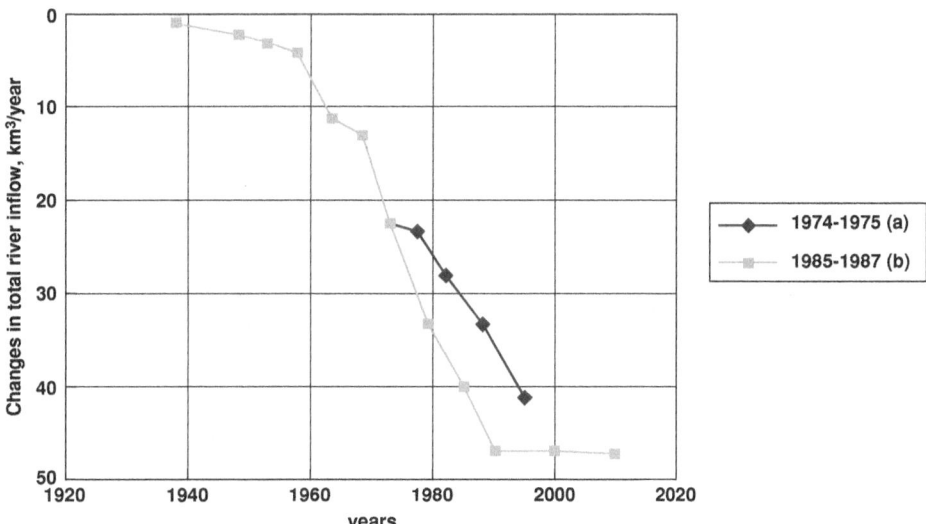

Fig. 5.26 Changes in total river water inflow to the Aral Sea due to economic activities for 1936–2010. Forecast was calculated at 1974–5 (a) and at 1985–7 (b).

5.7.4 The ecological consequences of the development of the Aral Sea Basin

The development of the Aral Sea Basin has changed the quantity and quality of the runoff to the sea and its surroundings and has caused an ecological disaster that has no parallel elsewhere in the world. No similar large area has suffered such an adverse impact to its environment. Furthermore, the living conditions of people in the basin, especially in the region immediately surrounding the Aral Sea (Israel *et al.*, 1988; Voropayev, 1992), have deteriorated dramatically. This deterioration of the environment is a result of the unprecedented reduction in the runoff; a reduction accompanied by serious pollution of the river water due to the discharge of mine drainage, drainage from the irrigated areas and discharges of untreated industrial and domestic waste. Mineralization of the water in the lower reaches of the Syr-Darya near Kazalinsk was 2.8 g/l, a four-fold increase over the upper reaches where it was 0.3–0.4 g/l. In the lower reaches of the Amu-Darya the concentration of salt and ions has doubled reaching 1.6 g/l (Ivanova, 1992). River waters in the lower reaches of the Amu-Darya and the Syr-Darya have become highly polluted with large concentrations of phenols, oil products and pesticides whose levels significantly exceed the maximum permissible concentration (MPC). According to the classification adopted, the waters of the Amu-Darya are "contaminated" water and waters of the Syr-Darya are "rather contaminated" water (Khomenko and Yemelyanova, 1991). The medical and epidemiological situation in the coastal regions of the Aral Sea is characterized by a high level of intestinal infection morbidity, particularly with liver and lung diseases. The morbidity of the hepatitis virus and typhoid fever has increased especially by 2 to 7 times. In some regions the rate of child mortality has reached 100 per 1000 births. Of serious concern is the state of the health of the female population: 70% of women suffer from anaemia (Elpiner, 1992).

The process of desertification is widespread in the coastal regions. The original vegetation has disappeared, salinization of soils has taken place and there has been a decrease in the area of farm land. But the ecological condition of the Aral Sea itself has undergone the largest changes. It no longer exists as one water body. All the features of the earlier marine ecosystem are gone. More than 200 species of flora and fauna have been eliminated (Kotlyakov, 1991). The Aral Sea has completely lost its importance as a fishery and for transportation. The continental character of the climate in the adjoining regions has become more prevalent and the dried-out part of the sea floor (more than 30 000 km^2) has become a source of dust storms. The prevailing transfer of aerosol particles according to Bortnik and Chistyayeva (1990) takes place in the westerly (Fig. 5.27) and southeasterly directions (27%). Some 75 million tonnes of dust and salt has been removed and up to 0.5 tonnes/ha has been washed out by precipitation. This process has been detrimental to the health of the inhabitants and to the state of the vegetation and soil cover in the Karakalpak and Kzyl-Orda areas.

5.7.5 Ways and means of addressing the problem

Was the actual disappearance of such a unique natural water body unexpected? Probably not, but the rapidity of the changes and the severity of the ecological disaster they generated were not expected, nor were their wider ramifications. It must be mentioned however that in the 1950s to 1960s, neither science nor practice could anticipate the scale of the possible consequences as the development of irrigation took place. This development was also accompanied by significant errors in the design,

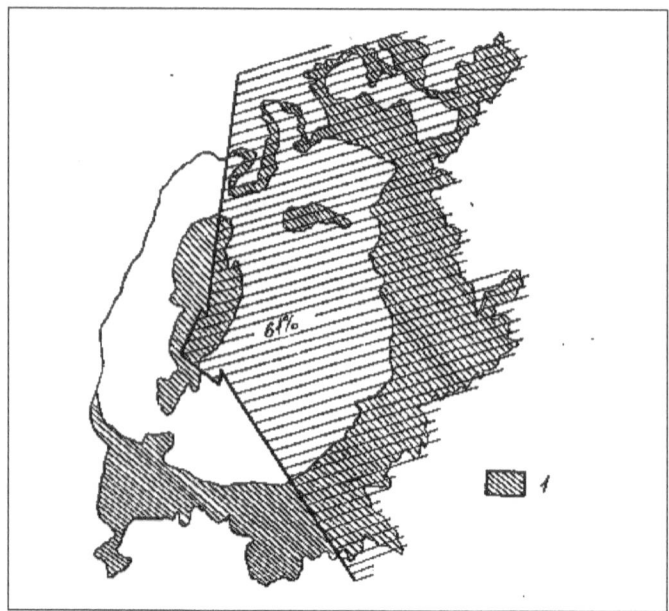

Fig. 5.27 The largest repetition of wind direction during the dust–salt storms in the Aral Sea basin (from Bortnik and Chistyayeva, 1990).

construction and management of the different water systems. Indeed the fate of the Aral Sea was determined more than 30 years ago at the out-of-town session of the Terrestrial Science Branch of the USSR Academy of Science in Tashkent in 1964. At this session the standpoint (although with some reservations) was confirmed, that the death of the Aral Sea was unavoidable and justified in terms of the economy. Everyone quoted Voyeykov who had said in 1882 that the Aral Sea is a useless evaporator of the water of the Amu-Darya and Syr-Darya Rivers and it would be better for this water to be evaporated from irrigated lands rather than from the sea. At that time the terminology "productive" and "non-productive" evaporation was popular. The latter meant the evaporation from natural lands that reduced the runoff.

However it is not true that the different reclamation projects did not consider the problems related to the need for environmental conservation. But this was done in a peculiar way, since the sections devoted to nature protection measures followed the results of water management calculations and the totalling of the water requirements of all water users. Thus the ecology had to be satisfied with the so-called remainder principle. This situation continued for a long time. It was not even altered by the consequences of the 1974–5 dry period. At that time the decrease in the runoff and the fall in level of the Aral Sea were considered fully but serious problems occurred in the supply of water for the existing irrigation system. The need for urgent measures to be undertaken only became apparent when it was obvious that the degradation of the

Aral Sea and its basin was out of control and affecting vast areas.

The hydroecological situation in the Aral Sea Basin has been used (Glantz, 1994, 1999) as a typical example in studies of the problem of gradually accumulating ecological changes. Glantz considers that each change (deterioration in the drinking water quality and the health of the inhabitants, the decrease in the flow to the sea, the drop in sea level, desertification, etc.) has its threshold level, and on reaching this level there is complete destruction of the ecosystem. Although such thresholds are revealed in most cases retrospectively, the signals of the forthcoming disaster in the Aral Sea were observed in advance, but no effort was made to stem and control the adverse ecological changes. The sharp deterioration in the ecology of the Aral Sea and the surrounding region has however made it necessary to undertake actions to reverse the decline. They were started in 1987 to 1988 as a result of the work of the Governmental Commission (Israel *et al.*, 1988). The proposed measures were primarily aimed at creating the required sanitary–epidemiological conditions in the region and providing a guaranteed flow of surface water to the sea (with collector–drainage waters), increasing from 9 km^3 in 1990 to more than 20 km^3 by 2000.

However the proposals and recommendations developed by the Commission did not solve the problem of restoring the Aral Sea to its former state. About 1 km^3/year of water is required per 1 km^2 of surface area for the existence of the sea, at an annual evaporation of 1 m (Bortnik and Chistyayeva, 1990). Hence the search for the ways and means to resolve this problem has continued and to this end a staged programme has been planned. For the initial stage it was proposed to stabilize the sea level and in the later stages the aim was to raise the level close to the natural one. To bring this about it was necessary to increase the inflow to the sea to about 45–50 km^3/year. The proposals were mainly aimed at solving the hydrological and water management problems: where and how much water can be taken. The numerous proposals can be conventionally divided into two main groups. The first direction consists of attempts to directly increase the volume of water resources by:

* regulation of water flushes from mountainous lakes
* increased melting of glaciers
* artificial enhancement of precipitation
* water use from ground water
* water transfers from other regions (pumping water from the Caspian and the Black Seas and from the Siberian rivers.

However these measures incur substantial costs and need much time. Also, those proposals which are still at the theoretical stage require experimental testing. The experience of regulating the water flushes from mountainous lakes (for example, Lake Sevan) show failures; and these procedures also generated new problems. More realistic are those measured for the rational use of the available water resources. They include:

- increasing the efficiency of irrigation systems
- replacing crops needing large amounts of water with those requiring less
- introducing scientifically based standards for irrigation and modern irrigation technologies to save water.

According to Kotlyakov (1991), increasing the efficiency of the irrigation systems over the area of 5.2 million ha will yield about 14 km^3/year of water. Decreasing the area of rice crops in the lower reaches of the Amu-Darya and Syr-Darya and replacing cotton on salinized areas where production is low (not more than 5% of the entire area is irrigated) with fruit and vegetables can save about 4 km^3/year. The introduction of new water-saving irrigation techniques will provide a further saving of 5 km^3/year. The approximate volume of water saved by these measures is estimated to be 26 km^3/year and from data in Sumarokova and Tsytsenko (1991) and Tsytsenko and Sumarokova (1999) some 20–25 km^3/year. The authors (Kotlyakov, 1991) believe that in addition, due to water charges (4 km^3/year) and reduced losses of agricultural products during transportation, processing and storage (15 km^3/year), the total volume of saved water can be 45 km^3/year.

The measures proposed are intended to be implemented over a period of not less than two decades and without any noticeable decrease in the irrigated area. In order to save the Aral Sea over a shorter period, it is recommended to remove from production those areas where the soil has been salinized (about 30% of the irrigated area), to introduce strict limits for water use and to discharge the water from the reservoirs into the rivers (Altunin *et al.*, 1991; Mirzayev and Rachinsky, 1991). Along with the recommendations (Kotlyakov, 1991) for the preservation of the Aral Sea, some studies of the Institute of Water Problems of the Russian Academy of Sciences (Voropayev *et al.*, 1992) propose a different approach. According to this approach, the unused reserves of water resources of the basin are quite limited and do not exceed 10–12 km^3/year (Antonov, 2000). Hence the goal of saving the sea is unreal and efforts should be focused primarily on the partial regeneration of the ecosystems in the Amu-Darya and Syr-Darya deltas to provide more favourable living and working conditions for the population in the lower reaches of these rivers. As to the sea itself, it is advisable to create several artificially maintained water bodies in it and to expand them if there is an increased flow into the sea. Serious doubts about the possibility of saving the Aral Sea were also contained in the studies of GOIN (Bortnik *et al.*, 1991; Tsytsarin and Bortnik, 1991). It was considered that under the conditions of extreme stress in the water balances of the Amu-Darya and Syr-Darya Basins, it would hardly be possible to allocate the necessary water volumes even for sea level stabi-

lization. Provided that it is possible to raise the level of the sea to 53 m (BS), such a water body would not possess the ecosystem that the Aral Sea had before 1961 and it would be impossible to reconstruct the earlier hydrochemical and hydrobiological regimes in the sea. Events in the Aral Sea region during the first half of the 1990s showed that the ideas of the scientists of GOIN and the Institute of Water Problems were closer to reality than those of the developers (Kotlyakov, 1991).

Following the break-up of the Soviet Union, the Concept Plan for overcoming the Aral Sea crisis adopted by the USSR Supreme Soviet in 1991 was not realized. The independent states of Middle Asia had to deal with the problem themselves (Polad-Zade, 1994). In spite of the decisions of numerous congresses and symposia, very few practical moves were made to rectify the situation indicated by the continued fall in sea level. For example, in Uzbekistan the real rates of reconstruction of the irrigation system in 1991 to 1993 turned out to be extremely low, some 40 000 ha against a planned 100 000 ha. The water saved was mainly used for increasing the irrigated areas (Anonymous, 1993). A very disturbing feature of this situation is the termination of observations of the state of the sea. This fact testifies to the refusal of attempts to conserve the Aral Sea and it postpones the solution of the problem to the distant future.

Thus it is clear that under present conditions, the goal of rehabilitating the Aral Sea can hardly be achieved. Now it is necessary to focus efforts on creating normal living conditions for the people in the coastal regions of the Aral Sea and reviving and restoring the environment in this part of the basin. As to conservation within certain limits (an area of water of not more than 25 000 km^2) of the existing water bodies (Large and Small Seas), it will require, according to the estimates from SOGI, a regular flow of mainly fresh river water of not less than 15–18 km^3/year from the Amu-Darya and 3 km^3/year from the Syr-Darya (Bortnik *et al.*, 1991). It will also be necessary to restrict the discharges of the highly polluted collector–drainage waters (up to 9–12 g/l) and the large amount of toxic chemicals they contain.

Analysis of the water management situation indicates that the hydroecological crisis of the Aral Sea basin cannot be overcome without addressing the cardinal question: whether abstractions for irrigation and industry can be restricted and decreased so that river flows can be restored. Such a move will require a significant revision of the existing concept of development based on the prevailing use of the land-water resources of the basin for irrigation, mainly for cotton-growing. Another very complicated problem is the level of the permissible pollution loads in the runoff and this is a problem which should be investigated, and one requiring political support from the nations surrounding the Aral Sea. For this purpose it would be necessary to create and test a system of mutually related socio-economic, hydroecological and water-management models, in order to determine the optimal conditions for overcoming this situation considering the basin and sea as a whole. Development of such models must be based on data provided by the monitoring of the water resources of the Aral Sea basin. This monitoring should include a basin-wide system of observations for the hydrological variables and for the use of water, so that the water resources and their use can be quantitatively assessed on an operational basis and the changes in runoff due to natural and human factors can be determined on a continuous basis. Such a system should include a capability for forecasting the changes in the hydrological and water-management situation. The close dependence of these monitoring and forecasting objectives on information support has resulted in the planning and organization of a common automated information system for the Aral Basin (Dukhovny and Sokolov, 1994; Dukhovny *et al.*, 2000).

5.8 CONCLUSIONS

Asia has a considerable volume of annually renewable water resources, some 13 500 km^3/year. However, their distribution over the continent is extremely varied. South East Asia (6650 km^3/year) and Siberia and the Far East (3100 km^3/year) are the rich regions. Much smaller volumes are available in Transcaucasia (67.9 km^3/year), in Middle Asia and Kazakhstan (181 km^3/year), in Western Asia (490 km^3/year) and an especially small volume in the countries of the Arabian Peninsula (less than 7 km^3/year). The runoff from the continent to the World Ocean is 12 740 km^3/year. Of this water 51% flows to the Pacific Ocean, 28% to the Indian, 19% to the Arctic Ocean and only about 2% to the Atlantic Ocean and its adjoining seas. The water resources in the region of inland drainage in the centre of the continent are 406 km^3/year. These resources are not distributed uniformly in time. During the eight months from April to November, 80% of the total runoff reaches the oceans. However, the runoff from some regions (Northern China and Mongolia, Middle Asia and Kazakhstan, and Western Asia) is very unevenly distributed within the year with up to 50% of the runoff occurring in three months of high water. This pattern is typical of Siberia and the Far East and of Southern Asia, where more than 60% of the annual runoff takes place in the space of three months. Southeast Asia and Transcaucasia have the most uniform distribution of river flows during the year. Because of the continued growth of the population there is a downward trend in the available water resources per head. The variations in the water

resources of Western Asia, South East Asia, the Islands and the mainland regions draining to the Pacific and Indian Oceans, as well as India, China and Pakistan, show this decrease starting in the 1950 to 1960s. No patterns were found in the variations in the rivers discharging to the Atlantic, Pacific, Arctic and Indian Oceans.

The water resources of Asia are employed by the different sectors: industry, agriculture, energy production, and domestic and community purposes. The main user of water is agriculture. The total area under irrigation across the continent was assessed as 175 million ha in 1995, but by 2025 it is likely to increase to 230 million ha. Abstraction and consumption in Asia are 2157 km^3/year and 1565 km^3/year, respectively. By 2025 they are expected to be 3100 km^3/year and 1970 km^3/year. The largest percentage use of the available water resources is in Middle Asia and Kazakhstan (more than 77%), but in those areas it is expected to increase by only 9–10% by 2025. A small increase (less than 20%) is also possible in Siberia and the Far East. This is the region where currently the smallest proportion of the available water resources (less than 1%) is used. South East Asia is placed second among the regions for the abundance of its water resources: it has a water availability of 4500 m^3/year per head. By 2025 this figure is expected to decrease to 4000 m^3/year per head. About 8% of the available water resource is currently used, but by 2025 it is expected to exceed 11%. Throughout Asia water availability is expected to decrease during the twenty-first century with a number of regions and countries experiencing serious water deficits.

If the present trends in water use persist in Southern Asia, Western Asia, and Middle Asia and Kazakhstan, the future availability of water in these areas will decline to the very low values of 590 to 2300 m^3/year per head, which are likely to decrease even further, possibly to one-tenth of these levels during dry years. However, it should be noted that the estimates of water resources and water use in the regions and countries of Asia are uncertain, despite the large amounts of observational data available in some areas. This uncertainty is attributed mainly to the poor hydrological knowledge of the continent, due to errors in the hydrometry such as the location of hydrometric stations relative to the borders of the regions, lack of measurements in many areas and gaps in the records in others. The possible alterations in hydrological regimes due to future global climate change have also not been taken into account. New data on water resources, water use, and water availability are badly needed for a more reliable and complete assessment for Asia for the twenty-first century.

6 Water resources, water use and water availability in Africa

6.1 INTRODUCTION

Africa, with its adjoining islands, occupies an area of 30.1 million km^2. Some 23.11 million km^2 drains to the Atlantic and about 7.01 million km^2 to the Indian Ocean basin, while the runoff from several large areas of the continent does not reach the ocean at all. Africa is the hottest continent on Earth, since much of it lies between the tropics of both hemispheres. It is characterized by sharp climatic contrasts: from dry deserts to equatorial jungles due mainly to the differing amounts and timing of the precipitation. The mean annual precipitation varies considerably: from 20 to 50 mm in the Sahara, to more than 5000 mm in the lower reaches of the Niger River. High mean annual temperatures result in high rates of evaporation. The river network is most dense in the region with an equatorial climate. There are five major rivers: the Congo, Nile, Niger, Zambezi and Orange whose basins together cover one-third of the area of the continent. Of these rivers the Congo is the World's third largest river in terms of the annual runoff (1320 km^3/year), while the Nile is the longest river on Earth (6670 km). African rivers have a vast hydropower potential but are characterized by very variable regimes over time-scales shorter than a year and also over longer periods. Desert areas that have only a sparse network of temporary water courses occupy around one-third of the continent. However, these areas frequently possess significant ground water resources in extensive artesian basins, for example in the Algerian and Libyan Sahara. Geographical position, orography and the climatic features of the continent govern the distribution of runoff and the different hydrological regimes of its regions. By reason of its physical and geographical characteristics and its economic development the continent can be divided into five regions: North, Eastern, Western, Central and Southern Africa.

Knowledge of African hydrology is poor. The total number of hydrological stations operating is about 2000. However, information from most stations is fragmentary and often difficult to acquire. The Nile Basin has been best investigated in terms of the number of stations and length of records. There are at present around 200 observation points in the Nile Basin whereas the Congo Basin is characterized by an extremely sparse observational network. This assessment of the water resources of Africa in this Monograph is based on observational data from 330 key hydrological sites, but data are also used from other stations along with certain meteorological information. The annually renewable water resources of Africa, estimated for the period 1921 to 1985, are about 4050 km^3/year, on average. The water resources of rivers draining to the Atlantic Ocean are 4.5 times greater than those which drain to the Indian Ocean. The water resources of the areas of inland drainage are about 150 km^3/year. The largest water resources (1770 km^3/year) exist in the Central region and the least (41 km^3/year) in the North. The average variability of the water resources of the continent is 0.10. The largest variability, namely 0.34, is typical of the North while the least variability (0.09) is characteristic of the Central region. The variation in water resources displays a weak decreasing trend over the period 1921 to 1985. This is especially obvious during recent decades in the North and Western regions, including the Sahel.

There are 53 countries in Africa. Its population is rapidly increasing and has reached about 710 million. Nigeria has the largest number of inhabitants (109 million), followed by Egypt (61 million), Ethiopia (53 million), Zaire (43 million) and South Africa (41 million). There are 13 countries with fewer than 1 million people. The average population density is 24 persons/km^2. The highest is typical of the island states (200–550/km^2) but it varies from 5 to 70/km^2 for the mainland countries. The gross national product (GNP) per capita in the northern countries varies from \$US500 to 2000. The Republic of South Africa is the most developed country (\$US2600). Some countries of Central and Eastern Africa are the poorest in the world with an income per capita of \$US80–500.

Per head of population, the available water resources are decreasing: they are now one-third of what they were in 1900. For 1995 there were 5400 m^3/year per head, but by 2025 this figure will decrease to 2400 m^3/year per head. In the North there is now only 220 m^3/year per head and by 2025 this figure will have decreased to 60 m^3 for the average year. Non-renewable water resources are used extensively in this region. In the Maghreb countries alone, the available water resources can be increased from this source by 13 to 14 km^3 over the period 1980 to 2025. Water availability in the Central region is extremely high (28 300 m^3/year per head). The differences in water availability reflect the high level of water use in the regions with the smallest water resources (78 km^3/year in the North and 41 km^3/year in the Eastern region) and low water use in the region with the best resources (1.4 km^3/year in the Central region).

The major water user in Africa is agriculture, predominantly for irrigation – 134 km^3/year (62% of the total water used for all needs). There have been high rates of growth in the irrigated area: from 1950 to 1995 this increased three-fold, to some 12 million ha. By 2025 the irrigated area is expected to be 16 million ha with water use in the agricultural sector increasing to 175 km^3/year (1.3 times), for industry and thermal power generation to 18.8 km^3 (a two-fold increase) and in the domestic sector to 59.9 km^3 (up by 3.5 times). There are many water management problems in Africa caused by the lack of water resources, the need for their rational use and protection from depletion and pollution. The Aswan High Dam hydropower complex is widely known and has contributed to the rapid economic development of some countries, but it has also resulted in serious hydroecological and social consequences.

6.2 PHYSICAL CONDITIONS

6.2.1 Relief

The surface of Africa, as compared to the other continents, is only slightly dissected. In its geological past the continent has experienced slow earth movements. As a result the elevated areas were formed: the Atlas Mountains (the highest point is 4165 m), Fouta-Djallon (1537 m), Mount Kilimanjaro (5895 m), the Drakensberg Mountains (3482 m) and the Ethiopian Highlands (4620 m). The plains and plateaux are located mainly in the centre of the continent occupying extensive tectonic depressions: Lake Chad and the White Nile in Sudan, the Congo in Central Africa and the Kalahari in South Africa. Deserts occupy a significant area in Africa (Bromley, 1982). The greatest desert on Earth is the Sahara with an area of about 7 million km^2 extending from the Atlantic coast to the Red Sea. Ahaggar and the Tibesti Mountains rise in its centre and there is the Darfur Plateau in Sudan. The second arid area, the Kalahari Desert (0.9 million km^2), is situated in the south of the continent (Gromyko, 1987). Only the northern ranges of the Atlas Mountains were created by Alpine folding. The remainder belongs almost entirely to the African platform and is composed of Precambrian crystalline and metamorphic rocks. Its eastern margin of over 6000 km is crossed by a belt of fractures, the most extensive on Earth. It is expressed in the relief by narrow very deep depressions most of which are occupied by lakes.

Dormant and active volcanoes occupy the line of fractures, including Kilimanjaro and Mount Kenya (5199 m). The mean height of Africa is 750 m above sea level.

6.2.2 Climate and vegetation

Africa is located in three climatic belts: much of it is in tropical latitudes, the smaller part is in the equatorial belt, and the northwestern and southeastern margins belong to the subtropics. A positive radiation balance throughout the year causes high air temperatures. Africa is the continent with the highest mean annual temperature (about 20 °C). The continent is situated on both sides of equator, hence the seasons of the year in its northern and southern parts are in opposition: the winter period in the Southern Hemisphere corresponds to the summer period in the Northern Hemisphere and vice versa. The temperature differences are comparatively small; however, the character of precipitation and its amount varies considerably within and between years. Much of the continent experiences a moisture deficit and there are pronounced wet and dry seasons of differing duration over large areas (FAO, 1982; UNDP, 1997).

The compactness of the continent and its areal extent increases the continental character of climate, especially in the Northern Hemisphere where permanent high-pressure areas noticeably influence the climate. These are the Azores and South Atlantic Highs over the Atlantic Ocean and the South Indian High over the Indian Ocean. The main circulation transfers tropical air through the tradewinds (northeastwards in the Northern Hemisphere and southeastwards in the Southern Hemisphere). The warm Mozambique and Agulhas currents washing the eastern shores south of the equator raise the temperature of the coast. The cold Canaries and Benguela Currents decrease the temperature and increase the aridity of the western shores in the tropics.

In equatorial latitudes the annual precipitation mostly exceeds 1000 mm per year, decreasing slightly only in eastern regions where the westerly winds become depleted. More than 1500 mm falls on the coast of the Gulf of Guinea and in the basin of the River Congo. Here is the wettest place in Africa – Debundja near the Cameroon Mountains (9655 mm per year). In other regions the annual precipitation is more than 1500 mm. On the windward slopes, in Liberia and Sierra Leone, as well as along the eastern coast of Madagascar, the annual precipitation is more than 3000 mm. A narrow strip with heavy precipitation throughout the year extends from the Niger Delta across the middle part of the Congo Basin to the Rift Valley in Eastern Africa. The driest place in the equatorial belt is the Horn of Africa (Korzun, 1974a). On the coast of the Gulf of Guinea and in the Congo, a temperature of 25–26 °C is observed throughout the year with a range from 1–2 up to 4–5 °C. The Highlands of East Africa have a cooler climate and the highest mountains (Kilimanjaro and Ruwenzori) extend above the snow line which is situated at a height of 4400–5000 m.

Polewards the equatorial climate gives way to subequatorial areas (equatorial monsoon) with humid summers and dry winters. The dry season lengthens from 2 to 10 months with annual precipitation decreasing from 1800 mm to 300 mm. In the Northern Hemisphere the desert is very dry, the annual precipitation decreasing to 100 mm and less in the Sahara. The driest region is in the east where 10–20 mm falls at Aswan and rain does not occur every year. Along the western coast the climate is oceanic desert with a high relative humidity. In the Southern Hemisphere there are three dry areas: the oceanic desert in the west, the moderately dry centre and the east with a summer precipitation maximum. The continent's highest temperatures occur in North Africa. In the summer in the south of Sudan and in the Sahara mean temperatures vary from 30 to 32 °C with daytime temperatures rising to 40 °C. The highest temperature of 57.8 °C was recorded in Al-Azizia (Libya). The dryness of the air and lack of clouds contribute to the fierce heating of the underlying surface (in deserts the sand is heated to 70 °C). There are winter frosts north and south of the tropics, as well as in some mountain regions. Along the western shores of the Sahara and in Namibia there are subocean deserts with lower air temperatures than inland (the Namib Desert is the driest place in Africa in the Southern Hemisphere, with only 25 mm of precipitation per year in some places). The main source of moisture in these deserts is dew and fog. The temperature in the summer months around the Mediterranean coast is usually 20–25 °C and 10–15 °C in the winter months. Snow occurs on the Drakensberg Mountains during winter frontal cyclones and temperatures can fall to −6 °C.

The margins of Africa belong to the subtropical belt. The climate in the north is Mediterranean with precipitation in the colder part of the year. There is snow on the Atlas Mountains and in the Sinai Peninsula and the temperature decreases to below zero (to −8 °C in the Atlas Mountains). The humidity decreases eastwards with distance from the Atlantic Ocean. The amount of precipitation decreases from 500 to 1000 mm per year in Morocco and Algeria to 100–250 mm in Egypt. In the Southern Hemisphere, the Mediterranean climate is pronounced only in the southwest, with a summer precipitation maximum to the east. Normally the climate is sufficiently humid, large amounts of precipitation occurring in the Drakensberg Mountains (up to 2000 mm per year) and in Madagascar (Gromyko, 1987).

Aridity restricts agriculture in many regions of Africa. Where precipitation is less than 600 mm per year, irrigation is necessary. Many regions suffer from drought. In the Sahel and along the fringes of the Kalahari Desert serious droughts are quite frequent, with the recent drought in the Sahel persisting for more than 20 years. Today, savannahs occupy about 33% of the area of Africa, deserts and semi-deserts 40%, with the remaining areas (around 27%) in forests and open woodland. Tropical vegetation dominates the whole of Africa. Extratropical species cover very small areas in the extreme north and south and in high mountains. Where the tropical climate is hot and wet, there are evergreen and mixed forests such as around the Gulf of Guinea and in the Congo Basin. More than 3000 species of trees grow in these permanently humid forests, the trees reaching heights of 40–50 m. Shrubs are usually absent and the herbaceous cover is open.

To the south and north of the humid equatorial forests with decreasing precipitation and an increasing dry period (up to 2–3 months), the vegetation becomes more open and there are alternating humid mixed tropical forests. In the mountain regions of the tropics above 1300 m, there are humid tropical "moss" forests. Dry open woodland with communities of shrub occupies large areas in southern Africa (to 20 °S), predominantly on the high plateau where the dry period lasts for 5–7 months. The trees are 7–25 m high and there is evergreen shrub underbrush with baobabs being typical. Depending on the duration of the rainy season and the amount of precipitation, moving from the equator towards the poles, the grasslands are subdivided into moist high herbaceous areas, dry low herbaceous areas and desert. Dry woods, open woods, shrub communities and desert savannah are represented in the Southern Hemisphere.

The vegetation cover in the Sahara, the Namibian semi-deserts and in the Karroo Plateau deserts depends strongly on the soil characteristics. In the Sahara, the rocky deserts have an irregular tree–dwarf shrub vegetation cover with a thick deep root system. The pebble deserts and mobile barkhans are almost devoid of vegetation. In the deserts of southern Africa plants with pulpy stems and leaves prevail. There is Mediterranean vegetation in North Africa especially within the Atlas Mountains, as well as along the narrow coastal strip from Tunisia to the mouth of the Nile. Subtropical hard-leaf evergreen forests are widespread here, consisting of oaks (rocky oak, cork oak) and maritime pine. In Africa generally the type and distribution of vegetation is highly correlated with precipitation. Humid evergreen tropical and equatorial forests are located where there are large amounts of precipitation (more than 2000 mm) and evaporation about 1000 mm a year. These are the areas where runoff occurs because of the surplus in the water balance.

6.2.3 The river network

The hydrographic network of the continent is very unevenly developed. The highest density occurs in the western and central parts of equatorial Africa where the extensive river system of the Congo (Zaire) is located. North and south of this area the density of river channels decreases. During the dry season the runoff decreases sharply with some of the rivers drying up. In the deserts there are only ephemeral streams except where several large rivers cross

Table 6.1. *Major morphometric characteristics of the principal rivers of Africa*

River	Drainage area, km$^2 \times 10^3$	Length, km
Congo	3680	4370
Nile	2870	6670
Niger	2090	4160
Zambezi	1330	2660
Orange	1020	1860
Chari	880	1400
Juba	750	1600
Senegal	441	1430
Limpopo	440	1600
Volta	394	1600
Ogowe	203	850
Rufiji	178	1400
Cuanza	149	630
Sanga	135	860
Tana	91	720
Kuilu	62	600

Source: Korzun (1974b).

them, the Nile for example. Runoff takes place only during the rare and brief showers or when a cloudburst causes a flash flood. By way of contrast, the density of the river network increases in humid mountain areas with numerous small streams, some emptying to the oceans and others flowing from the leeward slopes to the inner regions.

Areas of inland drainage cover about one-third of Africa, including the Sahara, Danakil and Namibian Deserts, the semi-deserts and desert savannahs of the Kalahari, together with the basins of Lakes Chad, Turkana and several others. The rivers of these areas flow predominantly to the shallow, often saline lakes that periodically dry up to form playas. The rivers of the remainder of Africa flow to the Atlantic or the Indian Oceans. The main watershed is displaced eastwards and only 18.5% of Africa drains to the Indian Ocean through such rivers as the Zambezi, Limpopo, Ruvuma, Rufiji and Juba and those of Madagascar. About 50% of the continent drains to the Atlantic Ocean including all the large rivers such as the Nile, Congo, Niger, Orange and Senegal and the largest lakes (Table 6.1). The largest rivers originate in the areas with more than 1000 mm of precipitation per annum.

The relatively uniform character of the geology and relief reinforce the latitudinal distribution of the physical zones over much of Africa, including the distribution of runoff, which is changed only around the mountain margins. In the Atlas Mountains the largest runoff (100–200 mm) is observed in the upper parts of the northwestern slopes of the High and Middle Atlas Ranges (Korzun, 1974a). Towards the Atlantic Ocean coast the runoff decreases sharply to as little as 10 mm. The regimes of the rivers

flowing from the Atlas Mountains vary dramatically from season to season. During the dry season many river channels are dry. Most of the rivers on the southern slopes draining to the Sahara carry water only during the rainy season. In the karst regions of the Rif and the Middle Atlas there is flow from the limestone strata near the base of the plateau during dry summers, for example, in the Oum er Rbia River.

In the Sahara Desert there is occasional local runoff, particularly during and after the localized cloud bursts which transform the dry channels of wadis and which drain to depressions where short-lived saline lakes known as shotts are formed. More abundant runoff occurs in the northern submontane Sahara where floodwater flows from the mountains along the wadi channels into the desert. Many of these channels are buried under the sands of the Great Western Erg Desert where there are chains of spring sand wells. In the Egyptian Sahara a narrow strip of land is irrigated from the Nile whose discharge originates in precipitation falling in the Highlands of Eastern Africa (the White Nile) and in Ethiopia (the Blue Nile). About one-third of the runoff reaches the sea, the remainder evaporating and being used for irrigation. Ground water recharge occurs along the irrigation canals. Some of this water is returned to the river but a significant portion is lost due to evaporation from the soil and from small sink holes (sebhas). Except for the Nile and short temporary streams along the desert margin, no river reaches the sea from the Sahara. Settlements in the Sahara are situated where ground water reaches the surface in several artesian basins or in oases.

In the Sudan runoff increases from north to south and the percentage of ground water flow in rivers also increases. The rivers that cross this region – the Niger and the White Nile and the Chari, which empties into Lake Chad – originate where precipitation is more abundant. A large amount of runoff is lost by evaporation and infiltration to ground water in the dry zone. For example, the Niger loses more than 50% of its runoff by evaporation in the inner delta and by infiltration in the Sahara. The White Nile loses about half its runoff due to evaporation in the swamps of southern Sudan. Between August and October up to 70–80% of the annual runoff occurs in Sudanese rivers. During the spring most of the rivers dry up. African runoff reaches a maximum (2000 mm) from the mountain massifs of Fouta-Djallon and Cameroon. Runoff increases in the Congo Basin from the marginal plateau to the centre, the river descending from the plateau through a series of waterfalls to the lowest level. Then the river flows in a wide meandering channel, crossing the equator twice and through the Crystal Mountains in a narrow ravine and the series of waterfalls. The river meets the Atlantic in a deep estuary which can take ocean-going ships. The Congo's exceptionally large volume of runoff results from its position in the equatorial and subequatorial climatic zones where precipitation is abundant (Czaya, 1981; Gromyko, 1987).

The Somali Peninsula is very dry and there are few perennial rivers. Only the Juba River reaches the Indian Ocean; the Wabi-Shebelle River is lost in swamps near the coast. The distribution of runoff over the East African Highlands, an important hydrological unit, is very uneven because of the diverse and variable climate and the complicated relief. Here are the sources of several large rivers, for example the Nile, Congo and Zambezi and a number of smaller ones, which flow from the mountains of Kenya and Uganda. Over the remainder of the region, where precipitation is less than 1000 mm, the annual runoff varies from 10 to 100 mm and changes abruptly depending on the character of the relief and geology.

In Southern Africa where the Zambezi, Limpopo, Orange and many smaller rivers are to be found, the runoff gradually decreases from east to west. On the slopes of the Great Terrace the runoff is about 50 mm, increasing to 100 mm on the humid slopes of the Drakensberg Mountains. Runoff decreases westwards towards the dry plains of the Kalahari where less than 20 mm occurs. Here temporary channels, like the wadis of the Sahara, lose water due to infiltration through channel deposits. Closer to the Namibian Desert runoff is almost absent, flows occurring occasionally after showers. Water is lost partly by evaporation and partly by seepage through the bottom of the channels into water bearing horizons. The situation is similar on the Orange River. On Madagascar runoff occurs on the narrow humid coast where the rivers flow down from high steep slopes of the Central Highlands.

There are many lakes in Africa, especially where tectonic fractures and faults occurred such as where Lakes Tanganyika, Nyasa and Victoria are located. Table 6.2 presents data on the largest lakes of Africa (ILEC/UNEP, 1993). The total area of the East African lakes is about 170 000 km². In areas of internal drainage, the lakes are, as a rule, shallow and most are saline. During the dry season many of them transform to solonchaks. The levels and areas of such lakes differ considerably during the year, increasing during the period of rains, decreasing during droughts. The largest of them is Lake Chad whose area varies from 10 000 to 26 000 km² with a mean depth of 2 m. The largest swamps and wetlands are situated in the inner delta of the Niger River, around Lake Chad, as well as around the lakes of East Africa and in the basins of the Upper Nile and Congo. Their total area is about 340 000 km².

Reservoirs form an important component of the network of rivers and waterways. Up to 1954 when Lake Victoria was turned into a reservoir by the construction of the Owen Falls Dam on the Victoria-Nile, there were no large artificial water bodies in Africa. Now there are about 20 large reservoirs in operation or under construction with a total volume of more than 5 km³. There are more than 100 reservoirs with a volume exceeding 100 million m³ and hundreds of smaller ones. Their total volume is more than 1000 km³, approximately 20% of the volume of all the world's reservoirs. Most of these reservoirs produce hydroelectricity, since

Table 6.2. *Major morphometric characteristics of the principal lakes of Africa*

Lake	Volume, km³	Surface area, km²	Maximum depth, m
Tanganyika	17 800	32 000	1471
Nyasa	8 400	30 900	706
Victoria	2 750	68 800	84
Kivu	569	2 370	496
Albert	280	5 300	58
Edward	78.2	2 325	112
Chad	72.0	10 000–25 000	10–11
Shirva	45.0	1 040	2.6
Shalla	37.0	409	266
Mweru	32.0	5 100	15
Tana	28.0	3 150	14
Katnit	14.0	1 270	60
Abaya	8.20	1 160	13
Bangweulu	5.00	4 920	5
Langana	3.82	230	46.2
Fagibin	3.72	620	14
Hora-Abjyata	1.56	205	14.2
Avusa	1.34	130	21
Zwai	1.10	434	7
Upemba	0.90	530	3.5
Gjyer	0.64	213	7
Turkana	–	8 660	73
Rukwa	–	4 500	–
Leopold II	–	2 325	6
Tumba	–	765	–
Gabel-Aulia	–	600	12
Chamo	–	551	12.7
Naivasha	–	140	–

the African rivers possess immense hydropower potential (Africa is second only to Asia in this respect). The largest production of hydropower is on the Congo (380 GW) and Zambezi (137 GW) Rivers. In addition to hydropower the reservoirs are also used for irrigation, flood control, water supply and fisheries (Table 6.3).

6.3 SOCIO-ECONOMIC CONDITIONS AND THE USE OF WATER RESOURCES

Africa is a continent of great potential. It possesses all known mineral resources and large fertile areas. In some regions, in spite of "demographic pressure", there are still areas which are sparsely populated, for example in the Congo, Central African Republic, Zaire, Cameroon, Côte d'Ivoire, Liberia, Madagascar and Zambia. Algeria is third in the world in terms of the size of its unused land area after Russia and Canada (Gorshkov *et al.*, 1994). Nevertheless

Table 6.3. *Principal reservoirs of Africa*

Reservoir	Country	Basin	Year of filling up	Dam backwater, m	Useful volume, km³	Type of Use[a]
Owen Falls (Lake Victoria)	Uganda, Kenya, Tanzania	Victoria-Nile	1954	31	204.8	H F I
Nasser	Egypt	Nile	1970	95	169	I H N F A
Kariba	Zambia, Zimbabwe	Zambezi	1959	100	160.3	H N I F A
Volta	Ghana	Volta	1965	70	148	H N I F
Cabora Bassa	Mozambique	Zambezi	1977	127	62	H I N F
Kossou	Côte d'Ivoire	Bandama	1972	57	28	H I F
Suanity	Guinea	Konkure	1961	125	17.2	H
Kainji	Nigeria	Niger	1968	41	15	H I N F

[a] S, water storage for different purposes; B, water supply; I, irrigation; T, timber rafting; A, flood control; R, recreation; F, fishery; N, navigation; H, hydropower engineering.

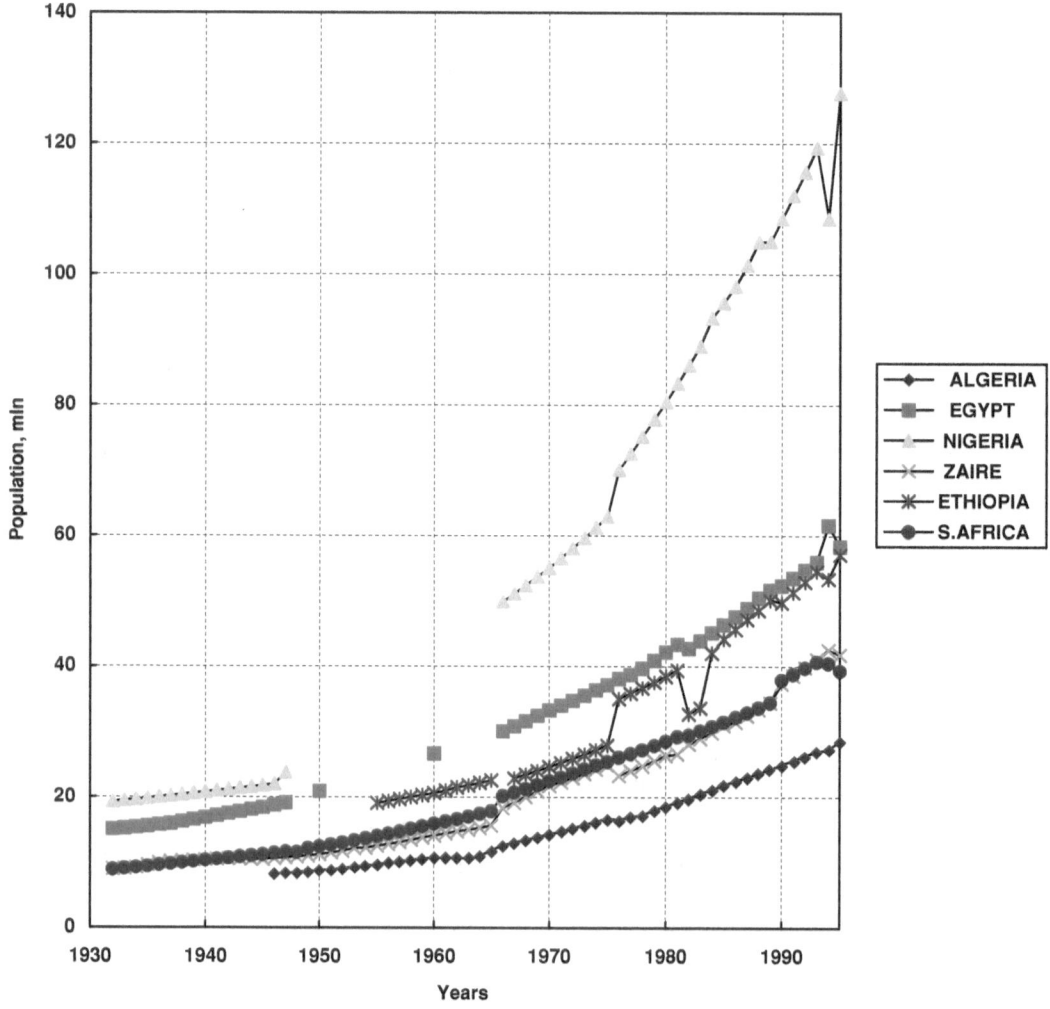

Fig. 6.1 Dynamics of population in the largest countries of Africa (million people).

most African countries possess considerable manpower resources and have great potential as tourist attractions. The growth of population is one of the most significant factors governing social and economic processes and water use in African countries. Africa differs from the other regions of the developing world in that its demographic expansion has been delayed. For much of the world the maximum population growth rates occurred in the late 1960s, but in Africa the most rapid rise in population took place in the late 1970s and 1980s (Fig. 6.1). According to UN estimates, during the period 1950 to 1955 Africa had 11% of the world population growth, but from 1958 to 1990 this increased to 20% (Bessonov, 1994). Both the North African (Arab) and traditionally sub-Saharan African types of population reproduction are characterized by high birth rates. Lower birth rates are typical of some island states and of some Central African countries.

Diversity is a common characteristic of African countries, from the standpoint of their social and economic structures, their natural resources and their demographic potential and geographical locations. Amidst the nations that the United Nations refers to as belonging to the least developed world, there is the Republic of South Africa – a highly developed country. Population density is extremely uneven across the continent. In the Chad Republic, for example, it is 4/km² whereas in adjoining Nigeria it is 60/km² (Sagoyan, 1993; Anonymous, 1995a). There are many nations with a small population – 36 of them have fewer than 10 million inhabitants, nine have fewer than 1 million. Only 10 countries have more than 20 million people. High population growth rates are combined with its increased concentrations in towns and cities (Pulyarkin and Lipetz, 1991). Where in 1960 the urban population ranged from 1.5–1.8% (Lesotho, Botswana) to 45–50% (South Africa, Djibouti), by 1990 it varied from 7–10% (Burundi, Rwanda, Uganda) to 60–80% (South Africa, Tunisia, Libya, Djibouti and Equatorial Guinea).

Industrial growth and urban development in some regions have contributed to the increased demand for water, especially for the mining and processing industries, as well as for the construction of large hydropower stations. In 1990 most water was used by industry in Egypt (4.6 km³), South Africa (1.45 km³), Algeria (0.7 km³), Nigeria (0.6 km³), Morocco (0.3 km³), Sudan (0.2 km³), Libya (0.1 km³) and Zimbabwe (0.1 km³). Domestic water use varies considerably across the continent (from 10 to 180–200 l/day per head) and depending on GNP (Fig. 6.2). The diagram is based on data contained in the UN Yearbook (UN, 1993a) and from the World Resources Institute (WRI, 1990, 1992) supplemented for some countries by data published by French and Arab investigators (Margat, 1990, 1994, 1995; ROSTAS, 1995; UNESCO, 1995a). More water is used per person in the arid northern countries and less in countries with a humid equatorial climate (Congo, Gabon).

One of the main prerequisites for the development of Africa is the building of an industrial infrastructure in energy, irrigation and transport. However, up to the present development has been random in its distribution, both across the continent and within each country. In the first stage of industrial development, small power plants were constructed in several coastal regions, particularly in South Africa using imported fuel. In 1929 in Morocco the first significant hydro-electric station was built on the Oum er Rbia

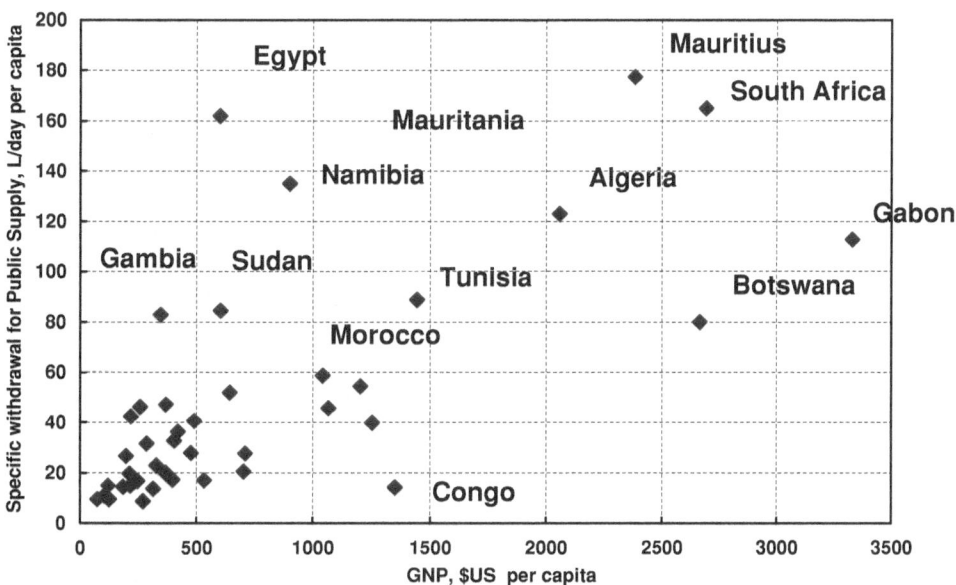

Fig. 6.2 The relationship between specific public and domestic water use (l/day per head) and gross national product per head (US$) in countries of Africa (1990).

River (Pulyarkin and Lipetz, 1991). During the pre-war period more sources of electrical energy were developed, for example in the Maghreb (Morocco and partly Algeria) and in the Belgian Congo. Some hydropower was developed on Madagascar and in Angola. After the Second World War, the construction of thermal and hydropower stations continued in the Maghreb, in Zaire and Angola. Important stations were built in Uganda, Guinea, Mozambique, Congo, Cameroon, Côte d'Ivoire, Egypt, on the border of Zimbabwe and Zambia, in Ethiopia, as well as in Ghana and Nigeria (Dmitriyevsky, 1967).

The 1970s were characterized by the rapid development of power stations. The percentage of energy produced by hydropower stations rose from 13% to 16% in the early 1960s, to 30–32% in the early 1980s and to 60–65% if South Africa is not included (Arkhipova, 1986). By the mid-1980s hydropower was in the lead in 28 countries of Africa. There are however, countries where hydropower is actually undeveloped (Libya, Western Sahara, Benin, Burkina Faso, Gambia, Guinea-Bissau, Mauritania, Niger, Senegal, Somali, Chad) (Zair-Bek, 1988), some of these being arid countries. The highest hydropower potential in Africa is in Zaire (530 TWh), followed by Ethiopia (162 TWh), Cameroon (11.5 TWh) and Angola (100 TWh). Construction continues in a number of countries. For example, in Angola the capacity of stations under construction was 130% of the total capacity of existing power stations as of 1991. In Cameroon, Ethiopia, Morocco and Niger these values are 23%, 45%, 50% and 45%, respectively (Gleick, 1993, 1998). An important feature of the modern geography of African hydropower is its movement inland, to support the expansion of the mining industry and to stimulate the development of new areas of the economy.

As a result of the interaction of natural, economic and social systems and the consequences of the industrial exploitation of water resources, several Hydroeconomic–Ecological Systems (HEES) were formed in Africa (Zair-Bek, 1988). Figure 6.3 illustrates in a general way the present location of the main types of HEES. Historically the first HEESs were formed in the old irrigation regions of the Maghreb, in the lower middle reaches of the Nile in Egypt and in the upper middle reaches of the Niger. Development of natural resources and mining began in the coastal regions (the coast of the Gulf of Guinea) extending to southern and inland regions. Large multi-purpose reservoirs and the distribution systems for water supply for domestic and industrial purposes (Mamilton and Maizels, 1989; Dmitriyevsky, 1990) have various ecological, economic, and socio-geographical (settlements, preservation of traditions, etc.) impacts. Of the projects implemented during recent decades, the dam on the Kru River in Morocco for water for Rabat, the treatment plant and reservoirs on the Elil and Caceb Rivers and also on the Merjerda River for water for Tunis are noteworthy. In 1982 in South Africa the new water supply of Cape Province began operating. The largest Southern African system was built on

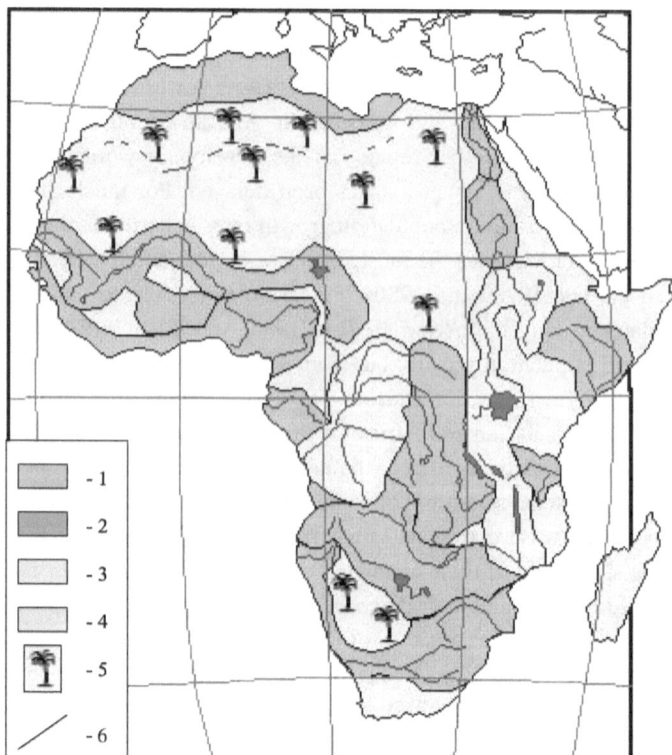

Fig. 6.3 Distribution of main types and subtypes of Hydroeconomic and Ecological Systems (HEES) in Africa. 1, HEES for a complex use of water resources (hydropower engineering, irrigation, industrial and municipal water use, fisheries, transportation); 2, HEES for traditionally irrigated land use, where modern irrigation systems and small hydropower engineering were developed (fisheries, industrial and municipal water use, small-size fleets); 3, HEES with the predominance of industrial and municipal water use (fisheries, water transportation); 4, HEES of commercial fisheries and water transportation developed on the basis of hydropower reservoirs; 5, HEES forming on the basis of oasis irrigation (traditional and modern), oasis breeding farms; 6, Boundaries of HEES now being formed.

the Orange River. There is also a system on the Vaal River supplying water to the main industrial regions; an installation on the Usutu River supplying thermal power stations in the Camden, Kril, Kendal and Matla regions with water; a hydropower unit at the Palmit River providing Cape Town with 30 million m^3 of water per year. In 1988 the governments of Lesotho and South Africa signed an agreement for the construction of a large hydropower complex in the mountains of Lesotho (South African Information Service, 1993). The influence of large reservoirs on water resources and climate in the tropics is more significant than in the temperate and cold zones. The high rates of evaporation from natural water bodies, reservoirs, irrigation canals and irrigated areas can result in a changed precipitation regime so reducing the total water resources (Maganza, 1996, 2000; Talling and Lemoalle, 1998). According to Zair-Bek (1988), the annual evaporation from six African reservoirs alone – Volta, Cabora Bassa, Kainji, Kariba, Kossou and Nasser – is about 14% of their total effective volume.

Irrigation in Africa is not so widespread as in Asia. From recent FAO data for the mid-1990s (FAO, 1995a) the largest irrigated areas are in Egypt (more than 3.2 million ha), in Sudan (1.9 million ha), in South Africa (1.3 million ha) and in Morocco and Madagascar (1.1 million ha each). Of other countries, those in the Maghreb are noteworthy: Algeria (0.45 million ha), Libya (0.47 million ha), Tunisia (0.36 million ha), as well as Nigeria (0.2 million ha). Elsewhere the areas are smaller: Ethiopia (about 0.2 million ha), Tanzania (0.15 million ha), Kenya (0.07 million ha), Zimbabwe (0.12 million ha) and Mozambique (0.11 million ha). In 1995 the total for Africa was about 12 million ha. There has been rapid growth in the development of irrigation in Africa with an almost three-fold increase over the last 30 years (FAO, 1995a, b). In some countries the area irrigated has increased between 10 and 18 times (Burundi, Zambia, Sierra Leone, Malawi, Niger and Chad). The largest growth in absolute terms in the irrigated area over this period was in Morocco (970 000 ha), Sudan (850 000 ha) and Madagascar (615 000 ha). Irrigation has been developed in some agricultural regions in Algeria, Morocco, Tunisia and Egypt, in the basins of the Nile in Egypt and the Sudan, along the Niger in Mali, and on the Juba in Somalia, in the upland regions of southwestern Kenya and northwestern Tanzania, as well as in desert oases.

Some countries already show a quite high level of use of water resources and have considerable problems with water supplies. The largest water resources deficit is in Libya where surface waters are extremely limited and mainly ground water is used for irrigation. There are around 100 000 wells and artesian boreholes in the country (Abramova, 1993); the total abstraction of ground water in 1995 (UNESCO, 1995a) was 1.74 km^3 for agriculture and 0.37 km^3/year for industry. Irrigation is based on ground water in some other countries (FAO, 1995b) such as: Djibouti (100%), Tunisia (61%), Botswana (44%), Morocco (31%) and South Africa (18%). Desalination plants contribute to the available resources; for example in Libya in 1989 there were 386 large desalination plants operating with a total capacity of 619 000 m^3/day, the largest capacity in Africa. There are 123 large plants in Algeria with a total capacity of 176 000 m^3/day, 110 (68 000 m^3/day) in Egypt and 39 (23 000 m^3/day) in Tunisia. Smaller plants are operating in Morocco, Sudan, Mauritania, Nigeria and other countries of Africa (Gleick, 1993, 1998; Pearce, 1996; Shahin, 1996).

At the present time in Egypt, a country where irrigation has been practised since the time of the Pharaohs, there are large problems with water supply for irrigation, in spite of experience of regulating and using the Nile runoff. These problems are mainly related to the long dry period which started in the early 1980s and the increased water use in the countries located in the upper part of the basin (Smith and Al-Rawahy, 1990). The main irrigation areas in Egypt are situated in the Nile valley and delta, but there are some small areas along the coast of the Mediterranean and on the Sinai Peninsula where irrigation uses the seasonal rainwater. Here are to be found the most ancient methods for the collection and use of rainwater (ROSTAS, 1995; UNESCO, 1995a). Agriculture in Egypt has undergone profound changes over the last 100 years due to the move to year-round irrigation, mechanization, application of pesticides and fertilizers and the introduction of new varieties of cultivars. In the 1980s Egyptian farmers began to use drip irrigation and the plastic hothouses which had become widespread in the other Arab countries from the mid-1970s. By 1990 around 3000 such hothouses had been set up (Library of Congress, 1991). Since the government had made investment in irrigation for many decades without parallel investment in drainage, more than 70% of irrigated lands in Egypt were subject to waterlogging and salinization. From 1974 government finances were available for the construction of drainage schemes in the Nile Delta and in Upper Egypt; between 1981 and 1986 between 50% and 60% of the irrigated area was provided with drains and major schemes were continued in the 1990s. Comprehensive land-use planning began in Egypt in 1966. From that time, due to the increase in the planting of crops with a high water demand, water use in irrigation has increased markedly. In 1977 Egypt commenced work on the General Plan for Development and Use of Water Resources. Efforts were focussed mainly on the analysis of the possibilities for using all types of water resources including agricultural drainage, treated sewage and industrial waste water (Arar, 1987). In 1989 the amount of recirculated irrigation water in use in the Nile Delta was about 2.7 km^3/year (Abu-Zeid and Abdel-Dayem, 1991). For 1990 the volume of recycled water in the industrial and domestic sectors was 4.6 km^3. It was estimated by FAO and the Ministry of Agriculture that by 2000 this figure could reach 7.6 km^3/year (Biswas, 1991). Egypt has also been implementing one of the largest projects (the El-Salam Canal) for irrigating 205 000 ha in the eastern delta and in the Sinai Peninsula using 2.3 km^3/year Nile runoff and drainage discharge (FAO, 1995).

The droughts of the 1980s have drawn the attention of the world to the countries of the Nile Basin, especially to Sudan and Ethiopia. Although climatic conditions have improved a little in recent years, the countries of the Basin are still concerned with the possibility of periodic droughts, the deterioration of water quality and famine (Smith and Al-Rawahy, 1990; Sircoulon et al., 1999; Maganza, 2000). Considerable difficulties surround the solution of water problems in Ethiopia, where about 60% of the Nile runoff originates; while an increasing amount of water is required for expanding irrigation in the Sudan (Dmitriyevsky, 1990). Lack of water resources in a number of countries has prompted the introduction of new water-saving irrigation techniques. Morocco is one successful example. Work on the optimization of the use of the restricted water and land resources has been carried out in South

Africa (Conley and Hansman, 1985). In terms of water availability per head, South Africa is approaching the group of countries which face the most serious shortages. In addition the development of irrigation is restricted by the reduction of funding for irrigation. The main factor for most African countries, which limits the further development of irrigation, is their restricted economies and limited financial possibilities (Gavrilov, 1989). An analysis of recent data indicates a close relationship between the GNP (per capita) and the rate of growth of the area irrigated. This is typical of both the developed countries of the continent (South Africa, Morocco, Libya) and the countries with the lowest income (Madagascar, Tanzania, Kenya, Malawi, etc.). These circumstances are of great importance for forecasting the demand for water for irrigation. The reliability of these forecasts depends, in many respects, on the verisimilitude of the scenarios for the economic development of the countries and regions.

Over the last 30 years a series of severe and widespread droughts and serious floods have played an important part in the development of African water resources and of the economy based on them. Especially pronounced were the droughts and related "hungry years" in the Sudan–Sahel zone[1] which were repeated through the twentieth century (1900–03, 1913–14, 1926–7, 1929–31, 1944–5, 1955, 1969–73, 1984–8, 1988–90). The worst drought (1969–73) contributed to the change in the vegetation zones (the conversion of forest zones and perennial grasslands to savannahs, the desertification of the savannahs) and to the southern borders of the Sahara advancing over 350 to 500 km (Neronov, 1990; Gonzalez, in press). Ecological disasters have been superimposed on a number of degrading effects induced by human activities, such as the rapid growth of population, pre-industrial management methods, incorrect use of irrigated lands, transfer of dirty industries and their waste from other countries to Africa. At the present time around three-quarters of the African population lives in these disaster zones: for example in the Sudan–Sahel zone, and in the deserts and semi-deserts of the south (Palmer and van Rooyen, 1998). The pasture lands and dry farming areas of the north are 80–90% affected by desertification (UNEP, 1997). Large-scale deforestation has also been undertaken. Over several decades the tree–shrub vegetation complex has retreated from Khartoum by over 100 km (Vol'skaya and Yagya, 1991). Woodlands have been almost cleared within 70 km of the capitals of Niger and Burkina Faso. It is predicted that by 2000 to 2010 quite a number of countries could be completely deforested (Gambia, Guinea-Bissau, Kenya, Lesotho, Côte d'Ivoire) (Bessonov, 1994).

In the Sahel unplanned urban development and growth of industry causes considerable damage to the environment. In some regions it brings a greater concentration of population and a sharp increase in pollution and a rise in the rate of depletion of water resources, especially the deep non-renewable ground water sources.

In order to avoid the worst effects of urbanization in arid and semi-arid regions, it is necessary to carry out a wide range of measures for water supply, drainage and sewage disposal. However, the implementation of such methods is impossible due to the economic difficulties of the region. Suffice to say that even such large states as Chad, Niger, Mali and the Central African Republic belong to the group of the poorest countries of the world.

6.4 HYDROLOGICAL DATA

Less is known about African hydrology than about the hydrology of other continents. The total number of stations in the continent's hydrological network was about 2000 in the 1970s (Korzun, 1974b), but the number has declined since then (WMO/UNESCO, 1991). The distribution of stations is and has been highly irregular. The Nile Basin has been best studied with respect to hydrology. In this basin the observations have been carried out from ancient times: the Nilometer at Aswan existed from about 2000 BC. Systematic measurements of the discharges of the Nile were started in Egypt in 1871 because of the need to develop irrigation and for navigation. In 1950 the Egypt Irrigation Administration and the Hydrological Service of Uganda began regular measurements of water levels and discharges in the Nile and its main tributaries from the mouth to the Kagera River. At the present time the hydrometric network in the basin consists of about 200 stations, the densest network being in the upstream of the basin. Downstream from Juba the number of stations decreases and they are located mainly along the Nile.

The Congo Basin has an extremely sparse hydrological network. The longest records of flow in the Congo River are for Kinshasa and date from 1903. Measurements in the basin have been made by more than 200 gauges. However, many of these records are fragmentary because the measurements of discharge were made sporadically (mainly using floats). Some parts of Western Africa have been the subject of a number of hydrological studies, such as Senegal, Mali and Burkina Faso, but there are other countries which are poorly covered by hydrological observations such as Nigeria and Niger and not so well known in hydrological terms. In the northern part of Western Africa from 12° N to the boundary with the Sahara Desert, hydrological studies have been made from time to time. The large rivers, namely the Niger and Senegal, have quite long records but there are numerous small rivers, flowing from the slopes of Fouta-Djallon to the Atlantic Ocean which have not been studied sufficiently.

1 The UN Permanent Interstate Committee for Drought Control in the Sahel (CILSS) includes 15 countries in the Sudan–Sahel zone: Mauritania, Senegal, Mali, Niger, Burkina Faso, Chad, Gambia, Cape Verde Islands, Sudan, Nigeria, Cameroon, Ethiopia, Somali, Kenya and Uganda.

Some of the continent's mountains have a relatively dense network of hydrological stations, but they are sited mainly in the valleys at low altitudes (1000 to 1500 m). There are many rivers in Morocco, Tunisia and Algeria where there is a dense network of well-equipped gauging stations because of the need to regulate the use of river water for irrigation, domestic supplies and for power production. Zimbabwe, South Africa and other Southern African countries have been comparatively well studied because of the demand for water for the mining and other industries, because of domestic demand and the need of water for agriculture. Recent hydrology in Southern Africa has benefited from the UNESCO Friend Project and the different studies that have been conducted under its aegis (WMO/UNESCO, 1991). The rivers of Madagascar have not been uniformly studied: the eastern slope of the Central Ridge has fewer stations than the western.

The publications devoted to the questions of hydrology of Africa are quite numerous. Of significant interest are the studies performed by L'vovich (1972), Dmitriyevsky (1967), Karasik (1970) and Baumgartner and Reichel (1975) containing detailed analyses and summaries of the hydrological data. The UNESCO publication *Discharges of Selected Rivers of the World* (UNESCO, 1993) is a significant contribution to the studies of African rivers. A map of the mean annual runoff of African rivers was first compiled by L'vovich (1945). Then a map of the runoff of Africa was constructed by Karasik (1970) who used the records from over 100 gauging stations but without the reduction of the data to a single period. For regions without runoff records, he deduced the runoff from precipitation. There are also the maps of runoff, precipitation and evaporation for the continent compiled by Baumgartner and Reichel (1975). The water resources evaluated using these maps are likely to be underestimates due to omission of the errors of precipitation measurement. The map of mean annual runoff of Africa compiled by Nemaltsev (1969) is even more detailed. The *Atlas of World Water Balance* (Korzun, 1974a) also presents the map of runoff of Africa. This map was based largely on the data from 800 hydrological and meteorological stations with a single period being assumed (1945–65). For some regions the runoff was related to a runoff coefficient and to an aridity index.

The publication *Water Resources of African Countries* (FAO, 1995b) presents the first assessment of the long-term mean resources for the various African countries, including data on runoff and ground water. The following attributes were used to characterize the water resources of the different countries: the runoff in those local areas where the runoff originates, the cross-border flows and the sum total of these resources. The local runoff and the inflow were estimated for both surface water and ground water runoff and assessments were made of the cross-border runoff. Assessments of water resources are given for different observational periods and are quite approximate.

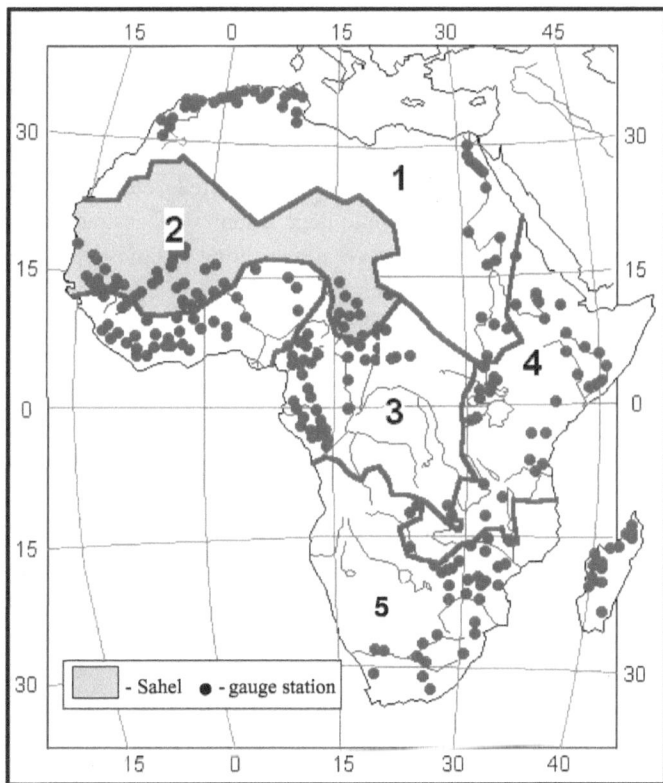

Fig. 6.4 Zoning and the scheme of location of main runoff gauge stations in the natural regions of Africa. 1, North; 2, Western; 3, Central; 4, Eastern; 5, Southern.

Due to the diversity of the continents's physical geography and the range of socio-economic conditions found over African, the assessments of water resources and the study of the trends and changes in water use and water availability were carried out for large natural regions. The countries with a similar physical geography, with populations from the same racial background and with like economic and political status were combined. On this basis, the continent was divided into five regions: North, Western, Central, Eastern and Southern Africa (Fig. 6.4). In general these regions coincide more or less with the normal geographical regions of Africa.

NORTH AFRICA

The region of North Africa is located predominantly in the subtropical zone and it includes: Egypt, Sudan, Libya, Tunisia, Algeria, Morocco and Western Sahara. Most of the vast Sahara Desert, the Atlas Mountains and the lower valley and delta of the Nile River are within its limits. Its area is 8.78 million km², about one-third of the continent, and it contains about 160 million people. The largest country is Sudan (2.5 million km²), and the country with the largest population is Egypt, with 62 million people. Lake Nasser is located in Egypt and is one of the world's largest reservoirs. There is a volume of ground water, part of which is intensively used (in the Ahaggar, Dahla, Faraff, etc.). The region has a relatively high level of economic development and the presence of large areas of irrigation with ancient irrigation traditions. Egypt and Sudan are among the 20 countries with the largest areas of irrigation. The hydrological knowledge of this area is relatively good, there are high absolute levels of domestic and industrial water use, while large volumes are abstracted and consumed by the agricultural sector resulting in almost complete depletion of the surface water resources (99% of the runoff). The main water management problems are related to saving water and increasing the volume of freshwater resources.

WESTERN AFRICA

The region of Western Africa, with a total area of 6.96 million km^2, includes: Mauritania, Senegal, Mali, Gambia, Cape Verde, Guinea-Bissau, Guinea, Sierra Leone, Liberia, Côte d'Ivoire, Burkina Faso, Ghana, Togo, Benin, Niger, Nigeria and Chad. It covers tropical deserts, savannahs and equatorial forests between the Sahara and the Gulf of Guinea. Western Africa is very rich in water resources with a number of large rivers. The largest of them is the Niger which is more than 4200 km long. The world's largest reservoir, the Volta reservoir, is situated on the Volta River. Nigeria has the largest population of any African country. It occupies one-fifth of West Africa and has more than 50% of the population of the region, some 111 million people. Nigeria also has the largest area of irrigation (0.9 million ha) but across the whole region the irrigated areas approach 1.5 million ha (four times less than in North Africa).

CENTRAL AFRICA

The region of Central Africa, with an area of 4.08 million km^2, includes Cameroon, the Central African Republic, Equatorial Guinea, São Tomé and Principe, Gabon, Congo and Zaire. This natural area occupying western equatorial Africa includes the extensive flat, wet, Congo Basin. The only highland areas are the Adamaoua Mountains in Cameroon which reach 3008 m and the volcanic massif Mount Cameroon, up to 4070 m above sea level. The Congo Basin has the densest river network in Africa, the Congo (Zaire) being the largest of them. The population of the region is about 63 million. Due to the abundant water resources and the relatively low population density, the region has the least stress on its water resources of any in the continent.

EASTERN AFRICA

The region of Eastern Africa, with a total area of 5.17 million km^2, includes Ethiopia, Djibouti, Somalia, Kenya, Uganda, Tanzania, Rwanda, Burundi, Zambia, Malawi and Madagascar and covers the elevated margin of the continent, namely the Ethiopian Highlands, the Somali Peninsula and the East African Highlands. Torrid deserts and wet forests, diverse savannah and open woods all oc-

cur in this region. The highest mountains are covered by perennial snow and ice caps. The largest African lake, Lake Victoria, is situated in the central part of the East African Highlands in a trough. The White Nile and the Blue Nile originate in this region and the highest mountain of the continent, Mount Kilimanjaro (5895 m), is also located here. The population of the region exceeds 194 million and by 2025 it is expected to increase 2.5 times. There has been rapid expansion of irrigation in the majority of countries, the irrigated area reaching 1.8 million ha by the mid-1990s, and there is quite a high use of surface water resources.

SOUTHERN AFRICA

The region of Southern Africa, with an area of 5.11 million km^2, includes Angola, Mozambique, South Africa, Namibia, Botswana, Zimbabwe, Swaziland and Lesotho. The region is located in subtropical and tropical latitudes with various landscapes from forest to desert. Its centre is occupied by the sand deserts of the Kalahari surrounded by uplands and mountains. The Drakensberg Mountains are located in the southeast. The largest rivers are the Limpopo and Zambezi flowing to the Indian Ocean and the Orange and Cunene which flow to the Atlantic Ocean. The rivers provide large amounts of hydropower. The Kariba reservoir with an area of 4450 km^2 is located on the Zambezi River. The population is about 84 million people and it is the region with the highest percentage of industrial water use (about 15%). A great deal of attention is paid to irrigation and it is most developed in South Africa with 1.2 million ha, but the main problems are related to the lack of available water resources.

THE SAHEL

The Sahel, which is part of North and Western Africa, is a narrow (up to 400 km) transient strip of semi-deserts and desertified savannas, extending from the Atlantic coast in Mauritania, Gambia and Senegal across Mali, Burkina Faso and Niger to Chad. In the north the precipitation is 100–300 mm per year and in the south 300–600 mm. Rain is a rare event and 80–90% of the moisture is evaporated. The zone is subject to frequent prolonged droughts occurring on average every 30 years. In 1984 a very severe drought spread across much of the area, which turned into a national disaster for many countries bringing famine, death and destruction.

6.5 DISTRIBUTION OF WATER RESOURCES IN TIME AND SPACE

6.5.1 Data and methodology

The number of stations in the hydrological network, their distribution, the duration and quality of the records and similar factors mean that the assessment of water resources for African countries and regions is not adequate, even with additional hydrometeorological data for precipitation and temperature. The publications *Discharges of Selected Rivers of the World* (UNESCO, 1993), and *Discharge of Selected Rivers of Africa* (UNESCO, 1995b) provided data, as did the Global Runoff Centre (Koblenz, Germany), as well as certain countries such as Morocco, Mali and Côte d'Ivoire. Using this information an African database was set up at the State Hydrology Institute (SHI). This database contains monthly runoffs over the period of observation, the location of the stations and the areas of the basins above the stations. For assessing the water resources, more than 330 hydrometric sites were selected from the available observation points, operating over various time periods (Fig. 6.4, Table 6.4). Table 6.4 shows that the largest number of stations are located in Burkina Faso, Cameroon, Madagascar, the Central African Republic, Uganda and Ghana. The Nile in Egypt has the longest record (116 years) and the Congo River in Zaire has 82 years of record.

The observations of runoff are often discontinuous and short. To make them longer and to reduce them to one period (1921 to 1985), both for the monthly and the annual periods the correlation methods (single and multiple correlation) were widely used, as presented in Chapter 2 (Shiklomanov, 1997, 1998a, 2000a, b; Babkin, 1998; Shiklomanov et al., 2000). The analogue stations were selected on the basis of the similarity in physical-geographical conditions in the basin in question and between basins. Frequently it was impossible to establish a close correlation between stations. Hence a full correlation relation between the runoff of the river in question and the runoff of several analogue rivers was considered. The gaps in the monthly runoffs were eliminated by using the regression equations between the runoff values of neighbouring months for the same river. In case of low correlation ($r \leq 0.75$), a graphical analysis of the regression relationship was performed. Data on precipitation and temperature both averaged over the basin and data from separate stations were used. For these studies regression equations between the runoff of one river and the runoff of several rivers were quite frequently replaced by simple equations. In this case the runoff of all the independent rivers was summed up and this sum was represented as one independent variable from which the required regression equation was calculated.

The rivers of Africa were found to have peculiar regimes. The runoff in even adjoining basins was often poorly correlated. Due to the large percentage of evaporation and water resources changes in the basin water balance for some years, runoff is often poorly correlated with precipitation. Hence for gap filling and lengthening short series, a method of interrelation of water balance elements was applied (Chapter 2, equation 2.6). In this event the runoff values calculated for separate years are approximate values, since in using equation 2.6 one has to neglect an important element – the change in water resources in the river basin. In connection with the fact that the change in the water resources in the basin over a long

Table 6.4. *Observations of river runoff in the countries of Africa*

Country	Area, km² × 10³	Number of observation stations	Area per station, km² × 10³	Number of years of observations
Algeria	2382	12	199	3–4
Benin	113	5	23	17–40
Burkina Faso	274	22	13	9–39
Cameroon	475	18	26	10–51
Central African Republic	623	15	42	20–65
Chad	1284	9	143	20–58
Congo	342	12	29	10–35
Côte d'Ivoire	323	11	29	2–3
Egypt	1001	6	167	12–116
Ethiopia	1222	10	122	1–6
Gabon	268	10	27	6–46
Ghana	239	14	17	9–49
Guinea	246	7	35	3–33
Kenya	583	2	292	22–42
Lesotho	30	4	8	10–20
Liberia	111	6	19	3–7
Madagascar	587	16	37	3–35
Malawi	119	4	30	9–29
Mali	1240	13	95	19–83
Mauritius	1031	6	172	4–8
Morocco	447	8	56	4–39
Mozambique	802	8	100	2–4
Niger	1267	5	253	20–61
Nigeria	924	3	308	1
Rwanda	26	3	9	27–29
São Tomé and Principe	1	7	0.1	2–9
Senegal	196	13	15	12–81
Sierra Leone	72	7	10	2–4
Somalia	638	6	106	28
South Africa	1221	11	111	13–23
Sudan	2506	8	313	9–72
Tanzania	945	6	158	18–25
Togo	57	4	14	10–16
Tunisia	164	6	27	3–4
Uganda	263	15	18	4–32
Zaire	2345	1	2345	82
Zambia	753	8	94	5–30
Zimbabwe	391	12	33	11–30
Africa as a whole		333		

period is close to zero, determining the average runoff for the site in question using the particular method is usually quite reliable. The runoff of the Zambezi River was assessed in this way for the years without data.

Table 6.5. *Renewable water resources and water availability by natural–economic regions of Africa for 1921–85*

Region	Area, $km^2 \times 10^6$	Population (1994), $\times 10^6$	Water resources, km^3/year				Coefficient of variation, C_v	Potential water availability, $m^3 \times 10^3$/year	
			Inflow	Local				Per km^2	Per head
				Average	Maximum	Minimum			
North	8.78	157	140	41	96	19	0.34	4.7	0.71
Western	6.96	211	30	1088	1948	581	0.28	156	5.22
including Sahel	5.30	46.9	77	104	175	52.3	0.29	19.6	3.04
Central	4.08	62.9	80	1770	2263	1453	0.09	434	28.80
Eastern	5.17	193	29	749	940	504	0.11	145	3.94
Southern	5.11	83.5	86	399	549	270	0.14	78.1	5.29
Africa as a whole	30.1	708		4047	5082	3073	0.10	134	5.72

The runoff regime of the typical river varies across the continent. In the equatorial zone and around the Gulf of Guinea, where precipitation is heavy, the volume of runoff increases with the increase in basin area. For rivers draining from the mountains, e.g. the Zambezi and Nile, the runoff disperses on the surrounding plains and large losses are observed. Especially large losses are typical of the largest rivers, for example the Congo, Nile, Niger, Zambezi and Orange. The water resources of the Nile Basin were determined as follows.

1. A coefficient which took into account the runoff from the Nile tributaries (the Baro, Tekeze and Atbara Rivers), was introduced to the runoff of the Blue Nile at Khartoum.
2. The runoff in the left-bank tributaries without observations was determined from the runoff of the Bahr-el-Jebel River, with a correction coefficient assessed using the runoff map (Korzun, 1974a).
3. The calculated runoff values were summed and to this sum was added the runoff of the Bahr-el-Jebel River at Mongalla for the period up to 1962. For the period 1962 to 1985 the runoff of the Bahr-el-Jebel River at Mongolla was assessed using precipitation data for that basin.

The water resources of the Congo Basin were determined from the runoff at Brazzaville. The runoff series at this site was reduced to the standard period from the flow at Kinshasa. The difference in the Congo runoff at these sites exceeds $30 \, km^3$/year over the period of observation. The runoff at Brazzaville was used to estimate the discharge at the mouth of the Congo by taking account of the additional runoff from the basin, located below this site (see Chapter 2). The water resources of Niger were estimated as the sum of the runoff of the Niger River at Koulikoro, and those of the Bani River at Douna, together with that of the Benue River at its mouth multiplied by the coefficient taking into account the runoff in its tributaries (Kaduna, Sokoto, Dallol and the other tributaries

of the Niger along the segment from Koulikoro to Niamey). The water resources of the Zambezi River were evaluated in two stages.

1. The runoff for years without observations was estimated from precipitation and temperature data from Livingstone using equation 2.6.
2. The water resources of the entire basin were determined by means of a coefficient taking into account changes in runoff along the length of the river (see Chapter 2).

The water resources of the other large rivers of Africa were assessed in a similar manner. Assessment of the water resources of the natural regions and countries of Africa was carried out in the following manner: runoff originating in the country plus the flow into the region (or country) gave the total water resources. As few hydrometric stations were located at the frontiers, the characteristics of the water resources were determined using the methods described in Chapter 2. The water resources of those areas draining to the oceans were assessed from the water resources of the rivers concerned. The runoff from interfluve areas and from those without areas of interior drainage was estimated using the same methods as for the assessment of water resources of the river basins. In cases of extreme difficulty, the runoff map was used from the *Atlas of the World Water Balance* (Korzun, 1974a). The flow into the oceans was calculated from data recorded at the stations nearest to the ocean. These calculations took into account changes in flow over the length of the river using, in some cases, the runoff map and data about possible losses.

6.5.2 Water resources of natural regions and individual countries

The annually renewable water resources of the continent averaged over the period 1921 to 1985 are estimated to be 4047 km^3/year (Table 6.5), only Europe and Australia having smaller volumes.

Due to the presence of large arid areas (the Sahara and Kalahari Deserts and the Somali Peninsula) Africa's specific runoff is the least of all the continents (Riebsame *et al.*, 1995). African water resources have been assessed a number of times, for example L'vovich (1972) – 4657 km³/year, Zubenok (1970) – 7826 km³/year and Karasik (1970) – 4225 km³/year. The *World Water Balance and Water Resources of the Earth* (Korzun, 1974b) gives the figure of 4570 km³/year for the period 1918 to 1967, while Baumgartner and Reichel (1975) estimated 3400 km³/year. In the publications from the World Resources Institute (WRI/IIED, 1986, 1988; IIED/WRI, 1987; WRI, 1990, 1992, 1996, 2000) and in Gleick (1993, 1998) the estimates of the water resources of the continents are based on the L'vovich (1972) data, but the water resources of the separate regions of Africa were rarely evaluated. Studies are more often devoted to assessing the available water of particular parts of the continent (Margat, 1990, 1994; Pearce, 1996; Shahin, 1996; FAO, 1997; Margat and Vallée, 2000), revealing, for example, the long-term droughts in the Sahel (Folland *et al.*, 1986; WMO, 1986; Hubert and Carbonnel, 1987) and the changing levels of Lake Chad. The most complete assessment of water resources and water availability in the regions is presented by Shiklomanov and Markova (1987) and Shiklomanov (1988) where the runoff was assessed using the data published in *World Water Balance* (Korzun, 1974b). *Water Resources of African Countries* (FAO, 1995b) contains data on the total runoff of African rivers which are close to the values given in the present study. According to data in Table 6.5, the volume of water flowing from each square kilometre of Africa (including Madagascar) is 134 000 m³/year on average. The potential water availability per head in 1994 was 5720 m³/year, on average. The values presented indicate the average water availability in space and time.

The distribution of the water resources over Africa is very uneven. The areas around the equator are well supplied with water throughout the year. As the distance from the equator increases, the water resources decrease to minimums in the deserts and semideserts of the Northern and Southern Hemispheres. In areas where water resources originate, such as the Central Region (Table 6.5) astride the equator, they reach 1770 km³/year, in the Southern Region 399 km³/year, and only 41 km³/year in the Northern Hemisphere. The North has the smallest specific runoff. Table 6.5 shows the value of 5720 m³/year per head for the continent as a whole and this is predominantly due to the large resources in the Central Region with 28 800 m³/year per head while in the other four regions it is much lower.

From year to year and decade to decade the water resources of the continent and its regions can vary quite significantly. For the continent as a whole these variations range from 102 000 to 169 000 m³/km² per year and from 4340 to 7180 m³/year per head. For the North the average available water resource per head is very small at 710 m³/year and much less in dry years. The coefficient of variation of runoff for the continent as a whole is 0.10. The Central Region has the smallest coefficient (0.09) and the North the largest (0.34). Table 6.6 contains data on water resources and water availability for some countries estimated for this Monograph and some data taken from *Water Resources of African Countries* (FAO, 1995b). In this study water resources were evaluated taking account of renewable groundwater resources not related to the runoff. For Zaire and Nigeria, water resources were determined for each year during the period from 1921 to 1985. These countries have the largest water resources among African countries. It should be noted that data on the water resources for Gambia, Mali and Niger that were estimated by SHI and FAO separately differ little, but for Senegal and the Sudan the differences are rather large. These differences are due to the differences in the initial data including the lengths of the periods of observation and the different methods used.

The year-to-year variations in water resources reveal the underlying hydrological regimes and demonstrate the availability of water in the areas concerned. Figure 6.5 shows the changes in water resources in time for all the regions, for the whole of Africa and Nigeria and Zaire. It displays the periods of increased and decreased runoff and the trends and variations. To investigate the yearly variations of water resources an analysis was performed for the period 1921 to 1985. The various runoff series were divided into two periods: 1921 to 1953 and 1954 to 1985. Using the methods described in Chapter 2, tests were made for the presence or absence of a stationary state (uniformity) for the complete period.

For the North and Western regions including the Sahel and for the whole continent, a decreasing trend in water resources was revealed over the period with a probability of 95%. In the Western region and in the Sahel these trends were confirmed by studies of precipitation variations (Balek, 1977). According to these studies, there was a summer drought in these regions in the period 1920 to 1931. From 1940 to 1949 there was a dry period in all Sahel countries. From 1950 to 1967 there was a wet period in the region and from 1968 a pronounced dry period produced a precipitation deficit of 20–30%. A weak continent-wide downward trend was found for the period 1921 to 1985. In contrast no changes were found over the period in the water resources for North, Eastern, Central and Southern Africa, using the Kolmogorov–Smirnov and Student's *t* tests. Mean values for the Western region and the Sahel were found to be non-stationary by the Student's *t* test. However other tests confirmed the stationary character of these series (Fisher and Kolmogorov–Smirnov).

It appears that the water resources of all regions of Africa are, in general, stationary and their variations are of a cyclic character. The periods of increased, decreased and average resources are quite lengthy as Fig. 6.6 shows, but the most recent period appears to be dry. At first (from 1964) it covered Western Africa, including the Sahel, from 1970, the Central Region from 1971,

Table 6.6. *Water resources of some countries of Africa*

| Country | Area, km² × 10⁶ | Population (1994), ×10⁶ | Water resources, km³/year[a] | | | Potential water availability, m³ × 10³/year | | | |
| | | | Inflow | Local | Local and ground water | Without ground water | | With ground water | |
						Per km²	Per head	Per km²	Per head
Algeria	2.38	27.3	0.4	13.2	13.9	5.55	0.48	5.84	0.51
Burkina Faso	0.27	10.1	0.0 / 2.0	13.0 / 14.7	17.5	48.1 / 54.4	1.29 / 1.46	64.8	1.73
Chad	1.28	6.2	28.0 / 28.3	13.5 / 15.8	15.0	10.5 / 12.3	2.18 / 2.55	11.7	2.42
Egypt	1.00	61.6	65.6	0.5	1.8	0.5	0.01	1.8	0.03
Gambia	0.01	1.1	5.0 / 4.7	3.0 / 3.2	3.0	300 / 320	2.73 / 2.91	300	2.73
Libya	1.76	5.2	0.0	0.1	0.6	0.06	0.02	0.34	0.12
Madagascar	0.59	14.3	0.0	332.0	337.0	563	23.20	571	23.60
Malawi	1.24	10.5	40.0 / 44.4	50.0 / 50.0	60.0	40.3 / 40.3	4.76 / 4.76	48.4	5.71
Mauritania	1.03	2.2	11.0 / 11.0	0.1 / 0.4	0.4	0.10 / 0.39	0.05 / 0.18	0.39	0.18
Morocco	0.45	26.5	0.0	22.5	30.5	50.0	0.85	66.7	1.13
Niger	1.27	8.8	29.0 / 30.4	1.0 / 3.0	3.5	0.79 / 2.36	0.11 / 0.34	2.76	0.40
Nigeria	0.92	108.0	59.0 / 43.6	214.0 / 274.0	221.0	233 / 298	1.98 / 2.54	240	2.04
Senegal	0.20	8.1	13.0 / 17.4	23.8 / 17.4	26.4	119.0 / 87.0	2.93 / 2.15	132	3.26
South Africa	1.22	40.6	5.2 / 5.2	40.0 / 47.4	44.8	32.8 / 38.9	0.98 / 1.17	36.7	1.10
Sudan	2.51	27.4	119.0 / 140.0	28.0 / 22.0	35.0	11.20 / 8.76	1.02 / 0.80	13.9	1.28
Zaire	2.34	42.6	84.0 / 313.0	934.0 / 987.0	935.0	399 / 422	21.90 / 23.20	400	21.90

[a] Nominator, data of FAO (1995b); denominator, data of SHI (Shiklomanov, 1997).

the Southern region from 1979, the Eastern region from 1981 and the North from 1982. From 1971 this drier phase had a marked effect on the water resources of the entire continent. Table 6.7 presents fuller information about the cyclic variations of the water resources: it shows that the duration of the drier phases can reach 40 years (North and Central Regions), while wetter phases last for 20 and 40 years (Eastern and North). Various factors were important in influencing the location of the intertropical convergence zone (ITCZ), the Azores and South Indian Highs, and the resulting climatic patterns, particularly that of the Sahel, and these have been investigated (Katz, 1964; Borisenkov and Borisov, 1987). However it has been impossible to determine precisely how these factors operate and the way they control the water resources of the continent. An analysis of the year-by-year changes in the annual water budget reveals that the patterns of the groups of dry and wet years are important regulators of the continent's water resources. The results are given in Table 6.8, which shows that the mean duration of dry (less than average) groups of years by regions varies from 1 (Eastern region) to 4.29 years (North), average resources from 1.25 (North) to 10.42 years (Central) and wet (higher than average) from 1.14 (Eastern) to 2.67 years (Central).

In areas other than the Central region the probability of occurrence of groups of years of the duration under consideration varies from 58% (Eastern) to 82.6% (Southern). In the Central region

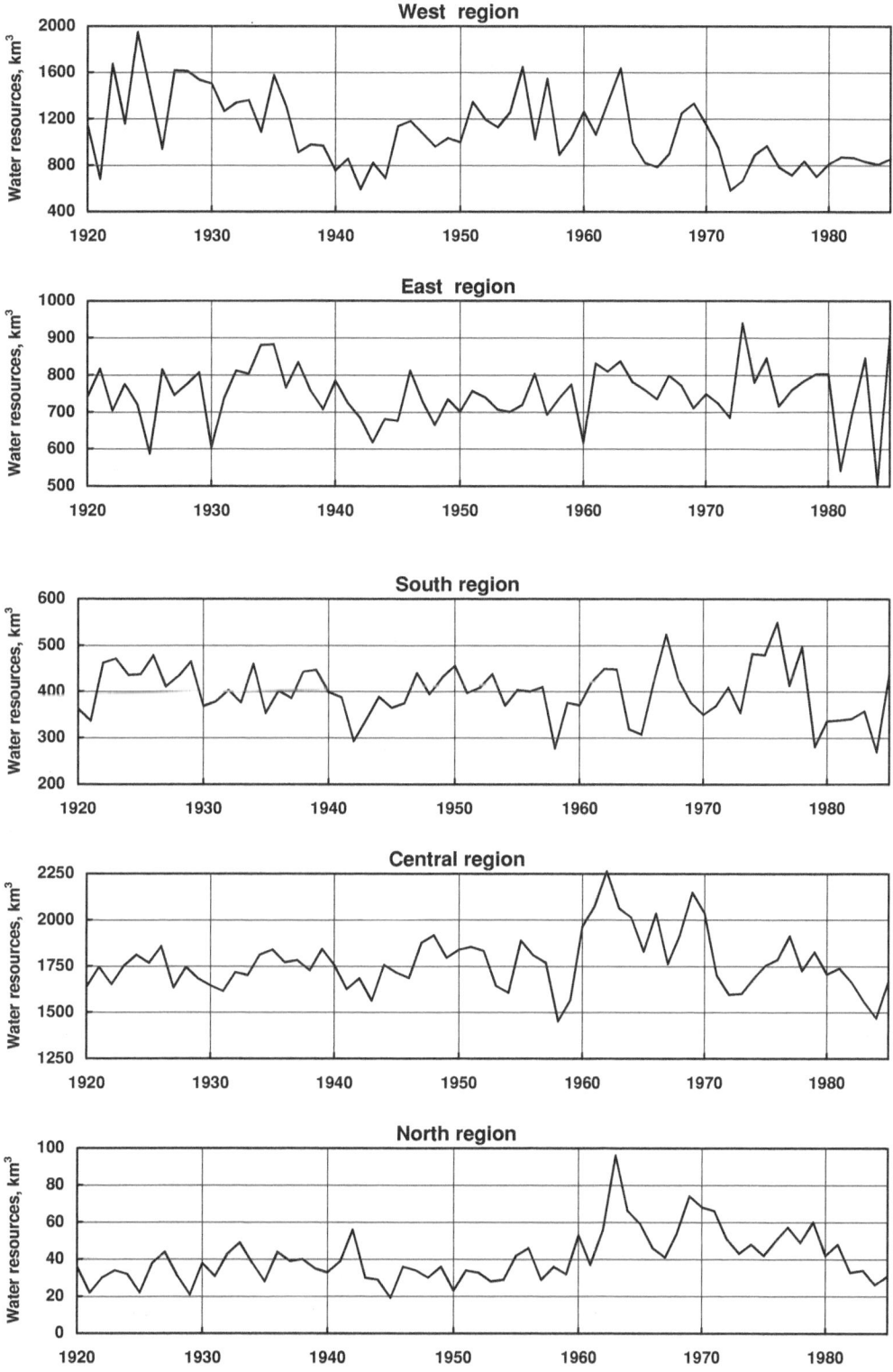

Fig. 6.5 Mean annual renewable water resources by African regions, the continent as a whole, selected countries and the Sahel region.

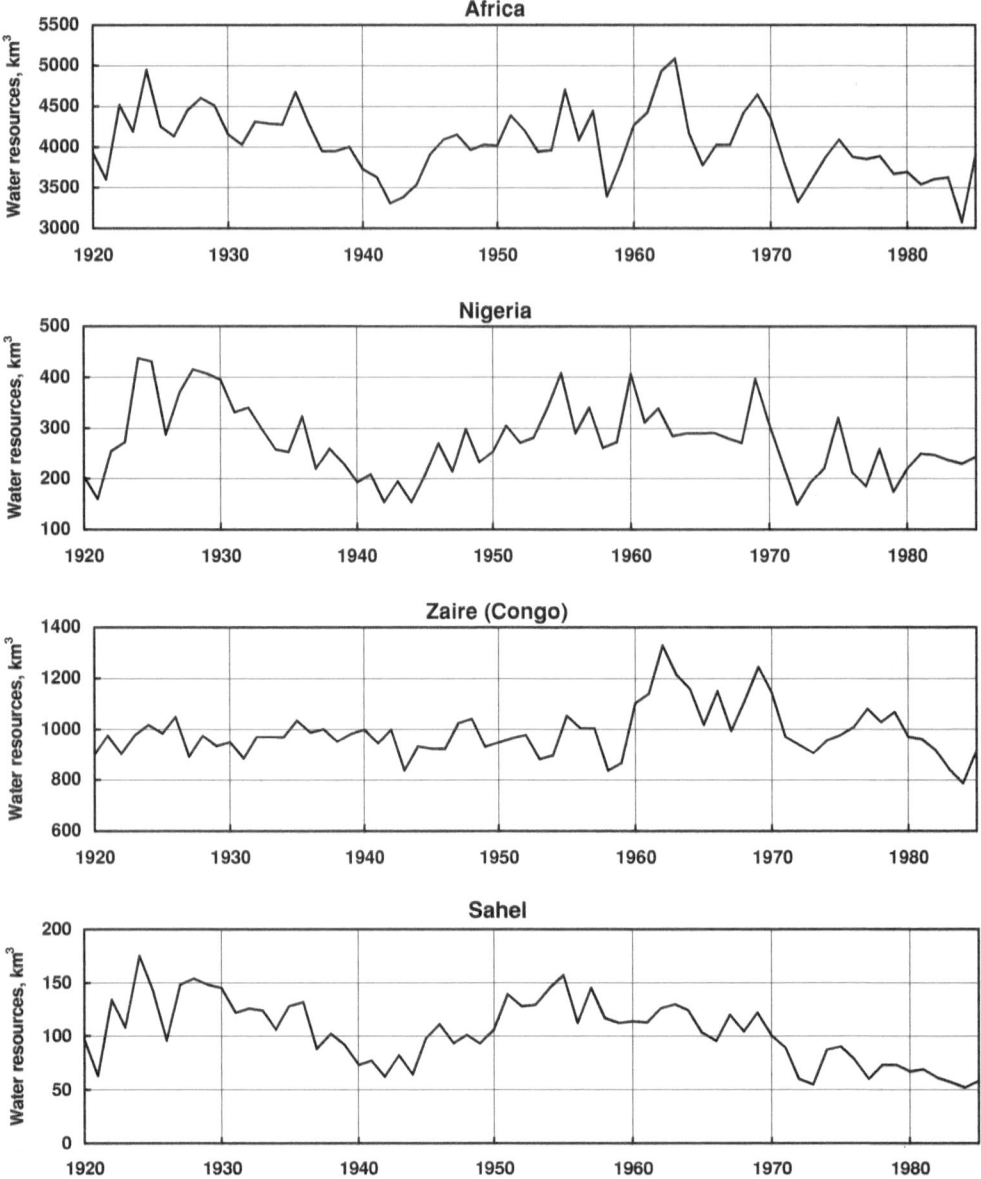

Fig. 6.5 (*cont.*).

this probability is small (25.1%), indicating significant probabilities of the occurrence of groups of wet, dry or average years of long durations. Figure 6.7 shows the probabilities of the change of years with these different water resources for the regions. As is seen, each region has individual regular features of changes of years of different water content. Only the Central region possesses the regular features of preserving same water content as observed in the previous year the next year. The probabilities for preserving this water content are the largest: low water year $P = 40\%$; average water year $P = 90\%$; high water year $P = 62\%$.

In the North, Western and Central regions, in more than 50% of cases high water years are again followed by high water years and

for the other two regions, the Eastern and Southern regions, this is unlikely. Low water in the North and Western regions in more than 75% of cases follows low water again. For the other regions it is more typical that low water years are replaced by average years in water content. For the continent on the whole the highest probabilities are typical for preserving the low and average water content. The years with high water content are replaced by years of average water content with greater probability ($P = 78\%$). The synchronicity and asynchronicity in the between-region variations of water resources have been assessed using correlation coefficients for the period (1921 to 1985). These studies showed that, except for some correlation between the resources of the North

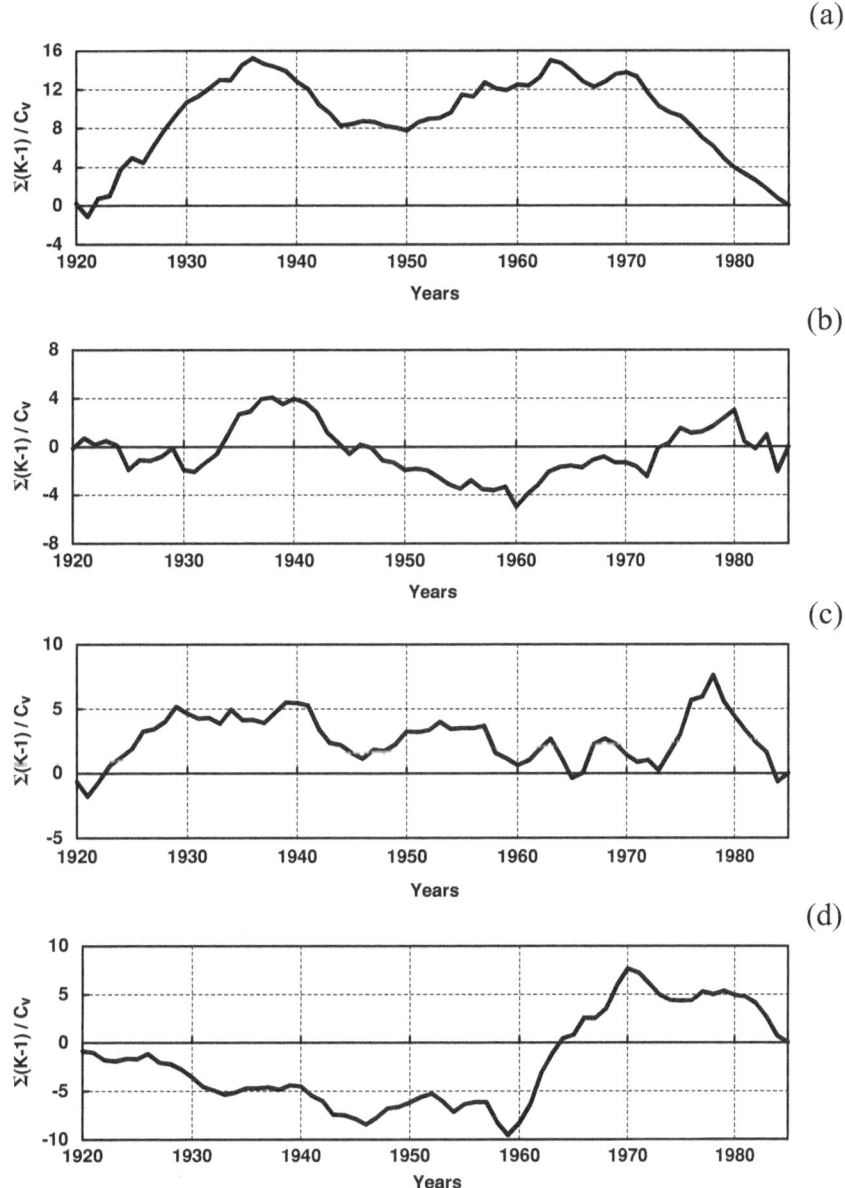

Fig. 6.6 Normalized difference integral curves of water resources of natural–economic regions and of Africa as a whole. (a) Western region; (b) Eastern region; (c) Southern region; (d) Central region; (e) North region; (f) Africa as a whole; (g) the Sahel.

and Central regions, one region's resources show no correlation with those of another. However significant correlations were found between the runoff of Western Africa and the Sahel ($r = 0.90$) and Nigeria ($r = 0.78$), between those water resources of the Sahel and Nigeria ($r = 0.82$), as well as between the water resources of the Central Region and Zaire ($r = 0.95$) and between the water resources of the North and Zaire ($r = 0.65$). Naturally the water resources of the continent are significantly correlated with the water resources of all the regions: with the

Western region ($r = 0.86$), with the Sahel ($r = 0.78$), with Nigeria ($r = 0.71$), and with the Central region ($r = 0.61$). The weakest correlations are with North ($r = 0.20$) and the Eastern regions ($r = 0.27$)

6.5.3 Major rivers and flows to the oceans

Africa has four large river systems. The Nile is the longest river on Earth. Its length is 6671 km and its basin area is 2 870 000 km². The Nile flows from the equatorial plateau of East Africa across the tropical plains of the Sudan and the deserts of Sudan and Egypt to the Mediterranean Sea. Lake Victoria is the Nile's main river source but some consider the source of the Nile to be the source of the Kagera River, the largest river flowing into Lake

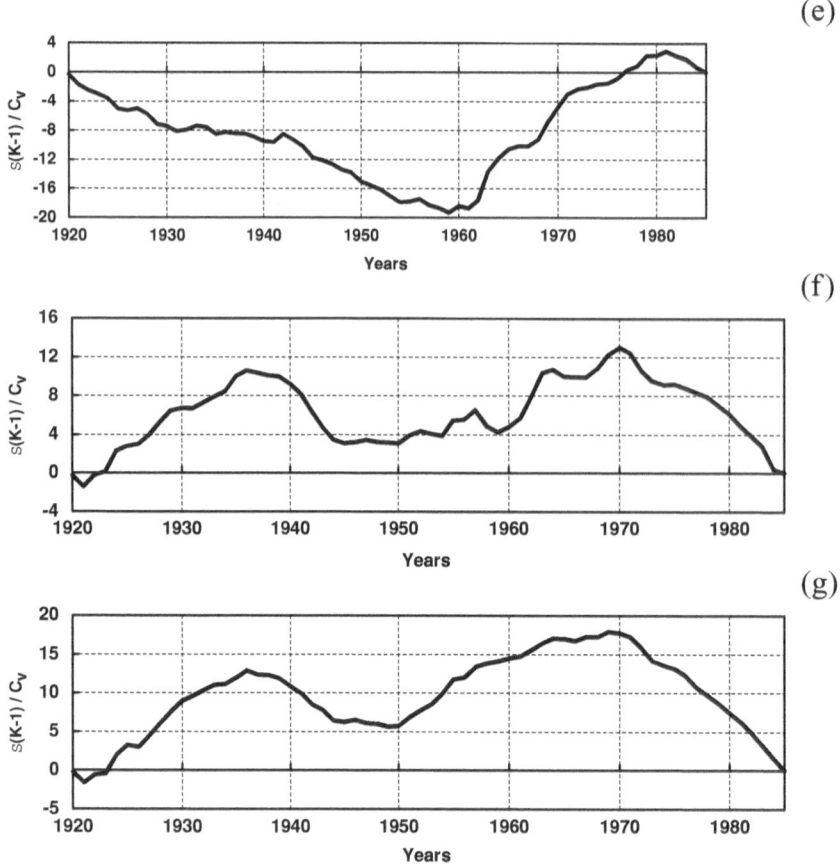

Fig. 6.6 (cont.).

Victoria, or the source of the Rukarara River. In the northern part of Lake Victoria there is a small bay from which the river called the Victoria-Nile flows. The flow from the lake is currently regulated by a dam which is located on the Ripon Falls. The Victoria-Nile carries flow into Lake Kyoga leaving the equatorial plateau in the vicinity of the Murchison Falls 32 km above Lake Albert. Here the river is called the Albert-Nile and downstream it is known as the Bahr-el-Jebel. Much of the Bahr-el-Jebel flows along quite a flat plain with a very gentle gradient. The velocities in this section are very low and due to the intense evaporation, large volumes of water are lost here. After the confluences with the Bahr-el-Ghazal and Sobat Rivers, the river is called the White Nile. During periods of high flow in the Sobat the flow in the White Nile sometimes reverses creating a temporary reservoir.

A similar phenomenon is observed at the confluence of the Blue Nile with the White Nile at Khartoum where the waters of the Blue Nile during high flows dam the waters of the White Nile. The basin of the Blue Nile receives abundant rain for several months (May–September) resulting in increased discharges from July to October. Near Khartoum during this period the mean daily discharge of the Blue Nile is approximately five times larger than that of the White Nile in flood. Numerous small rivers flow from the Congo–Sudan Upland but are mostly lost in sand and swamps without reaching the Nile. Hence the losses here are quite large. Along the course of the Nile from the Sobat to Wadi Halfa, its flow increases due to the input of large tributaries such as the Blue Nile and the Atbara. The losses here are relatively small. The Atbara flows into the Nile in an area of desert and for almost half the year this river is often only a chain of shallow ponds. Although the Atbara during the flood period carries much less water than the Blue Nile, the influence of its high flows on the regime of the Nile is considerable, since its flood reaches the main river less smoothed than the flood of the Blue Nile. The discharge of the White Nile is fairly constant whereas the discharges of the Blue Nile and Atbara are subject to sharp variations. During the high water period most of the flow of the Nile is formed by the waters of the Blue Nile and the Atbara and during the low water period by the White Nile. Downstream of the entry of the Atbara the Nile is controlled by the systems of reservoirs and irrigation canals located in that section. A remarkable feature of the Nile is its delta. There are two main arms: the Rashid (239 km long) and the Damietta (245 km). The area of the delta is $22\,000\,\text{km}^2$, containing several lakes and lagoons separated by spits from the sea. The total water resources of the Nile Basin exceed $161\,\text{km}^3$/year. The White

Table 6.7. *Periods with different water content by natural–economic regions, selected countries and Africa as a whole*

Territory	Periods of low water content			Periods of average water content			Periods of high water content		
	Years	Number of years	Ratio of mean for period to long-term mean, K_{av}	Years	Number of years	Ratio of mean for period to long-term mean, K_{av}	Years	Number of years	Ratio of mean for period to long-term mean, K_{av}
North	1920–59	40	0.84				1960–81	22	1.31
	1982–85	4	0.76						
Western	1937–50	14	0.85				1921–36	16	1.27
including	1964–85	22	0.80				1951–63	13	1.16
Sahel	1937–50	14	0.85				1922–36	15	1.28
	1970–85	16	0.68				1951–69	19	1.18
Central	1920–59	40	0.98				1960–70	11	1.14
	1971–85	15	0.96						
Eastern	1939–60	22	0.96	1920–33	14	1.00	1934–38	5	1.10
	1981–85	5	0.94				1961–80	20	1.04
Southern	1942–46	5	0.88	1930–41	12	1.00	1921–29	9	1.09
	1979–85	7	0.84	1958–74	17	0.99	1947–57	11	1.04
							1975–78	4	1.12
Nigeria	1937–50	14	0.80				1924–36	13	1.27
	1981–85	15	0.82				1951–70	20	1.13
Zaire	1920–59	40	0.97				1960–70	11	1.16
	1971–75	5	0.96				1976–79	4	1.06
	1980–85	6	0.91						
Africa as	1937–45	9	0.92				1922–36	15	1.07
a whole	1971–85	15	0.91				1946–70	25	1.04

Nile provides about 29% of this figure, the Blue Nile about 57% and the Atbara some 14%. The discharge to the Mediterranean is only one-third of its water resources. (The Nile Basin is described in more detail in Section 6.7.)

The River Niger is 4160 km long and its basin is 2 090 000 km^2. The Djoliba River is considered to be its source, rising at a height of about 800 m in the Fouta-Djallon Plateau. In its upper reaches the Niger is joined by its right-bank tributaries the Milo and Sankarani in the North Guinea Uplands. In this part of the river there are many rapids. In the middle section the Niger receives one significant tributary, the Bani River. In the Sansanding region, the so-called inner delta of Niger begins, consisting of two parts. The larger part is the "living" delta which is subject to annual river floods and has numerous arms and lakes. The length of the flood zone is around 400 km with its largest width (in the region of confluence of Niger and Bani) about 100 km. The "dead" delta is situated approximately north of Sansanding. It is separated from the "living" delta by a 30-km strip of sand deposits. The "dead" delta is much smaller than the "living" delta. The Niger Delta proper is one of the largest in the world and begins approximately 180 km from the sea. It is formed by sediments from the Niger and some small rivers of the Guinean coast. The main river in the delta is divided into 14 larger arms and many smaller ones. The Niger is fed predominantly in its upper and the lower reaches, whereas in the middle of its course it is a transit river. Since much of the precipitation in the basin falls in the summer in both the upper and lower Niger, high flows occur in the summer and autumn months. The flows in the middle Niger are not particularly strong. Furthest downstream the second flood often merges with the main flood. The highest floods occur below the confluence with the Bani River. The floods in the inner delta occur in the summertime. During inundation the Niger water mixes with some small lakes. Extensive flooding lasts for 1 to 3 months. The water resources of the Niger are 302 km^3/year and the flow of the river is about 134 km^3/year, the difference being due to evaporation and irrigation.

The Congo (Zaire) is Africa's largest river in terms of discharge and the third largest in the world after the Amazon and the Ganges–Brahamputra. Its mean runoff is 1320 km^3/year which is almost one-third of the total runoff of African rivers. The Congo has a basin of 3 680 000 km^2 and a channel 4370 km long. The Chambeshi River south of Lake Tanganyika is considered the source of the River Congo. Lake Tanganyika is connected to the Congo by the Lukuga River. Its middle course begins below

Table 6.8. *Water content transition from one grouping to another by natural–economic regions, selected countries and Africa as a whole*

Territory	Water content characteristic[a]	Probability (%) of appearance of year groups of 1 to 5 years in duration					Average duration of groups, years
		1	2	3	4	5	
North	1	2.53	1.94	1.49	1.14	0.87	4.29
	2	14.74	2.95	0.59	0.12	0.02	1.25
	3	6.18	3.40	1.87	1.03	0.57	2.22
Western including Sahel	1	2.90	2.18	1.63	1.22	0.92	4.00
	2	12.70	3.17	0.79	0.20	0.05	1.33
	3	7.03	3.68	1.93	1.01	0.53	2.10
	1	1.80	1.43	1.13	0.89	0.71	4.80
	2	14.05	3.31	0.78	0.18	0.04	1.31
	3	4.84	3.03	1.89	1.18	0.74	2.67
Central	1	2.73	1.09	0.44	0.17	0.07	1.67
	2	0.74	0.66	0.60	0.54	0.49	10.42
	3	1.81	1.13	0.71	0.44	0.28	2.67
Eastern	1	10.60	0	0	0	0	1.00
	2	4.34	3.30	2.51	1.90	1.45	4.17
	3	10.77	1.35	0.17	0.02	0	1.14
Southern	1	6.74	3.15	1.47	0.69	0.32	1.88
	2	7.18	4.64	2.92	1.83	1.15	2.69
	3	10.12	3.37	1.12	0.37	0.12	1.50
Nigeria	1	4.33	2.82	1.84	1.20	0.78	2.88
	2	11.3	4.91	2.14	0.93	0.40	1.77
	3	6.47	3.40	1.79	0.94	0.50	2.11
Zaire	1	5.49	1.57	0.45	0.13	0.04	1.40
	2	1.54	1.32	1.13	0.97	0.83	7.00
	3	1.54	1.03	0.69	0.46	0.31	3.00
Africa as a whole	1	3.30	1.92	1.12	0.65	0.38	2.40
	2	4.20	3.15	2.36	1.77	1.33	4.00
	3	8.33	1.05	0.41	0.09	0.02	1.29

[a] 1, Low; 2, average; 3, high.

Kisangani where the river is called the Congo proper. Here flow is steady and there are numerous islands dividing the river into arms and branches. Its width reaches 800 m and the valley of the river is up to 14 km wide in some places. The banks are poorly defined because of the swamps and lakes; and in this middle section the Congo is joined by some important tributaries. The lower reaches of the Congo begin at the Livingstone Falls extending over 360 km. The river here runs over 32 waterfalls and rapids with a fall of 220 m and the width here does not exceed 400 m. Near Matadi the rapids and waterfalls end and the river becomes sluggish. About 100 km from the ocean the river is about 5 km wide and its depth varies from 20 to 100 m. Further downstream the Congo is divided into arms and it reaches depths between 100 and 200 m and up to 400 m in some places. The Congo Basin is situated near to the equator and has high temperatures throughout the

year and small changes from month to month. In the equatorial part of the basin the average temperature is close to +27 °C. The precipitation is more than 1000 mm per year across most of the basin with totals exceeding 1500 mm in the north. The hydrological regime reflects the distribution of precipitation which is fairly uniform through the year but with maxima in autumn and spring.

The Zambezi is the largest African river discharging to the Indian Ocean. It is 2660 km long and its basin is 1 330 000 km^2 in area. The source of the Zambezi is in the flat Congo–Zambezi Plateau. The upper reaches of the river have a small gradient, then a series of rapids and waterfalls begin. In the 75 km above the Victoria Falls the Zambezi River receives the Chobe River flowing from the Northern Kalahari. This system complicates the hydrology of the Zambezi since the connection between the Chobe and Zambezi waters can be in either direction. As a result, the

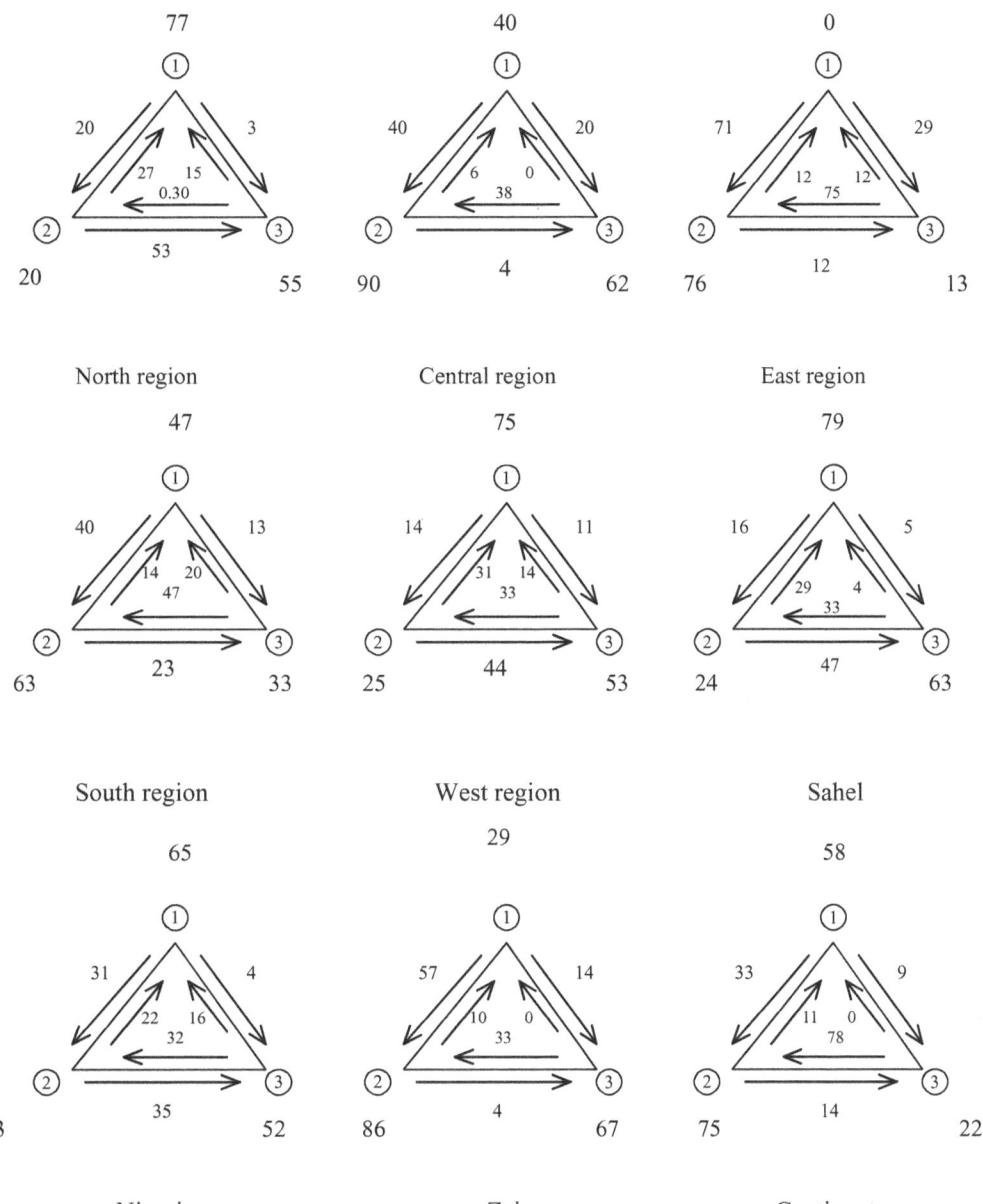

Fig. 6.7 Probability of change of year groups with different water content by natural–economic regions and selected countries of Africa. The water content of years is given in the circles (1, low; 2, average; 3, high). Near the circles the probabilities (%) for the continuation of this water content are shown. The arrows show the directions of the change in water content and the figures above and below them the probabilities of a wet dry or average year being replaced by the other.

area of the basin of the Zambezi can vary from 500 000 km² to 1 200 000 km². Upstream of the Victoria Falls the river is 1650 to 1800 m wide. Further downstream the Zambezi is joined by a number of large tributaries and the river is from 5 to 8 km wide. In the lower reaches of the river the regime is influenced by releases from the Kariba and Cabora Bassa reservoirs. The Zambezi flows

in part of the continent with a subtropical climate which is characterized by a summer precipitation maximum. Hence the rivers of the Zambezi Basin have a summer flood season extending in many cases into autumn. Water levels rise in December and maximum discharges are observed in March and April. The flow of the Zambezi is 154 km³/year and about 72% of this volume reaches the mouth of the river, the rest being lost by direct evaporation and by irrigation.

It is estimated that the water resources of the four largest rivers of Africa are almost half (48%) of the total runoff of all the rivers of the continent. Together their basins cover some 34% of Africa. For comparison, Table 6.9 presents data on the basin areas and discharges of the largest rivers of Africa taken from various studies

Table 6.9. *Area of catchment and water discharge of principal rivers of Africa by the data of different authors*

River	Baumgartner and Reichel (1975) Area of catchment, km² × 10³	Water discharge, m³/s	Czaya (1981) Area of catchment, km² × 10³	Water discharge, m³/s	Szestay (1982) Area of catchment, km² × 10³	Water discharge, m³/s	Meybeck (1988) Area of catchment, km² × 10³	Water discharge, m³/s	Shiklomanov (1997)[a] Area of catchment, km² × 10³	Water discharge, m³/s
Congo	3822	42000	3822	42 000	4015	40 000	4000	39 200	3680	41 250
Niger			2092	5 700	1114	6 100	1125	6 100	2090	4 217
Zambezi			1330	2 500	1295	7 000	1340	7 100	1330	3 519
Nile			2881	1 584	2980	2 800	3000	2 830	2870	1 696
Chari			700				600	1 320	880	1 252
Senegal					338	700			441	545

[a] The runoff at river mouth, 1921–85.

beginning in 1975. There are large differences even in the estimates of the areas of the basins of the Congo, Nile, Zambezi and other rivers. The differences in the data for the Zambezi runoff are especially large ranging from 2400 m³/s to 7000 m³/s. The Monograph data show the runoff at the mouths of the rivers. They are based on the most complete records and are the most reliable. In addition to the evaluation of the water resources of the largest river systems, the present study contains an analysis of the time series of discharge from 40 gauging stations. Table 6.10 contains the main variables derived from their flow data, and shows that the runoff of African rivers varies over a wide range of values. The largest coefficients of variation ($C_v = 0.60$ to -0.85) were found for the Orange–Vioolsdrift, Ouergha–Ourtzagh, Sebou–Azib Soltane and Volta–Sendi-Halcrow stations and the least ($C_v = 0.10$ to 0.11) for Congo–Kinshasa and Ouham–Batangafo.

Runoff from Africa flows to the Atlantic (77%) and to the Indian Oceans (23%) (Fig. 6.8). There are a number of parts of Africa, which are areas of inland drainage, and they total 9.6 million km² in area. About 85% of these areas are within the parts of the continent draining to the Atlantic and only 15% to the Indian Ocean. Included in the 85% is the Sahara Desert and the adjoining Lake Chad basin, as well as the basins of Lakes Eyashi and Rukwa. The Indian Ocean areas are the the basins of Lake Turkana and the Awash River in East Africa, and the Kalahari Desert and the basin of the Okovango River in the west. Table 6.11 shows the water resources of Africa divided between drainage to the Atlantic and Indian Oceans. The former receives 4.6 times the flow of the latter, but the coefficients of variation of the annual flows are very similar – 0.12 and 0.14, respectively; some 54% of the waters received by the Atlantic come from the Congo, Niger and Nile. The Zambezi provides 12% of the total volume flowing to the Indian Ocean; the remaining part comes from numerous small and average-size rivers. Table 6.12 contains data on flows from the continent to the oceans by latitudinal zone. The inflow to

the Atlantic Ocean is 2982 km³/year and to the Indian Ocean 674 km³/year. The water used is estimated to be 338 km³/year from runoff to the Atlantic Ocean and 53 km³/year from the Indian Ocean runoff.

The year-to-year variations in water resources are given in Figs. 6.9 and 6.10. Checks were made on the discharge time series for 59 gauging stations for stationarity (uniformity and the presence of a trend). For 30 records the trends in runoff changes were determined to the 95% probability level. The results of the analysis showed 20 series to be stationary both by mean values and dispersion, 2 series non-stationary by dispersion and 11 only by mean value, while 9 series were completely non-stationary. Mean discharges over the period from the late 1960s to the early 1980s in all the rivers studied in North and Western Africa decreased significantly, as compared to the preceding period (by 20–30%). This is of course the result of the severe droughts in this region from the late 1960s to the mid-1980s. The studies indicated that the series for the Congo River are non-stationary while the Nile series and flows to the Atlantic are stationary only at the first approximation. The series for the Niger and the runoff to the Indian Ocean are completely stationary. A trend is absent in the Nile time series and in discharges to the Indian Ocean, but is present in the flows to the Atlantic. Only a weak trend can be detected in the flows of the Congo and the Niger.

In order to investigate variations in the annual runoff six series were selected differing in the length of observations. Figure 6.11 presents the cumulative flow curves for these rivers: the Congo, Nile, Senegal, Ogooue, Niger and White Nile. Similar studies were carried out for a further 42 rivers. These studies show that in the different phases of the runoff of the Niger, Senegal, Nile and Ogooue Rivers two low water phases existed, one from 1938 to 1950 and the other from 1967 to 1983. There were two high water phases from 1920 to 1938 and from 1950 to 1967. The duration of the low periods varies from 7 to 12 years (module coefficients

Table 6.10. *Major parameters of river runoff in Africa*

River	Station	Area of catchment, km^2	Observation period	Surface water resources, Q, m^3/s	Coefficient of variation, C_v
North Africa					
White Nile	Dongola	–	1912–85	2 622	0.18
Blue Nile	Khartoum	325 000	1912–85	1 532	0.22
Atbara	Kilo 3	69 000	1920–85	358	0.31
Sebou	Azib Soltane	17 200	1951–85	56.8	0.59
Ouergha	M'Jara	6 190	1951–89	90.8	0.60
	Ourtzagh	4 404	1951–89	62.2	0.62
Oum er Rbia	Dechra	3 300	1951–88	31.0	0.43
Western Africa					
Niger	Niamey	700 000	1929–91	905	0.26
Chari	N'Jamena	600 000	1933–90	1 180	0.19
Volta	Sendi-Halcrow	394 000	1937–85	1 120	0.59
Senegal	Dagana	268 000	1903–85	655	0.25
Black Volta	Bomboi	134 200	1951–85	265	0.37
Niger	Koulikoro	120 000	1907–90	1 414	0.28
Bani	Douna	101 600	1951–80	533	0.45
Faleme	Kadira	28 900	1930–85	183	0.41
Pra	Daboasi	22 700	1951–85	242	0.49
Tano	Alenda	15 800	1951–85	145	0.15
Central Africa					
Congo	Kinshasa	3 475 000	1903–85	40 251	0.10
Oubangui	Bangui	500 000	1936–90	4 012	0.21
Ogooue	Lambarene	205 000	1930–85	4 689	0.15
Sanaga	Edea	131 520	1943–80	1 972	0.15
Kotto	Kembe	78 400	1951–85	449	0.20
Benoue	Garoua	64 000	1949–85	346	0.27
Kouilou	Sounda	55 010	1951–85	899	0.16
Chinko	Rafai	52 500	1951–85	407	0.16
Ouham	Batangafo	44 700	1951–85	354	0.11
Nyong	Dehane	26 400	1951–80	434	0.21
Ntem	Ngoazik	18 100	1951–85	276	0.18
Wouri	Yabassi	8 026	1951–85	302	0.15
Dja	Somalomo	5 150	1955–85	66.6	0.20
Eastern Africa					
Shebelle	Belet Uen	211 800	1951–80	66.1	0.32
Juba	Luch Ganana	179 520	1951–85	193	0.27
Tsiribihina	Betomba	45 000	1958–85	1 033	0.32
Tana	Garissa	42 220	1934–85	157	0.48
Pangani	Korogwe	25 100	1959–85	28.3	0.24
Ikora	Ansatrana	18 600	1951–85	436	0.16
Ivondro	Ringaringa	2 545	1951–85	106	0.28
Mandrare	Amboasary	2 435	1951–85	61.5	0.35
Betsiboka	Ambodiroka	1 800	1951–85	277	0.22
Southern Africa					
Orange	Vioolsdrif	851 000	1964–85	159	0.85

Table 6.11. *Renewable water resources of oceanic basin slopes and endorheic regions of Africa*

River basin, ocean slope	Area, km² × 10⁶	Water resources, km³/year			Coefficient of variation, C_v
		Average[a]	Maximum	Minimum	
Niger Basin	2.09	302	482	163	0.26
Congo Basin	3.68	1320	1775	1050	0.10
Nile Basin	2.87	161	248	95	0.16
Atlantic Ocean slope	23.1	3320	4281	2530	0.12
endorheic runoff region	8.2	(90)			–
Indian Ocean slope	7.01	727	987	496	0.14
endorheic runoff region	1.4	(60)			–

[a] Values in parentheses are approximate data.

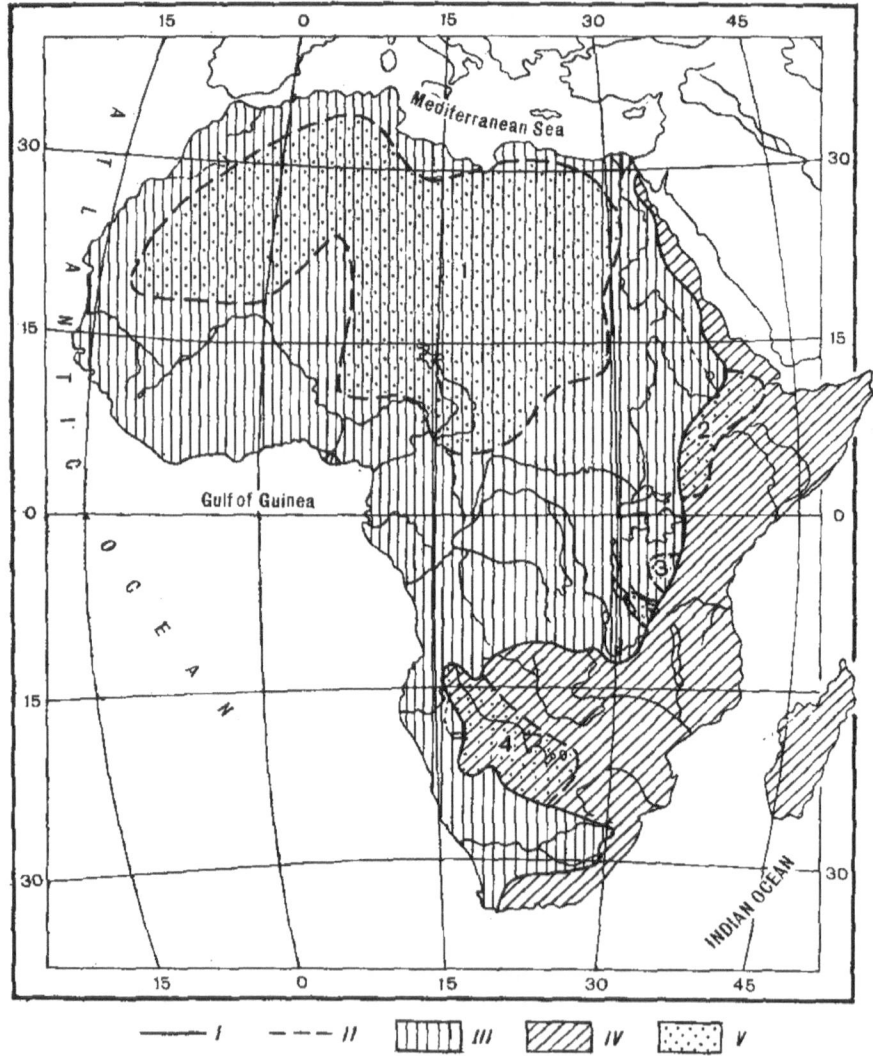

Fig. 6.8 Distribution of the drainage basins of the oceans on the territory of Africa. Boundaries: I, drainage basins; II, areas of internal runoff. Drainage basins: III, Atlantic Ocean; IV, Indian Ocean; V, endorheic regions. 1, Sahara Desert with the basin of Lake Chad; 2, basin of Lake Turkana; 3, basin of Lake Nyasa; 4, Kalahari semi-desert with the basin of the Okovango River.

Table 6.12. *Distribution of ocean basin inflow*
(km³/year) by latitudinal zones in Africa

Latitude	Atlantic Ocean	Indian Ocean
30–40° N	89	–
20–30° N	1	0.2
10–20° N	149	4
0–10° N	1074	4
0–10° S	1580	113
10–20° S	35	393
20–30° S	25	147
30–40° S	6	13
Total	2982	674

of 0.82 to 0.88) and of high periods from 12 to 23 years (module coefficients of 1.07 to 1.11). The results obtained are consistent with the variations in the annual precipitation. Similar conclusions about the phases of runoff can be reached for the Congo and White Nile. They had periods of low water from 1920 to 1959 (39 years) and from 1920 to 1961 (41 year) and periods of increased flow from 1959 to 1983 (25 years) and from 1961 to 1985 (24 years). The module coefficients of the runoff for the low water and high water periods are 0.96 to 0.90 and 1.08.

Despite the sparse hydrometric network and the short periods of observation, it was possible to determine that the runoff in 42 rivers show cyclic variations which are regional in character (Table 6.13). From this Table the Congo is seen to have the longest period of low flows, which lasted 40 years, as well as shorter periods when the runoff was reduced by 10%. During high flows the runoff increased by more than 15%. In the Nile low flows can decrease flow by 18% whereas high flows increase it by only 5–10%. The length of period of high flows and low flows is about the same in the Niger; but the discharge is reduced by 20% in dry periods and increased by 25–30% in wet ones. Discharge to the Atlantic can be reduced by 9–12% during periods of low flow and increased by 6–8% in wet periods, while variations in the flows to the Indian Ocean are slightly greater.

Table 6.14 presents data on the probability of occurrence of groups of years of differing flows with durations of 1 to 5 years for the largest rivers and the flows to the oceans. The total probability of the differing groups of years varies from 76% to 92.5% for the Niger, Nile, Atlantic Ocean and Indian Ocean flows. It is smaller for the Congo Basin, reaching 35.4%. An assessment of synchronicity and asynchronicity of the flows of 31 rivers and the discharges to the oceans was performed by correlation analysis. Of 465 correlation coefficients, 63 turned out to be the most significant ($r = 0.50$). The most closely correlated were the following: Niger–Gaya and Nakambe–Yakala ($r = 0.99$), Ouham–Batangafo and Mou Houn–Dopola ($r = 0.80$), Oubangui–Bangui

and Congo–Kinshasa ($r = 0.80$), Volta–Sendi-Halcrow and Pra–Daboasi ($r = 0.76$), Niger–Gaya and Niger–Mopti ($r = 0.89$). There were 189 significant correlation coefficients ($r > 0.30$), i.e. 41%. Of them, 32% indicate synchronous runoff variations and about 9% asynchronous variations. These studies showed that the flows of the Congo, Niger and Nile are highly related to the total discharge to the Atlantic. No significant correlation was found to exist with discharge to the Indian Ocean, however.

6.5.4 The distribution of runoff within the year

The distribution of the runoff of African rivers is characterized by great variety. For most rivers the flow regime is closely related to the rainfall regime of their basins. Snow has an influence on runoff only in the mountains of Uganda and Kenya and in the northwestern Atlas Mountains. Most rivers have a marked regime with a ground water component which controls flow during the dry season, but with little or no significant influence during the wet season. Ground-water-fed rivers occur only in karst regions, such as the Atlas Mountains and in Tanzania. Large lakes and reservoirs as well as swamps and wetlands have a regulating effect on runoff.

Using the methodology presented in Chapter 2, an assessment was made of the monthly distribution of runoff for the largest rivers, the regions, selected countries, for drainage to the two oceans, and for the continent as a whole (Fig. 6.12, Table 6.15). It can be seen that in the North region the runoff is unevenly distributed with 56% occurring from August to October and only 44% during the remaining 9 months. In the Central region the monthly distribution is very uniform while in the Eastern region it is much more variable, particularly because of the Zambezi River. More than half of the total discharge takes place from December to March in the Southern region while in Western Africa some 64% of the flows occur from August to October. Only 10.8% of the discharge takes place from January to May. In the Sahel the monthly distribution is close to that for the Western region with 75% of flows from August to October and only 2% from February to June. Zaire has a uniform monthly distribution of runoff while in Nigeria only 7% of the flows occur from May to July and 21% in September, i.e. the distribution within the year is extremely varied. For the whole continent the distribution is similar to that of the Central Region, although it is more uniform.

During a dry year distribution of runoff differs significantly from its distribution during an average year (Table 6.16). In North Africa more than 60% of the total discharge occurs between August and October while in Western Africa a similar volume of discharge takes place from August to November. In Central and the Southern Africa runoff is fairly uniform during a dry year. More than 35% of the discharge occurs from April to May in Eastern Africa. Zaire, as compared to Nigeria, has a more uniform

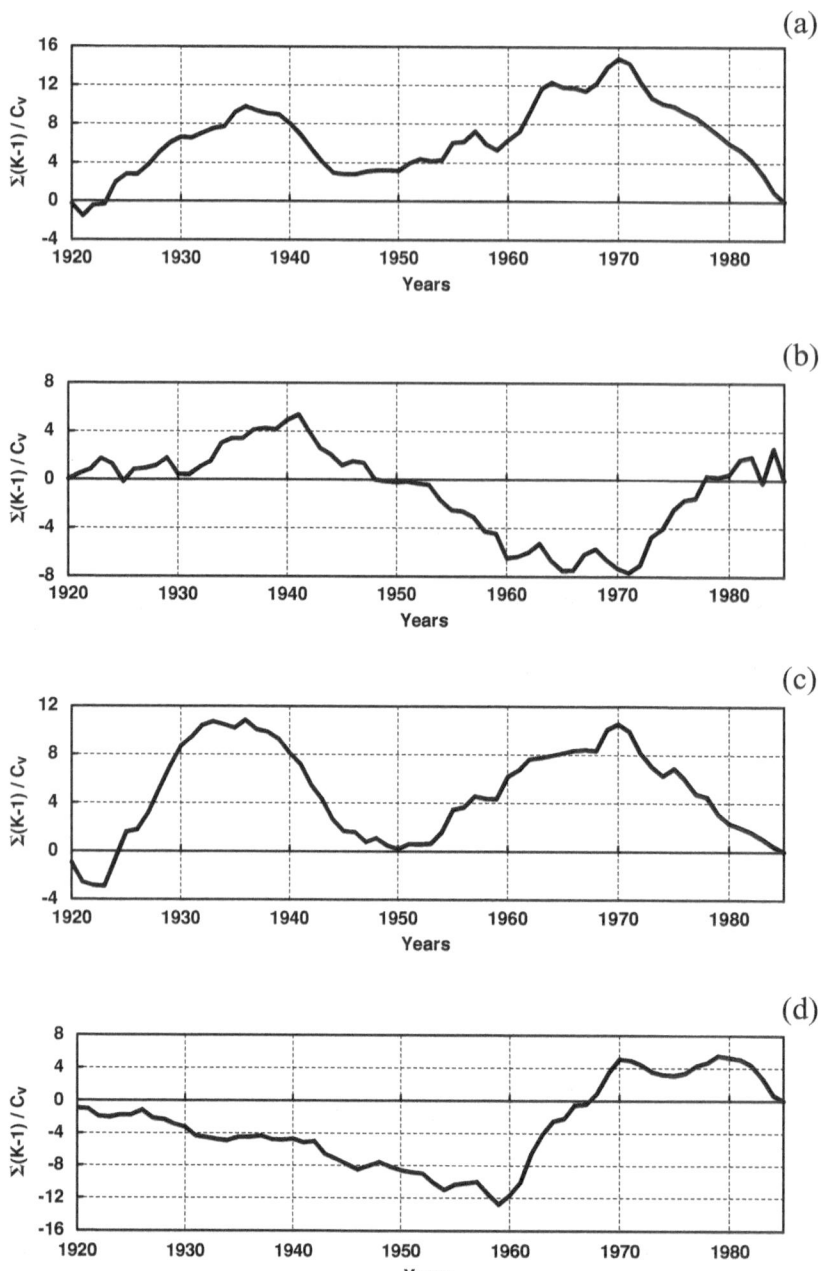

Fig. 6.9 Normalized difference integral curves of selected countries and
ocean slopes of Africa. (a) Nigeria; (b) Zaire; (c) Atlantic Ocean;
(d) Indian Ocean.

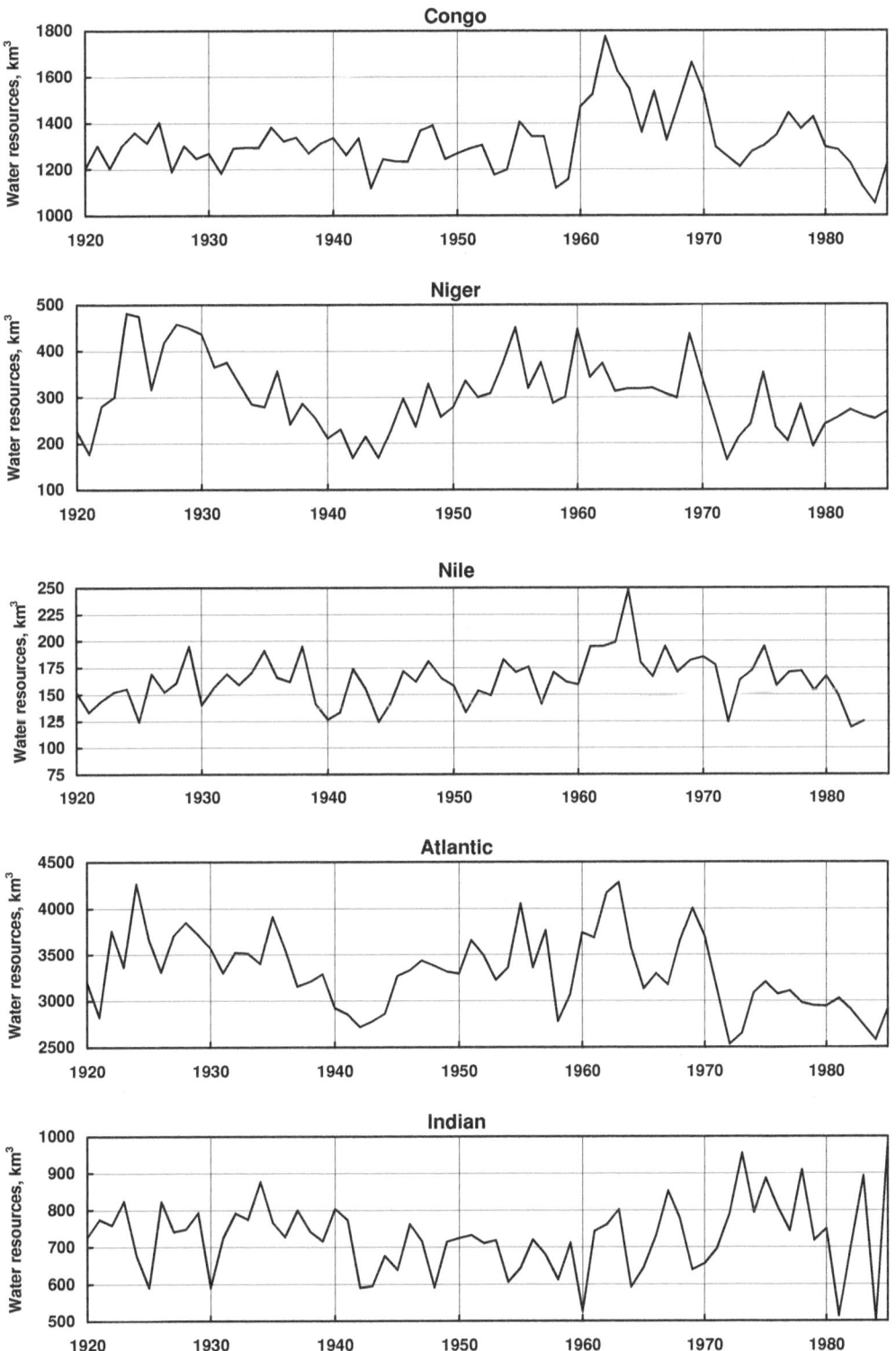

Fig. 6.10 Mean annual water resources of the Rivers Congo, Niger and
Nile and of the continental slopes of Africa.

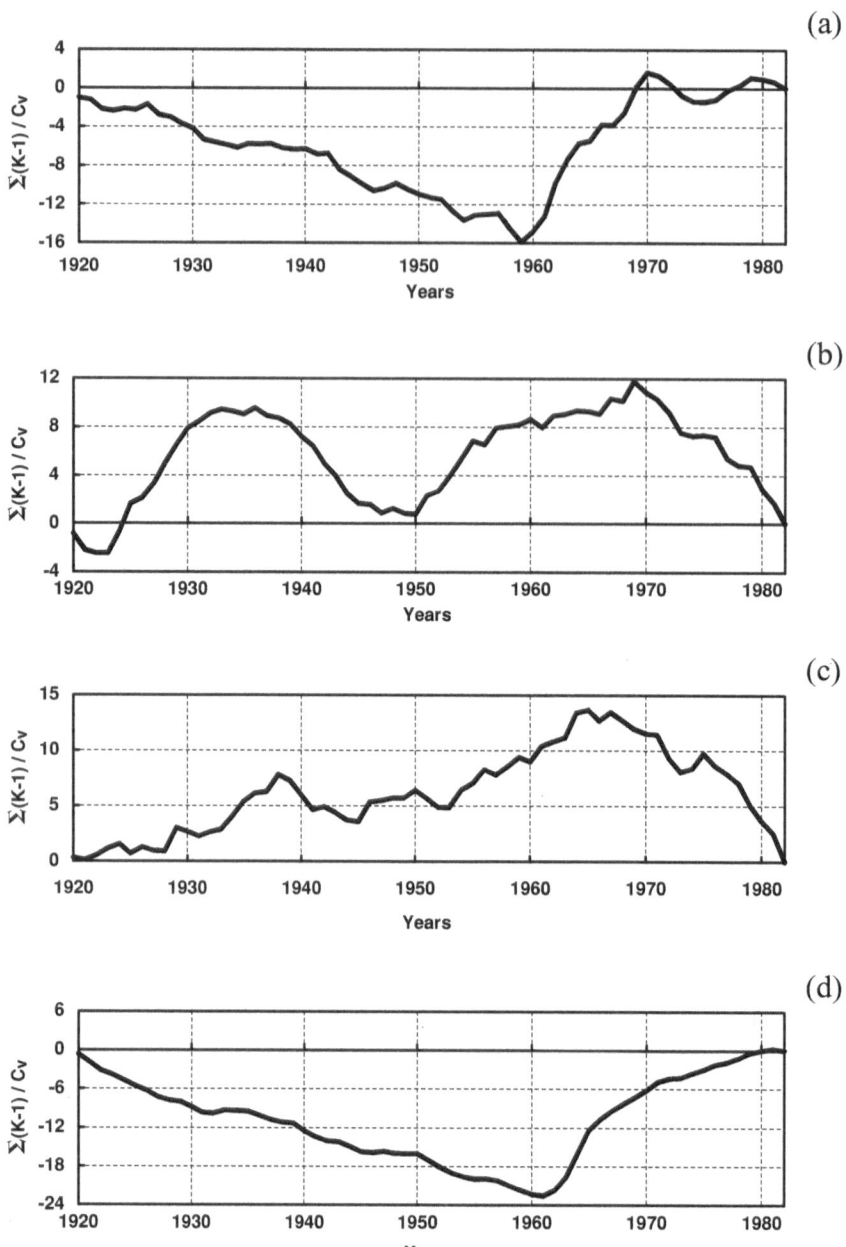

Fig. 6.11 Normalized difference integral curves of module coefficients
of runoff for individual African rivers. (a), Congo–Kinshasa;
(b) Niger–Koulikoro; (c) Nile–Dongola; (d) White Nile–Malakal;
(e) Senegal–Dagana; (f) Ogooue–Lambarene.

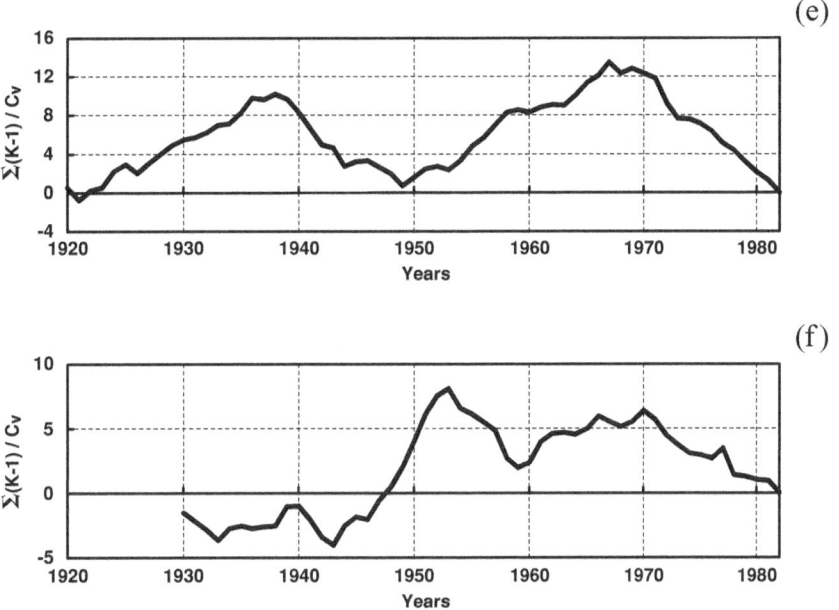

Fig. 6.11 (cont.).

Table 6.13. *Periods of different water content by principal river basins and continental slopes of Africa for period 1920–85*

River basin, continental slope	Periods of low water content			Periods of average water content			Periods of high water content		
	Years	Number of years	Ratio of mean for period to long-term mean, K_{av}	Years	Number of years	Ratio of mean for period to long-term mean, K_{av}	Years	Number of years	Ratio of mean for period to long-term mean, K_{av}
Congo	1920–59	40	0.97	–	–	–	1960–70	11	1.16
	1971–75	5	0.96	–	–	–	1976–79	4	1.06
	1980–85	6	0.91	–	–	–			
Niger	1937–52	16	0.83	–	–	–	1924–36	13	1.28
	1971–85	15	0.81				1953–70	18	1.14
Nile	1920–27	8	0.92	–	–	–	1928–38	11	1.05
	1939–53	15	0.94	–	–	–	1954–78	25	1.09
	1979–85	7	0.82						
Atlantic Ocean slope	1937–45	9	0.91	–	–	–	1920–36	17	1.08
	1971–85	15	0.88				1946–70	25	1.06
Indian Ocean slope	1942–60	19	0.92	1920–31	12	1.00	1932–41	10	1.07
				1961–73	13	1.02	1974–76	3	1.14
				1977–85	9	1.02			

Table 6.14. *Probability of appearance of year groups with different water content and their duration for principal river basins and continental slopes of Africa*

River basin, continental slope	Water content characteristic[a]	Probability (%) of appearance of year groups of 1 to 5 years in duration					Average duration of groups (years)
		1	2	3	4	5	
Congo	1	5.49	1.57	0.45	0.13	0.04	1.40
	2	1.54	1.32	1.13	0.97	0.83	7.00
	3	1.54	1.03	0.69	0.46	0.31	3.00
Niger	1	3.18	2.25	1.60	1.13	0.50	3.43
	2	10.05	4.57	2.00	0.94	0.43	1.83
	3	6.47	3.40	1.79	0.94	0.50	2.11
Nile	1	7.20	3.36	1.57	0.73	0.34	1.88
	2	9.63	5.51	3.15	1.80	1.03	2.33
	3	8.00	3.20	1.28	0.51	0.20	1.67
Atlantic Ocean slope	1	2.98	1.98	1.32	0.88	0.59	3.00
	2	7.23	4.47	2.76	1.70	1.05	2.62
	3	7.51	3.29	1.44	0.63	0.28	1.78
Indian Ocean slope	1	14.64	2.93	0.50	0.12	0.02	1.25
	2	14.01	7.18	3.63	1.89	0.97	2.05
	3	14.95	1.36	0.12	0.01	0.00	1.10

[a] 1, Low; 2, average; 3, high.

monthly distribution of runoff during a dry year and the continent as a whole has an even distribution of monthly runoff. Table 6.17 contains data for the three largest rivers and for discharges to the Atlantic and Indian Oceans during an average year. The Congo and the Zambezi have a more regular regime than the Niger and the Nile. Monthly discharges to the Atlantic Ocean are similarly distributed to those of the Congo and are more evenly distributed than flows to the Indian Ocean (Fig. 6.13).

6.6 CHANGES IN WATER USE AND WATER AVAILABILITY

6.6.1 Data

To make a reliable assessment of changes and trends in water use and water availability for Africa by region and by country is very difficult. In some African countries there is no water legislation that controls the use of water. Some countries have organizations responsible for overseeing the use of water and technical personnel for data acquisition (Alekseyevsky, 1974; Tuffuor, 1987). To generalize the data obtained at some key sites is also difficult, because of the high temporal and spatial variability of the climatic and hydrological regimes, the socio-economic conditions and the indicators derived from them (Ojo, 1987). The available estimates of African water use made by different researchers are only very

approximate and contain a number of assumptions which could produce serious errors.

The most complete and systematic data on the water use were obtained by SHI in the mid-1980s and published by Shiklomanov and Markova (1987) and by Shiklomanov (1988). These publications show changes in water use for all regions of Africa during the twentieth century, with forecasts for 2000. For most countries and all regions the volumes of water used were calculated from indirect indicators. (A brief description of these methods is given in Section 3.3.) This was quite justified, since at that time complete data on water use, including future estimates, were available only for two countries – Egypt and South Africa (Fyurin, 1966; Samaha, 1979; Bekker, 1983). However, practically all countries and regions had extensive literature covering some problem or other concerned with the development of the water economy and the water supply for the different sectors and these were used in these studies such as Grammatikati (1974), Briese (1981), Regional Report ECA (ECA, 1978), Barney (1980) and Van der Leeden (1975). Articles were also used about the current state and future developments of irrigation, industrial and domestic water use for the countries of North Africa (Ivanov and Luzhetsky, 1973; Schliephake and Deparage 1977; Salih, 1978; El Hares and Aswed, 1979; Samaha, 1979), for West Africa, and the Sahel (Anonymous, 1983; Maiga, 1984; Senghor, 1977), for Nigeria, Ghana and Togo (Van der Leeden, 1975; Barney, 1980), for Central Africa – Equatorial Africa, Cameroon and Zaire (Van der Leeden,

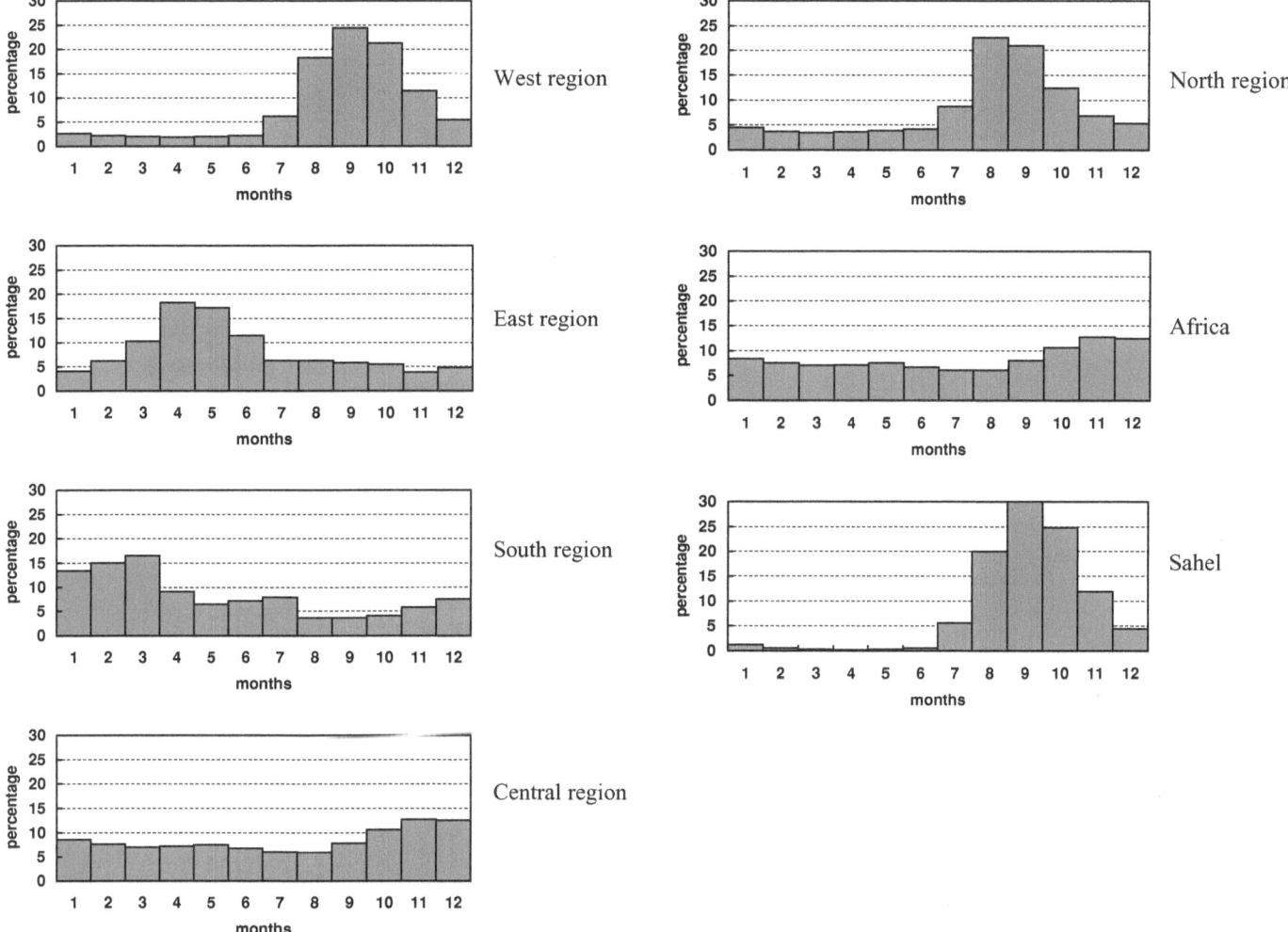

Fig. 6.12 Annual distribution of water resources by natural–economic regions of Africa and the continent as a whole with an average water content.

1975; ECA, 1978), for East Africa – Kenya, Somali, Tanzania (Van Alphen, 1979; Finn, 1983), and for Southern Africa – South Africa, Mozambique, Botswana, Namibia (Van der Leeden, 1975; ECA, 1978; Briese, 1981; Bekker, 1983).

The irrigation areas were based on FAO data (FAO, 1995a) in these accorded with the estimates available for some countries. The specific water use for irrigation was determined from analysis of the publications listed above. For all regions the return were assumed to be 20% of the intake volume. In calculating the water used in domestic water supply the following values of specific water use were assumed: for North Africa 170 and 40 l/day per head for 1980 and 300 and 60 l/day for 2000; for West Africa 170 and 50 and 300 and 70 l/day, respectively; for Central Africa 250 and 40, 400 and 60 l/day. Water consumption for agricultural was 70–90% of water intake for 1980 and up to 40–70% for the future; in community water use from 15% to 20% and 8% to 15%,

respectively. The industrial water use was calculated only approximately taking into account the data available for some countries, as well as the changes and trends in industrial production in the developing countries and the growth of the urban population.

Many new international and national publications have been published in recent years devoted to some extent to the water use problems of African countries. Some are general publications for the whole world (IIED/WRI, 1987; WRI/IIED, 1988; Mamilton and Maizels, 1989; WRI, 1990, 1992, 1996, 2000; Kulshreshtha, 1992; Andressian, 1993; Gleick, 1993, 1998; UN, 1993; Margat, 1994; Strzepek and Bowling, 1995; UN, 1995; Seckler *et al.*, 1998; Rijsberman, 2000), some are studies of the Arab region (ROSTAS, 1995; FAO, 1995c; UN/DPCSD, 1995c; UNESCO, 1995a; Shahin, 1996), while some deal with the countries of the Mediterranean and the Maghreb (Margat, 1990, 1995; Ali Ayub and Kuffner, 1994; Pearce, 1996; Margat and Vallée, 2000; Ragab, 2001). There are some detailed studies on the changes in water use for certain countries and in large river basins (Oniango Ogembo, 1980; Bekker, 1983; Russel, 1984; Conley and Hansmann, 1985;

Table 6.15. *Average monthly distribution of water resources during a year by natural–economic regions and selected countries of Africa in the year with an average water content*

Territory	Local water resources, km³/year	Monthly runoff distribution, % of mean annual value											
		1	2	3	4	5	6	7	8	9	10	11	12
North	41	4.5	3.7	3.4	3.6	3.9	4.2	8.7	22.6	20.9	12.4	6.8	5.3
Western including Sahel	1088	2.7	2.2	2.0	1.9	2.0	2.2	6.2	18.3	24.4	21.3	11.4	5.4
	104	1.2	0.6	0.3	0.2	0.3	0.6	5.6	20.0	30.0	24.8	11.9	4.5
Central	1770	8.5	7.6	7.0	7.2	7.5	6.7	6.0	5.9	7.8	10.6	12.7	12.5
Eastern	749	4.1	6.2	10.3	18.2	17.2	11.4	6.3	6.3	5.8	5.5	3.9	4.8
Southern	399	13.4	15.0	16.5	9.1	6.5	7.1	7.9	3.6	3.6	4.1	5.8	7.5
Nigeria	274	9.6	10.2	9.8	6.7	2.8	1.4	2.8	9.4	20.5	11.2	7.5	8.2
Zaire	987	8.3	7.2	6.5	7.5	8.4	8.6	7.8	6.9	8.4	9.4	10.2	10.7
Africa as a whole	4047	8.4	7.5	7.0	7.1	7.5	6.6	6.1	6.1	8.0	10.6	12.7	12.4

Table 6.16. *Water resources distribution during dry years by natural–economic regions and selected countries of Africa*

Territory	Local water resources, km³/year	Monthly runoff distribution, % of mean annual value											
		1	2	3	4	5	6	7	8	9	10	11	12
North	41	3.1	2.6	2.4	2.6	2.5	3.5	8.1	27.3	20.7	14.3	7.8	5.1
Western including Sahel	1088	4.0	2.8	2.2	6.7	4.3	5.0	8.3	14.2	20.4	16.1	10.1	5.9
	104	3.2	1.6	0.8	3.9	1.1	3.9	8.2	16.8	25.6	19.1	10.7	5.1
Central	1770	9.3	7.5	7.3	7.8	8.2	7.5	6.4	6.3	7.4	8.7	11.5	12.1
Eastern	749	4.0	6.3	8.3	17.5	19.0	8.2	6.6	6.4	5.9	6.6	6.8	4.4
Southern	399	10.4	6.0	10.6	6.2	5.3	9.0	12.3	6.4	6.9	6.9	6.7	13.3
Nigeria	274	13.2	6.4	2.3	0.7	0.4	0.3	2.2	6.4	15.6	16.1	16.8	19.6
Zaire	987	9.5	7.8	7.1	7.3	7.7	7.7	6.6	6.8	8.1	9.9	11.0	10.5
Africa as a whole	4047	9.2	7.5	7.3	7.8	8.1	7.5	6.5	6.4	7.4	8.7	11.5	12.1

Table 6.17. *Average monthly distribution of water resources during a year by principal rivers and continental slopes of Africa*

River basin, continental slope	Water resources, km³/year	Monthly runoff distribution, % of mean annual value											
		1	2	3	4	5	6	7	8	9	10	11	12
Congo	1320	9.8	7.8	7.3	7.7	8.0	7.5	6.5	6.5	7.6	9.1	10.8	11.4
Niger	302	14.3	14.3	11.7	6.9	2.7	1.0	1.0	4.0	9.0	10.4	11.6	13.1
Nile	161	3.5	3.3	2.7	2.6	2.2	4.3	6.0	21.0	26.1	16.6	7.2	4.5
Atlantic Ocean	3320	9.2	7.6	7.0	7.4	7.7	6.9	6.2	7.7	9.2	9.6	10.8	10.7
Indian Ocean	727	9.0	9.5	12.3	11.8	9.8	9.0	9.2	4.6	4.5	5.3	6.4	8.6

Schmidt, 1985; Arar, 1987; Smith and Al-Rawahy, 1990; Abu-Zeid and Abdel-Dayem, 1991; Biswas, 1991; Conway and Hulme, 1993; Ongweni, *et al.*, 1993; Sagoyan, 1993; Sircoulon *et al.*, 1999). There are also numerous articles on the different aspects of water and development for a number of countries.

An analysis of the material presented in these publications fully confirmed the 1987 estimates made by SHI (Shiklomanov and Markova, 1987) and, as a consequence, the SHI data for 1900 to 1980 were used here without changes. The estimates for 1990 (this was the forecast date used in the 1987 study), were overestimated significantly for most regions. Hence for the present time (1990 to 1995) and for the forecasts for 2000, 2010 and 2025, all the estimates were newly calculated using the methodology presented in Section 3.3 (Shiklomanov, 1997, 1998, 2000a, b; Penkova and

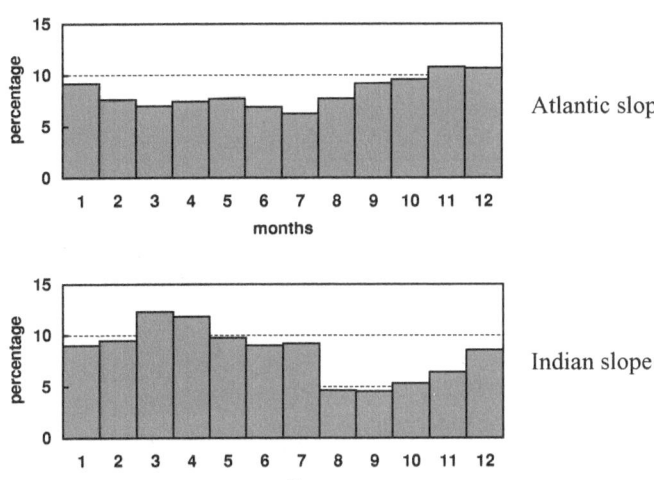

Fig. 6.13 Annual distribution of water resources by the territories of the Atlantic and Indian Ocean slopes of Africa.

Shiklomanov, 1998; Shiklomanov, *et al.*, 2000). The further development of water will be governed by the trends in the development of agriculture and industry and by the policies on water use and water in general. In the second half of the 1980s and in the early 1990s it became evident that the most difficult stage in the "crisis of Africa" was in the past and that the rates of economic growth would gradually increase (Bessonov, 1994). The most immediate pattern and the distant perspective of agricultural development will depend on the extension of the agricultural fringe, development of new crop species and other factors. These developments will be important to the development of trade within the continent and with other continents.

Most African countries have a high irrigation potential (FAO, 1995b). In particular, the Sahel countries (Mauritania, Senegal, Gambia, Mali, Burkina Faso, Niger and Chad) have more than 2.6 million ha in total. Integration of the plans for two or more countries based on basin-wide development of water resources, such as for the Senegal and Niger Rivers, should play a large role in the future development of irrigation. During the next decade in West Africa inter-basin transfers will become very significant. For example, in Nigeria five such schemes are planned. Two of them have already been developed: Zaire–Chad–Niger and the Transaqua scheme for using part of the flow of the Zaire in the Sudan–Sahel zone which will significantly increase its irrigation potential (Ojo, 1987; Nizskaya, 1989; Vitukhina and Onuchko, 1989; Tempelman, 1994; Oyebande, 1991; Sagoyan, 1993; Venema *et al.*, 1997).

North Africa is of great interest in the development of irrigation. The countries of this region have reached a high level of economic development and have achieved success in harnessing water resources over the last 20 years. However, within 10 to 15 years the available water resources will be depleted. Ways of

resolving the water problems in this region are connected with the regulation of water distribution, measures for pollution control and revision of plans for irrigation. A significant increase in the irrigated areas is unlikely in the near future. The increased cost of water will require production of cereals to be replaced by more valuable products for export purposes. Land degradation and the decrease in the agri-resource potential will also be of concern (Medvedev, 1989; Abramova, 1993; Ali Ayub and Kuffner, 1994; Margat, 1995; USAID, 2000). In the countries of East Africa attention is given to the introduction of new irrigation methods and the careful selection of sites for irrigation systems. In particular, in Zimbabwe there has been an efficient industry for manufacturing modern irrigation equipment and its sale to large farms in the private sector (Higgins *et al.*, 1988). Experiments at field stations in Zimbabwe initiated in 1991 have shown that irrigation efficiency can be increased by 40% by applying irrigation from a closed subsurface network (Institute of Hydrology, 1995). Construction of irrigation schemes continue in the basins of the Sabi and Lundi Rivers in the southeast where the irrigation is expected to reach 126 000 ha (Dmitriyevsky, 1990). Large areas of lands suitable for irrigation are also available in the western plains and the Ethiopian Highlands (Shahin, 1985; Pyatigorsky, 1990). In South Africa the development of irrigation is limited by the almost complete depletion of the irrigation budget. However, in accordance with national estimates (Bekker, 1983), in spite of decreasing rates of water use for irrigation from 2.3% in 1970 to 1975 to 0% in 2010 to 2020, the increased intensity of water use will result in the growth of production from irrigated lands by 2.9% a year.

Forecasting water use for irrigation in the African countries is based in the study on the generalization of data for recent years and on the analysis of trends in the areas irrigated in both absolute terms (thousand hectares) and in specific terms (hectares/1000 person). The study also takes into account population growth, GNP trends, the extent of development of the irrigation fund, the available water resources and national policies for the use of water employing the methodology presented in Section 3.3. Mean weighted gross irrigation norms for each country were determined as the ratio of the actual water use for irrigation to the area of the irrigated lands. These water use norms for most countries of North Africa vary from 7000 to 8000 m^3/ha, increasing for the countries with a rice crop rotation (Niger, Burkina Faso, Egypt) to 12 000 to 20 000 m^3/ha. The smallest norms were found for Cameroon, Congo and Zaire where they range from 5000 to 8000 m^3/ha. In Eastern African countries the largest norms occur in Madagascar (>20 000 m^3/ha). In the other countries the values range from 5000 to 12 000 m^3/ha and similar norms are typical of Southern Africa.

Water consumption in agriculture is largest (75–90% of the water intake) in the developed countries of the North and Southern regions of the continent and in the exceptionally dry countries of

Table 6.18. *Dynamics of fresh water use in Africa by type of economic activity (km³/year)[a]*

Water use	Assessment								Forecast		
	1900	1940	1950	1960	1970	1980	1990	1995	2000	2010	2025
Population, ×10⁶			220	273	354	481	642	708	871	1158	1596
Irrigation area, ha × 10⁶	2.8	3.5	4.0	6.0	7.1	8.4	11.4	12.0	12.6	14.1	16.1
Agriculture	40.8	47.7	53.5	79.4	89.0	106	127	134	140	156	175
	33.1	38.4	43.6	63.3	71.3	85.4	98.0	102	106	118	131
Industry	0.4	0.8	1.4	2.7	5.8	9.7	9.0	9.6	10	12	19
	0.1	0.1	0.2	0.5	0.8	1.4	1.6	1.7	2	2	3
Domestic	0.3	0.7	1.3	3.1	5.8	11.4	12.8	17.2	22	35	60
	0.2	0.3	0.5	0.9	1.2	1.8	1.7	2.1	3.5	4	6
Reservoirs	0	0	0	1.0	15.0	40.0	50.0	55.0	58.0	66.0	77.0
Total	41.0	49.0	56.0	86.0	116	168	199	215	230	270	331
	34.0	39.0	44.0	66.0	88.0	129	151	160	169	190	216

[a] Nominator, water withdrawal; denominator, consumption.

Western Africa and the Sahel. The smallest values are characteristic of countries in the Central region (60–75%). For forecasts, the calculated water use norms are decreased to take into account the more efficient use of water. African domestic and industrial water use is likely to increase when the economic downturn of the 1990s reverses and there is increased interest in the purchase of raw materials (Bessonov, 1994). In most countries the infrastructure is poor but the population is growing and become increasingly urban, while production will increase (Pulyarkin and Lipetz, 1991).

The forecasting of water use in the community and domestic sectors is based on likely changes in specific water use and forecasts of population numbers (ROSTAS, 1995). The specific water use (l/day per head) was calculated from actual 1990 data. Future water abstractions were forecast taking into account the plans for providing the population with a piped water supply while the expected water consumption was slightly decreased (by 10–20%). The largest domestic water use norms occur in North Africa (85 to 230 l/day per head), in Southern Africa (170 l/day per head). In the countries of Western and Central Africa they range from 10–15 to 130–140 l/day increasing only in the most developed countries (Mauritius, 180 l/day per head).

The volume of current industrial water use is based on generalizing the data for several countries. For forecasting the corrected UNIDO studies were used (Strzepek and Bowling, 1995). They contain the industrial water use growth coefficients for 2025 relative to the 1990 level. The largest coefficients were obtained for Nigeria, Ghana, Cameroon, Congo and Gabon, where they reach 3.6 to 4.8. Similar coefficients were obtained for Senegal and the Sudan. The highest growth in industrial water use should be found in Zambia, Zimbabwe and Kenya (5.1–5.3), while lowest coefficients are for Mozambique, Zaire, Egypt and Libya (2.2–2.3). For

the majority of countries they fall between 2.7 and 2.8. Based on the data for certain countries and the trends in global industrial water use for the last decades, the UNIDO coefficients for African countries appear to be slightly overestimated. Hence for this Monograph the 2020 figures were reduced by 25%. For 2000 and 2010 industrial water use was assumed to be proportional to the forecasts of GNP for these years. Water consumption by industry in the relatively developed countries was assumed to be constant at 8–10% of the total water intake or slightly above these percentages. For less developed countries, it was taken to be 15–20% but slightly decreasing into the future. The predicted GNP values are based on the so-called "Balanced Global Growth" scenario (Strzepek and Bowling, 1995). According to this scenario, the average annual rates of growth in the coming decades is likely to be 4.9%. The initial increase in income was assumed from data from the World Bank and the International Monetary Fund (World Bank, 1993, 1995; IMF, 1994), initial and projected population number from UN (UN/DPCSD, 1995).

6.6.2 Changes and trends in water use

The estimated volumes of water used at present and in the future are presented in Table 6.18 by country, region and for Africa as a whole. Figure 6.14 illustrates that the largest volumes of water (134 km³/year in 1995) are used in agriculture where a large volume is lost during irrigation. The other water uses consume relatively small amounts of water, if losses by evaporation from reservoirs are excluded. The latter is very significant for Africa and greatly exceeds the total water used by industry and in the domestic sector. By 2025 the volume of water used for domestic needs is expected to reach 60 km³/year. The actual water consumption

a)

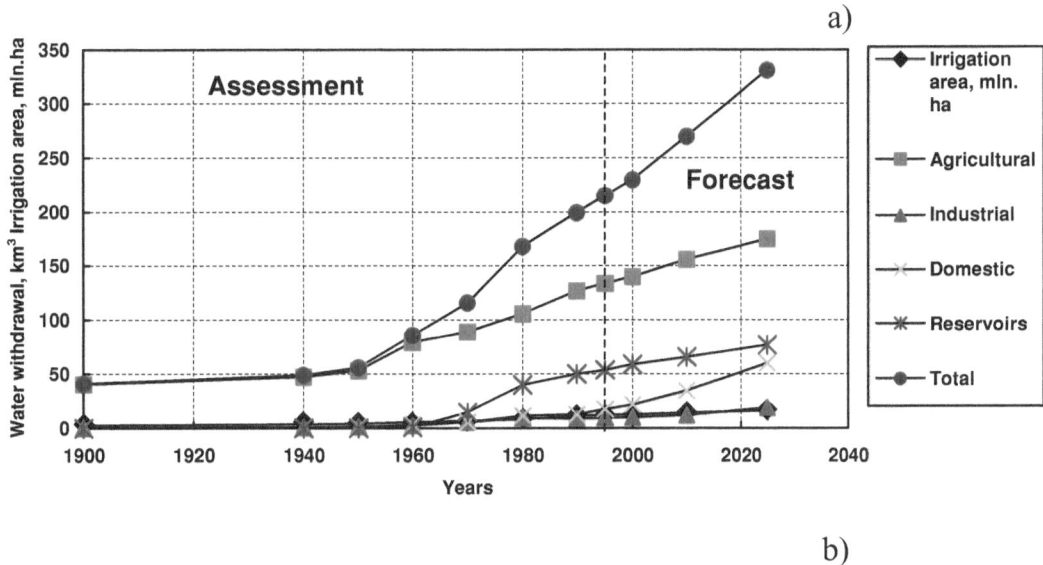

b)

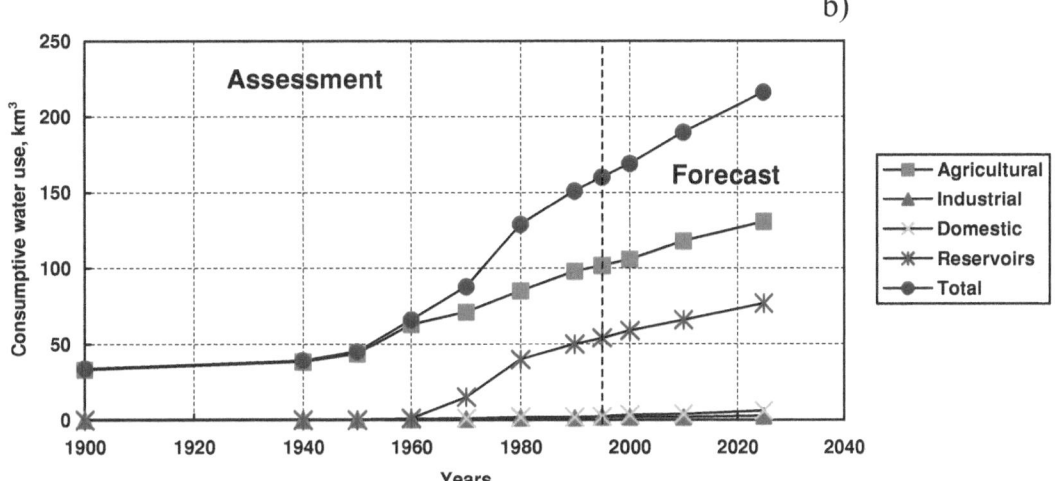

Fig. 6.14 Dynamics of water use in Africa over the kinds of economic activities. (a) Water withdrawal; (b) actual water consumption.

will, however, be only about 10% of the total water intake, whereas the losses in agriculture are likely to be about 75%.

Table 6.19 and Fig. 6.15 present the changes in the volume of water abstracted and consumed by region. In 1995 almost half of the 160 km³/year of the water consumed in Africa was used in the North. The least water use is used in the Central region (abstraction of about 2.5 km³/year and consumption of about 1.5 km³/year). The spatial distribution of water use is in good agreement with the distribution of water resources (Korzun, 1974a). The deficit is 2500 mm in North Africa and 1500 mm in the east and the south of the continent. This is in contrast to Central and Western Africa where resources exceed demand by between 1000 and 3000 mm. The changes in water use over time differ significantly over the continent. A three-fold increase (73 km³) in the total volume of water abstracted was observed in North Africa between 1900 and 1995. However rates of growth are almost as high in

Eastern Africa. The current trends in water use are expected to continue to 2025 when the water abstracted may exceed the 1995 level by 55%. The largest growth will still be in the countries of the North and Eastern Africa in spite of the possible rapid growth (3.6 times by 2025) in the Central region.

Table 6.20 shows the changes in the estimated volumes of water used by country and for the continent. Although the water used in Nigeria, the most densely populated country, may double by 2025, this total will be less than those for the developed Mediterranean countries. In the poorest countries of the Central region although water use will increase by more than 3 times, its total volume will not exceed the water use of the Sahel countries; in particular in Gambia and Zaire water use is small, only 0.1 to 0.3 km³/year. It is interesting to compare future changes in water use for certain countries predicted by different authors using different scenarios. For example Fig. 6.16 compares estimates the possible future water use in Egypt and Sudan made by Smith and Al-Rawahy (1990) and by Biswas (1991) with the values calculated for this Monograph. As might be anticipated, these forecasts differ considerably,

Table 6.19. *Dynamics of fresh water use by natural–economic regions of Africa (km³/year)*[a]

Region	Assessment								Forecast		
	1900	1940	1950	1960	1970	1980	1990	1995	2000	2010	2025
North	37.0 / 30.4	41.0 / 32.5	43.0 / 34.6	65.0 / 51.0	78.0 / 61.0	100 / 80.0	105 / 75.6	110 / 78.0	114 / 80	127 / 88	144 / 94
Western	1.0 / 0.7	1.5 / 1.1	2.3 / 1.7	3.8 / 2.6	8.4 / 6.3	19.0 / 14.0	22.7 / 18.2	26.0 / 20.1	29 / 22	37 / 26	52 / 32
Sahel							5.71 / 4.30	6.49 / 4.75	7.5 / 5.0	9.5 / 6.5	14 / 9.0
Central	0.1 / 0.05	0.3 / 0.12	0.5 / 0.18	1.0 / 0.29	1.6 / 0.6	2.8 / 1.3	2.0 / 1.3	2.5 / 1.4	3.0 / 1.5	4.5 / 2.0	9.0 / 3.0
Eastern	1.0 / 0.8	2.1 / 1.6	3.7 / 2.8	6.1 / 4.6	12.0 / 9.3	23.0 / 18.0	44.7 / 37.3	50.4 / 41.0	56 / 45	68 / 52	83 / 59
Southern	1.9 / 1.5	4.4 / 3.5	6.5 / 5.0	10.0 / 7.2	16.0 / 11.0	23.0 / 16.0	24.5 / 18.2	26.4 / 19.1	28 / 20	33 / 22	43 / 28
Africa as a whole	41.0 / 34.0	49.0 / 39.0	56.0 / 45.0	86.0 / 66.0	116 / 88.0	168 / 129	199 / 151	215 / 160	230 / 169	270 / 190	331 / 216

[a] Nominator, water withdrawal; denominator, consumption.

a)

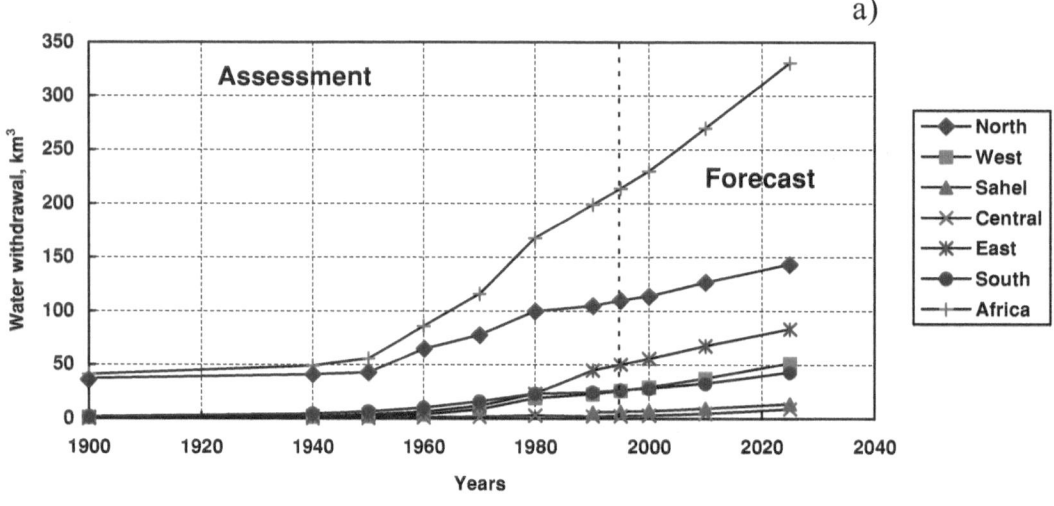

b)

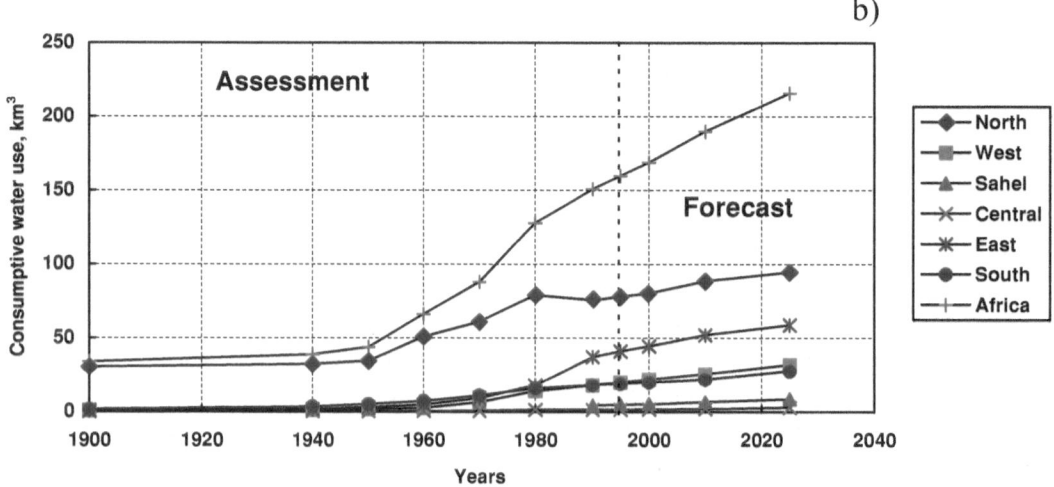

Fig. 6.15 Dynamics of water use in the regions of Africa. (a) Water withdrawal; (b) actual water consumption.

Table 6.20. *Dynamics of water use by selected countries of Africa (km³/year)[a]*

Country	Assessment			Forecast		
	1980	1990	1995	2000	2010	2025
Maghreb countries	17.60	24.90	26.80	28.60	32.40	42.50
	11.90	17.00	18.20	19.50	22.40	29.90
Morocco	10.30	11.00	11.40	11.70	12.40	15.10
	7.14	7.64	7.78	7.91	8.07	9.51
Algeria	2.99	4.50	5.20	5.91	7.06	9.90
	1.78	2.63	2.92	3.21	3.87	5.24
Mauritania	0.75	1.63	1.76	1.90	2.58	3.27
	0.56	1.22	1.32	1.41	1.91	2.42
Tunisia	1.89	3.08	3.11	3.14	3.20	3.50
	1.26	1.94	1.92	1.89	1.89	1.93
Libya	1.70	4.69	5.30	5.92	7.14	10.70
	1.30	3.46	4.19	3.82	4.88	7.07
Burkina Faso		0.38	0.51	0.64	1.04	2.06
		0.26	0.34	0.41	0.62	1.16
Chad		0.19	0.29	0.39	0.61	1.07
		0.13	0.20	0.26	0.38	0.64
Egypt	50.50	53.10	53.70	54.30	61.60	67.70
	35.10	36.90	37.40	37.80	38.90	39.90
Gambia		0.03	0.04	0.06	0.09	0.30
		0.02	0.03	0.04	0.06	0.10
Madagascar		19.70	22.40	25.20	28.00	31.60
		15.40	17.20	19.10	21.00	22.90
Mali		1.49	1.66	1.82	2.18	2.69
		1.19	1.28	1.37	1.55	1.74
Niger		0.50	0.64	0.79	1.32	2.54
		0.35	0.42	0.50	0.78	1.46
Nigeria	2.80	4.02	4.70	5.38	7.82	10.80
	1.70	2.03	2.16	2.29	2.69	3.18
Senegal		1.50	1.58	1.67	1.83	2.02
		0.97	1.08	1.20	1.22	1.23
South Africa	12.00	13.30	13.90	14.50	16.00	18.60
	8.00	8.10	8.10	8.00	8.00	8.00
Sudan	18.70	17.40	18.00	18.60	19.70	22.90
	13.40	12.40	12.60	12.90	13.20	14.70
Zaire		0.04	0.05	0.06	0.12	0.23
		—	—			

[a] Nominator, water withdrawal; denominator, water consumption.

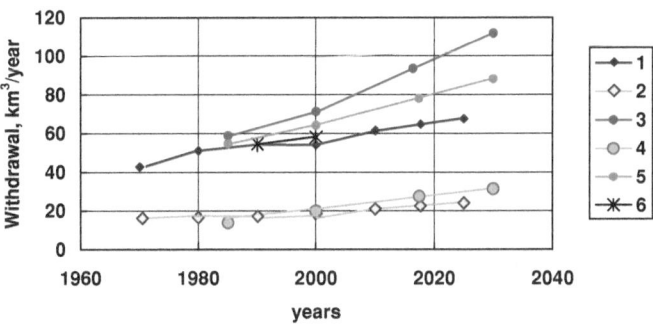

Fig. 6.16 Dynamics of water use (km³/year) in Egypt and Sudan from data of different sources. 1, Egypt (SHI, 1997); 2, Sudan (SHI, 1997); 3, Egypt (Smith and Al-Rawahy, 1990); 4, Sudan (Smith and Al-Rawahy, 1990); 5, Egypt (Smith and Al-Rawahy, assessment 1988); 6, Egypt (Biswas, 1991).

especially for Egypt, indicating the difficulties of making such forecasts. Irrigation uses about 85% of all the total water used, but this percentage could change during the coming years because of changes in government policy in agricultural production and because of large demands from the industrial and domestic sectors. However figures for 1990 and 1991 appear to be overestimated, since they were based on the extrapolation of growth in recent decades, and this also applies to the forecasts by Shiklomanov and Markova (1987). These forecasts suggested that if existing trends in the 1980s had been continued, the total water use in the North re-

gion in 1990 would be 125 km³/year, whereas the actual value was 105 km³/year (Table 6.14), i.e. some 20 km³ less. There were similar differences elsewhere: 10 km³/year for Western Africa, about 2 km³/year for the Central region and 12 km³/year for Southern Africa.

It is valuable to compare forecasts of African water use by sector (Table 6.18) with data from other studies. The most complete of recently published reports is by Margat (1994), although no account is taken of the water used for thermal power production which is not so important for Africa. For the 1980s, and for 2025, the estimates of industrial water use by Margat (8.6 and 18.0 km³/year, respectively) turned out to be quite close to the estimates obtained in this study. Margat assumed high levels of water use in for agriculture (270 km³/year) and in the domestic sector (54 km³/year), for 2025, and 165 km³/year and 39 km³/year if a low water use scenario was adopted. This study obtained the values of 175 km²/year and 60 km³/year for 2025. Taking into account the additional losses due to evaporation from reservoirs, this Monograph's forecast of total water use for Africa for 2025 (331 km³/year) is almost the same as Margat's forecast, assuming the high scenario (335 km³).

A comparison of water use with the renewable surface water resources is presented in Table 6.21. This table shows that the water use in 1995 accounted for only 5.3% (4% water consumption) of surface water resources, a figure almost half the world mean. This low figure is not only indicative of how poor are many African countries and of their low rate of growth, but it also demonstrates the variations in demand for water across the continent. In the North, the region with the smallest volume of runoff, almost all the renewable water resources (99%) are used, compared to the Central region where the resources are the largest and the demand is lowest. This situation will probably remain in 2025, but by then the water intake in the North will significantly exceed the renewable surface water resources (by 30%). This shortfall will be made

Table 6.21. *Water use as a percentage of water resources by natural–economic regions of Africa*

Region	Water resources, km³/year		Water use, km³/year				Water use as percentage of water resources			
			1995		2025		1995		2025	
	Inflow	Local runoff	Withdrawal	Consumption	Withdrawal	Consumption	Withdrawal	Consumption	Withdrawal	Consumption
Northern	140	41.0	110	78.0	144	94.0	99.0	70.2	130.0	84.6
Western including Sahel	30	1088	26.0	20.1	51.7	32.0	2.4	1.8	4.7	2.9
	77	104	6.49	4.75	14.0	9.0	4.6	3.2	9.8	6.3
Central	80	1770	2.50	1.40	9.0	3.0	0.14	0.08	0.5	0.2
Eastern	26	749	50.4	41.0	83.0	59.0	6.7	5.4	10.8	7.7
Southern	86	399	26.4	19.1	43.0	28.0	6.0	4.3	9.7	6.3
Africa as a whole		4047	215	160	331	216	5.3	4.0	8.2	5.3

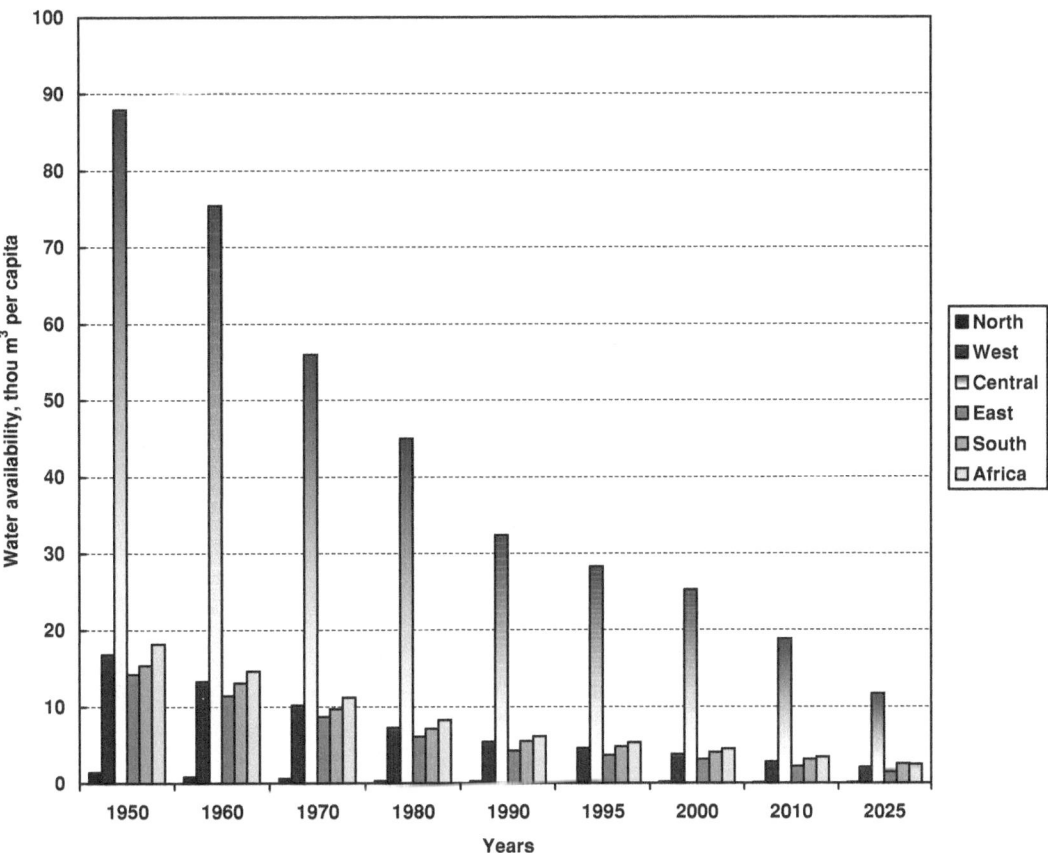

Fig. 6.17 Dynamics of water availability (m³ × 10³/year per head) by natural–economic regions of Africa.

up by the use of ground water and non-traditional sources. In the Central region no significant changes are predicted.

6.6.3 Water availability by region and country

The changes and trends in the specific availability of water (m³/km² per head) over the period 1950 to 2025 are shown in Table 6.22. In 1950 in all regions (except for the North) water was readily available (>10 000 m³/year per head). In 1995 only the Central region was in this category (>20 000 m³/year per head). The North declined sharply from the very low category to the catastrophically low (<1000 m³/year per year), while the other regions had low levels of water availability (2000–5000 m³/year per year). By 2025 Eastern Africa is expected to be a region with very low specific water availability, while other regions will be similar except for the Central region. Water availability has decreased all over Africa during the last 45 years and this decrease is expected to continue up to 2025 (Fig. 6.17). As this decrease occurs and the limits of the available resource are reached and surpassed, policies on water use will have to be revised. In some

countries water availability is much lower than at the regional level. For example, in 1990 in Niger, Nigeria and Senegal it was almost 50% less than the average in Western Africa and by 2025 it is expected to be at least 67% lower in Niger (Table 6.23). This situation will be exacerbated during droughts (Table 6.24) and even for the Central Region the mean specific water availability is expected to fall to 27 400 and 23 800 m³/year per head.

In North Africa, even in 1960, the demand for water (52 km³/year) exceeded the surface water resources (41 km³/year). By 1980 this deficit reached 38 km³/year, but, of course, this is fed by flows from outside (140 km³/year). However the use of ground water is traditional in Africa and is a resource which will continue to be developed (Ali Ayub and Kuffner, 1994; ROSTAS, 1994; UNESCO, 1995a). In Libya, for example, the available water resources over the period 1980 to 2025 can be increased due to ground sources by up to 9 km³/year. This will allow water supplies to be maintained at the 1980 level (330 m³/year per head) (Table 6.25). For the Maghreb countries, to increase the available water resources by 13 km³/year will require substantial capital investment. Table 6.25 is based on data of ground water use published in ROSTAS (1995); UNESCO (1995a) and FAO (1995b).

Table 6.22. *Dynamics of population and water availability by natural–economic regions of Africa*

Region	Area, km² × 10⁶	Year	Population, ×10⁶	Water availability, m³ × 10³/year per head
North	8.78	1950	51.4	1.49[a]
		1960	65.7	0.91[a]
		1970	82.9	0.60[a]
		1980	109.3	0.29[a]
		1990	140.4	0.25[a]
		1995	157.0	0.22[a]
		2000	175.6	0.18[a]
		2010	210.9	0.11[a]
		2025	274.4	0.06[a]
Western	6.96	1950	65.7	16.8
		1960	82.9	13.3
		1970	108.0	10.2
		1980	149.4	7.29
		1990	199.8	5.43
		1995	211.3	5.12
		2000	285.4	3.79
		2010	392.8	2.74
		2025	521.5	2.05
Central	4.08	1950	20.6	87.9
		1960	24.0	75.4
		1970	32.3	56.0
		1980	40.2	45.0
		1990	55.8	32.4
		1995	64.0	28.3
		2000	71.5	25.3
		2010	96.4	18.8
		2025	154.4	11.7
Eastern	5.17	1950	53.6	14.2
		1960	66.7	11.4
		1970	86.6	8.69
		1980	122.0	6.10
		1990	169.4	4.28
		1995	193.5	3.73
		2000	234.4	3.08
		2010	322.2	2.20
		2025	481.8	1.46
Southern	5.11	1950	28.3	15.4
		1960	33.3	13.1
		1970	44.5	9.69
		1980	59.7	7.14
		1990	76.5	5.54
		1995	83.5	5.06
		2000	104.0	4.06
		2010	135.2	3.11
		2025	162.9	2.54
Africa as a whole	30.1	1950	220.0	18.2
		1960	273.0	14.6
		1970	354.0	11.2
		1980	481.0	8.15
		1990	642.0	6.07
		1995	708.0	5.49
		2000	871.0	4.45
		2010	1158.0	3.33
		2025	1596.0	2.40

[a] With no account of water withdrawal from ground water storage.

Table 6.23. *Water availability of population with surface water resources in individual countries of Africa*

Country	Population (1994), $\times 10^6$	Water resources, m³/year		Water Consumption (1990), km³/year		Water availability, m³ $\times 10^3$/year per head						
						Assessment					Forecast	
		Local runoff	Inflow	Withdrawal	Consumption	1970	1980	1990	1995	2000	2010	2025
Sudan	27.36	22.0	140	17.4	12.4	5.02	4.21	3.16	2.78	2.40	1.88	1.30
Senegal	8.102	17.4	17.4	1.50	0.97			2.68	2.34	2.00	1.52	0.97
Gambia	1.081	3.20	4.70	0.03	0.02			6.36	6.25	6.14	4.86	2.93
Mali	10.46	50.0	44.4	1.49	1.19			7.71	6.65	5.60	4.16	2.84
Burkina Faso	10.05	14.7	2.00	0.38	0.26			1.73	1.59	1.45	1.07	0.60
Niger	8.846	3.00	30.4	0.50	0.35			2.31	2.07	1.82	1.31	0.78
Chad	6.183	15.8	28.3	0.19	0.13			5.36	4.71	4.06	3.16	2.21
Nigeria	108	274	43.6	4.00	2.00			2.71	2.26	1.81	1.29	1.04
Zaire	42.6	987	313	0.04	0.01			30.6	27.3	24.0	17.9	11.4
South Africa	40.6	47.4	5.2	13.3	8.10	1.84	1.38	1.03	0.94	0.84	0.67	0.60

Table 6.24. *Water availability throughout long-term dry period and the driest year by natural–economic regions and selected countries of Africa*

| Territory | Population, ×10^6 | | Water resources during dry periods and years | | | | | | Water consumption, km^3 | | Water availability, m^3 × 10^3/year per head | | | |
| | | | | | | | | | | | Period | | Year | |
	1994	2025	Period	Inflow	Local runoff	Year	Inflow	Local runoff	1994	2025	1994	2025	1994	2025
North region	157	274	1982–85	91.6	31.0	1984	76.8	26.0	77.6	94.5	0	0	0	0
Western region	211	521	1964–85	34.7	879	1972	15.6	581	19.7	32.0	4.15	1.66	2.70	1.07
Sahel countries	46.9	116	1970–85	52.4	70.7	1984	38.5	58.0	4.7	8.7	1.96	0.76	1.42	0.54
Central region	62.9	154	1971–85	71.5	1692	1984	60.4	1469	1.4	2.8	27.40	11.20	23.80	9.72
Eastern region	193	482	1981–85	23.0	701	1984	15.6	504	40.2	58.8	3.48	1.36	2.44	0.94
Southern region	83.5	163	1979–85	74.6	337	1984	52.9	270	18.9	27.8	4.26	2.13	3.32	1.65
Nigeria	108	281	1937–50	34.7	219	1972	23.5	148	2.1	3.2	2.17	0.83	1.46	0.56
Zaire	42.6	99.4	1980–85	285	898	1984	249	786	0	0.1	24.40	10.50	21.40	9.16
Africa as a whole	708	1596	1971–85	–	3684	1984	–	3073	156.0	216.0	4.98	2.26	4.04	1.79

Table 6.25. *Dynamics of water withdrawal, water consumption and water availability by the countries of Maghreb*

Country	Parameter[a]	1980	1990	2000	2010	2025
Algeria	1	16.3	16.4	17.4	18.4	17.4
	2	2.99	4.50	5.91	7.06	9.90
	3	1.76	2.63	3.21	3.87	5.24
	4	0.779	0.552	0.424	0.358	0.272
Libya	1	2.30	5.29	6.52	7.74	11.3
	2	1.70	4.69	5.92	7.14	10.7
	3	1.30	3.46	4.19	4.88	7.07
	4	0.329	0.402	0.383	0.354	0.330
Mauritania	1	6.47	6.47	6.47	6.47	6.47
	2	0.75	1.63	1.90	2.58	3.27
	3	0.56	1.22	1.41	1.91	2.42
	4	3.624	2.596	1.687	1.132	0.791
Morocco	1	32.8	33.0	33.2	33.4	34.1
	2	10.3	11.0	11.7	12.4	15.1
	3	7.14	7.64	7.91	8.07	9.51
	4	1.280	1.012	0.857	0.749	0.539
Tunisia	1	4.80	4.81	4.83	4.85	4.95
	2	1.89	3.08	3.14	3.20	3.50
	3	1.26	1.94	1.89	1.89	1.93
	4	0.554	0.356	0.311	0.272	0.221
Maghreb as a whole	1	62.7	65.9	68.4	70.8	76.2
	2	17.6	24.9	28.6	32.4	42.5
	3	12.0	17.0	19.1	22.4	29.9
	4	1.019	0.757	0.600	0.497	0.358

[a] 1, Total water resources available (km^3/year); 2, water withdrawal (km^3/year); 3, water consumption (km^3/year); 4, water availability (thou m^3/year per head).

6.7 PROBLEMS OF WATER MANAGEMENT AT THE CONTINENTAL SCALE

6.7.1 A brief review of the problem

The African continent, with vast areas of desert and an extremely varied distribution of water resources and considerable differences in the density of population, has a large number of countries with very different levels of economic and social development. There is a range of quite acute water management problems which concern the interests of both individual countries and countries sharing a river basin. These problems require international cooperation for their solution.

The following problems are typical of the continent:

• water supplies in desert and urban areas
• contamination of surface and ground water
• impact of large hydraulic structures, primarily reservoirs.

A brief description of these problems and their possible solution are given below. Lake Nasser on the Nile – a large reservoir in an arid area – is considered in more detail.

PROBLEMS OF WATER SUPPLY IN DESERT REGIONS AND IN URBAN AREAS

The presence of the humid regions of Central and Western Africa close to the Sahara and to the areas in the Sahel, Eastern and Southern Africa with limited water resources offers the possibility of transferring water resources to the adjoining arid regions, primarily northwards for irrigation in the Sahel and the Sahara. Several diversion projects have been proposed. One of these envisages the creation of a large reservoir in the middle section of the Congo (on the site of an ancient lake) with an area of several hundred thousand square kilometres. Water from the Congo will be discharged from this reservoir into a canal to the Shali River which flows into Lake Chad and then across the Sahara northward to the Gulf of Gabes. This would create an artery similar to the Nile (60–100 km^3/year). It would pass across the central Sahara and solve the water-supply problem of the adjoining regions of Chad, Niger, Libya, Algeria and Tunisia.

At present the lack of cohesion between African states and their low level of development, as well as difficulties of reliably assessing the ecological consequences of these projects, prevent their implementation. But such projects could become the basis for the water supply for vast areas. Only small-scale transfers have been developed and implemented in Africa; this has happened in South Africa (Robbroec, 1979), in the Nile Basin and some in other regions. The Transvaal in South Africa where the large mines and the most extensive urban centres are to be found has the most severe water problems. Water shortages exist in the west of the Cape Province, in Cape Town and in its hinterland. A radical solution of the water-supply problem in Cape Town is planned, namely desalinization of sea water. However its high cost means that it will possibly be introduced later in the twenty-first century. Several small water management systems which capture the runoff of nearby mountain streams provide an intermediate solution. The last system was introduced in 1982; it was one of the most complicated hydraulic complexes of the 1980s and a good engineering solution. At the present time some 3–4 km^3/year is diverted from South African rivers, comprising 10–12% of the total runoff of the country (Robbroec, 1979), and it is likely more projects will be developed.

POLLUTION OF SURFACE AND GROUND WATER

Pollution of African surface and ground water is due to natural factors and to human activities. There are naturally occurring high levels of soluble salts both in surface and ground water in Tunisia, Morocco, Algeria and Ethiopia, large concentrations of fluorine in Morocco, Senegal and Kenya and high levels of suspended matter in Tunisia, Morocco, Algeria, and in a number of other countries. The increased salinity is due to a combination of acid rocks, imperfect drainage and intense evaporation causing salt accumulation at the soil surface and increased concentrations in surface and

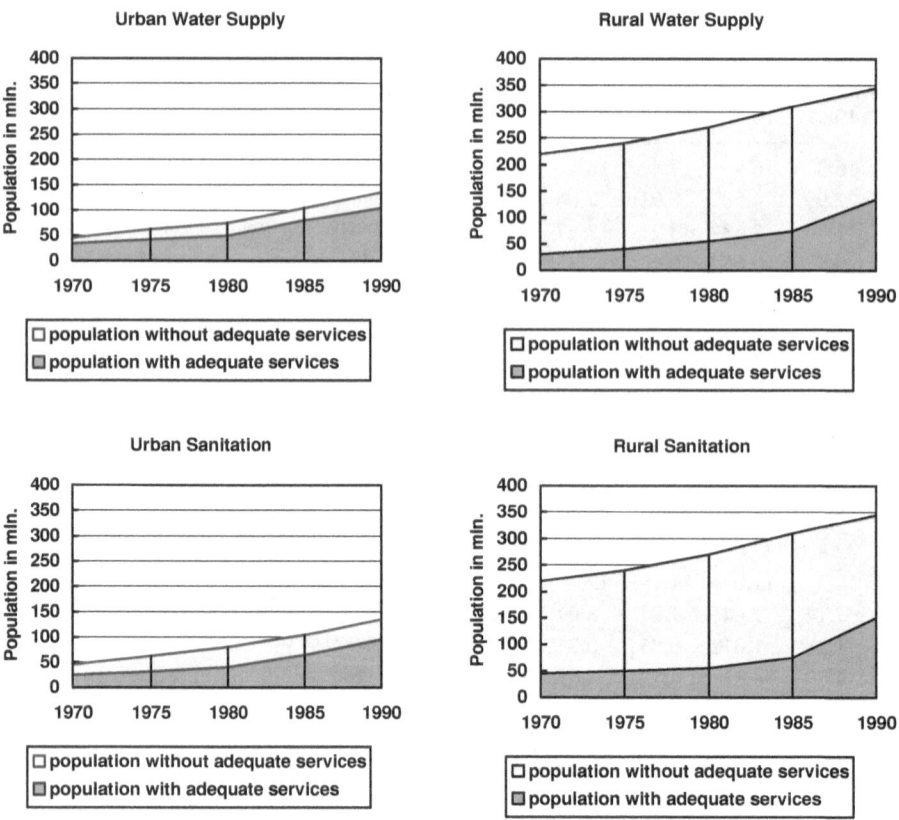

Fig. 6.18 Dynamics of water supply and sanitation in urban and rural Africa.

ground water. This occurs in most African countries. The problem is also aggravated by the process of sea water intrusion into aquifers in coastal areas such as the Gulf of Guinea countries, the Nile Delta and Senegal. Here, the total concentration of dissolved salts in ground water may reach 7500–16 000 mg/l (WHO/UNEP, 1991).

The characteristics of the main water quality variables are determined by the level of agricultural development, the activities of the mining industry and other branches of industry and by the degree, if any, of sewage treatment and the systems available to maintain standards of sanitation. The problems caused are even more difficult in most African countries, since except for Algeria, Morocco and Egypt, the systems for monitoring, control and management of water resources are virtually absent. Nine countries out of ten do not possess information on the quality of their rivers, lakes and aquifers, or specially equipped laboratories and trained technical personnel; there is no public awareness for the need and importance of ecological programmes. According to the report of 48th World Health Assembly (1995), only 259 million of the 563 million people in Africa have access to a water-supply system and only 208 million have sanitation systems, indicators which place Africa last among the continents. The trends and changes in the development of sanitation and water supplies in ur-

ban and agricultural areas between 1970 and 1990 are presented in Fig. 6.18 (WHO/UNEP, 1991).

These factors lead to unsatisfactory water quality levels across the continent with regard to organic matter, nitrate compounds and pesticides, as well as to high levels of bacterial contamination. The main sources of organic pollution and nutrients are industrial, domestic and agricultural wastes. For example, nitrate concentrations in basins around the Gulf of Guinea averages 200 mgN/l (WHO/UNEP, 1991), well above the WMO recommended limits for drinking water of 50 mgN/l. Direct infiltration of domestic waste and waste from cattle farms to aquifers produce the high level of nitrate in ground water. In some cases, for example, in the Sahel, the concentration of nitrates reaches 400 mgN/l in Burkina Faso and up to 800 mgN/l in Senegal. High loads of organic substances are also produced by the discharges from coffee producing plants with extremely high levels of biological oxygen demand (BOD) – 509 to 9000 mgO/l (WHO/UNEP, 1991). The discharge of these wastes causes sharp BOD increases in river water up to more than 100 mgO/l, resulting in fish death. The development of traditional industries in Africa – sugar refining, brewing, the paper and textile industries and the mining of gold, aluminium and phosphate (often without treatment of wastes) – has resulted in surface water contamination by heavy metals. This problem also occurs in the Nile Delta and in some regions of southern Nigeria, Zaire and Zambia.

Table 6.26. *Description of general state of water quality in Africa*[a]

Type of pollution	Northwestern Africa	Sahel	Gulf of Guinea	Congo Basin	Nile Basin[b]	Eastern and southern lakes	Great Lakes
Pathogenic substances	1–2	1–3	1–3	1–2	1–2	1–3	1
Organic substances	1–2	1–2	1–2	0–1	1–2	1–2	0
Salinity	0–2	0–2	0–1	0	0–1	1–2	0
Nitrates	0–2	1–2	1–2	0–1	0–1	1	0
Eutrophication	0–2	0–1	1	0	0–2	0	1[d]
Heavy metals	0–1	0–1	0–1	0–2[c]	0–1	0–1	0
Pesticides	0–1	1	1–2	1	0–1	0–1	1[d]
Industrial organics	0–1	0	0–1	0–1	0–1	0	0
Sediments	1–3	0–2	0	0	0–2	0–2	0
Fluoride	0–3	0–2	0	0	0	1–3	0

[a] 0, Water is not polluted; 1, moderate pollution, water can be used after appropriate purification; 2, strong pollution; 3, strong pollution, water cannot be used.
[b] Maximum pollution is observed in Nile Delta.
[c] Upper reaches of the Zaire River.
[d] Pollution of local scale.

Of special importance among the ecological problems is bacterial pollution, due to absence of treatment of sewage and because of the different microbiological processes occurring in artificial water bodies leading to the formation of pathogenic and carcinogenic organisms. In Abidjan (Côte d'Ivoire) and Lagos (Nigeria) sewage discharges contain up to 50 000 coliforms/100 ml. The concentrations of pathogenic bacteria in drinking water across the continent depend on the drinking water source. According to the World Health Organization, 3–50 coliforms/100 ml is found in rainwater and boreholes, 250–1500 coliforms/100 ml in ponds and small water bodies and 60–850 coliforms/100 ml in wells. The maximum concentrations of 500–100 000 coliforms/100 ml are observed in large rivers which are the main recipients of waste waters (WHO/UNEP, 1991). Eutrophication of water bodies is also a widespread African phenomenon. It is most typical in Morocco where several reservoirs have been built in recent years, in Egypt in Lake Nasser, as well as in the countries of the Gulf of Guinea. These countries mainly specialize in the food industry – production from canneries, palm oil and sugar (Ghana), coffee and cacao (Ghana, Togo) – and these are the source of nutrients. Table 6.26 presents water quality data for various parts of Africa (WHO/UNEP, 1991).

THE PROBLEMS CAUSED BY LARGE HYDRAULIC STRUCTURES
A number of large dams have been built for hydropower generation, for river regulation, flood control and reservoir storage for year-round irrigation. Some of the most important power stations have also been built to support mining and its associated industries. For example, in the Congo, the Katanga power stations were constructed for an expanding mining industry and for the new non-ferrous metals and chemical industries. In other cases power stations were built to provide electricity for new expanding industries. After the Second World War energy production increased in Algeria, Angola, Guinea, Zambia, Cameroon, Kenya, Congo, Liberia, Morocco, Nigeria, Tanzania, Tunisia, Uganda, Ethiopia and Egypt (Dmitriyevsky, 1967). The reservoirs built for hydropower plants also provide a source of water for water supply and for irrigation. However, these large water bodies have certain negative consequences due to changes in the water balance and the appearance of serious ecological problems, not only in the water body, but over the basin concerned. The impact of the Aswan Dam and Lake Nasser is considered below. However, it should be noted that the main cause of Africa's ecological problems is the low level of economic development in most countries, the absence of systems for water supply, sewerage and sewage treatment, as well as the lack of national programmes for water resource assessment, use and management.

6.7.2 The Aswan High Dam and Lake – the hydroecological and social consequences

BACKGROUND TO THE NILE
From its source to its mouth, the Nile River flows through nine countries: Burundi, Rwanda, Tanzania, Kenya, Zaire, Uganda, Ethiopia, Sudan and Egypt. As the Nile has been extremely important in the development of these countries, its regime has been studied from ancient times. Nile water has been used over the

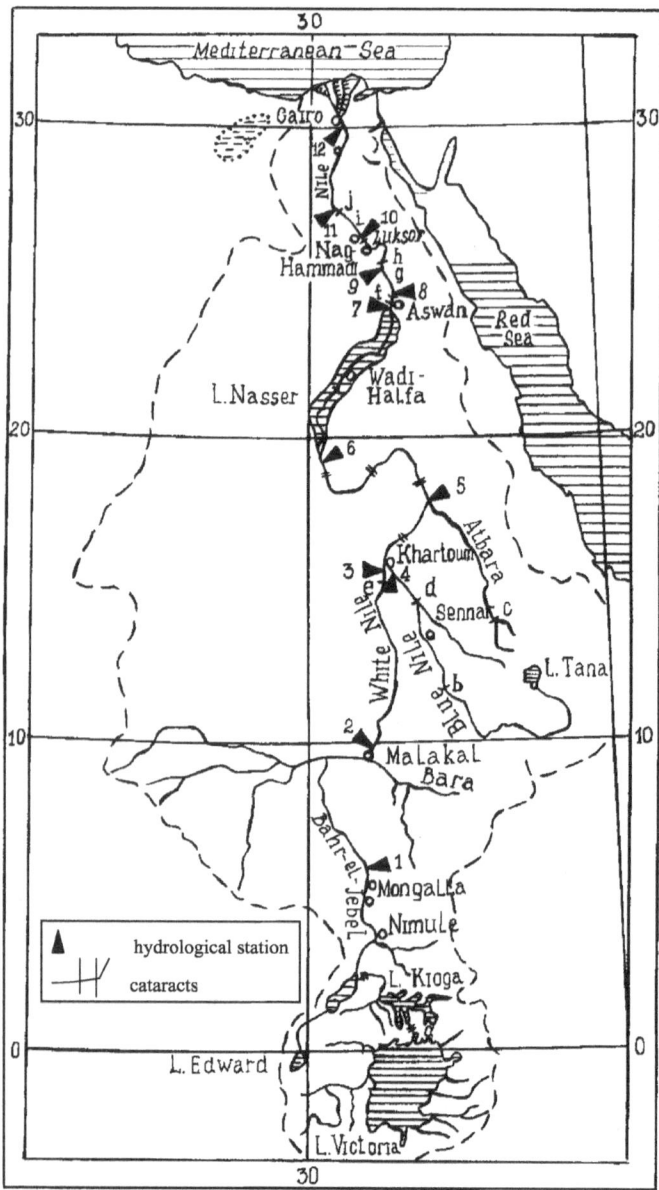

Fig. 6.19 The Nile Basin. a, Owen Falls Dam; b, Roseires Dam; c, Khashm el Girba Dam; d, Sennar Dam; e, Jebel-Aulia Dam; f, High Aswan Dam; g, Old Aswan Dam; h, Esna Barrage; i, Nag-Hammadi Barrage; j, Asyut Barrage.

Table 6.27. *Principal hydrological stations in the Nile Basin, in the territory of Sudan and Egypt*

River	Station	Period of observations
1. Bahr-el-Jebel	Mongalla	1912–82
2. White Nile	Malakal	1912–82
3. White Nile	Jebel-Aulia (Dam)	1973–82
4. Blue Nile	Khartoum	1912–82
5. Atbara	3 km	1912–82
6. Nile	Dongola	1912–84
7. Nile	Aswan Dam	1869–84
8. Nile	Gafra	1973–84
9. Nile	Isna	1973–84
10. Nile	Nag-Hamadi	1973–84
11. Nile	Asyut	1973–84
12. Nile	El-Ekzs	1973–84

The uneven development of the economies of the Nile Basin countries, as well as the consequences of the preceding colonial policy, has however resulted in the absence of an efficient monitoring system for the planning and management of water resources over much of the basin. The lower reaches of the Nile and its delta were the only exceptions. Observations of the river level were made there from the time of ancient Egypt, and during the colonial period they were extended by Great Britain, for the development of cotton which played a considerable role in the development of Egypt. The control of Nile waters was based on a series of agreements, first, in 1929 between Great Britain and Egypt, then in 1959 between the independent states of Sudan and Egypt. The latter agreement was connected with the design of the Aswan High Dam and it resolved the distribution of Nile waters between these countries. According to this agreement Egypt's share was increased, as compared to previously, by up to 55 km^3/year and that of Sudan by up to 18.5 km^3/year. The next important stage in the co-operation between the Nile Basin states was the agreement signed in 1967 by Kenya, Tanzania, Uganda, Sudan and Egypt for the acquisition and management of hydrometeorological observations in the basins of Lakes Victoria, Kijoga and Albert. In recent years the demand for water has increased in step with growth of population and the development of the economy, especially in Egypt and Ethiopia, whereas the water resources remained constant or have declined because of pollution. The equitable distribution of Nile waters is probably the most complicated problem facing the Nile Basin states. A full description of the Nile, the world's longest river, and its course can be found in Section 6.5.3, while Fig. 6.19 and Table 6.27 show its hydrological features.

DAM CONSTRUCTION IN THE NILE BASIN
The history of the use of the Nile is a history of the introduction of perennial irrigation in Egypt by creating a system of dams. The

centuries for irrigation. However it is only during recent decades and after they became independent that the economies of African countries have been changed from ones based on exploitation of resources, particularly minerals and agriculture, to more mixed economies. The socio-economic and political reforms occurring in these countries, have offered new possibilities for the use of the waters of the Nile. In addition to the traditional irrigation, hydropower production, mining and metallurgy began to develop. In assessing the current use of the water resources of the Nile, it is important to note the development of navigation, domestic water supply and fisheries.

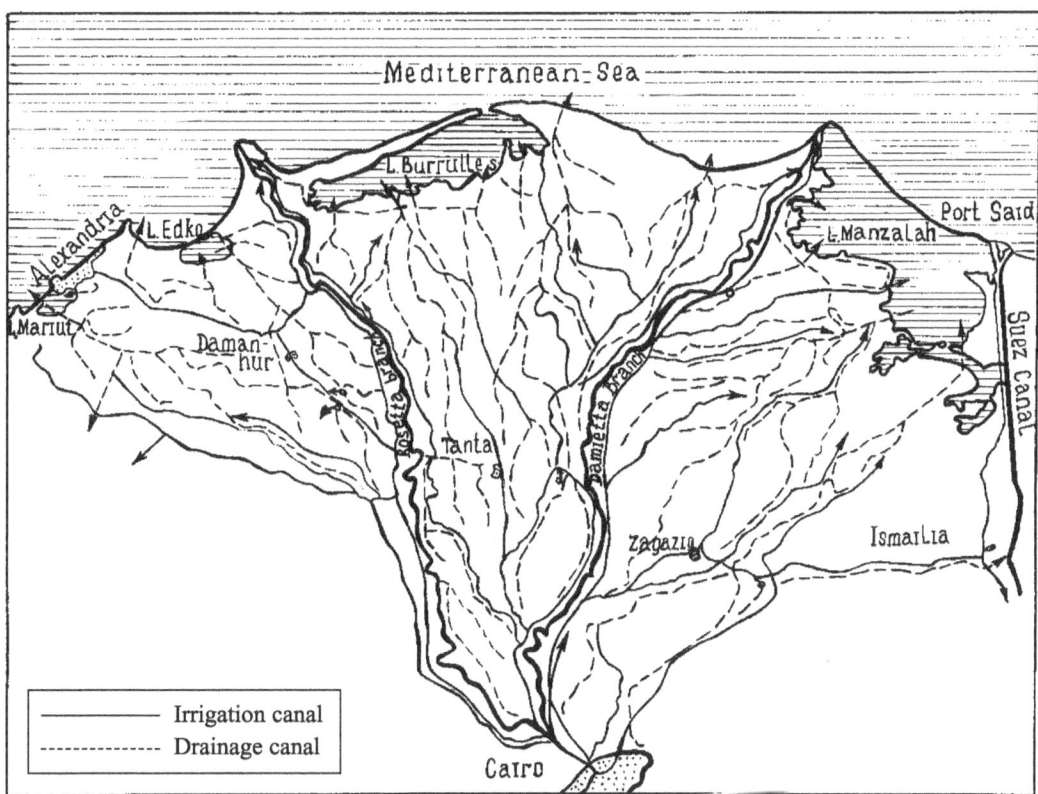

Fig. 6.20 The Nile Delta.

first of the Nile's dams, the Delta Dam, was constructed near Cairo in 1883 (Fig. 6.20). Then to expand the irrigated area, the Aswan dam was built (1889 to 1906) with a volume of 1 km³. It was twice reconstructed, increasing its volume to 5 km³. In 1902 the Asyut Barrage was constructed in Middle Egypt and the Zifta Dam in the Nile delta on the Damietta arm. In 1908 the construction of the Esna Barrage followed and from 1928 to 1930 the Nag-Hammadi Barrage was built in Upper Egypt. In 1950 the Idfina Dam was built not far from the mouth of the Rosetta. In Sudan the construction of dams began in 1922 with the Sennar Dam and the Roseires Dam on the Blue Nile. Then in 1937 the Jebel-Aulia Dam was built on the White Nile. The Khashm el Girba Dam was constructed on the Atbara. In Uganda the first dam on the Owen Falls was built in 1954. Table 6.28 presents details of the main dams and barrages on the Nile (Kashef, 1981b).

The rapid rise in the population, the need to extend the irrigated area and the increased industrial production created a demand for a much larger reservoir in order to produce more and cheaper electrical power for Egypt and the Nile Basin. To meet these needs the Aswan High Dam was erected between 1960 and 1968 taking Egypt into a new stage in the development of the country and allowing the solution of many water management and economic problems. On the other hand, it gave rise to serious ecological problems. The dam was constructed on granite outcropping in the bottom of the river and along its shores. The length of the Aswan High Dam at the top is 3600 m and 520 m at the bottom. The dam

width in the lower section is 980 m and 40 m in the upper section. The total height of the dam is 110 m. Kashef (1981b) gives details of the dimensions of the dam, its sluice gates and other features. The volume of the reservoir formed by the Aswan High Dam is 164 km³ when the water level reaches 182 m above sea level or 97 m above the bottom level. However after the dam was constructed, its actual volume turned out to be less than preliminary estimates. The maximum volume of 115 km³ was reached in 1975 with the water level at 175 m. The dead storage below the level of the spillway at 147 m is 31.5 km³ (Kashef, 1981b). The northern part of the reservoir is known as Lake Nasser and in the south where it is in the Sudan it is known as Lake Nubia. It has a length of 350 km along the centre line in Egypt and 150 km in Sudan. The total shoreline length of the reservoir is 8803 km at the 180 m level and 5960 km at 160 m. The length of the eastern side of the reservoir is almost twice the western side. At the 180 m level the maximum and mean depths are 81 m and 25 m, respectively, with maximum and mean widths of 25 and 17.9 km. The reservoir has numerous lagoons, some of them being more than 50 km long. (Table 6.29) (Kashef, 1981b).

THE HYDROLOGICAL AND HYDROCHEMICAL REGIMES OF THE NILE RIVER BEFORE THE CONSTRUCTION OF THE ASWAN HIGH DAM

The discharge of the Nile is governed primarily by the amount of precipitation which falls in the upper part of its basin, particularly in the Ethiopian Highlands and in the region of Lakes

Table 6.28. *Dams and barrages in the Nile Basin*

Dam or barrage	River	Year of construction	Area of reservoir, km^2	Volume of reservoir, km^3	Evaporation, km^3/year
Jebel-Aulia	White Nile	1937	–	3.575	2.8
Roseires	Blue Nile	–	315	2.700	0.45
Sennar	Blue Nile	1922	100	0.930	0.28
Khashm el Girba	Atbara	–	–	1.100	0.06
High Aswan	Nile	1968	5168	115.000	14.0
Old Aswan	Nile	1906	–	5.000	–
Esna	Nile	1908	–	0	–
Nag-Hammadi	Nile	1930	–	0	–
Asyut	Nile	1902	–	0	–
Delta	Nile Delta	1883	–	0	–
Zifta	Nile Delta	1902	–	0	–
Edfina	Nile Delta	1950	–	0	–

Table 6.29. *Main water balance characteristics and elements of the Nasser Reservoir*

Period	Maximum level of reservoir, m	Inflow,[a] km^3	Runoff plus accumulation in reservoir,[a] km^3	Actual water losses,[a] km^3	Surface area[b], km^2	Volume of reservoir,[b] km^3	Designed losses by evaporation,[b] km^3
1964–65	127.60	119.60	119.082	0.348	550	9.8	1.4
1965–66	132.70	79.550	78.497	1.053	–	–	–
1966–67	142.45	69.120	67.760	1.360	–	–	–
1967–68	151.25	82.370	87.290	5.080	2200	41.1	5.9
1968–69	156.55	71.102	62.740	8.362	–	–	–
1969–70	161.30	69.539	61.263	8.276	–	–	–
1970–71	164.88	81.790	70.371	11.417	3500	78.5	9.4
October 1975	175.00	–	–	–	5168	115.0	13.9
Not reached	182.00	–	–	–	6200	157.4	15.3–16.7

[a] Data of Bafa and Labib (1973), in Kashef (1981b).
[b] Data of Waterburi (1979), in Kashef (1981b).

Victoria and Albert. The mean annual precipitation in these areas is 1300 mm, but in some years and certain regions it can reach 2300 and even 3500 mm a year. In the highest mountains some precipitation occurs in the form of snow. However the losses over the basin are considerable due to evaporation from the large lakes and reservoirs, swamps and irrigated areas and because of transpiration. Evaporation and precipitation values for the Nile Basin are presented in Table 6.30 (Kashef, 1981a). From the source to the mouth the waters of the Nile follow a complex redistribution pattern which is governed by the diverse natural conditions in the basin. In the Upper Nile the regulating influence of the large lakes Victoria, Kjoga and Albert is apparent, as well as that due to the vast swampy Sudd. As a result, the annual variations in the discharge of the White Nile are quite small (Fig. 6.21).

The Blue Nile has a regime which contrasts with the regime of the White Nile. It is characterized by large variations in discharge throughout the year. This is because there are few lakes and wetlands apart from Lake Tana which influences only its upper reaches. The role of the Blue Nile varies through the year. During the flood season the Blue Nile yields about 5230 m^3/s which is about 70% of the total runoff. The Blue Nile floods earlier than the White Nile and causes the White Nile to back up (Fig. 6.21). During dry periods the discharge of the Blue Nile is about 180 m^3/s whereas the discharge of the White Nile at this time is of the order of 750–1000 m^3/s (Table 6.31). The Atbara, a lower tributary of Nile, has almost no discharge during the dry season, but during the wet season its maximum discharge is about 1640 m^3/s and is a significant addition to the Nile. The Blue Nile provides on average about 51% of the flow of the Nile, the White Nile about

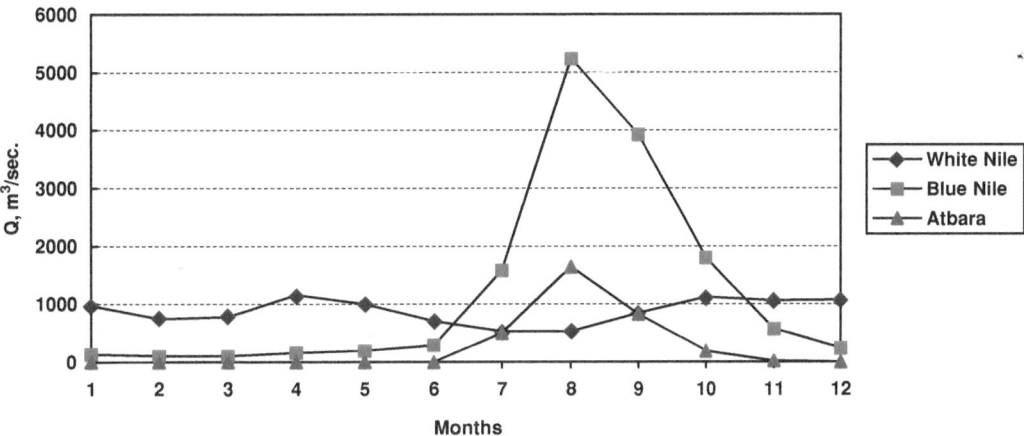

Fig. 6.21 Annual distribution of runoff of the Nile main components from data averaged over the period 1973 to 1982.

Table 6.30. *Mean annual precipitation and evaporation from water surface in different regions of the Nile River basin*

Territory	Evaporation, mm/year	Precipitation, mm/year
Lake Victoria	1400	1400
Lake Edward	1420	1400
Lake Albert	1420	1000
Lake Tana	–	1300
Sudd Region	1860	1000
South Sudan	1240	1000
Central Sudan (up to Roseires)	2300	600
Khartoum	2840	200
Northern Sudan (Halfa–Atbara)	2780	0–25
Aswan	1640	0–5
Cairo	1020	25
Nile Delta	840	25–200
Mediterranean sea coast	1100	200

38% and the Atbara 11%, although these figure differ depending on the year (Fig. 6.22). For example, during floods the Blue Nile gives 70%, the White Nile 10% and the Atbara 20%.

During periods of low water the White Nile provides 83% of the total Nile runoff, and it plays the dominant role against the Blue Nile's 13%. According to flow records from a gauging station operating since 1869, the natural regime of the Nile at Aswan, before regulation by the old Aswan Dam, was characterized by a high prolonged flood with a slow rise and even slower fall (Popov, 1958). The flood began in late May, sometimes in late June, with the water rising rapidly during the first 2 to 3 weeks, followed by a slower rise. The highest water levels were observed in late August or September. After that a comparatively rapid fall took place then a slower fall began. The rise in water levels near Aswan

was 8–10 m, on average. The rise began first in Wadi Halfa and some time later it was observed near Aswan. The hydrograph at this point is presented in Fig. 6.23. The flood peak moved from Aswan to Asyut over a period of 10 to 11 days with a mean speed of about 2.1 km/h. Near Asyut the highest level was usually observed in late September or early October, the rise reaching 7 m. Near Cairo the maximum level was usually observed in September to early October, some 3.5 to 4 m above normal. If all the sluices on all dams were open, the Nile's regime differed little from the natural one. It is changed after the passage of the flood peak, i.e. at the beginning of the recession when the old Aswan reservoir begins to be replenished.

The hydrological regime of Nile depends on a number of factors. In addition to climatic conditions and relief, they include:

- regulation of the runoff by the changes in the character of the channel and the flood plain and any resulting redistribution of the runoff in the different segments of the river
- differences in the timing of the runoff from the tributaries into the main river
- the influence of the lakes through which the White Nile and the Blue Nile flow and the smoothing of the discharge that they provide
- the influence of areas of wetland and their influence on runoff and evaporation
- the regulation and redistribution of runoff due to human activities and its subsequent use for irrigation and water supply.

Under natural conditions, the river transported up to 100 million m^3 of sediment a year, with 70% of this volume being carried in August and September. This total was made up of 30 million tonnes of fine sand, 40 million tonnes of silt and 30 million tonnes of clay. Of the suspended sediment transported during the flood season, up to 40% was deposited in the flood plain forming layers of silty sediments, the remainder was carried to the delta. Figure 6.24 shows the distribution within the year of the

Table 6.31. *Average monthly discharge (m³/s) of the River Nile and its tributaries during 1973–82*

River	Months												Mean annual discharge
	1	2	3	4	5	6	7	8	9	10	11	12	
White Nile	977	749	781	1162	1002	710	531	535	844	1120	1058	1068	878
Blue Nile	138	111	106	168	196	294	1588	5229	3927	1803	571	232	1197
Atbara	–	–	–	–	–	–	498	1639	827	191	17	4	264

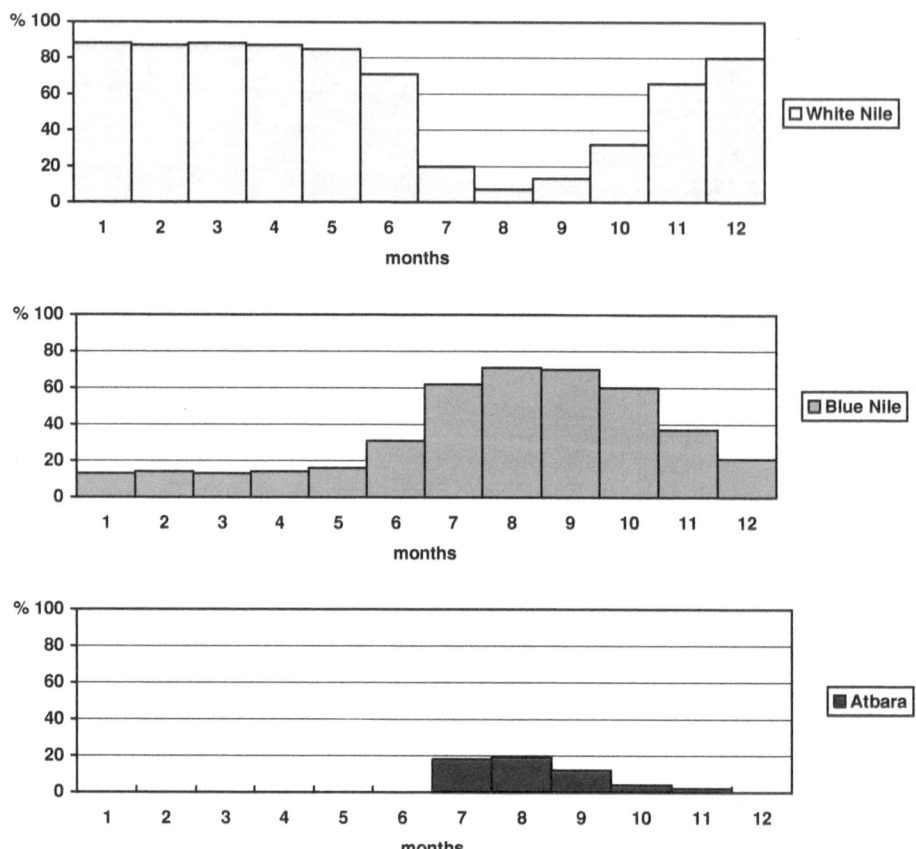

Fig. 6.22 Proportion of White Nile, Blue Nile and Atbara in the runoff of the main Nile for the period 1973 to 1982.

sediment discharge of the Nile prior to the construction of the Aswan High Dam.

The hydrochemical regime of the Nile is governed primarily by the variations in water chemistry along the length of the river. These variations are due to the changing physical-geographical conditions from the source to the mouth and the inputs of the tributaries. Mean values of the main water chemistry variables at different points in the Nile Basin are given in Table 6.32. The water chemistry of the Nile is mainly dependent on the chemistry of its main tributaries the White Nile and the Blue Nile which are very different. Waters of the Blue Nile belong to hydrocarbonate calcium-rich water which drains from water-bearing volcanic

measures that do not contain carbonates. As a result of carbon dioxide weathering of calcium minerals in the volcanic rocks, water is enriched with hydrocarbonates of calcium. Ca^{2+} is taken from the rock and HCO_3 is formed from carbon dioxide of atmospheric and biochemical origin. The chemistry of the White Nile is governed by the mixing processes in Lake Victoria, which has a mineral concentration of 49 mg/l, and in Lake Kijoga (61.2 mg/l), and in the highly mineralized waters of Lake Albert (393 mg/l). In view of the fact that only 15.1% of the runoff from Lake Albert is discharged to the White Nile, water mineralizing in the mixing area has a concentration of about 100 mg/l. As the White Nile approaches the Blue Nile, its mineralizing increases to 172 mg/l (WHO/UNEP, 1990). The waters of the Blue Nile have a lower (by 15%) mineralization and they belong to the alkaline-earth

Table 6.32. *Mean values of selected indicators by the data of GEMS/WATER stations within the Nile Basin above Aswan*

	Lake Victoria	Lake Kioga	Lake Albert	White Nile	Blue Nile
GEMS Station number	110001	110002	110003	078001	078002
Period of observations	1/4/79 to 12/15/80	1/15/80 to 12/15/80	1/29/79 to 9/25/80	1/2/80 to 6/60/80	2/1/80 to 6/60/81
Number of samples	~50	~10	~20	~18	~17
Electric conductivity, μS/cm	95 ± 12	97 ± 7	613 ± 17	2577	255
Temperature, °C	25.3 ± 1.7	26.1 ± 1	28.1 ± 3.5	24.7 ± 4.7	24.2 ± 4.0
pH	8.1 ± 0.4	8.0 ± 0.7	8.8 ± 0.1	8.1 ± 0.2	7.8 ± 0.2
Na^+, mg/l	10 ± 5	11 ± 2	76 ± 2	29 ± 8	14 ± 4
K^+, mg/l	4 ± 0.6	4.6 ± 1.4	51.2 ± 16.4	11.7 ± 3	6.5 ± 3.8
Ca^{2+}, mg/l	3.4 ± 1.3	4.2 ± 1.2	8.0 ± 1.7	15 ± 3.0	24.4 ± 5.6
SO_4^{2-}, mg/l	2.0 ± 2.0	2.0 ± 0	19 ± 4	26 ± 7	27 ± 8
CI, mg/l	4.0 ± 2	5.0 ± 1	23 ± 8	18 ± 5	18 ± 5

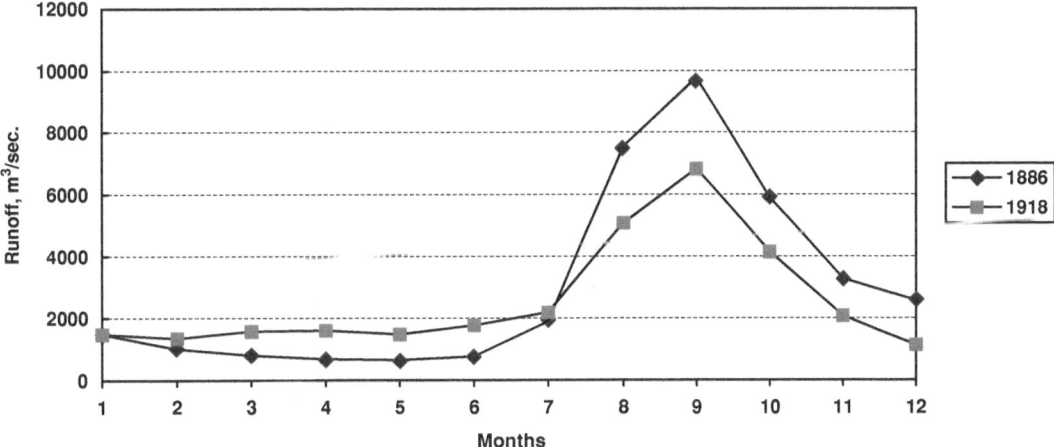

Fig. 6.23 Changes in Nile water discharge at Aswan for representative
years before (1886) and after (1918) the construction of the old
Aswan Dam.

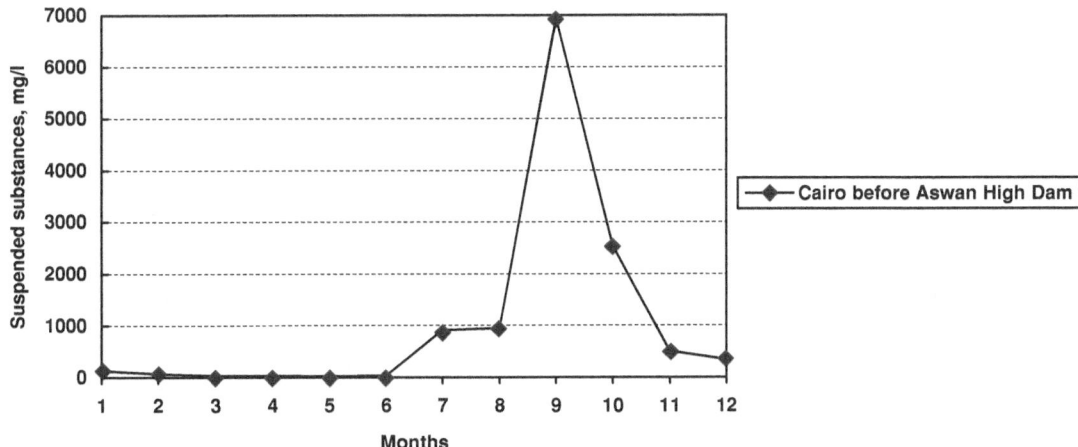

Fig. 6.24 Annual distribution of suspended substances (mg/l) near
Cairo before the construction of the Aswan High Dam (1963).

Table 6.33. *Annual change in water chemical composition (mg/l) of the Nile (Cairo) before the construction of the Aswan High Dam*

Month	Turbidity, silica scall	Total content of dissolved substances	Total hardness of CaCO₃	Cl⁻	SO₄²⁻	SiO₂	F⁻	Alkalinity
January	125	190	103	13.5	10.4	20	0.40	121
February	65	199	114	16.5	12.8	16	0.34	120
March	26	200	112	18.5	11.0	24	0.30	120
April	25	191	111	21.5	12.3	24	0.25	130
May	20	192	111	19.0	9.9	24	0.30	188
June	26	182	112	19.5	10.0	18	0.35	137
July	910	210	112	21.5	9.6	5	0.45	162
August	950	167	117	13.6	10.1	10	0.25	130
September	6935	172	88	9.5	10.2	20	0.20	62
October	2550	167	80	8.5	10.4	13	0.20	57
November	500	175	100	10.5	10.8	20	0.20	64
December	340	170	90	10.0	9.5	18	0.20	118

carbonate type. The calcium level is almost twice as high as in the White Nile. On the whole the main Nile belongs to a group of rivers with a small mineralization, up to 200 mg/l. The prevailing anion is hydrocarbonate and among the cations sodium and potassium ions dominate.

Regular observations of water chemistry are lacking before the construction of the Aswan High Dam, so it is rather difficult to assess reliably the hydrochemical regime of the Nile and its tributaries. However the published data indicate a pronounced seasonal variability in hardness, alkalinity and the concentration of suspended matter. The concentration of total dissolved solids varies, from 210 mg/l during the period of low water to 172 mg/l during the flood season. The changes in the surface water chemistry of Nile within the year before the construction of the Aswan High Dam are presented in Table 6.33 (Shalash, 1980).

THE INFLUENCE OF THE ASWAN HIGH DAM ON NATURAL AND SOCIAL PROCESSES

The regulation of flow by the Aswan High Dam has enabled the waters of the Nile to be used more efficiently, primarily for the irrigation of the fertile lands in the Nile flood plain and delta, but also for drinking water supply and for producing cheap electrical power. The construction and management of the Aswan High Dam allowed many problems in the basin to be addressed, in particular:

- storing about 90 km³ of fresh water: a guaranteed volume of water for the supply of Sudan and Egypt
- providing for the complete control of the water resources of the Nile

- offering a basis for the implementation of the project in Upper Egypt for the western desert
- increasing the area irrigated by 30%
- increasing the volume of irrigation water by 7.5 km³ to allow perennial irrigation with several harvests a year
- producing about 10 billion W/h of electrical power annually and, in addition, 120 MW/h from the old Aswan Dam
- providing flood protection during wet years (1967, 1975) and drinking water supplies during dry years (1972, 1979, 1982, 1984)
- facilitating the cultivation of all the islands in the Nile flood plain that were flooded earlier
- developing the urban areas
- improving the efficiency and quality of drainage systems
- maintaining the necessary depths in the river channel and thus providing a permanent navigation route for shipping.

These are the positive effects of the construction of the Aswan High Dam. Whereas all dams that were constructed earlier influenced relatively small areas and were as a rule of a seasonal character, the Aswan High Dam has radically affected natural processes. It has drastically changed the hydrological regime in the basin, so causing a number of adverse ecological and social changes.

INFLUENCE OF THE ASWAN HIGH DAM ON THE HYDROLOGICAL AND HYDROCHEMICAL REGIME OF THE NILE BETWEEN ASWAN AND CAIRO

The result of constructing the Aswan High Dam has been to change completely the hydrological regime of the River Nile.

Table 6.34. *Mean monthly water discharge (m^3/s) of the Nile at Aswan for different periods*

Period	Months												Annual mean discharge
	1	2	3	4	5	6	7	8	9	10	11	12	
1881–1890	1770	1295	954	721	630	704	2150	8101	9402	6272	3060	2445	3125
1941–1950	1192	1109	905	898	1019	1110	1767	6474	8020	5015	2339	1635	2623
1971–1980	1340	1597	1674	1625	1938	2264	2578	2251	1640	1493	1517	1420	1778

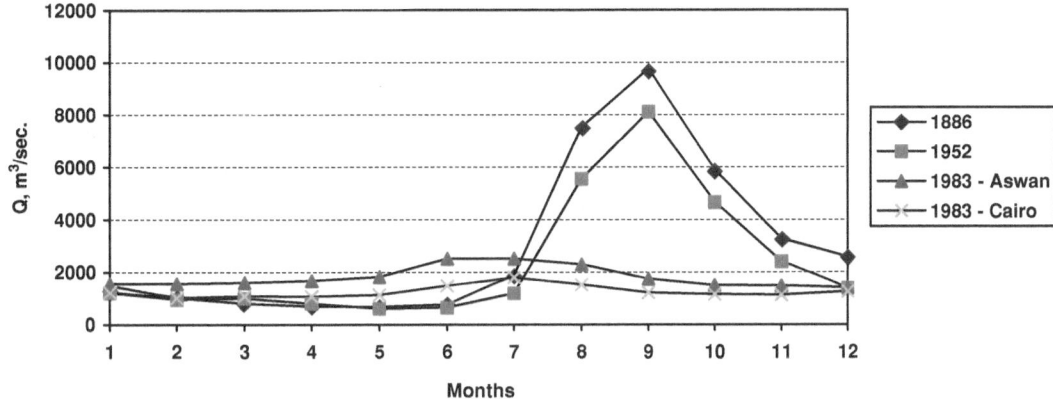

Fig. 6.25 Annual distribution of discharge of the River Nile during years with an average water content at Aswan and Cairo.

Discharges during the year vary within a smaller range, namely from 1340 m^3/s to 2580 m^3/s (Table 6.34) and the hydrograph is very smooth. The hydrograph near Aswan is similar to the hydrograph near Cairo but with differences due to direct evaporation losses and those due to abstractions for irrigation (Fig. 6.25). Such conclusions follow from the Table 6.34 and Fig. 6.25 showing changes in the mean monthly discharges near Aswan for different periods. They correspond to the natural regime (1881 to 1890), a regime slightly changed by the old Aswan Dam (1941 to 1950) and the regime due to the construction of the Aswan High Dam (1971 to 1980). The construction of the old Aswan Dam led to a 14% decrease in the mean annual runoff. The Aswan High Dam and the reservoir it created have resulted in a radical change to the distribution of runoff and a reduction in its volume by almost half (43%).

The impact of the Aswan High Dam is especially important for sediment discharge. On average the Nile transports 110–130 million tonnes per year of suspended matter, predominantly very small particles (Shalash, 1986), more than half the entire volume being carried during the flood. The Blue Nile and the Atbara are the source of these sediments which mainly consist of particles of apatites and clay shales. At the present time only 1–2 million tonnes per year (1–2%) of suspended matter is exported to the tailrace canal, the rest being deposited in Lake Nasser (Fig. 6.26). Below the reservoir there is the development of erosion in the

channel of the Nile which produces 2–3 million tonnes per year, on average (Shalash, 1986). This quantity is directly dependent on the volume of water released from the reservoir.

After the construction of the Aswan High Dam, Egypt began to implement a monitoring programme including observations of sewage discharges and their composition and the sampling and chemical analysis of river water (Abdel-Khalik *et al.*, 1998). An analysis of these results indicates that after the construction of the dam there was deterioration in the chemistry of the surface waters from Aswan to the mouth of the Nile. This is governed primarily by the fact that there was a decrease in the self-purification capabilities of the river during the flood period. In addition, before the construction of the dam, the seasonal changes in the physical and chemical variables were quite pronounced, but after the construction of the dam they changed little (WHO/UNEP, 1990) (Fig. 6.26). It is also important to note that the seasonal variations of mineralization change over the periods 1933 to 1936 and 1959 to 1960. Comparison of the level of dissolved substances from 1959 to 1960 and from 1969 to 1970 shows a trend towards increasing mineralizing (a 1.55 times increase) (WHO/UNEP, 1990) (Table 6.35). Increased concentrations are observed for all determinands: for chlorine and sodium ions, the most conservative ions, by factors of 3.6 and 2.14, respectively; for calcium, magnesium and bicarbonates by 1.3; and for sulphates by 2.6.

Table 6.35. *Comparison of mean weighted concentration before and after the construction of the Aswan High Dam*

Period	pH	Alkalinity	SO_4^{2-}, mg/l	Cl^-, mg/l	Ca^{2+}, mg/l	mg^{2+}, mg/l	Na^+, mg/l	K^+, mg/l	Total dissolved solids, mg/l
1959–60	7.74	2.02	0.22	0.25	1.14	0.68	0.69	0.09	193
1969–70	8.07	2.58	0.58	0.90	1.44	0.93	1.48	0.17	299
Factor of increase of concentration	1.04	1.28	2.64	3.60	1.26	1.37	2.14	1.80	1.55

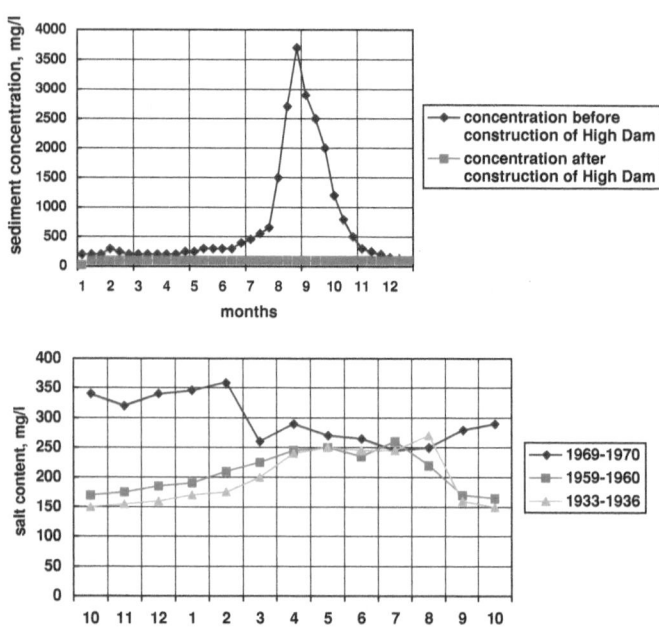

Fig. 6.26 Annual variations in sediment content and concentration of chemicals at Aswan Dam during different periods of time (WHO/UNEP, 1990).

INFLUENCES ON THE WATER QUALITY OF THE RIVER NILE

The natural water quality of the Nile Basin is governed largely by the conditions of climate, relief, soils and lithology that are typical of the arid region under consideration. New problems are caused by:

• the high levels of salts mainly in ground waters
• the intrusion of sea water into coastal aquifers
• the increased level of fluorine in surface and ground water.

At the present time the annual water use in the industrial sector is 3.4 km³ in Egypt and 0.24 km³ in Sudan. Most large towns and the industries of sugar refining, brewing, textiles, wood-processing and cellulose production as well as mining (gold, copper, aluminium and phosphorus) have no waste-water treatment facilities. The situation is similar for domestic sewage.

In addition to point sources the non-point agricultural discharges produce a considerable load for the Nile. For the years 1951 and 1952 the total amount of nitrate and phosphate was 0.88 million tonnes, but for 1987 and 1988 the total had increased to 6.19 million tonnes (WHO/UNEP, 1990). This amount exceeds the concentration of nitrogen and phosphorus in the bottom sediments that were deposited during the construction period of the Aswan High Dam, which is estimated to be 12 000 tonnes/year and 6000 tonnes/year, respectively. Although the quality of the Nile along the length to Cairo, taking into account industrial and agricultural discharges and return irrigation water, remains within maximum permissible concentrations, its deterioration since the mid-1970s is obvious. The continued increase in sewage discharges leads to considerable pollution, which is a serious obstacle to improving its quality. The BOD values along the Nile vary with the level of the domestic sewage load and are presented in Fig. 6.27 (Shalash, 1980). The maximum BOD values correspond to the positions of the sewage discharges and reflect the extent of pollution of the entire Nile.

Changes in the quality of the river water connected with the concentration of suspended sediment, the concentration of dissolved mineral substances and phytoplankton depend directly on the physical, chemical and biological changes occurring in Lake Nasser. According to data from the Laboratory for Water Contamination Control (Shalash, 1980), the trend towards increased levels of the different water quality determinands is observed along the entire river from Aswan to Cairo (Table 6.36). These trends are confirmed by other assessments. According to GEMS/WATER (UNEP/WHO, 1995) (Fig. 6.28), the levels of chlorides from Aswan to Idfina have increased 2.5 times. Alkalinity and specific electrical conductivity have increased by 1.4 due to the high concentrations of bicarbonates. The levels of phosphorus compounds in this part of the river have shown a small increase but increase sharply in the delta, while the concentration of nitrogen compounds decrease to Cairo. These changes in nitrogen and phosphorus compounds are explained by the high level of phytoplankton, predominantly of diatom algae and cyanobacteria, which develop in the reservoir and consume nutrients over the entire river length to Cairo.

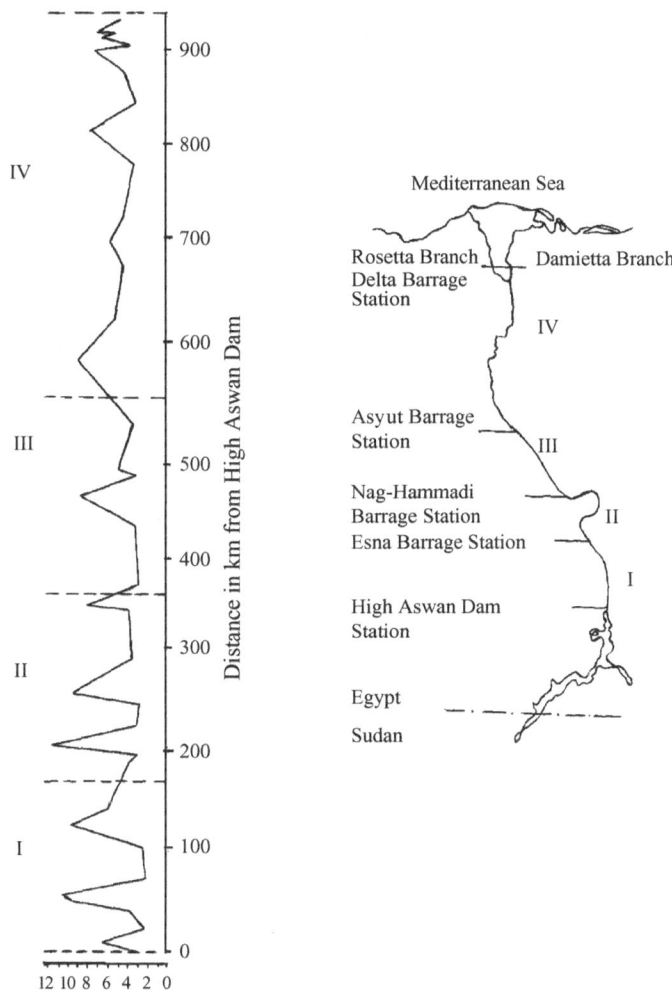

Fig. 6.27 Changes in biological oxygen demand (BOD) (mg/l) along the River Nile.

POLLUTION PROCESSES IN THE NILE DELTA

The situation is quite different in the Nile Delta. It is divided by two natural branches, the Rosetta and Damietta, into three parts: eastern, central and western. Each of these parts is dissected by systems of canals, used for irrigation and water supplies to settlements. The arid climate, the highly developed irrigation and the regulation of the runoff by numerous dams have had a considerable impact on the water quality in the delta. Soil salinization is an additional problem in the delta, to which the construction of the Aswan High Dam has contributed. Here two factors are important. On the one hand, drainage of irrigation water transports a large amount of dissolved salts to the Nile Delta. On the other hand, the intrusion of sea water has increased, due to the low volume of discharge of the Nile and the year-round irrigation. Additionally, as a result of the development of more irrigation after the building of the Aswan High Dam, the ground water level in the Nile valley and in the delta has increased by 2 m on average and in Upper Egypt by 1.5 m. Ground water quality has deteriorated and the higher

water table has caused waterlogging. Intense evaporation and soil salinization have contributed to the increased salinity of ground water. The mineralization of ground water in the delta has changed from 500 to 4000 mg/l and it has reached 7500–16 000 mg/l in the western part of the delta (WHO/UNEP, 1991).

A serious problem also arises because of the intense use of fertilizers, causing an excess of nutrients in the soil. Until they are actually bonded in the plant cells, these materials can be washed out, the process when the dissolved nitrates are removed being often coupled with transport of phosphates and nitrates on the surface by erosion. These nutrients discharged into the waters of the Nile Delta have a serious influence on their trophic state. Stagnant zones without currents, the low level of assimilation of untreated sewage, the high organic content and the large volumes of nutrients lead to eutrophication of the delta. The concentration of nitrates at the present time in the right Nile branch, Damietta, is 1.8 mgN/l, and for the left branch, Rosetta, it is 1.3 mgN/l. The concentration of mineral phosphorus forms in the Damietta reach 0.98 mg/l and in the Rosetta 0.18 mg/l. The BOD values

Table 6.36. *Changing water chemical composition along the River Nile from Aswan to Cairo*

Distance from the Aswan High Dam	pH	O$_2$ dissolved, mg/l	NH$_3$, mg/l	BOD, mg/l	Salt concentration, mg/l	Weighted substances, mg/l	F$^-$, mg/l	PO$_4^{3-}$, mg/l	Cl$^-$, mg/l	Coli, per 100 ml
0	8.20	8.2	0	3.0	160	70	0.2	0.13	10	35
10.1	8.20	9.2	0	6.6	188	58	0.3	0.13	15	45
24.0	8.20	9.0	0	2.0	168	48	0.2	0.13	12	25
42.4	8.25	9.9	0	3.6	200	56	0.2	0.26	10	275
44.0	8.25	8.7	0	3.1	186	28	0.4	0.26	12	170
49.4	8.00	8.0	0	9.6	280	40	0.4	0.26	14	900
55.2	7.80	8.5	0	10.6	286	26	0.2	1.21	12	140
518.0	7.68	9.0	0	3.4	216	16	0.4	0.27	20	900
580.0	7.68	9.1	0	6.0	204	48	0.5	0.13	16	900
620.0	7.40	9.5	0	6.0	210	40	0.5	0.13	20	450
676.0	7.40	9.8	0	5.0	210	48	0.5	0.50	16	450
912.0	8.30	6.8	0	7.2	224	68	0.5	0.27	26	350
914.8	7.80	7.8	0	5.6	224	36	0.4	0.27	18	250
930.0	7.95	8.0	0	5.0	226	58	0.4	0.81	24	250

significantly exceed the permissible concentrations: 10 mgO/l in Damietta and 6.4 mgO/l in Rosetta which points to a high organic load (Fig. 6.28). The solution of the pollution problem depends directly on the economic state of the country, and it becomes more difficult to solve when the three factors are combined: where many sources of pollution are present, where water use is high, and where water resources are stressed, so that there is a water deficit. This is the situation in the Nile Basin and especially in the delta since the construction of the Aswan High Dam.

INFLUENCE ON SOCIAL AND ECONOMIC CONDITIONS

Along with the negative effect of the Aswan High Dam on natural processes, there are other and not less important implications for society and the economy. These are connected with the decreased productivity of soils, economic losses in fisheries, the spread of infectious diseases, the deterioration in the state of holiday resorts and the migration of indigenous people (Kashef, 1981b; Shalash, 1986). The expansion of irrigation, the intensification of agricultural use and the absence of the silt previously brought by the flood have resulted in the decreased productivity of soils. Some 10% of agricultural production is lost annually due to the decrease in the natural fertility of the soil requiring the use of artificial fertilizers. At the present time Egypt is a leader in using chemical fertilizers (175 kg/ha per year) (WHO/UNEP, 1990) and this amount is increasing.

Another drawback of the High Dam is the decreased fish catch. Before its construction the annual fish catch in the Nile, the lakes of the Nile Basin and in the Mediterranean and Red Seas was estimated to be 125 thousand tonnes but it decreased sharply after

the dam was built. For example, the catch of sardines fell from 150 thousand tonnes in 1964 to 4600 tonnes in 1965 due to the absence of silt in the river. In 1966 it was only 554 tonnes (Kashef, 1981b). However, favourable climatic conditions and its remoteness provides the possibility of stocking Lake Nasser with fish to make up the shortfall. Another serious effect of the change in natural processes is spread of malaria and some other infectious diseases. This effect was especially pronounced when the High Dam was first built. As a result of these changes ground water levels have increased, leading to waterlogging of soils and further social consequences.

As a result of the increase in erosion due to the absence of sediment, the coastline of the Mediterranean Sea known for its summer resorts has been destroyed. This phenomenon is very evident near the mouths of the Damietta and the Rosetta and has touched Alexandria, the popular resort. The economic damage from these various processes has been quite important in a number of localities. With the building of the Aswan High Dam and the flooding of the area upstream near Wadi Halfa and the adjoining settlements, about 600 000 people had to move. About 100 000 of the inhabitants of Wadi Halfa moved 46 km north of Aswan to 43 newly built villages while about 500 000 people moved in Sudan to Khashm el Girba where 20 000 new villages were built for them.

SUMMARY

In summarizing the problems caused by the construction of the multi-purpose Aswan High Dam, it is important to stress its vital role for the development of the economy of Egypt and other countries. In particular, these countries were transformed from feudal

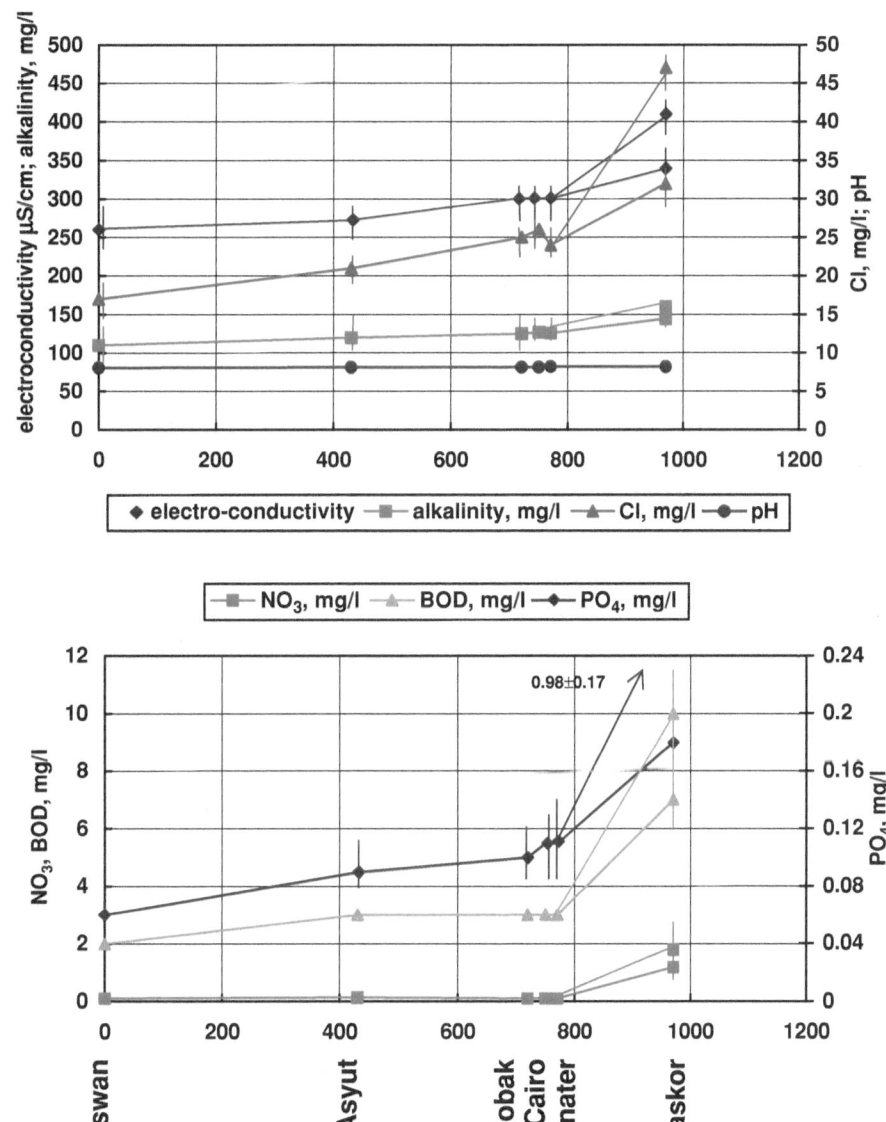

Fig. 6.28 Changes in some chemical indicators along the length of the River Nile from Aswan.

backward-looking states to countries which are developing and in some cases countries which have advanced further than some developed countries. With the dam it became possible to solve some of the water management and economic problems that had appeared to be insoluble. Hence the influence of the Aswan High Dam has to be evaluated in accordance with the common purposes of the state and the strategy for the continued development of the country, on the results achieved and on the social consequences.

Obviously from the economic and political viewpoints the transformation of Egypt from an agrarian country to an industrially developed country is of immense importance for the entire Arab world. From the ecological viewpoint, the drastic change in the

natural processes has resulted in serious problems. On the one hand, they were predicted in the design of the dam, and on the other hand, they were not properly taken into account or not accurately calculated. In some cases they were unexpected. All these problems require special attention and need to be solved.

Further studies should be aimed primarily at investigating the hydraulic regime under these changed conditions, particularly the balance between the fresh and sea water in the intrusion area, taking into account the changed water levels in the river channel and in the aquifers. The earlier erroneous estimates of evaporation and infiltration losses require updating. According to some authorities, they were caused because the location of the dam was chosen incorrectly. The health problems require serious attention by the government. For this purpose large-scale state projects on changing the drainage system and decreasing groundwater levels need to be developed. Their cost is estimated to be US$600 million. The government should also develop measures for preventing the spread of infectious diseases. The problems of the pollution of the Nile also require comprehensive studies. For this purpose it is necessary to expand and develop the monitoring system, carry out hydrochemical and hydrobiological studies, equip modern laboratories and provide professional training for their personnel.

These studies can best be undertaken within the framework of effective co-operation between the countries of the Nile Basin. Joint studies and the construction and management of structures for further utilizing the water resources of the Nile Basin are the main avenues for the future. For the operation of the basin's water resources system, especially the control of floods, flow regulation and navigation, it is necessary to undertake certain measures. It is also necessary to create the corresponding hydrological services, especially for forecasting, involving the national and regional weather services and those concerned with water resources (Sehmi, 1996).

6.8 CONCLUSIONS

This chapter indicates that the annually renewable water resources of Africa are quite small, namely 4050 km^3/year. Most of this, some 1770 km^3/year, originates within the Central region in the basin of the River Congo, but the water resources of the remainder of the continent are very unevenly distributed. About one-third of the continent is an area of inland drainage with only a small volume (150 km^3/year) of runoff. The distribution of runoff within the year is characterized by a very irregular pattern strongly correlated with the occurrence of the wet and dry seasons. In North and Western Africa the monthly distribution of runoff is rather skewed with 56% and 64% of the discharge taking place from August to October. In Southern Africa more than half the discharge occurs from December to March and in Eastern Africa 57% of the runoff takes place from March to June. Only in Central Africa is the distribution fairly uniform throughout the year and here it is most favourable for the development and use of water resources. The changes and trend in the available water resources show a decrease in all regions of the continent per head of population. In 1995 high water availability (more than 10 000 m^3/year per head) was typical only of the Central region. The water availability in North Africa has become extremely low (less than 1000 m^3/year per head). The other parts of Africa are characterized by low water availability (2000–5000 m^3/year per year).

From the overall estimates of the annually renewable water resources and the volume of water used the continent has an acute water problem. There are both insufficient water resources in many countries and regions and a threatening growth of pollution of these resources to hinder development. In addition these problems have to be set against a background of a rapidly rising population. Of course the estimates of water resources and water use are very poor for most regions and basins. This is due to the sorry state of the network of hydrological stations – a network which has declined over the last 20 years in terms of number of stations, their maintenance and their representativeness. The lack of stations at mouths of many rivers is one problem, but there are many others such as poor quality control and numerous gaps in the observation series. This creates the problem that the use of gap-filling techniques leads to the occurrence of errors. The estimation of the volume of water used is an even bigger problem.

Assessment of water resources, water use and water availability in the African perspective is error-prone and must be regarded in this way. In addition there are the possible changes due to the shifts in the global climate, to which Africa appears the most vulnerable of continents in terms of water resources. However, in spite of these drawbacks the data presented here on water resources, water availability and water use appear to be the most reliable of those available for Africa.

7 Water resources, water use and water availability in North America

7.1 INTRODUCTION

North America is the third largest continent (after Eurasia and Africa); its area is 20.36 million. km^2, or 24.25 million. km^2, if the adjacent islands are included. In the shape of a gigantic triangle, it projects from the Arctic Ocean southwards to connect with South America at the Panama Isthmus. Greenland, Newfoundland, Vancouver, the Canadian Arctic Islands, the Antilles and Aleutians, as well as a number of smaller islands in the Pacific and Atlantic Oceans belong to North America.

The continent has a highly indented coastline. The large peninsulas of Labrador, Alaska, California, Yucatan and Florida jut out far into the oceans. On its northern and southern coasts Hudson Bay and the Gulf of Mexico cut deep into the mainland, causing a pronounced effect on its climate. The west of the continent is occupied by the Cordilleras mountain system, while the east contains vast plains, plateaux and medium-size mountains. The climate ranges from arctic in the far north to tropical in Central America and the Caribbean region, oceanic along the coast and continental inland. Precipitation amounts are highest on the Pacific coast of Alaska and Canada, in the northwest of the USA and in the mountains of Central America. In the north, there are many lake which are glacial in origin. The largest of them are the Great Lakes, Winnipeg, Great Slave, and Great Bear Lakes. The principal river system is the Mississippi–Missouri which has a basin 2 980 000 km^2 in area and a length of 6420 km. Among the other large rivers are the St Lawrence, Mackenzie, Nelson, Yukon, Rio Grande, Columbia and Colorado.

There are 34 countries within North America: 10 are on the mainland and 24 on the islands of the Caribbean. The largest are Canada (9 976 000 km^2), the USA (9 363 000 km^2) and Mexico (1 973 000 km^2). By 1995 the population of the continent amounted to 453 million people. More than half of them, 262 million, live in the USA, about 28 million in Canada, 95 million in Mexico, and 68 million in the Central American countries. The USA and Canada are the most developed countries with high living standards. Mexico is a big industrial and agricultural country where the economy is based on the exploitation of natural resources. The rest of the Central American countries, with few exceptions, are agricultural, specializing in growing tropical crops; the role played by industry is of minor importance.

The total renewable water resources of the North America continent, as estimated for the period 1921 to 1985, is 7870 km^3/year; 4100 km^3 flows to the Atlantic Ocean, 2590 km^3 to the Pacific Ocean and 1170 km^3 to the Arctic Ocean, while 15 km^3 remains in areas of inland drainage. The largest volume of runoff (8820 km^3) took place in 1958, and the least (6660 km^3) in 1953. A major part of the continent's water resources originate on the Pacific rim, especially in southern Alaska and the Pacific coast of Canada. There are also regions of higher rainfall around the Gulf of St Lawrence, in the Appalachian Mountains, on the eastern part of the Gulf Plain, the Bay of Campeche and in Central America. There are dry and very dry areas in the inland plateaux and tablelands of the Southern Cordilleras, in Lower California, and in a number of Mexican plateaux and bolsons. Statistical analyses showed that, for the principal regions of North America, almost all runoff remained stationary in character during the period in question, with only a few major rivers showing some insignificant trends.

The distribution of runoff varies considerably during the year across North America. For most of the continent the maximum runoff is in spring due to the meltwaters from the snow cover, from glaciers and due to heavy rains. The exceptions are the Pacific coast of the USA and Florida where runoff is at a maximum in the winter.

Human activities throughout the twentieth century have considerably changed the natural pattern of runoff in many drainage basins. This is especially obvious in the centre of the continent. The flow in many rivers is regulated while numerous abstractions and discharges are made for a variety of different purposes. In 1995 annual abstractions were 672 km^3 and water consumption 241 km^3. By 2025 total abstractions are expected to increase to 790 km^3 and consumption to 290 km^3. Some 76% of the total or 503 km^3 was abstracted by the USA in 1995. In Canada water use was 55 km^3 in 1995, and in Mexico about 83 km^3. Most water is used by agriculture and this accounts for 45% of the total water abstracted; industry takes 40%, while abstractions for domestic purposes account for 10%. At present the available water resources total is 17 100 m^3/year per head but by 2025 this figure is expected to decrease to 13 000 m^3/year per head. However, the available water resources are very unevenly distributed. The north of the continent is better supplied with water, while in some regions, particularly in Mexico, more water is abstracted than is available in surface water resources and use is made of ground water. However, in many of the arid areas, ground water levels are falling rapidly because of over-abstraction. By 2025 these areas will expand. In addition, there will be a great expansion in the areas where water availability will be extremely low (less than 1000 m^3/year per head). These problems may be solved by economizing in the use of the available water resources, the conjunctive use of surface and ground waters and other measures. Inter-basin transfers can further assist the management of water resources, as is the practice in Canada.

7.2 PHYSICAL CONDITIONS

7.2.1 Geology, relief and landscape

North America is characterized by a great variety of relief although it is less mountainous than Eurasia; the mountain: plain ratio is approximately 1:2. The highest mountainous areas are located mainly in the west near the Pacific Ocean, while the centre and east contain extensive plains and lower mountains. The northeast is occupied by the Laurentian Shield. This is the oldest part of the continent formed by Precambrian crystalline rocks. Its relief is a response to glaciation. Permafrost covers most of the area; the landscape is a patchwork of low hills and numerous lakes connected by a dense but poorly developed river network. The predominant taiga landscape changes northward into tundra (Fig. 7.1). The Canadian Arctic Islands also originated from the recent glacial period and the subsequent rise in sea level. They are dominated by a plateau-like relief. The eastern edge of the archipelago is uplifted and dissected by fiords. The region is characterized by glacial, desert–arctic and tundra landscapes (Isachenko and Shlyapnikov, 1989).

South and west of the Laurentian Shield a plain runs south towards the Gulf of Mexico bounded by the Cordilleras on the west and the Appalachians to the east. There is a crystalline rock basement deep beneath the thick and varied sedimentary cover. This vast plain is divided into the three regions: the Central Plains, the Great Plains and the Gulf Plains (Bromley, 1980).

The Central Plains represent uplifted horizontal beds laid down over the Palaeozoic–Precambrian platform, at some places with cuesta relief and karstic features. In the north, there are glacial and lake–glacial relief forms, while there are erosional features in the south. The natural landscapes ranges from taiga and subtaiga

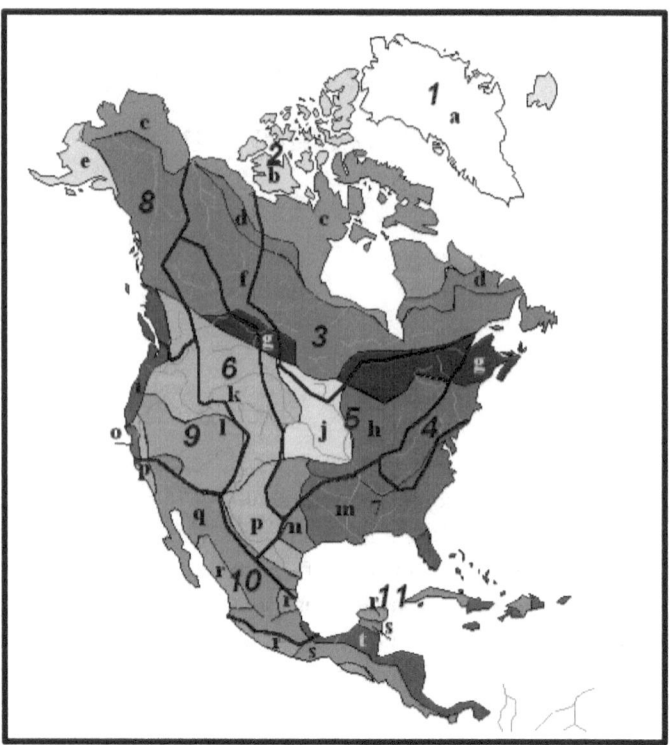

Fig. 7.1 Types of landscapes and physiographic zones of North America. Types of landscapes: a, glacial; b, desert–arctic; c, tundra; d, forest–tundra; e, suboceanic meadow; f, taiga; g, subtaiga; h, subboreal broadleaved forest; i, subboreal humid forest, subpacific; g, subboreal forest–steppe; k, subboreal steppe; l, subboreal desert; m, subboreal tropical rainforest; n, subtropical forest–steppe; o, Mediterranean; p, subtropical dry steppe and desert–steppe; q, subtropical and tropical desert; r, tropical–savannah; s, tropical variable rainforest; t, tropical permanent rainforest. Physiographic zones: 1, Greenland; 2, Canadian Arctic Archipelago; 3, Laurentian Shield (Canadian Shield); 4, Appalachians; 5, Central Valley; 6, Great Plains; 7, Coastal lowlands: subatlantic and submexican; 8, Northern Cordilleras; 9, Central Cordilleras; 10, Mexican Plateau; 11, Central America. (From Isachenko and Shlyapnikov, 1989.)

in the north to broadleaved forest and forest–steppe in the centre, and subtropical forest in the south. Due to human activites this natural pattern has been changed, especially by urban landscapes.

The Great Plains are stepwise uplifted stratified beds lying on the Mesozoic and Cenozoic sediments of the North American platform extending in a broad band along the eastern slope of the Cordilleras. The Great Plains are largely flat, but they are dissected by rivers which rise in the mountains to the west. Steppe landscapes are predominant here combined with continental taiga and narrow transition bands of subtaiga and forest–steppe.

The Gulf Plain is underlain at depth by part of the Palaeozoic folded basement. Its surface is flat terminating in swamps in the coastal belt but there are cuestas on the Mesozoic and Cenozoic sandstones and limestones inland. The humid subtropical climate gives rise to forest landscapes but away from the coast to the west, they gradually change to forest–steppe and steppe. In the south of Florida, there are swamps, the Everglades, with tropical forests which make a unique landscape, threatened by urban development.

The geological structure and relief of the east of the continent result from the Appalachian orogeny. The Appalachians are a system of Hercynian fold ridges and tablelands. The north was glacier-reworked, in the east the formations plunge below the Atlantic Ocean to form the continental shelf. In the southeast, between the mountains and the coast, there are two zones – the crystalline tableland of the Piedmont, and the flat Atlantic Coast Plain characteristic of the Appalachians. A landscape of boreal, subboreal and subtropical forest, varying with altitude, is present.

The main mountain system of North America, the Cordilleras, occupies the western part of the continent. It was formed as a result of the development of the Pacific Ocean basin, so the mountain ranges took the form of huge arcs framing the ocean depression, while the ridges and tablelands are cut with a system of tectonic faults originating on an ocean slope far away from the continent. The Cordilleras mountain range may be divided into the Northern Cordilleras, Central Cordilleras, the Mexican Plateau, and Central America.

The Northern Cordilleras consist of several longitudinal bands. The western one is formed of high ridges, such as the Alaska Range and the Canada Coast Range, exhibiting the effects of old glaciers and still highly glaciated, which turn northwestward into the volcanic arc of the Aleutian Islands. Eastwards there is a system of folded ridges (Brooks Range, Mackenzie Mountains, Rocky Mountains), with a succession of plateaux surfaces in between, such as the Yukon and Fraser. The Northern Cordilleras are situated within subarctic boreal latitudes. The Pacific rim is covered with coniferous forest, while the far northwest has a meadow-like oceanic landscape. The eastern slope shows a succession of tundra, taiga, subtaiga, forest–steppe and steppe landscapes, each within its characteristic elevation belt. The most dry and continental landscapes are those of the internal plateau.

The Central Cordilleras form the widest part of the mountain belt. As in the Northern Cordilleras, there are three longitudinal bands. In the west there are two low ridge systems divided by a deep longitudinal depression, whereas the other ridge system, including the Cascade Range and Sierra Nevada, is high and covered with snow and ice. The internal tableland region is very broad, embracing the Columbia Plateau, the Great Basin and the Colorado Plateau. The eastern margin is formed by branched range systems providing the Rocky Mountains with dramatic alpine landscapes. The side to the Pacific Ocean is characterized by subboreal forest and Mediterranean landscapes, while there are steppes along all the eastern slope and arid landscapes on the internal highlands.

The Mexican Highland is an extension south of the Central Cordilleras; Lower California is an extension of the coast ridges while the Gulf of California is a longitudinal depression. The internal highland is framed by the Sierra Madre Occidental in the west, the Sierra Madre Oriental in the east, and in the south by the Transverse Volcanic Sierra. The relief of the Mexican Highland is subdued because it is composed of sedimentary rock in the north and lava sheets and other volcanic products in the south. The internal highland has an extremely arid climate which results in tropical deserts; the slopes of the peripheral ridges are more humid giving rise to tropical mountain forest landscapes.

Central America including Central America proper – the isthmus between North and South America – represents an extension of the Cordilleras, and the Antilles. Its relief is most complicated. Numerous faults and fractures, non-uniform weathering of different rock types under hot and humid tropical conditions and deep fluvial incisions distort the surface of the ancient massifs. This has resulted in a mosaic containing ridges, peaks and plateaux with their depressions and canyons in juxtaposition. A typical relief forms here in the dozens of volcanic domes making an almost unbroken chain from the Mexico–Guatemala boundary to Panama. The character of the landscape is rather variable: savannahs occur on the plains, but in the mountains the distribution of savannahs, tropical and subequatorial and permanent rainforests depends on the exposure to wind.

7.2.2 Climate

North America extends from 80° N to 7° N and covers almost all the climatic belts from the Arctic to the subequatorial. However its largest extent is within the subarctic, temperate and subtropical belts. The amount of heat from solar radiation increases southwards from belt to belt accompanied by changes in the dominant air masses.

North Greenland and the northernmost islands of the Arctic Archipelago have a negative annual radiation balance which has resulted in the widespread recent glaciation. Greenland has 1 803 000 km^2 of ice and the Arctic Archipelago are 150 000 km^2. The supply of water in the Greenland ice cap and glaciers is 2 340 000 km^3. This huge mass affects not only the climate of North America but also that of the entire Northern Hemisphere. The annual radiation balance of the rest of the continent is positive, ranging from 15 kcal/cm^2 per year in the north to 80 kcal/cm^2 per year in the south. The radiation balance for most of Canada is 20–45 kcal/cm^2 per year, for the USA 50–70 kcal/cm^2 per year, and for Mexico and Central America 60–80 kcal/cm^2 per year.

The major moisture sources for North America are the air masses coming from the Pacific Ocean, the Caribbean and the Gulf of Mexico. The constant westerlies are characteristic of most of the continent, except for the far north and the southernmost regions, where meridional air circulation prevails. The air mass circulation over the continent depends on the location of the high and low atmospheric pressure centres near the coasts: the Aleutian Low (60° N) and Hawaiian High (30° N) in the Pacific Ocean, the Azores High (40° N) and the Icelandic Low (60° N) in the Atlantic Ocean.

The varied relief of North America determines the extent of the effects of air masses of different origins. The high Cordilleras restrict the inflow of humid air from the Pacific Ocean into the continent, and most rainfall takes place on its windward slopes. Warm humid air masses from the Gulf of Mexico and the Caribbean Sea can penetrate the eastern part of the continent where there are no large orographic barriers. Cold dry Arctic air masses meet no significant barriers and can reach the centre of the continent, the area of interior plains. There they encounter the moist air masses from the Pacific and the Atlantic to form active frontal zones which cause rainfall (Lins, 1985a).

These features of the atmospheric circulation persist throughout the year, slightly varying with the season. In winter, there are two areas with temperature anomalies: a warm northwest and a cold northeast, while the constant westerlies bring warm rainy weather to the Pacific coast. Even in January, the temperature over most of the Pacific coast of Canada is about 0°. Moving to the east the marine air is transformed into a continental type, resulting in frosty and dry winters in the interior of the Cordilleras. The continental northeast is characterized by low winter temperatures: from −9° C (an average January temperature) around the Great Lakes to −24° C to −32° C in the north of Labrador due to strong winds from the Arctic. Warm humid air masses from the Atlantic Ocean warm the Atlantic coast of the USA and the Appalachians. In the northeast of the Appalachians and in the southern part of Labrador, snow cover reaches 1.5–2 m in depth by the late winter. The snow cover is rather thick in Canada north of the Great Lakes, but away from the Atlantic it grows much thinner. In the Central Plains, accessible both to the arctic and the tropical air masses, the weather in winter is extremely changeable. Hard frosts are frequently followed by thaws (since the snow cover is rather thin the ground is frozen to considerable depth during these periods). South of 30° N tradewinds and dry monsoons from the continent prevail. The winter is sunny, dry and warm in most areas, with temperatures ranging from +8° C (in the northern part of Mexico) to +24° C in Central American countries (Bromley, 1981).

The summer weather in most parts of the continent is hot with rainfalls in the east. The exception is the Arctic region of Canada which is affected by cold Arctic air masses as well as by the Labrador Current. Low temperatures are also characteristic of the western coasts of North America: the cold California Current significantly decreases average temperatures in July. Summers are dry over most of the west of the continent because of the cold dry air masses arriving from the northwest. Rain and snow are abundant only in the south of Alaska and British Columbia, where air masses have to cross pass the mountain ranges. The cold California Current is also an important factor in the climate of the west. California and its surroundings has hot dry summers and cool humid winters. The climate becomes drier and more continental away from the Pacific Ocean. In depressions of the Great Basin and in the southwestern regions of the USA, such as Death Valley, summer temperatures reach extreme values that have yet to be observed elsewhere in the world. However, the temperature drops rapidly with altitude so that large areas of the Rocky Mountain, the Sierra Nevada, the Colorado Plateau and the eastern parts of the Great Basin have pleasantly warm summers.

South of 30° N the summer is predominantly hot and humid. The Pacific air mass encounters the humid Atlantic air to form a tropical front whose movement is accompanied by considerable falls of precipitation. In the southeast of the continent and on the adjacent islands, the summer weather is mainly warm, with abundant precipitation which reaches a maximum in the hurricanes which hit the Caribbean with great destructive force in summer and autumn. The amount of summer precipitation decreases sharply from east to west throughout the Central and Great Plains. There may be no rain for several weeks in the Cordilleras piedmont plateau, and then it falls in heavy showers which produce torrents flowing down from the mountains charged with sediment. Tornadoes with great destructive force are typical of the Great Plains in summer. Dust storms raised by winds in the Great Plains often reach the Central Plains and, sometimes, the Atlantic coast. The plains, in general, are well endowed with heat and moisture so that agriculture is highly productive on the fertile soils.

The average annual precipitation for North America is 805 mm (Korzun, 1974b). However variations in annual precipitation over the continent are large: from 30–50 mm in Death Valley and in the south of California to 6000–7000 mm in the mountains of the Central American isthmus. For a large part of the continent, precipitation is above 1500 mm, and almost all of Central America has over 2500 mm. The general tendency is for annual precipitation to increase, from north to south, from 180 mm (80°–90° N) to 900 mm (50°–60° N) and 2600 mm (0°–10° N), except for the area between 20° and 30° N (Mexico), where it drops to 700 mm (Fig. 7.2).

7.2.3 Hydrology

North American water resources are very unevenly distributed. The annual runoff ranges from 0–20 mm on the inland plateaux to 3000–4000 mm on the Pacific Coast (Fig. 7.3). There are a few areas with higher runoff values: the Pacific Coast (north of 40° N), the Laramian region of the Canadian Cordilleras, the coast of the Gulf of St Lawrence, the Appalachians, the eastern part of the Gulf Plain, the Bay of Campeche and the Central American region. Lower runoff values are characteristic of the

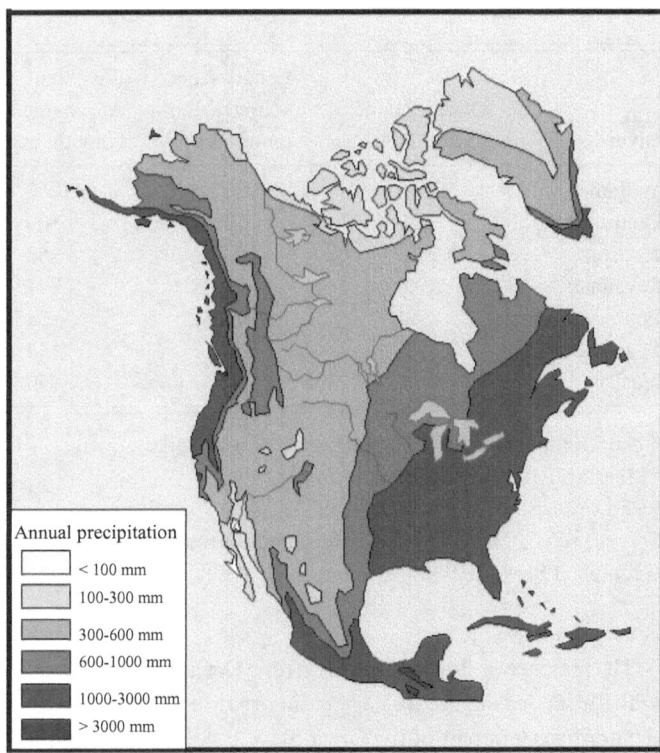

Fig. 7.2 Mean annual precipitation for North America.

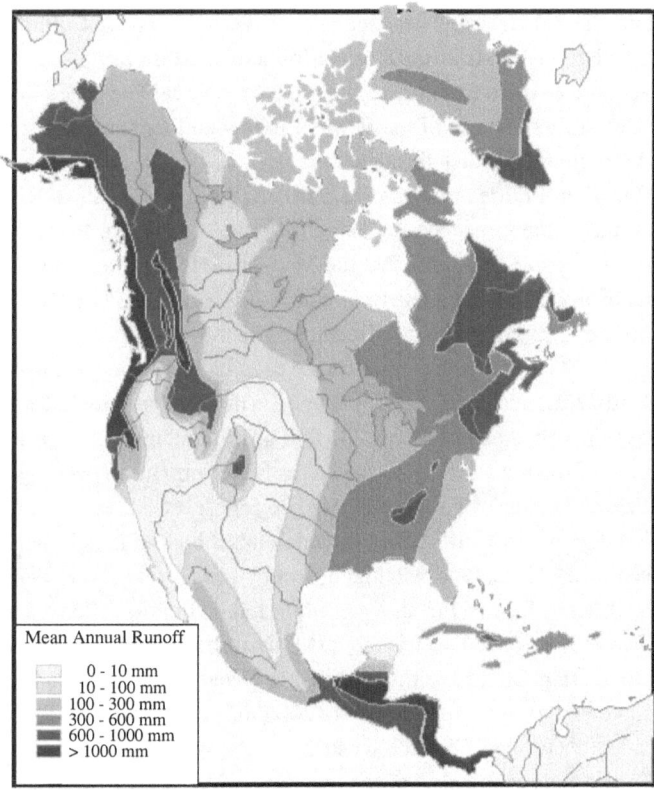

Fig. 7.3 Mean annual runoff for North America.

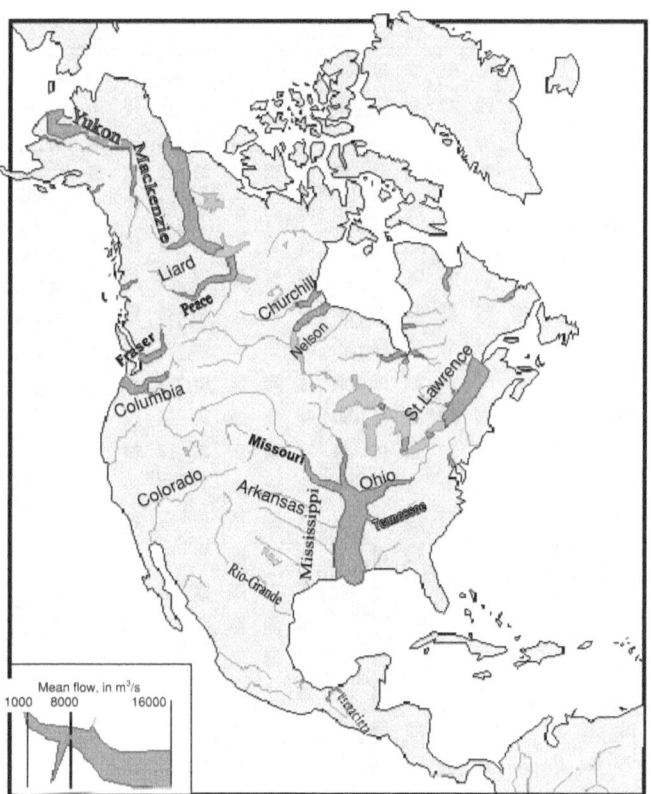

Fig. 7.4 Principal river systems of North America and those channels carrying mean flows of more than 1000 m³/s.

inland plateaux and highlands of the Southern Cordilleras, Lower California, Mexico Bolsons de Mapini, Saint-Louis-Potosi and the plateaux of Lianos–Estacado and Edward.

The major North American rivers are the Mississippi, Mackenzie, St Lawrence, Rio Grande, Yukon and Colorado (Fig. 7.4, Table 7.1). The largest in terms of the drainage area and by the volume of discharge is the Mississippi. It rises in the hilly area of Minnesota west of Lake Superior and drains to the Gulf of Mexico. In its upstream section, the Mississippi runs through a number of lakes and is full of rapids. Before the Ohio River enters the Mississippi, its valley is not more than 15 km wide; downstream it gets much wider, reaching 120 km in some places. Finally, the Mississippi Delta has many channels forming a vast area of swamps on the edge of the Gulf of Mexico. The principal tributary of the Mississippi is the Missouri (right bank). In its upstream section, the Missouri traverses the Rocky Mountains to flow through a plateau and the Great Plains. The left bank tributary of the Mississippi, the Ohio River, is a full-flowing river. The upstream section of the Ohio forms deep gorges in the Appalachian Plateau, its downstream section flows along a wide flat valley. One other large tributary of the Mississippi is the Arkansas River, rising in the Rocky Mountains and flowing across the Great and Central Plains.

Table 7.1. *Principal rivers of North America*

River	Drainage area, km² × 10³	Length from source to mouth, km	Average discharge at mouth, m³/s
Mississippi	2980	3780[a]	16 280
Mackenzie	1787	5472	10 300
Missouri	1370	4090	2 160
Nelson	1132	2574	2 830
St Lawrence	1026	3057	10 100
Rio Grande	870	3033	270[b]
Yukon	850	2897	6 220
Columbia	668	1953	7 500
Colorado	637	2333	505[b]
Ohio	526	2110	7 950
Arkansas	417	2350	1 160
Churchill (Hudson Bay)	298	1609	1 270
Snake	280	1790	1 610
Atchafalaya	246	2280	1 640
Red	241	2070	1 585
Fraser	233	1370	3 620
Platte	220	1590	169
Yellowstone	181	1110	371
Kansas	154	1200	198
Gila	151	1040	65
Albany	134	975	1 100
Koksoak	133	1300	2 420
Kuskokwin	124	1160	1 900
Canadian	121	1460	30
Brazos	118	2060	40
Porcupine	117	920	651
Mobile	116	1250	1 900
Pecos	115	1490	2.4
Tanana	115	1060	1 160
Colorado (Texas)	110	1390	77
Back	107	960	560
Tennessee	106	1420	1 920
Severn	101	976	722
Churchill (Atlantic coast)	79.8	560	1 620
Susquehanna	70.4	720	1 180
Copper	63.2	460	1 670
Sacramento	55.0	610	642
Susitna	51.8	500	1 440
Stikine	51.8	610	1 580
Attawapiskat	50.2	810	626
Papalaopan	47.0	350	1 300
Eastmain	46.4	680	909
Manicouagan	45.8	520	852
Apalachicola	45.3	880	636
George	41.7	550	881
San Joaquin	35.2	560	136

Table 7.1. (*cont.*)

River	Drainage area, km² × 10³	Length from source to mouth, km	Average discharge at mouth, m³/s
Willamett	29.5	500	1060
Connecticut	26.9	550	500
Hudson	21.0	490	388
Savannah	19.6	510	291
Penobscot	17.4	340	34
St Johns	15.3	640	133
Nushagak	13.4	460	950

[a] The length of the Mississippi River is without Missouri River; with Missouri River it is 6420 km.
[b] Mid-reaches.
Source: CNC/IND, 1978; Rosenberg and Barton, 1986; US Geological Survey, 1986.

The major river flowing into the Arctic Ocean is the Mackenzie, with the second largest drainage basin in North America. It rises in the northwestern part of the Great Slave Lake and flows directly northwards. However, it becomes a huge river system rising in the Rocky Mountains, to include seven major rivers, three lakes and three deltas. The longest tributaries in the south are the Peace and the Athabasca. The Athabasca River rises in the Columbia Glacier, running from the Rocky Mountains, to flow northeastward across the Albert Plains, and into Lake Athabasca. The Peace River rises at the junction of the Rivers Findlay and Parsnip, and from the mountains flows through the Province of Alberta across the Canadian Prairies towards the northern forests. The Slave River connects the Great Slave Lake with Lake Athabasca. It is north of the Great Slave Lake that the Mackenzie River proper rises. It takes in many tributaries, the longest of which is the Liard River, before reaching the Arctic Ocean in a vast delta.

The St Lawrence River forms the third largest drainage basin in North America. Its distinctive feature is that the basin includes the five Great Lakes (total area 245 000 km²) containing a large volume of fresh water. These are Lakes Superior, Huron, Michigan, Erie and Ontario. They are located on four levels. The uppermost is Lake Superior, 184 m a.s.l., which has a bottom mostly below sea level. Huron and Michigan occupy the second level, 177 m a.s.l. Lake Erie is 175 m a.s.l, while Lake Ontario is 75 m a.s.l. The St Lawrence River is the last link of the Great Lakes system connecting Ontario with the Atlantic Ocean. Its length from the Lake Ontario to Cape Gaspe is 1200 km, or 3057 km if it is considered as rising from the River St Louis, with its source not far from the upper reaches of the Mississippi, before it reaches Lake Superior. The St Lawrence River flows between the Appalachian and Laurentian Mountains along a broad and fertile valley. The streambed is very broad, and beyond Orleans Island, it becomes an estuary. Because of the Great Lake system, the water is very clean although farther downstream there is salt water and the flow is affected by the tides.

One other river flowing to the Atlantic is the Rio Grande which is part of the boundary between the USA and Mexico in its downstream reach. It rises in the Rocky Mountains, in the northwestern part of the State of Colorado, and flows south and southeastward. Almost all the flow in the Rio Grande is used for irrigation, so that when it reaches the Gulf of Mexico the total discharge is about 10 m^3/s. The River Yukon rises in the Coast Range in Canada (the Lewis River is considered to be its source), and flows across Alaska to the Pacific. Its valley varies in width from 3–5 km to 80–100 km at the widest and it enters the Bering Sea in a branched delta.

In the southwestern part of the USA, there is the Colorado River which rises in the Rocky Mountains and flows from the Front Range towards the Pacific Ocean. Cutting across the Rocky Mountain ridges and the Colorado Plateau, the river forms a series of picturesque canyons, the Grand Canyon being the most spectacular. The upper part of the Colorado is joined by its tributaries: the Green River, the San Juan, the Little Colorado and the Sevier. Downstream, a considerable volume of water is abstracted for irrigation and domestic water supply. When the Colorado enters the Gulf of California in Mexico, its discharge is only a few cubic metres per second.

The River Columbia rises in the Rocky Mountains in Canada cutting across the Columbia Plateau forming a deep gorge. Downstream from Rigland the Columbia turns west through the Cascade Range where it reaches the Pacific Ocean in an estuary. There is one more large river in North America, the Fraser, which flows through the Canadian part of the Rocky Mountains. It rises a little east of the main watershed with two small glacier-fed streams running from Mount Robson. Near Mount Hope it leaves the mountains and turns eastward to flow along the extensive fertile valleys of British Columbia.

The geological history of the continent gave rise to a great number of lakes, especially in the north where in Canada they occupy about 8% of the area (700 000 km^2) (Wolman, 1990). They are most numerous in the Lower Arctic, in the Canadian Shield, in the Atlantic Provinces, in the northeast and central northern parts of the United States, and in Florida. In some parts of the Laurentian Shield they occupy 60% of the area. These lakes are extremely varied in size: over 20 exceed 1000 km^2 (the morphometric characteristics of the major lakes are given in Table 7.2). They are mostly glacial or glacial–tectonic in origin. Wolman (1990) distinguished nine lake types in North America: arctic; subarctic; Canadian Shield (temperate); mountains south of the subarctic; glacial terrain, (eastern; glacial terrain, prairie); riverine; desertic; and sinkhole lakes (Fig. 7.5).

Table 7.2. *Major morphometric characteristics of principal lakes of North America*

Lake	Location	Surface area, km^2	Maximum depth, m	Volume, km^3
Superior	Canada, USA	84 500	406	11 600
Huron	Canada, USA	63 500	229	3 580
Michigan	USA	58 000	281	4 680
Great Bear	Canada	31 400	137	1 010
Great Slave	Canada	28 600	156	1 070
Erie	Canada, USA	25 800	64	545
Winnipeg	Canada	24 400	19	127
Ontario	Canada, USA	19 300	236	1 710
Nicaragua	Nicaragua	8 030	70	108
Athabasca	Canada	7 940	60	110
Reindeer	Canada	6 640		
Winnipegosis	Canada	5 360	12	16
Netilling	Canada	5 530		
Wipigon	Canada	4 850	162	
Manitoba	Canada	4 700	28	17
Great Salt	USA	4 660	14	19
Lake of the Woods	Canada, USA	4 410	21	
Dubawnt	Canada	3 830		
Amadjuak	Canada	3 120		
Iliamna	Alaska	2 590		
Mistassini	Canada	2 190	120	
Managua	Nicaragua	1 490	80	
St Clair	Canada	1 200	7.2	5.3
Lesser Slave	Canada	1 190	3	
Chapola	Mexico	1 080	10	10.2
Winnebago	USA	818	6	4.1
Marion	USA	465		2.8
Winnipegosis	USA	181	55	3.8

Sources: Wolman, 1990; Korzun, 1974b.

The largest on the continent are the lakes included in the Great Lakes system, consisting of Lakes Superior, Michigan, Huron, Erie and Ontario, as well as the small Lake St Clair. Lake Superior is the world's largest freshwater body. All the basins of Great Lakes are glacial–tectonic in origin. The lakes themselves were formed in the Quaternary due to extensive ice-sheet melt. The total water volume of the Great Lake system is estimated at 24 100 km^3 (Gusakov and Petrova, 1987), and they cover 245 000 km^2. The water level varies seasonally by 25–50 cm. Among the other principal lakes of the continent are the Great Bear, the Great Slave, Lake Winnipeg, and Reindeer Lake. There are a lot of small mineral water lakes in

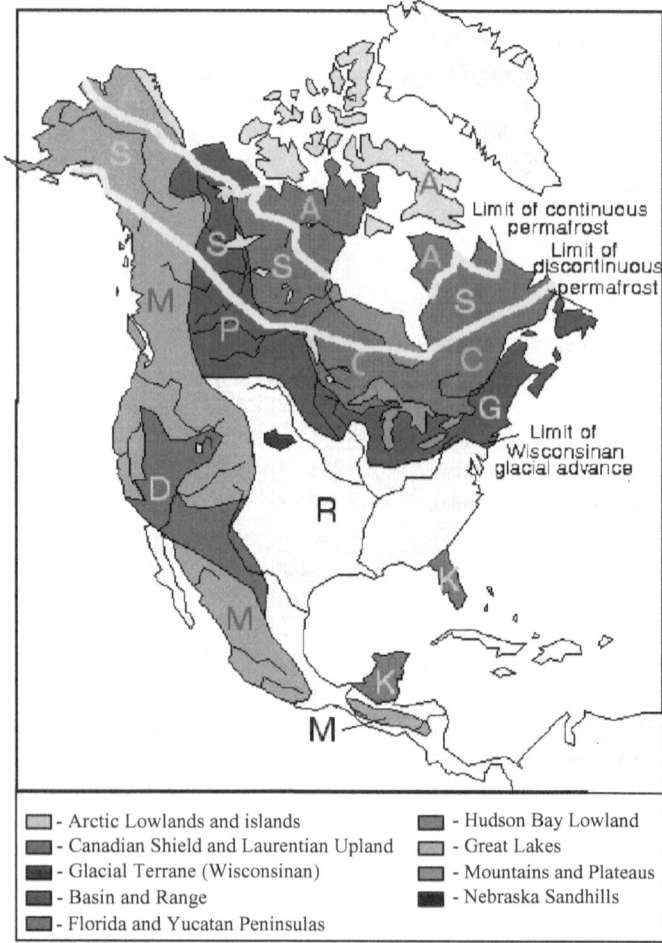

Fig. 7.5 Physiography of North America and distribution of general hydrogeologic types of lakes and wetlands (Wolman, 1990). Lakes types: A, arctic; S, subarctic; C, Canadian Shield (temperate); M, mountains, south of subarctic; G, glacial terrain (eastern); P, glacial terrain (prairie); R, riverine; D, desertic; K, sinkhole.

the Great Basin kettles; the largest of them is the Great Salt Lake, whose salinity is 270 ppm.

7.3 SOCIO-ECONOMIC CONDITIONS AND THE USE OF WATER RESOURCES

7.3.1 Population

The total population of North America was 453 million in 1995. More than half (about 262 million) live in the USA, where the average population density is 27.8/km². In Canada with 28.0 million people in 1995, the population density is not very high: 3.0/km². The highest population density is in Central America: 57.8/km², with a total population of 163 million people in 1995. The greatest population growth is also in Central America. During the last

decade of the twentieth century this growth was about 30 per 1000 inhabitants, while in Canada and the US it was three times less. Population figures for the principal North American countries are given in Fig. 7.6.

North America is one of most urbanized parts of the world, with four-fifths of its population living in towns. The urban inhabitants of Canada and the USA make up 97% of the population (Bromley, 1980), while in Central American countries the figure is 70%. Typical of North America urbanization is the rapid rise in the number and extent of suburban settlements. These are vast suburban areas that develop into agglomerations; over two-thirds of the inhabitants of the continent live in such agglomerations.

7.3.2 Economic characteristics

North America countries are far from uniform in their economic development. The USA and Canada are the most highly developed nations in the world. Mexico is a large agricultural–industrial country rich in natural resources. Most Central American countries are developing, being largely agricultural with little industry and poor water management systems. Canada ranks among the states with the most highly developed economies. It is a large industrial–agricultural country with a high economic potential. With a gross national product of US$21 900 per capita in 1995 (Fig. 7.7) it is far ahead of many developed countries, and in its life comfort index, it is among the countries at the top of the list of the world's best. Canada is far from being evenly populated. About 95% of people live along the southern border in a long narrow band occupying only 20% of the total area. The population of the northern regions is clustered around sites where minerals and energy are produced.

As Canada is rich in natural resources, much of its industry is in primary production and export of raw materials. Primary production gives 25% of the material production value (Prokhorov, 1987). Almost all the main useful minerals are found in large quantities in Canada. The leading manufacturing industries are machine tools, metal fabricating, chemicals, wood products, food and consumer goods. Most plants are located in the southern part of the country. Canada is extremely rich in sources of power, and water power amounts to more than two-thirds of all power production. With vast areas covered by forest, Canada became a world leader in the paper and wood products industries.

Agriculture is a highly developed part of the Canadian economy. Not only does it meet the home needs but it provides 15% of the total value of all exports (Bromley, 1980). Extensive fertile soils and the variety of climate favour different types of farming. Farms now cover 170 million ha of which 46 million ha are cultivated (FAO, 1995) and 0.96 million hectares are irrigated. There

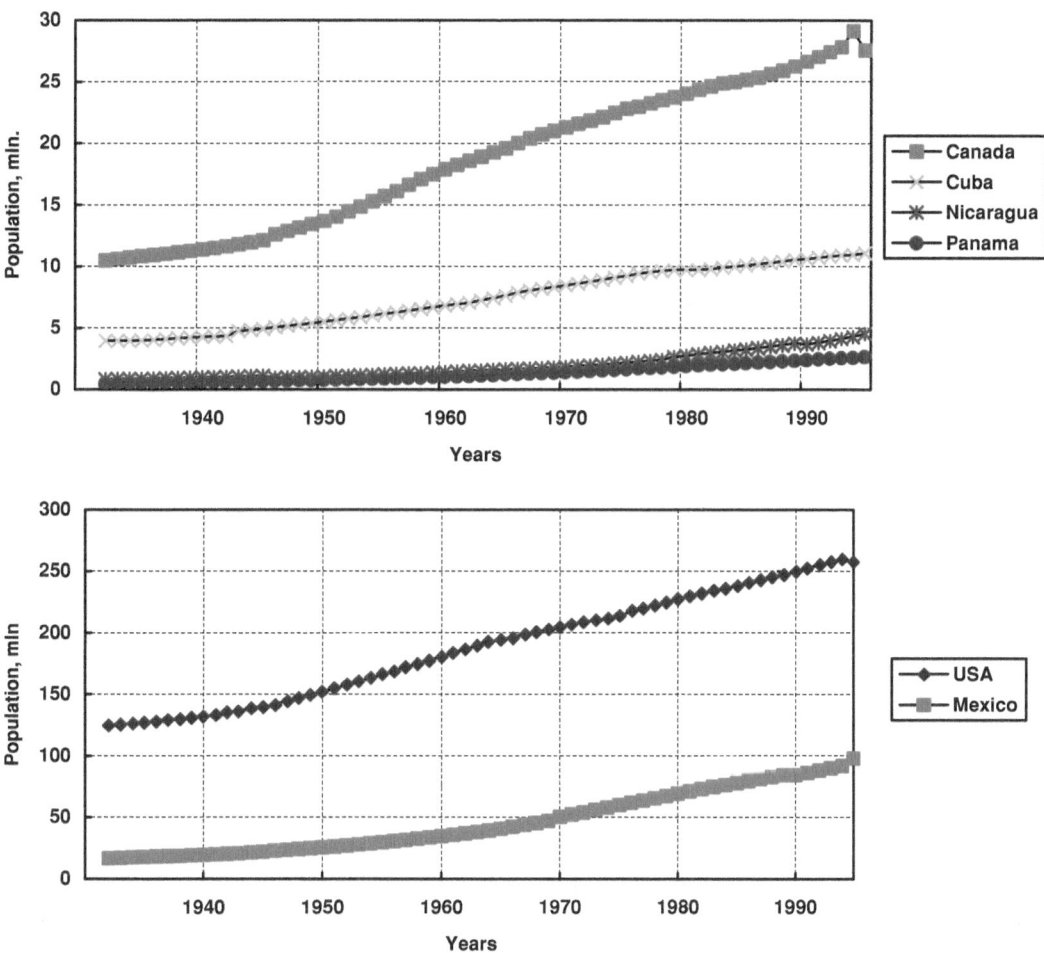

Fig. 7.6 Dynamics of population in selected countries of North America.

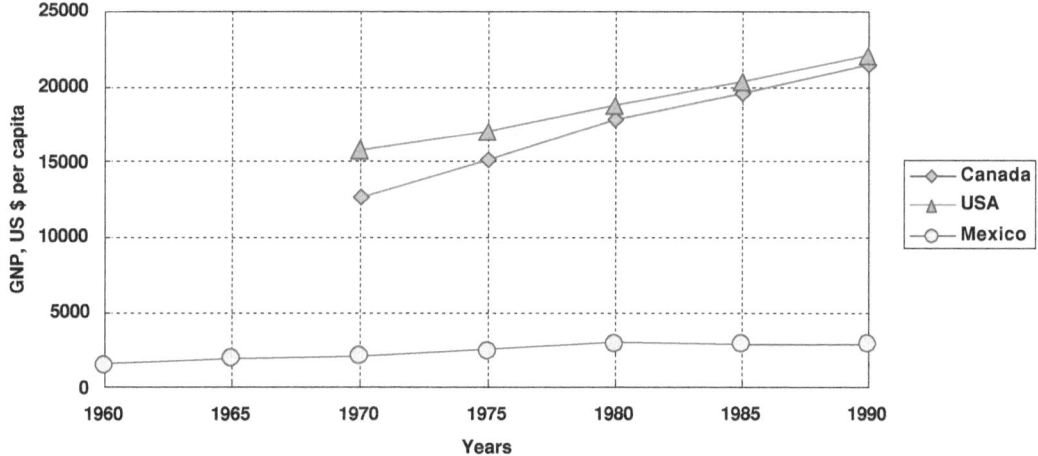

Fig. 7.7 Dynamics of GNP in selected countries of North America, 1960–90 (IMF, 1994).

are significant reserves for further agricultural development, by cultivating virgin lands and irrigating arid areas.

The USA has the world's largest economy. It surpasses all other countries in its gross national product with an annual income of US$24 500 per capita (Fig. 7.7) and in its industrial and agricultural production, foreign trade turnover, export of capital, labour productivity and in most of the other major fields of the economy. About 30% of the US national income is provided by industry (Prokhorov, 1987). Industry is very diverse, self-sufficient in raw materials, and supported by high technology, the most important characteristic of technological progress. The USA is a leader in the world of mining but it also imports most important minerals. There are advanced fuel and power industries, and massive oil refineries. Electric power production is a fast-growing part of the US economy, with hydroelectric power accounting for little more than 10% of all power. The ferrous and non-ferrous metals, machine tools, consumer goods and food industries are well developed.

Agriculture in the USA not only meets the home demand for basic foods and raw materials but it also provides excess for export. The US agricultural area of about 770 million ha includes about 190 million ha of arable land with 20.4 million ha being irrigated. In the eastern more humid areas, the land consists mainly of arable land and forests, while in the arid west it is mainly pasture. The proportion of arable land is especially high in the prairies. In the West, most arable land is irrigated. Cattle production provides 55% of all produce (Moore et al., 1990). Since the 1980s the economy of the USA has been changing. The proportion of agriculture and forestry, primary production and manufactured goods has been steadily declining (Vasilevsky, 1994). But in 1992 the economy entered a new cycle (Bogacheva, 1995).

Mexico is a rather large industrial–agricultural country. Its gross national product is US$2930 per capita. The economy is based on the country's raw materials. The mineral wealth includes ample reserves of various ores, oil, natural gas and sulphur. The power industries are fed on oil and gas. The petro-chemicals chemistry is a comparatively new part of the economy. About 75% of all electric power is generated thermally and about 25% by water. Mexico is relatively rich in sources of hydropower, but the easily accessible sites have already been exploited. Industry includes metallurgy, machine tools and food production.

The climatic conditions in Mexico are not favorable for agriculture. About 40% of the area is desert or semi-desert, and another 40% mountains and forests. Only about 15% can be used for agriculture (Bromley, 1981). By 1995 about 25.0 million ha were cultivated, of which 6.1 million ha were irrigated. The latter are situated mainly in the northwestern and northern parts of the country. Crops provide over 60% of farm commodity values, while cattle-rearing produces about 35% (Bromley, 1980). The latter is located in the semi-arid northern and northeastern districts where there are vast natural pastures.

The economies of almost all the smaller countries of Central America are agricultural. The only exceptions are Cuba, whose industry provides about 40% of the gross national product, and Jamaica with 30% of its GNP produced by industry (Prokhorov, 1987). In Cuba there is production of ferrous and non-ferrous metals, and machine tools, chemical and petro-chemical industries and power generation. The basis for Jamaican industry is bauxite production. Costa Rica, Salvador, Guatemala, Honduras, Nicaragua and the Dominican Republic have oil refining, while there is a petro-chemical industry in Guatemala, as well as textiles and food production. Central American countries produce good crops of coffee, banana, citrus fruits, cotton and sugar cane. They also grow maize, kidney beans, sorghum and rice. About 15% of the arable lands are irrigated.

7.3.3 Assessments of water use

Since the beginning of the twentieth century the rapid growth in industrial and agricultural production has resulted in a sharp increase in the use of water resources and in their depletion. Irrational use of water has caused surface water and ground water pollution and has degraded the overall environment of North America. By the late twentieth century the problem had become so severe that restriction of water use was put on the agenda of the Federal Government of the USA. It entailed obtaining objective information on the main water users, the amount of water used, abstracted, consumed and returned to rivers, as well as on the most water-intensive production technologies.

The USA was the first to start collecting quantitative information on water use by various industries on a regular basis. Since 1950 the US Geological Survey has produced a series of national assessments of water use every 5 years (Estimated Use of Water in the United States for 1950, 1955, 1960, 1965, 1970, 1975, 1980, 1985, 1990 and 1995). Data were published on water use in industry, agriculture (for irrigation and livestock) and public supply, as well as data on in-stream water use for hydropower production. Assessments were made of abstractions of surface and ground water, both fresh and salt. Data on water consumption have been regularly published since 1960.

The data on water use in the USA before 1975 have been summarized in a number of publications (MacKichan, 1951, 1957; MacKichan and Kammerer, 1961; Murray, 1968; Murray and Reeves, 1972, 1977). However these data were based on information differing in reliability. Emphasizing the necessity for more homogeneous and reliable data on water use, the Congress required the Geological Survey to follow the National Water Use Information Program which is part of the Geological Survey's Federal–State Cooperative Program and differs from the previous one in that more reliable data on the quality and quantity of water were acquired. Later on, the data on water use were summarized (Solley *et al.*, 1983, 1988, 1993). The 1987 *National Water Summary: Hydrologic Events and Water Supply and Use* was published (US Geological Survey, 1990), where the changes in use by the different sectors was analysed, and the pattern of abstractions and consumption was described in detail for all the states.

Thomas (1959) of the Department of Natural Resources and Technical Surveys was the first to carry out systematic studies on abstractions in Canada. The author focussed his studies on the qualitative aspects of water use and made no efforts to give any quantitative information. Systematic estimations of abstractions were started in Canada in the 1970s with the collection and analysis of data on abstractions by industry. The first survey of industrial water use was made in 1972, and was continued in 1976, 1981, 1986 and 1989 (Tate, 1977, 1984; Tate and Scharf, 1985; Tate *et al.*, 1992). The analysis of municipal water use was carried out in 1976, 1983 and 1986 (Tate and Lacelle, 1978, 1987, 1992), taking into account data to determine the cost of municipal water. No detailed studies were carried out on the changes of agricultural water use in Canada, although some estimates were produced of national and regional water use in agriculture (PPWB, 1982).

The data on municipal water use in Canada were generalized in Grima (1972), Sewell and Rouelche (1974), Kitchen (1975), Tate and Lacelle (1987, 1992), Tate (1989) and McNeil and Tate (1991). The results of comprehensive investigations on industrial water use are given in Tate and Scharf (1985), Tate (1986), Renzetti (1987), and Tate *et al.* (1992). The data on patterns of water use for certain river basins and regions are given in Earmme (1979), Tate (1984, 1990) and Pearse *et al.* (1985). Data on ground water abstractions are given in Hess (1986).

No regular observations of water use have been carried out in the countries of Central America. Only a few estimates have been made for major sectors for individual years. As to assessing the use of water resources, the situation is satisfactory in Mexico, Barbados, Costa Rica, the Dominican Republic and Panama. However, there are no programmes to assess regularly the water used in any Central American country or those of the Caribbean. Nevertheless, abstractions were always recorded in the countries where irrigation has been practised. A review of water use in Mexico is given by Anaya (1967), who collected data on domestic, industrial and agricultural abstractions for 1965 and made a forecast for periods to 1980. In the National Hydrologic Plan of Mexico (*Plan Nacional Hidraulico de Mexico*, 1981) there are estimates of the amount of the intake, diversion, transfer and redistribution of water over regions for the current year and for future years. Calvo Jullo (1990) analysed sectoral water use data in Costa Rica for 1970 and 1987 and gave a forecast for 2000. Some tentative data on water use for certain countries are given by L'vovich (1974), Gleick (1993, 1998), Seckler *et al.* (1998), Rijsberman (2000) and in WRI (1990, 1992, 1996, 2000).

7.3.4 Use of water resources

A greater part of the North American continent is located in arid and semi-arid areas, so a principal concern in water management is irrigation. The amounts of water used for irrigation are affected by many factors: climate, the basic physiological needs of crops, the application technology, water costs, crop and livestock prices, non-farm competition for water resources and public policies. Water-supply costs depend on access to and location of surface water storage and aqueducts, the stability of ground water levels and energy costs (Moore *et al.*, 1990). To smooth out the natural variations in stream flow and thus to make irrigation water readily available throughout the growing season, runoff is controlled; and systems of canals, channels and pipes are used to convey and distribute the water. The building of irrigation systems in North America reached a peak in the 1960s, but since the 1980s the focus has shifted towards improving existing systems and making them more efficient.

The continent's irrigated area has grown by 6 times from the beginning of the twentieth century to 1990: from 4.2 million ha to 26.3 million ha. However between 1980 and 1990 there was some reduction in the area irrigated. Likewise the amount of water abstracted for irrigation had been rising to 1980, but it declined to 1990.

The USA has the most extensive irrigation, as well as the most water abstracted. As early as 1900, there were 3.1 million ha of irrigation (about three-quarters of the total for the continent). By 1950 the area had expanded to 10.4 million ha, by 1980 to 20.6 million ha (according to FAO) or to 23.5 million ha (Solley *et al.*, 1983). The Federal Government provided subsidies for this expansion. The Bureau of Reclamation supplied water to farmers at a reduced price, which resulted in a Federal budget loss. Irrigation costs were partly paid from the gains obtained from hydroelectric power generation. In general, between 1949 and 1977 users paid only 26% of the total cost of irrigation. In 1977 the Administration initiated a shift in irrigation policy by halting the planning and the construction of many prospective water projects. Since then Federal funding of irrigation has continued, but at a very low level, and only one new major project has been started (Moore *et al.*, 1990). As a result, the irrigated area contracted between 1980 and 1990. Another factor was the cyclic recession in production, including agricultural production, which occurred in the USA in the 1980s. At the same time, US legislative policy turned to the environment and protection of water resources, which made irrigation more costly. Some states have taken measures to halt falling ground water levels by limiting pumping. So, water became valuable but the cost of its use for irrigation failed to compete with that of the water used in industry or for public supply. In general, the irrigated area has remained more or less the same since 1982 with some variation depending

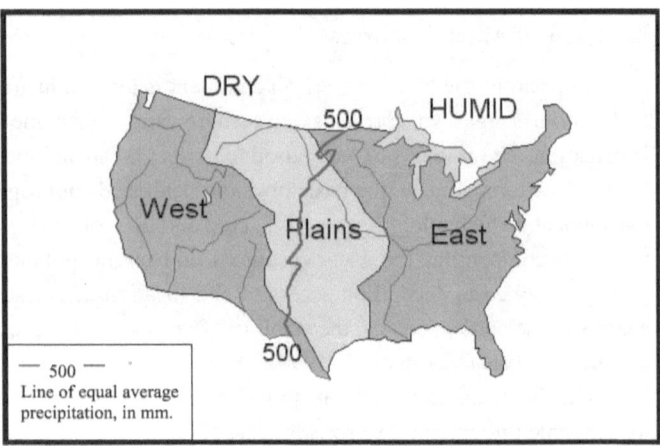

Fig. 7.8 USA regions, as used in this chapter. The line of 500 mm average annual rainfall is the boundary between the humid region and the dry region. (From US Geological Survey, 1990, fig. 49.)

on the national programme for agriculture and for environmental protection.

The 500 mm isohyet of mean annual precipitation stretches from southwest Texas to the extreme northwestern corner of Minnesota dividing the USA almost equally into a humid east and a dry west (Moore *et al.*, 1990). The dry region is then subdivided into the Centre (Great Plains) and the West (Fig. 7.8). Irrigation in the dry regions developed when the price of water was low. While use of surface water resources dominated in the West, in the Great Plains, the main source for irrigation was ground water. In the humid East, irrigation is used as a supplement to precipitation to raise yields and reduce the year-to-year variability in yields and product quality.

Irrigated areas are located mostly in the Central (plains) regions and in the West (Fig. 7.9). In the 1980s, the Central regions were a leader in this field. In the 1990s, due to the reducing extent of irrigation, the Central and Western regions were about equal in area irrigated in the Great Plains, especially in Texas and Oklahoma, but there was the lowering of ground water levels, as well as rising power costs and the temporary recession in farming. Smaller areas are irrigated in the Eastern USA. The main irrigated areas are the Lower Mississippi (45% of the irrigation in the East) and the South Atlantic–Gulf (36%). However, the growth in irrigation is faster here than in other regions of the USA: between 1960 and 1990 the area irrigated expanded 5.2 times in these regions. In the Central part of the USA, the irrigated area increased by 59% over the same period, while that in the West increased by only 4% because of shortages of water.

The amount of water abstracted for irrigation in the USA in 1950 was 125 km^3; by 1980 it had increased to 206 km^3. Between 1980 and 1990 the total dropped to 192 km^3, mainly due to reductions in the West and Centre. In the West, the driest area, abstractions were greatest (119 km^3) in 1990, but the increase was slower than

in other regions. Between 1960 and 1990 the figure increased by 10%, and water use per hectare increased by 6%, with a small decrease in abstractions during the last decade. Abstractions in the Centre were 55 km^3 in 1990. This represents an increase of 28% compared to 1960, while the irrigated land area grew by 59%, with the specific water use decreasing by about 20% over the same time. In the East, water for irrigation was only 18.2 km^3 in 1990, with the most (16.6 km^3) being used in the Lower Mississippi and South Atlantic–Gulf areas. The rate of abstraction in the East was much higher than in the rest of the country. Between 1960 and 1990 abstractions increased 7 times, and specific water use rose by 35%.

The water used for irrigation in the USA comes both from surface (63%), and ground water (37%) (Solley *et al.*, 1993), although a small quantity of treated sewage is also employed (Moore *et al.*, 1990). Surface water is used mostly in the West, ground water in the Lower Colorado basin, California, in the Pacific Northwest, and on the plains. Water pumped from the well-known Ogallala aquifer is used in the Centre, in Arkansas, Kansas, Mississippi, Missouri, Nebraska, Oklahoma and Texas. Ground water is used as much as surface water in the East (Moore *et al.*, 1990). Irrigation is, of course, largely a consumptive use of water, amounting to 56% of the total abstracted in the USA, varying from 40% to 100% by region.

Canada has a humid climate, so irrigation occupies a comparatively small area of the country, mostly in the prairies. Irrigation is widespread in the Southern Saskatchewan basin; there is some in the Lower Fraser, the Upper Mainland, and Okanagan (Pearse *et al.*, 1985), as well as in the Great Lakes region. Early in the twentieth century less than 0.1 million ha were irrigated, but by 1990 this area had expanded to 0.86 million ha. Since the 1980s the rate of growth has increased. Inter-basin transfers could be used for further development of irrigation in the prairies. In 1990 4 km^3 of water was used, of which about 75% went to the prairies, 17% to British Columbia, 5% to Ontario and 2.5% to Quebec (Pearse *et al.*, 1985).

Mexico has an arid climate, so agriculture cannot develop without irrigation. About 30% of farms are irrigated and provide 50% of the gross value of agriculture, producing a wide assortment of foods. Over the greater part of Mexico there is no relation between the available water resource and the demands of agriculture. For example, there are about 1.5 million ha of fertile land in the north of the Pacific Coast not cultivated because of the lack of water, while there is plenty of water in the Central Pacific region. The National Plan for Water Resource Development (Gordunio *et al.*, 1984) made allowance for transfers from other regions to irrigate the northwest. However economic problems in the 1980s delayed most of the projects envisaged and because of insufficient funds during the 1980s irrigation remained at 5 million ha. Only in the early 1990s did growth start again, to reach

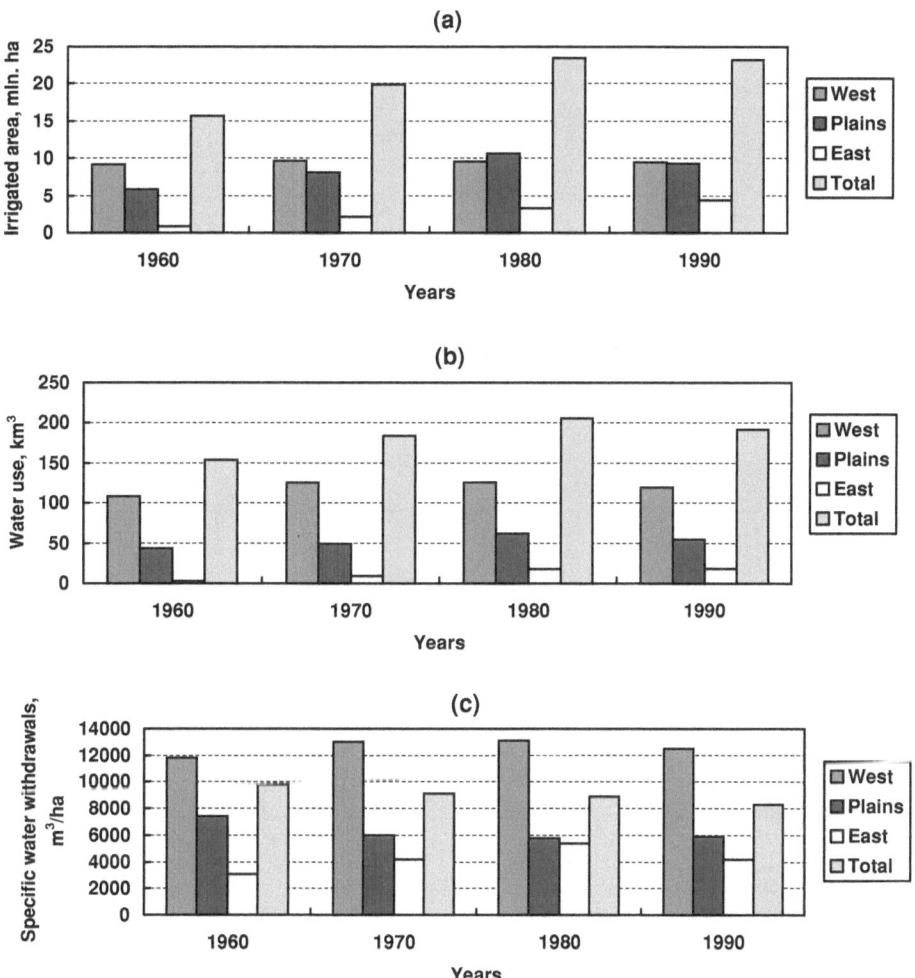

Fig. 7.9 Dynamics of water use in the USA. (a) Irrigated area, ha $\times$ 10^6; (b) water use, km^3; (c) specific water withdrawals, m^3/ha.

6.1 million ha by 1992 with water use exceeding 60 km^3/year. Further development of irrigation is of great importance for Mexico because of its rapidly growing population, accompanied by an increasing per capita income and a rising demand for agricultural products. Further development of irrigation is possible by inter-basin transfers and by better use of local water resources; an important resource remains the more efficient use of fresh water. In a number of regions of Mexico, especially in the north, because of restricted water resources and the high rate of abstraction, the ground water level is being lowered, causing adverse effects on the environment.

Much land is irrigated in the Central American countries. The greatest areas of irrigation are in Cuba (0.9 million ha), in the Dominican Republic (0.225 million ha) and Costa Rica (0.12 million ha). In 1990, water used for irrigation was 7.5, 2.6 and 1.8 km^3/year, respectively. During the past decades, the rate of irrigation in most of Central American countries has risen rapidly. According to FAO (1995), from 1970 to 1990 the area irri-

gated increased 3 times in Nicaragua, 4 times in Costa Rica and 6 times in El Salvador. In most other countries it grew by between 30% and 80%. However, this significant expansion of the irrigated areas was a result of completing previous projects, while new irrigation systems are being developed only slowly because of the lack of direct governmental investments (ECLAC, 1995).

In addition to water for irrigation, considerable amounts of water are abstracted in the North American continent for other types of rural use. In the USA, this sector includes aquaculture. In the USA, more water was abstracted for rural use than in the other countries of the continent. From 1900 to 1990 this sector grew from 3 km^3 to 11 km^3/year. Water used for livestock has increased since the mid twentieth century from 2 km^3 (1950) to 3.08 km^3/year (1980). Between 1980 and 1985 the demand rose approximately twice as a result of a wide spread of aquaculture. In 1990 6.3 km^3/year was used for livestock. The amount of water used to meet rural needs in Canada has increased gradually since the beginning of the twentieth century, and by 1990 it was 0.5 km^3/year. At the same time the rural population has decreased

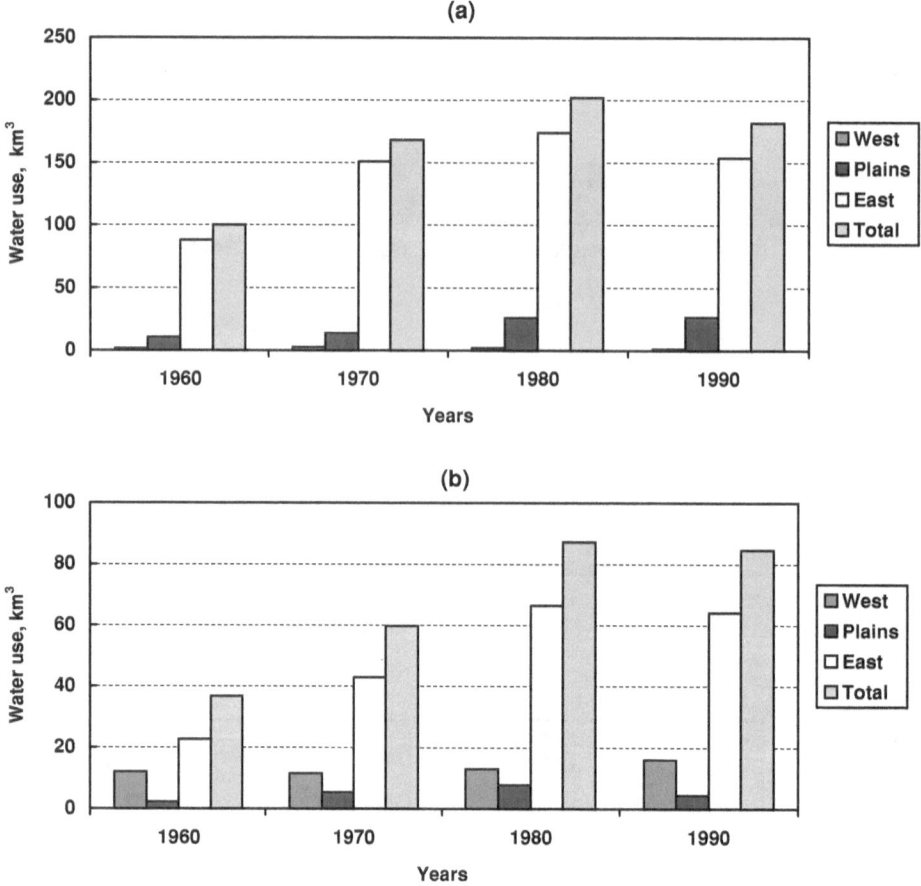

Fig. 7.10 Dynamics of thermoelectric power water use in the USA by regions, 1960–90. (a) Fresh water; (b) salt water.

considerably, while the specific water use (per capita) has increased sharply. In Mexico, like in the other Central American countries, the proportion of rural population is high while the specific water use is much lower than in the USA and Canada.

Considerable volumes of water are abstracted in North America for use by industry. Industrial needs increased from 1900 to 1980. In the 1980s new technologies were introduced; at the same time industrial production decreased slightly in the USA and Canada; as a result the increase in industrial water use halted. Industrial abstractions in the overall figure rose from 31% in 1900 to 44% in 1970, then reduced to 40% in 1990. Water use by industry is grouped as: thermal power generation, manufacturing, and mining. Thermal power is the largest water user in North America, mainly for cooling. As early as in 1950, this figure was as large as that for manufacturing and mining together. In 1990, it abstracted 5 times as much as all other industries together. In the USA, thermal power provided 90% of the electricity generated (Solley *et al.*, 1993), in Canada, which is rich in hydroelectric power resources, about 40%, and in Mexico about 75% (Tryoshnikov, 1988).

Manufacturing also needs much water. The major sectors requiring water are: chemical and associated products, paper and associated products, petroleum refining, steel processing and food processing (David, 1990). These are well-developed industries in the USA and Canada; some of them are also present in some of the other countries. Mining usually uses smaller volumes of water compared to other industries. Unlike manufacturing, mining makes use of salt water and this amounts to half of all abstractions. All countries in North America have well-developed mining industries.

In the USA and Canada, the water use in the three industrial groups is usually assessed separately. Data for the USA starts from 1950, and for Canada from 1972. In the USA, abstractions for thermal power generation as compared to other industrial sectors predominated throughout the second part of the twentieth century, growing rapidly in the middle. After 1980 there was a marked decrease from 202 km³/year in 1980 to 182 km³/year in 1990. This resulted from the economic slowdown as well as from increasing use of sea water. Thermal plants use 73% and nuclear 27% of abstractions for thermal power production in the USA. Changes in abstractions are given in Fig. 7.10.

Abstractions for manufacturing and mining in the USA grew until 1970 when they reached 66 km³/year including about 11 km³ of

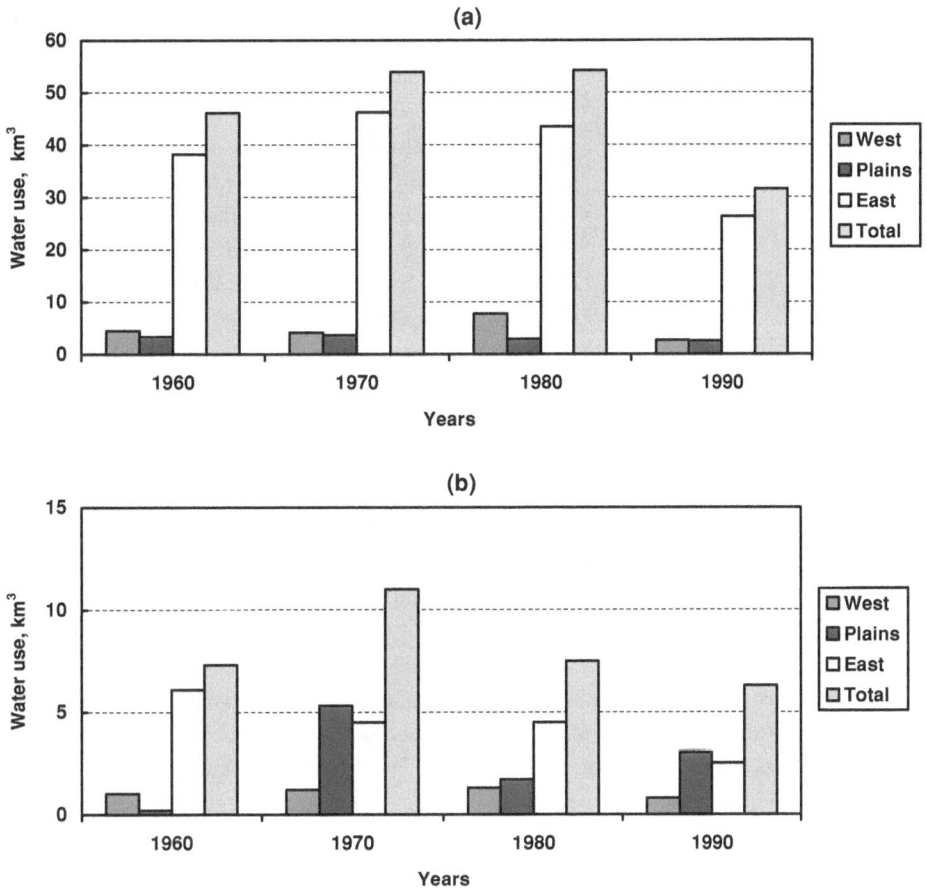

Fig. 7.11 Dynamics of industrial and mining water use in the USA by regions, 1960–90. (a) Fresh water; (b) salt water.

Table 7.3. *Water discharge per unit of production for five major water-using manufacturing sectors in the USA*

Manufacturing sector	1954	1960	1970	1980	1983
Chemical and allied products	830	590	340	190	150
Paper and allied products	330	270	180	140	130
Petroleum refining	210	170	120	90	80
Steel processing	580	560	450	270	260
Food processing	100	90	80	60	50

Source: David (1990).

sea water (Fig. 7.11), but after 1980 these volumes began decreasing. Water use by industry depends on the following: gross output, the amount of water used per unit of production, and the number of reuse cycles (the water intake). Gross output in US manufacturing grew until the mid-1970s. The slowdown in the economy resulted in a decrease in steel processing, and in branches of the chemical industry and in petroleum refining. Gross water use per unit of production (abstraction + reuse) in the 1950s remained constant and began to decrease from the early 1960s (David, 1990). Abstraction per unit of production steadily reduced from the 1950s to the 1990s (Table 7.3). Changing products, technology and legislation caused this trend. Notably the chemical industry reduced use of water while, after 1965, most water was used in the production of steel. In most industries the number of reuse cycles has risen significantly since the middle of the twentieth century (Table 7.4). The greatest number of such cycles is in the petroleum industry, followed by the paper and chemical industries. According to David (1990), the largest volume was abstracted in 1968, some 59.4 km^3/year, by 1990 this figure had fallen to 27.0 km^3/year. The mining industry used the most water in 1973 (6.4 km^3/year),

the total falling to 4.6 km^3/year in 1990. The amount used per unit of production also dropped from 400 to 300 million l/day (David, 1990), while the number of reuse cycles increased from 2.5 to 3.0. Most of the water used for manufacturing and mining is abstracted in the eastern states, but the proportion used in the Plains has increased during recent decades. The West uses salt water.

Canada is rich in hydroelectric resources and uses them efficiently. In 1950, thermal power accounted for only about 4% of all the country's electricity; the proportion started growing to reach 23% by 1970, and was about 40% by 1995. The highest rate

Table 7.4. *Ratio of gross water use to water withdrawal for the manufacturing industry in the USA*

Manufacturing sector	1954	1960	1970	1980	1983
Chemical and allied products	1.6	1.7	2.4	2.8	2.8
Paper and allied products	2.4	3.0	3.1	4.8	3.9
Petroleum refining	3.3	4.4	5.6	7.2	7.5
Steel processing	1.3	1.5	1.7	2.1	2.5
Food processing	2.1	2.0	1.8	2.0	2.2
Total	1.8	2.2	2.6	3.4	3.4

Source: David (1990).

of abstraction for power production took place between 1960 and 1970, then the rate began to drop; in 1990 it was 28.5 km^3/year. In Canada the maximum demand for water for manufacturing occurred in 1981. The total was 10 km^3/year (Tate and Lacelle, 1992), but this was followed by a decline for the same reasons that occurred in the USA. The number of reuse cycles in Canadian manufacturing was about two. The major water user is the paper industry (about 40% of all water abstracted for manufacturing), followed by the chemical industry (about 20%) and metallurgy (about 20%) (Tate and Lacelle, 1992). Use by mining is rather small as compared to other Canadian industries; it reached a maximum of 0.7 km^3/year in 1976 (Tate and Lacelle, 1992) and since the 1980s the volume has remained constant at about 0.6 km^3/year. The number of reuse cycles in Canadian mining is 4–5 which is much higher than in the USA. Total industrial abstractions in Canada as a whole and in the regions rose until 1986, then leveled out and later slowly declined (Tate and Lacelle, 1987, 1992). Industrial use is greatest in Ontario, and this province accounted for 70% of all abstractions.

Mexico and the Central American countries abstract much less water for industry than the USA and Canada. In Mexico the total was 5.9 km^3 in 1990. However the rate of increase was rather high in recent decades and this pattern is likely to pertain in the near future. There is little thermal power production; most water is used in mining, petroleum refining and food production. Hydroelectric power is well developed in a number of the Central American countries, the most being in Mexico (IWPDC, 1991).

North American industry consumes about 4–5% of the water abstracted, this proportion being lowest (1–2%) in thermal power production. It is somewhat higher in manufacturing (10% in the USA). In mining the water drained from the mine often exceeds the volume abstracted for use.

The continent's large urban population provides a high demand for public services which include both municipal water use in general, and domestic use in particular.

The USA has the highest demand for domestic water. In 1950, 111 million people were served by the public supply system and by 1990 this number had risen to 207 million. Abstractions grew from 19.6 km^3 to 53.9 km^3/year, i.e. 2.7 times, between 1950 and 1990. The specific domestic water use increased from 485 to 700 l/day per head between 1950 and 1980, and this figure has remained relatively constant. Figure 7.12 shows data on public water supply trends in the West, Central and East regions of the USA. In Canada, the volume of the domestic water increased 3.5 times from 1950 to 1990 when it reached 5.9 km^3/year. The number of people served by the public supply grew from 10 million to 23 million people, while the specific domestic water use rose from 460 to 700 l/day per head.

Mexico is characterized by a high rate of population growth especially in the cities. In 1950, only 40% of the population was urban, but in 1995 it was about 70%. Urban agglomerations are denser in the central and northern parts of the country. In the north, their growth was caused by industrial development but restricted by a significant water resources deficit. There are two major metropolitan areas in the centre of the country, Mexico City and Guadalajara, and their total population exceeds 50 million people. The problem of meeting the water needs of the urban population is severe all over the country. The present day rate of specific domestic use is about 200 l/day per head.

In most countries of Central America, the percentage of urban population is much lower than in Mexico, but its growth rate is very high. The specific domestic use in the towns and cities of most Caribbean countries does not exceed 100 l/day per head; but on the continent it is 100 to 300 l/day per head. In fact, the rate of growth of specific water use in all the Central American countries has been low during the recent decades because of the rapid population growth. One of the main problems of public supply in the Central American and Caribbean countries is the poor quality of drinking water and the poor sanitary conditions. Less than 10% of the sewage systems in Latin America are equipped with treatment plant and this gives rise to frequent epidemics. There are programmes in most Central American countries to improve drinking water quality and to build efficient sewage treatment systems. However, according to ECLAC (1995), it is unlikely that adequate sewerage systems will be provided for most of the towns and cities before 2020. Improving public water supplies remains an urgent problem in most Central American countries.

In order to make North America water resource management more efficient, a great number of reservoirs have been built, for seasonal and long-term storage (Table 7.5). The earliest dams and reservoirs were built in the latter half of the nineteenth century. First, they were designed to perform one or, at most, two functions; but later, multi-purpose designs became viable. Different functions are now fulfilled by reservoirs such as hydroelectric power production, flood control, fish and wildlife conservation, augmenting low flows, meeting recreational needs, irrigation, municipal

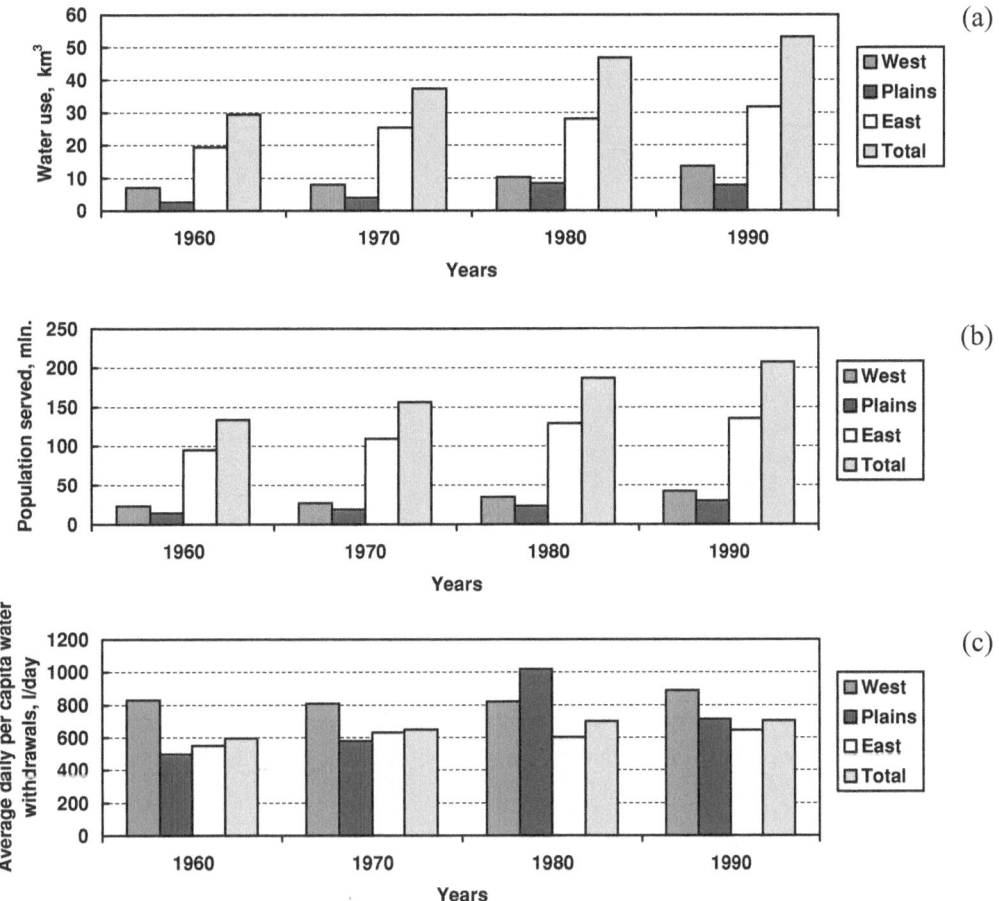

Fig. 7.12 Dynamics of (a) domestic water use, (b) population served and (c) average daily per capita water withdrawals in the USA.

and industrial supply (Foxworthy and Moody, 1986). The total number of reservoirs in North America now is about 60 000. However, 95% of them are small, with a capacity of less than 5000 m^3. There are about 800 large and medium-size reservoirs with capacities over 100 million m^3, and their combined normal storage capacity approaches 1600 km^3, with a total surface area of 200 000 km^2. Most of the reservoirs are located in the USA and Canada.

In the USA, reservoir construction started in the second half of the nineteenth century and was particularly rapid after 1930. Figure 7.13 shows the cumulative normal storage for reservoirs in the USA for the period 1880 to 1985. The most rapid growth took place in the 1960s, but after 1977 it slowed down. This was caused by lack of suitable sites and by the change in the Federal water policy to give more emphasis to protecting the natural environment (Beard, 1994). By the 1990s the total reservoir volume, including storage capacities, was over 900 km^3. Approximately 2500 reservoirs and controlled natural lakes with capacities over 0.006 km^3 provide a combined normal storage capacity of more than 590 km^3, with around 600 of the largest reservoirs ac-

counting for almost 90% of the total storage. There are also about 50 000 smaller reservoirs with capacities ranging from 0.000 06 to 0.006 km^3, and about 2 million smaller farm ponds used for storage (Foxworthy and Moody, 1986). Large and medium-size reservoirs are located mainly in the mountains or highlands. A characteristic of the US reservoirs is their high control ability, due to a significant excess of the useful storage capacity over the clearance volume. Most reservoirs in the USA with capacities over 100 000 m^3 are equipped for seasonal and long-term flow control.

Numerous reservoirs in the USA (up to 70%) are used for flood control, hydroelectric power generation and water supply. For power production there are systems of reservoirs in series, particularly on the Tennessee, Colorado and Columbia Rivers. Other reservoirs are used for irrigation, recreation, navigation and fisheries, and during the last 20–25 years a number of important reservoirs have been built. The largest reservoir in the USA is held back by the Hoover Dam on the River Colorado, which was built in 1936 with a capacity of 34.8 km^3; however, it is far down on the list of the world's largest reservoirs. Figure 7.14 shows the principal reservoirs and the ratio of the storage capacity to water resources by water management regions (Wolman, 1990).

Table 7.5. *Principal reservoirs of North America*

Reservoir	Country	Basin	Year of filling up	Backwater height, m	Reservoir capacity, km³	Type of use[a]
Daniel Johnson	Canada	Manicouagan	1968	214	141.8	E N F
WAC Bennet	Canada	Peace	1967	183	70.3	E F N
La Grande 2	Canada	La Grande	1978	168	61.7	E
La Grande 3	Canada	La Grande	1981	93	60.0	E
Caniapiscau	Canada	Caniapiscau	1981	56	53.8	E
Hoover	USA	Colorado	1936	221	34.8	I E F A
Glen Canyon	USA	Colorado	1966	216	33.3	E A F R
Kemano	Canada	Nechako	1952	104	32.7	E
Churchill Falls	Canada	Churchill	1971	32	32.3	E
Nechaco	Canada	Nechaco	1953	25	32.2	E
Jenped	Canada	Wilson	1975	30	31.8	N E
Garrison	USA	Missouri	1956	62	30.1	F I P E A
Iroquois	USA, Canada	St Lawrence	1958	20	30.0	E N A
Oahe	USA	Missouri	1962	75	28.8	E F N R
Mica	Canada	Columbia	1976	175	24.7	E A
Kenney	Canada	Nechako	1952	104	23.7	E
Fort Peck	USA	Missouri	1937	76	23.0	F E I N
La Grande 4	Canada	La Grande	1984	125	19.4	E

[a] E, Hydroelectric power; N, navigation; F, flood control; I, irrigation; A, accumulation; R, recreation; P, public supply.

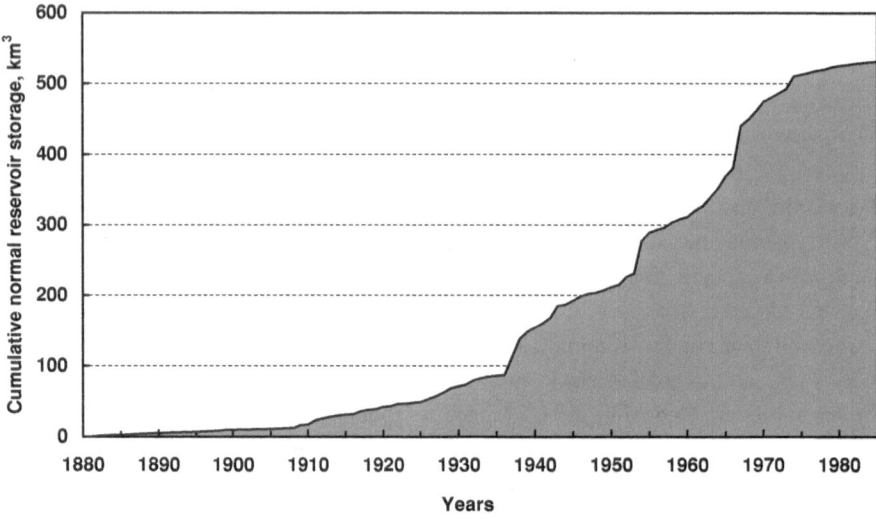

Fig. 7.13 Cumulative normal storage of reservoirs with at least 0.006 km³ capacity in the USA, 1880–1985. (From US Geological Survey, 1990, fig. 75.)

Canada is second to the USA in number of reservoirs, but they are similar in their combined capacity. The latter is due to not only the presence of very large reservoirs in Canada (with capacities over 50 km³) but also the use of controlled lakes. There are about 100 reservoirs with capacities of 0.1–1.0 km³ and more than 50 with capacities over 1 km³ in Canada (Pearse *et al.*, 1985). Most

Canadian reservoirs were built in the 1950s. They are most numerous in Quebec, British Columbia and Newfoundland–Labrador. The largest reservoir is the Daniel Johnson, which was filled in 1968 and has a capacity of 141.8 km³. About 80% of Canadian reservoirs are used to generate hydroelectric power, a little under 10% are used for irrigation, still fewer reservoirs are used to meet the needs of domestic and industrial water supply. Canadian reservoirs are largely single-purpose; multi-purpose ones are fewer in number.

Among the Central American countries, Mexico has most reservoirs: there are more than 1000 with a combined capacity of

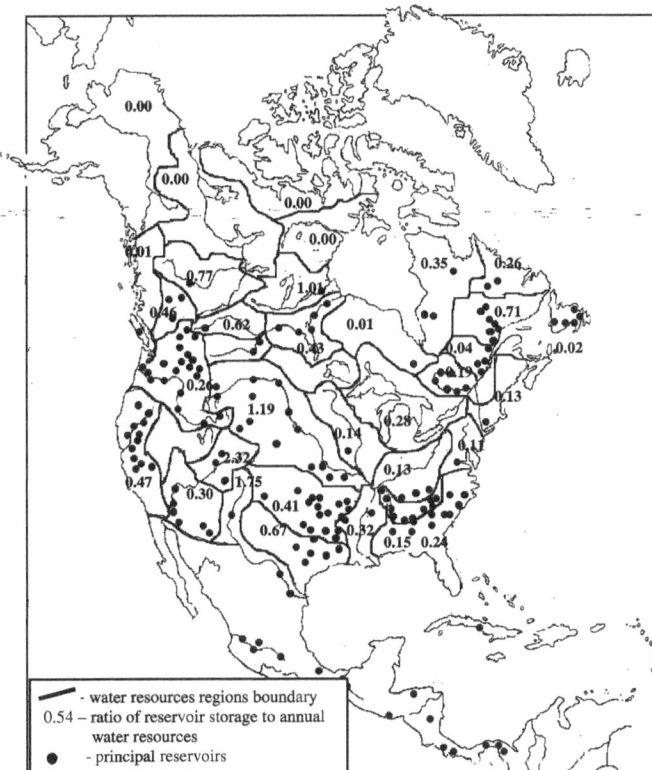

Fig. 7.14 Principal reservoirs of North America and ratio of reservoir storage to annual water resources.

140 km³. They are widespread and used predominantly for irrigation. During the last 20–25 years a number of multi-purpose reservoirs have been built in Mexico, primarily for power generation, water supply and flood control. There are about 100 reservoirs in Cuba mainly used for irrigation, with a few of 100 million m³ capacity. In other Central American countries the reservoirs are much fewer and most have capacities under 100 million m³. Guatemala, Honduras, El Salvador, Nicaragua, Costa Rica, Panama and the Dominican Republic each have a few reservoirs used for power generation or irrigation as well as several multi-purpose reservoirs.

Of course water is lost by evaporation from the surfaces of these reservoirs. The amount of water evaporated depends on water surface area and on the local climate. The annual value of the loss due to evaporation from reservoirs was estimated to be 27 km³ in 1980.

7.4 HYDROLOGICAL DATA

The hydrological network of North America contains about 25 000 stations, with various lengths of records. About 80% of them are in USA and the network is also well developed in the southern provinces of Canada (Manitoba, Ontario, Quebec, Saskatchewan

and Alberta). In the central part of the continent, there is one station on average for each 500–1000 km². The network is especially dense on the Atlantic coast of the USA, where there is one station per 200–500 km². In Mexico, where the network is sparse and the length of record is shorter, the density is one station is per 2000–5000 km². Short-term and intermittent records are available for some rivers in Central America and the Caribbean; however, the density of the network is generally higher than in Mexico with one gauge station is per 500–2000 km². The northern regions of Canada have very few stations, with one station per 20000–50 000 km².

The earliest records for Canadian rivers date from 1854 (St Mary's – Sauf St Mary's, Niagara – Queenston and others). Since 1904 the Department of the Interior of Canada has been carrying out surveys of river flow and of depths of lakes, harbours and reservoirs to meet the needs of navigation, irrigation and fisheries. In 1908, the Parliament of Canada made a decision to measure discharges in the provinces of Alberta and Saskatchewan, and in 1911, the Hydrological Service was founded (Pearse *et al.*, 1985). Most of the Canadian network is in the south of the country. Observations in the centre and north were started in 1951. There are quite a lot of sites with data for 10–20 years, but the density of the network is far from adequate in the north. As a matter of fact, there are no permanent stations in the entire Arctic Ocean coastal region and on the vast area of the Canadian Arctic Islands.

Four times a year Canada publishes runoff records. Observations of surface water quality have been made since 1934. In 1905 the International Commission on the Canada–USA waterways worked out the basic principles of joint use of the border water resources of the Great Lakes–St Lawrence River system. In subsequent years the two countries started close co-operation in the fields of hydrology, water use, water protection and water quality recovery in the Great Lakes Basin. General assessments of the water resources of Canada have been published by Pearse *et al.* (1985), Prowse and Ommaney (1990) and in the *Hydrological Atlas of Canada* (CNC/IHD, 1978). In Canada, much emphasis is given to the socio-economic value of water resources. Of particular interest are the publications of Tenant (1976) and Mitchell and McBean (1985). The impact of human activities on water and water resources has been considered by Hare (1984), and the problems of water transfers and related water resources changes have been analysed by Clark (1987) and by the Saskatoon Hydrological Congress (held in 1987), and others. Assessments of the water resources of Canada have been given by Bruce (1974), Dreier (1978), Pearse *et al.* (1985) and Wolman (1990).

In the USA, the Geological Survey is the main body responsible for collecting and publishing hydrological data. However, data on discharges and surface water condition are also collected by several other Federal agencies. In 1840 the US Federal Government started the systematic collection of quantitative data on water bodies for navigation purposes, commencing with the Ohio and

Mississippi Rivers. Water resources have been determined since 1860 in Eastern states and since 1978 in the West, principally for irrigation. The intensive use of water for industrial and agricultural purposes since the early twentieth century has depleted water resources. Therefore the estimation of water use was recognized an urgent problem in the 1950s. In 1968, the Commission on Water Policy was set up and in 1978, National System of Hydrological Information Exchange was established.

The average period of record for the principal rivers of the USA (Mississippi, St Lawrence, Colorado and Columbia) including rivers on the Atlantic coast is 60–120 years. Runoff series for rivers in inland regions of the USA is shorter, and the network density is much less. Every year there are 16 publications with information on runoff in the USA. Since 1983, the US Geological Survey has published annually the *National Water Survey* giving the condition of water resources for the whole country and for the regions and states, as well as summaries of surface and ground water data, their quality quantity and use. The USA water resource estimations are summarized in the *National Water Summary* (US Geological Survey, 1986, etc.) and in the US Water Resource Council (1968, 1978). The water resources of the continent were estimated by Wolman (1990) and the methodology employed for water resources assessment is described by Fiering (1983). Besides these publications, long-term runoff variations were produced by Langbein *et al.* (1949), Busby (1963), Hardison (1972), Langbein (1982), Meko and Stockton (1984), and Lins (1985a, b). Some of these deal with the human impact on water resources. During recent decades much emphasis has been given to the effects of climate change on water resources. The following publications deal with this problem: Matalas and Fiering (1977), Stockton and Boggess (1979), Matalas *et al.* (1982), Nemec and Schaake (1982), Revelle and Waggoner (1983), USEPA (1984), Klemes (1985), Beran (1986), Frederick (1986), Gleick (1986b, 1988), Riebsame and Smith (1986), Hanchey *et al.* (1987), Peterson *et al.* (1987), Waggoner (1990), Gleick *et al.* (1995) and Izmailova, (1999). Consideration is given both to the general aspects of the impact on water resources of global climate effects and to the particular problems of alterations in national and regional water resources. In addition, the *Hydrological Atlas of the USA* provides an overall survey of water resources. Assessments of US water resources are also given in: Dreier (1978), US Water Resources Council (1978), Timashev (1983), Mann (1985) and Wolman (1990).

In Mexico, observations of runoff were first started in the 1930s and 1940s. They were carried out predominantly in the major basins of central and northeastern Mexico. In 1972, intensive studies of runoff were started, when a working group was created, with UN participation, to formulate a plan on using and developing national water resources (Gordunio *et al.*, 1984). Most time series used in that study were rather short.

In most Central American countries observations of runoff were only started in the 1960s. From 1967 to 1976 a WMO/UNDP Central American Hydrometeorological Project was carried out in six countries to develop a network of hydrometeorological stations. As a result, each of these countries has a well-developed hydrometeorological network, support for operation and servicing, as well as facilities for data processing, analysis and application (Dengo, 1982). However the recession in the 1980s led to a general deterioration in data-collection programmes with reductions in basic networks. In practice, no monitoring is undertaken in many countries in the region. There are many river basins where even the basic hydrological variables are not measured and the water balance remains unknown in most of the countries. No observations of ground water are made in most countries. According to WMO/UNESCO (1991), hydrometeorological agencies generally have a low priority in governments' programmes and a general deterioration in data-collection programmes has resulted, particularly at the level of ongoing national rather than project-based monitoring. According to the ECLAC (1995) Central American countries can be classified according to the recognition given to water resources assessment as between high (Mexico, Barbados, Costa Rica, Panama), medium (the Dominican Republic, Honduras) and low (El Salvador).

The network of stations for monitoring water quality and the records are poorly developed in most Central American countries. There are few laboratories for the control of drinking water quality and the GEMS network has ceased to function in this region, despite UNEP efforts to resuscitate it. Mexico has the best surface water quality system. Water resource estimations for the Central America region are given in: Garcia-Quintero (1950), Anaya (1967), Dreier (1978), *Plan Nacional Hidraulico de Mexico* (1981), Calvo Jullo (1990), WRI (1990, 1992, 1996, 2000), UN (1993), Babkin (1998), Izmailova (1998, 1999), Izmailova and Moiseenkov (1998), Reboucas (1998), Seckler *et al.* (1998) and Rijsberman (2000).

Significant difference in its nature and economic development enabled the continent to be divided into the three natural–economic regions: northern, central and southern (Shiklomanov and Markova, 1987) (Fig. 7.15). The Northern region occupies the northern half of the continent and the numerous adjacent islands. Its constituents are Canada – 9976 thousand km^2, Alaska (the US state) – 1519 thousand km^2, Greenland – 2176 thousand km^2. The Northern region is the largest in area (13 670 thousand km^2) and the least in population (29.0 million people in 1995). The Central region includes 48 conterminous states of the USA. Its area is 7844 thousand km^2, and the population was 261 million people in 1995. This is the most economically developed natural–economic region in the world. The Southern region unites Mexico and the continental and island countries of Central America. Its total area is 2735 thousand km^2, and the population was 163 million people in 1995.

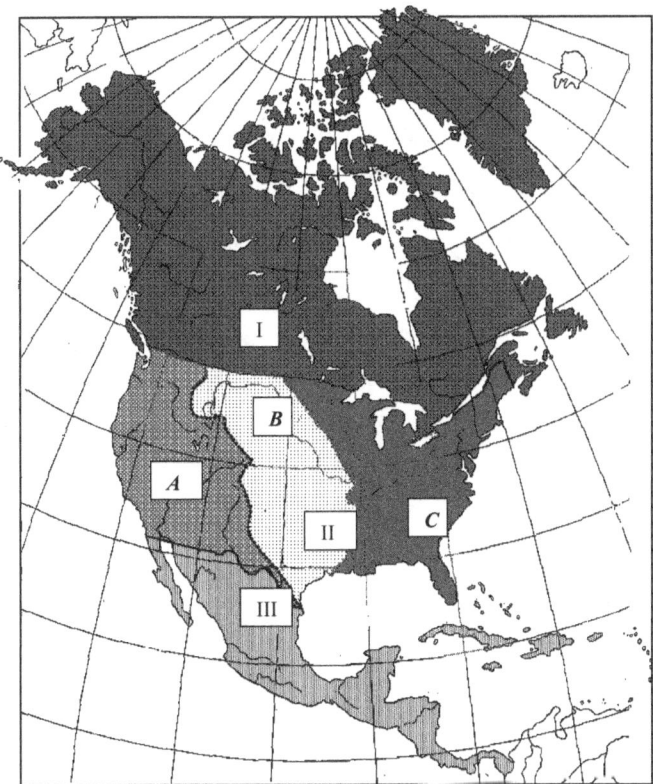

Fig. 7.15 The natural–economic regions of North America. I, Northern region; II, Central region; III, Southern region. A, Western USA; B, Plains; C, Eastern USA.

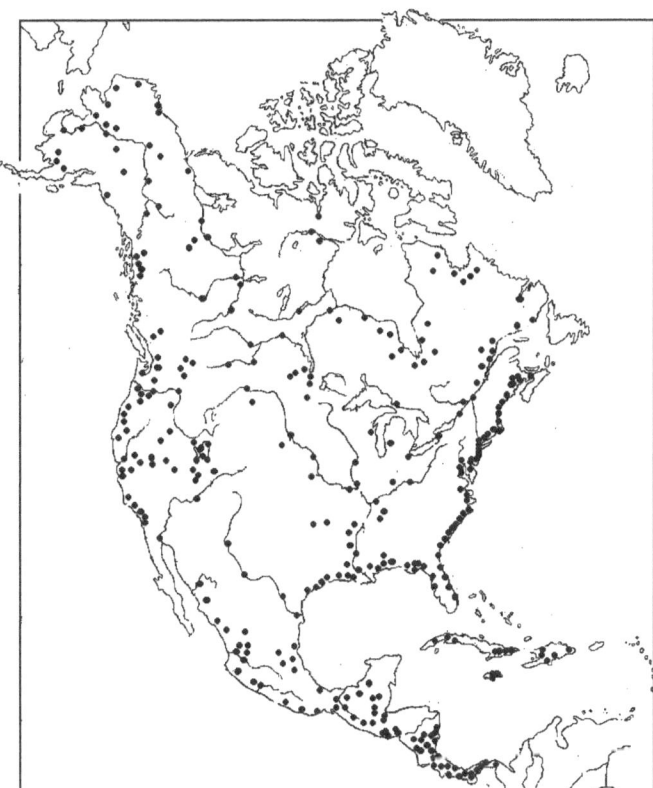

Fig. 7.16 The location of main runoff gauge stations of North America.

7.5 DISTRIBUTION OF WATER RESOURCES IN TIME AND SPACE

7.5.1 Data and methods

In this Monograph the annual runoff for the period 1921 to 1985 has been calculated for the North American continent, for the basins draining to the surrounding seas and oceans, for the natural regions, and for principal countries. Data from about 300 river gauging stations were used to assess these renewable water resources, including the 156 most reliable long-term series of flows. The distribution of these stations is shown in Fig. 7.16. Using the measured discharges estimates were made of flows at river mouths. The annual runoff values were estimated at sites where averages were available; most of the series were extended to the period 1921 to 1985. This was impossible at some sites because of the absence of nearby gauges to use as proxies and with a sufficient length of observations to calculate runoff. At these sites meteorological data were used and the runoff was determined as a residual in the water-balance equation or from precipitation coefficients. The entire continent was divided between the various river basins. The runoff from islands with no stations was determined either by using data from the continent or from meteorological data (see Chapter 2).

To compute drainage to the Atlantic, assessments were made separately of the runoff to Hudson Bay, to the Atlantic, to the Gulf of Mexico, to the Caribbean, and from the islands of the West Indies. As these regions were not uniformly covered by stations (Table 7.6), different methods had to be employed. Data for Hudson Bay were first collected after the 1950s, so the previous runoff was computed from meteorological data. Some authorities consider Hudson Bay to be a part of the Atlantic, others the Arctic. In Korzun (1974a), the basin was classified as part of the Arctic Ocean. However, an analysis of currents showed that almost all the water from Hudson Bay runs into the Atlantic Ocean. The volume of water transported through the Hudson Strait is estimated to be approximately equal to the drainage to Hudson Bay plus the flow east through the Fury and Hecla Straits (Fairbridge, 1966). Therefore, in this Monograph Hudson Bay is considered to be part of the Atlantic.

The flow from the eastern United States into the Atlantic and into the Gulf of Mexico was assessed from data for the period 1921 to 1985 with assessments of the runoff from Newfoundland and Baffin Island. Records of discharge records from Mexico began as late as the 1940s and 1950s, so estimates of runoff back to 1921 were estimated from meteorological data. The discharges of certain rivers in the Caribbean have been recorded for about

Table 7.6. *Ratio of measured to total runoff by ocean basins of North America*

Basin	Area, km² × 10³	Measured runoff area, km² × 10³	Ratio of measured runoff area to basin area, %	Total runoff, km³/year	Measured runoff, km³/year	Ratio of measured runoff to total, %
Atlantic Ocean	12 620	9400	75	4100	2525	62
Hudson Bay	3 600	2840	79	970	700	72
Central part of slope	2 480	1770	71	1485	920	62
Mexico Bay	5 160	4670	91	965	830	86
Caribbean Sea	350	60	17	390	60	15
Central American Island	330	60	18	100	15	15
Greenland	700	–	–	190	–	–
Arctic Ocean	5 450	1900	35	1170	390	33
Continental part	2 630	1900	72	580	390	67
Canadian Arctic Archipelago	1 340	–	–	190	–	–
Greenland	1 480	–	–	400	–	–
Pacific Ocean	5 300	3520	66	2590	1120	43
Alaska	1330	1130	87	980	350	36
Canada	990	690	70	750	380	51
USA	1960	1400	71	490	290	59
Mexico	850	250	29	120	35	29
Central America	170	50	29	250	65	26
Endorheic basins	880	40	5	15	2.5	17
USA	520	40	8	10	2.5	25
Mexico	360	–	–	5	–	–

30 years; however, data are sparse, so the assessments given are only approximate.

Drainage to the Arctic Ocean includes: the Mackenzie River basin; the Arctic coast rivers bounded in the east by the ranges of the Melville Peninsula; the Arctic Archipelago and a part of Greenland. Certain Canadian authors regard the Arctic coast and the Arctic Archipelago as a part of the Arctic Basin (Pearse *et al.*, 1985; CNC/IHD, 1978). However, the greater portion of the water draining into the gulfs and straits of the Canadian Arctic flows subsequently into the Atlantic Ocean.

Records of flow for rivers draining the continental part of the Arctic basin have only been made since 1943 (Mackenzie River – Norman Wells). For the rest of the basin they were started even later, in the 1950 to 1970s. To estimate the runoff to the Arctic Ocean, data were only available for four rivers draining an area of 1.9 million km². For the period without data, runoff was determined from meteorological data. The estimate of runoff from Greenland and the Arctic Islands before 1968 was taken from Korzun (1974a) as since then no more information has been collected. For the subsequent period, the runoff was assessed from meteorological data. Estimates of the average runoff to the Arctic made by Canadian authors (CNC/IHD, 1978; Pearse *et al.*, 1985) were used in this study.

The runoff to the Pacific Ocean, like that to the Atlantic, was estimated by summarizing the runoff from Alaska, Canada, USA, Mexico and Central America. The inflow from Alaska and Canada mostly occurs in a narrow coastal zone with few measurements. Flow from southern Alaska is determined at only a few gauging stations, including a number on small rivers but this very small area of about 120 000 km² produces approximately 450 km³/year of runoff. There are many stations on the rivers draining from the USA to the Pacific. Flows from Mexico have been recorded since about 1950 and from Central America since the 1960s. Earlier runoff estimates were determined from meteorological data. The water resources of the areas of inland drainage within the USA were assessed from data for the entire design period.

There is a sharp contrast between the excellent state of knowledge of the hydrology of the USA and southern Canada and the poor state of knowledge for most of the countries of Central America (Table 7.7). However for much of the USA and southern Canada there are very few basins with natural flow regimes, i.e. basins which could be referred to as analogues. Here water resources had to be determined as the sum of the observed flow data and the volume of water consumed averaged over the entire design period. This approach was used for the USA as a whole and for the water-management regions of the USA and Canada.

Table 7.7. *Ratio of measured to total runoff by regions and selected countries of North America*

Teritories	Area, km^2 × 10^3	Measured runoff area, km^2 × 10^3	Ratio of measured runoff area to basin area, %	Total runoff, km^3/year	Measured runoff, km^3/year	Ratio of measured runoff to total, %
Regions						
Northern	13 668	7520	55	4980	2230	45
Central	7 844	6460	82	1800	1470	82
Southern	2 735	940	35	1090	390	36
Countries						
Canada	9 976	6350	64	3290	1860	57
USA	9 363	7630	81	2930	1840	63
Mexico	1 973	790	40	346	210	61

Table 7.8. *Renewable water resources by natural–economic regions of North America*

Region	Area, km^2 × 10^3	Population, ×10^6	Water resources, km^3/year Inflow	Local Average	Maximum	Minimum	Potential water availability, m^3 × 10^3/year Per 1 km^2	Per head
Northern	13 668	29	130	4980	5830 (1957)	4360 (1953)	364	174
Central	7 844	261	70	1800	2480 (1972)	950 (1931)	230	7.03
Southern	2 735	163	2.5	1088	2000 (1958)	530 (1949)	398	6.69
North America as a whole	24 247	453		7870	8820 (1958)	6660 (1953)	324	17.4

The records of all of the rivers employed to estimate the continent's water resources were checked for stationarity, homogeneity and trends using the methods described in Chapter 2. The same test was carried out on the series of the total runoff. Then, to show up long-term variations an analysis was made of the runoff series for most of the major river basins and the values calculated for drainage to the different oceans, from regions and selected countries. In addition an analysis was made of the transition probabilities from one group of years to another. The runoff for individual years was classified using the following coefficients for the different river basins:

$K < 0.9$ – low water content year
$0.9 < K < 1.1$ – average water content year
$K > 1.1$ – high water content year.

It is acknowledged that estimates for regions and countries are lower than those for individual river basins and so the classification adopted was:

$K < 0.95$ – low water content year
$0.95 < K < 1.05$ – average water content year
$K > 1.05$ – high water content year.

In addition, an assessment was made of the distribution in time of the variations in the distribution of water resources across the

regions and river basins. The distribution of the average monthly runoff was computed for selected rivers and those locations with a similar distribution of streamflow within the year were identified. The distribution of the mean average monthly runoff, taking into account a weighting coefficient in proportion to area, was used to determine the average value of the water resource within the year for the regions and for selected countries (Shiklomanov, 1997, 1998a, 2000a, b; Babkin, 1998; Izmailova, 1998; Shiklomanov *et al.*, 2000).

7.5.2 Regional and national water resources

The volumes of the renewable water resources are given in Tables 7.8 and 7.9 by region and country and for areas draining to the oceans. The runoffs from the Northern region were estimated to be 4980 km^3/year (local) and 5110 km^3/year (total), making up 64% of the continent's water resources. The richest areas are the southern part of Alaska and the parts of Canada draining to the Pacific; the poorest are the Prairie provinces of Canada. The Central region's water resources are 1800 km^3/year (local) and 1870 km^3/year (total), or 22% of the continent's resources; runoff from the Southern region is 1090 km^3/year, both local and total, which makes up 14% of the continental resources (Fig. 7.17).

Table 7.9. *Renewable water resources by countries of North America*

Country	Area, km² × 10³	Population, ×10⁶	Water resources, km³/year				Potential water availability, m³ × 10³/year	
			Inflow	Local			Per 1 km²	Per head
				Average	Maximum	Minimum		
Canada	9976	28.0	130	3290	3760 (1957)	2910 (1953)	330	120
USA	9363	262.0	146	2930	3680 (1972)	1960 (1931)	313	11.5
Mexico	1973	94.8	2.5	346	645 (1958)	229 (1949)	176	3.66

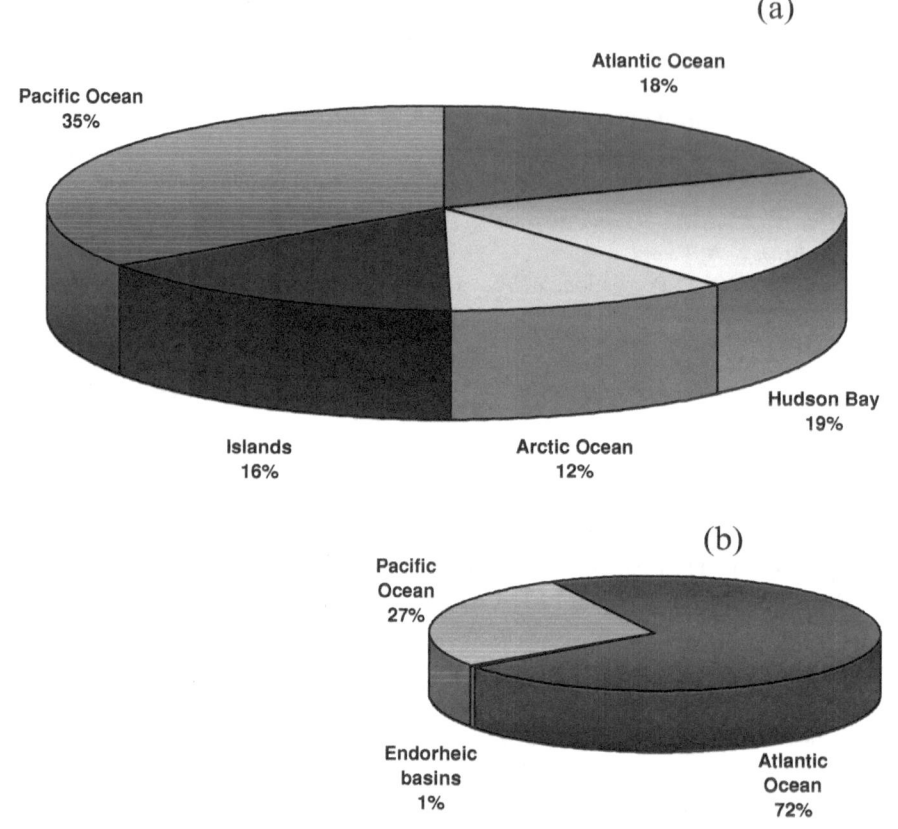

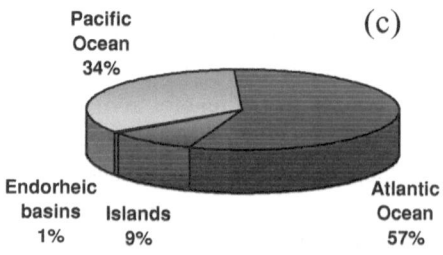

Fig. 7.17 Natural water resources by natural–economic regions of North America. (a) Northern region; (b) Central region; (c) Southern region.

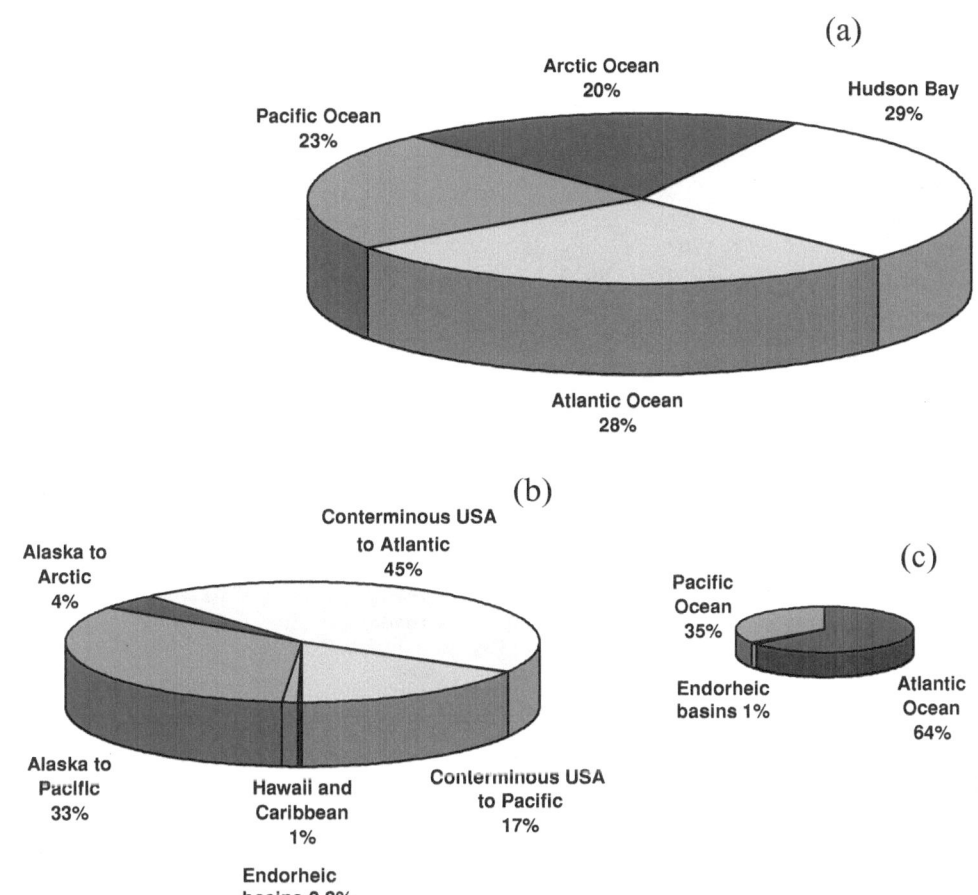

Fig. 7.18 Natural water resources by selected countries of North America. (a) Canada; (b) USA; (c) Mexico.

The water resources of Canada are assessed to be 3290 km³/year (Fig. 7.18a, Table 7.9), most draining to Pacific Ocean, the least to the Arctic. In addition to its own water resources, Canada receives some 130 km³/year from rivers originating in the USA, making a total for Canada of 3420 km³/year. Flow to the USA from Canada amounts to 146 km³/year, largely from the Great Lakes into the St Lawrence. The US water resources are 2930 km³/year (local) and 3076 km³/year (total) (Fig.7.18b, Table 7.9); 1800 km³/year (61%) flows from the conterminous USA, 1100 km³/year (39%) from Alaska and 25 km³/year from the Hawaiian Islands and the Caribbean islands belonging to the USA. Alaska has the richest resources, especially the south; the Great Basin has the least. The runoff from Mexico is 346 km³/year (local) and 348 km³/year (total) (Fig. 7.18c). Most water drains to the Gulf of Mexico. Water resources are poorest in the Mexican Highlands and Lower California; but most of the country is deficient in water resources. Several assessments of the renewable water resources of Canada, the USA and Mexico are given in Table 7.10.

Trends in the continent's renewable water resources, for the areas draining to the different oceans, for the regions and for selected countries are given in Fig. 7.19. Analyses showed that records of runoff are statistically stationary for both averages and distributions about the average. Flows from the Northern and Central regions showed very small trends with mean values for the former increasing by approximately 200 km³/year, and for the latter by 120 km³/year. In general, runoff from the continent is somewhat higher during the second part of the period (1955 to 1985) than for the first part (1921 to 1954). Different statistics of the water resources of Canada, USA and Mexico and for the various regions is given in Tables 7.11 and 7.12. Over the long term, variations in water resources show changes between wet and dry periods. The year-to-year streamflow variations are caused mainly by changes in climate and in human activities. Such variations are greatest in arid and semi-arid regions where a small change in precipitation has a large effect on runoff.

In the Northern region the average conditions from 1921 to 1956 changed in 1957 to a more humid regime. In the Central region a humid period followed a dry one from 1943 with another dry period from 1953 to 1967. In the Southern region the average conditions from 1921 to 1932 became dry in 1933 and humid from 1954. Most records are characterized by periods of average conditions, particularly in the north of the continent. The probability of a year with low water is from 25% (in the Northern region) to 45%

Table 7.10. *Water resources of selected countries of North America by the data of different authors*

Source	Year	Water resources, km³/year[a]				
		Canada	Conterminous USA	Alaska	USA, total	Mexico
Garcia-Quintero	1950					181
Anaya	1967					360
Doxiadis	1967		1660			
Bruce	1974	2900			2478	
Dreier	1978	2767	1735	645	2380	354
US Water Resources Council	1978		1830[b]			
US Water Resources Council	1980		1704	1250	2954	
Plan Nacional Hidraulico de Mexico	1981					394
Timashev	1983		1848	1251	3099	
UN *Statistical Yearbook*	1985					330
WRI (1992)	1985					357
Pearse *et al.*	1985	3300				
US Geological Survey (1986)[c]			1770	1270	3040	
US Geological Survey (1986)[c]	1986		1930*		3200*	
Prowse and Ommaney	1990	3279				
Wolman	1990	3300	1900*			

[a] Total value of water resources calculated basing on resources data of water resources regions.

[b] Restored natural runoff.

[c] US Geological Survey (1986), Foxworthy and Moody (1986) and Wolman (1990) cite the data on water budget of water resources regions of the USA. Calculations of water budget in these studies were based on runoff data for 1951–80; for water resources regions a ground water flow was also taken into account.

Table 7.11. *Statistical characteristics of water resources by regions and selected countries of North Amerrica*

Territory	Area, km² × 10³	Average runoff, km³	Coefficient of variation, C_v	Maximum (year)	Minimum (year)
Northern region	13 668	4980	0.10	5830 (1957)	4360 (1953)
Central region	7 844	1800	0.17	2480 (1972)	950 (1931)
Southern region	2 735	1090	0.20	2000 (1958)	530 (1949)
Canada	9 976	3290	0.10	3760 (1957)	2910 (1953)
USA	9 363	2930	0.11	3680 (1972)	1960 (1931)
Mexico	1 973	346	0.18	645 (1958)	229 (1949)

(in the Central region), while that of a wet year is from 20% (in the Northern region) to 40% (in the Central region). The average duration of the dry periods varies from 1.6 years (Southern region) to 2.7 (Central region), and average conditions from 1.2 (Central region) to 1.8 (Northern region); groups of wet years range from 1.1 (Northern region) to 2.3 (Central region) (Table 7.13). The longest durations of groups of years were recorded in the Central region.

7.5.3 Drainage to the oceans and flows in the major rivers

Discharges to the oceans are given in Table 7.14, and for selected countries and regions in Fig. 7.20. Flow to the Atlantic is 4100 km³/year, or 52% of the continent's total water resources (Fig. 7.20a); 3740 km³/year comes from the continent itself and 360 km³/year from the islands. Less water drains to the Atlantic than to the Pacific Ocean. The central part of the continent

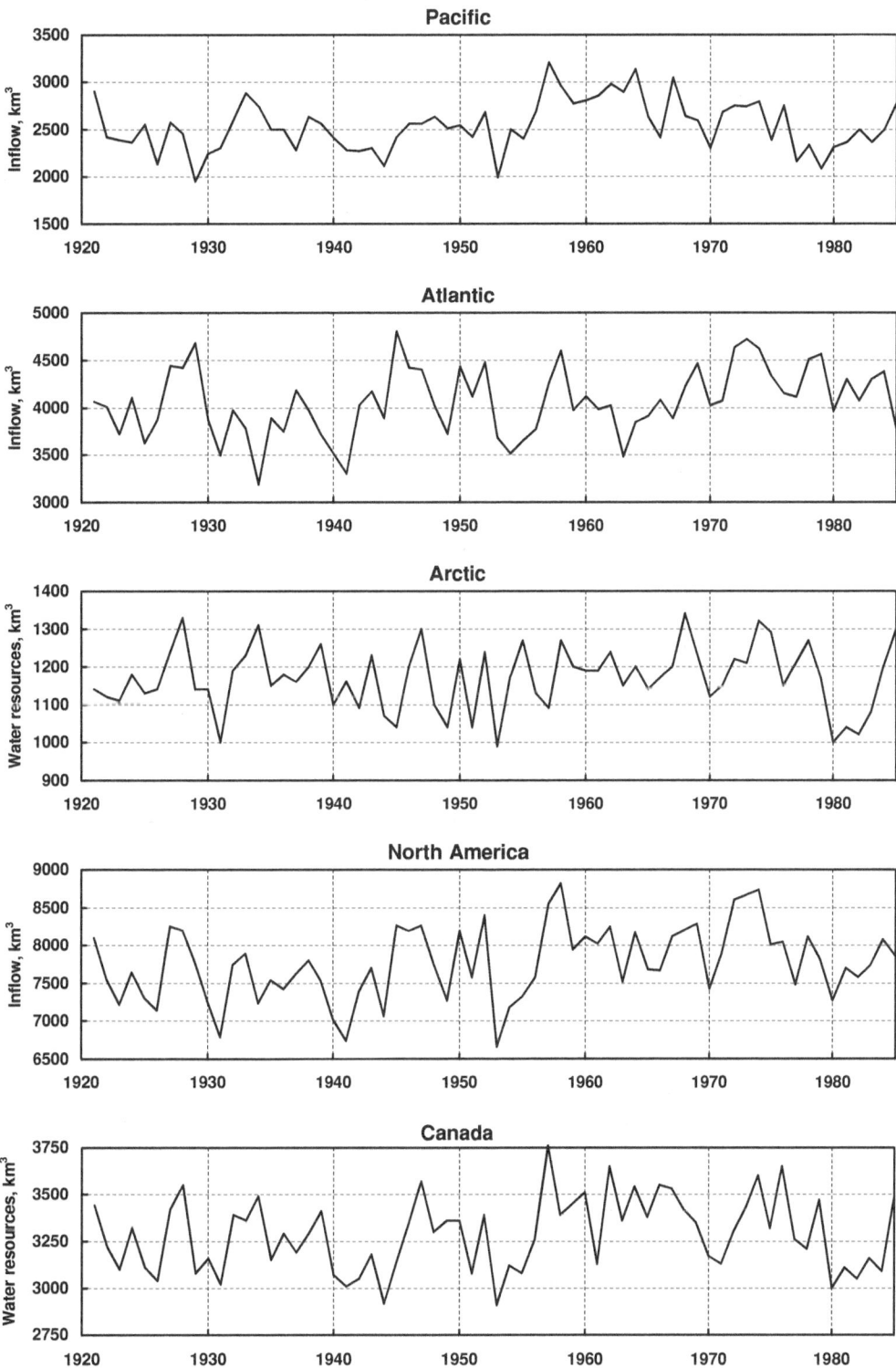

Fig. 7.19 Variations in renewable water resources (km^3/year) by ocean slopes, selected countries and natural–economic regions of North America.

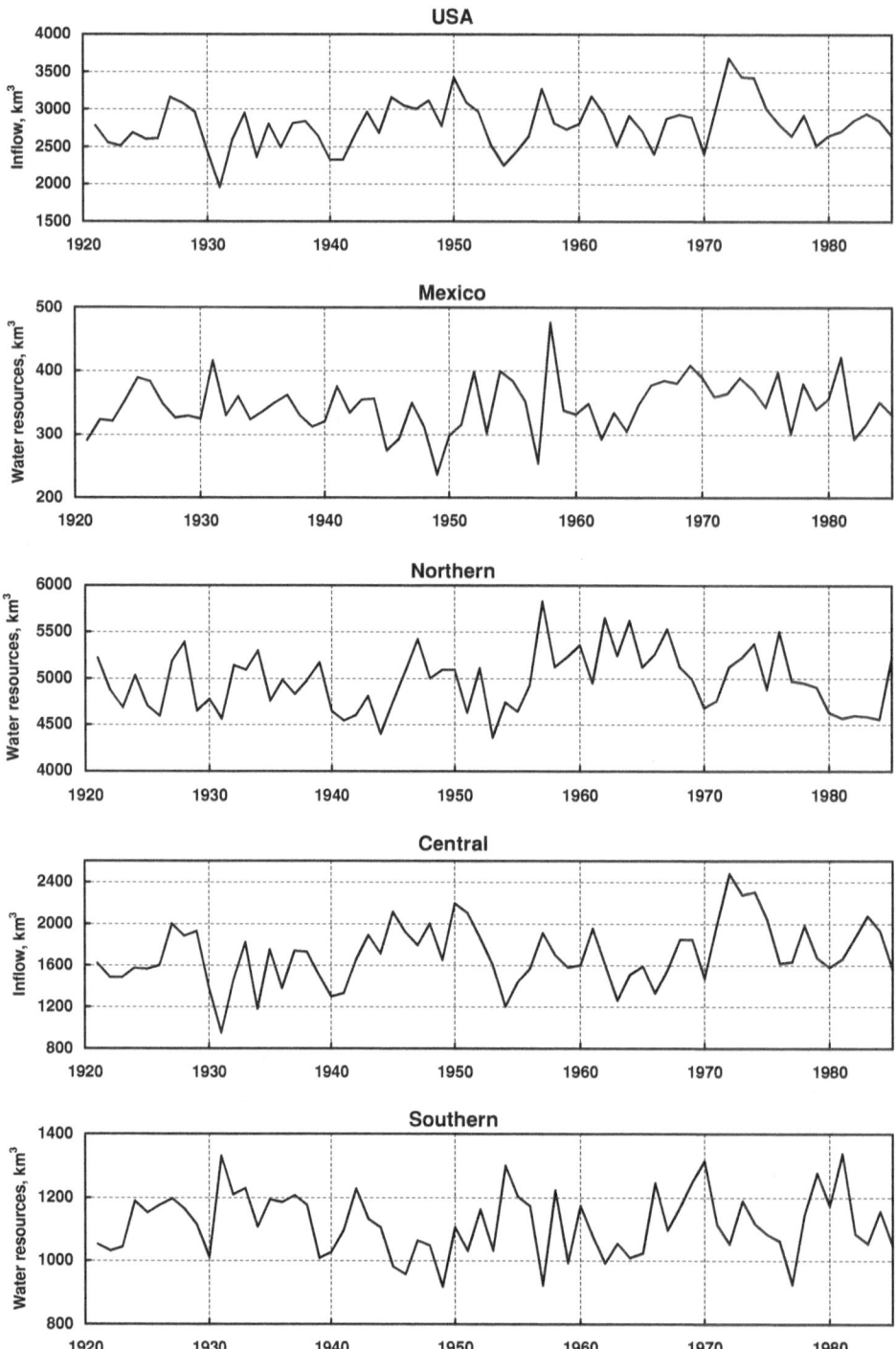

Fig. 7.19 (*cont.*).

Table 7.12. *Periods of different water content by regions and selected countries of North America*

Territory	Wet periods			Dry periods			Periods of average water content		
	Years	Ratio of mean for period to long-term mean, K_{av}	Number of years	Years	Ratio of mean for period to long-term mean, K_{av}	Number of years	Years	Ratio of mean for period to long-term mean, K_{av}	Number of years
Northern region	1957–78	1.05	22	1979–85	0.95	7	1921–56	0.99	36
Central region	1943–52	1.13	10	1921–42	0.91	22			
	1968–85	1.10	18	1953–67	0.91	15			
Southern region	1954–85	1.06	32	1933–53	0.90	21	1921–32	1.01	12
	1979–84	1.04	6						
Canada	1957–79	1.04	23	1980–85	0.94	6	1921–56	0.99	36
USA	1943–52	1.13	10	1921–42	0.95	22	1953–70	0.98	18
	1971–85	1.05	15						
Mexico	1954–61	1.11	8	1933–53	0.93	21	1921–32	1.01	12
	1979–85	1.10	7				1962–78	0.99	17

Table 7.13. *Probability of water content transition from one grouping to another by regions and selected countries of North America*

Territory	Character of water content[a]	Probability of character transition from one grouping to another, %			Average duration of water grouping, years
		1	2	3	
Northern region	1	39	61	0	1.64
	2	30	42	28	1.74
	3	8	62	30	1.44
Central region	1	63	15	22	2.70
	2	42	17	41	1.20
	3	24	20	56	2.27
Southern region	1	38	29	33	1.62
	2	45	32	23	1.47
	3	14	48	38	1.62
Canada	1	41	53	5	1.70
	2	23	46	31	1.84
	3	17	75	8	1.09
USA	1	46	36	18	1.83
	2	45	32	23	1.47
	3	15	30	55	2.22
Mexico	1	39	33	28	1.64
	2	33	41	26	1.69
	3	16	53	31	1.46
North America as a whole	1	36	43	21	1.56
	2	24	59	17	2.43
	3	6	50	44	1.78

[a] 1, Low water content grouping; 2, average water content grouping; 3, high water content grouping.

Table 7.14. *Renewable water resources by ocean slopes of North America*

Basin	Area, $km^2 \times 10^3$	Water resources, km^3/year		
		Average	Maximum	Minimum
Atlantic	12 620	4100	4800 (1945)	3190 (1934)
Arctic	5 450	1170	1340 (1968)	990 (1953)
Pacific	5 300	2590	3200 (1957)	1950 (1929)
Endorheic regions	880	15		

produces small volumes of runoff ranging from 10 to 50 mm. Flow to the Arctic Ocean is 1170 km^3/year (Fig. 7.20b), and to the Pacific is 2590 km^3/year (Fig. 7.20c), the largest volume. The southwestern part of the USA and adjoining areas of Mexico produce the least amounts of runoff.

There are some areas of inland drainage in North America, notably the Great Basin (USA) and some parts of California. These areas extend over 520 000 km^2 and there is a further 360 000 km^2 in Mexico. Precipitation is rare here and many rivers flow only after heavy rain. These are the driest regions of the continent. The Great Basin produces 10 km^3/year of runoff, 5.2 km^3/year being used. The runoff from the Mexican Plateau amounts to 5 km^3/year (Fig. 7.20d). About 3 km^3/year of runoff is utilized in the Basin of the Great Salt Lake while Nevada uses about 1.5 km^3/year, and the State of Utah some 3.5 km^3/year. Most runoff occurs in the basins of the Rivers Nasas, Aguanavaal and Santa Maria in Mexico.

The estimate of the total renewable water resources of the North America continent made for this Monograph is 7870 km^3/year; other estimates are shown in Table 7.15. According to Korzun

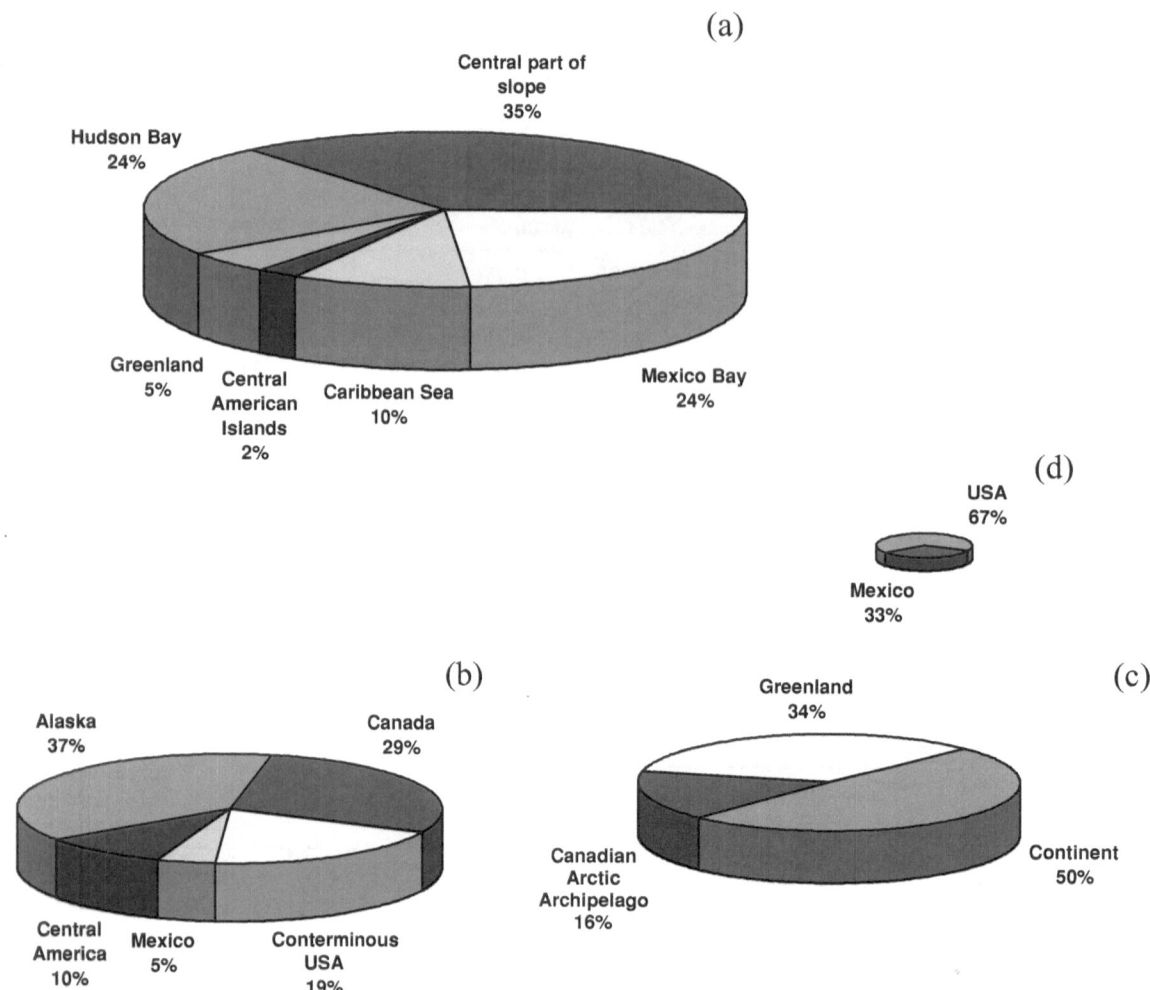

Fig. 7.20 Natural water resources by ocean slopes of North America. (a) Atlantic slope; (b) Arctic slope; (c) Pacific slope; (d) endorheic basins.

Table 7.15. *Water resources of North America by data of different authors*

Author	Year	Water resources, km³/year
Korzun	1974	8200
Baumgartner and Reichel	1975	5840
L'vovich	1974	6940
Budyko	1986	7980
Shiklomanov	1997	7870

Table 7.16. *Distribution of ocean basin inflow (km³/year) in North America by latitudinal zones*

Latitude	Atlantic Ocean	Pacific Ocean	Arctic Ocean	Total
80–90° N			50	50
70–80° N			410	410
60–70° N	370	630	710	1710
50–60° N	1410	1020		2430
40–50° N	620	480		1100
30–40° N	230	90		320
20–30° N	930	55		985
10–20° N	455	185		640
0–10° N	85	130		215
Total	4100	2590	1170	7860

(1974), the renewable water resources of North America were 8200 km³/year, with runoff to the Atlantic Ocean 3380 km³/year, 2040 km³/year to the Arctic and 2780 km³/year to the Pacific. At that time Hudson Bay was considered as part of the Arctic drainage.

The distribution of flows into the oceans by latitude is shown in Table 7.16 and Fig. 7.21, with a maximum between 40° N and 70° N. Most flow enters the Atlantic and Pacific Oceans between 50° N and 60° N, and the Arctic Ocean between 60° N and 70° N. The Mississippi makes an enormous contribution to the discharges

Table 7.17. *Statistical characteristics of oceanic basin resources and runoff of selected rivers in North America*

Basin, river	Area, km² × 10³	Average runoff, km³/year	Coefficient of variation, C_v	Maximum (year)	Minimum (year)
Atlantic Ocean	12 620	4100	0.10	4800 (1945)	3190 (1934)
Arctic Ocean	5 450	1170	0.10	1340 (1968)	990 (1953)
Pacific Ocean	5 300	2590	0.11	3200 (1957)	1950 (1929)
Endorheic regions	880	15			
Mississippi River	2 980	515	0.24	881 (1972)	281 (1931)
Mackenzie River	1 787	325	0.12	427 (1962)	284 (1970)
St Lawrence River	1 026	320	0.10	405 (1973)	242 (1935)
Yukon River	850	196	0.26	335 (1962)	122 (1954)
Columbia River	668	237	0.18	331 (1887)	144 (1977)
Colorado River (middle reaches)	637	16.0	0.38	32 (1959)	4 (1959)
Fraser River	233	115	0.13	155 (1976)	82 (1929)
North America as a whole	24 247	7870	0.10	8820 (1958)	6660 (1953)

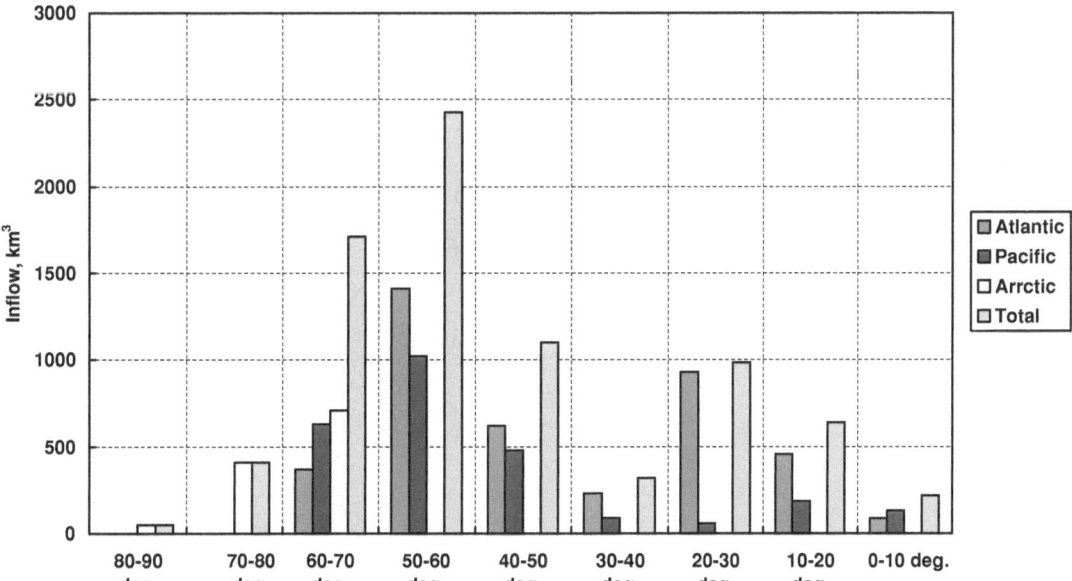

Fig. 7.21 Ocean basin inflows of North America by latitudinal zones.

to the oceans. An analysis of the variability of long-term runoff showed that almost all flow series for major North American rivers are statistically stationary, for both averages and variances. The only exception is the Rio Grande basin where the increasing abstractions have caused an ever-decreasing flow. Some negligible trends were revealed for 12 basins including the Colorado, Columbia and Rio Grande. All subsequent estimates were based on the assumption that all the runoff series being used are stationary. The same statistical analysis showed that drainage to the oceans is also stationary; a negligible trend was found in the runoff

series for the continent: the average runoff from the continent during the second half of the period (1955 to 1985) was 400 km³/year above that for the first half (1921 to 1954). Statistics for the flows in the major rivers and to the oceans are given in Table 7.17.

Long-term runoff series show long cycles of wet and dry periods, and short-term series show more drastic variations in the average runoff caused by the climatic conditions in certain years. The latter include periodically recurring droughts lasting several years that deplete ground water resources. There is also the opposite situation when several years of heavy rainfalls and high

Table 7.18. *Periods of different water content of selected rivers of North America*

Region	Period of analysis	High water content period	Years	Low water content period	Years	Average water content period	Years
Pacific Ocean							
Alaska rivers (to 60° N)	1945–85	1945–67	23	1968–85	18		
Canada Coast Ridge rivers	1928–85	1952–76	25	1928–51	24		
				1977–85	9		
Columbia River basin	1878–90	1878–04	27	1919–45	27	1905–18	14
		1946–59	14	1977–89	13	1960–76	17
US northwest rivers	1900–85	1900–21	22	1922–45	24		
		1946–85	40				
US southwest rivers	1910–85	1935–58	24	1910–34	25		
		1971–85	15	1959–70	12		
Colorado River basin	1911–85	1911–29	19	1960–85	26	1930–59	30
Lerma River basin	1948–85	1965–77	13	1948–64	17		
				1978–85	8		
Mexican and Central American rivers[a]							
Endorheic regions							
Cordilleras intermountain plateaux	1921–85	1942–53	12	1925–41	17		
		1965–85	21	1954–64	11		
Interior regions of Mexico	1942–85	1942–49	8	1950–57	8		
		1958–68	11	1969–85	17		
Arctic Ocean							
Mackenzie River basin	1945–85	1957–77	21	1978–85	8	1945–56	12
Atlantic Ocean							
Saskatchewan–Nelson River basin	1910–85	1910–28	19	1929–46	18		
		1947–56	10	1957–64	8		
		1965–76	12	1977–85	9		
Hudson Bay basin[b]							
Great Lakes basin	1860–90	1860–90	31	1891–40	50	1941–63	23
		1964–90	27				
Rivers of northeast part of Atlantic slope and Sargasso Sea basin[c]	1913–90	1933–50	18	1913–32	20		
		1971–80	10	1951–70	20		
				1981–90	10		
		1942–60	19	1913–41	29		
		1972–84	13	1961–71	11		
				1985–90	6		
Rivers of northern Great Plains	1910–85	1943–52	10	1910–42	33		
		1965–85	21	1953–64	12		
Rivers of central Great Plains	1913–85	1950–72	23	1940–49	10	1913–39	27
				1973–84			
Mississippi River basin	1913–88	1941–52	12	1930–40	11	1913–29	17
		1973–86	14	1953–72	20		
US southeast rivers	1913–90	1943–90	7	1930–42	13	1913–29	
		1958–80	22	1950–57	8		
				1981–90	10		
Rio Grande basin and northwestern Mexico Bay basin	1910–90	1919–42	24	1910–18	9		
				1943–90	48		
Rivers of southwestern part of Mexico Bay (Usumchinta, Panuko, Grehalva)	1948–85	1965–85	20	1948–64	17		
Central American rivers[d]							

[a] On other rivers of Mexico no large water content phases have been revealed; duration of dry and wet groupings is about 10 years here. It is mostly determined by local factors. For Central American rivers the available observational period is insufficient to analyse water content phases.

[b] On Hudson Bay basin rivers the observational period starts at 1951. No large water content phases have been revealed here, only short waves (of less than 10 years) determined by local climatic processes.

[c] For rivers of region 15 two types of oscillations have been revealed.

[d] On Central American rivers the observational period is insufficient to reveal large water content phases.

summer temperatures occur, and snow cover is thick in winter. A wet period is usually caused by snow and rain, or rain and glacier-melt, or snow and glacier-melt. The diagrams of long-term annual runoff variations usually show rather complicated patterns: during periods of 10–30 years there are short-term wet and dry cycles (3–6 years). Analysis of long-term data revealed 22 regions where the variations in water resources were synchronous (Table 7.18). The annual variations in the mean annual discharge of the principal North American rivers are given in Fig. 7.22.

In general, low flows were recorded on most rivers during the first half of the twentieth century, but from the 1940s to the 1950s flows have been larger. After 1980, most rivers draining to the Atlantic showed decreased flows but a number of rivers continued to experience average or high flows up to 1984, then flows decreased. The Mississippi was an exception as high flows lasted to 1986. Because the period of record finished in 1985, it was impossible to reveal later variations in runoff. An average duration of either high or low flows periods is 15–25 years. Periods when flows are about the average are less frequent; but they last for 20–30 years. Sharp variations in the regime are typical of a number of rivers, for example, the rivers flowing into the Gulf of Mexico from the west. Here the length of the cycles is about 20 years, dry and wet phases included.

The coefficient of variation for the long-term runoff of North American rivers calculated for this Monograph is from 0.10 to 0.90, compared to that of Riggs and Harvey who found values from 0.10 to 1.10 (quoted in Wolman, 1990). The coefficient of variation is higher for rivers in arid and semi-arid regions: Trinity – Romayor, 0.50; Rio Grande – Laredo, 0.52; Colorado (Texas) – Wharton, 0.56; Arkansas – Tulsa, 0.56; Red River (North) – Grand Forks, 0.62; Green River – Green, 0.66; Humboldt – Imlay, 0.80; Kansas – Desoto, 0.85; San Joaquin – Vernalis, 0.87. Figure 7.23 shows variations in the annual mean runoff for selected rivers using data collected for this Monograph and from Wolman (1990). According to Table 7.19, the flows for all rivers in the northern part of the continent are grouped closely around the mean. For arid-region rivers the amplitude of fluctuations is considerably higher, and the probability of average years is significantly less than that of years with low flows and high flows. For example, on the Colorado River, the probability of low flow years is the highest, at 47%. On the rivers with small variations in flows, dry years can be followed both by dry and average flow years, whereas for rivers with large variations, a dry year is usually followed by a dry year. On rivers with small variations in flow, years of average flow usually follow similar years while high flows follow years with high flows. The Columbia, Fraser and downstream sections of the Mississippi are the exceptions as high flows are frequently followed by an average year. The duration of groups of wet or dry years can reach 30 years. The longest groups occur in the records of the St Lawrence, Mackenzie and Colorado Rivers.

The analysis of river regimes gave the following results: the rivers draining to the Atlantic, especially short rivers, showed correlation coefficients with neighbouring rivers from 0.7 to 0.9, and from 0.3 to 0.6 for rivers farther away. No significant correlation coefficients were found for Canadian rivers. For rivers draining to the Pacific, significant correlation coefficients were found only for the flows of some small rivers with adjoining catchments. For most of these rivers the correlation coefficients were in the range of -0.3 to 0.5 (Table 7.20).

7.5.4 The distribution of runoff within the year

The within-year distribution of runoff for most rivers can be very different. Most have marked periods of high flow followed by periods of low flow within a given year. High flows usually result from heavy precipitation and snowmelt, while dry weather and relatively high evaporation cause low flows. The timing of high and low flows differs from basin to basin and depends on the climatic conditions. However, there are certain regions where the distribution of runoff during the year is similar.

In regions where the temperature is rarely lower than $0\,^\circ$C the seasonal distribution of runoff corresponds closely with the monthly distribution of precipitation (the Pacific coast of the USA and Mexico, the Gulf of Mexico coast and Central America). For example, there is a winter maximum of precipitation and runoff in streams along the Pacific coast of the USA and in the south of Canada. Conversely, most precipitation and runoff occur in summer in Arizona, Florida and the central part of Mexico. In the east of the USA runoff is usually higher in winter and early spring when there is the lowest evaporation. Figure 7.24 shows these different patterns of runoff for selected rivers. For Arctic and mountain rivers the runoff depends on snowmelt and the timing of this period governs the monthly distribution of runoff. For basins in the central and southern states the snow usually lasts for a brief period but the melting may be concentrated into a short period. For example in the Goose River in North Dakota more than half of the annual runoff usually occurs during April (Foxworthy and Moody, 1986). Some mountain rivers are fed not only by snow and rain but also by meltwater from glaciers. Glacier-fed rivers have less variable flows and longer periods of highwater. Wetlands and lakes also regulate streamflow, the extent of the regulations depending on the number and sizes of lakes and wetlands, their location in the basin and the amount of water they retain. There are many lakes in Saskatchewan, Manitoba, Ontario and Quebec, and in the states of New York and Minnesota. They produce a rather uniform distribution of streamflow particularly into Hudson Bay.

Ground water also plays an important part in streamflow regulation. The geological structure of the continent results in a ground water component of the flow of most rivers, from 15% to 30%

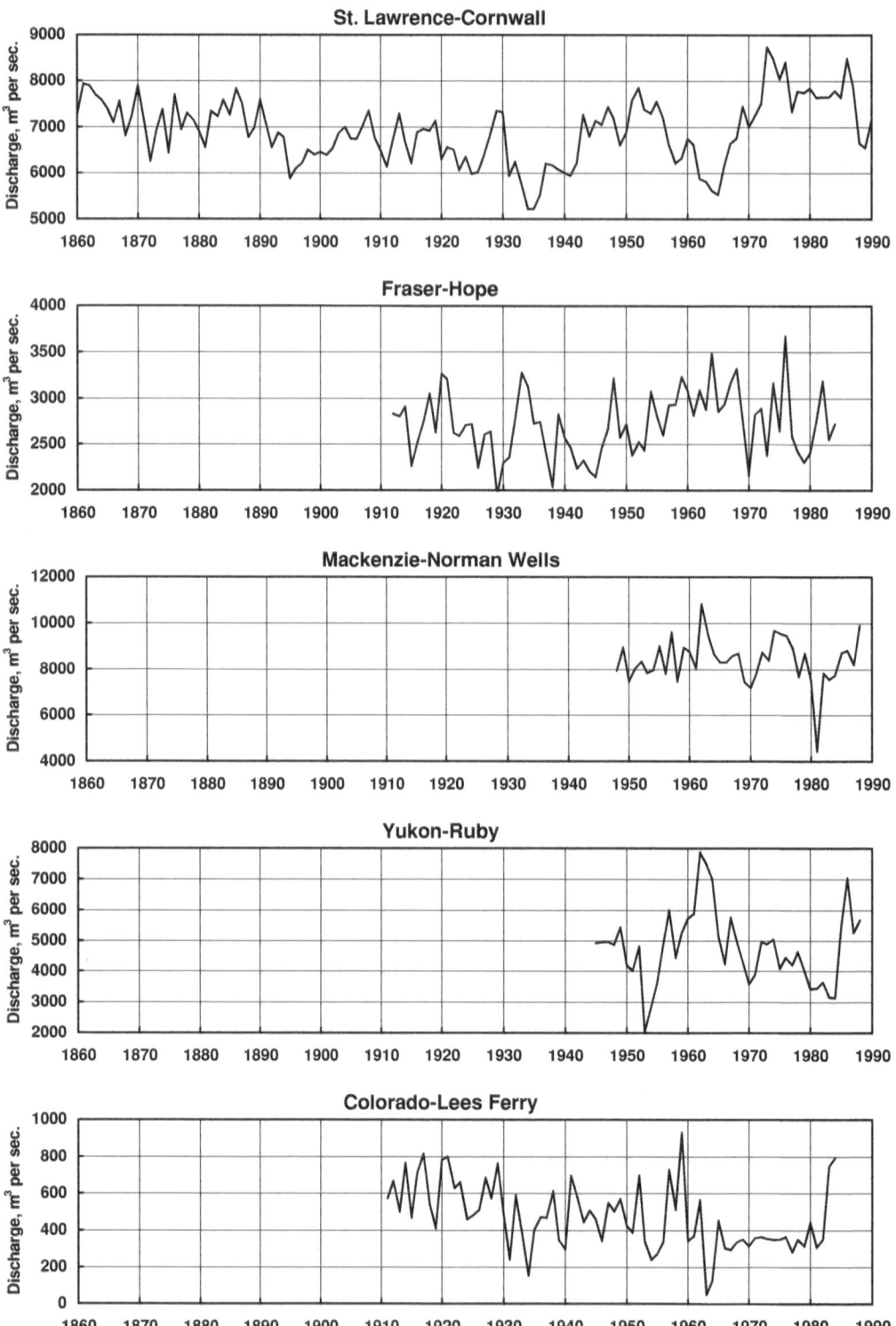

Fig. 7.22 Average discharge for selected rivers of North America.

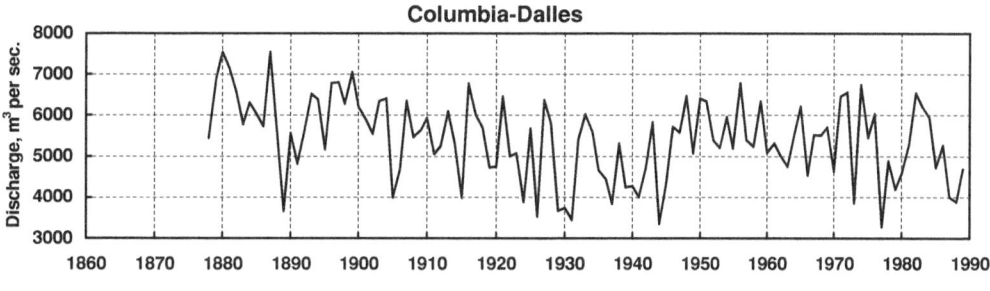

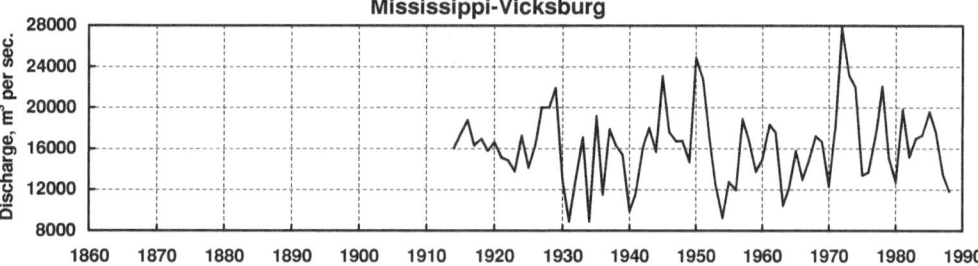

Fig. 7.22 (*cont.*).

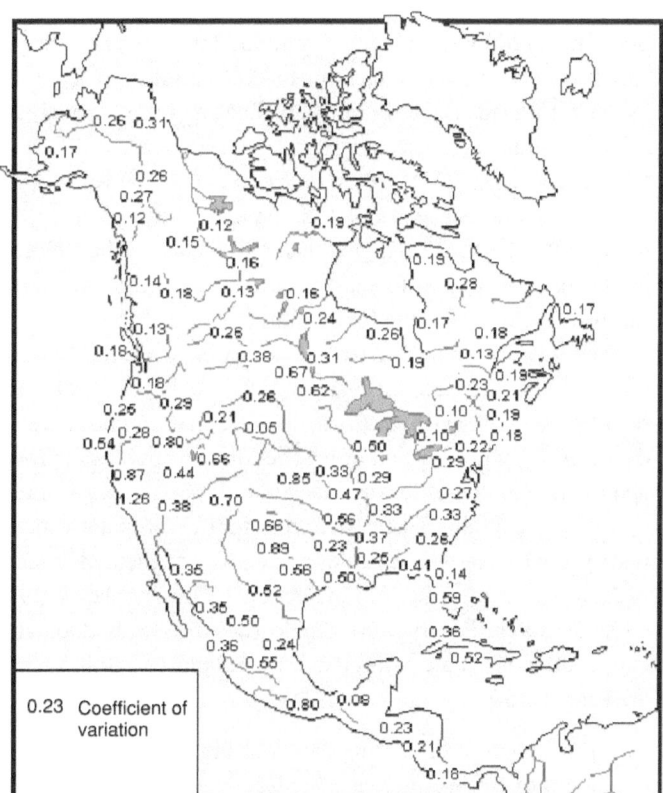

Fig. 7.23 Coefficient of variation of annual mean flows for selected rivers of North America.

(Dreier, 1978), but during dry periods it can amount to 40% of the annual flow (Foxworthy and Moody, 1986). Regions where limestone is widespread have either a wide seasonal variation in flows or an unusual distribution pattern. Limestone is most widespread in Florida, the lake regions of Missouri and Arkansas, Yucatan, Cuba and on some other islands. The rivers of Florida have a fairly uniform distribution of flow, while the pattern of flow in the Missouri responds to widespread heavy rainfalls (Foxworthy and Moody, 1986). Rivers in volcanic regions with permeable soils, such as those in Idaho, Oregon, California and Hawaii, exhibit a long delay between precipitation and flow (Foxworthy and Moody, 1986). Flows in arid regions, except for large rivers crossing these regions, are often fed from mountains or large ground water systems.

Of course the flows of many rivers in the USA and Canada are controlled by reservoirs (see Fig. 7.14), the natural variations in the volume and distribution of runoff being modified to a greater or less extent. Much water is abstracted for irrigation especially in the southwestern states of the USA and in Mexico. In these regions, flow patterns have been changed by human activities to a large extent.

Across the continent the higher flows occur from April to October, with maxima in June and low flows from November to March (Fig. 7.25, Table 7.21). The highest flows in Canadian rivers as well as in rivers of the Northern region take place from May to July coincident with the seasonal snowmelt maximum in June. Low flows last from December to March when most rivers are covered with ice. US rivers also reach a maximum between May and July, but the variations in the monthly runoffs are much lower here. Minimum flow values are recorded in November to January. In central areas high flows last from

Table 7.19. *Probability of water content transition from one grouping to another by ocean basins and by selected rivers of North America*

River, basin	Character of water content[a]	Probability of parameter transition from one grouping to another, %			Average duration of water grouping, years
		1	2	3	
Mississippi– Vicksburg	1	38	38	24	1.62
	2	31	47	22	1.88
	3	17	50	33	1.50
Mississippi– Alton	1	60	20	20	2.50
	2	29	21	50	1.27
	3	23	17	60	2.50
Missouri	1	56	25	19	2.29
	2	33	25	42	1.33
	3	15	32	52	2.07
Mackenzie	1	33	67	0	1.50
	2	11	75	14	4.00
	3	17	33	50	2.00
St Lawrence	1	60	40	0	2.50
	2	9	81	10	5.31
	3	0	37	63	2.71
Yukon	1	40	30	30	1.67
	2	15	60	25	2.50
	3	23	31	46	1.86
Columbia	1	45	34	21	1.81
	2	28	37	35	1.59
	3	12	50	38	1.62
Colorado	1	68	9	23	3.09
	2	25	31	44	1.45
	3	30	30	40	1.64
Fraser	1	44	45	11	1.80
	2	28	46	26	1.86
	3	0	75	25	1.33
Atlantic slope	1	44	50	6	1.78
	2	28	41	31	1.71
	3	11	42	47	1.90
Arctic slope	1	33	47	20	1.50
	2	18	52	30	2.06
	3	25	50	25	
Pacific slope	1	48	38	14	1.91
	2	33	50	17	2.00
	3	16	21	63	2.71

[a] 1, Low water content groupings; 2, average water content groupings; 3, high water content groupings.

March to June, snowmelt (in the north) combining with heavy rainfalls particularly along the Atlantic coast. Low flows occur from August to December. In Mexico, the variations of monthly streamflows are much higher than in the remainder of the continent. The highest flow occurs between June and October with a maximum in September, when the rainfall maximum takes place and evaporation is less. The lowest flows are recorded between February and May.

7.6 CHANGES IN WATER USE AND WATER AVAILABILITY

7.6.1 Data

The assessment and forecasting of water use and water availability has to be approached in different ways across the continent. For Canada and the USA detailed nationwide information exists, but data are scarce for Mexico and the countries to the south. For the Southern region as a whole, comprehensive data on trends and changes in water use are given by Shiklomanov and Markova (1987). The same publication overviews data for the Northern and Central regions taken from MacKichan (1951, 1957); MacKichan and Kammerer (1961); Murray (1968); Murray and Reeves (1972, 1977); Solley *et al.* (1983) and Pearse *et al.* (1985). Data on water use before 1980 were obtained from Shiklomanov and Markova (1987) and from later studies for the USA (Solley *et al.*, 1988, 1993), for Canada (Tate, 1986, 1990; Tate and Lacelle, 1992) and for the Central American countries (WRI, 1990, 1992, 1996, 2000; Gleick, 1993, 1998; UN, 1993; Seckler *et al.*, 1998; Rijsberman, 2000). The forecasts for 1990 (Shiklomanov and Markova, 1987) turned out to be 10–17% higher than actual data for the 1990s, so the latter were used to forecast water use and water availability for the present (1995) and for the future.

The forecasts for 2000, 2010 and 2025 were made for the major sectors taking into account changes in the extent of irrigation and rates of use, industrial growth, the increase in recycling as well as the decrease in use per unit of production, the population growth and the rate of domestic water use. Earlier data for water use by sectors was carefully analysed, except for the sea water which is widely used along the coasts of the USA. The methods used to make forecasts are described in Section 3.3 (Shiklomanov, 1997, 1998, 2000a, b; Penkova and Shiklomanov, 1998; Shiklomanov *et al.*, 2000); but some points that are particular to North America are listed below.

- Population forecasts are shown in Fig. 7.26 for natural regions of North America using UN data (UN, 1995a).
- GNP forecasts in US dollars per capita are based on the sustainable global development scenario (Strzepek and Bowling, 1995). The initial GNP values are those of the

Table 7.20. *Periods with different water content by ocean basins of North America*

Basin	Web periods			Dry periods			Periods with average water content		
	Years	Ratio of mean for period to long-term mean, K_{av}	Number of years	Years	Ratio of mean for period to long-term mean, K_{av}	Number of years	Years	Ratio of mean for period to long-term mean, K_{av}	Number of years
Atlantic	1945–52	1.08	8	1930–44	0.93	15	1921–29	1.01	9
	1968–85	1.06	18	1953–67	0.95	15			
Arctic	1958–79	1.03	22	1940–57	0.97	18	1921–39	1.00	19
				1980–85	0.94	6			
Pacific	1956–76	1.09	21	1921–45	0.95	25	1946–55	0.99	10
				1977–85	0.94	9			
North America as a whole	1957–76	1.03	20	1921–44	0.96	23	1977–85	0.98	9
							1945–56	0.99	12

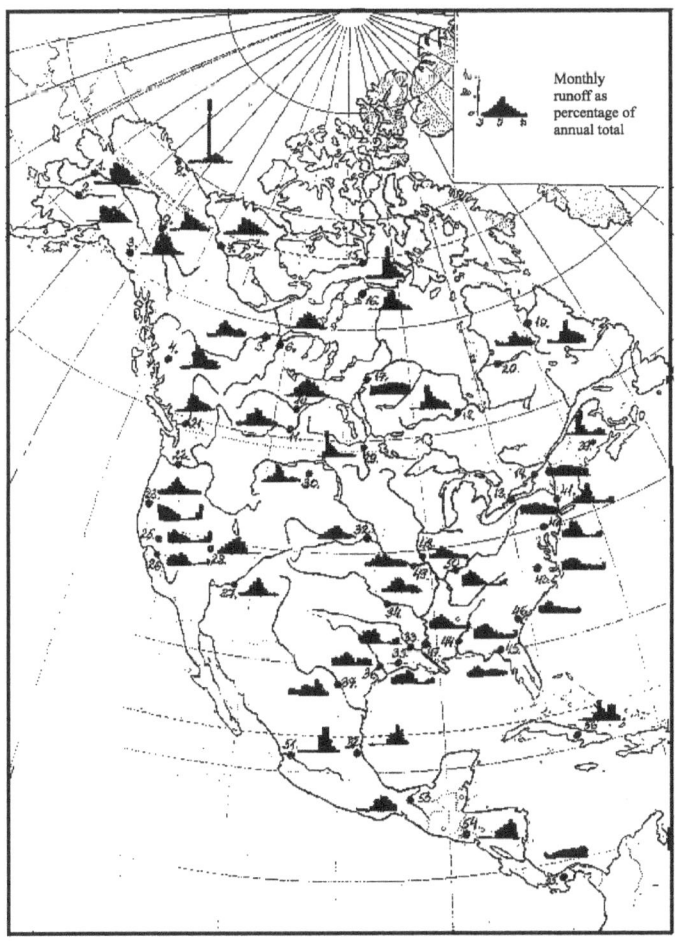

Fig. 7.24 Average monthly distribution of runoff for selected rivers of North America.

UN and International Monetary Fund (UN, 1993; IMF, 1994).

• UNIDO data on trends in industrial water use (Strzepek and Bowling, 1995) are used, where industrial water use growth rates for 2025 are compared to those for 1990. For Canada and the USA water use in 2025 is forecast to be 1.4 times usage in 1990, for Mexico 5.1, and for other Central American countries the values range from 2.8 to 6.2. For this Monograph the UNIDO rates are significantly decreased for a number of countries as a result of analyses of trends in recent years and the possibility of rationalization of future industrial water use.

The forecasts for all levels were made for every country, then summed for the entire region. Of course, the forecasts for the Central and Northern regions differ from those for the South as they are based on detailed data. Furthermore, in those countries, the growth in use of water has ceased and abstractions are decreasing, making for more reliable forecasts. For the Southern region data are extremely scarce and estimates have to be relied upon. In addition there is no evidence that the growth in use is slowing and this makes it more difficult to produce forecasts.

To determine abstractions for agriculture, changes in the area irrigated were estimated for the different periods. The changes in irrigated areas were extrapolated both in terms of the absolute (thousands of hectares) and in specific (hectares/1000 people) indices, with population and GNP changes taken into account (see Section 3.3). Account was also taken of the proportion of the area which was already irrigated against the potential area. The values of specific water use for irrigation m³/ha) were forecast against the possibilities for improvements in irrigation systems. If specific water use in the Northern, Central and Southern regions is now 4650 m³/ha, 8300 m³/ha (Solley *et al.*, 1993) and 11 200 m³/ha, respectively, the forecast figures are estimated to be 4650, 7240

Table 7.21. *Average monthly water resources distribution by regions and selected countries of North America*

| Territory | Monthly runoff distribution, % of mean annual value | | | | | | | | | | | |
	1	2	3	4	5	6	7	8	9	10	11	12
Northern region	3.8	3.6	3.5	6.1	13.1	17.2	13.8	10.3	8.9	8.6	6.4	4.7
Central region	7.7	9.4	10.3	12.1	12.1	11.5	8.6	5.8	5.2	5.1	5.7	6.5
Southern region	3.6	2.7	2.2	2.4	4.8	12.8	13.8	14.6	19.8	14.9	4.8	3.6
Canada	3.9	3.6	3.6	6.7	13.8	17.6	14.1	9.7	8.3	7.9	6.2	4.6
USA	6.2	7.4	7.7	9.3	11.7	13.2	10.2	8.1	7.1	7.1	6.1	5.9
Mexico	3.0	2.1	2.1	1.8	1.8	6.3	15.6	18.2	23.6	15.2	6.2	4.1
North America as a whole	4.7	4.9	5.0	7.0	11.6	15.2	12.6	9.9	9.6	8.6	5.9	5.0

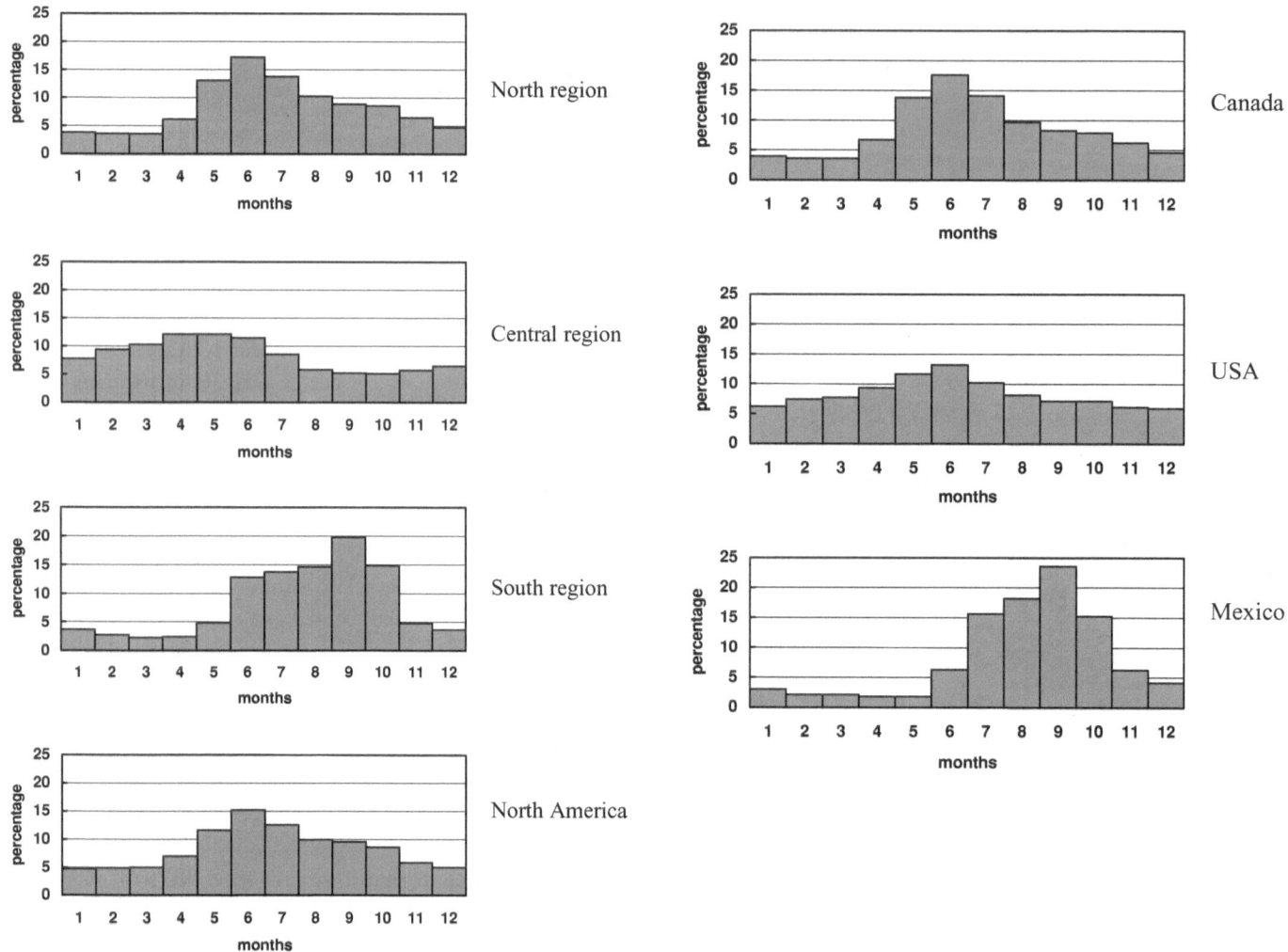

Fig. 7.25 Average monthly runoff distribution by regions and selected countries of North America.

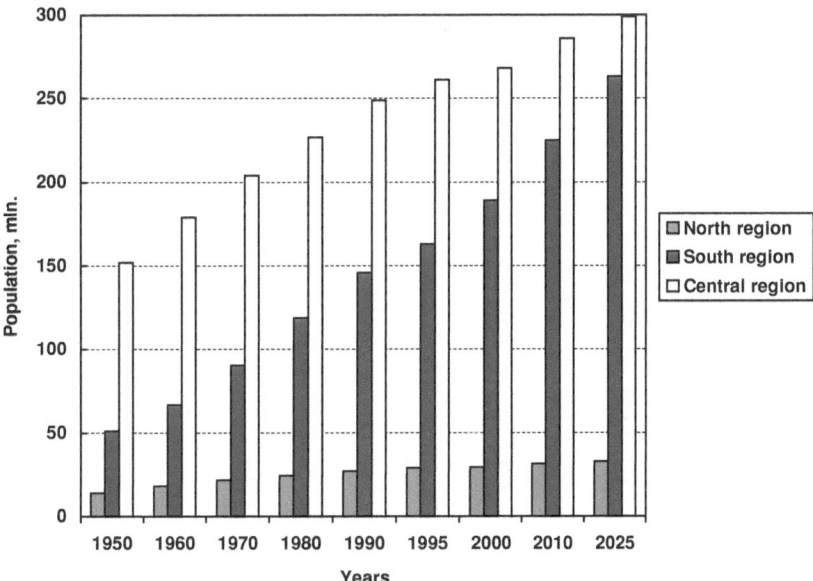

Fig. 7.26 Dynamics of population by natural–economic regions of North America.

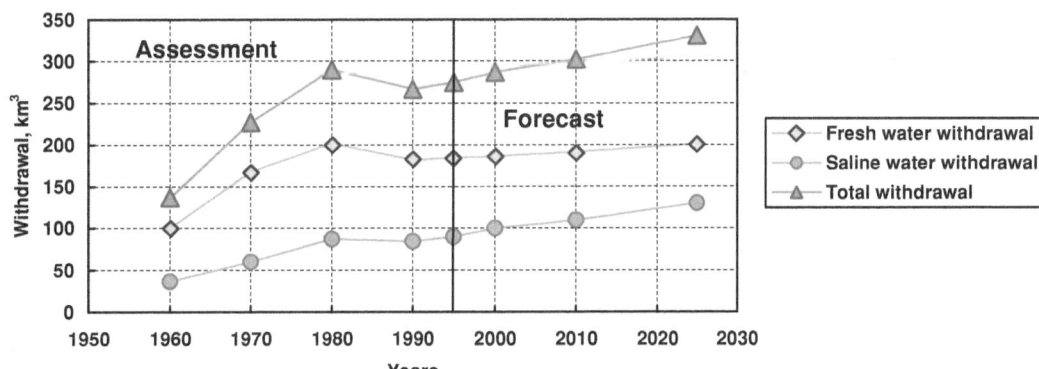

Fig. 7.27 Dynamics of water withdrawal for thermoelectric power in the USA.

and 9500 m³/ha, respectively. Irrigation water use forecast for the Central region (conterminous USA) was made separately for Western, Plains and Eastern regions and then summed using national data on changes in the area irrigated (MacKichan, 1951, 1957; MacKichan and Kammerer, 1961; Murray, 1968; Murray and Reeves, 1972, 1977; Solley *et al.*, 1983, 1988, 1993).

In the Northern, Central and Southern regions, agriculture currently consumes 77%, 56% and 71% of the total volume of water used. These percentages were assumed to apply to the future agricultural use. Industrial water use by region was estimated including use in thermal power production. Forecasts of water use took into account growth rates to 2025 using the 1990 figures as a base, as employed by UNIDO. The UNIDO figure of 1.4 for Canada was used in this Monograph, but the UNIDO figure for

the USA was reduced to 1.1, which was considered more realistic in the light of the reductions in the use of water seen during recent decades in the USA.

The UNIDO growth rates for the Central American countries range from 2.8 to 6.2, but it is likely that industrial development in these countries will be accompanied by greater economy in the use of water, so rates can be expected to be lower. Therefore, the UNIDO rates were decreased for a number of countries. For the Southern region as a whole by 2025 growth in abstractions by industry was estimated to be by a factor of 2.5. For 2000 and 2010 the figures were taken to be proportional to the estimated growth in the GNP. Estimates of future use of water were also made for certain industries in Canada and USA, such as power production, manufacturing and mining, using the trend for previous years and taking account of changes in abstractions of fresh and salt water particularly in power production (Fig. 7.27). In the case of manufacturing and mining, account was taken of decreasing abstractions per unit of production, and the increase in the number

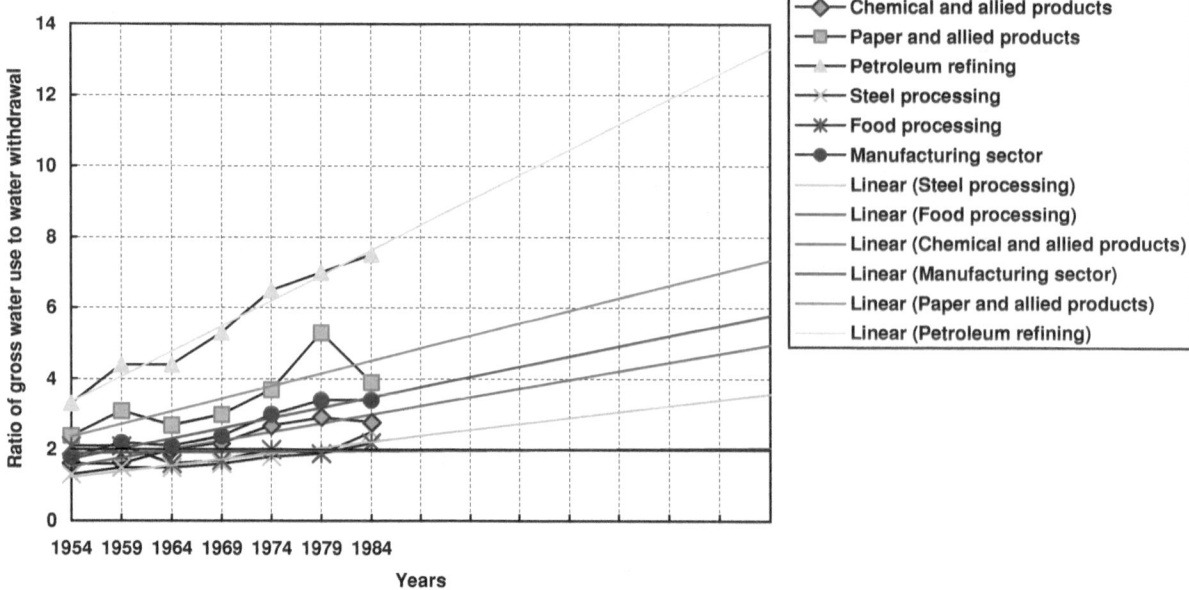

Fig. 7.28 Ratio of gross water use to water withdrawal for five major water-using manufacturing sectors in the USA.

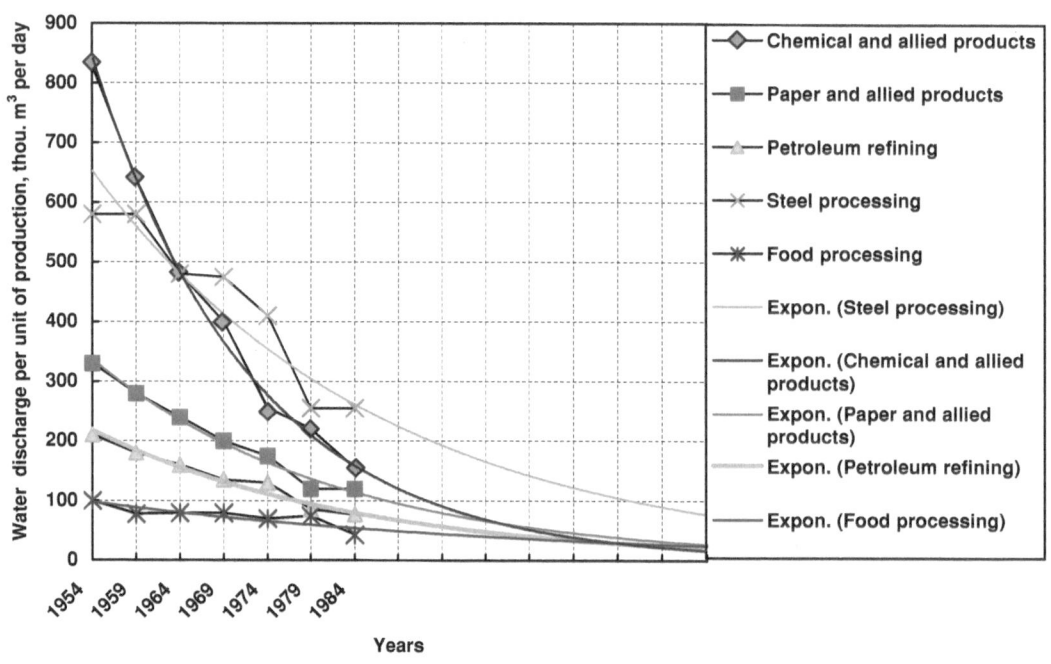

Fig. 7.29 Water discharge per unit of production for five major water-using manufacturing sectors in the USA.

of re-use cycles. David (1990) showed these changes and the results are contained in Tables 7.3 and 7.4 and in Figs. 7.28 and 7.29. The current industrial water consumption in the Northern region is about 4% of the total volume used, about 5% in the Central region and about 14% in the southern region. The Northern and Southern percentages were assumed to apply in 2025, but the Central region figure was increased to 6%.

The forecasts of abstractions for public supply water were based on UN forecasts of population growth, taking into account the change in specific water use per head. At the present time the amount of water used in the Northern and Southern regions is 700–710 l/day per head. The figure of 710 l/day per head was taken to apply to these regions for the near future. These forecasts took account of the growth in the population served by public supply as given in MacKichan (1951, 1957); MacKichan and Kammerer (1961); Murray (1968); Murray and Reeves (1972, 1977); Solley *et al.* (1983, 1988, 1993) and Tate (1986, 1989). For the Southern region, because there was no data on the population served by public supply, use was made of the estimates of the growth of the urban population. Rates of water use for the countries of the Southern region were assessed from actual data for 1990 where this was available. Plans for providing the population with piped water supplies and waste disposal systems were taken into account. For the Southern region as a whole the water use in 1990 was 205 l/day per head; the figures for 2000, 2010 and 2025 were taken to be 215, 225, 240 l/day per head, respectively.

In the Central region, during the 1970s to 1980s, abstractions for public water supply were 21–22% of the total, falling to 13% by 1990. The future percentage was taken to be 15%. In the Northern region, in 1990, water consumption was about 15% of total abstractions and the same value was assumed to apply in the future. In the Southern region, according to Shiklomanov and Markova (1987), before 1980 water consumption was 25–30% of the total, but this percentage has decreased during the last decades and a figure of 18% was assumed to be a reasonable estimate for the future. To forecast the additional losses from reservoir evaporation, data on existing reservoirs was used with the plans for new reservoirs.

Water availability was estimated both per unit area and per head according to the recommendations given in Section 3.3, for regions, countries, the water resources regions of Canada and the USA, and for the zones of Mexico. The water resources regions of Canada and the USA were assumed correspond to the administrative regions, with the water resources of these regions being estimated from data collected for this Monograph and from Foxworthy and Moody (1986) and Pearse *et al.* (1985). The zones of Mexico were those used in the *Plan National Hidraulico de Mexico* (1981).

When forecasting water availability for the individual regions of Canada, the rates of population growth and the rise in water consumption were taken as the average for the country. For the USA, the population growth and the rise in consumption of water were assessed for every region individually, based on trends for the previous years. For Mexico, population growth was estimated for every zone on the basis of earlier trends, while abstractions were estimated from data in the *Plan Nacional Hidraulico de Mexico* (1981).

7.6.2 Changes and trends in water use

Tables 7.22 to 7.26 and Figs. 7.30 to 7.35 show the volumes of water used in North America. Table 7.22 demonstrates that from 1900 to 1995, irrigation areas expanded from 4.2 to 29.2 mil-

lion ha, while agricultural (irrigation and rural) water use rose from 43 to 304 km^3/year. However the water used in agriculture decreased from 62% in 1900 to 45% in the period 1980 to 1995. By 2025 about 37.8 million ha are expected to be irrigated, about 30% more than at present. This increase will be about 20% in the Central region while in the Southern and Northern regions it will be about 50%. Some 75% (6.1 million ha in 1995) of the irrigation in the Southern region is in Mexico. However, because of water scarcity, by 2025 the Mexican irrigated area is expected to be a smaller percentage of the total area in the Southern region, and will make up about 60%. For most of the countries of this region the forecast increase in the area of irrigation is likely to be about 40%. Higher rates of growth (50–75%) are expected in Guatemala, Honduras and Panama. For the continent as a whole the area irrigated is expected to expand slightly but where irrigation is well established the increase will be very small.

The amount of water used in agriculture in North America is expected to rise to 350 km^3/year by 2025 with a consumption of 215 km^3/year. Agricultural abstractions will account for about 45% of the total. The increase in use is expected to occur mainly in the Southern (29 km^3)/year and Central (16 km^3/year) regions (Tables 7.25 and 7.26).

Over the period 1900 to 1980 use of water by industry in North America grew steadily from 21.8 to 293 km^3/year. Then it stopped rising and by 1995 it had fallen to 266 km^3/year. Industrial water rose from 31% of all abstractions in 1900 to 45% in 1970, then it fell to 40% in 1995. The continent's total water use by industry is forecasted to rise to 306 km^3/year by 2025, mainly in the Northern and Southern regions. The largest volume of abstraction for industry is the Central region (conterminous USA) and this is expected to remain so in the future. In 1995 this region took 82% of the total industrial water, but earlier it was about between 88% and 90%; by 2025 it is expected to decrease to 77%. Actual consumption of water by industry was less than 1.0 km^3/year early in the twentieth century, then it increased to reach 14.6 km^3/year in 1995, and this trend is expected to continue (Table 7.25).

The total amount of water used in North America for domestic purposes was 4.8 km^3/year early in the twentieth century. By 1950 it had grown to 22 km^3/year and to 72.5 km^3/year by 1995, and this total is expected to reach 94 km^3/year in 2025. The greater proportion will continue to be utilized in the Central region (67 km^3/year). The water used for domestic purposes in the Northern region is expected to be 7.1 km^3/year by 2010 and 7.6 km^3/year by 2025, and between 4.0 and 8.5 km^3/year in 2010 (Pearse *et al.*, 1985). In the near future abstractions for public supply water in the Southern region are expected to double to 19.0 km^3/year by 2025. In North America as a whole, water consumption in public water supply grew from 1 km^3/year to 11 km^3/year between 1900 and 1995, and it is expected to rise to 15 km^3/year by 2025.

Table 7.22. *Dynamics of freshwater use (km³/year) in North America*

	Assessment							Forecast		
	1900	1950	1960	1970	1980	1990	1995	2000	2010	2025
Population, $\times 10^6$										
Urban[a]		142	178	222	282	329	358	391	450	504
Rural		75.5	84.9	94.3	88.9	94.5	95.0	95.0	93.0	90.8
Total		217	263	316	371	423	453	486	543	595
Irrigated land, ha $\times 10^3$ Water use	4200	12 700	18 100	21 400	28 100	26 300	29 200	30 600	33 800	37 800
Irrigation										
withdrawal	39.3	149	198	244	286	274	286	295	309	327
consumption	24.6	86.3	114	143	171	166	173	179	188	200
Rural use										
withdrawal	3.5	6.1	7.3	9.6	12.4	16.8	17.7	18.7	21.0	23.1
consumption	2.7	4.7	5.4	7.0	8.6	11.5	12.0	12.6	14.2	15.5
Public supply										
withdrawal	4.8	22.0	33.0	44.0	56.3	67.1	72.5	76.8	84.9	94.0
consumption	1.0	4.8	5.8	9.8	12.0	9.1	10.9	12.4	13.8	14.9
l/day per head		425	505	545	555	560	550	540	520	510
Industry										
withdrawal	21.8	104	165	246	293	259	266	273	286	306
consumption	0.7	3.9	6.4	10.2	13.4	13.8	14.6	15.4	17.4	19.1
Reservoirs	0.2	8.3	14.3	22.7	27.0	29.1	30.1	31.0	34.0	37.0
Total withdrawal	69.6	289	418	566	675	646	672	695	734	788
Total consumption	29.2	108	146	193	232	229	241	250	267	286

[a] Public supply population served.

Table 7.23. *Dynamics of water withdrawal and water consumption (km³/year) by natural–economic regions of North America*

	Assessment							Forecast		
Region	1900	1950	1960	1970	1980	1990	1995	2000	2010	2025
Northern region										
withdrawal	2.6	13.2	19.2	25.9	41.1	52.0	55.6	58.5	65.4	74.1
consumption	0.5	2.3	3.3	4.8	7.8	10.3	11.1	11.6	13.9	15.9
Central region										
withdrawal	54.2	247	347	470	538	492	503	512	530	550
consumption	19.9	86.6	109	143	163	154	159	163	170	175
Southern region										
withdrawal	12.8	28.5	51.5	69.9	95.4	102	113	124	139	164
consumption	8.8	18.8	34.2	45.5	61.2	64.8	70.4	75.5	83.4	95.3
North America as a whole										
withdrawal	69.6	289	418	566	675	646	672	695	734	788
consumption	29.2	108	146	193	232	229	241	250	267	286

Table 7.24. *Dynamics of freshwater use (km³/year) in the Northern natural–economic region of North America*

			Assessment						Forecast	
	1900	1950	1960	1970	1980	1990	1995	2000	2010	2025
Population, $\times 10^6$										
Urban[a]		10.1	13.6	17.2	20.4	23.0	24.8	25.6	27.5	29.0
Rural		3.8	4.5	4.5	4.2	4.2	4.2	4.0	4.0	3.8
Total		13.9	18.1	21.7	24.6	27.2	29.0	29.6	31.5	32.8
Irrigated lands, ha $\times 10^3$ Water use		300	360	421	580	860	960	1070	1290	1460
Irrigation										
withdrawal	0.2	0.9	1.2	1.6	3.1	4.0	4.5	5.0	6.0	6.8
consumption	0.15	0.65	0.8	1.1	2.4	3.1	3.4	3.7	4.6	5.3
Rural use										
withdrawal	0.2	0.2	0.3	0.3	0.4	0.5	0.5	0.5	0.6	0.6
consumption	0.15	0.15	0.2	0.2	0.3	0.3	0.3	0.3	0.4	0.4
Public supply										
withdrawal	0.5	1.7	2.3	3.1	4.3	5.9	6.3	6.6	7.1	7.6
consumption	0.1	0.5	0.5	0.5	0.6	0.9	1.0	1.0	1.1	1.1
l/day per head		460	460	495	580	700	705	710	710	710
Industry										
withdrawal	1.7	9.8	14.2	18.5	29.7	37.1	39.5	41.4	45.7	52.1
consumption	0.1	0.4	0.6	0.6	0.9	1.5	1.6	1.6	1.8	2.1
Withdrawal										
Thermoelectric power		0.8	2.6	9.3	19.1	28.5	29.7	30.7	32.6	37.0
Manufacturing		8.6	11.2	8.8	10.0	8.0	9.0	9.7	11.7	13.3
Mining		0.4	0.4	0.4	0.6	0.6	0.8	1.0	1.4	1.8
Reservoirs	–	0.6	1.2	2.4	3.6	4.5	4.8	5.0	6.0	7.0
Total withdrawal	2.6	13.2	19.2	25.9	41.1	52.0	55.6	58.5	65.4	74.1
Total consumption	0.5	2.3	3.3	4.8	7.8	10.3	11.1	11.6	13.9	15.9

[a] Public supply population served.

The increase in reservoir construction since the start of the twentieth century has resulted in a considerable loss of water from the surfaces of the reservoirs. Starting from 0.2 km³/year, the water loss reached 8.3 km³/year in 1950, and by 1980 reservoir evaporation accounted for 27.0 km³/year. By 2025 this figure may reach 37 km³/year. The largest losses occur in the Central region (21.7 km³/year at present), about two-thirds of the total losses across the continent. The Northern region has most reservoirs, but the evaporation from their surfaces is much lower than in other regions, comprising about 15% of losses. In the Southern region, where there are few large reservoirs, the losses due to evaporation are small (3.6 km³/year). They are, however, expected to increase in number in the future. Recently the ecological implications are carefully considered in planning new reservoirs and a few existing reservoirs have been removed. Hence for the future reservoir evaporation is expected to experience a very small increase in the USA.

As estimates of water use have been carried out in the USA and in Canada on reliable data, the changes and trends in water use can be considered in some detail. Tables 7.24 and 7.25 contain these values for the period 1950 to 1995 and forecast volumes to 2025, not only for the main sectors as in Tables 7.22 and 7.26, but also for irrigation and rural supplies, thermal power production, manufacturing and mining, for Mexico and some countries of Central America. More general data are presented for Mexico and some countries of Central America in Table 7.27 and in Fig. 7.36.

Most water used in the USA is for irrigation and for thermal power production. Since 1980 the total volume used in power production has been a little smaller than for irrigation, but if salt water is included it is far greater. Most water in Canada is used by industry. Up to the 1970s processing took most water but in subsequent years it was thermal power production. Chapter 7.3 presents a detailed analysis of trends in the use of water in the USA and Canada and the results of forecasting water use are now considered. Between 1990 (the last year when published national estimates were available) and 2025, total abstraction in the USA is expected to increase by 58 km³/year (12%). The largest growth, some 25%, will be for domestic purposes, followed by 20% for rural water

Table 7.25. *Dynamics of freshwater use (km³/year) in the Central natural–economic region of North America*

	Assessment							Forecast		
	1900	1950	1960	1970	1980	1990	1995	2000	2010	2025
Population, × 10⁶										
Urban[a]		111	133	156	187	207	219	228	246	259
Rural		41	45	48	40	43	42	40	40	40
Total		152	178	204	227	250	261	268	286	299
Irrigated lands,[b] ha × 10³	3100	10 400	13 800	15 900	20 600	18 400	20 400	21 050	23 300	24 740
Irrigation										
withdrawal	27.0	125	154	184	206	192	196	199	205	210
consumption	16.0	68.0	82.7	102	115	107	110	112	115	117
Rural use										
withdrawal	3.0	5.0	5.0	6.3	7.8	11.0	11.4	11.8	12.6	13.2
consumption	2.3	3.9	3.9	4.8	5.4	7.3	7.6	7.8	8.3	8.6
Public supply										
withdrawal	4.0	19.6	29.4	37.4	46.8	53.9	57.0	59.0	63.6	67.4
consumption	0.8	4.1	4.8	8.3	10.0	7.0	8.3	9.3	10.0	10.6
l/day per head		485	595	650	700	705	710	710	710	710
Industry										
withdrawal	20.0	91.0	146	223	256	214	217	220	226	235
consumption	0.6	3.1	4.9	8.6	11.5	11.2	11.7	12.2	13.6	14.3
Withdrawal										
Thermoelectric power		47.5	100	168	202	182	185	187	192	201
Manufacturing		40.8	43.0	50.0	50.1	27.0	27.5	28.0	29.0	29.0
Mining		2.7	3.2	4.7	4.1	4.6	4.7	4.7	4.8	5.0
Reservoirs	0.2	7.5	12.5	19.0	21.0	21.5	21.7	22.0	23.0	24.0
Total withdrawal	54.2	247	347	470	538	492	503	512	530	550
Total consumption	19.9	86.6	109	143	163	154	159	163	170	175

[a] Public supply population served.
[a] Irrigated area by FAO data.

supplies, 10% for thermal power production and a little less than 10% in the manufacturing and mining industries. By 2025, the forecast increase for irrigation is 9%, totalling 210 km³/year, occurring largely in the east of the country and in the prairie states. In the west, due to the lack of water resources, special attention is being paid to water-saving measures, and the expected small increase in the irrigated area will not result in increased water use. In some prairie states there may be a slight increase in abstractions because of expansion of irrigation, but there is likely to be a decrease in the southwest and in Texas (Young and Bennett, 1992). According to a forecast for 2025, rural water use will total about 13.0 km³/year provided the current rate of increase continues.

The freshwater abstraction for thermal power production in the USA is likely to be around 200 km³/year by 2025, i.e. it will remain at the level of 1980. With increasing electrical production one may expect an increase in the saline water intake. By 2025 the abstractions for the manufacturing and mining industries in the

USA are likely to increase by 5–10% in line with the predicted growth of industrial production and the high level of success in utilizing technology for achieving measures to save water.

By 2025 water use in Canada will increase by 40%, as compared to 1990, comprising 73.0 km³/year (the last national water use estimates for Canada are dated to 1989). The main growth will take place in the mining industry (three-fold), in manufacturing and in agriculture (by 70%). In the other sectors the increase will be 20–30%. According to estimates for this Monograph, by 2010 water used for irrigation will increase to 6.0 km³/year (i.e. by between 2.8 and 7.1 from Canadian estimates) and to 6.8 km³/year by 2025. Water for thermal power production will increase by some 30% reaching 37 km³/year, but past rates of growth for power production were higher. The growth in the mining industry is explained by the overall growth of industrial production and very high use of recycled water, which significantly exceeds the level in the USA.

Table 7.26. *Dynamics of freshwater use (km³/year) in the Southern natural–economic region of North America*

	Assessment							Forecast		
	1900	1950	1960	1970	1980	1990	1995	2000	2010	2025
Population, × 10⁶										
Urban		20.4	31.3	48.6	74.3	98.7	114	137	176	216
Rural		30.7	35.4	41.8	44.7	47.3	49.0	51.0	49.0	47.0
Total		51.1	66.7	90.4	119	146	163	188	225	263
Irrigated lands, ha × 10³	1050	2000	3900	5100	6900	7000	7800	8500	9200	11 600
Irrigation										
withdrawal	12.1	23.0	43.0	56.8	76.8	78.3	85.0	91.0	98.0	110
consumption	8.4	17.3	30.9	40.2	53.7	55.6	59.8	63.4	68.3	76.9
Rural use										
withdrawal	0.3	0.9	2.0	3.0	4.2	5.3	5.8	6.4	7.8	9.3
consumption	0.3	0.7	1.3	2.0	2.9	3.8	4.1	4.5	5.5	6.5
Public supply										
withdrawal	0.3	0.7	1.3	3.5	5.2	7.3	9.2	11.2	14.2	19.0
consumption	0.1	0.2	0.5	1.0	1.4	1.2	1.6	2.0	2.6	3.2
l/day per head		94	115	180	190	205	210	215	225	240
Industrial										
withdrawal	0.1	3.7	4.6	5.3	7.0	7.8	9.5	11.3	14.4	19.2
consumption	0	0.4	0.9	1.0	1.0	1.1	1.3	1.6	2.0	2.7
Reservoirs	0	0.2	0.6	1.3	2.2	3.1	3.6	4.0	5.0	6.0
Total withdrawal	12.8	28.5	51.5	69.9	95.4	102	113	124	139	164
Total consumption	8.8	18.8	34.2	45.5	61.2	64.8	70.4	75.5	83.4	95.3

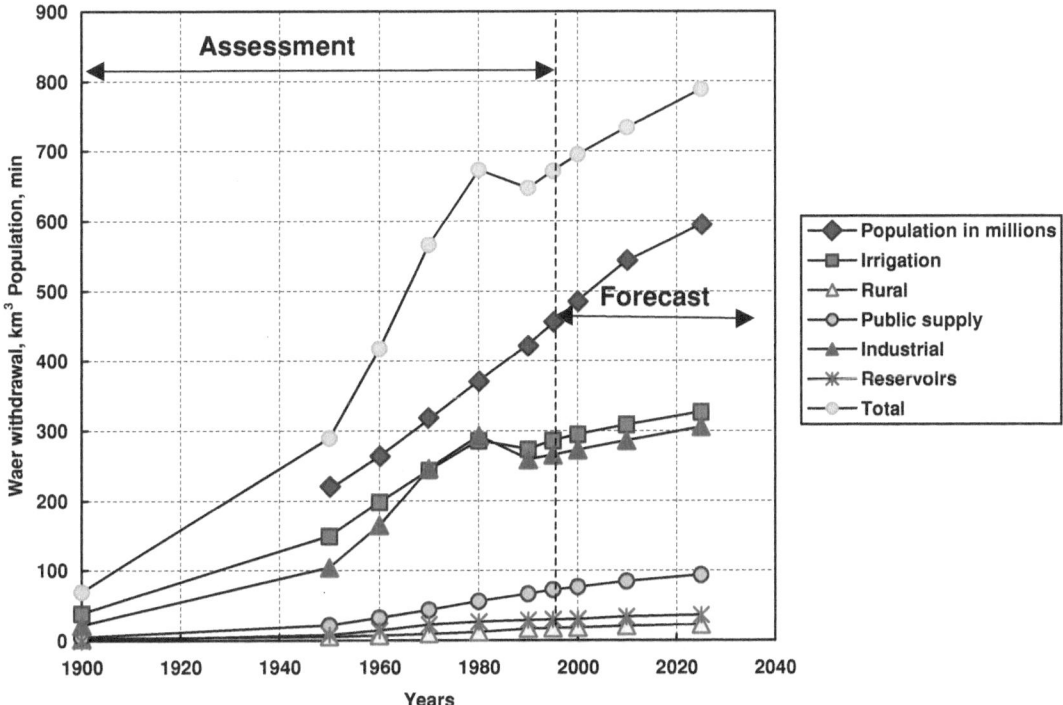

Fig. 7.30 Dynamics of freshwater withdrawal in North America.

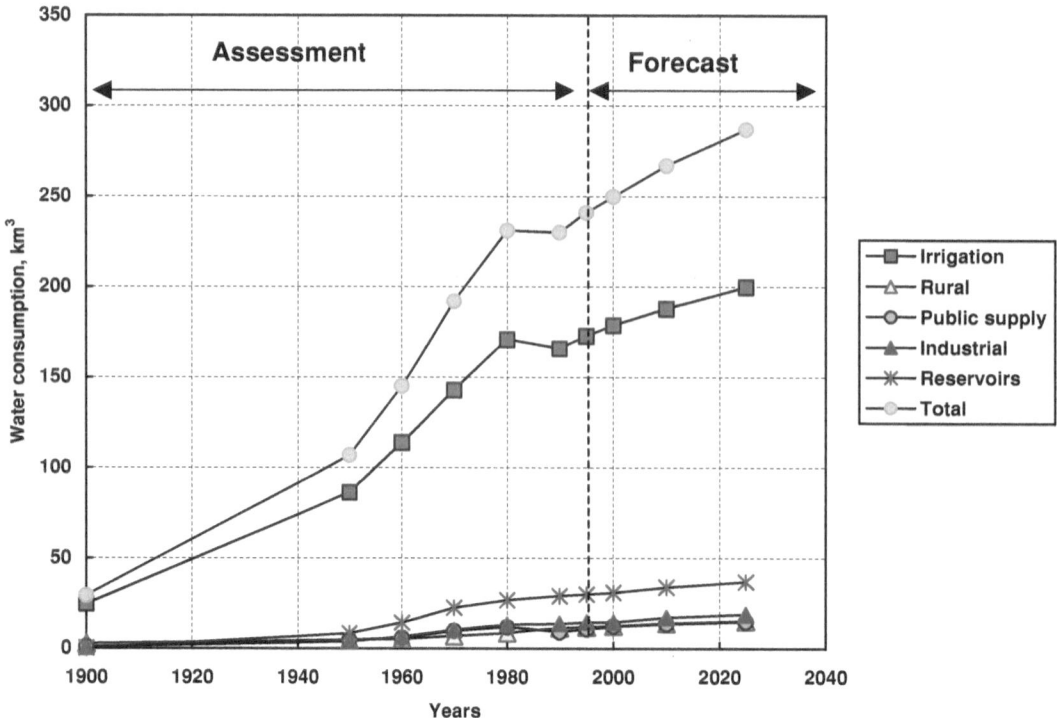

Fig. 7.31 Dynamics of water consumption in North America.

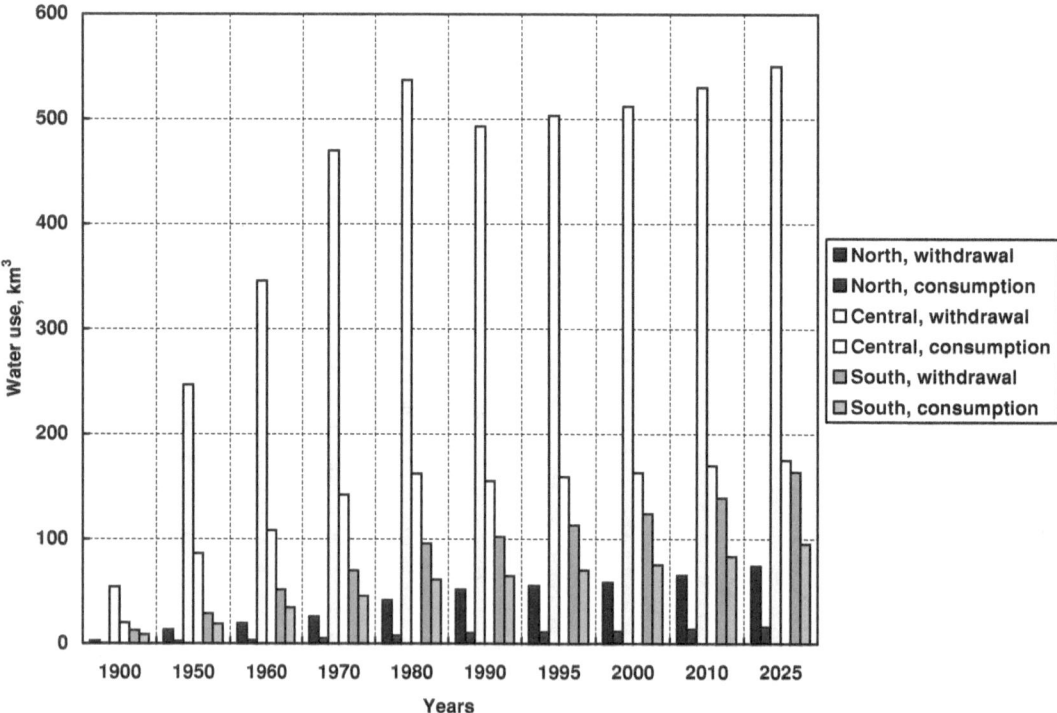

Fig. 7.32 Dynamics of water withdrawal and water consumption by natural–economic regions of North America.

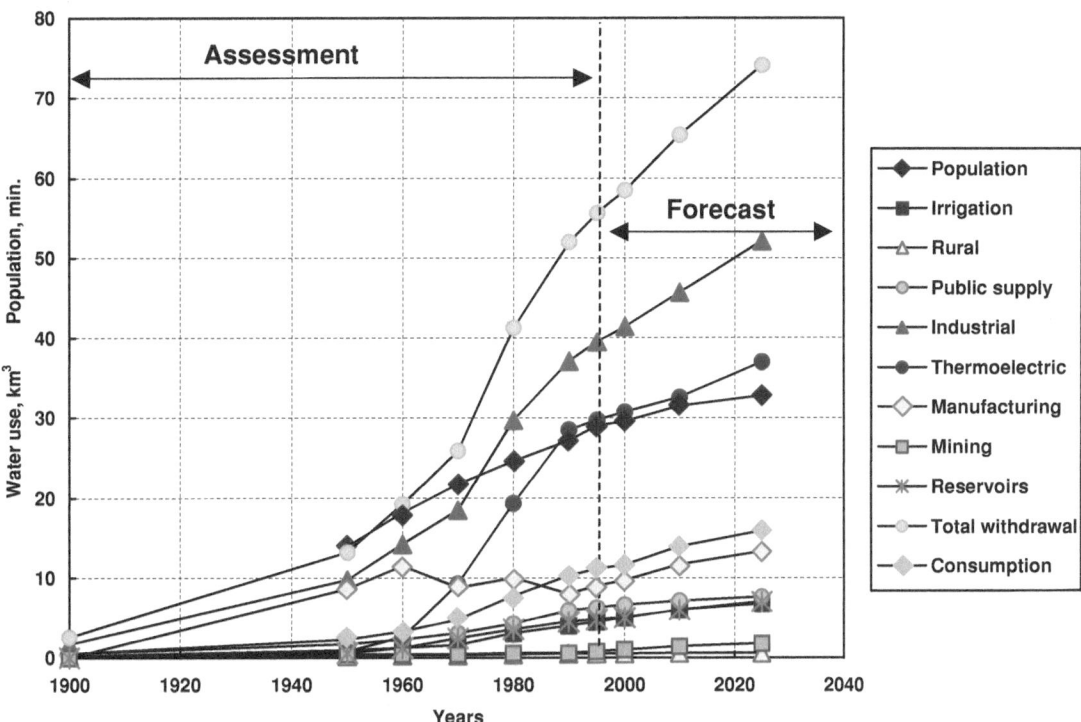

Fig. 7.33 Dynamics of water use in the Northern natural–economic region of North America.

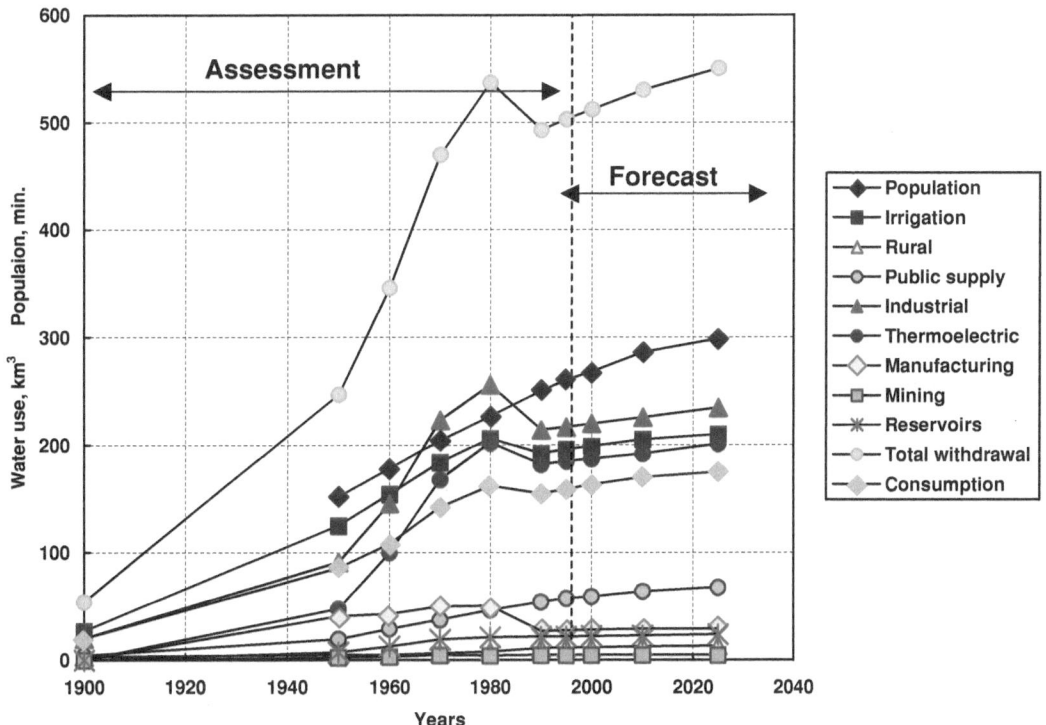

Fig. 7.34 Dynamics of water use in the Central natural–economic region of North America.

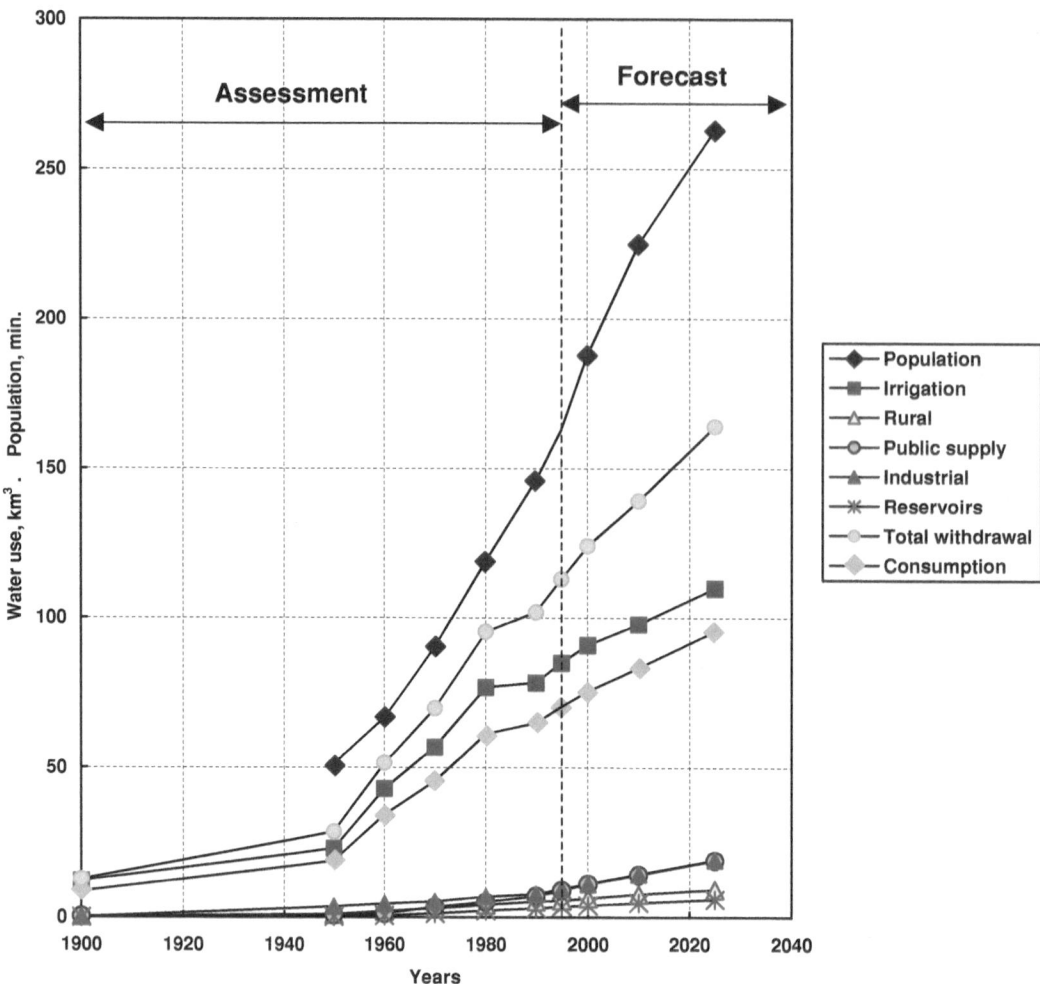

Fig. 7.35 Dynamics of water use in the Southern natural–economic region of North America.

In most countries of Central America, except for Mexico, Cuba, the Dominican Republic and Barbados, the current abstractions do not exceed 10% of water resources and this percentage is expected to be maintained. However, the rates of growth of abstractions will be much higher here than in the USA and Canada. Thus between 1995 and 2025 a 35% increase is expected in Mexico and Cuba, and in most other countries of Central America it is likely to be between 70% and 200%. Table 7.27 shows changes in water use from 1950 to 1995 and forecasts to 2025.

A comparison of the use renewable water resources is given in Table 7.28. It shows that the total average water use is some 8.5% of surface water resources while water consumption is 3%. The Central region has figures of 27% and 8.7%, respectively. By 2025, these values will increase and are expected to be 10% and 3.6% on average for the whole continent. The most significant increase will occur in the Southern region. For the present and in the near future the total water use in the Northern region is likely to be slightly more than 1% of water resources while water consumption is less than this.

7.6.3 The availability of water in the regions and in certain countries

Table 7.29 shows that mean specific water availability for the continent in 1995 is now 324 000 m³/year per km² or 17 100 m³/year per head. The available water was 398 000 m³/year per km² in the Southern Region, some 364 000 m³/year per km² some in the Northern region, and only 230 000 m³/year per km² in the Central region. Per head the largest volume was available in the Northern region at 174 000 m³/year per head, while the figures for the Southern and Central regions were 6600 m³/year and 6400 m³/year per head, respectively (Fig. 7.37). By 2025 availability per head for the Continent as a whole will decrease by about 1.5 times to 12 900 m³/year consisting of 154 000 m³/year in the Northern region, 5500 m³/year in the Central region and 4000 m³/year in

Table 7.27. *Dynamics of water withdrawal, water consumption and water availability for selected countries of North America*

Country	Area, km² × 10³	Assessment						Forecast		
		1950	1960	1970	1980	1990	1995	2000	2010	2025
Canada	9976									
Population, × 10⁶		13.7	17.9	21.3	24.0	26.7	28.0	28.9	30.7	31.9
Water use, km³/year										
withdrawal		13.0	18.9	25.6	40.8	51.6	55.0	58.0	65.0	74.1
consumption		2.3	3.3	4.8	7.8	10.3	10.9	11.5	13.8	16.0
Water availability, m³ × 10³/year per head		240	184	154	137	123	119	113	107	103
USA	9363									
Population, × 10⁶		152	181	205	228	250	262	268	286	300
Water use, km³/year										
withdrawal		247	346	470	537	493	502	512	530	550
consumption		86	108	142	162	155	159	163	170	175
Water availability, m³ × 10³/year per head		18.6	15.5	13.6	12.1	11.1	10.6	10.2	9.5	9.0
Mexico	1973									
Population, × 10⁶		25.8	35.0	50.7	69.4	84.5	94.8	109	128	150
Water use, km³/year										
withdrawal		18.2	33.0	45.0	60.8	73.9	83.0	92.1	102	111
consumption		12.0	21.9	29.3	39.0	47.0	52.0	56.4	61.9	64.5
Water availability, m³ × 10³/year per head		13.0	9.3	6.3	4.4	3.6	3.1	2.7	2.2	1.9
Nicaragua	130									
Population, × 10⁶		1.1	1.4	1.8	2.7	3.7	4.5	5.3	6.8	9.2
Water use, km³/year										
withdrawal		0.6	0.6	0.7	1.2	1.4	1.6	1.9	2.4	3.0
consumption		0.2	0.2	0.3	0.6	0.7	0.8	0.9	1.0	1.2
Water availability, m³ × 10³/year per head		170	120	96.0	64.0	47.0	38.7	33.0	26.0	19.0
Cuba	111									
Population, × 10⁶		5.5	6.8	8.5	9.7	10.6	11.1	11.7	12.6	13.0
Water use, km³/year										
withdrawal		3.6	4.6	5.5	8.8	9.2	10.0	10.8	12.4	13.7
consumption		2.3	2.9	3.4	5.6	5.8	6.3	6.8	7.6	8.4
Water availability, m³ × 10³/year per head		6.0	4.7	3.8	3.0	2.7	2.6	2.4	2.2	2.0
Guatemala	109									
Population, × 10⁶		2.8	3.8	5.1	6.9	9.2	10.6	12.2	15.8	21.7
Water use, km³/year										
withdrawal		0.5	0.6	0.7	1.1	1.3	1.4	1.6	2.1	2.7
consumption		0.3	0.4	0.5	0.7	0.7	0.8	0.9	1.2	1.5
Water availability, m³ × 10³/year per head		41.0	30.0	23.0	16.6	12.5	10.8	9.4	7.3	5.3
Panama	77									
Population, × 10⁶		0.8	1.1	1.4	2.0	2.4	2.6	2.9	3.3	3.9

Table 7.27. *(cont.)*

Country	Area, km² × 10³	Assessment						Forecast		
		1950	1960	1970	1980	1990	1995	2000	2010	2025
Water use, km³/year										
withdrawal		0.8	0.9	1.1	1.5	1.8	2.0	2.1	2.5	3.0
consumption		0.5	0.5	0.6	0.9	1.1	1.2	1.2	1.4	1.7
Water availability, m³ × 10³/year per head		176	135	100	73.0	59.0	54.2	49.0	43.0	37.0
Costa Rica	51									
Population, × 10⁶		0.8	1.2	1.7	2.2	3.0	3.4	3.6	4.2	5.2
Water use, km³/year										
withdrawal		0.9	1.1	1.4	1.8	2.4	2.7	3.1	4.1	5.4
consumption		0.5	0.6	0.7	1.0	1.4	1.5	1.6	2.4	3.2
Water availability, m³ × 10³/year per head		120	80.0	54.0	42.0	31.0	27.1	26.0	22.0	17.5

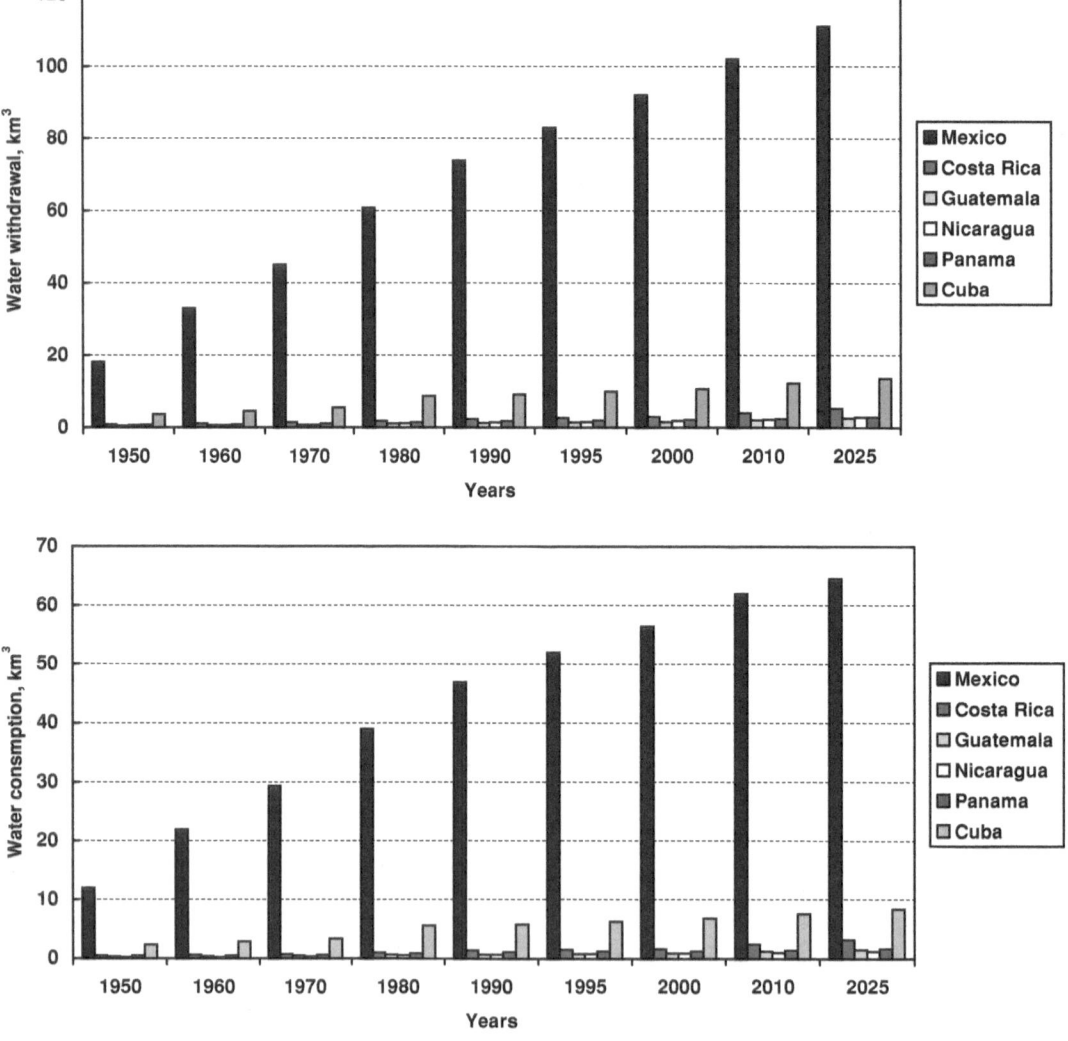

Fig. 7.36 Dynamics of water withdrawal and water consumption by selected countries of Central America.

Table 7.28. *Water use as a percentage of water resources by natural–economic regions of North America*

| Region | Water resources, km^3/year | | Water use, km^3/year | | | | Water use as a percentage of water resources | | | |
| | Inflow | Local resources | Withdrawal | | Consumption | | Withdrawal | | Consumption | |
			1995	2025	1995	2025	1995	2025	1995	2025
Northern	130	4980	55.6	74.1	11.1	15.9	1.2	1.5	0.21	0.31
Central	70	1800	503	550	159	175	27.0	30.0	8.70	9.60
Southern	2.5	1090	113	164	70.4	95.3	10.4	15.0	6.50	8.80
North America as a whole		7870	672	788	241	286	8.5	10.0	3.00	3.60

the Southern region where the population growth will be greatest. Some parts of each of the three regions have large volumes of available water, others have shortages. The Northern and the Southern regions are at the extremes in this respect. Tables 7.30 to 7.32 present the specific water availability for 1990 and 2025 for the regions of the USA and Canada, as well as the availability for some parts of Central America and Mexico, while Fig. 7.38 maps this information.

Overall North America has large volumes of available water, although there are even now some areas with pronounced shortages of water resources. If a specific water availability below 1000 m^3/year per head is considered as catastrophically low (Shiklomanov and Markova, 1987), this level has been reached in the Assiniboine–Red River Basin (Canada) and in two regions of the USA, as well as in the northern and central zones of Mexico. In some basins of Mexico abstractions exceed surface water resources and ground water is used. The Caribbean region (USA) and Haiti have low values of available water and in the future with the increased population and the growth in demand, water supply will become an even more acute problem. Water availability will decrease in the Northern and Central zones of Mexico; in the Northern zone it is likely to become less than 100 m^3/year per head. In the South Saskatchewan (Canada), in Texas, in the Great Basin, California and on some Caribbean islands, as well as in Mexico as a whole, in the Dominican Republic, on Cuba and in El Salvador, specific water availability will reduce to between 1000 and 2000 m^3/year per head. This decrease will be less for the USA and Canada where by 2025 it will be 1.2 times less than in the Central America countries, where it will decrease by 1.5 to 2.5 times. While in Canada, Mexico, Cuba and especially in the USA the decreased water availability will be induced both by population growth and a significant increase in water consumption, in most countries of Central America the main cause for the decrease in available water will be the rapid growth of population. However, in most of these countries water consumption is negligible as compared with water resources. The values in Tables 7.30 to 7.32 refer to mean values of water resources, but during dry

years the specific water availability becomes especially low. As an example, Table 7.33 presents the specific values in dry years for some regions that are characterized by low water availability. As can be seen, dry years can cause grave problems for supplies of water and this emphasizes the need to develop efficient measures for their solution.

7.7 WATER MANAGEMENT PROBLEMS IN NORTH AMERICA: CANADIAN TRANSFERS

7.7.1 The background

The North American continent extends from 82° N to 7° N and covers practically all climatic types. It is extremely diverse in terms of physical geography and economic development. There are unexplored areas with a severe climate, and large cities with a very dense population. The Arctic Archipelago has only 0.005 persons per km^2, and the entire northern part of the continent above 50° N on the whole has a very low population density. The most densely populated areas are the centre and the south of North America; in the Central region the mean density is 33 persons per km^2 and in the Southern region some 69 persons per km^2. The Caribbean islands and the Mexican plain are densely populated, with 150 to 400 inhabitants per km^2 and 700 to 800 inhabitants per km^2, respectively.

Canada and the USA have economies which are amongst the most advanced in the world. Northern Canada and Alaska are regions rich in natural resources. Regions of the USA that are arid and semi-arid have been developed because of the availability of resources of ground and surface water and because of water transfers from other regions. Mexico is a large country mainly with an arid climate. Lack of water resources is a factor restricting economic development. Where there are water resources there is development but generally limited financial resources prevent solutions to the problems of shortage water resources. Some countries of Central America and the Caribbean are amongst the least

Table 7.29. *Dynamics of population and water availability by natural–economic regions of North America*

Region	Area, km^2 × 10^3	Year	Population, ×10^6	Water availability, m^3 × 10^3/year	
				Per km^2	Per head
Northern	13 668	1950	13.9	364	364
		1960	18.1		280
		1970	21.7		232
		1980	24.6		205
		1990	27.2		186
		1995	29.0		174
		2000	29.6		171
		2010	31.5		160
		2025	32.8		154
Central	7 844	1950	152	230	11.4
		1960	178		9.6
		1970	204		8.3
		1980	227		7.4
		1990	250		6.7
		1995	261		6.4
		2000	268		6.2
		2010	286		5.8
		2025	299		5.5
Southern	2 735	1950	51.1	398	21.3
		1960	66.7		16.3
		1970	90.4		12.0
		1980	119		9.2
		1990	146		7.5
		1995	163		6.6
		2000	188		5.7
		2010	225		4.7
		2025	263		4.0
North America as a whole	24 250	1950	217	324	35.8
		1960	263		29.4
		1970	316		24.6
		1980	371		20.9
		1990	423		18.4
		1995	453		17.1
		2000	486		15.7
		2010	543		14.2
		2025	595		12.9

developed in the world. Most rely on the products of tropical agriculture.

These exceptionally diverse conditions across the continent result in all kind of water management problems and the different levels of development govern the ways that can be explored to overcome them. The main water management problems of North America include:

1. Problems caused by lack of water resources because of the climate conditions and because growth of population and the demands of industry and agriculture have resulted in increased requirements which exceed the natural resource.

2. Problems connected with too much water which require measures for the control of runoff, drainage and protection from floods.

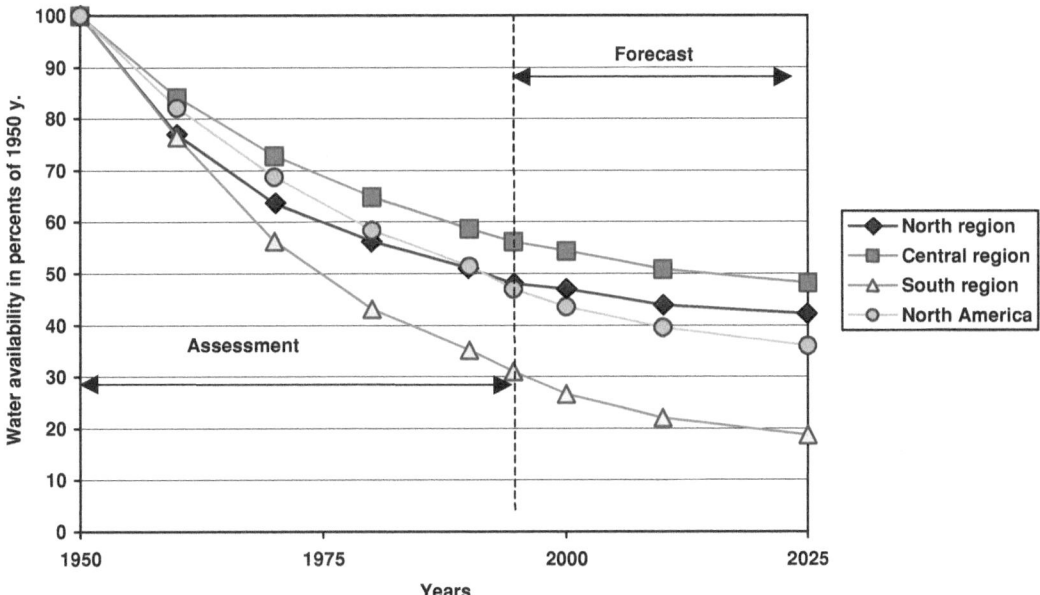

Fig. 7.37 Dynamics of water availability by natural-economic regions of North America, as percentage of water availability in 1950.

3. Problems of the rehabilitation of surface and ground water which have resulted from deterioration in their quality because of pollution.
4. Problems related to the decline in ground water levels in some regions as a result of over-pumping.
5. Problems of the spatial redistribution of water resources.

Since regions with a lack of water resources occupy a considerable part of North America, the problems of water shortages have received most attention. Much of the USA and the Canadian Prairies lack water resources, particularly surface water resources, which have been overcome by the use of ground water and especially by inter-basin transfers. To date, there are many complex water management systems consisting of large dams and reservoirs, small reservoirs and many aqueducts and canals. Increased attention to the ecological problems these cause has resulted in numerous publications devoted to the more efficient use of available water resources taking into account environmental protection. In recent years, various measures have been undertaken for the protection of water resources in the driest regions such as California (Kennedy, 1992; DWR, 1993, 1994; Gleick *et al.*, 1995; Maddock and Hines, 1995) and in the south of the USA (Loh, 1994; Rush, 1988).

Mexico is characterized by a clear discrepancy between the availability of water resources and the existing demand. The main demand is centred in the north and in the centre of the country (Gordunio *et al.*, 1984) whereas the water resources are in the south. Rapid population growth requires increasing agricultural production but this is impossible without irrigation. To date,

there are sufficient reserves of land suitable for irrigation, but they need water which can, in the main, only be obtained by inter-basin transfers. The *Plan Nacional Hidraulico de Mexico* (1981) contains a regional strategy for the development and use of water resources, but the implementation of the plan has been put back because of lack of financial resources. Due to the decrease in capital investment in water management in the 1980s, the area of irrigation declined in the early 1990s (Gleick *et al.*, 1995).

The increasing demand in the northeast of the USA, a humid region, has produced an excess of demand for domestic and industrial purposes, especially in dry years, over the resource. Such a situation requires careful management of water resources.

The global warming predicted by climatologists may result in more water shortages in certain regions of the USA and Canada and this is of great concern to both hydrologists and water managers. An increasing amount of attention has been given recently to estimates of future water resources and to water supply problems with a non-stationary climate (Waggoner, 1990; Izmailova and Moiseenkov, 1998). These studies concern the available resource and the expected changes in agriculture and demands for energy, against the background of changes in future monthly and annual temperatures, precipitation and soil moisture. Ways of stabilizing the situation are considered including the greater regulation of runoff and its redistribution in space and time. Too much water is a problem in the southern part of the continent, particularly in tropical areas. Nevertheless in the USA, in spite of a high degree of river regulation, extreme floods can inflict significant damage (US Geological Survey, 1991). Measures for flood protection are practised in many regions and in some of the wetter areas land drainage is important.

Table 7.30. *Water availability by water resources regions of Canada*

Number of region	Water resources region	Area, $km^2 \times 10^3$	Population, $\times 10^6$		Water resources,[b] km^3/year	Water availability, $m^3 \times 10^3$/year		
			1990	2025[a]		Per km^2	Per head 1990	Per head 2025
1	Pacific Coastal	352	0.69	0.81	520	1480	750	640
2	Fraser–Low Mainland	234	1.92	2.26	126	538	65	56
3	Okanagan–Similkameen	14	0.21	0.25	2.35	168	10	8
4	Columbia	90	0.18	0.21	63.7	708	350	300
5	Yukon	328	0.02	0.03	79.5	242	3 460	2 840
6	Peace–Athabasca	487	0.32	0.37	92.0	189	290	250
7	Lower Mackenzie	1300	0.04	0.05	233	179	5 420	4 480
8	Arctic Coast Islands	2025	0.01	0.02	325	160	25 000	20 300
9	Missouri	26	0.01	0.02	0.38	13.5	25	19
10	North Saskatchewan[c]	146	1.21	1.42	7.70	50.1	6.1	5.1
11	South Saskatchewan[c]	170	1.43	1.68	8.30	36.2	3.7	2.0
12	Assiniboine–Red River[c]	190	1.45	1.70	1.53	7.5	0.8	0.5
13	Winnipeg	107	0.09	0.10	24.0	224	260	240
14	Lower Saskatchewan–Nelson	363	0.25	0.29	60.6	167	240	205
15	Churchill	298	0.08	0.09	22.2	74.4	275	240
16	Keewatin	689	0.01	0.01	123	179	24 600	20 500
17	Northern Ontario	694	0.18	0.21	190	274	1 050	900
18	Northern Quebec	950	0.12	0.14	534	562	4 450	3 800
19	Great Lakes	319	8.47	9.93	97.0	304	11	10
20	Ottawa	146	1.42	1.66	63.1	432	44	37
21	St Lawrence	116	5.80	6.81	67.9	585	11	10
22	North Shore–Gaspe	403	0.73	0.85	276	685	370	320
23	St John–St Croix	37	0.44	0.51	24.7	668	56	48
24	Maritime Coastal	114	1.46	1.73	97.7	857	66	56
25	Newfoundland–Labrador	376	0.64	0.74	296	787	460	390
	Canada as a whole	9976	26.7	31.9	3290	330	123	103

[a] Population by water resources regions of Canada is given with the assumption that it will be uniformly increasing all over the territory of the country.

[b] Water resources by provinces of Canada are cited by the data of Pearce (1985), except for the regions marked[c], where they are given by the data of our calculations.

The intense economic development of the North American continent has caused significant deterioration of the ecological situation in many regions. This deterioration became very pronounced in the second half of the twentieth century. Domestic, industrial and agricultural discharges all contributed to the decline in the quality of surface and ground water, especially in the highly developed industrial regions of the USA and Canada. For example, the consequences of this economic activity were observed in the Great Lakes, the greatest freshwater reservoir in the world. Along with an abrupt change in the chemistry of the lake water and increased levels of toxic substances, intense eutrophication took place. From the late 1970s both the USA and Canada have embarked on programmes to solve the problem of the Great Lakes, making large financial investments. These measures have stopped the degradation and significantly improved the ecological situation. The restoration of the water quality of the Great Lakes serves as an example of a good solution to an ecological problem. Similar projects have been carried out on other rivers in the USA and Canada, with funds being invested in the construction of treatment facilities and in the monitoring of surface and ground water quality. The decline of water quality is also typical of many areas of the Southern region. However, no large-scale measures for the protection of water resources have been undertaken there.

Table 7.31. *Water availability by water resources regions of the USA*

Number of region	Region	Area, km² × 10³	Population, ×10⁶		Water resources,[b] km³/year	Water availability, m³ × 10³/year		
			1990	2025[a]		Per km²	Per head 1990	2025
26	Alaska	1520	0.6	0.7	1100	725	2000	1670
27	New England	153	12.8	14.2	108	706	8.4	7.5
28	Mid Atlantic[c]	228	41.5	44.0	100	592	2.4	2.3
29	South Atlantic, Gulf	722	34.7	44.2	298	413	8.4	6.5
30	Great Lakes	332	21.4	23.0	107	322	4.9	4.5
31	Ohio[d]	419	21.9	23.0	196	468	8.8	8.3
32	Tennessi	110	3.9	4.5	60.6	497	15.4	13.3
33	Upper Mississippi[e]	471	21.3	24.5	101	236	4.6	4.0
34	Lower Mississippi[f]	3212	72.6	81.5	658	205	8.3	7.3
35	Souris-Red-Rainy	155	0.7	0.8	10.8	69.6	14.7	13.3
36	Missouri[c]	1347	10.0	12.0	85.0	53.4	6.8	4.8
37	Arkansas–White–Red	699	8.2	9.4	89.0	127	9.2	7.8
38	Texas–Gulf[c]	477	15.2	19.0	50.3	105	2.4	1.8
39	Rio Grande	406	2.2	3.1	7.0	17.2	1.0	0.3
40	Upper Colorado	308	0.6	0.9	17.2	55.8	18.0	12.4
41	Lower Colorado[c,g]	668	5.4	7.7	12.5	23.5	0.3	0.2
42	Great Basin	518	2.2	3.1	11.6	22.4	3.1	1.8
43	Pacific Northwest	666	8.9	11.0	407	611	43.7	35.3
44	California[c]	290	29.4	45.0	95.0	421	2.2	1.4
45	Hawaii	16.7	1.1	1.7	20.0	1200	17.1	11.2
46	Caribbean	9.0	3.6	4.8	7.1	790	1.9	1.5
	USA as a whole	9363	250	300	2930	313	11.1	9.0

[a] Population in 1990 by water resources regions of the USA from Solley *et al.* (1993).

[b] Water resources of US regions are given by Foxworthy and Moody (1986), except for the regions marked[c], where they are given by the data of our calculations.

[d] Except for outflows from the region of Tennessee.

[e] Except for outflows from the region of Missouri.

[f] Represents the conditions in the regions of Ohio, Tennessee, Upper and Lower Mississippi, Missouri, Arkansas–Red–Rainy Rivers.

[g] Represents the conditions in the regions of Upper and Lower Colorado.

Over-abstraction of ground water, primarily for agriculture, has significantly reduced levels in some regions of North America. Gleick (1993) contends that more than 4 million ha (about 20% of the national irrigation fund) are irrigated by ground water at rates which exceed its replenishment. The largest aquifer in the country – the Ogallala, which is the main source of water supply for the Great Plains – has been over-abstracted, particularly in its southern part. Over much of northern Mexico and in some central parts of the country the ground water levels have declined causing a number of problems, such as land subsidence (Bertoldi, 1992) and salt water intrusion in coastal regions. Use of ground water, especially for irrigation, is often accompanied with salination of the soil, primarily in regions with poor drainage. To prevent the decline of ground water levels measures need to be implemented for the rational use of water, including reductions in abstractions and recharge. Numerous publications discuss these problems in the USA (Jacobs, 1992; Pope, 1992). There are proposals for restricting the pumping of ground water in some Prairie states, and in California a programme for the comprehensive use of surface and ground water has been developed and implemented. As many parts of Mexico rely on ground water, restriction of abstractions appears unlikely at the present time. However, in the future with depleting aquifers this would appear to be inevitable.

Table 7.32. *Water availability by selected countries of Central America*

Number of region	Country, Zone	Area, km² × 10³	Population, ×10⁶ 1990	2025ᵃ	Water resources, km³/year	Water availability, m³ × 10³/year Per km²	Per head 1990	2025
	Mexico:	1973	84.5	150.0	346	176	3.6	1.9
47	North zone	516	14.0	28.0	11.7	23	0.2	0.03
48	North and Central Pacific	630	9.5	15.0	49.4	78	3.4	1.9
49	Central zone	263	43.0	79.0	55.5	211	0.9	0.45
50	Mexico Bay and south-east	562	18.0	28.0	278	495	15.1	9.4
53	Honduras	112	5.1	11.5	95.0	864	19.0	8.6
54	El Salvador	21	5.2	11.3	19.0	950	3.6	1.6
59	Haiti	28	6.5	13.2	11.0	367	1.7	0.8
60	Dominican Republic	49	7.2	11.4	20.0	400	2.8	1.7
61	Jamaica	11	2.4	3.5	8.3	830	3.4	2.3
62	Trinidad and Tobago	5.1	1.2	2.0	5.1	1020	4.1	2.5

ᵃ Population in 1990 and 2025 by the zones of Mexico is calculated by the data for the 1960–80s with the assumption that it will be increasing in every zone at the same rates that have been observed before.

7.7.2 Problems of inter-basin transfers in Canada

There are many transfers in Canada aimed at increasing the hydropower potential and some of the problems they have produced and Canadian experiences in overcoming them may also be useful for other countries.

TYPICAL FEATURES OF CANADIAN TRANSFERS

There are some features of these transfers which identify a special "Canadian" type of inter-basin transfer (Quinn, 1981). In most cases the diversions are characterized by relatively short routes not exceeding 20 or 30 km where the water flows under gravity. The nature of the relief formed by the Quaternary glaciation is an important factor, characterized by a dense network of river valleys and troughs, as well as by a large number of lakes situated at different levels. The watersheds are, as a rule, low and the interfluves are narrow. The lakes often play the role of reservoirs in regulating the runoff, eliminating the need to construct reservoirs. Another important feature of most Canadian transfers is that they are carried out in sparsely populated regions and thus do not run counter to the interests of other water-users. Such projects are low cost with a high economic efficiency for increasing power production with a rapid repayment of the capital involved and a comparatively small environmental impact.

THE MAIN TRANSFER SYSTEMS

In order to exclude small local abstractions for municipal, industrial and irrigation uses, Day and Quinn (1987) suggest the

following criteria for defining transfers:

• diversion of the runoff over a distance of not less than 25 km from the intake point
• mean annual volume of diverted water should not be less than 0.016 km³/year (0.5 m³/s).

At the present time there are 59 large diversions in Canada in nine Provinces with a total volume of about 140 km³/year (Fig. 7.39, Table 7.34). Canada is the world leader in inter-basin transfers. Of these diversions 94% are for hydropower and only a fraction for shipping and irrigation. The Welland Canal constructed in 1829 to bypass the Niagara Falls can be considered as the earliest diversion in Canada. Its aim was to create a continuous waterway from the Great Lakes to the St Lawrence River and the Gulf of St Lawrence, as well as to obtain cheap electrical power. On average, 8 km³/year of water is carried along the canal from Lake Erie to Lake Ontario. Of the largest diversions undertaken in the first half of the twentieth century, the Bridge–Seton and the Coquitlam Lake–Buntzen in British Columbia are noteworthy with volumes of 2.9 and 0.9 km³, respectively. The diversion of part of the runoff from Albany River (Hudson Bay Basin) to Lake Superior was made with the aim of stabilizing the level in the Great Lakes and maintaining a high flow in the Nipigon, Niagara and St Lawrence Rivers. This diversion follows two routes: from the Ogoki River (a tributary of the Albany River) across Lake Nipigon to Lake Superior, and from the Kenogami River (a tributary of the Albany River) to Long Lake and then along the canal to the Agusabon River and Lake Superior (Fig. 7.40). The total volume of water transferred by the two routes is about

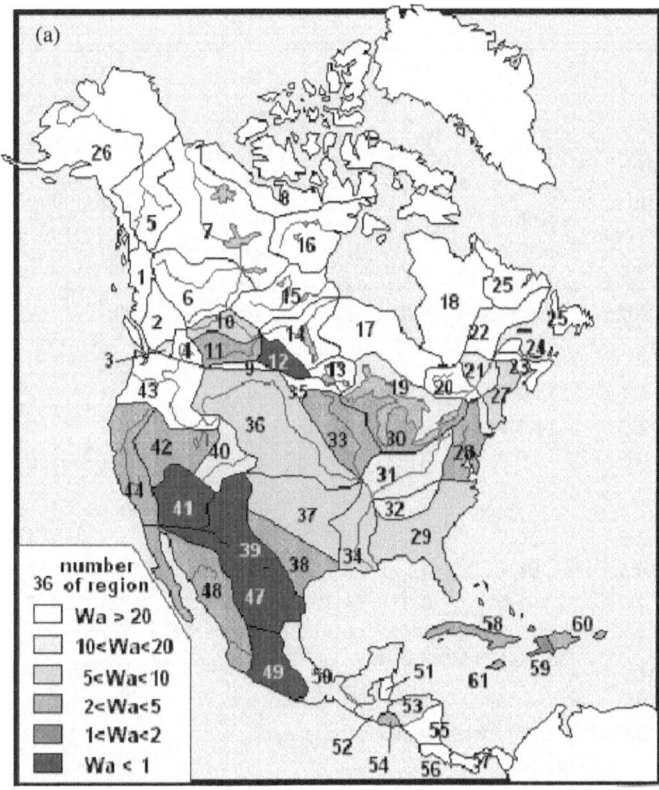

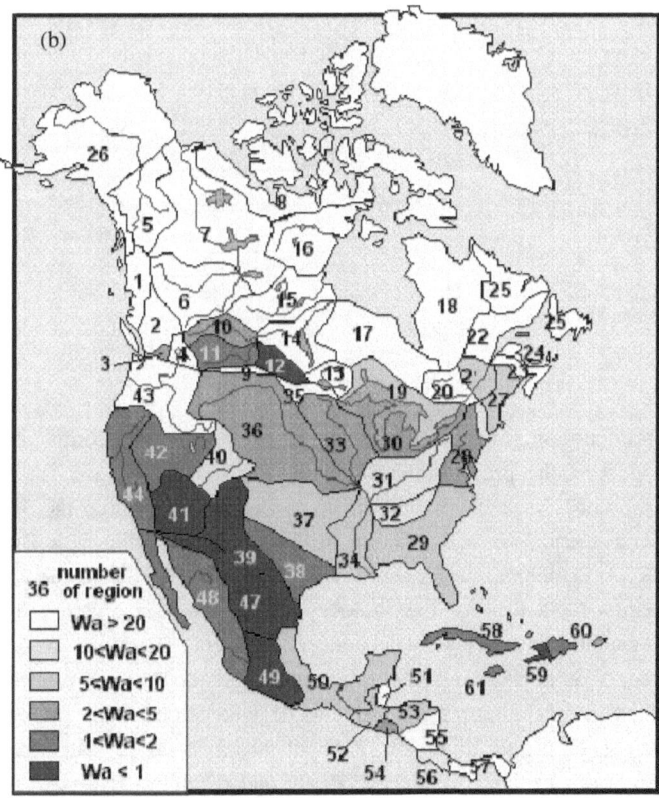

Fig. 7.38 Water availability (m³ × 10³/year per head) by water resources regions of North America: (a) at 1990, (b) at 2025. See Tables 7.30 to 7.32 for key to numbers of regions.

5 km³/year (Peet and Day, 1980), including 3.6 km³/year from the Ogoki River. This flow causes a rise in water level of 10–20 cm, compensating for current abstractions, in particular that of 2.9 km³/year from Lake Superior for the water supply of Chicago and its subsequent discharge to the Illinois River. The decreased flow of the Albany River results in a decrease in the maximum levels of the Great Lakes. Thus the management of the water resources including the abstractions reduces the range of variations in the levels of the Great Lakes. Other examples of important transfers are at Churchill in Manitoba, James Bay in Quebec and Churchill Falls in Labrador. The seven diversions included in these three projects account for about two-thirds of all diversions in Canada.

The James Bay project in Quebec (Fig. 7.41) takes the runoff from the Eastmain, La Grande and Caniapiscau Rivers, some 36% of all water diverted in Canada (51 km³/year out of 140 km³/year). Exploitation of the La Grande River was the first part of an ambitious plan for developing James Bay. For the next stage it is planned to include the Great Whale, Nottaway, Broadback and Rupert Rivers which can increase the runoff by up to about 50%. According to the plan, the area of the basin of the La Grande River is increased from 97 000 km² to 176 000 km². About 90% of the runoff of the Eastmain River and two of its tributaries, the Opinaca and Lower Opinaca (840 m³/s), is directed through Lakes Boyd and Sacami to the lower reaches of the La Grande (reservoir LG-2). There are a series of other diversions and after completing the second phase of the plan, the discharge will be increased to 6000 m³/s for power generation.

The Churchill project in Manitoba (Fig. 7.42) which was completed in 1977, is also important in terms of hydropower generation. The feasibility study showed that although the Churchill River has quite a high hydropower potential, it is more profitable to divert 75% of its runoff (24 km³/year) to the Nelson River and exploit this system rather than each of these rivers separately. The diversion began with the construction of the Missi Falls Dam across the northern exit of the Churchill River from Southern Indian Lake. During the second phase, a canal connecting Southern Indian Lake with the Red River of the Nelson River system was constructed. This diversion was accompanied by the regulation of Lake Winnipeg discharging into the upper reaches of the Nelson River. The level of Lake Winnipeg is controlled with the aim of creating the winter water supply for the hydroelectric station Jenpeg. For this purpose two canals were constructed. One of the canals connects Lake Winnipeg and Lake Playgreen and the second canal Lake Playgreen and Lake Kiskittogisu (Playle et al., 1987).

The third largest scheme is at Churchill Falls in Newfoundland, which was built in 1971. Like the system on the La Grande River, it consists of three large diversions: 196 m³/s from the Julian River,

Table 7.33. *Water availability by selected water-resources regions and selected countries of North America during low water years*

Territory	Period of analysis	Water resources[a] km³/year	Coefficient of variation C_v	Water availability, m³ × 10³/year per head		Minimum water resources during period, km³/year	Water availability during dry year, m³ × 10³/year per head		Water resources at 95% low water content year km³/year	Water availability during 95% low water content year, m³ × 10³/year per head	
				1990	2025		1990	2025		1990	2025
Canada											
North Saskatchewan	1921–85	7.7	0.38	6.1	5.1	4.3	3.3	2.7	3.9	3.0	2.4
South Saskatchewan	1921–85	8.3	0.26	3.7	2.0	2.7	0	0	3.9	0.6	0
Assiniboin–Red River	1921–85	1.5	0.68	0.8	0.5	0.2	0	0	0.1	0	0
USA											
Mid Atlantic	1921–85	100.0	0.21	2.4	2.3	56.7	1.3	1.2	66.0	1.5	1.4
Great Lakes	1921–85	107.0	0.10	4.9	4.5	91.8	4.2	3.9	82.7	3.8	3.4
Upper Mississippi	1921–85	101.0	0.29	4.6	4.0	45.0	2.0	1.7	55.3	2.5	2.1
Missouri	1921–85	85.0	0.33	6.8	4.8	28.2			44.4	2.7	1.5
Texas	1921–85	50.3	0.52	2.4	1.8	13.2	0.3	0.08	12.0	0.3	0.02
Rio Grande	1921–85	7.0	0.47	1.0	0.3		0	0	2.7	0	0
Lower Colorado	1921–85	12.5	0.50	0.3	0.2		0	0	4.9	0	0
Great Basin	1921–85	11.6	0.70	3.1	1.8		0	0	1.1	0	0
California	1921–85	95.0	0.40	2.2	1.4	32.0	0.1	0	49.1	0.7	0.4
Mexico											
North zone	1950–85	11.7	0.50	0.2	0.03				4.5	0	0
Pacific zone	1950–85	49.4	0.38	3.4	1.9				28.0	1.2	0.5
Central zone	1950–85	55.5	0.35	0.9	0.5				26.9	0.3	0.1
El Salvador	1965–85	19.0	0.23	3.6	1.6				12.5	2.3	1.1
Cuba	1969–79	35.0	0.20	2.7	2.0				23.7	1.7	1.2
Dominican Republic	1976–84	20.0	0.37	2.8	1.7				10.8	1.5	0.9
Jamaica	1969–79	8.3	0.26	3.4	2.3				4.7	1.9	1.3

[a] Water resources estimated by our data from the period.

200 m³/s from the Nascaupi River and 130 m³/s from the Kanairktock River which are brought into the basin of the Churchill River (Labrador Peninsula). This diversion has produced a 30% increase in the discharge of the Churchill River which was 1620 m³/s before the diversion. The largest diversion in western Canada is the Kemano project (1952) built for the aluminum industry in British Columbia. Some 3.6 km³/year (115 m³/s) is transferred from the Nechako River (a tributary of the Fraser River) to the Kemano River (Fig. 7.43). An increase in the flow of the Kemano River was engineered by further diversions of the Nechako and the Nanika Rivers.

There are some small water transfer systems in the Prairie provinces for hydropower, but most are for the irrigation needs, the total amount transferred for this purpose being about 2 km³/year. Most of the programmes for irrigation transfers in the Prairie provinces were developed before and immediately after the Second World War with Alberta in the leading position. The Bow River is the main water source in this province. In Saskatchewan, part of the flow of the St Mary River runoff is diverted into Montana but part goes to southern Alberta. There are a number of other examples of similar transfers in the Prairie provinces. In view of the future requirements for water, some other projects envisage a total diversion of about 50 km³/year of water from the basins of the Athabasca, Peace and Churchill Rivers to the basin of the Saskatchewan River, the main river system of the Prairies.

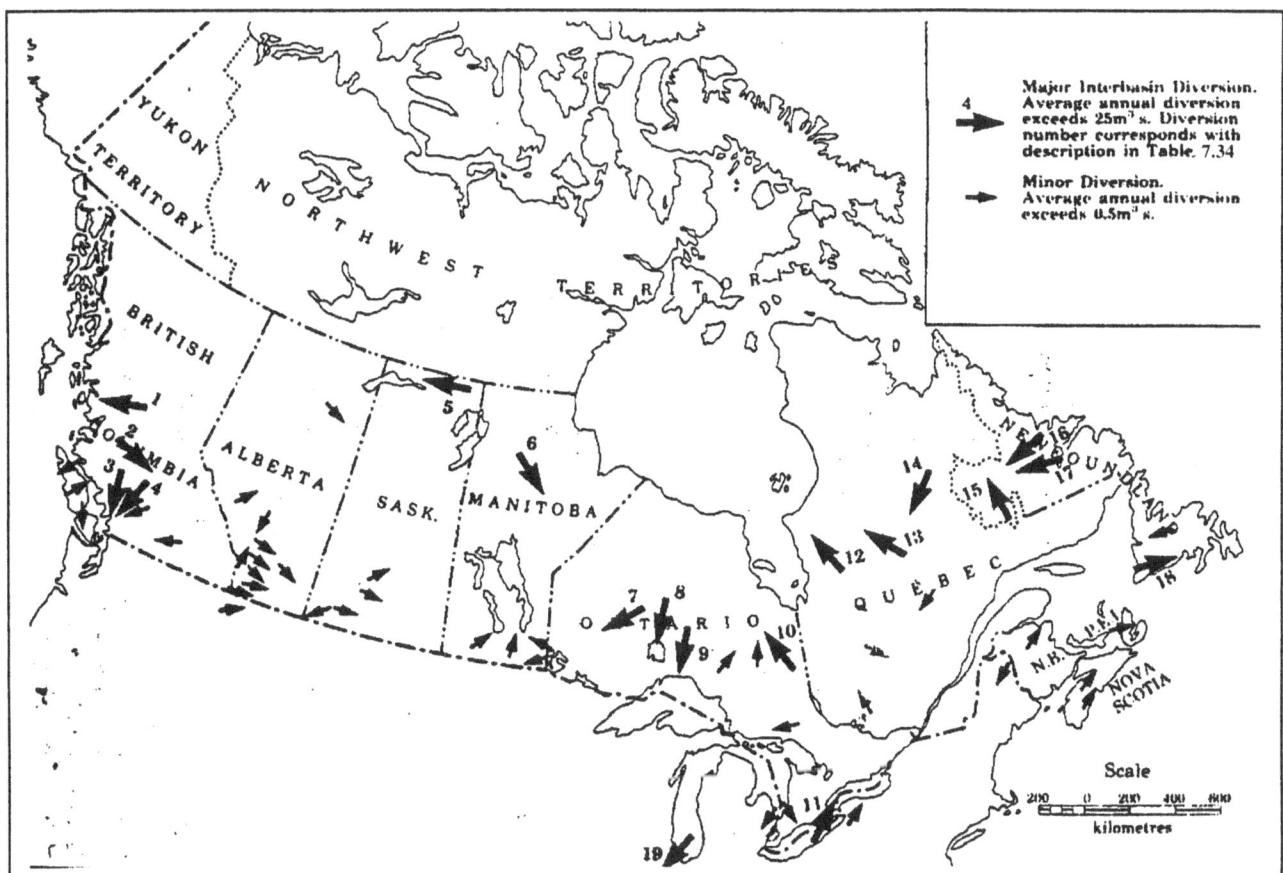

Fig. 7.39 Water diversions in and affecting Canada. (From Day and Quinn, 1987.)

Up to the 1950s and 1960s diversions of water were strongly supported by the Federal Government, which provided funds, but by the 1970s proposals for further schemes were being criticized and soon most of the proposed diversions within Canada were being shelved. Irrigation and hydropower projects were not favoured, because of economic problems but there were also other problems some socio-economic and then in 1980s environmental questions were raised. In 1987, a symposium on the diversion problems was held in Saskatoon. It considered the ecological and socio-economical consequences caused the export of water and discussed the prospects for further transfers. A great deal of attention was devoted to the Grand Canal scheme. A number of problems were identified to be studied before making decisions about the new diversions. It was recognized that the diversion of large volumes of water from one river system to another would result in some changes in the runoff regime, in the river morphology, in water chemistry, and in the life depending on the water, leading to changes in number and diversity of species. Such diversions may also influence the socio-economical sphere and may lead to climatic change at the local and larger scales.

INFLUENCES ON THE RIVER REGIMES AND MORPHOLOGY

Gomes-Amaral and Day (1987), MacLean and Beckstead (1987), Roy and Messier (1987) and Woo and Waylen (1987) have considered the impacts of water transfers, such as the increased channel and shore erosion and the increased sedimentation in reservoirs and at the mouths of rivers. There are however, a number of positive consequences. For example the Kemano project has reduced flood peaks and has decreased the frequency of floods in the Fraser River. However in the lower reaches of the Nechako River, erosion has increased, making the mouth of the Nautley River deeper by 1.8 m. The increase in the flow of the Cheslata River from 5 m^3/s to 138 m^3/s has stimulated erosion: the channel has expanded from 5 m up to 75–150 m and it has become entrenched by some 5–10 m. It has induced erosion of the rocky channel around the Cheslatta waterfall and encouraged silting in Lake Cheslatta (Gomes-Amaral and Day, 1987). The transfer from the Churchill River to the Nelson River has resulted in intense erosion of the shore of Southern Indian Lake and bed erosion in the Barntwood River (Playle et al., 1987). The construction of the Bennett Dam

Table 7.34. *Major river diversions in and affecting Canada*

Number in Fig. 7.39	Province	Project name	Contributing basin(s)	Receiving basin	Average annual diversion, km³	Operational date	Type of use
1	British Columbia	Kemano 1	Nechako (Fraser)	Kemano	3.6	1952	Hydropower
2	British Columbia		Bridge	Seton Lake	2.9	1934 (1959)	Hydropower
3	British Columbia		Cheakamus	Squamish	1.2	1957	Hydropower
4	British Columbia		Coquitlam Lake	Buntzen Lake	0.9	1902 (1912)	Hydropower
5	Saskatchewan		Tazin Lake	Charlot (Lake Athabasca)	0.8	1958	Hydropower
6	Manitoba	Churchill Diversion	Churchill (Southern Indian Lake)	Rat–Burntwood (Nelson)	24.6	1976	Hydropower
7	Ontario		Lake St Joseph (Albany)	Root (Winnipeg)	2.7	1957	Hydropower
8	Ontario		Ogoki (Albany)	Lake Nipigon (Superior)	3.6	1943	Hydropower
9	Ontario		Long Lake (Albany)	Lake Superior	1.3	1939	Hydropower, logging
10	Ontario		Little Abitibi (Moose)	Abitibi (Moose)	1.3	1963	Hydropower
11	Ontario	Welland Canal	Lake Erie	Lake Ontario	7.9	1829 (1951)	Hydropower, navigatation
12	Quebec	James Bay	Eastmain–Opinaca	La Grande	26.8	1980	Hydropower
13	Quebec	James Bay	Fregate	La Grande	1.0	1982	Hydropower
14	Quebec	James Bay	Caniapiscau	La Grande	25.1	1983	Hydropower
15	Newfoundland	Churchill Falls	Julian–Unknown	Churchill	6.2	1971	Hydropower
16	Newfoundland	Churchill Falls	Naskaupi	Churchill	6.3	1971	Hydropower
17	Newfoundland	Churchill Falls	Kanairiktok	Churchill	4.1	1971	Hydropower
18	Newfoundland	Bay d'Espoir	Victoria, White Bear, Grey, Salmon Lakes	Northwest Brook (Bay d'Espoir)	5.9	1969	Hydropower
19	Illinois	Chicago Diversion	Lake Michigan	Illinois Mississippi	2.9	1848 (1900)	Municipal supply, sanitation

Source: Day and Quinn (1987).

in British Columbia has caused considerable changes in morphology of the Peace River both upstream and downstream of the dam; about 98% of the solid discharge is deposited in the reservoir. Due to the decreased transport in some rivers, deposition takes place and deltas form (Van Kooten *et al.*, 1987). As a result of the diversion of part of the flow of the Eastmain River to the La Grande River its velocity has significantly decreased and sedimentation has started, because transport has decreased by a factor of 10 and movement of suspended matter from the estuary to the bay has experienced a 25-fold decrease (Drinkwater, 1987).

CHANGES IN WATER CHEMISTRY
The problems of water chemistry changes as a result of diversions, were considered by Playle *et al.* (1987) using the Churchill project as an example and by Remenda and Davis (1987) con-

sidering projects implemented in the Prairie provinces. Monitoring, analysis and archiving of data on water chemistry are necessary for a better understanding of the influence of hydropower construction on the quality of river water in northern Canada. It should be noted that no significant changes in water quality were revealed in the north, whereas in the Prairie provinces there were significant changes in the chemistry of the diverted water.

The lakes and rivers of the Churchill diversion region are located on post-glacial deposits lying over the Precambrian basement where water does not contain much suspended matter. The Saskatchewan–Nelson River in the Prairies has greater turbidity. The total concentration of ions and suspended matter in the Nelson River is much higher than in the Churchill River. Measurements of water chemistry before and after the diversion at some points in

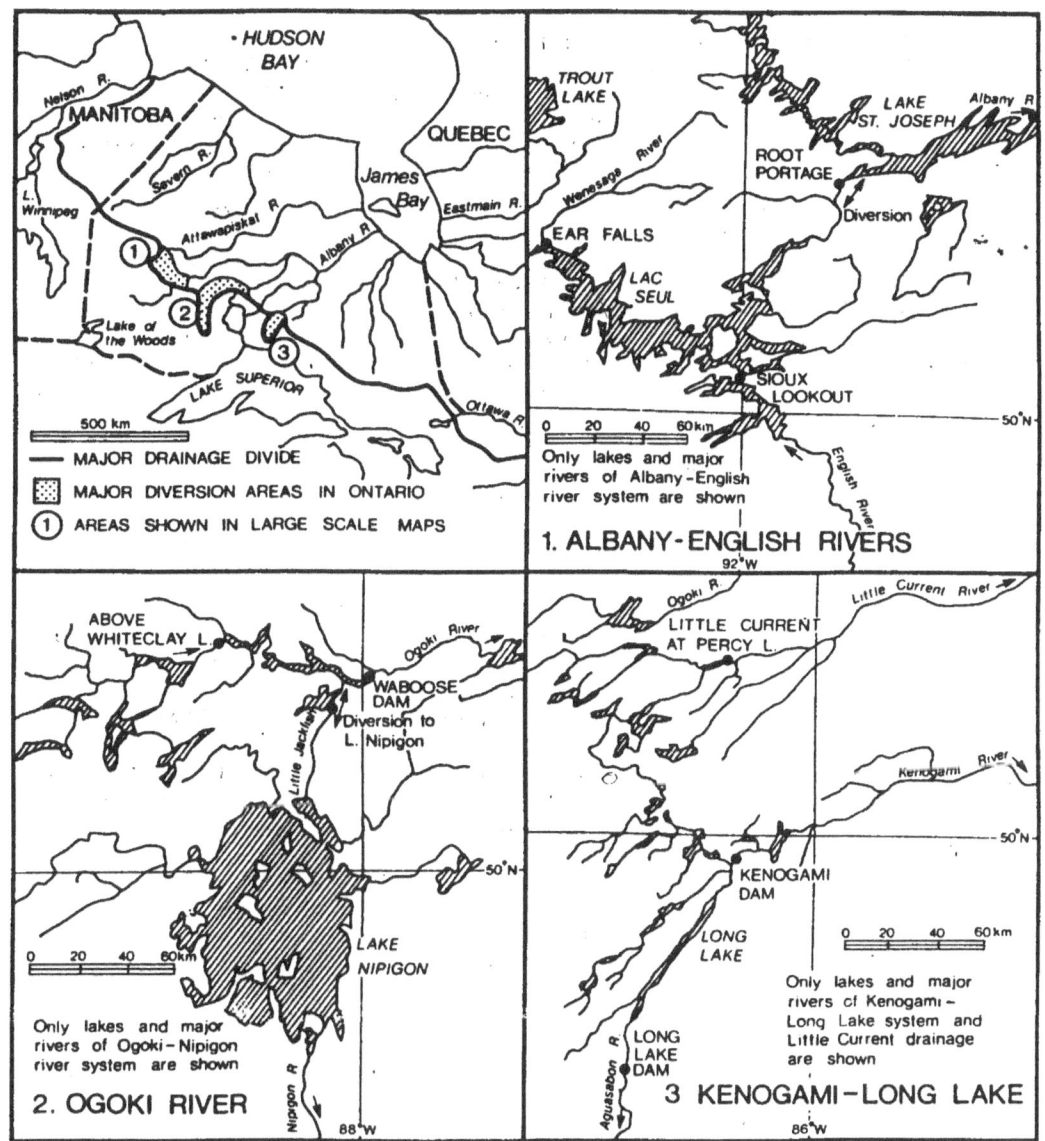

Fig. 7.40 Location of inter-basin water transfer sites in Ontario. (From Woo and Waylen, 1987).

the Churchill and Nelson Rivers have revealed some changes in a number of determinands. For example, the phosphorus concentration has actually increased in all stretches of the Churchill transfer system; turbidity has increased and nitrogen concentration has decreased, the total inorganic carbon decreased at three out of five observation points, the electrical conductivity and alkalinity decreased at practically all the other points. In Lake Winnipeg in the Nelson River the total inorganic carbon increased at three out of four points. The electrical conductivity increased at one point and alkalinity remained unchanged (Playle *et al.*, 1987).

In the Prairies where the evaporation sometimes exceeds precipitation, these conditions promote a high concentration of mineral salts in ground water. Hence during diversions special attention

should be given to the route to reveal possible sources of strongly mineralized ground water. Thus according to Remenda and Davis (1987), the diversion from Lake Diefenbaker has resulted in a significant deterioration in water quality south and east of Saskatoon. Water is pumped from Lake Diefenbaker passes through five reservoirs and a significant change was revealed in the third reservoir. A calcium bicarbonate water type entering the lake changes to a sodium sulphate type at the exit. Deterioration of water quality, an increase of concentration of sodium and sulphate ions and in total dissolved solids is observed over the entire lake length. Studies have shown these changes to be caused by the discharge of groundwater to the lake from the horizons that are characterized by a high level of pollution. Hence, as stressed in Remenda and

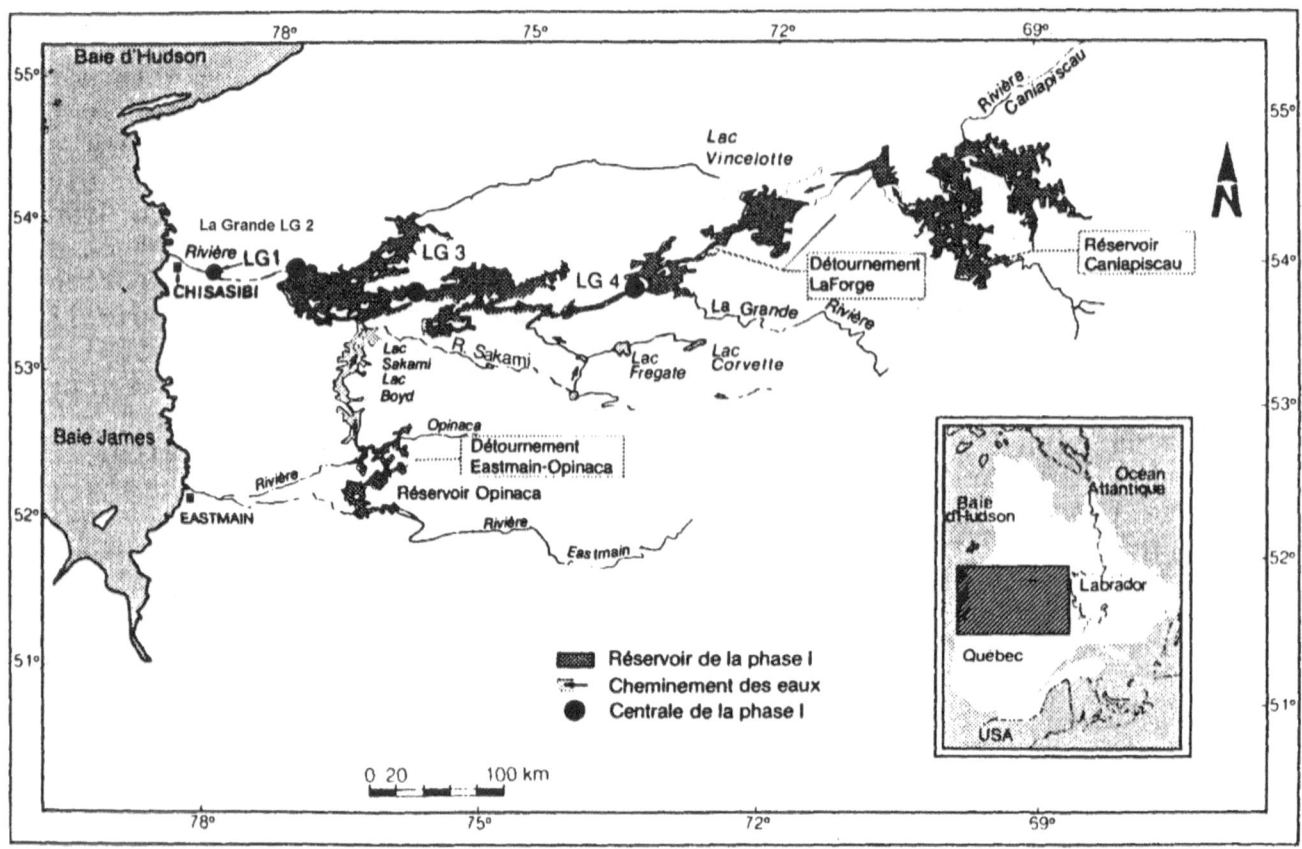

Fig. 7.41 The hydroelectric complex at La Grande River, Quebec, Canada. (From Roy and Messier, 1987.)

Davis (1987), for planning diversions in the Prairies a careful preliminary hydrogeological and hydrochemical analysis should be undertaken along possible routes. The authors have also assessed the possible changes in water quality in other reservoirs and lakes of the region which can be used in the future.

HYDROBIOLOGICAL AND ECOLOGICAL CONSEQUENCES

The biological changes induced by transfers appear to be very significant (Drinkwater, 1987; Gomes-Amaral and Day, 1987; Milko, 1987; Penn, 1987; Seagel, 1987; Thompson, 1987, etc.). They may concern both aquatic and coastal flora and fauna. Any diversion affects the primary productivity in the water body governing the development of the entire trophic web. Diversions of large water volumes significantly change the regimes of rivers and can lead to the change of biological species. Thus a diversion of 90% of the runoff of the Eastmain River has caused the significant changes in its estuary. Before the diversion the river estuary was brackish fresh water, but after the diversion, due to a significant decrease in the current speeds, brackish water began penetrating 8 km upstream. Intrusions of saline water are accompanied by phytoplankton blooming and a change from freshwater to ma-

rine species (Drinkwater, 1987). The diversion of the Churchill River has changed the processes of ice cover formation and ice melting, decreasing summer temperatures and reducing the flood discharge from the Churchill River to the northern part of Southern Indian Lake. Along the diversion route to the Notigi reservoir, an increase in primary production has taken place with a simultaneous increase in the concentration of naturally occurring inorganic mercury in the water and in the fauna and flora. The latter has resulted in the decrease in the market for fish products in the region, since many lakes now have a higher mercury concentration in fish (Hecky, 1987).

The transfers have quite serious implications for mammals changing their numbers and leading to the death of higher plants. Thus in Van Kooten *et al.* (1987) the construction of the Bennett Dam (British Columbia) had an extremely unfavourable environmental impact. As a result, a reservoir of 680 square miles was created flooding the lower reaches of the basins of three rivers (Finlay, Parsnip and Peace). The Peace–Athabasca Delta was one of the most productive in Canada before the construction of the dam. Construction of the dam has cut the spring floods and for two years the water level in Lake Athabasca and in the delta decreased by more than 1 m with several minor water bodies

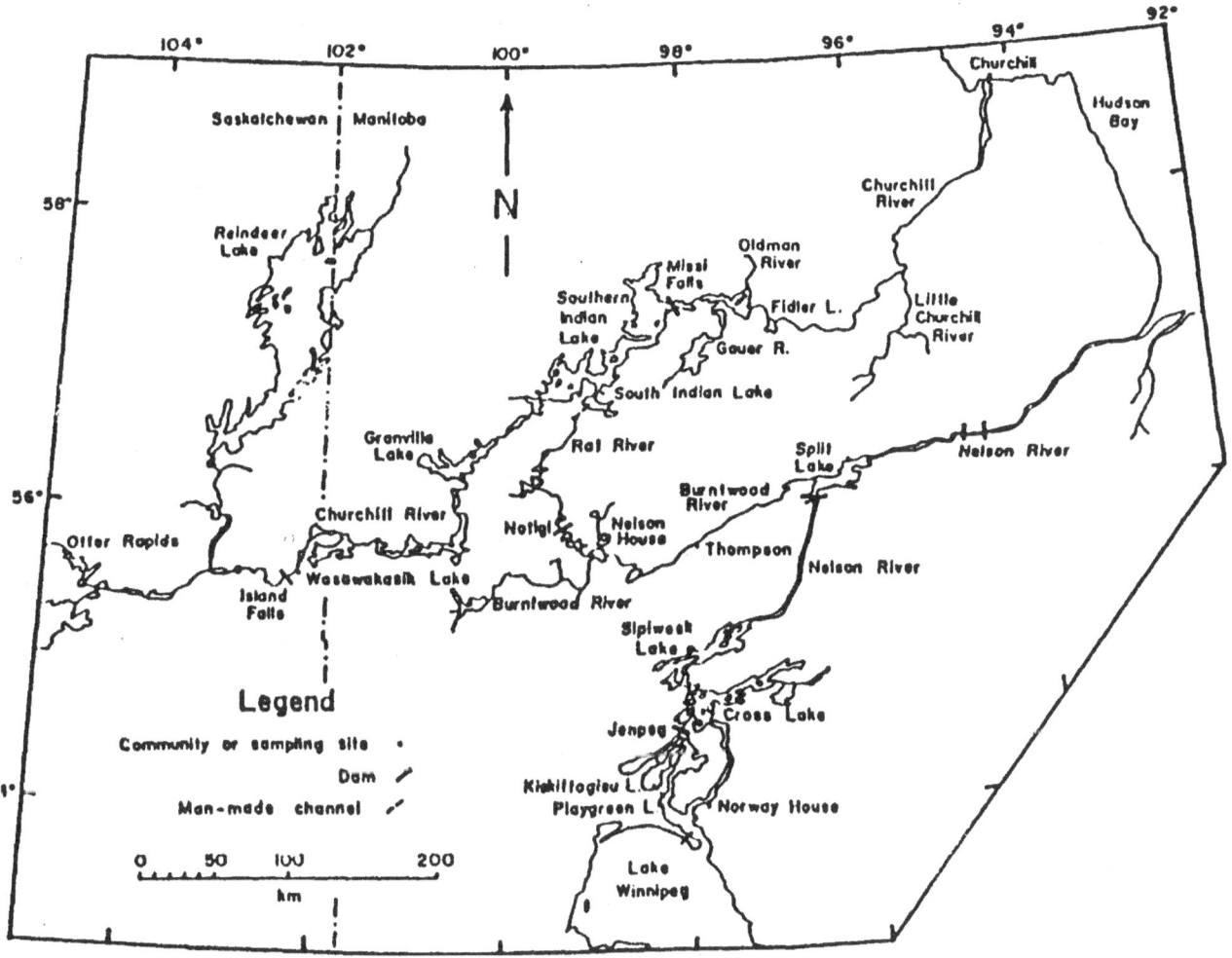

Fig. 7.42 The Churchill River diversion and Lake Winnipeg regulation project area, Manitoba, Canada. (From Playle *et al.*, 1987.)

in the delta completely drying up. As a result, the population of waterfowl has decreased. Due to the winter freezing in some lakes the population of muskrat has been reduced considerably. Many fish species were affected and the number of bison has decreased.

The implementation of the Kemano project caused a loss of trees (from 19% to 28%) due to the pollution by fluoride which has spread across an area of more than 7042 ha. The forest was completely destroyed over the 32 ha close to the aluminum plant. The level of fluorine in vegetation between 1976 and 1983 increased by more than 60% (Gomes-Amaral and Day, 1987). One more negative consequence of the Kemano project was the decrease in the spawning of salmon in the rivers of the region which was caused by both the water temperature increase and silting of Lake Cheslatta. As a result, the number of eggs in the salmon has decreased from 2500 to 100 (Canadian Fish and Wildlife Service, 1979). The influence of diversions on the spawning of salmon is a topic which is especially sensitive in Canada.

A detailed study of the possible ecological consequences is extremely important for the planning of future diversions. One of the best examples is the study of the effect of the transfer of water from the McGregor River, a tributary of the Fraser River (Pacific Basin) to the Parsnip River, a tributary of the Peace River (Arctic Basin). For this diversion, it was planned (1976) to construct a 140-m dam in the McGregor River to create a reservoir of 23 500 ha and to connect the Arctic and the Pacific Basins. However, a detailed study of the redistribution of the fauna between the Pacific and the Arctic Basins which was made for four groups (fish, bacteria, viruses and parasites) has shown the project to be not advisable and it was halted in 1978. In particular, it was revealed that as a result of the diversion, a parasite (*Ceratomyxa shasta*) would get in to the Parsnip River basin. It is distinguished by a high pathogenic ability and it can affect many valuable fish species of the Arctic Basin over quite a wide area, and probably over the entire Peace–Athabasca–Mackenzie system (Seagel, 1987).

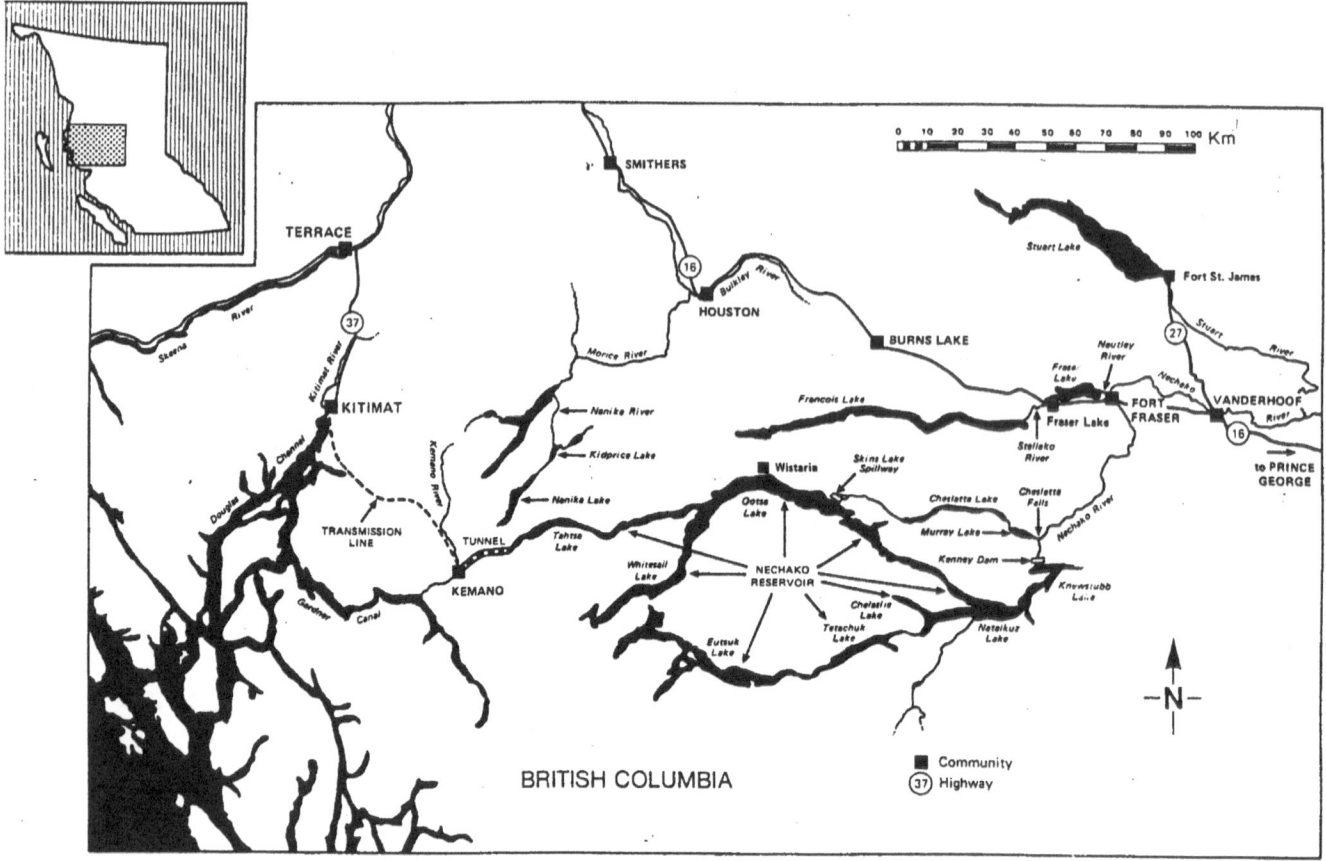

Fig. 7.43 The Kemano diversion, British Columbia, Canada. (From Aluminum Company of Canada Ltd., 1982.)

CLIMATIC CHANGE

Climatic changes caused by small runoff diversions are usually minor. It is generally believed that they do not result in serious consequences but the schemes at a larger scale can cause significant climate changes. Eley and Lawford (1987) have analyzed the climatic effect of diversions depending on their scale. For diversions of smaller scale (less than $10 \, km^3$/year) the main changes are observed in albedo, transpiration, evaporation and surface temperature. The climatic effect is evident at a distance of only a few kilometres from the source. For diversions at the medium scale ($10–100 \, km^3$) the climatic effect is evident at a distance of tens of kilometres. Usually changes relate to additional losses by evaporation which cannot be balanced by increased precipitation. Also, the diversions which change flows from one ocean to another can be the cause of the change in surface water temperatures in the ocean, together with changes in salinity and oceanic stability. An example of diversions of the medium scale is the Grand Canal project which is considered below. Large-scale projects (from 100 to $1000 \, km^3$) of the NAWAPA (North American Water and Power Alliance) type can cause significant climate change over large areas. For example, a diversion of $300 \, km^3$ is equiva-

lent to a 100 mm increase in evaporation from a basin of about 3 million km^3 and can cause the precipitation to increase over adjacent areas.

SOCIO-ECONOMIC CONSEQUENCES

Most Canadian diversions aim at the production of cheap hydropower. However, the predicted costs and market benefits both for energy and agriculture were sometimes significantly overestimated and the profit margin was much lower than that planned. Thus a social analysis of the Kemano project (Gomes-Amaral and Day, 1987) has shown the cost of the project to be much higher than the profit primarily as a result of large initial investments, but also because of the decrease in the world price of aluminum. The benefit of some diversions for irrigation schemes in the Prairies were also questionable.

The diversions in Canada served in the development of remote regions creating work places and improving living conditions. The Kemano project established of a large industrial centre in a sparsely populated region of British Columbia. As a result, a small fishing village has become a town with a large population with a well-developed aluminum industry and an increased standard of

Table 7.35. *Water diversion projects in and affecting Canada*

Province	Contributing basin(s)	Receiving basin	Average annual diversions, km^3	Type of use
British Columbia	Coutenay (Columbia River basin)	Columbia	5.4	Hydropower
British Columbia	McGregor (Peace River basin)	Parsnip (Peace River basin)	6.2	Hydropower
British Columbia	Yukon	Taku	23	Hydropower
Alberta	Apper–Athabasca	North Saskatchewan	3.6	Irrigation
Alberta	Peace, Smoky River, Lower Athabasca	North Saskatchewan	23	Irrigation, hydropower
Saskatchewan	North Saskatchewan	South Saskatchewan	8.9	Irrigation
Saskatchewan	Churchill	Saskatchewan	8.9	Irrigation
Manitoba	Saskatchewan	Assiniboine (Red River basin)	4.5	Irrigation
Quebec	Peace, Smoky River, Prodback	Rupert	31	Hydropower

living. However, the creation of large industrial centres sometimes conflicts with the interests of the local indigenous population. The life of many American Indians is based on hunting and fishing. The flooding of some areas inhabited by Indian tribes and the ecological consequences deprive the tribes of the sources of living. The Kemano project resulted in Cheslatta Band reservation being flooded. For some reason the Indians who had to move did not receive timely compensation for the losses incurred. The construction of the Bennett Dam has also caused losses to the Indians inhabiting that region. During the construction of the hydropower complex planned in the Great Whale River (Grand Vallin) the losses of 7000–9000 American Indians inhabiting the construction area were expected to be compensated (Dubeau, 1991). Experience has shown that in all cases, a careful analysis of the interests of all groups of the population affected by diversions is necessary, probably undertaken by commissions consisting of the representatives of these groups, together with widespread consultation. Careful control of the amount of compensation and of its payment is also required.

Day and Quinn (1981), in assessing the socio-economic influences of diversions, came to the conclusion that such diversions are a reasonable way of maintaining economic activity and employment. In this sense the potential of diverted water can be considered a resource to be used for stimulating economic activity, provided this does not afflict future generations. The projects should be approved only after a careful consideration of the possible implications and after signing flexible long-term contracts, so that the cost of energy can be indexed annually. Diversions at the smaller and medium scale are preferable to large-scale schemes both in terms of lower environmental damage and fewer adverse economic effects. Recently Canadians have adopted a rather negative attitude towards diversion projects; the inhabitants of the source regions are usually discontented with diversions. The position of the Canadian Government after the 1970s and 1980s was also generally negative. However it is quite probable that in the future the political climate may change with regard to diversions. In any case participation of the population affected in the planning of the project appears to be very important for eliminating many possible future conflicts.

A PERSPECTIVE ON DIVERSIONS

In the 1960s the Canadian Government was quite active in supporting water transfers. At that time a number of projects were being implemented and new ones planned. A list of these projects, mainly small-scale projects, based on Shiklomanov and Markova (1987) is given in Table 7.35. Some projects were developed jointly with the USA, addressing water supply problems at the continental level. Such large-scale projects include the well-known NAWAPA project, a scheme for reversing the Liard River and the CeNAWP (Central American Water Project) and Grand Canal Projects.

The NAWAPA project, developed by the Ralph M. Parsons Company of Los Angeles, envisaged the diversion of 135 km^3/year of the flow of the Yukon and the Mackenzie Rivers across British Columbia to the west coast of the USA and the north of Mexico. The project included six dams more than 450 m high, a large reservoir in the Rocky Mountains in British Columbia, a shipping canal across the Prairies from the Great Lakes, and many other dams, canals and reservoirs. The cost of this project was extremely large. The value to Canada of the project was much less than to the USA. It was planned that only 20% of diverted water would be used in Canada, allowing only a 15% increase in the area of irrigation and an annual profit of 9 billion dollars. Based on Canadian water, the USA could increase its irrigated area by 72% and obtain an annual profit of 30 billion dollars (at 1960s prices). It was believed that the construction of these large systems would have a positive influence on the development of the economy of Canada, revive remote northern and northwestern regions, increase the population and employment, and permit more efficient and multi-purpose use and protection of water resources

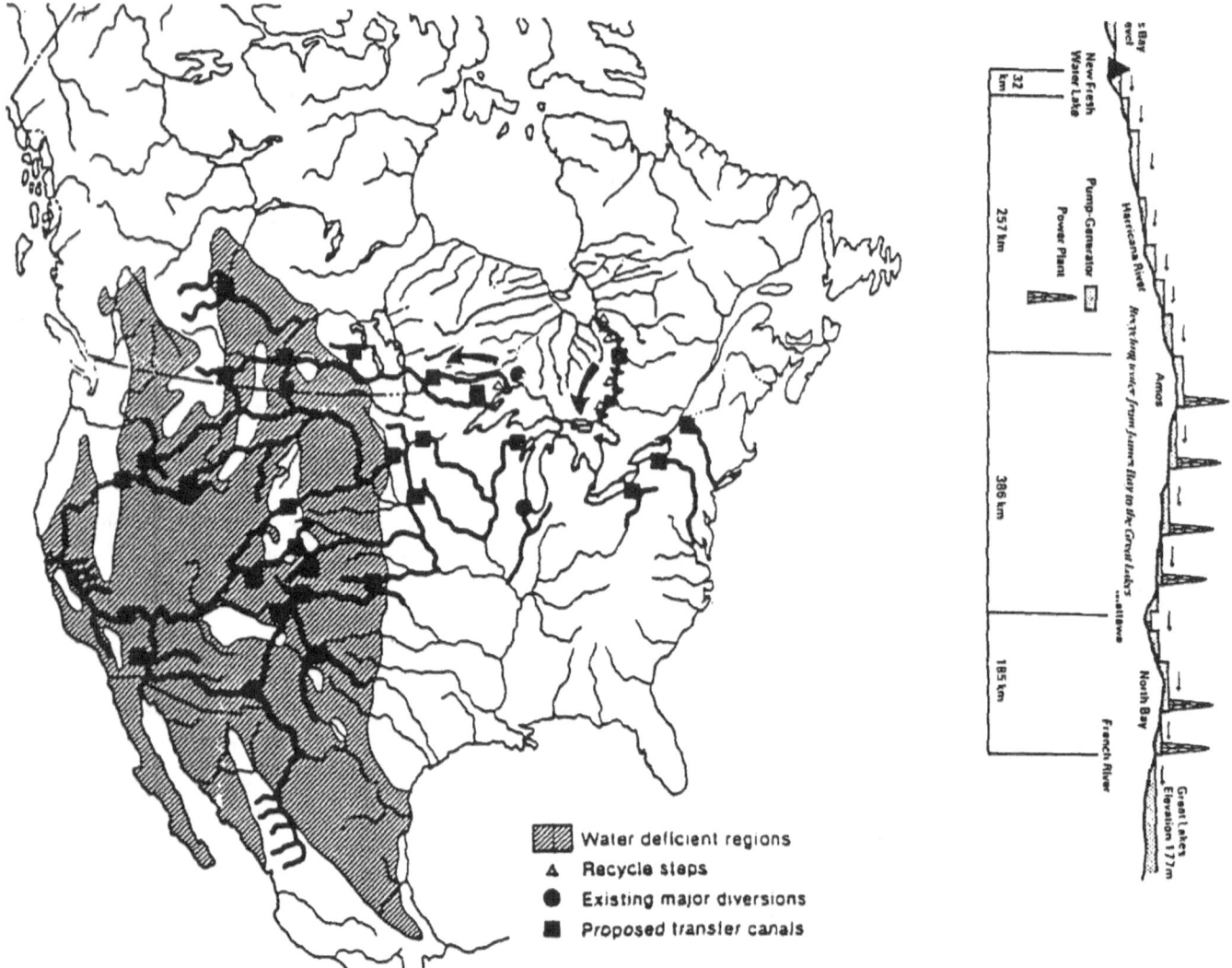

Fig. 7.44 Great Recycling And Northern Development (GRAND)
Canal Project, James Bay, Quebec, Canada. (From Kierans, 1987.)

of the country (Kierans, 1980). However, the NAWAPA project
was not approved, although the smaller-scale diversions within
the framework of the project seemed to be quite feasible and economic, but the problem centred on overcoming the political and
economical difficulties. The Western States Augmentation Concept project was developed by Lewis Gordy Smith. It included
reversal of the flow in the Liard River and the diversion of part
of the Mackenzie River across the Rocky Mountains to Williston
reservoir and then along the Fraser River south to the USA. Roy E.
Tinney developed the CeNAWP project, which was three times
less expensive than the NAWAPA. The project envisaged abstract-
ing water from the Mackenzie, Churchill and Nelson Rivers to
Lake Great Bear and then transporting it through the Great Slave
Lake and Lake Athabasca to Lake Winnipeg. From there the flow
was to be transferred southward to the USA and eastward to Lake
Superior.

The GRAND (Great Recycling and Northern Development)
Canal project was developed in 1959 and envisaged diversion of
part of the water from James Bay across the Great Lakes to the
regions with water shortages. The GRAND Canal project appears
to be the most promising of those outlined, and at the conference in
Saskatoon in 1987 special attention was given to this project, since
it was the only one of all other projects whose implementation was
possible to begin before the end of the twentieth century.

The main features of this project are (Kierans, 1987)
(Fig. 7.44):

1. A new enclosed freshwater lake in James Bay with a system
   for controlling the level.
2. A system for transferring water from James Bay to the Great
   Lakes, including interception systems, special aqueducts,
   reservoirs, pumps, power stations and flow control systems.

3.   A system for the regulation of the Great Lakes.
4.   A system to supply water from the Great Lakes to regions in Canada lacking resources, including intermediate reservoirs, aqueducts, pumps, power stations and distribution systems.
5.   A system to supply water from the Great Lakes to similar regions in the USA.

The GRAND Canal scheme envisaged the transfer of 60% of the freshwater balance of Hudson Bay which would inevitably lead to a number of ecological problems for northern Canada. Special attention was devoted to the study of these problems. Milko (1987) indicated that the transfer would change the onset of the formation of the ice cover and its thickness in Hudson Bay and that the ice might increase. In addition, vertical mixing in Hudson Bay would decrease contributing to pack ice formation. According to Milko, changes in the ice cover and increased salinity, caused by the restricted freshwater flow to Hudson Bay, could result in the reduced primary production, which in turn would cause decreased productivity at all levels in the trophic web.

According to Hackl and Mulamoottil (1987), for the enclosed part of James Bay, the thickness of the ice after dam construction would increase as a result of the decreased salinity and ice would be formed earlier. The attenuated circulation would contribute to the formation of strong thermal stratification and would practically stop complete vertical mixing. As a result, the concentration of suspended matter in James Bay would increase and transparency decrease resulting in a smaller depth of the eutrophic zone. In addition, the accumulation of organic allochthonous material would increase. The intrusion of salt water to James Bay and farther to freshwater lakes would influence the mixing system, and as a result water at the bottom of James Bay would become anaerobic. A series of chemical changes could be induced by the release of some chemical substances such as ammonium and hydrogen sulphide which would worsen the water quality significantly.

A study of the changes in productivity in James Bay has yielded favourable results (Duthie, 1979; Baxter, 1980; Hackl and Mulamoottil, 1987). The enrichment of water in James Bay would increase primary productivity leading to the increased productivity at all stages of the trophic web. The entire system would become eutrophic. As indicated by Hackl and Mulamoottil, most sea fish adapt well to the decreased salinity. However, for some fish migrating at different stages of their life cycle between James Bay and Hudson Bay, the construction of a dam would disturb their usual habitats.

As shown by Milko (1987), the construction of a dam would result in the increased salinity and decreased primary production in Hudson Bay. This would lead, on the whole, to significant depletion of fish stocks. A number of rich fishery regions such as the Grand Banks, Scatian Shelf and Gorge Banks would suffer. The enclosure of James Bay would change the situation in the estuaries of most rivers flowing to the bay. Salinity might significantly increase, especially in the southeastern part of Hudson Bay which would inevitably induce a change in species. The ability of fish to adapt to the increased salinity would play the main role in their survival under the new conditions.

Milko (1987) and Hackl and Mulamoottil (1987) showed that the construction of a dam would negatively influence the mammals inhabiting the basin. Thus it would be difficult for whales to adapt to the reduced living space resulting from enclosure of the bay. For seals and walruses the increased ice cover would reduce the open water area necessary for respiration. The decrease in the number of seals might also reduce the numbers of polar bears. Milko pointed out that in connection with water temperature and salinity changes the populations of marine and shorebirds could decrease. Hackl and Mulamoottil however, believed that in James Bay, there would be more food for migrating birds. Assessing the hydroclimatic consequences of the GRAND Canal project, Eley and Lawford (1987) indicated that the decreased freshwater inflow to Hudson Bay might have certain implications for the North Atlantic. Here, plankton numbers would decrease, salinity would rise, the vertical mixing layer in the ocean would increase and the ocean-to-atmosphere heat and moisture flux would increase. This effect would be more evident in the years with below-average runoff and there might also be implications in time for the climate of Europe.

Canadian experience shows that for planning new transfers, detailed hydroclimatic, ecological and socio-economic studies are necessary in order to avoid possible negative consequences. Adoption of projects should be based on economic principles such as an unbiased prediction of costs and benefits both for the basins which are the source of the water and for the receiving basins. In view of the possible ecological consequences, the social and environmental factors should be taken into account on an equal basis with engineering and economic indicators. The possible long-term changes resulting from the implementation of the project (climatic, ecological, technological, etc.), should also be considered (Fitzgibbon, 1987).

Since the early 1990s the economy of Canada has climbed out of recession and in the light of the predicted changes of global climate, the need for new transfer projects may again become acute. Many specialists (e.g. USEPA, 1984) predict a significant air temperature increase and a decrease in precipitation for the coming decades over much of the USA and Canada including the regions where water resources are under stress. In these regions water requirements are substantial, but the resources are small and water quality is a growing problem. Many scenarios of future climate show that in the middle latitudes of the USA and in some regions of Canada, the mid-summer will become more hot and dry, ground water recharge will decrease and water availability will be less. Analysis of the sensitivity of the river systems to climate changes (Stockton and Boggess, 1979; Flashka *et al.*, 1989; IPCC, 1990, 1992, 1995, 2001a, b; Shiklomanov and Lins, 1991) have shown that the increased temperature and decreased precipitation can have extremely adverse consequences for the water resources in arid regions, in particular in the basins of the Rio Grande and Colorado and also in California. Thus a temperature increase of 1–2 °C and a precipitation decrease of 10% can lead to the decrease of runoff in arid and semi-arid regions by up to 40–50%. All this may result in changes in agricultural production and create a need to expand irrigation. The implementation of the transfer projects, including large-scale projects in the interests of both the USA and Canada, could again become quite acceptable. According to available global circulation model scenarios, in many regions of the northern and northeastern Canada global warming could lead to a significant increase in the importance of water resources. These are, however, the problems of the future.

7.8 CONCLUSIONS

This chapter attempted to assess the water resources of North America and to estimate the changes and trends in the continent's water availability and its use of water over the twentieth century; forecasts were made for water availability and use of water to 2025. A brief overview of the continent's main water management problems were presented with a more detailed consideration of Canadian inter-basin transfers. As a result of these studies, the renewable water resources of North America were estimated to average 7870 km³/year for the period 1921 to 1985. This calculation is based on data on the annual runoff observed at 156 hydrometric stations taking into account the mean runoff at an additional 160 stations sited mostly at mouths of rivers. Compared to previous estimates (Australian Water Resources Council, 1974; Korzun 1974a; L'vovich, 1974; Baumgartner and Reichel, 1975), this estimate was made for a longer period and it took account of data from stations in Central America set up in the 1970s and 1980s.

Hydrological knowledge of the continent is extremely varied and consequently the estimates of renewable water resources differ in accuracy from one region to another. The most accurate estimates relate to the central region of the continent: the USA, the Pacific rim and the areas of Canada draining to the Atlantic. For some regions with scarce data, such as the Arctic Basin, the estimates were based on figures from national publications (CNC/IHD 1978; Pearse *et al.*, 1985). Calculations of runoff from Central America are based on meteorological data due to records of runoff being short and on estimates in L'vovich (1974) and WRI (1990, 1992,

1996, 2000). The current assessment of renewable water resources of North America is the most reliable up to the present, in spite of these drawbacks.

A statistical analysis of the discharge series for the largest rivers of the continent, as well as for the total runoff from the regions, from countries and for those parts draining to the different oceans, has not revealed any significant trends over the period of calculation. In some basins there were, however, slightly increased discharges during the second half of the period. All subsequent calculations were based on the assumption that all the time series were stationary.

The natural runoff patterns in most basins of North America have changed considerably during the twentieth century. The runoff in many rivers is highly regulated as it is harnessed for different needs. An assessment of the water used in the different sectors from 1900 to 1995 showed a significant growth in abstractions: the water intake over the period under consideration has increased 10-fold. At the present time it is 672 km³/year with the consumption of water standing at 241 km³/year. The most water is used in the Central region. Up to 1980 it was about 80% of the water used across the entire continent, but thereafter, it decreased to some 75% of the total. The largest rates of growth in abstraction occurred in the Southern region. They were also quite high in the Northern region. Agriculture uses most water (about 45%), then industry (about 40%), of which 10% was used for domestic purposes. Industrial water use was considerable during the second half of the twentieth century. This was due primarily to the extensive development of thermal power.

Forecasts of water use to 2025 show that total abstractions are likely to increase by 116 km³/year or by 17% over the 30 years. The largest rates of growth will still occur in the Southern region and the largest absolute values in the Central region. Growth in abstractions is likely to be largest for the domestic sector, while agriculture and industry will use slightly smaller percentages, 44% and 39%, respectively. Specific water availability per head remains quite high, on average. For 1995 it was estimated to be 17 100 m³/year per head, decreasing to 12 900 km³/year per head by 2025. However specific water availability varies enormously across North America. Water resources are already stressed in the basins of the Colorado and Rio Grande Rivers (USA), in the Assiniboine–Red Basin (Canada) and in a large number of basins in northern and central Mexico where specific water availability values are now less than 1000 m³/year per head. By 2025 this situation will be far worse in these regions. Haiti suffers from the same low levels currently. During dry years much of the continent will have water availability values of less than 5000 m³/year per head.

The problems of water shortages in North America are extremely important. They appear not only in regions where water resources are small, but also where initially resources were sufficient, but where the excessive growth of cities and towns has resulted in an increased demand. During the second half of the twentieth century there was a need to restore the quality of surface and ground water which had become polluted. In some regions the level of ground water has dropped appreciably as a result of over-pumping. Soil salination has occurred primarily in regions with poor drainage, while subsidence has taken place where over-pumping has occurred especially in seismically unstable regions while sea water has intruded many coastal aquifers. The ways of addressing these problems are well known, but the results obtained are extremely diverse because of the great differences in the level of economic development across the continent.

Canadian inter-basin transfers are a specific water-management activity in North America and Canada is a world leader in this area. For the last 30 years the scientific community has given a great deal of attention to the question of further diversions. The morphological, hydrochemical, hydrobiological, ecological, climatic and socio-economic changes resulting from such projects have been investigated. In spite of a current attitudes against such transfers, there is the possibility of new projects emerging at the beginning of the twenty-first century, especially to counter the effects of changes in the climate. To take account of previous experience on this problem will be of particular importance to progress.

8 Water resources, water use and water availability in South America

8.1 INTRODUCTION

South America, with its adjacent islands, is 17.9 million km^2 in area. In the north, the continent borders Central America and it extends southwards for 7150 km. From west to east it is 5150 km. That part draining to the Pacific Ocean occupies 7% of the area, and to the Atlantic Ocean 85%, and the remaining 8% (some 1.4 million km^2) has no drainage to the sea. About 14% of the area of South America is situated in the Northern Hemisphere, and more than 80% lies within the equatorial zone. The continent is characterized by a large input of solar heat and high humidity: the annual value of the radiation balance varies from 45 to 70 W/m^2, and the mean annual precipitation is about 1600 mm. The population of the continent is more than 326 million with an average density of about 18/km^2. The highest densities are in the coastal regions, there are some especially large urban centres, while there are also vast spaces (e.g., in the Amazon basin, Guyana) which are sparsely populated.

There are 13 countries occupying the continent. According to the International Bank for Reconstruction and Development's classification, the group of countries with incomes above the average (GNP above $ US 2500) includes Brazil, Uruguay and Venezuela; while Bolivia and Ecuador have the smallest incomes (less than $ US 1000). The highest rates of economic growth have taken place in Chile and Brazil. In 1995, the total irrigated area was 9.06 million ha. The major centres of irrigation are near the coasts (Chile, Peru, Argentina, Brazil). Most water is used for agriculture in Chile (about 19 km^3/year), the least in French Guyana (about 0.1 km^3/year). Across the continent, about 6000 hydrological stations, and gauges have been operating at different times, most of them for short periods. To assess water resources, runoff data from 240 stations were used with information from other stations, as well as meteorological data. The continent can be divided into four large regions: Northern, most humid (the principal countries are Colombia and Venezuela); Western, mostly mountainous (well explored – Chile, Peru, Ecuador); Eastern (the largest region for future development – Brazil); Central (Argentina, Bolivia).

Using data for the period 1921 to 1985, the renewable water resources of South America were estimated to be 12 030 km^3/year on average, of which the average available water was 674 000 m^3/year per km^2. There are considerable year-to-year fluctuations in water resources: 14 350 km^3/year was recorded in 1984, and 10 330 km^3/year in 1925 – the highest and lowest figures. The greatest volume of water resources occurs in the Eastern region (6220 km^3/year of available water, 731 000 m^3/year per km^2). In the Northern, Western and Central regions the quantities are: 3340, 1720 and 750 km^3/year, respectively. Three South American rivers are important globally: the Amazon, Orinoco and Rio de la Plata. The flow of these rivers to the Atlantic Ocean, including the Caribbean, is eight times greater than the total flow from the continent into the Pacific. The runoff in regions of inland drainage is estimated at 56 km^3/year.

A statistical analysis of discharge, beginning in the 1950s for the continent as a whole, for flows to the Atlantic and flows in the Northern and Eastern regions, showed an increasing trend. No trend was observed in the volumes of discharge into the Pacific Ocean. Dry years are most frequently followed by average years, and years when conditions are average by dry years. Wet years tend to be followed by wet years. The present-day total volumes of abstractions and consumption for the continent are 166 and 98 km^3/year, respectively; by 2025 they are expected to increase to 260 and 120 km^3/year. The highest water use relative to water resources is found in the Central and Western regions: 4.1% and 2.6%, respectively; by 2025 these figures are forecast to be 5.8% and 3.7%.

Due to the growth in population, between 1950 and 1995 the availability of water in the Northern, Eastern, Western and Central regions decreased per head by 71%, 69%, 66% and 55%, respectively. By 2025 water availability could decrease by a factor of 1.6 in the Northern region and by 1.5 in other regions, comprising 35 200, 29 300, 21 800 and 13 900 m^3/year per head, respectively. The need to solve water-management problems, especially in the Central region, requires the strengthening of institutions generally and an integrated approach to water especially in the Rio de La Plata countries (Argentina,

Brazil, Bolivia, Paraguay and Uruguay) to ensure equitable use of water.

8.2 PHYSICAL CONDITIONS

8.2.1 Relief, climate and vegetation

The major topographic features are the Brazilian Plateau in the east of South America and the Andes fold mountains in the west, separated by the lowlands that extend from the Orinoco in the north, through the Amazon, to the Paraná-Paraguay in the south. The plateau can be subdivided into the Guiana Highlands, the Brazilian Plateau and the Patagonia Highlands, while the lowlands and plains can be divided betweeen the Llanos–Orinoco, the Amazon, Beni Mamore, Gran Chaco, Entre Rios (Paraná and Uruguay) and Pampas. In the east, highlands are bounded by narrow broken bands of coastal plains. The Guiana Highlands and Brazilian Plateau are gently sloping hilly plains (of up to 1500 to 1700 m in height) within which there are cone-shaped peaks and ranges as well as sandstone uplands (Tryoshnikov, 1988). The depressions of the Brazilian Plateau are occupied by aggraded plains (the São Francisco valley, etc.) or lava plateaux (in the middle reaches of the Paraná).

The Andes in the west of the continent is subdivided into the Northern, Central and Southern Andes. The Northern Andes stretching to 5° S is a system of alternating highly folded block mountains and deep valleys. In Ecuador, they consist of the Eastern and Western Cordilleras, the valley between being filled with volcanic products. In Colombia, there are the three major systems of the Cordilleras (Eastern, Central and Western) separated by the valleys of the Magdalena and Cauca Rivers. In the north and west of the continent, there are the largest lowlands in the Andes system: the Caribbean and Pacific Coastal Plains. South of the Northern Andes (27–28° S), there are the Central Andes. They occupy a large area and are more monolithic than the Northern Andes. Inland plateaux are typical of them, elevated to 3800–4800 m a.s.l. and framed by edge ranges. The southern part is the Central Andes Highlands, the widest sector of the Andes (up to 750 km across); its major element is the Puno Plateau with the ancient lake plateau Altiplano in the southwest and a series of block ridges in the east and south. In the north of the Southern Andes (to 41° 30′ S), there is the double Major Cordilleras bordering in the east the Pre-Cordilleras massifs. In the southernmost section of the Andes, the Patagonia Andes, the Coastal Cordilleras change into an archipelago, the Longitudinal Valley into a system of straits, and the submerged glacial troughs into fiords.

The position of South America, predominantly in the low latitudes, is responsible for the large input of solar energy almost everywhere on the continent. The radiation balance is from 45 to 70 W/m^2 per year, and only in Patagonia does it reduce to 20–30 W/m^2 per year. North of the Tropic of Capricorn mean monthly temperatures vary from 20 °C to 28 °C (the maximum can reach 49 °C in the Gran Chaco), reducing in summer (January) to 10 °C in Patagonia, and in winter (July) to 12–16 °C on the Plateau of Brazil, to 6–10 °C in Pampas, and to 1 °C in the extreme south of the continent. In the tropical part of the continent, the long-term mean monthly temperature varies from 20 to 24 °C. In the Amazon basin, it is generally about 24 °C for most of the year.

Much of the continent experiences large amounts of precipitation, enhanced by the relief in many areas (Korzun, 1974a). On the other hand, there are a few areas where the rainfall is very low, such as the Atacama Desert in northern Chile where annual totals are less than 100 mm. In Colombia, precipitation varies with altitude from 2000 mm/year to 5000 mm/year, but totals decrease at the highest altitudes. In the south of Chile, there is abundant precipitation throughout the year due to the interaction between the western mid-latitude air flow and the Andes. Ascending the sharp slopes the air soon approaches saturation and precipitation totals range from 2000 mm to 5000 mm/year. Abundant precipitation occurs over the Amazon Basin due to the tradewinds carrying in moist air from the Atlantic. The ascending air becomes saturated with moisture and cooled in higher layers of the atmosphere, resulting in heavy convective showers. The total precipitation is above 2000 mm/year, across most of the basin, but to the west in the foothills of the Andes, precipitation reaches 3000 mm/year and more. Over the greater part of the Amazon Basin, there are the two maximums in the monthly distribution of the precipitation, one in the spring and the other in the autumn, as the sun passes overhead.

In the humid equatorial zone there are some precipitation anomalies associated with the circulation of the air masses. Depending on the position of the intertropical convergence zone, droughts can alternate with periods of extremely heavy rain. This variability is responsible for arid regions in the northeast of Brazil and in Venezuela, where the precipitation is, on average, about 1000 mm/year, but locally it may decline to 300–500 mm/year. As well as the dry areas in Patagonia and in the Atacama Desert, there is the Central Andes Highlands, where the precipitation is between 50 and 150 mm/year: only in the east does it increase to 250–500 mm/year. Over the entire continent the greatest amount of precipitation falls during the hottest season when there is the greatest evaporation. Evaporation varies from less than 100 mm/year in the coastal deserts to 1500 mm/year and more in the equatorial humid regions.

Depending on geographical position, the predominant vegetation is evergreen forest, open woodlands, scrub and savannah-type vegetation. Of greatest extent are the evergreen, mainly equatorial, rainforests occupying almost all the Amazon Basin and the adjoining slopes leading to the Andes, and western Colombia and the eastern slopes of the Brazilian Plateau. Altitude controls the distribution of soil and vegetation in the Andes. Between 1000 and

Table 8.1. *Principal rivers of South America*

River	Drainage basin, km$^2 \times 10^3$	Length, km^a
Amazon (with Ucayali)	6915	6280
La Plata (with Paraná and Uruguay)	3100	4700
Orinoco	1000	2740
São Francisco	600	2800
Uruguay	365	1600
Parnaíba	325	1450
Magdalena	260	1530
Essequibo	155	970
Chubut	138	850
Rio Negro	130	(1000)

^{a} Values in parentheses are approximate data.

1200 m a.s.l., the forest and soil are similar to forests and soils of the plains: from 1800 to 2200 m ferns and bamboo predominate, and from 3000 to 3200 m cloud forests of undersized trees and bushes occur. In eastern Amazonia, away from the equator, a mixture of deciduous forests is to be found, while in the north of the Guiana Highlands and in the north and east of the Brazil Plateau, there are deciduous–evergreen forests. In river valleys in the subequatorial and tropical belts, savannah and open woodlands with galeria forests occur. There are alpine steppes on the plateau of Peru and northern Bolivia, and desert and semi-desert-type soils and vegetation in the southern highlands of the Central Andes and western slopes in the tropics. In the subtropical belt, in the west, the semi-deserts give way to dry sclerophyllous forests and bushes. East of the Andes, in the northwest of Argentina, semi-desert vegetation predominates. In the middle course of the Upper Paraguay, there are subtropical and tropical wetlands. The islands and western slopes of the Patagonia Andes are covered with evergreen forests, while the eastern slopes of Andes have coniferous–deciduous forests.

8.2.2 Hydrology

South America has a dense river network, which is most dense in the wettest parts of the continent. It includes the basins of the Amazon and the Rio de la Plata, which are the largest in South America and in the world. Table 8.1 presents details of the principal rivers of South America (Korzun, 1974b; Tryoshnikov, 1988). In the regions with a variable climate and year-to-year changes in rainfall such as in the north and southwest of the Amazonia, the density of the river network decreases as the length of the dry season increases. About 10% of the continent is desert and semi-desert and some rivers do not reach the oceans, finishing in sands, solonchaks and lakes. Areas of inland drainage cover about

1.4 million km^2 of the continent. Among them is the Plateau Altiplano of more than 220 000 km^2 located at an elevation of 3500–4100 m. There are the alpine lakes Titicaca, Poopó and Coipasa fed by runoff from the greater part of the Puno. Water is mainly lost due to evaporation from these lakes and wetlands surrounding the saline lakes Poopó and Coipasa. The area of inland drainage includes the Pampas (on the left side of the Paraná between the Rio Salado in the north and the Rio Colorado in the south and from 64° W to the Atlantic coast south of the Rio de la Plata) as well as the regions of the Pre-Cordilleras and the Pampas sierras. The total area of this region is over 1 million km^2. A comparatively small area is located in the semi-desert northeast of Patagonia between the Rio Negro and the Chubut River. It occupies, in the main, the Patagonian table country composed of sands and sandstone, where the low precipitation percolates deep or drains into depressions forming shallow small lakes on the clays, or stagnates in saline bogs. The total area of this region is about 130 000 km^2.

Other South American rivers drain to the Atlantic and Pacific Oceans or to the Caribbean. The major rivers draining to the Atlantic are the Amazon, Rio de la Plata, Orinoco, São Francisco and Magdalena, with a total catchment area of about 12 million km^2, i.e. 70% of that of the continent (Korzun, 1974b) (Fig. 8.1).

Fig. 8.1 Distribution of the territory of South America by catchment basins of oceans. 1, Atlantic Ocean; 2, Pacific Ocean; 3, endorheic regions.

The Amazon river system, the world's greatest, flows eastwards from the Andes, but it starts by flowing north parallel to the coast in a deep depression as the River Marañón, on the eastern slopes of the Western Cordilleras in Peru, at an elevation of 4840 m. Then it turns east, breaching the Andes, and joining the Ucayali River where it becomes the Amazon. For the rest of its length the river flows over flatlands which are frequently swampy and covered with tropical rainforests. Below the confluence with the Rio Negro, the Amazon floodplain reaches 80–100 km in width and only at Óbidos and Santarém it gets narrower. The river is 6300 km long and the area of its basin is 6.9 million km². At a distance of 350 km from the ocean the Amazon forms a delta, one of the greatest in the world (about 100 000 km² in area). The major portion of the flow passes through the northeastern arms and passes Marajó, one of the largest deltaic islands.

The Amazon is fed by numerous tributaries. About 20 of them are large rivers of 1500 to 3500 km in length. The Amazon tributaries are not only diverse in their lengths and discharges but they also differ in water colour. The waters of the Rio Negro are dark while the Rio Blanca waters are the colour of milk. There are rivers that are yellow, grey, greenish and even red. The hydrology of the Amazon is very complicated. The right-bank tributaries with their basins located in the Southern Hemisphere and the tributaries with catchments in the Northern Hemisphere flood at different seasons. This is because rain falls on the basins of the right-bank tributaries from October to March (the summer in the Southern Hemisphere), and on the ones on the left between April and October (the summer in the Northern Hemisphere). Consequently the variations in the seasonal runoff are smoothed. The southern tributaries are full-flowing from May to July, but in August and September, flows are lower. The Amazon provides between 15% and 17% of the total annual runoff of all of the world's rivers. The maximum discharge reaches 300 000 m³/s and more; yellowish river waters are noticeable at this time in the Atlantic up to 300 km from the shore. When flows are low the discharge is between 70 000 and 80 000 m³/s. Downstream the flow is affected by tides which can reach up to 1400 km inland. At the mouth, the tides are accompanied by the so-called "roaring water" – a steep wave of up to 5 m in height.

The Rio de la Plata system (Paraná, Paraguay and Uruguay) is the second largest basin in South America, with a length of 4700 km and an area of 3.1 million km². In its upper reaches the principal river of this system, the Paraná, flows in a rocky gorge with many rapids and falls. On leaving the gorge, it joins its major tributary, the Paraguay River, and then it divides into several arms reaching 1.0–1.5 km in width. From there it flows across the flooded lowland. Then the river widens to 50 km, and after the confluence with the Uruguay River the Rio de la Plata estuary is up to 300 km wide and 3–5 m deep (Tryoshnikov, 1988). The rivers Paraná, Uruguay and Paraguay have different regimes because of the contrasting pattern of rainfalls over their basins. Their levels vary widely, especially the Paraná where floods are frequent and can cause great damage (Danilevsky, 1987; Arduino, 1990). The Paraná is of great importance to the economy of Brazil, Argentina and other countries, mainly for navigation but also as a source of water supply. The water management of the basin is important, especially along the lower reaches of the river, where the most important industrial centres are located. However the potential of the Paraná has yet to be fully utilized, especially for power generation.

The River Orinoco with a basin area of 1 million km² and length of 2740 km is the third river of South America. The Orinoco originates in the western slopes of the Sierra Parima, in the southwestern part of Guyana, and flows into the Atlantic. In the upper part of the basin a river carries a portion of its discharge into the Amazon.

Runoff varies considerably across South America. Over the greater part of the continent (Korzun, 1974a) the mean annual runoff is above 400 mm. Values of 1500 mm/year and more are observed in the wettest regions: on the western slopes of the Andes in southern Chile and in Colombia, in the western part of the Amazon and on the Caribbean coast. In the southern part of Chile, south of 40° S, the runoff to the Pacific is considerably above 1500 mm/year. In Patagonia glaciers and ice sheet cover an area of 21 000 km² while the total area of all the glaciers is 25 000 km² (Karasik, 1974). There are numerous and large rivers flowing into the Pacific in the western part of Colombia where the precipitation is 5000 mm/year and more. Along the Pacific coast and in the adjoining slopes of the western desert of the Andes, in the tropical Andes Highlands, on the Plateau of Patagonia and in the northeast of the Brazilian Plateau, discharges are very small, ephemeral, or non-existent. In Peru, of the 52 rivers flowing from the mountains, only 10 rivers reach the ocean, and in the Pampas and the Atacama Desert only one river, the Loa, crosses the desert. The plateaux in the Central Andes produce a runoff of 20–50 mm/year in the west and south and 200–400 mm/year in the east. In Patagonia, south to the Rio Colorado and the Rio Negro and to the Tierra del Fuego, the mean annual runoff is below 20 mm/year, but eastward it increases to 50–100 mm/year.

The rivers of South America are predominantly rain-fed, but in the extreme south there are the meltwaters from snow and glaciers. Ground water recharge varies from 20% to 25% in the basins of the Southern Andes, to 50% on the margins of the Brazilian Plateau and the Guiana Highlands as well as in the desert areas of the continent. The large lakes of the continent are mainly of glacial origin and are concentrated predominantly in the Patagonia Andes (Lago Argentina, Buenos Aires). In the Central Andes, there is the world's largest alpine lake – Titicaca. Also there are residual lakes (Poopó, etc.) and large solonchaks. Large lagoon lakes are located in the north (Maracaibo) and in the southeast. Information about

Table 8.2. *Principal lakes of South America*

Lake	Location	River basin	Volume, km³	Surface area, km²	Depth Maximum	Depth Average	Area of catchment, km²
Titicaca	Bolivia, Peru	Desaguadero	893	8 372	281	107	58 000
Todos los Santos	Chile	Petrohue	34.4	178	337	191	3 036
Valencia	Venezuela		6.3	350	39	18	2 646
Salto Grande	Uruguay, Argentina	Uruguay	5.0	783	33	6.4	224 000
Maracaibo	Venezuela			13 300	35	–	–
Poopó[a]	Bolivia		2.0	2 530	3	–	–
Buenos Aires	Argentina, Chile		–	2 400	–	–	–
Argentin	Argentina		–	1 400	300	–	–
Nahuel Huapi	Argentina	Limay–Rio Negro	–	646	300		2 758

[a] Salt lake.

the lakes of South America is cited in Table 8.2 (Korzun, 1974b; ILEC/UNEP, 1991).

The unique combination of historic and socio-economic factors and the wealth of natural resources determine the characteristics of water use in South America.

8.3 SOCIO-ECONOMIC CONDITIONS AND THE USE OF WATER RESOURCES

Many of the countries of South America are much ahead of other "third world" countries. Compared to countries in the newly industrializing states of eastern and southeastern Asia, the per capita values of their GNPs are several times higher (Vol'sky, 1990). However these values are, on average, 5 to 7 times less than those of most developed countries (IDB, 1999). In South America the main emphasis is on the exploitation of natural resources (Tarasov, 1993). Many countries have developed water legislation, which gives the ownership and management of rights to water resources to the state. Similar legislation is or has been developed for forests, national parks and the protection of natural landscapes (Collier, 1992).

During recent decades more emphasis has been given to ecology and to problems of the use of natural resources by national and international organizations, particularly within projects aimed at revealing and analysing major tendencies and key factors in development. The acute social problems of most of large cities of the continent have also received attention (Tarasov, 1986; WRI, 1990, 1992, 1996, 2000; Klochkowsky, 1991; Reboucas, 1998; UNDP, 1998). Over much of the continent, there has been a decrease in the forest-covered area. During the 1980s 2.4 million ha were deforested in Brazil, 0.9 million ha in Colombia, 0.3 million ha in Ecuador, 0.3 million ha in Peru and 0.2 million ha in Venezuela (but as a result of the measures undertaken dur-

ing the 1990s the scale of deforestation decreased considerably). Forest has been converted into farmland with about 40% of the deforestation in the Amazon basin apparently due to clearance for livestock. Expanding pastures and the incoming savannah-type vegetation prevent afforestation. As a result of deforestation, the water balance has changed in many basins (Nikitin, 1984; Breinzer, 1991; Shuttleworth, 1991; Shuttleworth *et al.*, 1991; Dengo, 1992; Tarasov, 1992; Matsuyama *et al.*, 1993) and soil erosion has increased. For instance, in the "wheat belt" of the Argentina Pampas, erosion has affected about a quarter of the land (Vol'sky, 1990). UNEP considers that serious erosion has spread over 0.4 million km², while over 3.9 million km² are less seriously affected. Data from the WRI (1992) indicates that between 1945 and 1990 the loss of fertile land was 243 million ha, or 14% of the total area.

Desertification was being observed as long ago as during the times of the great American Indian civilizations in Peru and Colombia. At the present time UNEP data shows that 22% of the arid lands on the continent are subject to intensive desertification (Gerasimov, 1986). Arid lands on which dryland farming is possible are available in Colombia (Guajira Peninsula), Bolivia (the southern part of the Altiplano), Argentina (Patagonia) and in other countries and regions. Semiarid lands for dryland farming and livestock production include the Atlantic coast of Colombia and Venezuela, the central (especially southeastern) part of Bolivia (Oriente), and the entire region of the Gran Chaco. In Brazil, major arid and semi-arid zones are in the northeast and in the "Campos Serrados", where shrub and low-tree savannah is located in the centre of the country. A special savannah type is in the valley of the middle and lower São Francisco River. In this dry region, the shortage of water raises complicated economic, demographic and social problems. As with the Amazon, there is intensive degradation of the natural environment: reducing soil fertility, spreading erosion and depleting water resources. The water shortage is increasing because its source is only the São Francisco, whereas the flow of the other big rivers in the region (primarily the Paranaíba) is under-used (Tarasov, 1993). The variability in climate also is normally associated with phenomena that produce impacts with important socio-economic and environmental consequences in many subregions of the continent (CEPAL, 1998; IPCC, 2001a).

Major centres of population are separated by vast underdeveloped areas, although the "homogenization" of the national economy has been one of the principal concerns of most states for a long time and is included in many development strategies (Sheremetyev, 1994). River basins often are the basis for planning and regional policy, often with shared river basins involving groups of countries. For example, Argentina, Brazil, Bolivia, Paraguay and Uruguay are concerned about the Rio de la Plata basin. Efforts have been made to assess and develop water resources and co-ordinate activities in developing hydroelectric power, improving navigation, mitigating flooding and helping farming, irrigation and recreation (Danilevsky, 1987; Arduino, 1990; Avakyan and Sidoruk, 1993; Reboucas, 1998).

The variability of the density of the population is more significant than in other continents. The least populated country is French Guiana (with a population density below 1 inhabitant/km^2), especially away from the coast. Over most of the continent (excluding Colombia, Ecuador, Peru and Bolivia), the major part of the population is concentrated in the coastal zone. For instance, in Brazil, about half the population lives along the coast, which comprises about 7% of the country, and the vast areas of equatorial forest in the Amazon basin are almost depopulated. In Guyana, more than 90% of people live on the coastal plain which lies some 2 m below sea level (Daniel, 1990). Most of the countries of South America are multi-racial and have experienced a. rapid growth of population since the 1940s due to decreasing mortality and a high birth rate. The mortality decrease was largely due to the advances in medicine and improvements in sanitation (Russian Academy of Science, 1989; WRI, 1990, 1996, 2000). Since the 1960s growth rates have begun to slow. During the 1980s the least growth took place in Argentina, Surinam and Uruguay, but growth was especially rapid in Bolivia, Venezuela, Peru, Paraguay and Ecuador (Fig. 8.2).

Since the 1940s the population of South American cities and towns has been increasing faster than the population of the continent as a whole, reflecting an intensive urbanization process. In Peru, Argentina and Uruguay, between 23% and 50% of the population is urban (UN, 1985a), while a number of cities are listed amongst the world's largest or are forecast to become so: for example, São Paulo was expected to become the second largest city with a population of 24 million but Greater Buenos Aires, Rio de Janeiro and Lima Callao are also very large (UN, 1990b). By country, the urban population is highest in Venezuela (91%), and in Argentina, Chile and Uruguay (all 85%). By 2010 it is expected to reach 95%, 91%, 91% and 89%, respectively (WRI, 1990). The smallest urban population is in French Guiana (35%), Paraguay and Surinam (each 48%) and Bolivia (51%). In other countries, it is close to the average for the continent (60–75%).

In the late nineteenth and early twentieth centuries, the increasing urban population was largely due to immigrants from Europe. During the last 50 years it has been caused by inner migration, namely the planned and sporadic movement of people from villages to towns and cities, often above the requirements of industry for manpower. This migration increases the difficulties of providing urban public services, including water supply (Tarasov, 1986). The lack of available water resources and the problems of water supply are aggravated by poor management, while the most severe problem is the removal and treatment of industrial and domestic waste water, because there are few sewerage systems and even fewer treatment works in most countries (Dourojeanni and Nelson, 1982; Gavrilova and Tarasov, 1985; Meangus, 1993). Recently, great progress has been achieved in developing domestic water supplies in Chile. More than 95% of the population now has tap water, between 70% and 80% are connected to a sewer and it is expected that soon all the urban population will have a complete water service. However, only 35% of the rural population is provided with a public water supply and only 5% of the waste water is treated (Frostell and Ramier, 1994).

On average, in 1990, less than 60% of the rural population of South America and about 80% of the urban were provided with water services, or 35% and 75%, in terms of the waste water (*Water International*, 1991). The available resource, the level of service provided and the climatic conditions largely determine the volume of abstraction in a particular country. The largest volume, some 450 l/day per head is abstracted in Venezuela. In the relatively highly developed countries of the temperate belt (Argentina, Chile and Uruguay) this value is 220–270 l/day per head, in Paraguay about 150 l/day per head, in Brazil about 300 l/day per head, and in the Pacific Rim countries of the equatorial Andes (Colombia, Ecuador and Peru) from 100 to 200 l/day per head. Lack of water resources is frequently one of the factors limiting the spread of industry, although for most countries of South America, there is little or no information about industrial water use and estimates have to be arrived at (Gavrilova and Tarasov, 1985). Industrial water use is concentrated either in capital regions, or in special regions with heavy and processing industries. For instance, in Chile, it is concentrated in the basins of the Maipo, Bío Bío and Aconcagua Rivers. Industrial water use occurs along the valley of the Paraíba River in the city of Rio de Janeiro, in Buenos Aires, and in Santa Fe in Argentina.

Studies in Mendoza (Argentina), where the water services system is comparatively reliable, showed that only 5% of industrial water is taken from the public water supply system. Small users

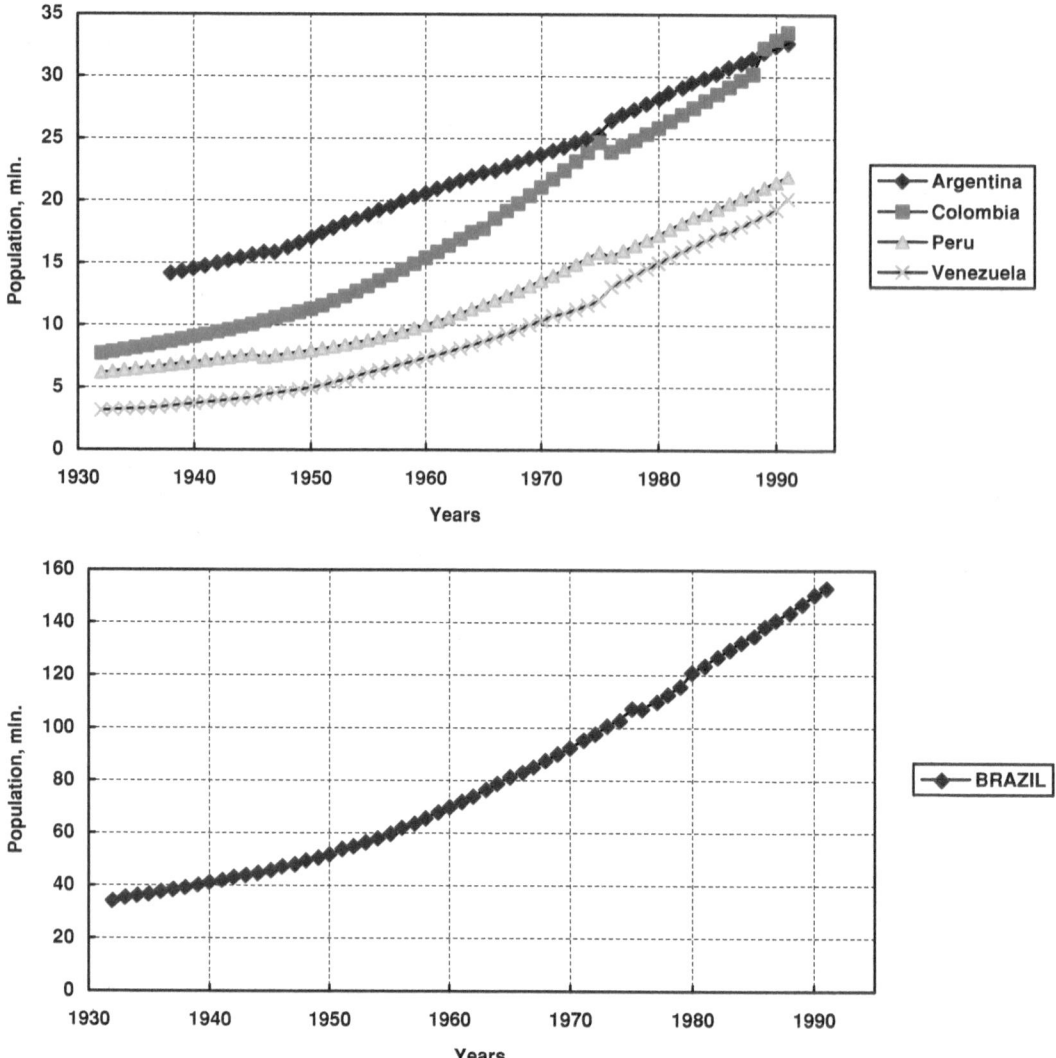

Fig. 8.2 Dynamics of population in some countries of South America.

usually employ ground water; however, if the user is large – oil refining, iron and steel production, etc. – water is normally taken directly from a surface source. In more arid regions, where water resources are restricted, sea water is used for cooling (Vasquer *et al.*, 1989; Waylen and Caviedes, 1990; Ca'ceres *et al.*, 1992). Determining the characteristics of changes in water use by industry and agriculture is complicated by the shortage of data. Some trends can be established only by proxy data, others by comparing the fragmentary literature, and also from information about the agricultural and industrial development in certain countries.

In the most general terms, several stages can be identified in modern economic development. The first stage is associated with the intensive development of natural resources and integration into the world's monetary economy (the first third of the twentieth century). During this period Argentina and Uruguay

became the largest exporters of wheat, meat and wool, Brazil became renowned for coffee and rubber, Bolivia for tin, Chile for copper and Peru for non-ferrous metals (Vol'sky, 1990). Growth slowed in South America during the "great depression" from 1929 to 1933, hitting the export of raw materials and causing a shortage of currency to pay for imported industrial goods. Since that time most nations have experienced growth and some economic stability helped by the activities of the UN Economic Commission for Latin America and the Caribbean (ECLAC).

Import substitution is often one of the important stages in industrial progress. In Brazil, for instance, between 1929 and 1962, the volume of industrial production grew 9.5 times (Karavayev, 1987). Great changes occurred in the structure of the processing industry, especially with the transition to the third stage of import substitution (1970 to 1990s), when in the most developed countries of the region (Argentina and Brazil) the most modern high-technology sectors started to develop. The importance of traditional sectors

Table 8.3. *Principal reservoirs of South America*

Reservoir	Country	Basin	Year of filling up	Backwater height, m	Full volume, km³	Type of use[a]
Guri	Venezuela	Caroni	1986	162	136.3	E
Serra da Mesa	Brazil	Tocantins	1993	144	54.4	E
Serros Coloradeso	Argentina	Neuguen	1977	35	43.4	E I
Tucurui	Brazil	Tocantins	1984	65	43.0	E
Sobradinho	Brazil	São Francisco	1979	43	34.2	N E W
Itaipu	Brazil, Paraguay	Paraná	1982	165	29.0	E
Loma de la Lata	Argentina	Neuguen	1977	16	25.1	E
Ilia Solteira	Brazil	Paraná	1974	85	21.2	E N
Yacyreta	Argentina, Paraguay	Paraná	1991	41	21.0	E N I
Furnas	Brazil	Rio Grande	1965	96	20.9	E A
El Chocón	Argentina	Limay	1975	65	20.2	E I A
Tres Marias	Brazil	São Francisco	1960	70	19.2	E A N
Emborcacão	Brazil	Paranaíba	1982	158	17.6	E
Itumbiara	Brazil	Paranaíba	1980	106	17.0	E
Tres Irmas	Brazil	Tiete	1990	62	13.8	E
São Simão	Brazil	Paranaíba	1978	110	12.5	E
Brocopondo	Surinam	Surinam	1971	42	12.4	E
Piedra del Aguila	Argentina	Limay	1993	163	11.3	E
Agua Vermelha	Brazil	Rio Grande	1979	85	11.0	E
Itaparica	Brazil	São Francisco	1988	105	10.8	E

[a] E, power engineering; N, navigation; W, water supply; I, irrigation; A, accumulation.

(mainly light industry) reduced and heavy industry increased. The chemical sector, oil refining and coal processing grew considerably in Argentina from 11.8% to 20.1%, in Peru from 10% to 30% and in Uruguay from 11.2% to 24.8%. In a number of countries, the wood pulp and paper and printing industries, metal processing and machine building began developing. In the 1970s, there was a shift to export-orientated industrialization. This was associated to a large extent with the expanding activities of multinational corporations and with structural shifts in labour. Developing and changing the structure of industry caused a considerable growth in water use. From estimates by the State Hydrology Institute (Shiklomanov and Markova, 1987; Shiklomanov, 1988), from the 1950s to 1980 the continent's total industrial water use increased by more than 10 km³/year and from 1970 to 1980 by almost 5 km³/year. Using the available data for 1990 the largest use of water by the industrial sector occurred in Brazil (6.3 km³/year) and in Argentina (about 6 km³/year), and the smallest in Chile, Venezuela, Colombia and Peru (1.1, 0.9, 0.8, and 0.6 km³/year, respectively).

Economic development was often led by the growth of the infrastructure across the continent, particularly in transport, power and irrigation. South America is the only continent where hydropower provides more than 50% of the energy, especially in Argentina, Brazil, Venezuela, Paraguay and Colombia. The total electrical power production in South America was 17 times greater in 1985 as compared with 1960. From 1975 to 1990 in Brazil, Colombia and Venezuela, production increased by 3 to 4 times, and for the continent as a whole it increased by a factor of 3. There is a vast hydropower potential: in Brazil some 1.19 million GWh, in Colombia 418 200 GWh, in Peru 412 000 GWh and in Argentina 390 000 GWh, and it continues to develop. In 1991, the power of all the hydropower plants in Argentina, Brazil, Colombia and Ecuador was only 3%, 18%, 6% and 4%, respectively, of the potential (Gleick, 1993, 1998). Some of the world's largest hydropower plants have been constructed in Brazil, such as the power complex Urubuiyunga of 4.6 million kW on the Paraná, the hydropower plants Marimbondu and Furnas of above 1 million kW on the Rio Grande, and several others. In 1981, the first nuclear power plant Angra dos Reis came into operation in the state of Rio de Janeiro. The construction of large hydropower plants is usually associated with the construction of reservoirs.

Reservoirs were being constructed at a high rate in South America in the late twentieth century. Their number and total volume have grown up by a factor of 5 since the early 1970s. At present the continent has about 2000 reservoirs with a total capacity of 750 km³ (Table 8.3). The total effective volume is estimated at 350 km³, but this value is only approximate because of insufficient data. Comparison of this figure with the total volume of water resources, namely 12 030 km³, shows just how small in percentage terms is this figure (8%) for the total volume, and only 3% for the total effective volume. The total surface area of reservoirs is assessed at 60 000 km², about 0.003% of the area of the continent (Avakyan *et al.*, 1987; Gleick, 1993, 1998). The total volume of the more than 500 reservoirs in Brazil is 450 km³, and in terms of reservoir construction rates Brazil ranks the first in the world. A considerably smaller number of reservoirs, about 100, have been constructed in Argentina with a total volume of about 130 km³. The remaining countries have significantly fewer. Venezuela, Colombia and Uruguay, have 20 to 30 reservoirs, including three to four reservoirs of 1 km³ capacity in each country. In Ecuador, Peru and Bolivia, there are numerous small and medium-size reservoirs; only one with a volume greater than 1 km³ is operating in Peru. In Paraguay, there are a few small reservoirs; however, Paraguay has agreements with Brazil and Argentina for the joint construction of several large hydropower plants with associated reservoirs on the River Paraná. These plants will turn Paraguay into the greatest exporter of electricity in South America. Medium-size and large reservoirs are constructed mostly (85%) on the plains and in the highlands, and only a few reservoirs are built in the mountains. Most of reservoirs are for runoff regulation on a seasonal or shorter-term basis. There are a few large and medium-size reservoirs used for irrigation; however, there is a large number of small reservoirs with a volume of 1–10 million m³. For the total area under of irrigation, reservoirs supply more than one-third of the water used, but irrigation produces a considerable part (20%) of food production of the continent (Borozdina, 1978).

Constructing dams to provide water for agriculture is especially widespread in the arid regions, in particular, in the northeast of Brazil where the coefficient of variation of the annual precipitation is 0.30–0.45 (the largest of tropical regions). Most dams were constructed after a long dry period at the beginning of the twentieth century. By the end of that century in this region, the total reservoir capacity (exclusive of the São Francisco River) was about 50% of the average annual discharge to the sea (Dubreuil, 1985). In Argentina, only in the provinces of Buenos Aires, Cordoba and Santa Fe, up to 25 million ha of land are periodically flooded; in Colombia about 7.5 million ha; in Chile some 3.0 million ha; and in Venezuela about 12 million ha. Reservoirs have been built to reduce flooding and they offer protection to about 10 million ha (Avakyan *et al.*, 1987; Richey *et al.*, 1989b). About 100 reservoirs are used to provide water for domestic supplies in combination with water for irrigation and for power. In spite of the considerable reserves of ground water, ground water is little used for drinking and for industry and agriculture (not more than 15 km³/year). For drinking water supply use is frequently made of surface sources.

The major user of water in South America is agriculture, mainly for irrigation. In 1990, the water use in this sector was 64% of the total use for all needs. The development of irrigation is associated with the modernization of the agricultural sector as a whole. In

the 1930s, agriculture passed from simply exporting traditional products (coffee, bananas, cocoa, etc.) to the diversification of agricultural production. Improvements in farming relied on intensive irrigation and better drainage. The major areas with ancient irrigation systems are in Chile and Peru and adjacent to the river basins on the Pacific Coast, and also in Argentina. To finance the construction of irrigation systems, credit was provided by international financial organizations and the USA (World Bank, 1995).

South American countries have experienced high rates of growth in irrigation over the last 30 to 35 years. Brazil saw the highest rates of increase in the area irrigated (6.8%) between 1965 and 1989. By the end of the period the areas irrigated comprised 1.1% of all the agricultural land in the country. In some states of the northeast (for instance, in Piauí) special programmes were developed for multi-purpose rational use of water with but predominantly for irrigation. As a result of the measures undertaken, by 1990 Brazil ranked first on the continent for the area irrigated (2.7 million ha) (World Bank, 1993; FAO, 1995a). In Uruguay, the rate of increase in irrigation reached 4.8% (by the beginning of the 1990s irrigation covered 0.7% of all farmlands); in Bolivia, 4.0% (0.6%); in Colombia, 3.8% (1.1%); and in Venezuela, 2.7% (1.2%). In the countries with developed irrigation systems the rates were lower: Chile 0.5% (7%); Peru 0.7% (4.1%); Ecuador 0.9% (7.1%); and Argentina 1.9% (1.0%) (World Bank, 1993). In Chile in the mid-1990s, about 1.265 million ha were irrigated, and in Peru about 1.3 million ha. In Ecuador, in the late 1970s the plan was for the irrigated area to increase to 600 000 ha (Alekseyewsky, 1974), but by 1990 it had only risen to 550 000 hectares. For all the countries of the Western region the specific irrigated area has gradually decreased since 1970: in Ecuador from 75 to 50 ha/1000 inhabitants, in Peru from 80 to 60/1000 inhabitants, and in Chile from 125 to 95 ha/1000 inhabitants. In Venezuela, this indicator has also decreased, although more slowly, since the late 1980s. In other countries of the Northern region (Colombia, Surinam), the rate of increase in irrigated areas has been above the rate of population growth.

In Uruguay and other countries of the Central region, the steady increase in irrigation was slowed somewhat during the economic downturn of the 1980s. During the last 30 years the area under irrigation has increased by 3 times. At the present time, the rate of increase is above that of population growth, because of investment by private business. Considerable assistance has been provided by international financial organizations and by certain countries (Australia, New Zealand, Canada and the USA). In other countries of the Central region there has been a steady decrease in the specific irrigated area during the last decade. The greatest changes occurred in Bolivia, because of great economic difficulties caused by the crisis in the mining industry. In Argentina in the 1990s, irrigation covered 1.7 million ha (Alekseyevsky, 1974; Direccion Nacional de Hidrografía, 1994).

A wide variety of crops are grown under irrigation: sugarcane, rice, vegetables, fruit crops, forage crops, grapes, cotton and tobacco. The mean weighted water application rates vary from 10 000 (Uruguay) to 12 000–17 000 m³/ha (Peru, Chile, Argentina, Venezuela) and even to 30 000–40 000 m³/ha (Uruguay, Guyana). The only exception is Surinam (7500–8000 m³/ha), where there are efficient rice field rotations with a mechanized water supply that provides irrigation along with drainage (Alekseyevsky, 1974). In Bolivia, Paraguay, Brazil and Colombia, where irrigation makes use of small and medium-size reservoirs, the mean weighted application rates vary from 5000 to 7000 m³/ha. In Brazil, application rates for rice are about 16 000 m³/ha, and for other crops between 1400 and 2000 m³/ha (Aoki, 1987). Consumption of water in irrigated areas varies widely: from 90–95% where drip or other modern systems are used to 25–30% on rice plantations with continuous flow checks. As a whole, agriculture on the continent, in spite of considerable reductions in its rate of expansion (from 3–4% in 1980 to 1.0–1.5% in the 1990s), was able to avoid the economic downturn of the 1980s because of national initiatives and also those of multinational corporations. In the 1980s the irrigated area increased by almost 2 million ha. Water consumption by agriculture rose by almost 19 km³/year and reached 97 km³/year, whereas in industry, water consumption increased by 2.5 km³/year (53% of the growth for the previous decade) (Sinelshchikova, 1989; Prebish, 1993).

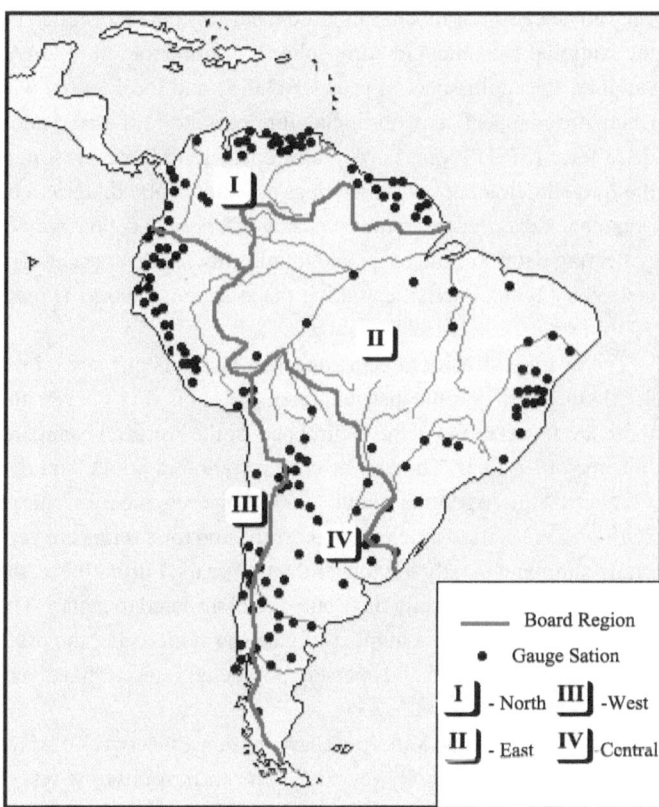

Fig. 8.3 Distribution of South America territory by natural–economic regions and location of main runoff gauge stations.

8.4 HYDROLOGICAL DATA

Less is known about the hydrology of South America than the other continents. River levels were first measured in the early twentieth century, but it was not until recent decades that discharges were measured. The development of the network of hydrological stations has been uneven and, in general, linked to projects (Fig. 8.3). A number of stations were opened then closed soon after hydrological observations were carried out by federal and private organizations in order to solve certain problems, for instance those associated with the design and construction of hydropower plants and new irrigation systems. On most of the navigable rivers – the Amazon, Orinoco, Paraná, Paraguay, Uruguay and others – water levels were not measured until the 1970s.

Brazil has a hydrological network which contains about one-third of stations in South America. Almost all of them are located on the rivers in the most developed southeastern part of the country (Dengo, 1992). The Amazon Basin occupies more than half the area of Brazil, but it only has a few permanent hydrological stations. The network of hydrological stations both in Brazil and over the entire continent is operated by a large number of federal, provincial and private organizations. Important amongst them in Brazil are the Department of Water Resources and Electric Power

(DNAEE) with its branches in the provinces; the State Sanitation Department; the State Department for Ports and Waterways; and the different parts of the Ministry of Agriculture.

Over the last 10 years a number of hydrometeorological data centres have been established in Brazil. For instance, the Centre of Hydroclimatology and Remote Sensing of Amazonia (CHCRSA), created within the Superintendencia para o Desenvolvimento da Amazonia (SUDAM) Project, is widely known. Hydrometeorological indicators of environmental conditions are being observed on a more frequent and systematic basis, and the institutions carrying out this research have been faced with the need for data management. Some institutions in the Amazon have gained much experience of databases, and noticeable progress has been achieved. From the 1980s the DNAEE with other bodies, both Brazilian and foreign, has been carrying out assessments of the water balance in certain catchments of the Amazon Basin. In addition the new UN/WMO project has the participation of all the countries of the basin. During the course of this project the water balance elements of the basin will be determined (Dengo, 1992), while the DNAEE has established 476 hydrological stations in the Amazon Basin (including the basin of Tocantins–Araguaya). Remote-sensing methods have been applied to study different features of the Amazon Basin. Use is made of satellite monitoring systems such as the very advanced high-resolution radiometer (AVHRR) installed on the National Oceanic and Atmospheric Administration (NOAA) satellite, the multi-spectral scanner (MSS) and the Landsat TM (thematic mapper), as well as Satellite probatoire d'observation de la terre (SPOT) images. A major obstacle to these sensors is the frequent cloud cover, which does not allow good images to be obtained. Nevertheless, remote sensing offers an excellent way of collecting data over the basin and monitoring the diverse changes caused by human activities taking place in the Amazon (Fread, 1976; Dengo, 1982, 1992).

The average density of rain gauges in Brazil is one station per 1500 km^2. However the distribution of gauges is very uneven and there are few stations in the central part of the Amazon basin and in some other areas. The first discharge station in South America was set up in Argentina about 1900. However, most discharge stations were started much more recently, and time series are generally short and mostly not longer than 10 years. Furthermore, the stations are not uniformly distributed or distributed to a plan. The rivers used most have a number of gauging stations but the other rivers, in undeveloped and sparsely populated regions, have very few, or no observations.

Chile, Uruguay and Venezuela have the densest network of stage and discharge stations. In Venezuela, the earliest observations of water level were started in 1911 on the Orinoco River, but were not carried out systematically until 1940. A few hydrological stations have been established on the flat lands on the left bank of the Orinoco basin. The rather high density (one station per 4500 km^2) does not reflect the actual hydrological knowledge of this country. The mean duration of discharge observations there is about 10 years. Only stage is measured in most of the stations of the hydrological network in Uruguay. Discharge is being measured on only a few rivers (e.g. the Rio Negro), particularly those used to generate electricity. In Bolivia and Paraguay, the network is less developed. For instance, in Paraguay, discharge is measured at only three stations. The hydrological network of Colombia covers only 40% of the country. Over the rest, there are only three stations measuring discharge and almost all the hydrological stations are located near the cities of Bogotá and Medellín, or in the basin of the Cauca River. There are no discharge stations in the Orinoco lowland. In Peru, the rivers on the western slopes of the Andes are comparatively well gauged; however, there are very few observations of discharge on rivers draining to the Atlantic. Peru has one of the sparsest rain-gauge networks in South America.

Information for the early 1980s for the South American hydrological network is presented in Table 8.4. The total number of stations and gauges operating across the continent is about 6000 with an average network density of about one gauge per 3000 km^2. While this density may be sufficient, most stations have only operated for a short period of time, and reliable station histories are usually not available. There is also the point that the number of

Table 8.4. *Hydrological stations in the countries of South America*

Country	Area, km$^2 \times 10^3$	Number of runoff stations	Density, km^2 per station
Argentina	2776.6	702	3 950
Bolivia	1098.6	63	17 400
Brazil	8512.0	2962	2 870
Chile	756.9	391	1 940
Colombia (1976)	1138.9	370	3 080
Ecuador (1981)	283.6	231	1 230
Guyana	215.0	46	4 670
Paraguay	406.8	10	40 700
Peru	1285.2	391	3 300
Surinam	160.0	98	1 390
Uruguay	186.9	86	2 170
Venezuela	912.1	700	1 300

Source: UN (1985a).

stations may have increased as compared with the early 1970s, but their distribution over the continent is extremely varied. Most stations are located on the small rivers draining to the Pacific, in Venezuela and on the rivers of the southeastern part of Brazil. In arid regions of the Central Andes Highlands and in Patagonia, where permanent rivers are absent, there are a few stations as is also the case in other sparsely populated regions, with a density of one gauge for 10 000 to 30 000 km^2. But there are even fewer gauges in remote parts of the Amazon Basin. The longest observations of discharge are available in Argentina and Uruguay: from 1898 on the Uruguay River near Salto (Concordia), and from 1901 on the Paraná River at Posados. Discharge has been recorded since 1910 on some rivers in Chile and Peru, and since the 1920s on certain rivers in Venezuela and Brazil. However, in most countries the majority of stations were installed at later dates.

The water resources of South America have been assessed a number of times (Karasik, 1974; Korzun, 1974; L'vovich, 1974; ECLAC, 1978a, b; WRI, 1990, 1992, 1996, 2000; Seckler et al., 1998; Rijsberman, 2000). Karasik (1974) L'vovich (1974) and Baumgartner and Reichel (1975) made estimates employing the water balance method. Korzun (1974) used discharge data where these were available and other observations and ratios to estimate the water resources of the whole continent. Table 8.5 shows the results obtained by the different authors, all the estimates falling between 10 000 and 11 770 km^3/year. Some values differ from each other by almost 15%, which is almost twice as much as the discharge of the Rio de la Plata.

South America can be divided into four regions: Northern, Eastern, Western and Central (Fig. 8.3). The Northern region includes Colombia, Venezuela, Surinam, French Guiana and

Table 8.5. *Water resources of South America by data of different authors*

Author	Water resources, km^3/yr
Korzun, 1974b	11 770
Karasik, 1974	10 380
L'vovitch, 1974	10 380
Baumgartner and Reichel, 1975	11 100
Dubreuil, 1985	10 000
WRI, 1992	10 377[a]

[a] Excluding Guyana.

Guyana with a total population of 57.3 million (UN, 1995a), of which 80% lives in cities and towns. The area of the region is 2.6 million km^2. The largest rivers are the Orinoco and the Magdalena. They receive large amounts of rainfall and consequently their soils are very wet and there is an excess of surface water, which results in the formation of large wetlands and vast areas which are flooded. This causes environmental problems, especially in the context of large-scale land reclamation for agriculture. Both Colombia and Surinam have large areas with irrigation, and most of the population of the region (except in Colombia) is concentrated along the coast. The municipal sector has the continent's highest demand for water and in Venezuela and Colombia, industry is increasing its use.

The Eastern region occupies the greater part of the Amazon Basin, a part of the basins of the Paraná and the Uruguay, as well as the São Francisco Basin. It has a hot humid climate which is only tempered in the São Francisco Basin. Its area is 8.5 million km^2, and it has 159.1 million people, of whom 78% are concentrated in the megalopolises of southeastern Brazil, the planet's largest urban area. The region has high rates of economic growth (the "Brazilian miracle"). There is the largest industrial water use, and the irrigated areas are increasing. The major water-management problems are related to pressure on the Amazon, particularly deforestation and the general deterioration of the environment stemming from development of industry and agriculture and the growth of population in certain areas coupled with lack of treatment of waste water.

The Western region is 2.3 million km^2 in area, stretching like a narrow stripe along the entire Pacific coast. It includes the three Andean countries, Ecuador, Peru and Chile, whose total population is 48.6 million with 73% of the people living in cities and towns. The hydrographic network contains the upper reaches of the Amazon and its tributaries, as well as the rivers flowing from the Andes into the Pacific Ocean. The entire coastal region (exclusive of the extreme south) has a water deficit of 1000 to 1500 mm/year. The region includes areas with ancient traditional irrigation systems and large areas under contemporary irrigation

(except for Ecuador), the total irrigated area being above 3.1 million ha. A number of countries have shared in the upsurge in economic development (the "Chilean miracle"), and a considerable improvement has taken place in domestic water supply, while industrial water use has risen to above 2 km^3/year. This region is hydrologically the best known in South America and there has been increased attention to the problems of management and protection of water resources, but the basic water management problem is rationalizing water use.

The Central region includes Argentina, Bolivia, Paraguay and Uruguay. It covers 4.5 million km^2, and in 1994 the population was 49.4 million, 82% classed as urban. Practically all the region (except for the lower reaches of the Paraná River) lacks water resources, especially in Argentina where there are poorly developed areas needing irrigation, and the large capital investment this requires. In Argentina, the industrial sector uses large quantities of water and public water services are well developed. In the most arid parts, ground water is used for water supply and for cooling, as well as sea water. Further development of the management of the shared water resources the Rio de la Plata's resources is very important to future progress, including their protection from pollution.

8.5 DISTRIBUTION OF WATER RESOURCES IN TIME AND SPACE

8.5.1 Data and methods

Although there are a few long records available for some rivers, as a whole South America lacks hydrological data. There are a few discharge stations on Amazon Basin rivers, but only in 1963, 1964 and 1967 was the flow of the Amazon measured near Óbidos (about two-thirds of the basin area) (Korzun, 1974b), to give an idea of its water resources. However, so far the flow has not been measured at the mouth because of a number of difficulties. In the present study, in order to assess the water resources of the continent, all the data on discharges from Korzun (1974b) were used. For subsequent years, these data were supplemented by discharge data from sources such as *Discharge of Selected Rivers of the World* (UNESCO, 1993). In addition, use was made of all the discharge data available in the Global Runoff Data Centre (Koblenz, Germany) and in national year books. In addition discharge data were requested from Venezuela, Guyana, Colombia, Chile, Peru and other countries. Data from all these sources were employed to create a database of monthly South American discharges. Data from 240 stations located near river mouths were selected from this database in order to assess water resources (Fig. 8.3, Table 8.6). To fill any gaps in the time series, analogue stations were selected and regression equations determined, excluding those where the correlation coefficient was less than 0.80. The remaining gaps were

Table 8.6. *Number of river runoff observation stations used to estimate water resources of South America*

Country	Area, $km^2 \times 10^6$	Number of stations	Number of years
Argentina	2.777	80	4–80
Bolivia	1.099	6	4
Brazil	8.512	28	4–66
Chile	0.757	29	10–50
Colombia	1.139	14	4–13
Ecuador	0.284	16	4–16
French Guiana	0.091	6	6–15
Guyana	0.215	2	4–11
Paraguay	0.407	1	8
Peru	1.285	16	4–61[a]
Surinam	0.163	4	4–7
Uruguay	0.178	3	15–70
Venezuela	0.912	35	3–39
South America as a whole	17.8	240	3–80

[a] Annual means.

filled by using relationships between rainfall and runoff (Babkin, 1998).

To assess the water resources of the principal rivers, of particular countries and of the four regions, use was made of data from the hydrological stations where the discharge values were the greatest, normally stations close to river mouths. Where records were short at particular stations, they were extended using the methods described in Chapter 2 (Shiklomanov, 1997, 1998a, 2000a, b; Babkin, 1998; Shiklomanov *et al.*, 2000). Hydrological analogues were employed for certain locations and geographical interpolation at others where data were absent. For the Amazon at Óbidos discharge values were obtained from UNESCO (1993, 1996b) for the years 1928 to 1947 and 1969 to 1983. Discharge data were available for 1976 to 1981 for the tributary Tocantins at Itupiranda and for the Xingu River at Altamira for 1976 to 1979. The discharge of the Amazon at its mouth was obtained for each year by multiplying the annual discharge at Óbidos by a coefficient which took account of the runoff from the area to the mouth and which was not measured. Discharges during years with no data were determined by a multiple regression equation which included rainfall. The correlation coefficient for this equation was 0.82.

The Global Runoff Data Centre supplied data for the Paraná at Corrientes from 1904 to 1983 and for 1984 and 1985 the discharges were taken from published data (Nikitin, 1984; Collier, 1992; Garcia and Vargas, 1996). For the Orinoco, the data for the station at Ciudad Bolívar (with a catchment of 1 million km^2) was supplemented by data from 1970 to 1992 for Musinacio which is located upstream (787 000 km^2). The water resources

of Argentina, Brazil and Bolivia were estimated for each year of the design period from 1921 to 1985. Those of Chile, Ecuador, Peru, Uruguay, Guyana, Surinam and Colombia were estimated as the means for this period. Account was taken of the runoff originating in the country concerned, inflows from outside and flows out of the country. From these figures the total resource was determined, where necessary, discharges were calculated from equations whose coefficients were changed for each length of every river.

The water resources of Argentina and Bolivia were assessed in this way and so also were those for Brazil, where the difference between the discharge to the Atlantic and the flow from Colombia, Peru, Ecuador and Bolivia was estimated (Korzun, 1974b). Also for Brazil, estimates were made of the runoff to the Atlantic from the area not covered by gauging stations. The water resources of the continent were estimated as the sum of the resources for the four regions and those for the regions in the same way as for the countries and the drainage to the oceans was determined in the same manner. In the case of the Northern region, they were estimated as the sum of the discharges of the Orinoco, Magdalena, Atrato, Cauca and Essequibo Rivers as measured at the lowest stations and multiplied by coefficients to take into account the runoff from areas not covered by measurements. These coefficients were determined by the methods described earlier.

The runoff from the Eastern region (which coincides largely with Brazil) and for the Western region (consisting of Chile, Peru and Ecuador) was estimated separately for each country and then summed. Where data were absent use was made of the methods described earlier for assessing resources. The resources of the Central region were estimated from the sum of the inflows and the runoff originating in the region. The volumes of the inflows were determined from equations whose variables were the discharges of the Rivers Madeira, Pilcomayo, Paraná, Paraguay, Uruguay and Negro at the corresponding sites taking account of the changes in discharge along the length of the river. The discharge originating in the Central region was estimated as the sum of that from Argentina, Bolivia, Uruguay and Paraguay. The runoff to the Pacific and Atlantic oceans (Fig. 8.1) was determined from equations which took account of the runoff from areas not covered by observations. Flows to the Caribbean and areas of inland drainage were determined in addition. Discharges to the Caribbean were reduced to the study period, excluding those of the Atrata River which had a low negative correlation with the discharges of the other rivers. The correlation coefficients between the discharges of the rest of the rivers flowing to the Caribbean (excluding the Magdalena) varied between 0.50 and 0.94. To determine resources in the areas of inland drainage, some 1.4 million km^2 in area, the methods described previously based on analogues (Salado, Dulce, Tercero, Lima) were applied. The discharges employed were reduced to the standard observation period, and corrections

Table 8.7. *Renewable water resources and water availability by natural–economic regions and individual countries of South America*

Territory	Area, $km^2 \times 10^6$	Population (1994), $\times 10^6$	Water resources, km^3/yr				Coefficient of variation, C_v	Potential water availability, $m^3 \times 10^3/yr$	
			Inflow	Local				Per 1 km^2	Per head
				Average	Maximum	Minimum			
Region									
Northern	2.55	57.3	0	3 340	4 670	2 390	0.15	1310	58.3
Eastern	8.51	159.1	1900	6 220	7 640	5 200	0.08	731	45.1
Western	2.33	48.6	0	1 720	2 380	992	0.18	738	35.4
Central	4.46	49.4	720	750	1 310	531	0.17	168	22.5
Countries									
Argentina	2.78	34.2	623	270	610	150	0.27	97.1	17.0
Bolivia	1.10	7.2	155	361	487	279	0.13	328	60.9
Chile	0.76	14.0	0	354				466	25.3
Colombia	1.14	34.3	0	1 200				1053	35.0
Ecuador	0.28	11.2	0	265				946	23.7
Guyana	0.22	0.8	0	270				1227	338
Peru	1.28	23.3	144	1 100				859	50.3
Surinam	0.16	0.4	0	230				1438	575
Uruguay	0.18	3.2	74	68				378	32.8
South America as a whole	17.9	314.5	0	12 030	14 350	10 330	0.07	672	38.3

were made in estimating the average long-term resources of these regions.

8.5.2 Estimates of the water resources of regions and countries

As in previous chapters, water resources are estimated as the sum of those originating in the area concerned and the inflows and outflows. Table 8.7 shows the details of these estimates including the figure of 12 030 km^3/year for the continent as a whole. Runoff from the Northern region was estimated to be 3340 km^3/year and from the Western region some 1720 km^3/year. The Eastern region (Brazil) occupies the greater part of the Amazon basin and the runoff from this region is 8120 km^3/year with the inflow from the Northern, Western and Central regions taken into account. Runoff originating in the region is 6220 km^3/year. The runoff originating in the Central region is 750 km^3/year and with the flows of the Paraná, Uruguay and Negro included the total amounts to 1470 km^3/year. Table 8.7 shows that the highest water availability is in the Northern region, where 1 km^2 produces 1.3 million m^3/year of water versus 168 000 m^3 in the Central region where water is least available. For the continent as a whole, water availability is 672 000 m^3/km^2. The resources per head are greatest in the Northern region at 58 300 m^3/year and least in the Central region, where they amount to 22 500 m^3/year.

Taking the whole of South America, the water availability is 38 000 m^3/year per head.

Detailed assessments of runoff from Brazil, Argentina and Bolivia were made for every year of the study period and for the long term for Chile, Ecuador, Peru, Uruguay, Guyana, Surinam and Colombia. The mean annual discharge of the Amazon was estimated to be 4600 km^3/year and that of the Paraná in Brazil some 270 km^3/year from an area of 802 000 km^2. The Uruguay River was estimated to have a mean flow of 151 km^3/year in Brazil from a basin of 244 000 km^2. The mean annual discharge of all Brazilian rivers was assessed at 8120 km^3/year for the 1921 to 1985 period, i.e. the same as the flow from the Eastern region (Table 8.7). The Rivers Paraná, Paraguay and Uruguay, which are navigable, flow from the eastern and northern parts of Argentina, and the Rivers Negro, Colorado, Chubut and Santa Croix and others originating in the Andes, which have potential for hydropower, produce a total runoff of 893 km^3/year. Of this 270 km^3/year originates in Argentina and 623 km^3/year is the inflow from Brazil, Paraguay, Uruguay and Bolivia. In Bolivia the runoff is 516 km^3/year, which includes an inflow of 155 km^3/year. Surinam has the largest available resource at 1.44 million m^3/year, which expressed per head is 575 000 m^3/year, and Argentina has the least, 97 100 m^3/year per km^2 or 17 000 m^3/year per head.

The estimates of the water resources of South America produced by Karasik (1974), L'vovich (1974), WRI (1990, 1992,

Table 8.8. *Water resources (km³/year) of some countries of South America by data of different authors*

Country	L'vovich (1974), Karasik (1974)		Gleick (1993), WRI (1990, 1992)		Engelman and LeRoy (1993)	Margat (1994)		Direccion Nacional de Hidrografía (1994)
	Local	Transit	Inflow	Outflow		Local	Inflow	
Argentina	289	300	694	300	994	694		718
Bolivia	300		300[a]		300			289
Brazil	5668	1760	5190	1760	6950	5610	2305	5363
Chile	168		468			927		483
Colombia	1113		1070		1070			1105
Ecuador	318		314		314	284		318
Guyana	241		241[a]					
Peru	750		40		40	1893	150	41
Surinam	200		200[a]		200			
Uruguay	59	65	59[a]	65	124	59	65	59

[a] Institute of Geography of Academy of Sciences of USSR (A.V. Belyayev).

1996, 2000), Gleick (1993, 1998) and Margat (1994) differ considerably. For instance, those for Peru are estimated as 40 km³/year by the World Resources Institute (WRI, 1990), but Karasik (1974) and L'vovich (1974) both give the figure of 750 km³/year. Margat (1994) assessed the runoff originating in Peru as 1893 km³/year, and the inflow as 150 km³/year, but SHI (Table 8.7) gives a figure of 1244 km³/year with an inflow of 144 km³/year (Shiklomanov, 1997, 1998a, 2000a, b; Babkin, 1998; Shiklomanov *et al.*, 2000; present Monograph) (Table 8.8). From the analysis of the long-term variations, the coefficient of variation (C_v) of the mean annual runoff for the 65-year period was found to be 0.07 (Table 8.7). Some of the lack of variability may be accounted for by more runoff in one part of the continent occurring when there was less in another. However, for particular river basins and certain regions, the runoff variations are more significant. The C_v for the Eastern region is 0.08, and in the Western and Central regions it is 0.18, while the highest variability ($C_v = 0.45$–0.51) occurs in the Southern region of the continent (Rapel, Maipo).

Tests of stationarity were conducted for the regions, for countries and for the continent as a whole for the period using the methods outlined in Chapter 2. It was found that the series for the Northern region and Bolivia are non-stationary both in terms of means and variances, while those of the East are non-stationary in terms of means but not for variances. There was no evidence of trends in the series for the Western and Central regions. The other time series were analysed for trends by the four methods discussed in Chapter 2. All four showed an upward trend in the runoff for the Northern and Eastern regions, as well as for Brazil and Bolivia. There were no trends for the Western region and Argentina, but for the Central region there was a tendency towards a trend. There are periods of years of differing durations when the runoff was

high or low in the different time series. These periods of runoff variations are shown in Fig. 8.4 and in Table 8.9. The duration of periods with high flows varies from 14 to 38 years, and for low flows from 27 to 51 years, when respectively the flows were 3% to 20% above the average for high flows and 4% to 7% below for dry periods.

The variations in the runoff from the Northern region are in phase with those of the whole continent, and the same applies to those for the Central region, Argentina and Bolivia. The longest period of high flows, some 38 years from 1948 to 1985, occurred in the Eastern and Central regions and in Bolivia and Argentina. Dry and wet years alternate; for example there is a 68% probability of dry years being followed by an average year and wet years are mostly followed by wet years. Table 8.10 shows no pattern to the regional variations in runoff, probably due to contrasts in the atmospheric circulation over the different parts of the continent. In Brazil, Argentina and Bolivia, patterns of runoff are very variable. Table 8.10 shows the groups of years of 5 years duration, with an 85% probability of occurrence. However, there are groups of years of longer duration, their probabilities being 11% and 9%. Studies were made correlating the runoff of the continent and the four regions and for the Northern and Eastern regions, the correlation coefficients being $r = 0.74$ and $r = 0.69$. Runoff from Argentina and Bolivia are closely related to the runoff of the Central region ($r = 0.95$ and $r = 0.87$), but for the rest of the regions the correlation coefficients are low.

8.5.3 River flow and drainage to the oceans

The characteristics of the continent's major rivers are shown in Table 8.11, where considerable differences can be seen in size of

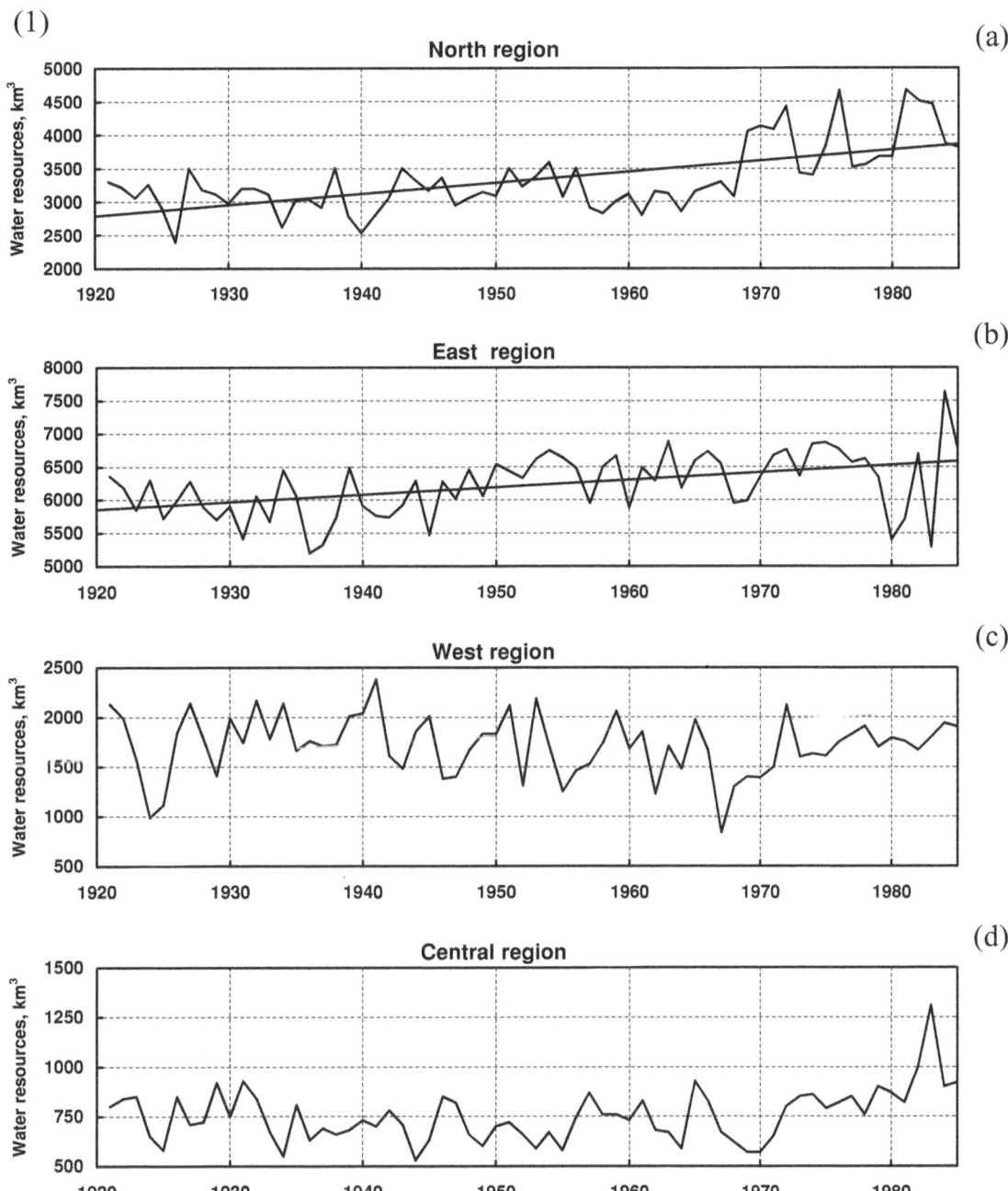

Fig. 8.4 Water resources variations on the continent of South America, in natural–economic regions and individual countries. (1) Year-to-year variations; (2) normalized difference integral curves. (a) Northern region; (b) Eastern region; (c) Western region; (d) Central region; (e) South America; (f) Brazil; (g) Argentina; (h) Bolivia.

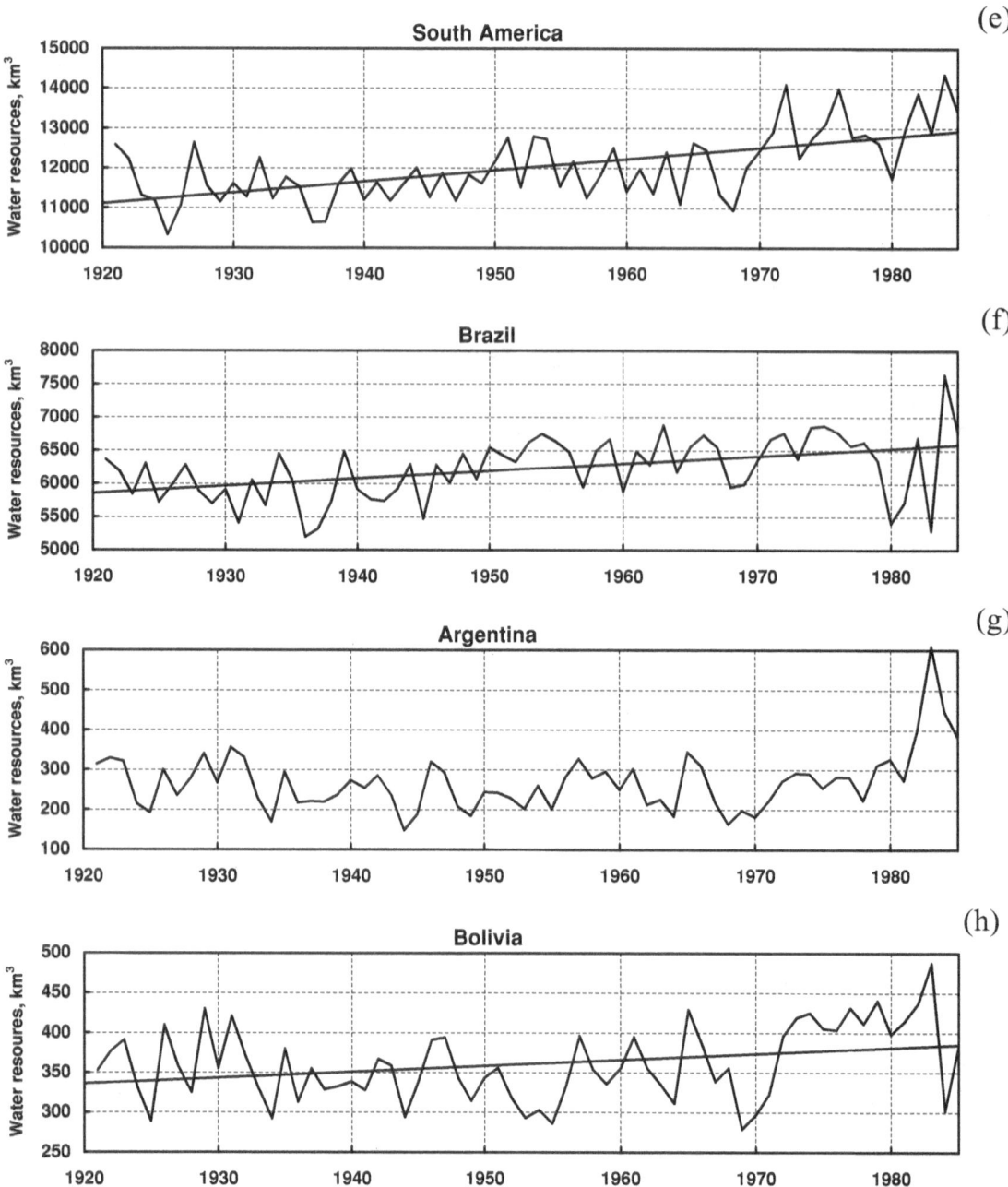

Fig. 8.4 *(cont.)*.

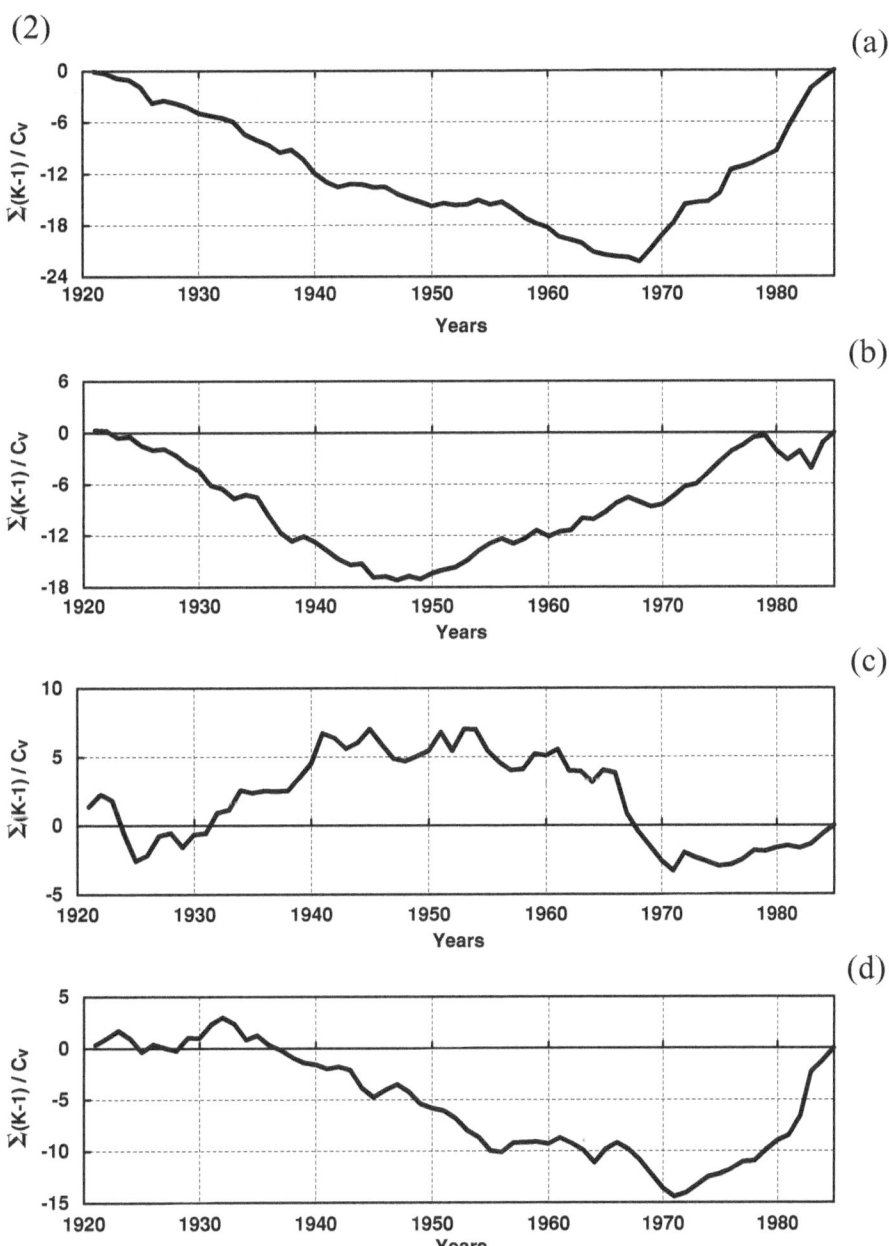

(2)

(a)

(b)

(c)

(d)

Fig. 8.4 (*cont.*).

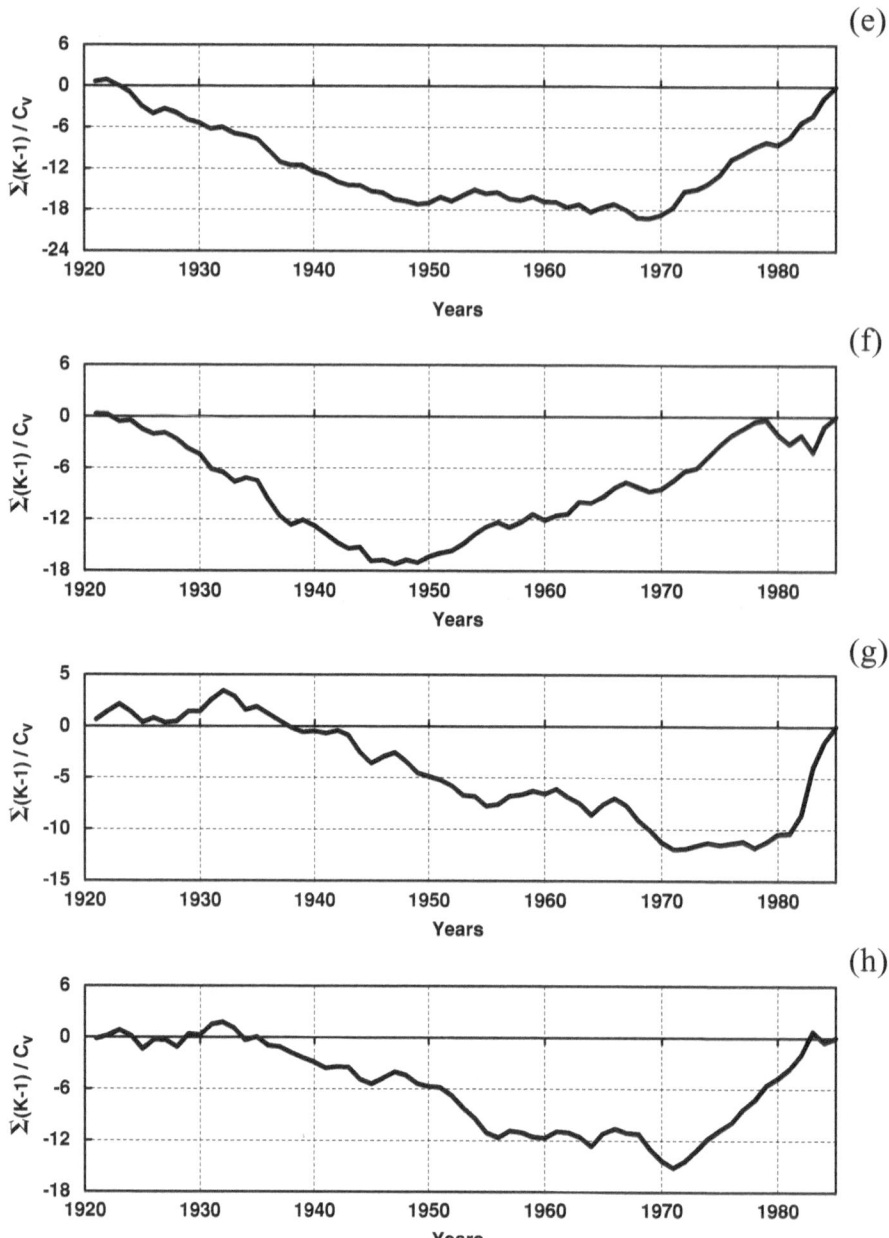

Fig. 8.4 (cont.).

Table 8.9. *Periods of different water content by natural–economic regions, selected countries and the continent of South America*

Territory	Periods of high water content			Periods of low water content			Periods of average water content		
	Years	Ratio of mean of period to long-term mean, K_{av}	Number of years	Years	Ratio of mean of period to long-term mean, K_{av}	Number of years	Years	Ratio of mean of period to long-term mean, K_{av}	Number of years
Regions									
Northern	1969–85	1.20	17	1921–68	0.93	48			
Eastern	1948–85	1.03	38	1921–47	0.95	27			
Western	1921–45	1.05	25	1946–85	0.97	40			
Central	1972–85	1.18	14	1921–71	0.95	51			
Countries									
Argentina	1972–85	1.23	14	1921–71	0.94	51			
Bolivia	1972–85	1.14	14	1921–71	0.96	51			
South America as a whole	1970–85	1.09	16	1921–49	0.96	29	1950–69	0.99	20

Table 8.10. *Water content transition from one grouping to another by natural–economic regions and individual countries of South America*

Territory	Water content characteristic[a]	Probability of characteristics transition from one grouping to another, %			Average duration of groups (years)
		$\overrightarrow{1}$	$\overrightarrow{2}$	$\overrightarrow{3}$	
Regions					
Northern	1	31	69	0	1.4
	2	23	69	8	3.2
	3	0	17	83	6.0
Eastern	1	44	45	11	1.8
	2	29	46	25	1.9
	3	11	33	56	2.2
Western	1	47	29	24	1.9
	2	23	47	30	1.9
	3	12	65	23	1.3
Central	1	57	29	14	2.3
	2	25	38	37	1.6
	3	16	42	42	1.7
Countries					
Argentina	1	58	27	15	2.4
	2	29	47	24	1.9
	3	29	24	47	1.9
Bolivia	1	36	50	14	1.6
	2	23	69	8	3.2
	3	7	27	66	3.0
South America as a whole	1	26	68	6	1.4
	2	47	37	16	1.6
	3	0	33	67	3.0

[a] 1, Low; 2, average; 3, high.

Table 8.11. *Runoff of individual rivers of South America and its variability*

River	Station	Area of catchment, km²	Study period	Runoff km³/year	Runoff mm	Coefficient of variation, C_v	Coefficient of asymmetry, C_s
Northern region							
Orinoco	Ciudad Bolívar	850 000	1924–92	788	927	0.15	−0.23
Magdalena	Galamar	257 000	1904–90	226	880	0.08	0.26
Apure	San Fernando	114 800	1963–91	69.9	609	0.22	0.41
Essequibo	Plantain Island	66 600	1917–84	67.6	1015	0.15	−0.82
Maroni	Langa Tabaki	60 930	1926–80	50.5	829	0.29	−0.60
Atrata	Tagachi	9 430	1926–90	47.0	4984	0.10	2.22
Tocuyo	Puente Torres	3 590	1914–87	0.38	105	0.36	0.09
Eastern region							
Amazon	Óbidos	4 640 000	1920–85	5016	1081	0.08	0.00
São Francisco	Juazeiro	510 800	1929–79	88.6	173	0.28	1.25
Parnaíba	Porto Formoso	282 000	1928–83	28.9	102	0.15	0.11
Jequitinhonha	Jacintop	63 900	1928–83	16.2	254	0.45	0.45
Paraíba	Dos Scampos	55 080	1928–83	27.0	490	0.33	0.26
Iguacu	Salto Ozorio	46 400	1926–80	30.1	649	0.28	0.66
Western region							
Baker	Colonia	23 740	1927–84	29.7	1251	0.06	−1.16
Bío Bío	Desembocadura	21 220	1917–91	35.6	1677	0.25	1.33
Rimac	Chosica	18 260	1921–72	0.89	49	0.19	0.04
Rapel	Las Balsas	13 190	1917–84	6.06	459	0.45	1.50
Puelo	Carrero De Basilio	8 620	1927–85	21.5	2494	0.18	1.18
Pastaza	Banos	7 694	1962–77	3.78	491	0.17	0.27
Mira	Lita	5 630	1927–84	4.70	835	0.08	0.27
Maule	Armerillo	5 454	1916–92	7.79	1428	0.27	−0.35
Chicama	Salinar	4 336	1911–72	0.90	208	0.62	2.20
Rio Bueno	Bueno	3 714	1929–79	12.1	3258	0.15	0.17
Huaura	Casa Blanca	3 380	1912–71	0.85	251	0.22	−0.09
Chancay	Lambayeque	2 380	1914–77	0.85	357	0.28	0.51
Maipo	Cabimbao	1 480	1912–91	3.34	2256	0.51	1.34
Central region							
Paraná	Corrientes	1 950 000	1904–83	520	266	0.23	0.91
Uruguay	Puerto Salto	244 000	1921–80	145	594	0.37	0.95
Negro	Paso Palmar (Uruguay)	63 000	1910–79	21.2	336	0.51	1.27
Negro	Primera Angostura						
Ter.Negro		40 600	1927–84	28.9	718	0.32	1.06
Salado	El Arenal						
Pr.Santigo Del Estero		40 000	1929–80	0.61	15.2	0.56	1.08
Chubut	Los Altares	20 000	1928–83	1.79	90	0.09	−2.06
Dulce-El Sauce, Pr.San-Tiago Del Estero		19 720	1926–80	2.91	148	0.49	0.76
Colorado	Pichi Mahuida	12 500	1927–84	3.44	275	0.33	1.00
Tercero	Embalse	3 300	1914–79	0.85	258	0.45	0.90

Table 8.12. *Renewable water resources of largest rivers and ocean slopes of South America*

Territory	Area, $km^2 \times 10^6$	Study period	Water resources, km^3/yr			Coefficient of variation, C_v
			Average	Maximum	Minimum	
River						
Amazon	6.92	1920–85	6 920	8 510	5790	0.076
La Plata	3.10	1904–85	811	1 860	450	0.26
Orinoco	1.00	1924–92	1 010	1 380	710	0.15
Ocean slopes						
Atlantic Ocean	16.6	1921–85	10 720	12 870	9280	0.075
Caribbean Sea	0.60	1904–85	453			0.07
Endorheic regions	1.41	1914–85	56	85.6	35.2	0.23
Pacific Ocean	1.24	1921–85	1 310	1 820	640	0.18

Table 8.13. *Comparison of assessments of average annual river runoff on principal rivers in South America by data of different authors*

Author, date	Amazon		La Plata		Orinoco	
	Area, $km^2 \times 10^3$	Runoff, $km^3/year$	Area, $km^2 \times 10^3$	Runoff, $km^3/year$	Area, $km^2 \times 10^3$	Runoff, $km^3/year$
Keller, 1952			3100	1330	1050	880
Korzun, 1974a	6915	6900	2970	725	1000	918
Karasik, 1974	7050	5992	3100	983	944	1043
Baumgartner and Reichel, 1975	7180	6000	2650	615	1086	915
UN, 1985a					982	1135
Dubreuil, 1985	4688	6300–7100				
Vorosmarty, 1989; Vorosmarty and Moore, 1992		7065				
Balek, 1983		6700[a]				
FAO, 1995		5520				1070
Meade *et al.*, 1991	6150	6300			900	1135
Vasquez *et al.*, 1989					1100	1135
Meybeck and Ragu, 1995	6112	6590			1100	1135
SHI (Shiklomanov, 1997)	6915	6920	3100	811	1000	1010

[a] Excluding Tocantins.

basin, discharge and other features. For example the smallest variations in runoff occur in the Amazon, Orinoco, Magdalena, Atrato, Baker, Mirra and Chubut Rivers. Across South America, there are considerable variations in the runoff of the different rivers and in the statistics derived from the time series of runoff, which in many cases are of limited value because records are short. As can be seen from Table 8.11, the series are not symmetrical in many instances, because of many more dry years than wet years. But there is a lack of knowledge of many South American rivers, including the Amazon, Orinoco and Rio de la Plata; while the discharge of the Amazon has been determined in a number of studies, records

are short for the main river and its tributaries. The measurements made at Óbidos are irregular, and models have been devised to estimate the flow from the data at Óbidos (Karasik, 1974; Korzun, 1974b; Fread, 1976; Sterling, 1980; WRI, 1990, 1992; Sioli, 1984; Richey *et al.*, 1989a; Vorosmarty, 1989; Matsuyama *et al.*, 1993; Russel and Miller, 1990; Meade *et al.*, 1991; Dengo, 1992; Vorosmarty and Moore, 1992; Meade, 1994).

The mean annual discharge at Óbidos is assessed at 159 000 m^3/s. The mean annual resource of the entire Amazon basin is 6920 $km^3/year$ (Table 8.12). Table 8.13 gives estimates of the discharge of the Amazon, Orinoco and Rio de la Plata obtained by a number of authors at different times. Assessments of Amazon water resources are in the range 6000 to 7100 km³/year. In the present study, more complete data for Amazon discharges have been used than those by the authors indicated in Table 8.13. The new figures differ from estimates made by other authors, over the range from 13% greater to −2% less. The water resources of the Rio de la Plata system are better known than those of the Amazon. There are more than 80 years of observations of the discharge of the Paraná at Corrientes; however, the estimates vary from 600 km³/year to 1330 km³/year depending on the basin area. This study found the water resources of the Rio de la Plata to be 811 km³/year (Table 8.12), which is 86 km³/year above the figure given by Korzun (1974). The difference may be caused by the wet period in the Rio de la Plata basin for the last 15 years and because a larger area for the basin was used in the present estimate. The Orinoco River has a smaller discharge than the Amazon, but larger than the Rio de la Plata. The Orinoco delta is divided into 36 arms but the stage and discharge considerably vary. The mean annual discharge at Cuidad Bolívar is 25 000 m³/s and the water resources of the Orinoco are estimated to be 1010 km³/year (Tables 8.12 and 8.13).

Data for the runoff to the oceans are given in Table 8.12. The renewable water resources of areas draining to the Atlantic, including the Caribbean Sea (453 km³/year) and areas of inland drainage (56 km³/year), are estimated at 10 720 km³/year for the period 1921 to 1985. The water resources of the areas draining to the Pacific amount to 1310 km³/year. The least variation in water resources ($C_v = 0.07$) occurs for drainage to the Caribbean and to the Atlantic, and is greatest for Pacific drainage ($C_v = 0.18$). The coefficient of variation in water resources of areas of inland drainage is 0.23. Human activities account for the use of 49 km³/year, on average, for the period 1921 to 1985. Table 8.14 presents data on flows to the oceans by latitude. From these data, the total inflow to the Atlantic Ocean is 10 620 km³/year and to the Pacific some 1300 km³/year.

Long-term variations in the flows of the principal rivers and in the drainage to the oceans are very significant (Fig. 8.5). Tests for stationarity were made by the Student, Kolmogorov–Smirnov and Fisher methods. As a result, it was found that flows to the Atlantic and those of the Orinoco River are non-stationary both by average values and by variances. The discharge series for the Amazon basin are non-stationary by average values but homogeneous by variances. Discharges to the Pacific Ocean and from the Rio de la Plata basins were found to be stationary. The non-stationarity of series for the Atlantic Ocean seems to be due to climate and was examined by dividing the series into equal parts; one of them was found to contain years that were similarly wet or dry. The results of analysing the runoff series for the presence of trend by the different tests (Kendall, Spearman, WMO and I.I. Polyak), show that runoff to the Atlantic and the continent as a whole increased. However,

Table 8.14. *Distribution of ocean basin inflow (km³/yr) by latitude zones in South America*

Latitude	Atlantic Ocean	Pacific Ocean
10–20° N	247	
0–10° N	1 351	157
0–10° S	7 446	25
10–20° S	194	3
20–30° S	123	11
30–40° S	773	182
40–50° S	444	494
50–60° S	43	432
Total	10 620	1300

there was no trend in the runoff series for the Pacific. Table 8.15 gives information and Fig. 8.5 shows differences in the runoff from South America about the wet and dry periods and those around the mean. It is noteworthy that figures for runoff to the Pacific are opposite in sign to those for runoff to the Atlantic Ocean. Variations in runoff to the Atlantic are in phase with those for the Northern region and the Orinoco. For Atlantic runoff, there is a 63% probability that average years will follow dry years and a probability of 39% that dry years occur after average years. Wet years follow wet years. Patterns of runoff to the Pacific are quite different. In particular, there is a 47% probability that dry years are followed by dry years and a probability of 29% that they will be followed again by an average year. Average years following average years occur with a probability of 47% and a probability of 30% for wet years.

Table 8.16 shows that the large river basins, the Amazon, Orinoco and Rio de la Plata, differ considerably in their patterns of runoff. However, those of the Amazon are very similar to the variations in the total runoff from the continent as the Amazon runoff represents more than 50% of the total. Groups of years with wet, dry and average conditions of 1 to 5 years duration appear in the Amazon runoff with a probability of 90.9%. Groups of dry years of 1 to 5 years duration have probabilities of occurrence of 32% and 26%, respectively, while for the continent and the Amazon, groups of average years have probabilities of 49% and 46%, and wet years 19% and 26%. The respective figures for groups of years of longer duration are 11% and 9%. An analysis was made of the timing of variations in runoff for areas draining to the oceans by determining correlation coefficients between the different time series. The runoff in most South American rivers was found not to be correlated. However, the total runoff for the continent was closely correlated with the total runoff of Atlantic-draining rivers ($r = 0.96$) as well as with runoff from the Amazon ($r = 0.69$) and Orinoco ($r = 0.74$), but the Rio de la Plata showed no correlation with the Amazon and Orinoco. The correlation coefficient between the Amazon and Orinoco was determined as 0.22. The conclusion

(1)

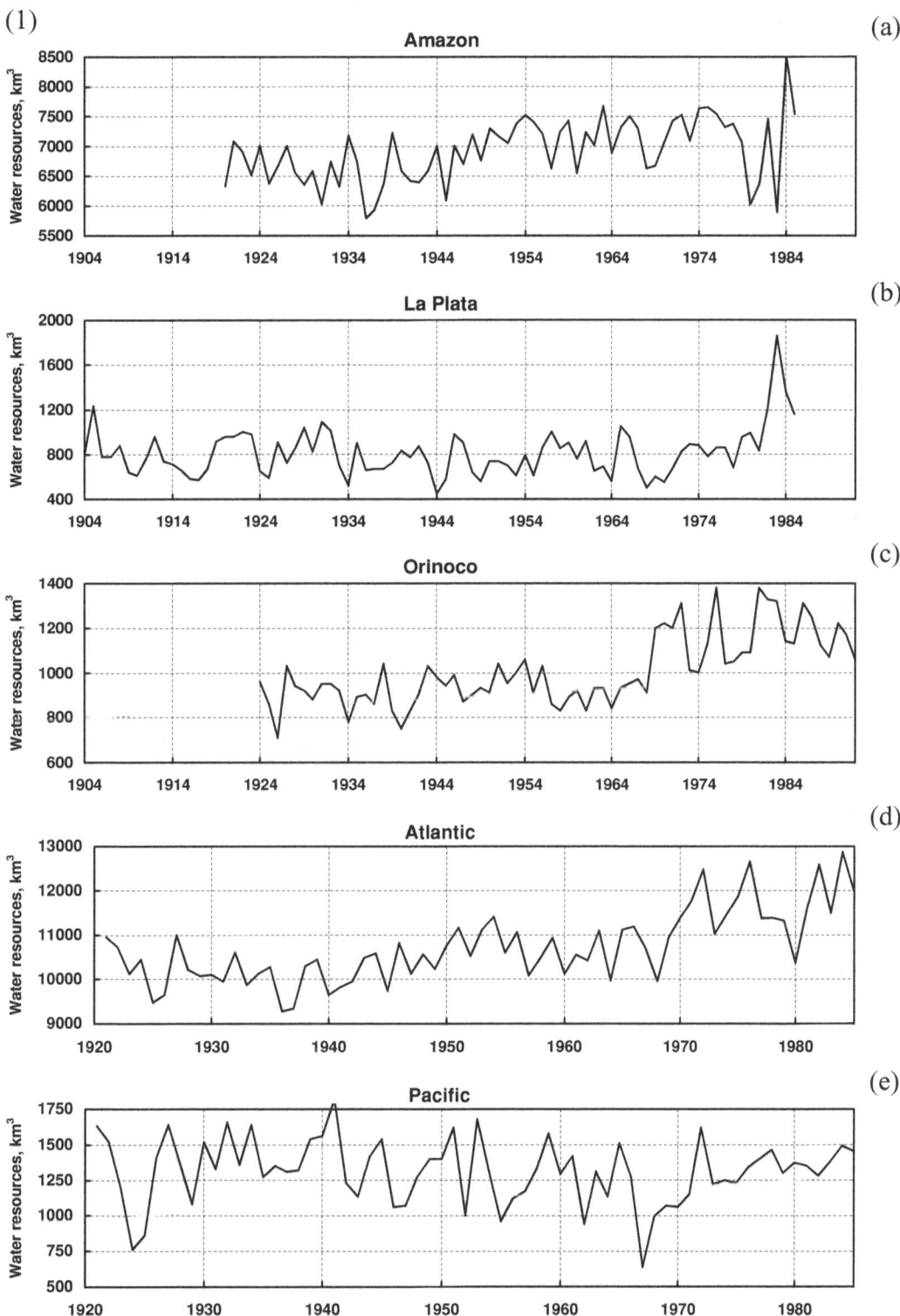

Fig. 8.5 Water resources variations (1) and normalized difference integral curves (2) of water resources for principal rivers and ocean slopes of South America. (a) Amazon; (b) La Plata; (c) Orinoco; (d) Atlantic Basin; (e) Pacific Basin.

(2)

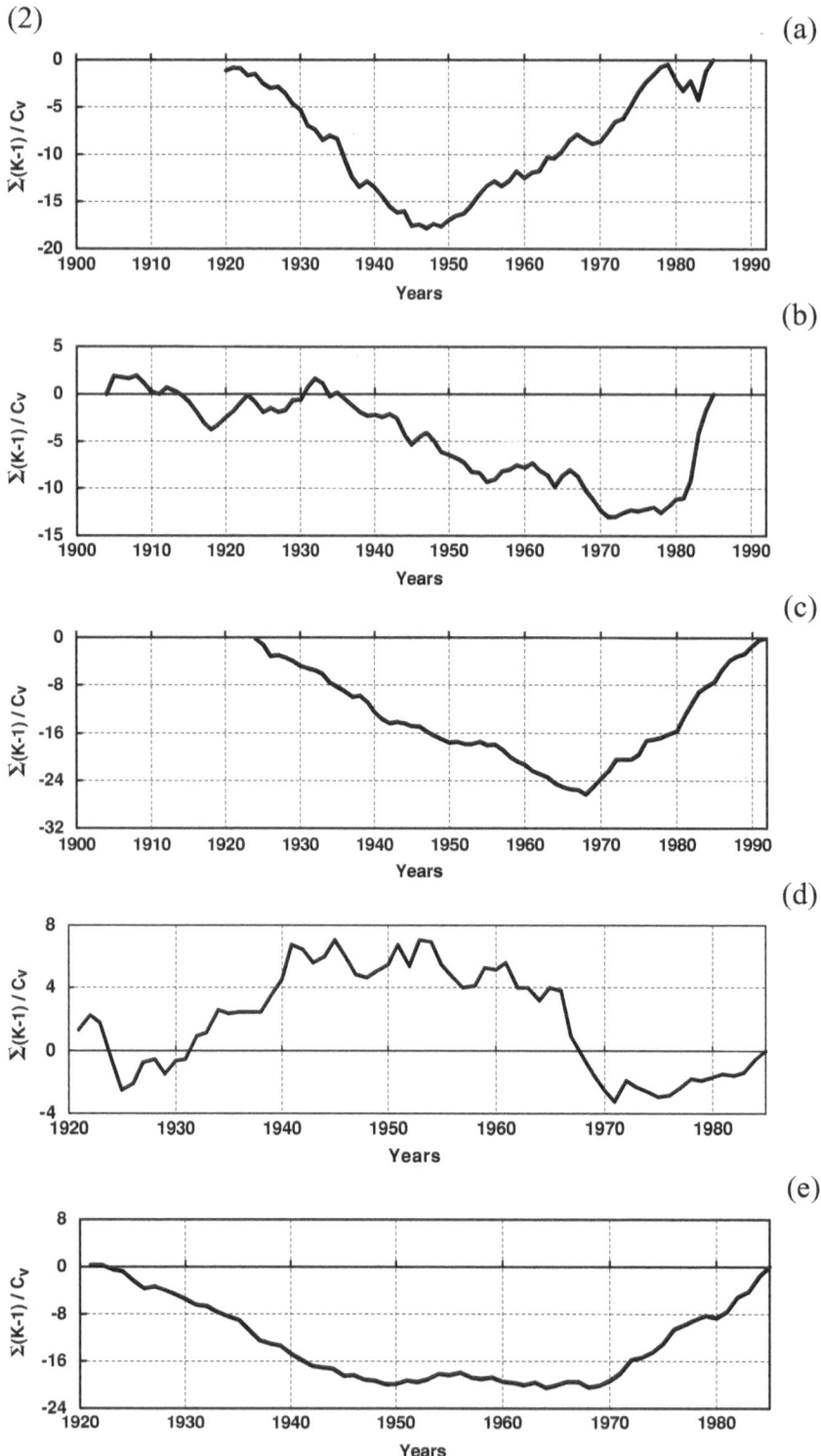

Fig. 8.5 (*cont.*).

Table 8.15. *Periods of different water content on major rivers and ocean slopes of South America*

Territory	Period of observations	Periods of high water content			Periods of low water content			Periods of average water content		
		Years	Ratio of mean for period to long-term mean, K_{av}	Number of years	Years	Ratio of mean for period to long-term mean, K_{av}	Number of years	Years	Ratio of mean for period to long-term mean, K_{av}	Number of years
Rivers										
Amazon	1920–85	1948–85	1.03	38	1920–47	0.95	28			
Orinoco	1924–92	1969–92	1.17	24	1924–68	0.91	45			
La Plata	1904–85	1972–85	1.25	14	1933–71	0.90	39	1904–32	1.01	29
Ocean slopes										
Atlantic Ocean	1921–85	1970–85	1.09	16	1921–49	0.95	29	1950–69	1.00	20
Pacific Ocean	1921–85	1921–45	1.05	25	1946–85	0.97	40			

Table 8.16. *Probability of appearance of groups of years with different water content on principal rivers and ocean slopes of South America*

Territory	Water content characteristic[a]	Probability of characteristics transition from one grouping to another, %			Average duration of groups, years
		$\rightarrow$ 1	$\rightarrow$ 2	$\rightarrow$ 3	
Rivers					
Amazon	1	29	59	12	1.4
	2	30	47	23	1.9
	3	11	33	56	2.3
Orinoco	1	56	44	0	2.5
	2	23	65	12	2.8
	3	0	25	75	4.0
La Plata	1	61	21	18	2.5
	2	25	42	33	1.7
	3	29	25	46	1.8
Ocean slopes					
Atlantic Ocean	1	37	63	0	1.6
	2	39	48	13	1.9
	3	0	21	79	4.7
Pacific Ocean	1	47	29	24	1.9
	2	23	47	30	1.9
	3	12	65	23	1.3

[a] 1, Low; 2, average; 3, high.

was that while there are many variations in the discharges of South American rivers, these variations are not synchronous.

8.5.4 The distribution of runoff within the year

Table 8.17 and Fig. 8.6 present patterns of runoff for South America. Both the Western and Eastern regions have a more even distribution than the other regions, including dry and wet years, but the Eastern runoff is more uniform. The mean long-term distribution for Brazil, Bolivia and Argentina is also depicted in Fig. 8.6, with the runoff for Brazil and Argentina being relatively constant. Compared to data in Korzun (1974b), the distribution of runoff during the course of a year shows few fluctuations (Fig. 8.7). The flows of the large rivers depend on the regimes of their principal tributaries. The discharges of the Amazon are maintained at a high level throughout the year due to the location of its tributaries in the different hemispheres. The right-bank tributaries, with basins in the Southern Hemisphere, are fed by rains from September to March, whereas the left-bank tributaries, located in the Northern Hemisphere, receive rain between April and October. Flow from the southern tributaries is greatest from May to July when the flow of the Amazon totals about 35% of the annual volume. The July monthly maximum is three or more times the November minimum (Table 8.18).

The flow of the Paraná River receives a large input from its right-bank tributary, the Paraguay. The summer maximum in the upper reaches of the Paraná becomes an autumn one downstream after the confluence of the Paraguay River. Most of the flow of the Orinoco River is located north of the equatorial belt; some 75% occurs over a period of 6 to 7 months with a maximum between July and September, due to the northeast tradewinds (Table 8.18). Runoff to the Atlantic (Table 8.18) from April to August is 54% of the annual volume, and the November to January runoff is 17%. Some 40% of the annual runoff to the Pacific occurs between February and April (May), and only 28% from September to December.

Table 8.17. *Renewable water resources distribution in years with different water content by natural–economic regions of South America*

Water content of year	Monthly runoff distribution, in % of mean annual value											
	1	2	3	4	5	6	7	8	9	10	11	12
Northern region												
Average	4.2	3.0	2.5	3.1	6.2	10.2	13.5	15.3	14.7	11.8	9.0	6.5
Dry	3.1	2.1	1.9	2.7	4.3	7.9	12.4	15.4	16.2	13.9	11.6	8.5
Wet	4.1	3.2	2.9	3.4	6.9	10.3	13.2	14.9	14.0	11.6	9.0	6.5
Eastern region												
Average	6.8	8.2	9.4	10.4	11.0	11.0	10.2	9.1	7.3	5.6	5.2	5.8
Dry	6.9	8.4	10.1	11.2	11.8	11.3	10.2	8.3	6.0	5.0	5.1	5.7
Wet	6.8	8.0	9.3	10.7	11.0	10.9	10.3	9.1	7.6	5.9	5.2	5.2
Western region												
Average	8.3	9.6	10.8	9.2	9.2	9.3	7.9	7.1	6.5	6.7	7.8	7.6
Dry	7.3	7.8	11.8	10.6	6.4	9.5	13.2	10.3	6.9	6.8	4.8	4.6
Wet	6.3	5.7	5.3	4.2	10.2	12.1	7.8	12.5	8.8	7.9	10.0	9.2
Central region												
Average	9.2	11.8	12.8	12.2	9.5	9.3	6.6	5.2	5.4	5.3	5.9	6.8
Dry	9.9	12.5	12.1	12.0	9.5	9.3	6.0	5.6	5.5	5.0	6.0	6.6
Wet	8.6	10.4	12.1	12.0	10.0	9.5	7.0	5.6	5.8	6.0	6.0	7.0

8.6 CHANGES IN WATER USE AND WATER AVAILABILITY

8.6.1 Data

The most comprehensive information about changes and trends in water use in South America for public supply, industry, power production and agriculture and about the additional losses due to evaporation from reservoirs is given in Shiklomanov and Markova (1987) and in Shiklomanov (1988). Because none of the countries concerned had data available on water use, especially data since the start of the twentieth century, the usage had to be estimated from the literature (Onuphriyev, 1971; Alekseyevsky, 1974; Van der Leeden, 1975; Yermolina and Kalinin, 1975; Borozdina, 1978; ECLAC, 1978a, b, 1984; Bromley, 1983). Some of these studies were concerned with the current status of irrigation and its future development and others with the development of industrial and municipal water use for certain countries: Colombia (Oranas, 1978), Venezuela (Eden, 1974), Peru (Nikolayev, 1971), Ecuador (Raumer, 1977), Chile (Hoyd, 1974), Argentina (Bertranou *et al.*, 1983) and Brazil (Pebayle, 1981).

FAO data were used for determining the size for the irrigated area at present and for the future for countries and the four regions. Specific water use for irrigation was assessed at 5000 to 12 000 m³/ha and return water at 20–50% of the abstraction. Spe-

cific water use for agriculture is estimated to average at 40–90 l/day per head and for domestic supplies 200–250 l/day per head. These values were expected to increase by the year 2000 to 70–110 and 250–400 l/day per head, respectively. Use of water by industry and for power generation for the regions was calculated from county data with the trends in industrial production taken into account along with the size of the urban population. Between 1980 and 1990, the period used by Shiklomanov (1988) and Shiklomanov and Markova (1987), and by the UN Economic Commission for Latin America and the Caribbean (ECLAC), in spite of economic problems, the water resources of most South American countries were developed rapidly, although not at the rate expected in the previous decade. ECLAC identified problems of investment for water management, lack of attention to agricultural water use in rainfed areas and to the protection of the environment, as well as to fragmentation of the water sector. To overcome the latter problem, in certain countries studies were started of integrated medium-term and long-term planning and managing water resources (UN, 1990a).

In Brazil, Peru, Chile and Argentina, strategies were designed for managing water for agriculture. They included both large-scale irrigation and drainage projects (Project São Francisco in Brazil) and irrigation projects on the medium and small scales. Attention was given to the servicing of available systems, the problems

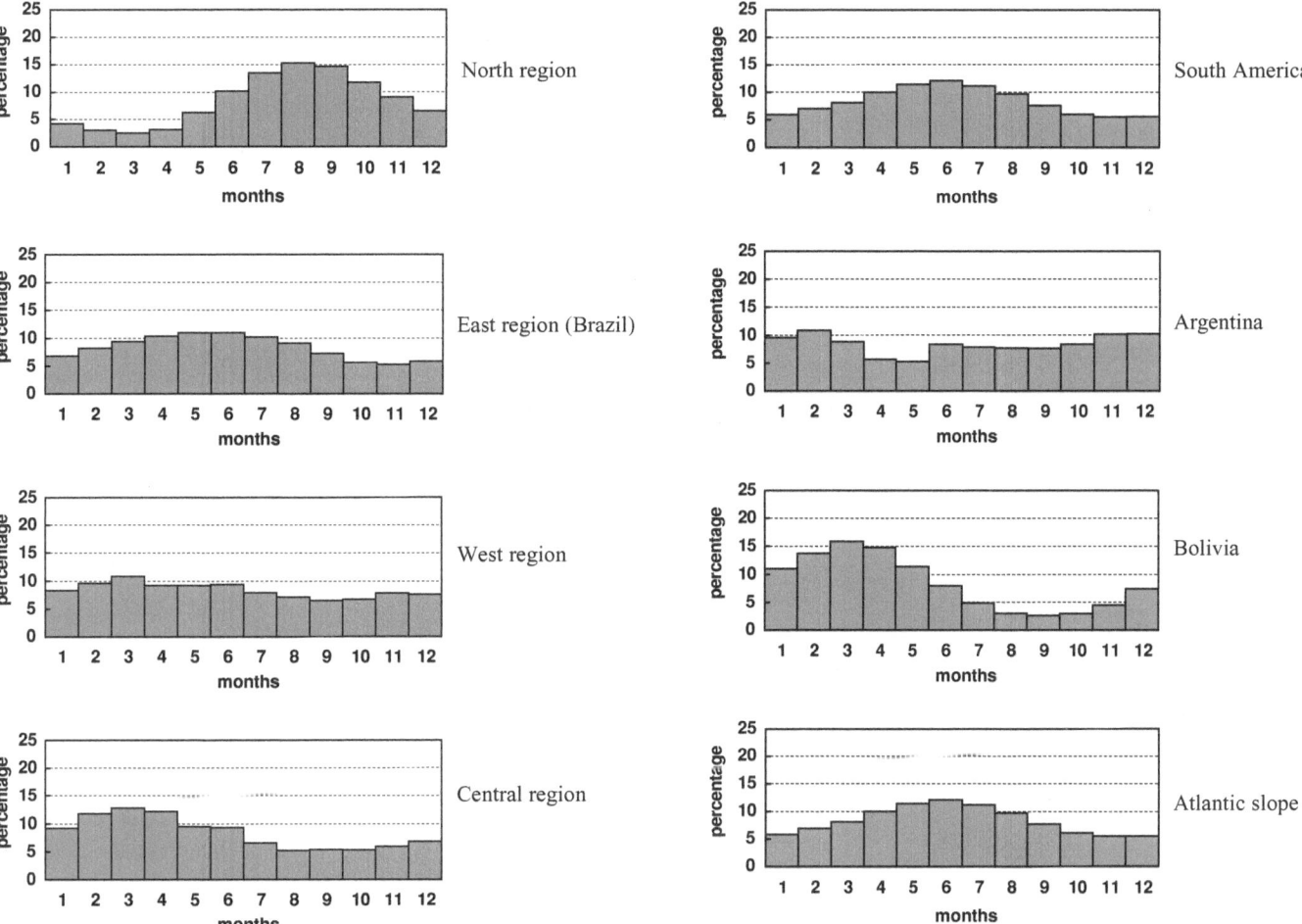

Fig. 8.6 Average monthly distribution of water resources of the natural–economic regions, continent, individual countries, ocean slopes and large river basins of South America.

of optimizing water management and the protection of the environment (UN, 1990). Experts in UNEP (1994) characterized the period 1980 to 1990 as the onset of the sixth stage in the evolution of the use of water in Latin America, together with the transition from river basin management to managing the natural environment as a whole. It was considered that water management plans should be included in the general national and regional development plans, but recognized that in practice this does not take place. The controlling factor in water management is the available capital for investment, mainly for hydropower engineering, drinking water supply and irrigation. Multi-purpose water use almost always is carried out retrospectively as problems and funds appear. To make forecasts of water use, normal trends were employed (Shiklomanov and Markova, 1987) in the current study.

Since the mid-1990s, these trends suggest that the past rates of growth in the use of water by agriculture will continue, but that there will be a considerable increase in water use by in-

dustry and for public supply. The forecasts of public use of water should take into account the national plans which have been produced by certain countries. The latest statistics for the continent confirmed water use estimates made by Shiklomanov for 1990 and earlier. For example the value forecast for 1990 was 150 km^3/year when the actual value was 152 km^3/year. However, there are changes in totals for particular sectors such as industry where the forecast was an overestimated by 9.1 km^3/year, whereas in the agricultural and domestic sectors, the forecasts underestimate by 6.7 and 6.0 km^3/year. Because of these differences which are due to particular features, new estimates were needed to be made for the 1990 totals and for later dates. In the present study for the period 1990 to 1995 and for 2000, 2010 and 2025 estimates of water use for the different sectors were made by the methods set out in Chapter 3 (Shiklomanov, 1987, 1988, 2000a, b; Penkova and Shiklomanov, 1998; Shiklomanov *et al.*, 2000). Data included in Gleick (1993, 1998, 2000), Kulshreshtha (1992), UN (1993a), Margat (1994), Strzepek and Bowling (1995), Seckler *et al.* (1998) and Rijsberman (2000) were used as well as the information contained in many published papers covering the different aspects of water in South America.

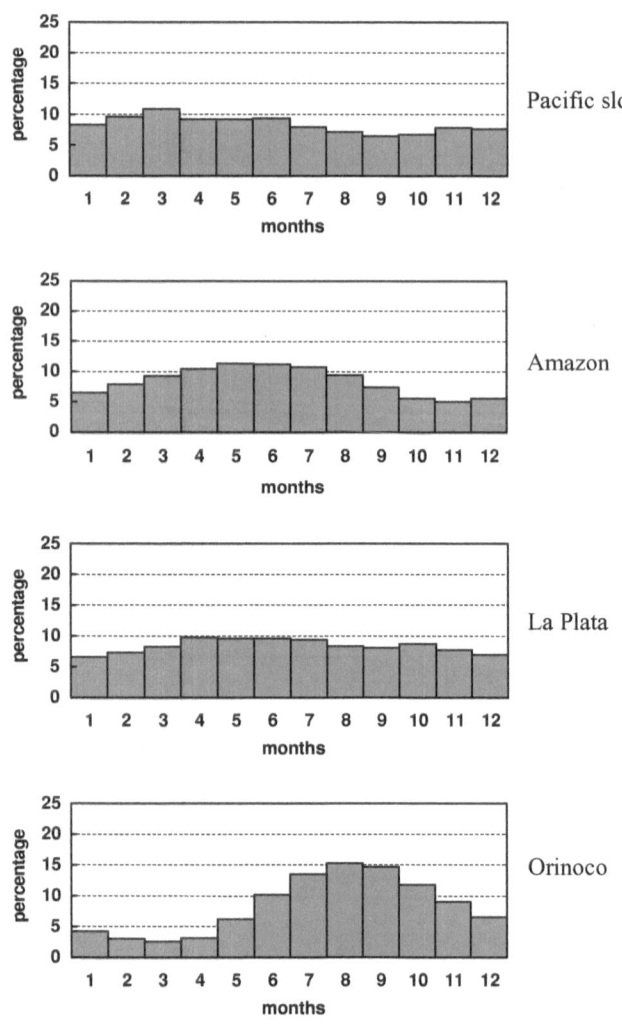

Fig. 8.6 (*cont.*).

Most South American countries have experienced an increasing demand for domestic water and this is considered as a continuing trend. For example, in Lima, by 2000 the water demand was expected to be 65–70% above the 1990 level. In this study, municipal water use was estimated separately for each country using forecasts of population growth and data on specific per capita water use (UN, 1995a). Rates of water use are foreseen to increase by 30–60%, but with a slight reduction in consumption, which is expected to be about 10–25% of the volume abstracted. Industrial water use was estimated country by country using the available statistics and by employing the UNIDO estimators for 2025 with the maximum values for Chile and Uruguay (6.6), and the minimum for Paraguay (2.2) and Argentina (3.2). For Brazil, Strzepek and Bowling (1995) give the value of 5.9 while for Ecuador, Peru, Colombia and Bolivia they give values of 4.5, 4.4, 4.2 and 4.0, respectively.

Taking into account recent changes in industrial water use, these estimators are recognized to be slightly overestimated, often be-

cause of the introduction of water-saving measures and pressures to conserve and protect the environment. These have caused more industrial water to be recycled and other means of saving to be followed. Taking these measures into account in the present study, the UNIDO coefficients were reduced by up to 25–30%. For industrially developed countries industrial water use was estimated to be 5–12% of abstractions (with the possibility of a slight growth), while for agrarian–industrial countries the percentage assumed was 15–20% with the possibility that this figure would lessen. For 2000 and 2010 use of water by industry was taken as proportional to the per capita income (GNP) using national financial data from publications of the International Monetary Fund (IMF, 1994) and the International Bank for Reconstruction and Development (IBRD) (World Bank, 1993, 1995). The growth in the GNP was estimated for groups of countries in accordance with UNIDO forecasts for the Global Balance Scenario. Strzepek and Bowling (1995) considered this growth to be about 5.6% per year and slightly more for more rapidly developing countries (e.g. Chile, Brazil).

There are possibilities for the further expansion of the area irrigated, the only exceptions being in Chile, in the Central Valley, in a strip of the coastal lowlands of Peru some 75–100 km wide (La Costa), in the Andes zone in Argentina and some central and southern regions of Brazil. At present in Chile about 97% of the potential area is irrigated, i.e. about 1.3 million ha (Dourojeanni and Nelson, 1982), but this is less than its previous extent which was between 2.0 and 2.5 million ha. However, there were plans to increase the irrigated area to 2.5 million ha by 1985 (Zonn and Nosenko, 1981; Dourojeanni and Nelson, 1982; FAO, 1995a). In Peru only 73%, of the potential area of 1.73 million ha has been exploited (Dourojeanni and Nelson, 1982), while in Ecuador the figure is 52% out of 1.15 million ha (Dourojeanni and Nelson, 1982). Drainage is more important than irrigation in Venezuela but there are plans for increasing irrigation to 323 000 ha, and for draining up to 1 22 7000 ha. Brazil has recently started large-scale drainage and irrigation programmes (National Irrigation Plan, 1982–86) increasing the area to 3 million ha (UN, 1990a). In Argentina, areas of low productivity occupy about 70% of the total area. About half (95 million ha) need some irrigation. However, estimates show that with modern techniques, irrigation is possible on about 5.5 million ha (Alekseyevsky, 1974; Zonn and Nosenko, 1981); by 2025 if present trends continue the area irrigated is expected to increase to 1.9–2.0 million ha.

Across South America irrigation is in different stages of development and trends indicate reductions in irrigation in a number of the countries. In these countries the 1990 level of irrigation is lower by 1.2 million ha than those forecast by Shiklomanov and Markova (1987). These changes and trends have been taken into account in the present study in developing forecasts for the coming decades along with other considerations such as those mentioned

Table 8.18. *Average monthly distribution of water resources for principal rivers and ocean slopes of South America*

Basin	Average water content km³/year	Monthly runoff distribution, in % of mean annual value											
		1	2	3	4	5	6	7	8	9	10	11	12
Rivers													
Amazon	6 920	6.5	7.9	9.2	10.4	11.3	11.2	10.7	9.4	7.4	5.5	5.0	5.5
La Plata	811	6.6	7.3	8.2	9.7	9.6	9.6	9.3	8.3	8.1	8.7	7.7	6.9
Orinoco	1 010	4.2	3.0	2.5	3.1	6.2	10.2	13.5	15.3	14.7	11.8	9.0	6.5
Ocean slopes													
Atlantic Ocean	10 720	5.8	6.9	8.1	10.0	11.4	12.1	11.2	9.7	7.7	6.1	5.5	5.5
Pacific Ocean	1 310	8.3	9.6	10.8	9.2	9.2	9.3	7.9	7.1	6.5	6.7	7.8	7.6
South America as a whole	12 030	5.9	7.0	8.1	10.0	11.4	12.1	11.1	9.7	7.6	6.0	5.5	5.6

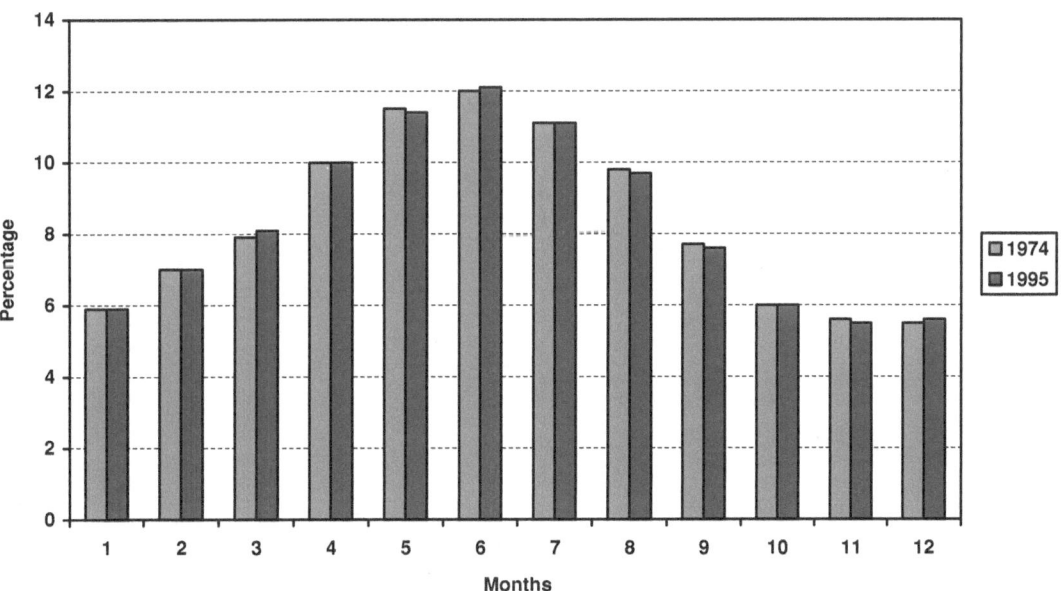

Fig. 8.7 Monthly distribution of water resources in South America in 1974 and in 1995.

by ECLAC (WRI, 1990, 1996, 2000). Accepting these points, then by 2000, 2010 and 2025 the total area irrigated is expected to be 9.3, 10.0 and 10.4 million ha, respectively.

8.6.2 Changes and trends in water use

Sectoral water use was estimated separately for each country and summed for the four regions and for the continent (Fig. 8.8, Table 8.19). From those data it appears that for 1995 the use of water for agriculture is about 100 km³/year, about 60% of all water used. Of the other sectors industry accounts for 11% and domestic supplies about 20%; some 14.5 km³/year or 9% is evaporated from

reservoirs. By 2025 industry is expected to use more water than the domestic sector, while agriculture will fall to 43% of the total. However, agriculture, primarily irrigation, will consume the greatest volume (about 70%) and about 20% will be spent in reservoir evaporation. Use of water is expected to increase in 2000, 2010 and 2025 by 8.4, 28.3 and 54.8%, respectively, as compared to water use in 1995. However rates of growth are expected to decrease from 95% for the period 1970 to 1995 to 55% for the period 1995 to 2025.

Use of water varies considerably across the regions (Fig. 8.9, Table 8.20). After 1970 the highest growth rates were recorded in the Eastern region, associated with the high rates of development in Brazil (WRI, 1992). Since the late 1970s Brazilian irrigation rates have began to increase. These different patterns of development will result in the redistribution of the volume used across the

Table 8.19. *Dynamics of fresh water use in South America by type of economic activity (km³/yr)[a]*

	Assessment								Forecast		
	1900	1940	1950	1960	1970	1980	1990	1995	2000	2010	2025
Population, ×10⁶			110	144	188	240	294	326	356	414	494
Irrigation area, ×10⁶ ha	1.2	2.2	4.81	4.97	6.24	7.9	8.76	9.06	9.35	10	10.4
Agriculture	13.6	24.6	54.3	58.6	65.9	77.3	96.7	99.9	103	110	112
	10.9	19.7	40.2	41.7	51.4	59.3	74.3	76.3	78	83	85
Industry	1.3	2.2	3.0	4.9	8.4	13.3	15.9	19.0	22	34	56
	0.26	0.4	0.6	0.8	0.9	1.1	1.2	1.6	2	3	6
Domestic	0.25	0.8	1.9	4.4	6.9	12.4	28.1	32.6	37	48	64
	0.14	0.4	0.7	1.2	1.5	2.5	4.9	5.3	6	7	8
Reservoirs	0	0.1	0.2	0.7	4.0	8.0	11.0	14.5	18	21	24
Total fresh water use	15.2	27.7	59.4	68.6	85.2	111	152	166	180	213	257
	11.3	20.6	41.7	44.4	57.8	71.0	91.4	97.7	104	113	123

[a] Nominator, withdrawal; denominator, consumption.

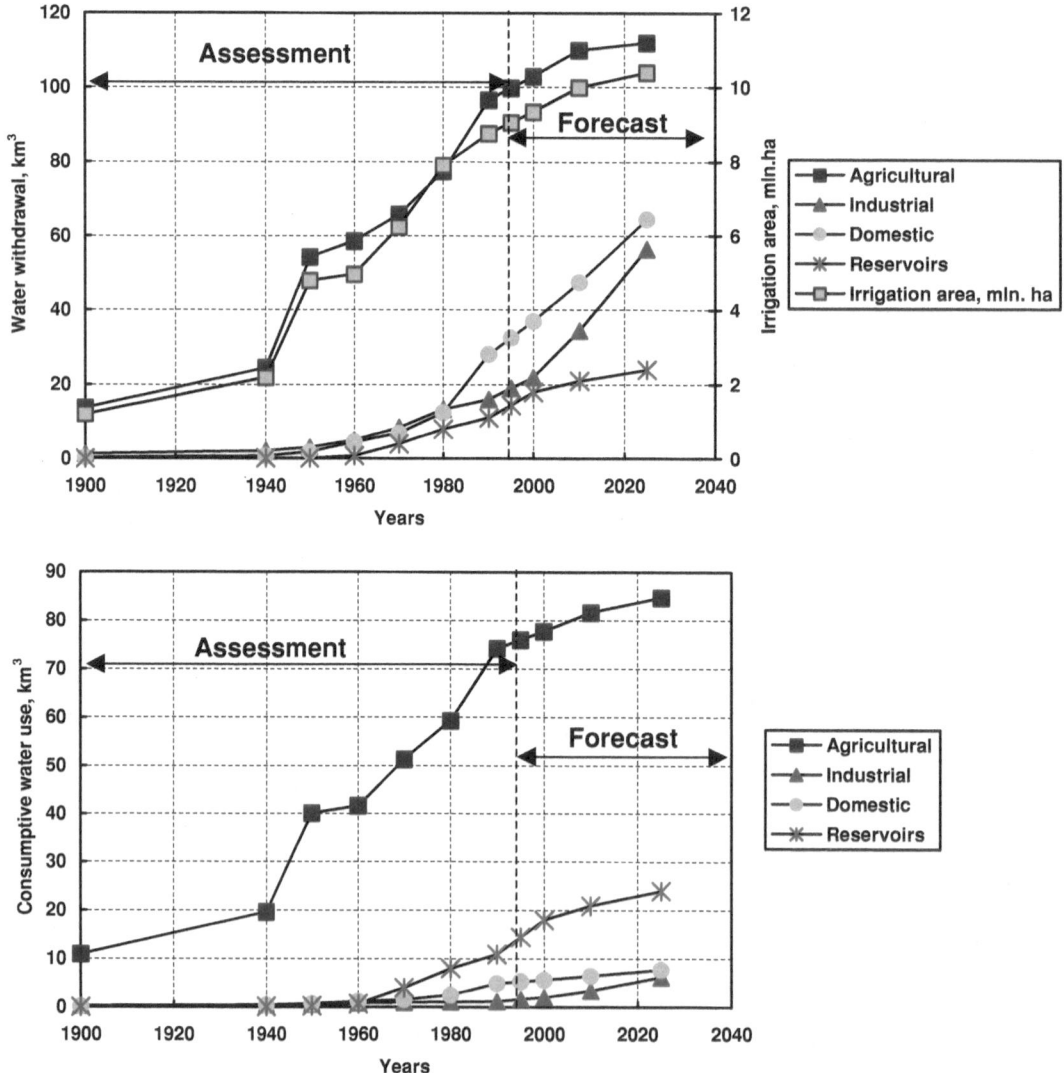

Fig. 8.8 Dynamics of water use by economic sectors on the continent of South America. (a) Water withdrawal; (b) water consumption.

Table 8.20. *Dynamics of water use by natural–economic regions of South America (km³/yr)*[a]

Region	Assessment								Forecast		
	1900	1940	1950	1960	1970	1980	1990	1995	2000	2010	2025
Northern	1.6	4.2	6.4	7.7	11.3	15.4	22.1	24.5	27	32	41
	1.3	3.4	5.0	5.8	8.3	11.0	14.4	15.6	17	18	21
Eastern	1.1	2.1	3.0	7.3	12.1	23.2	43.0	49.0	55	69	88
	0.52	0.96	1.15	3.0	5.6	10.3	18.0	20.4	23	26	28
Western	8.8	14.9	36.7	37.5	35.8	40.0	45.0	47.1	49	55	64
	6.9	11.7	26.0	25.1	27.3	29.9	32.7	33.6	34	37	40
Central	3.6	6.5	13.3	16.0	26.0	32.6	42.0	45.6	49	57	64
	2.6	4.5	9.6	10.5	16.6	19.7	26.3	28.0	30	32	34
South America as a whole	15.2	27.7	59.4	68.5	85.2	111	152	166	180	213	257
	11.3	20.6	41.7	44.4	57.8	71.0	91.4	97.7	104	113	123

[a] Nominator, withdrawal; denominator, consumption.

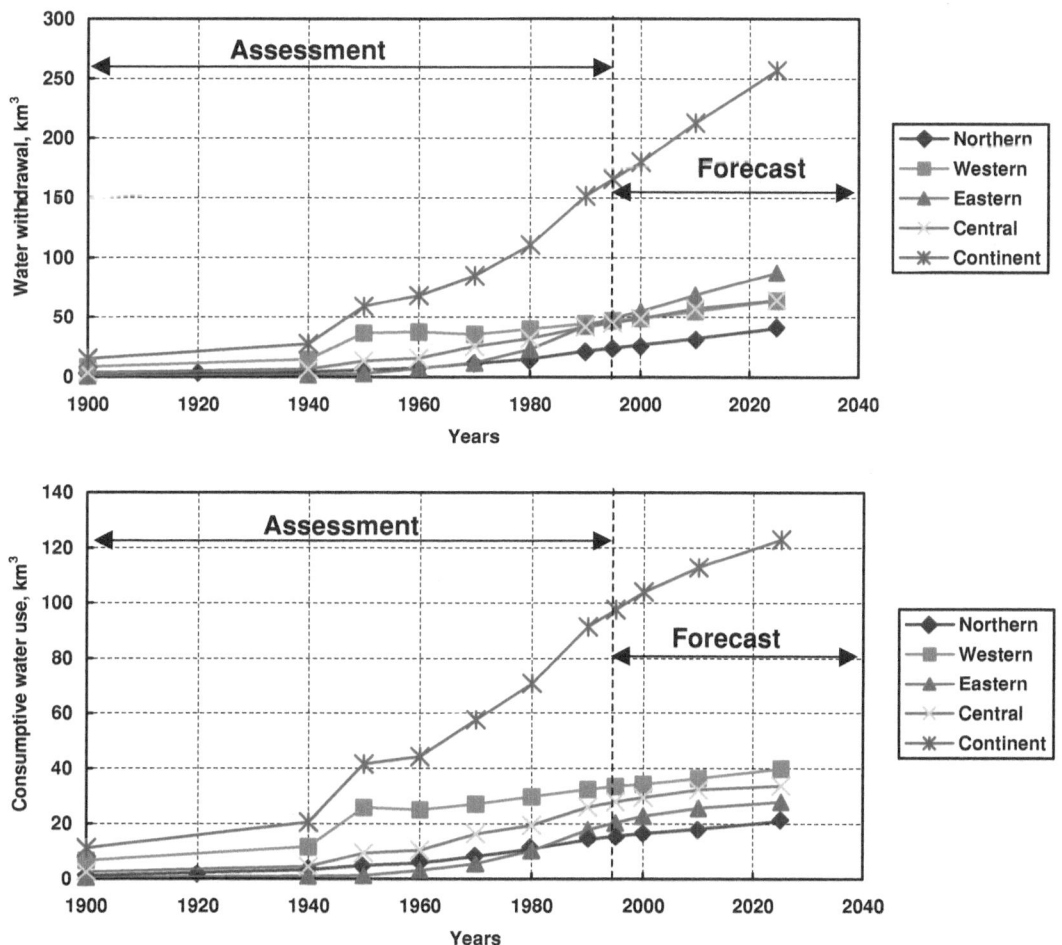

Fig. 8.9 Dynamics of water use in the regions of South America.
(a) Water withdrawal; (b) water consumption.

Table 8.21. *Dynamics of fresh water use by selected countries of South America (km³/yr)ᵃ*

Country	Assessment				Forecast		
	1970	1980	1990	1995	2000	2010	2025
Argentina	32.2	30.5	33.7	35.8	38	42	46
	20.8	19.2	20.1	20.6	21	22	20
Bolivia	0.70	1.15	1.38	1.68	2	4	4.5
	0.47	0.81	0.92	1.06	1.2	1.5	2
Brazil	12.1	23.2	43.0	49.0	55	68	88
	5.6	10.3	18.0	20.4	23	26	28
Chile	19.6	21.0	21.4	22.0	23	25	28
	14.6	15.6	15.4	15.4	15	16	16
Colombia			5.55	5.95	6.5	8	10
			3.62	3.72	4	4	5
Ecuador	4.95	5.56	5.98	6.23	6.5	7	8
	3.76	4.17	4.35	4.42	4.5	4.5	5
Peru	14.6	15.5	17.0	18.0	19	22	26
	11.0	11.5	12.3	12.8	13	15	17
Surinam			0.49	0.53	0.6	0.7	0.9
			0.31	0.32	0.3	0.4	0.4
Uruguay		3.50	4.18	4.78	5.5	6	7
		2.68	3.08	3.50	4	4.5	5

ᵃ Nominator, withdrawal; denominator, consumption.

continent. In 1995 the Eastern region used most water and forecasts indicated that by 2025 this difference would grow larger, even with the Central region, where the rate is expected to be 23 km³/year. More than 70% of the total volume (124 km³ in 1995) is used in Brazil, Argentina, Chile and Peru (Table 8.21). Ecuador, Colombia and Uruguay each use about 5–6 km³/year while the parallel figures are much lower in the remaining countries. Growth in these four countries, especially Brazil, is expected to be considerable in the future.

The 180 km³/year forecast for 2000 is 17% below the estimate obtained by earlier forecasts (Shiklomanov and Markova, 1987; Shiklomanov, 1988). From the perspective of the 1970s, it should have been 216 km³/year. Andressian (1993) and Margat (1994) forecast that by the late 1990s the total volume water used in South America (without thermal power production) would amount to 158 km³/year, while the volume consumed (including thermal power) would be 67.1 km³/year; for 2025 their "lower and higher scenarios" of water use showed that abstractions could amount to 123 and 170 km³/year, respectively, and consumption to 115 km³/year. By comparison with the values in this Monograph the Andressian and Margat figures are lower. The difference in the forecast for industry is 39.8 km³/year, and 26.6 km³/year for domestic supply, even with the "higher" water use scenario, and water losses from reservoirs could be a larger component of use in the future as more reservoirs are constructed. Today the use made of South America's water resources is rather low. Table 8.22 shows that at present it is about 1.3%, and that by 2025 this figure may rise to 2.1%. It is highest in the Central region but even there it only reaches 6% of the available water resource. The future will see an increase in this percentage, particularly as irrigation increases and the management of shared river basins improves, for example with the utilization of the Rio de la Plata basin (Avakyan and Sidoruk, 1993; Tucci et al., 1995).

8.6.3 Water availability

Table 8.23 contains population and water availability figures (in thousand m³/year per head) for the regions from 1950 to 2025. In 1950 all the regions had a high availability of water (>20 000 m³/year per head). In 1995 the Central region was in the category where 10 000–20 000 m³/year was available and by 2025 the Western region may also be in this category. The trends in specific water availability are also depicted in Fig. 8.10. The situation is likely to be worst in Argentina, where by 2025 12 000 m³/year per head will be available, whereas in Surinam and Guyana it will remain at a very high level (Table 8.24). During dry years and in droughts the water availability in Argentina is estimated to be 15 300 and 11 400 m³/year per head in 1994 and 2025, respectively (Table 8.25). In future it can be expected to decrease to 5.6 and 4.2 thou m³ (by 61%). Soon Argentina will join those

Table 8.22. *Water use as a percentage of water resources by natural–economic regions of South America*

Region	Water resources, km³/year		Water use, km³/year				Water use as a percentage of water resources			
	Inflow	Local runoff	1995		2025		1995		2025	
			Withdrawal	Consumption	Withdrawal	Consumption	Withdrawal	Consumption	Withdrawal	Consumption
Northern		3 340	24.5	15.6	41	21	0.73	0.5	1.2	0.6
Eastern	1900	6 220	49.0	20.4	88	28	0.6	0.2	1.2	0.4
Western		1 720	47.1	33.6	64	40	2.6	1.9	3.7	2.3
Central	720	750	45.6	28.0	64	34	4.1	2.4	5.8	3.1
South America as a whole		12 030	166	97.7	257	123	1.3	0.8	2.1	1.0

Table 8.23. *Dynamics of population and water availability by natural–economic regions of South America*

Region	Area, km$^2 \times 10^6$	Year	Population, $\times 10^6$	Water availability,[a] m$^3 \times 10^3$/year per head
Northern	2.55	1950	17.0	196.0
		1960	23.7	140.0
		1970	32.6	102.0
		1980	42.2	78.9
		1990	52.9	62.8
		1995	59.3	56.0
		2000	63.3	51.8
		2010	74.5	44.6
		2025	94.2	35.2
Eastern	8.51	1950	51.9	138.0
		1960	69.7	103.0
		1970	92.5	77.5
		1980	121.3	59.2
		1990	149.0	48.2
		1995	165.0	43.3
		2000	179.5	40.1
		2010	207.4	34.7
		2025	245.8	29.3
Western	2.33	1950	17.2	98.5
		1960	22.0	76.9
		1970	29.0	58.3
		1980	36.5	46.3
		1990	45.3	37.3
		1995	50.2	33.5
		2000	57.7	29.2
		2010	67.2	25.0
		2025	77.0	21.8
Central	4.46	1950	23.7	46.3
		1960	28.4	38.5
		1970	33.9	32.2
		1980	39.8	27.1
		1990	46.9	22.8
		1995	51.5	21.6
		2000	55.7	19.1
		2010	64.6	16.7
		2025	76.7	13.9
South America as a whole	17.86	1950	109.8	109.0
		1960	143.8	83.4
		1970	188.0	63.6
		1980	239.8	49.8
		1990	294.1	40.6
		1995	326.0	36.6
		2000	356.2	33.4
		2010	413.8	28.8
		2025	493.7	24.1

[a] With no account of withdrawal from ground water.

Table 8.24. *Water availability in selected countries of South America*

Country	Population (1994), $\times 10^6$	Water resources, km³/year		Water use (1990), km³/year		Water availability, m³ $\times 10^3$/year per head						
		Local runoff	Inflow	Withdrawal	Consumption	1970	1980	1990	1995	2000	2010	2025
Argentina	34.2	270	623	33.7	20.1	25.3	21.2	17.4	16.2	15.1	13.5	12.3
Bolivia	7.24	361	155	1.38	0.92			61.2	53.1	45.1	34.3	24.1
Brazil	159	6220	1900	43.0	18.0	77.5	59.2	48.2	43.3	40.1	34.7	29.3
Chile	14.0	354	0	21.4	15.4			25.7	24.3	22.9	20.7	17.1
Colombia	34.3	1200	0	5.55	3.62			37.0	34.3	31.5	27.3	22.1
Ecuador	11.2	265	0	5.98	4.35			24.7	21.0	17.4	15.0	13.1
Guyana	0.83	270	0	6.02	3.92			334	313	291	257	230
Peru	23.3	1100	144	17.0	12.3			55.0	49.5	42.4	35.4	31.8
Surinam	0.42	230	0	0.49	0.31			544	517	490	430	348
Uruguay	3.17	68	74	4.20	3.10			44.9	43.0	41.0	38.0	37.2

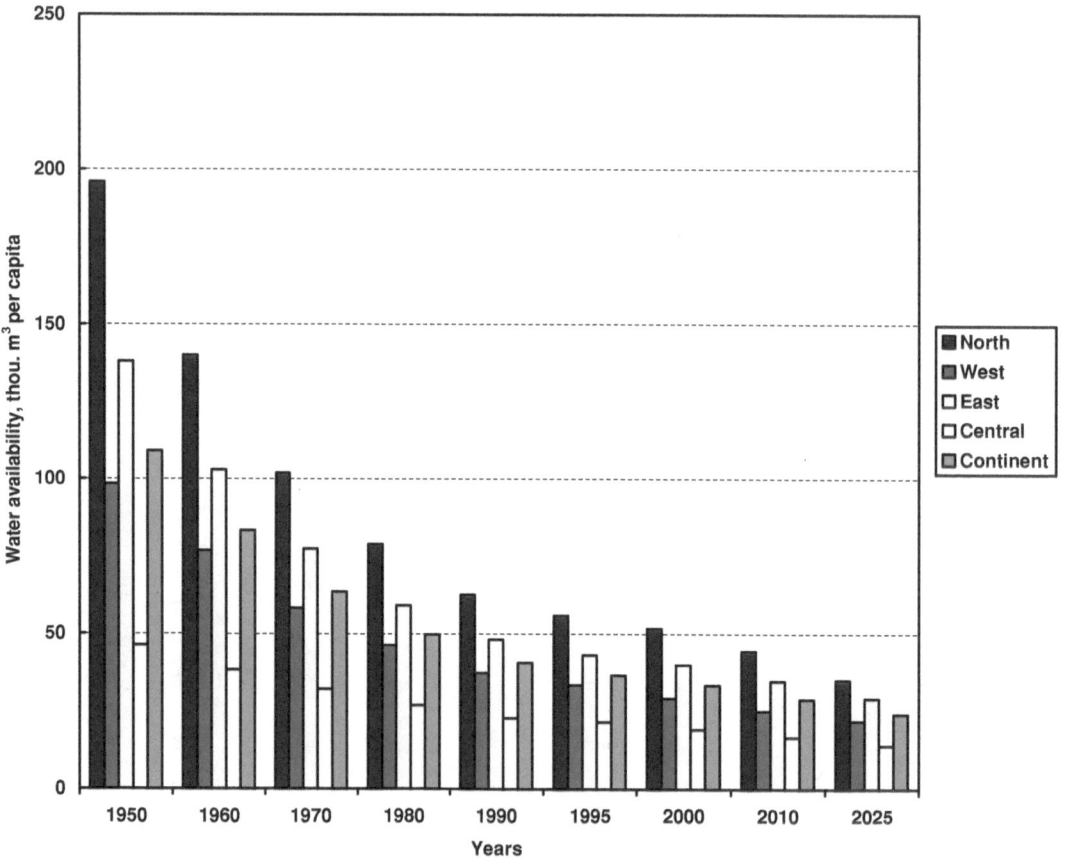

Fig. 8.10 Changes in water availability by natural–economic regions of
South America.

Table 8.25. *Water availability throughout long-term dry periods and the driest year by natural–economic regions and selected countries of South America*

Territory	Population, ×10⁶ 1994	2025	Period	Inflow	Local runoff	Year	Inflow	Local runoff	Water consumption, km³/year 1994	2025	Water availability, m³ × 10³/year per head — Dry period 1994	2025	Dry year 1994	2025
Regions														
Northern	57.3	94.2	1921–68		3 100	1926		2390	15.6	21	53.8	32.7	41.4	25.1
Eastern	159	246	1921–47	1817	5 910	1936	1596	5200	20.4	28	42.7	27.6	37.5	24.3
Western	48.6	77.0	1946–85		1 660	1924		990	33.6	40	33.4	21.0	19.7	12.3
Central	49.4	76.7	1921–71	634	716	1944	392	530	28.0	34	20.3	13.0	14.1	9.0
Countries														
Argentina	34.2	45.5	1921–71	581	253	1944	126	148	20.5	20	15.3	11.4	5.6	4.2
Bolivia	7.2	18.3	1921–71	149	346	1955	123	286	1.1	2.2	57.9	22.8	47.9	18.9
South America as a whole	314	494	1921–49		11 520	1925		10330	97.7	123	36.3	23.1	32.5	20.7

nations in the category with average and lower water availability (2-5,000 m³/year per head). During dry periods the Central and Western regions approach the average category (Table 8.25), but water will remain at high and very high levels of availability in the North and East.

8.7 CONCLUSIONS

South America has very large water resources: the runoff is some 12 030 km³/year. Most of this (6220 km³/year) originates in the Eastern region and flows into the Atlantic, representing about 89% of the total flow from the continent. Some 10% drains to the Pacific Ocean, while the remainder (about 56 km³/year) does not reach the oceans at all. The runoff is fairly evenly distributed through the year. Because of the growth in the population, the volume per head of population has reduced in all parts of the continent. Since the 1950s there has been an upward trend in runoff in the Northern and Eastern regions, in rivers draining to the Atlantic and for the continent as a whole. Flows to the Atlantic and to the Pacific show little or no correlation in time, either in phase or out of phase. The total volume abstracted is 166 km³/year of which 98 km³/year are consumed. By 2025 these figures are expected to increase to 257 and 123 km³/year. In the Central and Western regions, the water used currently comprises 3.8% and 2.6%, respectively, of the total volume of the water resource. By 2025 these volumes are expected to increase to 5.8% and 3.7%.

Despite the availability of a certain amount of data, these assessments of water resources and water use for countries, regions and the continent are only estimates. Lack of hydrological data due to poor instrument networks, short records and absence of systematic collection of data on the use of water are the chief impediments to improved understanding. Similarly, forecasts of the future positions are only approximate and do not account for the possibility of climate change. Nevertheless these estimates of water resources, water availability and water use at present are the most comprehensive and reliable statements for South America for the beginning of the twenty-first century.

9 Water resources, water use and water availability in Australia and Oceania

9.1 INTRODUCTION

Australia is the smallest continent, with an area of 7 615 000 km^2. It is an island located entirely in the Southern Hemisphere between 10° S and 39° S and between 113° E and 153° E. The continent of Australia is fringed by several islands, Tasmania being the largest covering an area of 68 400 km^2. Oceania is represented by numerous islands and groups of islands, located in the central and south-western Pacific between Australia and the Malaysian Archipelago in the west and extending over a vast area to the north, east and south. The islands of Oceania cover an area of 1 270 000 km^2, the three islands of New Guinea and the North and South Islands of New Zealand making up 83% of the land area. According to geography and ethnology the islands of Oceania can be divided into three groups: Melanesia, Polynesia and Micronesia.

The location of Australia and its adjacent islands near to the equator within a relatively warm ocean are important controls of its climate. Tropical and equatorial air masses dominate most of the continent and only the southern most parts of Australia, Tasmania and New Zealand are affected by temperate air masses. Australia and Oceania can be considered as the two regions of this part of the world. Estimates of water resources and water use have been carried out for these two regions while similar estimates have been made for Australia and New Zealand and for the drainage areas of Australia, including a separate estimate for the Murray–Darling River Basin.

Australia is well known as the driest continent. The availability of water is the lowest and it is the only continent where 50% of the runoff fails to reach the oceans. Almost the entire continent is dry, except a narrow fringe around the southern, eastern and northern coasts. The dryness increases towards the centre; in contrast the islands adjacent to Australia are very wet. The hydrology of Australia and even more so of Oceania needs more attention. The present publication presents estimates of water resources of the areas in question on the basis of the measurements at 190 permanent stations. The runoff from Australia was found to be to 352 km^3/year, for Oceania 2050 km^3/year and together the estimated water resources are 2400 km^3/year. Year to year variations

are great: the largest runoff from Australia and Oceania was observed in 1950 when it was 2880 km^3/year, while in 1937 it was only 1890 km^3/year.

As indicated by UN (1995a), the population of Australia in 1994 was 17.9 million. It is one of the most sparsely populated countries in the world (2.3 inhabitants/km^2). In New Zealand the population is 3.5 million (13 inhabitants/km^2), while Oceania as a whole has 10.8 million people (8.5 inhabitants/km^2). The potential water available in Australia is estimated to be 45 800 m^3/year per km^2, (19 700 m^3/year per head); in Oceania these figures are 1 614 000 m^3/year per km^2 and 190 000 m^3/year per head, while for Australia and Oceania they are 268 000 m^3/year per km^2 and 84 000 m^3/year per head.

Australian rivers are intensively used, particularly for irrigation. The irrigated area increased from 0.05 million ha in 1900 to 2.31 million ha in 1995. It is predicted that by 2025 the irrigated area will reach 2.8 million ha. Consumption of water in general is not high, consisting of 15.4 km^3/year in 1995. However considering the present trends, the total water used in Australia and Oceania was expected to increase by 7% to 2000 and by 16% in 2010, as compared to 1995. A similar increase is predicted for water consumption. In Australia the present water use amounts to 7.7% of the annual runoff, while for Oceania the same figure amounts to only 0.17%. Oceania uses the smallest proportion of runoff of any of the world's regions.

9.2 PHYSICAL CONDITIONS

9.2.1 Relief, climate and vegetation

The Australian continent is dominated by a vast plateau, concave in the centre and slightly raised at the periphery. The Great Dividing Range, extending along the eastern fringe of the continent, is the main mountain system, consisting of ridges and uplands alternating with lows and troughs. The eastern slopes of the mountains are steep and rugged, while their western edge slopes to foothills and then into a low plain. The southernmost part of the mountains contains a marked mountain ridge; this is the so-called Australian Alps with Mount Kosciuszko (2230 m) which is the highest point the Continent. A coastal lowland extends in a narrow band along the eastern periphery of the mountains. This is the most heavily developed and densely populated part of the country where the largest Australian cities are located. Western Australia is covered by a vast plateau, its mean elevation ranging from 300 to 500 m. The old, heavily weathered MacDonnell and Musgrave Ranges rise in the middle of the plateau (1300–1500 m). Extensive dry lands stretch to the northwest and southwest of these ridges; these are the Great Sandy Desert, the Gibson Desert and the Victoria Desert. There is a large low plain in the central part of the continent. Its level lands are covered with sand and pebbles and extend for hundreds of kilometres; in places there are lakes like Lake Eyre, Lake Torrens and others. This part of the continent is known as "the Dead Heart of Australia" (Grigorkina, 1976a).

The relief of the two main islands of New Zealand contrasts markedly with that of Australia. They are predominantly mountainous with high pointed summits covered by perennial snow and glaciers. The mountain ridges are broken up by deep valleys. The Southern Alps reach 3756 m in Mount Cook, while in New Guinea Mount Jaya (5029 m) is the highest point in Oceania. Low plains fringe the peripheries of most of the islands.

Most of Australia and particularly its northern coast have a tropical and subtropical climate. Only the southernmost part of Tasmania Island has a temperate climate. Temperatures in the north range from 23 to 28 °C, while in the south (the Melbourne area) the temperature in summer (January) is 20 °C, and in winter 9 °C. It snows only occasionally in the mountains, despite the name – the Snowy Mountains.

Over much of Australia the precipitation is extremely low. Summer monsoons bring precipitation to the tropical north. Southeastern coasts receive precipitation all year round; however, the winds from the sea do not reach far inland because of the mountains, hence the amount of precipitation decreases inland. Most of Australia receives less than 200–300 mm of precipitation annually, and this is the cause of the semi-desert climate. However there are no parts of the continent where rain never falls: although amounts are small, rain occurs throughout the year. The eastern and northern edges of the continent receive most rains, including the Kimberley Plateau (1400 mm/year), Cape York Peninsula (1600 mm/year), Eastern Queensland and south to the mountains of Eastern Australia (more than 1200 mm/year). Totals reduce to the southeast to 600–800 mm/year but increase to more than 2500 mm/year over Western Tasmania. The southwest corner of Australia receives more than 1000 mm/year. All these areas with high amounts of precipitation are located near the coast. Precipitation totals drop rapidly inland. The western part of the centre of Australia and the Western Australian Plateau receive only 300 mm/year. In the vicinity of Lake Eyre annual precipitation is about 100 mm. The average precipitation for Australia is 456 mm/year, Tasmania receives 1290 mm/year and Australia with Tasmania has an average of 460 mm/year.

In New Zealand tropical maritime air masses prevail in summer, while in winter they only penetrate the north. The South Island is dominated by polar maritime air masses and the Southern Alps lie almost perpendicular to the direction of the prevailing westerly winds, acting as an obstacle and causing orographic enhancement of precipitation amounts. Some coastal areas receive over 2000 mm/year; but as elevation increases totals rise to 5000 mm/year and more. Eastward from the crest, totals decrease rapidly to 1200–1300 mm/year, while in some areas in the rain shadow amounts decline to 700 mm/year. Most of the eastern side of the Island is drier, with totals in the region of 500 mm/year. The smallest totals in the South Island are found on the low plain in the vicinity of the Dunstan Range in central Otago. Here, at an elevation of 300 m, only 400 mm/year is recorded, the amount typical of a steppe zone. On average the island gets 1900 mm of precipitation annually, with the highest amount of 7180 mm/year recorded at Milford Sound.

Almost the entire island of New Guinea, with the exception of a narrow strip in the south, lies in the equatorial zone. This zone receives copious precipitation throughout the year, while the south is influenced by the monsoon and has a seasonal pattern of precipitation. A stable moist equatorial air mass alternates with a stable tropical air mass between one half of the year and the other but the mean precipitation on the island is 3080 mm/year. There are some areas where precipitation totals reach 5000 mm or even 5800 mm/year. Currents in the Pacific and Indian Oceans significantly affect Australia and Oceania making the climate milder. The Southern Tradewind Current brings warm waters (with temperatures higher than 28 °C) to the coasts of New Guinea. The cold West Australian Current carries water with temperatures of 10–12 °C in winter and 13–15 °C in summer.

Atmospheric circulation patterns play an important role in determining the climate of Australia. In the colder months (June to August) the continent is usually under the influence of high pressure, while in the warmer months (December to February) low pressure occurs. The climate of Australia, despite the relatively small size of the continent and its mid-oceanic location, is continental with a large number of sunny days. Total solar radiation in the south of Australia amounts to 130 kcal/year per cm^2, while in the northwest it does not exceed 180 kcal/year per cm^2. The intense heat of the central areas is induced by the continent's position near the equator and its east–west extent, while the fact that the coastline is not indented also contributes. The surface radiation balance is 70 kcal/year per cm^2. Evaporation accounts for 35% of this amount, the rest being utilized in the eddy exchange processes at the Earth's surface and in the atmosphere.

Australian vegetation is unique. More than 75% of the plant species are found only in Australia. They are well adapted to a dry climate having a strong and deep penetrating root system, narrow, dry and rigid leaves, covered in most cases by a dense wax layer. Eucalyptus trees are widespread. Eucalyptus wood is very hard and strong, while its bark and leaves are rich in ether oils used in medicine and in engineering. The forests, including shrubs, cover 26% of the country. Tropical evergreens grow in the areas bordering the northern and northeastern coasts (palms, rubber plants, eucalyptus trees). Large areas in the southeast of the country are covered with evergreen forests (mainly eucalyptus trees, tree ferns, horsetails (*Equisetum*), Dummar pines, and beeches). The vegetation in the southwest is typically represented by eucalyptus trees mixed with acacia. "Light" sparse eucalyptus and acacia forests are most widely distributed in Australia; these trees do not usually produce shade. The soil in these forests is covered by dense grass. "Light" forests are used as pastures for sheep and cattle. Dense, almost impassable prickly shrubs consisting of dwarf eucalyptus and acacia cover the inland deserts and semi-deserts. In places the scrub is replaced by thickets of high and rigid grass – spiniferous plants.

New Zealand is a country of forests and meadows. Forests cover 23.3% of the entire country. The primitive pristine forests are very peculiar with three-quarters of the species being endemic. These are huge kauri (*Agathis australis*) trees, tree ferns, palms, southern beeches and many other species.

9.2.2 Hydrology

Being a dry country, Australia has a rather sparse network of rivers and lakes. The distribution of runoff over the continent follows the precipitation, although it decreases more abruptly inland. Over most of Australia the runoff is small, i.e. less than 5–10 mm/year. This is typical of vast areas stretching from west to east (from the Indian Ocean to the Great Dividing Range), and from north to south (from the upper reaches of the Victoria River to the Great Australian Bight). Areas with a larger runoff (more than 100 mm/year) are limited, they are located in the north of the continent, in the east and southeast and in the extreme southwest. Table 9.1 presents the hydrological data for the major rivers of Australia and Oceania.

The smallest runoff (1–5 mm/year) is recorded in the Australian deserts and semi-deserts, while the largest (up to 1500 mm/year and higher) occur in basins along the Great Dividing Range, Liverpool Ridge, the Blue Mountains and the Snowy Mountains. The Australian rivers with the highest flows (the Burdekin, Fitzroy, Herbert, Clarence and Snowy) originate in the mountains of Victoria, New South Wales and Queensland and flow to the Pacific. These are short rapid rivers predominantly fed by rain. Most of the runoff from Australia reaches the Pacific such as in the Malgrave,

Table 9.1. *Principal rivers of Australia and Oceania*

River	Drainage area, km² × 10³	Length from source to mouth, km
Australia		
Murray	1072	3490
Fitzroy	143	960
Diamantina	115	896
Flinders	108	930
Gascoyne	79.0	770
Victoria	77.5	570
Burnett	33.4	
Hunter	22.0	465
Derwent (Tasmania Island)	9.55	
Warren	4.28	
New Guinea		
Sepik	81.0	1120
Mamberamo	77.6	900
Fly	64.4	1040
Purari	30.5	640
Degul	25.0	950
Ramu	15.5	650
New Zealand		
Clutha	22.0	338
Waikato	14.3	435
Waitaki	11.8	217
Wairau	8.13	215
Wanganui	7.20	290
Buller	6.50	177
Rangitaiki	3.19	153
Mohaka	2.35	150
New Caledonia		
Diao	0.30	65
Tiping	0.247	40
Cin	0.143	25
Hengen	0.114	30

Johnston and Tally rivers. Their mean annual runoff is between 1500 and 1700 mm/year. The Burdekin and Fitzroy Rivers are the largest of these rivers with a mean annual flow of 300 m³/s and 182 m³/s respectively. The runoff from areas along the northern coast such as the Kimberley Plateau and Arnhemland is between 100 and 300 mm/year while in the tip of the Cape York Peninsula it reaches 500 mm/year. This is the part of Australia with a monsoon climate where the rivers such as the Daly, Victoria and Mitchell contain large volumes of flow. The rivers of the west coast, such as the Murchison, Gascoyne and Ashburton flow to the Indian Ocean from the semi-desert plateaux and have smaller mean flows and a greater variability.

The Murray–Darling is the largest river system in Australia, including many rivers flowing from the inner slopes of the Great Dividing Range. In its upper reaches this is a typical mountain river, then it flows over a vast plain with an almost negligible slope where most of its water is either used for irrigation or evaporates. In the middle reaches between the Tokamole Peninsula and the Wakool River, the Murray often floods, forming a lake with a volume of 5 km^3. Much of this water infiltrates to recharge the ground water (Grigorkina, 1973). Because the Murray–Darling is intensively used, the discharge at its mouth is 10.0 km^3/year, which is less than half of the average flow (24 km^3/year) in the upper part of the basin. More than 80% of the irrigation in Australia is located in the Murray–Darling Basin (Crabb, 1988).

New Guinea, New Zealand, New Britain and other island in Oceania are relatively wet with annual runoffs almost everywhere exceeding 800 mm/year, particularly for rivers draining the highland areas. The southern slopes of New Guinea and the western coast of New Zealand's South Island have discharges close to 3000 mm/year. Lower areas have smaller discharges (Grigorkina, 1975). The largest river in New Guinea (by length and area of drainage basin) is the Sepik River which drains the northern slopes of the Central Mountains and flows to the Pacific Ocean. Its mean discharge is 3800 m^3/s. The Fly River is second in these terms, but it has the largest volume of flow, starting at the junction of the Hindenburg and Victor Emanuel Ridges and flowing to the Coral Sea. The Fly's mean discharge at its mouth is 4500 m^3/s (Grigorkina, 1980).

Almost all of New Zealand's rivers are short and steep. On the North Island they flow radially from the central volcanic area, while in the South Island they flow from the Southern Alps west to the Tasman Sea with courses 20 to 70 km in length. Some of the rivers that flow east connect a series of lakes and are heavily regulated. There are only 68 rivers longer than 50 km in New Zealand. The Waikato (435 km) is the longest, starting near Mount Ruapehu and flowing northward via Lake Taupo. Its mean annual discharge is 440 m^3/s. The Waitaki, Clutha and Wairau are important rivers in the South Island, each starting from a glacial lake. The Clutha is the largest in terms of flow (600 m^3/s) and the longest (338 km). Its lower reaches cross a dry valley and its waters are used as a source of drinking water and for irrigation.

Table 9.2 contains data on the large lakes of Australia and Oceania. Shallow salty lakes with no exits to the sea and which occasionally dry out predominate in Australia, Lakes Torrens and Eyre being the largest. These lakes are fed on average by flows of about 6.5 km^3/year, while evaporation from them is 13–15 km^3/year. Hence these lakes are ephemeral: they are filled with water in wet years and turn into salt marshes in dry years. New Zealand has the largest number of lakes. By origin they can be divided into three types: tectonic and volcanic in the North Island and glacial in the South Island. Lake Taupo in the centre of the

Table 9.2. *Principal lakes of Australia and Oceania*

Lake	Area, km^2	Max depth, m	Volume, km^3
Australia			
Eyre[a]	to 15 000	20	
Amadeus[a]	8 000		
Torrens[a]	5 800		
Gairdner[a]	4 780		
George	145	3	0.3
New Zealand			
Taupo	611	164	60
Te Anau	352	276	
Wakatipu	293	378	
Wanaka	194		
Manapouri	130		
Hawsa	119		

[a] Salt lake.

North Island is the largest lake in Oceania. It resulted from enormous volcanic explosions from about 50 000 to 22 000 years ago, and a further massive eruption only 2000 years ago (Soons and Selby, 1992). Lakes in the South Island are mostly in valleys such as Te Anau, Wakatipu and Manapouri. These lakes typically have depths of 200 to 380 m (Avakyan *et al.*, 1987). New Guinea has only a few lakes.

Coral atolls, unlike other islands, are devoid of rivers and lakes, and if there is no ground water it is difficult to find fresh water for drinking. Though there are copious rains in the tradewind zone of Oceania, rapid runoff and large rates of evaporation cause a deficit in the water balance.

The underlying geology often controls the water balance. For example in the Barkly Tableland and the Nullarbor Plain in Australia the fissured limestones allow no surface flow. Springs issue from the karst along the sea coast and probably on the ocean floor. In the deserts no runoff occurs even after showers because the rainfall that does not evaporate quickly seeps through the permeable soils to recharge the ground water. Ground water is abstracted in most of the arid areas of Australia from both shallow and deep aquifers. The volume of ground water in Australian rivers is estimated to be 10.6% of the total annual volume of runoff (Grigorkina, 1977).

Australian rivers are also fed by ground water from the artesian basins that cover about one-third of the area of the continent. Aquifers occur at various depths ranging from a few metres at the periphery of the basin to 2100 m in their centres. It is believed (Korzun, 1974) that the use of ground water for domestic purposes and for irrigation has a number of advantages over the construction of reservoirs to store the flow in ephemeral rivers and creeks. The major artesian basins of Australia are shown in Fig. 9.1 and

Table 9.3. *Principal artesian basins of Australia*

Basin	Area, thousand km^2	Depth of artesian water, m	Salt content, g/l
Great Artesian Basin	1751	0–2134	6.2
Desert	388	30–550	0.3
Murray	282	30–396	1.5–1.8
Eucla	191	90–610	6–37
Northwestern	77.5	60–1220	4–5
Coastal area (Perth plain)	54	60–760	
Ord–Victoria	31	60–300	

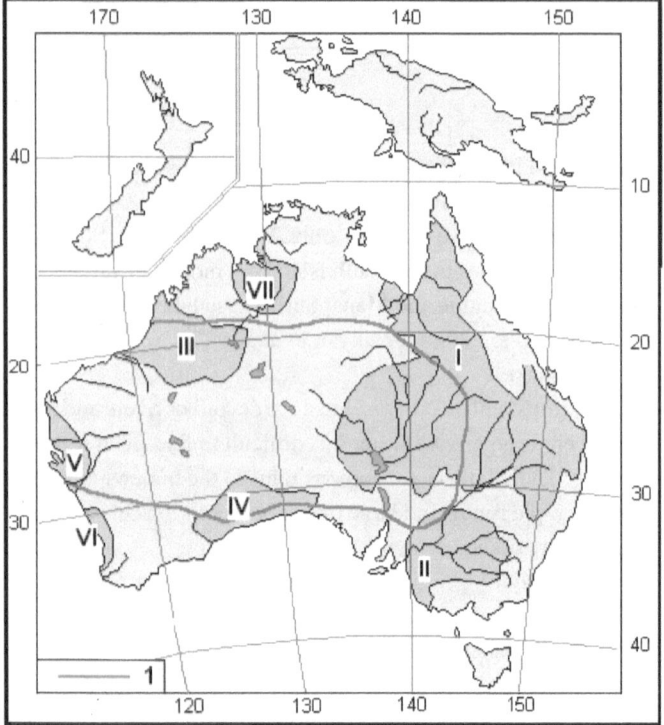

Fig. 9.1 Main artesian basins of Australia. I, Great Artesian Basin; II, Murray; III, Desert; IV, Eucla; V, North West; VI, Coastal Plain; VII, Ord–Victoria. 1, Boundary of the arid zone.

Table 9.3 presents their details (Australian Bureau of Statistics, 1971; Grigorkina, 1977). The Great Artesian Basin is the world's largest known source of ground water, extending south from the Gulf of Carpentaria across the continent. In general, the water in this basin is fresh or slightly mineralized. The Murray Artesian Basin mostly follows the topography of the Murray River Basin and is fed mainly by infiltration of river water and irrigation. It discharges at springs in the south where in the area of Hume, located south of the city of Gambier, there is a discharge of 270 000 m^3 per day. Altogether there are five major artesian basins in Western Australia, with high economic value. The survival of settlements in Nullarbor Plain depends on the availability of ground water, since there is no surface runoff.

9.3 SOCIO-ECONOMIC CONDITIONS AND THE USE OF WATER RESOURCES

The water problems and long-term trends in water use in Australia and Oceania have been conditioned by the history of colonization and development of this part of the world as well as by climate and distribution of water resources. Variations in climate are of considerable importance amongst these factors, particularly droughts, such as the severe drought of 1902 in Australia which put the problem of water in focus. In 1905 a special law was adopted and the Commission on Rivers and Water Supply was set up. It was the task of the Commission to deal with the nationwide water problem (Alekseyevsky, 1974).

Since then the problem of the optimum use of the limited water resources has become one of the priorities of national development and the subject of much research. In 1954 the Australian Academy of Sciences was set up and in 1956 the Academy established the National Committee for Hydrology. To ensure co-operation between scientists, engineers and economists the Australian Council on Water Resources was formed in 1962, with representation from all the states of the Commonwealth. The hydrological research carried out in laboratories and field stations of the Commonwealth Scientific and Industrial Research Organization contributed greatly to a better understanding of the overall water situation. Australia is an active participant in the programmes of the most important international organizations dealing with water resource and water use issues (Arthington *et al.*, 1998; Smith, 1998; Prime Minister's Science, Engineering and Innovation Council, 1999).

The prospects for the economic development of Australia and Oceania seem to be determined largely by the movement of capital, while the growth of population has been a result of immigration to Australia. However, there are areas in Australia and Oceania with a sparse population and a poorly developed economy and in these areas development is still going on (McDonald *et al.*, 1993) (Fig. 9.2). Australians mostly live in the cities and towns and not in the countryside. At the beginning of the twentieth century 50% of the population lived in urban areas and this number increased to 70% after the Second World War. Urbanization has continued steadily since then. In the mid-1960s the rural population was 16% of the total, while early in the 1980s it had dropped to 14%. It has been assumed that by 2000 this figure will be only 8% (Malakhovsky, 1988). Coastal areas will see most growth, compared to much slower development inland.

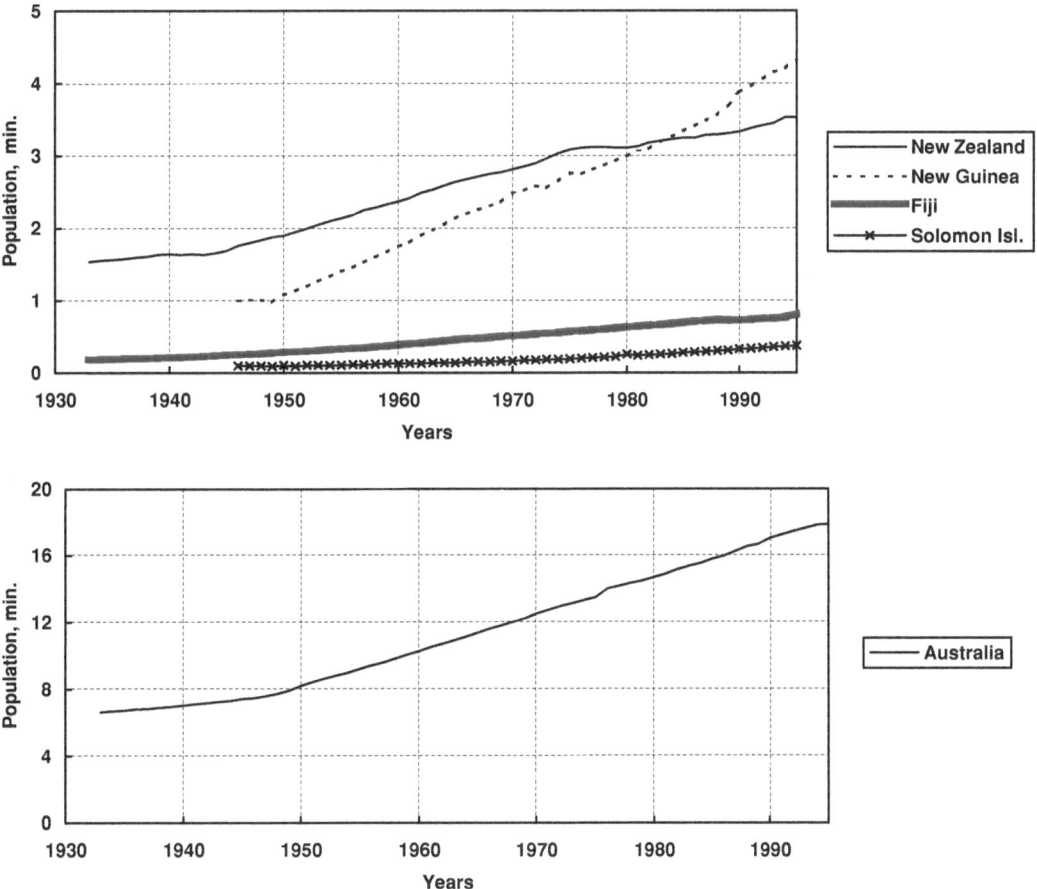

Fig. 9.2 Dynamics of population in Australia and selected countries of Oceania.

Australian Water Resources Council statistics from the mid-1980s indicate that the total urban and industrial water use from public sources was 2.58 km³/year and water use from local sources amounted to 0.484 km³/year. Use of water in Melbourne and Sydney, which accounts for 20% of Australia's total and specific domestic water use, is quite high, reaching more than 400 l/day per head in 1990. Government policy aims to decrease future water consumption as do the strategies for sustainable development and water resource management (Flemming and Daniel, 1994; Murray–Darling Basin Commission, 1999; Prime Minister's Science, Engineering and Innovation Council, 1999; Boer, 2000).

The semi-arid character of much of Australia and the specific features of its economy fostered the development of irrigation and this sector now consumes the most water. The dependence of the Australian economy on export-related income and the extremely rapid post-war development also contributed to this result. Australian water resources management for irrigation is facilitated by the promotion of a number of multi-purpose projects (Australian Academy of Science, 1964; Arthington *et al.*, 1998). Irrigation started to develop with the spread of European settlements, in the late eighteenth century (Bambrick, 1994). However, the rate of increase was highest after the Second World War, when the irrigated area increased from 0.5 to 2.1 million ha (Australian Bureau of Statistics, 1977; FAO, 1995). Today 70% of Australia's irrigation is in the Murray–Darling Basin, where water availability increased greatly after the completion of the Snowy Mountains project. In addition to its surface waters this basin is quite rich in ground water: the mean annual volume of available groundwater exceeds 488 million m³ (Prashchikin, 1986). The water resources of the Murray–Darling Basin are the most developed in Australia. Only 12% of the population of Australia lives in this area but the total value of the agricultural production is one-third of the Australian total (CSIRO, 1982; Crabb, 1988).

The domestic and industrial water use amounts to 4% of the total and several cities beyond the basin receive its water. The water resources of the Murray–Darling are highly regulated including the flows of the headwater streams: this regulation occurs along the 2200 km of the river from the Hume Dam to Lake Alexandrina near its mouth. Some 90% of the total volume of runoff is regulated; this is the largest percentage for any basin (Australian Water Resources Council, 1987), with 91% of the water being used for irrigation. There are other parts of Australia where irrigation occurs

Table 9.4. *Principal reservoirs of Australia and Oceania*

Reservoir	Location	Basin	Year of filling up	Dam back water, m	Reservoir capacity, km^3	Type of use[a]
Main Gordon	Australia	Gordon	1975	130	11.8	E
Ord River	Australia	Ord	1972	99	5.8	E
Pukaki	New Zealand	Waitaki	1992	–	5.8	E
Eucumbene	Australia	Eucumbene	1958	116	4.8	E I
Dartmouth	Australia	Mitta-Mitta	1979	180	4.0	P
Benmore	New Zealand	Waitaki	1965	118	2.2	E
Warragamba	Australia	Warragamba	1960	137	2.1	E
Blowering	Australia	Tumut	1968	112	1.6	I E
Copeton	Australia	Gwydir	1976	113	1.4	I
Wyangla	Australia	Lachlan	1971	85	1.2	I
Thomson	Australia	Thomson	1984	164	1.2	A I

[a] E, Hydroelectric power; P, public supply; I, irrigation; A, accumulation.

including areas along the coasts of Western Australia, in Tasmania and near the Gulf of Carpentaria (Prashchikin, 1986; Australian Water Resources Council, 1987). These account for about 25% of the total area under irrigation. The Fitzroy, Burdekin and Burnett Rivers are used for irrigation along the northeastern coast while the Pine, Brisbane and other rivers supply water to the large industrial area at Moreton. Ground water is abstracted for irrigation and other uses and amounts to 750 million m^3/year, including 320 million m^3/year in the Burdekin Basin (Prashchikin, 1986), used particularly for growing sugar-cane (Robinson, 1995).

Ground water is important in many parts of Australia because of the scarcity of the surface water. Ground water occurs in river alluvium, coastal dunes and deltaic deposits and in Queensland in the buried channels in the valley of the Fitzroy River. In New South Wales aquifers are recharged by spring floods, while water supplies in Newcastle and Botany Bay are based on ground water from sand dunes (Alekseyevsky, 1974; Shiklomanov and Markova, 1987). Water from artesian sources is often highly mineralized, and distribution losses were found to be high (up to 90% was lost from channels due to seepage). More artesian water has been used recently as surface water sources have become fully used for agricultural purposes even in the most humid areas. However abstraction of ground water has caused water tables to fall while there has been an increase in mineralization (Grigorkina, 1977).

By the mid-1980s there were about 400 reservoirs in Australia used for irrigation, the largest eighteen of them having a mean volume of between 1 and 10 km^3; there were 52 with capacities between 0.1 and 1 km^3, while the remainder were small and very small (Avakyan *et al.*, 1987) (Table 9.4). Twelve reservoirs with a surface area of 100 km^2 and with a total volume of 80 km^3,

hold 80% of the total volume of all the stored water in Australia (Australian Water Resources Council, 1987). Almost all the reservoirs are located on the periphery of the Continent, most being in the east, southeast and in Tasmania. Ord, the largest, was built on the Ord River in 1972; its total volume is 5.8 km^3 (useful volume 2.0 km^3), while its surface area is 100 km^2. This reservoir is used for flood control and it provides water for irrigation, some 180 000 ha of rich chernozem at the northwest of the country (Alekseyevsky, 1974). Some reservoirs have been constructed in Tasmania where the largest of them (Lake Gordon) was built in 1974 on the Gordon River. It has an area of 272 km^2 and a volume of 11.8 km^3 (Maclaine, 1981).

About 60 Australian reservoirs and lakes are used for hydropower generation, 14 solely for this purpose. Large hydropower plants are located in Tasmania and in the Snowy Mountains. Some 33 Tasmanian water bodies of the total of 43 are used for hydropower. Today half of the national hydropower potential has already been developed (Gleick, 1993, 1998), the total proven capacity of hydropower plants being 7.3 million kW. The largest hydropower system is in the Australian Alps, with a capacity of 3.7 million kW, and the second largest (more than 1.5 million kW) is in Tasmania, where more than 80% of the total useful volume is employed for power generation.

The increase of irrigation in Australia is primarily controlled by economics. Large reservoirs are needed to provide sufficient storage but there are large losses from these reservoirs because of evaporation. There is also a risk of mineralization and additional losses by infiltration (Frolova, 1974). Irrigation is, of course, a highly capital-intensive industry (Nimmo, 1949; Campbell, 1964). For many years alternatives to irrigation have been considered by Australians. The higher yields obtained from 1950 to 1962 were a result of the use of more intense fertilizers and better-quality seeds. Changes in the world market for raw materials and food required better breeding and higher wool exports, but the area under agriculture remained unchanged except for a small increase for some new commercial crops, such as cotton. Sheep rearing still remains economically successful over much of the country (Anonymous, 1992) and in the south it is being further developed (Campbell, 1964), although the structure of Australian exports has undergone large changes (Bambrick, 1994).

Today Australia relies on agriculture, especially in the southeast and southwest, and on mining where new areas are being developed in the west and northwest (Alayev and Kolosov, 1989). Since the 1960s Australia has become the world's largest exporter of mining products. The development of mining continued through the difficult 1970s when this energy-intensive mode of development seemed to slow down in many countries and when irrigation was affected. From 1970 to 1980 (FAO, 1995a) the irrigated area did not increase and water use decreased in much of the country. At the start of the 1980s there was a boom in the development of natural resources. The 1970s oil crisis put more emphasis on energy resources and attracted new investment. These factors stimulated irrigation and from 1980 to 1990 the irrigation growth rate exceeded the growth of population.

Most of the countries of Oceania, with their small populations and copious water resources, do not experience any problems with water availability as compared to Australia. New Zealand is the largest and most developed country in Oceania. Due to its favourable environment and climate as well as its social and economic situation, the country experienced continuous development in the first half of the twentieth century. In terms of per capita GNP New Zealand has held a record place in world ranking (between 1 and 3). The country's success was based on its agriculture, which produced 85–90% of its exports in the 1950s and 1960s. After the Second World War New Zealand started to develop its industry and from the mid-1950s to the mid-1980s agriculture, forestry, hunting and fishing contributed a decreased share of its GNP, falling from 22% to 9.1%. Direct foreign investments attracted to the country after the war had a significant impact on the economy (Rubtsov, 1987).

A number of economic crises in the 1970s and 1980s caused a relative decline in the quality of life, with large changes in demography and a serious effect on the development of various industries, particularly those that were related to water use. In general, the population of New Zealand grew at a higher rate than other countries, the highest growth occurring in the 1950s due to immigration (Government of New Zealand, 1988) (Fig. 9.2). In the nineteenth century immigrants made up 40% of the population. Changes in economic conditions are apparent in the growth of irrigation and the reduced income in the late 1970s and early 1980s, showing slow growth and no growth at all. At present the irrigated

area is 280 000 ha, no part of the country experiencing any serious water shortages. In some areas like the Canterbury and Heretunga Plains, Hawke Bay and in the vicinity of Nelson, ground water is the only source of water resources. Christchurch, Lower Hutt and Hastings use ground water (Government of New Zealand, 1988). About 2.3 million ha have copious water resources (Alekseyevsky, 1974), but some areas face the problems of land drainage, particularly the valleys with small gradients and coastal lands. For instance, development of the Hauraki Gulf and Bay of Plenty areas depends on effective land drainage.

Irrigation is necessary in the areas of the South Island which are in the rain shadow, e.g. the central Canterbury Plains, where irrigation started in 1878, and developed rapidly by the 1930s and 1940s (Alekseyevsky, 1974). By the mid-1980s about 60% of New Zealand's irrigation was located on the Canterbury Plains, 75% for lucerne, grass occupying the rest. By comparison the North Island had only 9% of the irrigation. Recently the rate of construction of new irrigation systems in New Zealand has fallen (*New Zealand Journal of Agriculture*, 1980; Taylor and Smith, 1997; Ministry for the Environment, 1999). New Zealand has a large number of sites suitable for the construction of reservoirs and considerable potential for generating power (74 000 GWh/year) (Gleick, 1993, 1998). More than 80% of the power is presently generated by water, the 15 largest reservoirs storing a total of 15.36 km^3. The largest reservoir is Lake Pukaki on the Waitaki River (volume 5.80 km^3 and surface area 150 km^2).

Most of the island states have large rural populations: 88% in Papua New Guinea, 60% in Fiji and Western Samoa, and 75% in Vanuatu. Agriculture produces most of the GNP and most of the export revenue, the only exception being in Nauru, where phosphate mining is important (Rubtsov, 1991). However Papua–New Guinea and New Caledonia also have significant mining (Ryan, 1972), while fishing and fish processing are important in the Solomon Islands, and Vanuatu has a growing service sector (Connell and Pritchard, 1990). Papua–New Guinea has 84% of the land area of Oceania (excluding New Zealand and the Hawaiian Islands) and has a large hydropower potential (98 000 GWh/year) (Gleick, 1993, 1998); however it has few consumers for this power. Fiji, after it became independent, successfully developed its power potential: in 1983 the Monasavu hydropower plant was built which today produces 90% of all power generated in the country.

Rapid population growth is characteristic of many of the states of Oceania (Fig. 9.2). In Papua–New Guinea the population has increased by 5 since 1950 and in Fiji by 4 since 1930. Estimates suggest that by 2025 the populations of Papua–New Guinea and the Solomon Islands will double, while Fiji's will grow by 15–20%. Statistics on the development of irrigation in Oceania are not easily available. It seems that in Fiji the area irrigated, about 1000 ha, has not changed since the late 1960s (FAO, 1995a). There are few reliable sources of data on water use in the countries of Oceania and virtually no data that can be used to assess changes. Estimates from the World Resources Institute (WRI, 1992, 1996, 2000) and in the *UN Statistical Yearbook 1992* (UN, 1993a) show the highest water use as 191 m^3/year per head for Nauru, while the Fiji figure was 41 m^3/year per head and that for Papua–New Guinea was 26 m^3/year per head. The total volume used was highest in Papua–New Guinea (0.09 km^3/year) and in Fiji (0.03 km^3/year); elsewhere it was very small (UN, 1993a).

9.4 HYDROLOGICAL DATA

The first systematic measurements of flow started in Victoria in 1865 on the initiative of the Department of Water Supply. Early measurements were carried out on the Murray River and the results have been published since 1887. Observations in other states started later: in New South Wales and South Australia (the Murray River Basin) in 1885, in Queensland in 1910, in Tasmania in 1925, in Western Australia in 1939 and in the Northern Territories between 1960 and 1965, although some rivers have been gauged since 1952. During the International Hydrological Decade (IHD) the Australian hydrological network was expanded considerably. Flow data for some states are published as monthly, annual and long-term average discharges or for the entire observation period (Victoria, New South Wales and South Australia), while for other states these data are published as annual reports for certain years (Queensland, Tasmania, Western Australia, Northern Territory). Daily flow data are published by the State Energy Commission for the Snowy Mountains (Snowy Mountains Hydroelectric Authority, 1971). In 1962 the Australian Council of Water Resources was established and since 1964 the Council has published flow records for each 5-year period, including the results of hydrological studies and long-term mean discharges (Australian Water Resources Council, 1965, 1971).

Hydrological and meteorological stations are unevenly located over the country (Table 9.5). The southeastern coast is best instrumented, particularly Victoria and New South Wales, where there is one gauge post per 353 km^2. Network densities are much less for the coastal areas of Western Australia (one gauge per 12 600 km^2), with the sparsest network of stations in inland areas of Western Australia (one gauge per 189 000 km^2). There are very few stations in the central and northern parts of the country. Over Australia on average there is one station per 2710 km^2.

Hydrological and meteorological observations in New Zealand started on the Waikato River in 1905 and by 1966 there were more than 500 stations across the country. The duration of observations usually ranges from 5 to10 years but there are some records of 50 to 60 years length. The stations are unevenly distributed over the country with an average of one gauge per 530 km^2. Stream-

Table 9.5. *Hydrological stations in Australia*

Hydrological region	Area, $km^2 \times 10^3$	Number of gauge stations in 1985	Mean area per gauge station, km^2
Northeast Coast	451	372	1 210
Southeast Coast	274	776	353
Tasmania	68	178	382
Murray–Darling Basin	1063	767	1 390
Great Australia Bight	82	55	1 490
Southwest Coast	314	244	1 290
Indian Ocean	519	41	12 660
Timor Sea	547	156	3 510
Gulf of Carpentaria	638	149	4 280
Lake Eyre	1170	59	19 830
Bulloo–Bankannia	101	21	4 810
Western Plateau	2455	13	188 850
Island Territories	–	no measurements	–
Australia	7682	2831	2 710

Source: Brown and Bergman (1987).

flow statistics are published in hydrological yearbooks and generalized statistics are published in the Hydrological Statistics series (Ministry of Works, 1970). Streamflow statistics for the small islands of Oceania are rarely published; however some data are to be found in French Yearbooks for Overseas Territories and in relevant US publications.

There have been numerous attempts by Australian scientists and others to estimate the country's water resources. The first attempt was made by Nimmo in 1949 (Nimmo, 1949), using all available observational data, and estimates for areas with no measurements based on precipitation and air temperature data. In 1963 an assessment of Australian surface water resources was made by Sergeant (1964), in which the discharge to the oceans was estimated as 250 km^3/year, 42% of this amount being measured and 58% estimated. It was assumed that flow occurred only from two-thirds of Australia, the rest having sporadic flow. The total surface flow from the continent was estimated as not likely to exceed 333 km^3/year.

The first official estimate of the surface water resources was presented in the *Review of Australia's Water Resources* in 1963 and published in 1965, by the Department of National Development and the Australian Water Research Council. The *Review* gave data from 197 gauging stations representative of each area and having long-term reliable data series. In 1963 there were 1439 gauging stations, 841 of them equipped with automatic water-level recorders. The runoff was calculated to be 299 km^3/year but only two-fifths of this amount was actually measured, the rest being estimated with various degrees of accuracy.

Table 9.6. *Renewable water resources of Australia and Oceania (km³/year) by data of different authors*

Territory	Korzun, 1974b	L'vovich, 1974	Department of National Development, 1968	WRI, 1990	Australian Water Resources Council, 1976	Baumgartner and Reichel, 1975	Australian Water Resources Council, 1987	Mosley, 1996
Australia (continent)	301	334	299				269	
Tasmania	47.2	48						
Australia and Tasmania	348	382		343	345		390	
Australia and Oceania	2390			2011		2394		2500

In 1968 a map of the surface water resources of Australia was published, the isopleths of mean annual runoff being shown as well as the annual hydrographs for selected gauging stations and data on their variations (Department of National Development, 1968). It demonstrates how Australian water resources (346 km³/year) are very unevenly distributed. Brown and Bergman (1987) summarized Australian studies and compared estimates of mean annual flow for the period 1949 to 1985. The Australian water resource estimates vary from 346 to 390 km³/year with 1982 standing out as a wet year when the flow to the Gulf of Carpentaria doubled. The Russian State Hydrological Institute (SHI) assessed the water resources of Australia and Oceania for this Monograph for the period of 1971 to 1973 using the latest data (Grigorkina, 1974). Table 9.6 presents estimates from different sources (Rozanov, 1967; L'vovich, 1974; Baumgartner and Reichel, 1975; Mosley, 1996).

9.5 DISTRIBUTION OF WATER RESOURCES IN TIME AND SPACE

9.5.1 Data and methods

A limited number of stations were selected for this study despite the fact that today in Australia there are 2800 (Brown and Bergman, 1987), and more than 500 in New Zealand. Observations were obtained from the Global Runoff Data Centre (Koblenz, Germany) and from the New Zealand National Institute of Water and Atmospheric Research Ltd. The database set up for this study included measurements from 186 hydrological stations from Australia and Oceania with periods of observation from 3 to 86 years (Fig. 9.3, Table 9.7). The stations that were selected have the most reliable data for the larger rivers of each region: for each of these stations monthly and annual data exist for the entire observational period. To estimate the total flow to the Pacific Ocean over the 65 years, discharge data were used for 14 rivers, their catchments covering 60% of the drainage area, while their discharges make up about 25% of the total flow.

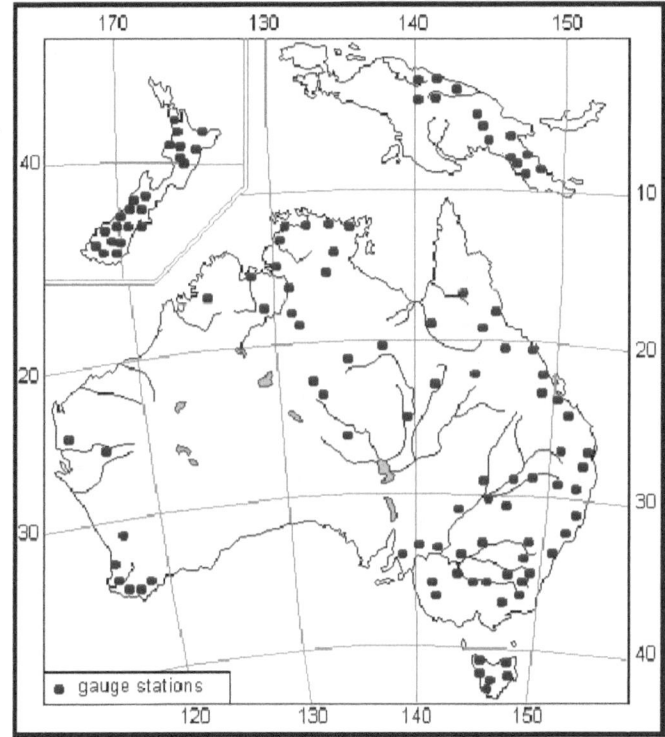

Fig. 9.3 The scheme of location of main hydrometric runoff gauge stations in Australia and Oceania.

For drainage to the Indian Ocean a 65-year series was devised by adding the runoff for the four regions, namely 1 – southeastern, 2 – southwestern, 3 – western, and 4 – northern (Fig. 9.4). For regions 1 and 2 actual runoff data were used, while for regions 3 and 4 due to the small number of observations the 65-year series was obtained from precipitation. The long term monthly discharge data for 72 stations which had observations for longer than 10 years were used to determine the seasonal distribution of runoff. If the observational period was short, then the hydrograph was selected which was the most similar to the average runoff from the area.

Table 9.7. *Number of river runoff observation stations used to estimate water resources of Australia and Oceania*

Country	Number of stations	Number of years of observations
Australia	94	5–80
New Zealand	42	10–85
Papua–New Guinea	26	5–17
New Caledonia	7	3–7
Samoa	3	7–8
Taiti	1	3
Marian Islands	12	12–15
Total	188	3–85

Table 9.8. *Statistical estimation of water resources by regions and selected countries of Australia and Oceania, 1921–1985*

Territory	Water resources, km³/year			Coefficient variation, C_v
	Average	Maximum (year)	Minimum (year)	
Australia	304	631 (1956)	164 (1966)	0.27
Australia and Tasmania	352	701 (1956)	228 (1961)	0.24
Oceania	2050	2570 (1962)	1515 (1928)	0.1
Australia and Oceania	2400	2880 (1950)	1890 (1937)	0.1
New Zealand	313	405 (1958)	246 (1950)	0.11

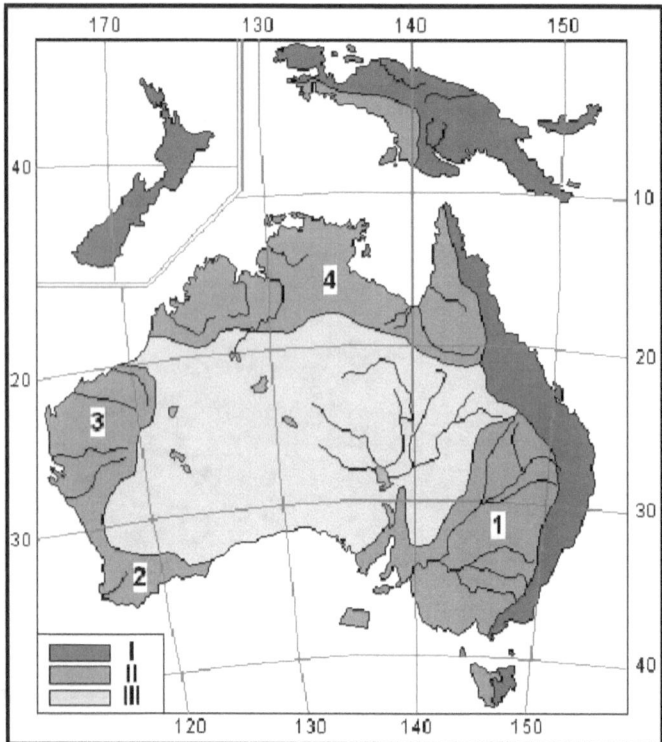

Fig. 9.4 Distribution of the territory of Australia and Oceania according to ocean basins. I, Pacific Ocean; II, Indian Ocean (1, southeastern area; 2, southwestern area; 3, western area; 4, northern area); III, endorheic basins.

The assessment of water resources and their variations over the long term was made for Australia and Oceania for the period of 65 years from 1921 to 1985. The runoff for each year for this period was estimated using discharges measured at stations located close to the mouths of rivers. The total surface water discharge to the different oceans was determined by introducing a coefficient for each ocean. Where there were no data for some rivers during the chosen period, monthly and annual runoff data were obtained

by the regression and analogy methods described in Chapter 2 (Shiklomanov, 1997, 1998b, 2000a, b; Babkin, 1998; Shiklomanov *et al.*, 2000). In some cases 65-year time series were constructed, while for less well-gauged areas discharge time series were developed from precipitation coefficients. Data from about 50 meteorological stations with long-term observations were used for this purpose using correlations between precipitation and runoff data.

9.5.2 Water resources for regions and countries

Australia is the world's driest continent. Its total annual runoff is less than 304 km³ (40 mm), which in turn is less than 1% of the global runoff. This is the only continent where regions of inland drainage cover 50% of the area. The water resources of Oceania and Tasmania far exceed those of Australia (Table 9.8). For example, the runoff from New Guinea is 1370 km³/year (1730 mm), from New Zealand some 313 km³/year (1180 mm), and from Tasmania about 48 km³/year (1610 mm), as presented in Table 9.9. Of course there is also infiltration to aquifers and evaporation, factors in the water balance of every basin. It is estimated (Grigorkina, 1976b) that 22 km³/year or 5.5% of the Australia's water resources does not drain to the oceans, but is lost in these natural processes and then in part pumped from aquifers for consumption by industry. Table 9.10 presents water resource of Australia.

Spatial and temporal variations are considerable in Australian water resources and generally increase from the periphery towards the centre of the continent. The coefficient of variation of annual runoff ranges from 0.30 to 1.70, the lowest being typical of rivers in the Australian Alps, while higher values (0.50–0.90) are characteristic of small rivers flowing along the slopes of the Great Dividing Range, the tributaries of the Murray, and northern and southwestern rivers (Grigorkina, 1976b; Australian Government Publishing Service, 1978). In arid areas the coefficients of variation increase

Table 9.9. *Water resources of individual large islands of Australia and Oceania, 1921–1985*

| Island | Area, km$^2 \times 10^3$ | Water resources | | | | Coef. of variation, C_v |
| | | Average | | Maximum, km^3/year | Minimum, km^3/year | |
		km^3/year	mm			
New Guinea	785	1370	1740	1770	912	0.12
Year				(1982)	(1928)	
New Zealand	265	313	1180	405	246	0.11
Year				(1958)	(1950)	
Other islands (New Caledonia, New Britain, etc.)	217	372	1710	373	164	0.13
Year				(1982)	(1923)	
Oceania	1267	2050	1620	2570	1515	0.1
Year				(1962)	(1928)	
Tasmania	68.4	48	700	76.5	29.3	0.24
Year				(1974)	(1979)	
All islands, total	1335	2100	1570	2590	1570	0.1
Year				(1982)	(1928)	

Table 9.10. *Water resources of Australia by ocean basins*

| Sea | Area, km$^2 \times 10^3$ | Water resources | | | Coefficient variation, C_v |
		Average, km^3/year	Maximum, km^3/year	Minimum, km^3/year	
Corall/Sea	420	74.4	179	19.5	0.41
Year			(1950)	(1967)	
Tasman Sea	262	48.3	159	14.8	0.54
Year			(1950)	(1940)	
Australia and Tasmania					
Indian Ocean slope	2083	100	198	63.1	0.23
Year			(1956)	(1972)	
Timor Sea	354	48.0	77.0	38.0	0.17
Year			(1974)	(1958)	
Arafura Sea	685	71.0	167	32.0	0.35
Year			(1974)	(1961)	

to 1.00 and to 1.30, and for rivers flowing from the Blue Mountains and those draining the Liverpool Range they are 1.6–1.7. Where runoff is sporadic and no hydrological observations are available, discharge variations may be estimated from precipitation, whose variations indicate the variations of flows. Records for several years at Birdsville (Diamantina River) and at Windore (Cooper's Creek) give $C_v = 1.00$ and 1.2.

The ephemeral rivers are also indicative of the variability of runoff, as also is Lake Eyre, the largest Australian lake, which is seldom filled to capacity. For instance during the period from 1880 to 1968 Cooper's Creek ran to Lake Eyre on only three occasions (Warren, 1963; Grigorkina, 1976b). The mean annual discharge

of Cooper's Creek at Windore in 1950 was 486 m^3/s, while in 1953 there was no discharge for the whole year. In arid areas the runoff decreases downstream, while its variability increases. Over the long term, dry periods are characteristic of Australia, but one or two years occur with large flows. Figure 9.5 shows the variation in renewable water resources for natural–economic regions in Australia, New Zealand and other parts of Oceania. The SHI estimates (Korzun, 1974b) are also shown for comparison and a fairly good agreement is apparent. Figure 9.6 demonstrates variations in runoff over the period from 1921 to 1985. These curves show patterns of dry and wet phases over different periods of time.

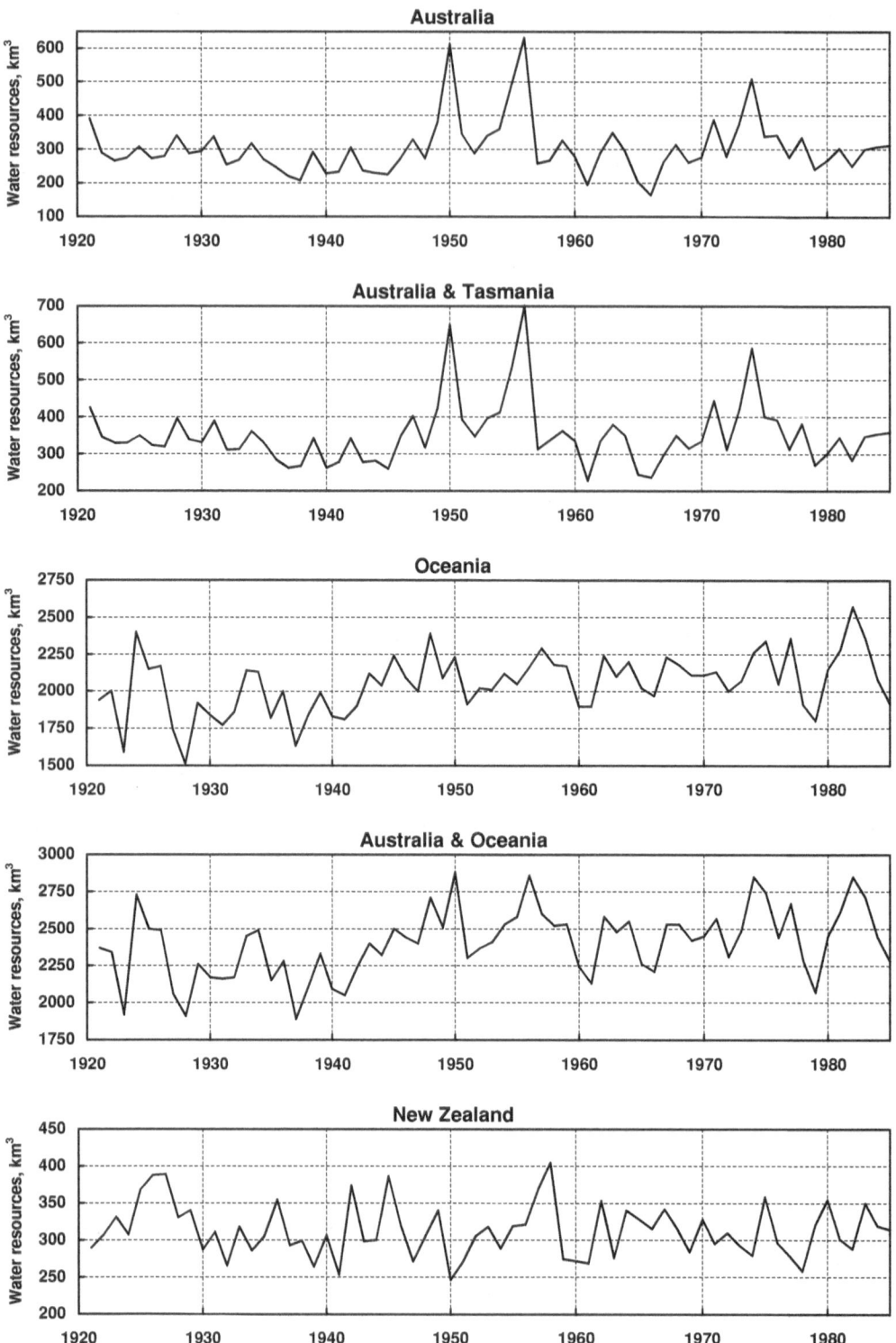

Fig. 9.5 Variations in renewable water resources (km³/year) by
natural–economic regions of Australia and Oceania.

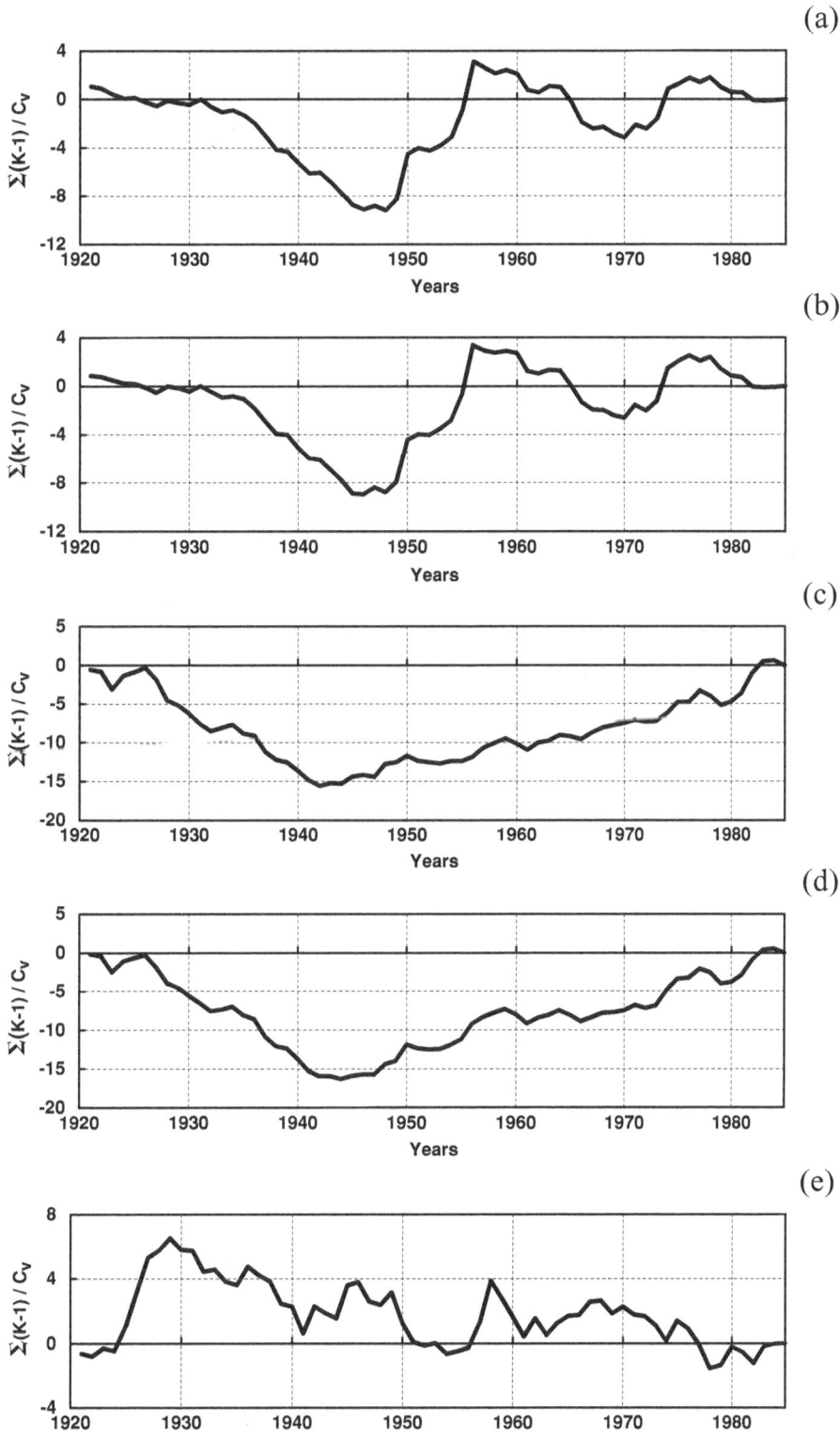

Fig. 9.6 Normalized difference integral curves of water resources by natural-economic regions of Australia and Oceania. (a) Australia; (b) Australia and Tasmania; (c) Oceania; (d) Australia and Oceania; (e) New Zealand.

Table 9.11. *Periods of different water content by natural–economic regions and selected countries of Australia and Oceania, 1921–1985*

Territory	Wet periods			Dry periods		
	Years	Ratio of mean of period to long-term mean, K_{av}	Number of years	Years	Ratio of mean of period to long-term mean, K_{av}	Number of years
Australia	1946–56	1.29	11	1921–45	0.88	25
	1971–76	1.22	6	1957–70	0.88	14
				1977–85	0.95	9
Australia and Tasmania	1946–56	1.27	11	1921–45	0.91	25
	1971–76	1.21	6	1957–70	0.90	14
				1977–85	0.93	9
Oceania	1943–85	1.04	43	1921–42	0.93	22
Australia and Oceania	1945–85	1.04	41	1921–42	0.94	24
New Zealand	1921–29	1.08	9	1930–54	0.96	25
	1955–58	1.13	4	1959–78	0.97	20
	1979–85	1.03	7			

At the beginning of the period the entire continent was dry. During the 1940s conditions were wetter and these continued until 1957. However, after 1957 a drier phase commenced (Table 9.11). During wet years the runoff in some areas increased from 80% to 200% by comparison with the average value, while in dry years it reduced by 50% to 75%. The wettest year was 1956 (210% increase) and the driest 1966 (63% lower than the mean value); similar variations were observed for the islands in 1982 (24%) and in 1928 (26%), respectively. The coefficient of variation for the total discharge from the entire continent is 0.27; this figure is higher than those of other continents. The flow variations for the Pacific Islands are much less pronounced, the coefficient of variation for the total discharge of the islands is 0.10, while for New Zealand it is 0.11, for New Guinea 0.12 and 0.24 for Tasmania.

Table 9.12 presents the probability of occurrence of groups of years of different types of climate for the continent and other areas. From this table it can be seen that following dry years, there is a 48% probability that the flow for the year will be either lower than average or average, and only a 4% probability that the flow will be even lower than the previous year. If the previous year had an average flow, there is a 45% probability that a year with low flow will follow and that average flows and high flows are less probable, their probabilities being 23% and 32%, respectively. There is a 47% probability that a wet year will follow a wet year, a 35% probability that an average year will occur and an 18% probability of a dry year.

The long-term flow variations are different in Oceania. Independent of the flow in the previous year there is a high probability

Table 9.12. *Probability of water content transition from one grouping to another by natural–economic regions and selected countries of Australia and Oceania*

Territory	Water content characteristic[a]	Probability of characteristics transition from one grouping to another, %			Average duration of water grouping, years
		$\overrightarrow{1}$	$\overrightarrow{2}$	$\overrightarrow{3}$	
Australia	1	48	48	4	1.90
	2	45	23	32	1.30
	3	18	35	47	1.90
Australia and Tasmania	1	43	52	5	1.75
	2	30	48	22	1.93
	3	25	25	50	2.00
Oceania	1	36	54	10	1.57
	2	16	73	11	3.67
	3	0	67	33	1.50
Australia and Oceania	1	27	64	9	1.38
	2	18	68	14	3.14
	3	0	78	22	1.28
New Zealand	1	31	46	23	1.44
	2	18	67	15	3.00
	3	17	58	25	1.33

[a] 1, Low; 2, average; 3, high.

Table 9.13. *Renewable water resources and potential water availability of Australia and Oceania*

Territory	Area, km² × 10⁶	Population (1994), ×10⁶	Water resources, km³/year			Coefficient of variation, C_v	Potential water availability, m³ × 10³/year	
			Average	Maximum (year)	Minimum (year)		Per km²	Per head
Regions								
Australia	7.68	17.9	352	701 (1956)	228 (1961)	0.24	45.8	19.7
Oceania	1.27	10.8	2050	2570 (1962)	1515 (1928)	0.10	1614	190
Australia and Oceania	8.95	28.7	2400	2880 (1950)	1890 (1937)	0.10	268	83.6
Countries								
Australia	7.68	17.9	352	701 (1956)	228 (1961)	0.24	45.8	19.7
New Zealand	0.27	3.50	313	405 (1958)	246 (1950)	0.11	1159	89.4
Principal river basin								
Murray	1.07	2.14	24	129 (1956)	1.16 (1972)	0.75	22.4	11.4

Table 9.14. *Renewable water resources by ocean slopes and principal river basins of Australia and Oceania*

Basin	Area, km² × 10³	Period of analysis, years	Water resources, km³/year			Coefficient of variation, C_v
			Average	Maximum	Minimum	
Murray River Basin	1072	1877–1988	24	129 (1956)	1.16 (1972)	0.75
Indian Ocean slope	3078	1921–1985	180	340 (1974)	118 (1961)	0.21
Pacific Ocean slope	612.3	1921–1985	144	294 (1950)	36.1 (1966)	0.42
Endorheic basins	3924	1921–1985	9.43	58.1 (1950)	1.18 (1965)	1.05

that the next year will have high flows. There is a probability of 54% that a dry year will be followed by an average year, and a probability of 73% that an average year will follow an average year and a probability of 67% that a wet year will be followed by an average year. The variations in the total discharge from Australia with Oceania are close to those of Oceania alone.

Statistical analyses of the flow data were undertaken using the methods described earlier such as those of Student, Kolmogorov–Smirnov (mean values), Fisher and Kolmogorov–Smirnov (variance), particularly for the presence of a trend. These analyses showed that from the tests of the two halves of the 1924 to 1985 record, the flow to the Indian Ocean and the total runoff from Australia and Oceania exhibited trends probably caused by climatic variations in the basins in question. The discharge from Australia is highly correlated with that to the Pacific Ocean ($r = 0.92$) and there is a slightly higher correlation ($r = 0.93$) between the sum of the discharges from Oceania and Australia and those from Oceania.

Table 9.13 presents data for the Murray Basin. Particularly high specific water availability exists for Oceania (1 614 000 m³/year) and New Zealand (1 159 000 m³/year), while the lowest amount is for Australia (45 800 m³/year). The water availability per head

is greatest in Oceania (190 000 m³/year) and in New Zealand (89 400 m³/year), the specific water availability in New Zealand being the world's highest.

The water resource estimates made for this Monograph for Australia and Oceania compare well with those obtained previously, such as by Baumgartner (269 km³/year) and L'vovich (334 km³/year) (L'vovich, 1974; Baumgartner and Reichel, 1975) as illustrated in Table 9.6. Estimates of Australian water resources have been made for 245 rivers and the mean annual runoff is estimated to range from 341 to 390 km³ (Australian Water Resources Council 1976, 1987).

9.5.3 Drainage to the oceans

Some 612 300 km² or 8% of Australia drains to the Pacific Ocean, about 3 078 000 km² or 40.5% to the Indian Ocean, while flow from the remaining 3 924 000 km² or 51.5% does not reach the oceans (Fig. 9.4). About 180 km³ (60%) of this discharge goes to the Indian Ocean with 144 km³ (37%) flowing to the Pacific Ocean and 9.4 km³ (3%) draining inland (Table 9.14). The centre of the continent is quite arid and there is a very poorly developed network of channels with a few mineralized (salty) lakes and short river courses and creeks that are usually dry. The largest rivers like Cooper Creek and the Georgina and Diamantina Rivers (Lake Eyre Basin) contain water after the short, though heavy, infrequent rains. In the upper reaches of these rivers the runoff is about 10 to 15 mm reducing to 1 mm in the lower parts of their basins. There is virtually no flow over much of Western Australia, such as the Victoria Desert, Gibson Desert and Nullarbor Plain. There are remnants of river channels and the beds of lakes and flows increase to 10 mm only along the northern and western fringes. These arid areas with ephemeral flows provide constitute only 3% of the total surface water resources of Australia, but they cover more than 50% of area of the continent. Lake Eyre and Bulloo-Bankannia represent about 2% of the total resource, while the arid areas contribute less than 1%. Major areas of inland drainage cover 16% of the continental area (1 244 000 km^2), and there are similar but smaller areas in the Murray–Darling Basin such as the Wimmera–Avoka Basin and some other northwestern areas. Lake Eyre is the largest lake in Australia. It turns into a chain of separate water bodies surrounded by areas with a layer of salt almost a metre thick, while in the wet season the lake covers an area of about 15 000 km^2. Rivers like Cooper Creek, and the river Diamantina and Fink, which contain water after the rains and sometimes reach the lake, usually disappear however into the sands. Lake Eyre is filled several times over a century either partially or to the brim.

Table 9.15 presents data for the major rivers using the stations with the longest period of observations, ranging from 46 to 100 years. These are the data which are used for the analysis of the variations of runoff. Figure 9.7 shows the variations in the annual runoff of the Murray–Darling River, while Fig. 9.8 presents the difference between the long-term average and the cumulative flows. Analysis of these curves shows that similar phases of water availability are to some extent shifted by duration and degree of deviation from the mean. In the basins draining to the different oceans, dry years prevail when the runoff values were below their means. The number of dry years for Pacific drainage is 43, with 22 wet years. There were 44 dry years for the Indian Ocean drainage and 21 wet years. The duration of dry phases ranges from 14 to 30 years, and from 8 to 13 years for wet phases.

Since the time series for Australian regions were not the same length, the variability of the annual runoff could not be determined accurately. The coefficient of variation of the total flow to the Pacific Ocean was 0.42, and 0.21 for flow to the Indian Ocean. The coefficient of variation for flow in the areas of internal drainage was much larger (1.05). Tables 9.16 and 9.17 show the probabilities of dry and wet periods for different areas. The Murray–Darling Basin has a high probability that a dry year with follow an average or wet year. For instance, a dry year with a 61% probability is likely to be followed by another dry year, while after an average year there is 67% probability that another dry year will

follow, and there is only a 35% probability that a wet year will follow a wet year.

Table 9.18 shows the flow from Australia and Oceania by latitude. The total flow to the Pacific Ocean is estimated to be 1800 km^3/year, with 574 km^3/year to the Indian Ocean. Taking account of the estimated losses due to evaporation and the industrial consumption (some 22 km^3/year) and the flow to areas of inland drainage (9.4 km^3/year) the total flow is 2400 km^3/year, i.e. this amounts to the total water resources of Australia and Oceania.

9.5.4 Distribution of runoff during the year

Discharges display not only year-to-year variations but also those from month to month. The seasonal distribution of flow is primarily controlled by the climate (precipitation, temperature, evaporation), the topography and geology of the basin. According to the pattern of runoff Australia can be divided into three zones (Fig 9.9): I, the arid zone covering the central areas; II, the semi-arid zone; III, the humid zone, including the eastern and southwestern coasts, and also Tasmania and Oceania. The typical monthly distribution of runoff over Australia is shown in Table 9.19. In the arid zone runoff is negligible. Very brief floods occur after periods of heavy rain. Precipitation amounts are small and the evaporation rates high. The network of stream channels is poor and this zone has no permanent flow, the exception being the Murray River and the short coastal rivers on the western periphery of the zone. Large floods occur in the Lake Eyre Basin with the heavy monsoonal rains, i.e. in March to April.

The area north of the arid zone extending to the dry tropics receives 500 to 600 mm of precipitation per year, but because the rates of evaporation are so high during the summer, there is no runoff. The only runoff occurs from December to March. In the central arid areas the very small volume of runoff seems to be evenly distributed between the summer and winter months. The annual precipitation here is between 150 and 300 mm and irregular. The desert areas in the south receive very little precipitation and the predominantly karstic soils preclude even brief periods of flow. In the Ashburton and Fortescue Basins with the prevailing summer precipitation, the runoff only occurs during that season, December to January. In the Gascoyne Basin it occurs both in summer and autumn, while in the south of Western Australia it takes place in winter.

In the coastal zone where precipitation is greater than evaporation all the year, rivers flows are large, for example in the Barron, Herbert, Clarence, Hunter and Snowy Rivers. In northeastern Queensland the major part of the flow (75%) occurs from January to April, with a maximum in March. Along the east coast the flow maximum shifts from north to south, occurring in early autumn in the Clarence River; in late autumn and winter in the

Table 9.15. *Runoff of individual rivers of Australia and Oceania*

River	Station	Area, km^2	Period of analysis, years	Average runoff, m^3/s	Coefficient of variation, C_v	Coefficient of assimetry, C_s
Australia						
Murray	Renmark	1 072 000	1877–84	320	0.75	2.60
Darling	Pooncarie	647 000	1886–85	54.2	0.63	1.94
Darling	Wilcannia	570 000	1886–85	4.79	0.98	2.29
Murray	Mildura Weir	238 200	1886–88	304	0.57	1.76
Murrumbidgee	Balranald	165 500	1886–88	77.2	0.67	1.62
Fitzroy	The Gap	135 900	1900–85	191	0.46	0.67
Barron	Walgett	132 000	1886–85	66.4	1.05	2.41
Diamantina	Birdsville	115 000	1898–83	18.6	1.00	3.05
Lachlan	Enabalong	44 000	1895–47	30.5	0.77	0.80
Burnett	Walla	32 220	1900–85	45.9	1.12	2.82
Namoi	Narrabri	31 100	1886–85	16.1	0.98	1.97
Murrumbidgee	Wagga Wagga	27 700	1877–88	112	0.74	2.66
Macintyre	Boggabilla	24 480	1886–85	31.8	0.79	2.01
Macquarie	Dubbo	19 710	1886–85	28.7	1.18	4.27
Murray	Albury	17 220	1877–88	145	0.61	2.36
Styx	Jeogla	16 800	1900–85	9.5	0.51	1.06
Clarence	Lilydale	16 693	1900–85	110	0.73	1.88
Hunter	Singleton	16 450	1900–85	27.8	0.94	2.81
Snowy	Jarrahmond	13 800	1900–85	66.0	0.35	1.07
Goulburn	Murchison	10 800	1877–88	67.9	0.61	0.87
Brisbane	Savage's Crossing	10 200	1900–85	27.1	0.81	2.00
Herbert	Ingem	8 806	1900–85	104	0.37	0.36
Macley	Lower Cruk	8 030	1900–85	33.6	0.66	2.04
Manning	Killawarra	6 555	1900–85	62.0	0.57	1.82
Mary	Miva	4 840	1900–85	35.3	0.68	1.57
Mitchell	Glenaladale	3 900	1882–85	34.5	0.43	0.68
Shoalhaven	Welcome Reef	2 770	1910–85	18.5	0.80	1.70
Barron	Myola	1 937	1915–70	25.6	0.57	0.95
Pioneer	Playstowe Mill	1 393	1900–85	21.9	0.70	1.53
Kiewa	Kiewa	1 145	1877–88	20.6	0.55	2.67
King	Crotty	449	1925–70	88.0	0.19	−0.30
Onkapringa	Clarendon Weir	445	1877–88	1.23	0.73	1.33
Oceania						
Clutha	Lake Hawea Outflow	15 470	1918–84	563	0.18	0.4
Waitaki	Mouth	11 820	1918–84	330	0.16	0.44
Wairau	Lake Manapouri Outflow	7 091	1918–84	478	0.13	0.45
Waikato	Lake Taupo Control	3 430	1905–67	128	0.13	0.38
Kaituna	Lake Rotoiti Outflow	632	1906–90	21.8	0.16	0.68
Dumbea	Mouth		1918–84	3.49	0.23	0.77
Yate	Mouth		1918–84	5.27	0.26	1.03

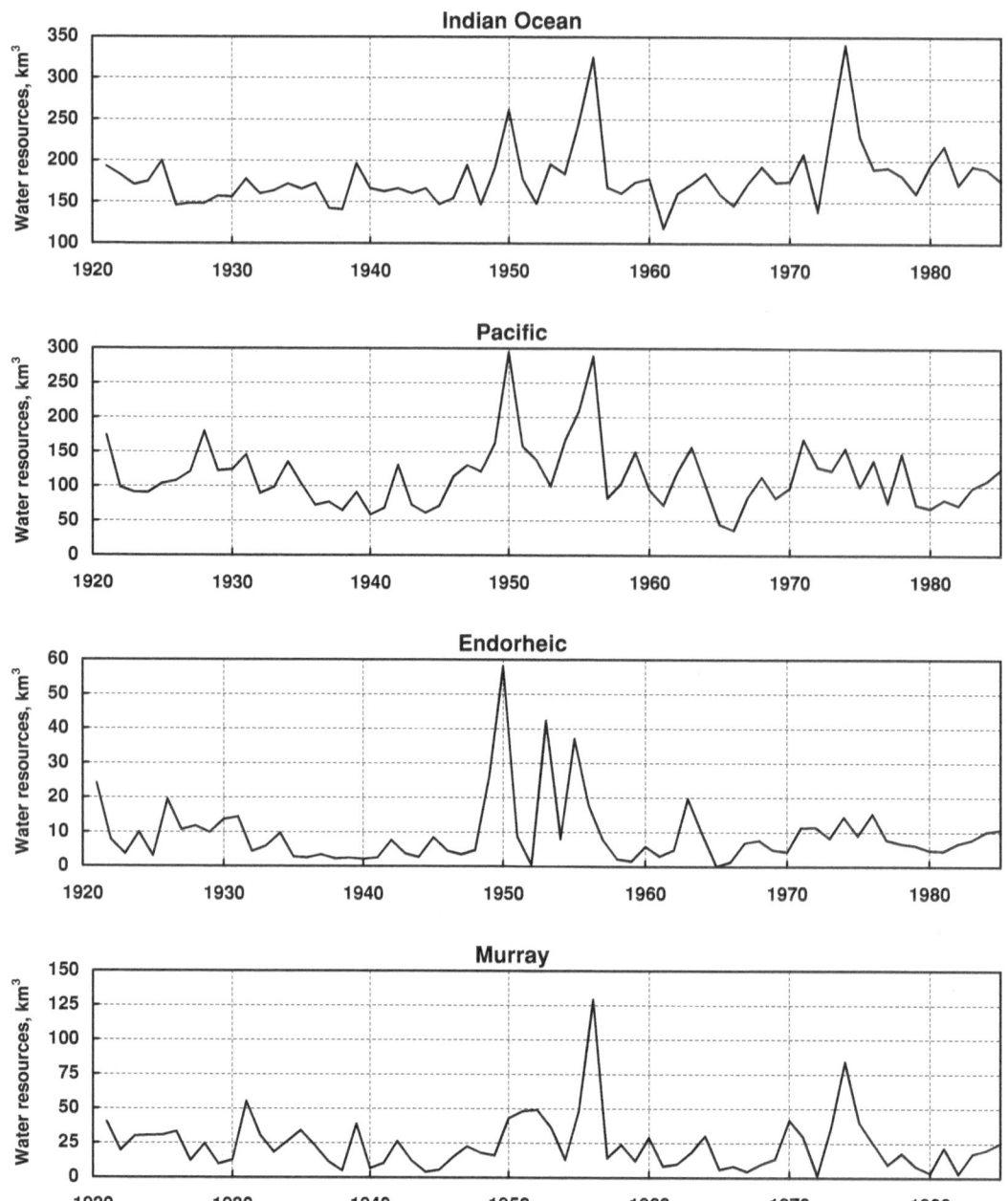

Fig. 9.7 Variations in water resources on continental slopes of the basins of the Indian and the Pacific Oceans, the endorheic region and the Murray River.

Shoalhaven River and in early spring in the Snowy River. In the Australian Alps the highest flows are reached in spring due to snowmelt and rains. In southwestern Australia 90–95% of the annual flow takes place in the winter and spring with the humid southwestern winds.

The extent of the semi-arid zone is determined by the climate and how it differs from year to year (Ceplecha, 1971). Runoff and infiltration to ground water is very variable and does not occur during the dry season. The discharges of the northern rivers (the Fitzroy, Ord, Victoria, Daly and others) are largest in the summer

and autumn due to the monsoon, when between 80% and 95% of the runoff occurs. During the winter and spring even the largest rivers dry up, and most of northern Australia suffers from drought between May and October; irrigation is practised even in areas where precipitation reaches 1000 mm/year. Seasonal variability of river flows is high, such as in the Burdekin, Fitzroy (Queensland) and Darling Rivers. The Fitzroy, with a catchment area of about 143 000 km², can record flows as high as 28 500 m³/s, while in a dry season there is no flow. In the north of the Murray Basin the rivers reach their highest discharges in summer, for example, the Balony, Kalgua and Warren, while those in the south, such as the Goulburn and Murrumbidgee, have a winter maximum

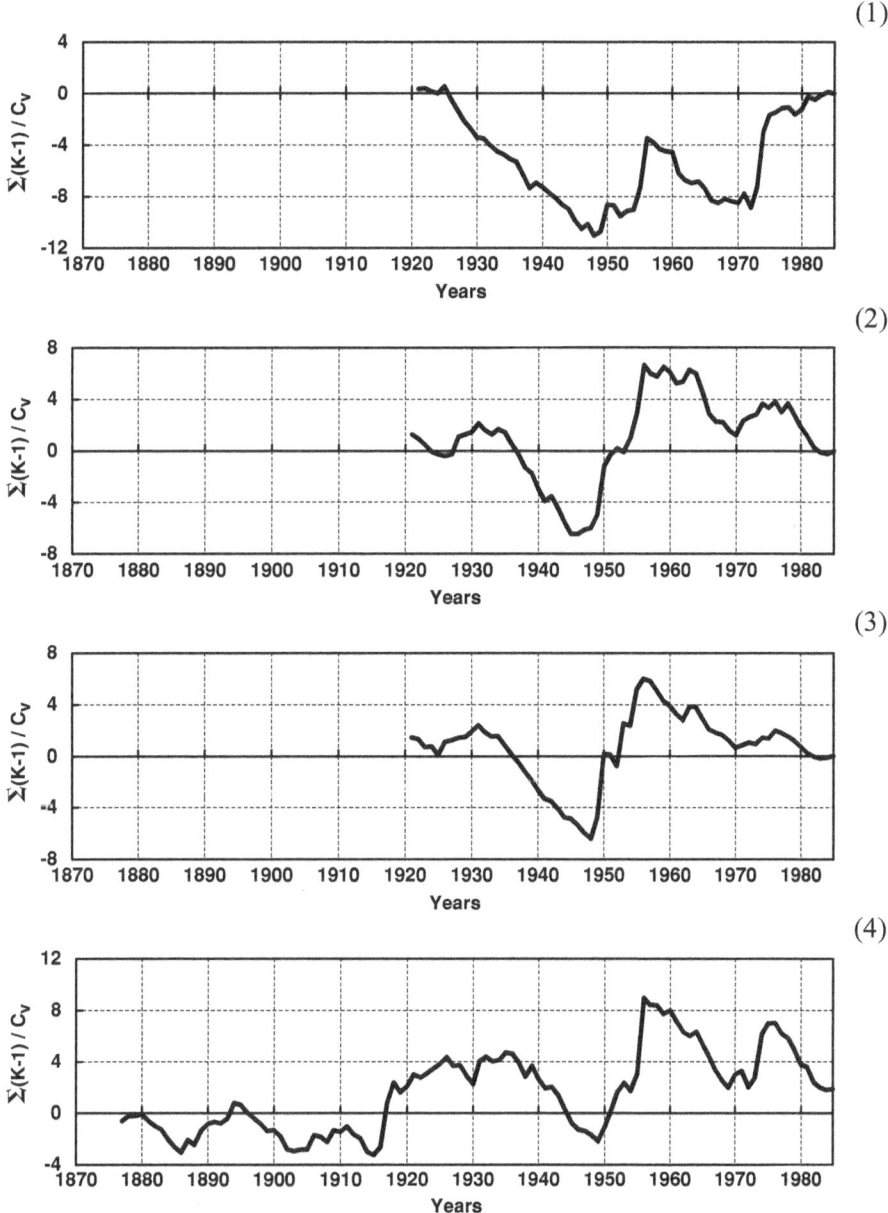

Fig. 9.8 Normalized difference integral curves of water resources of Australia and Oceania. (1) Indian Ocean slope; (2) Pacific Ocean slope; (3) interior runoff region; (4) Murray–Darling River Basin.

(Fig. 9.10, Table 9.20). Some 68% of the flow to the oceans occurs betweeen January and March while the rest of the year experiences low flows and no flow at all.

The flows from the Pacific Islands near Australia exhibit a pattern. New Guinea has its largest flows in the summer and early autumn and smallest in winter. In New Zealand where the flow is regulated by lakes, the runoff is evenly distributed through the year: the North Island has a winter maximum while flows in the South Island peak during summer. The Tasmanian rivers have winter and spring maxima and flows are low during the summer.

Thus over most of the continent the scanty water resources are unevenly distributed during the year, which is why efficient management of water resources is one of the major requirements in Australia. Estimates of the monthly variation of the water resources for Australia as a percentage of the annual total are shown in Fig. 9.11; they are also compared with estimates made in 1974 (Korzun, 1974b) and quite good agreement is reached.

9.6 CHANGES IN WATER USE AND WATER AVAILABILITY

9.6.1 Background

According to United Nations forecasts, the population of Australia will reach 23 million by 2025 and will constitute a major

Table 9.16. *Periods of different water content by ocean slopes and principal river basins of Australia and Oceania*

Basin	Period of observations	Periods of high water content			Periods of low water content		
		Years	Ratio of mean for period to long-term mean, K_{av}	Number of years	Years	Ratio of mean for period to long-term mean, K_{av}	Number of years
Murray–Darling River	1877–1988	1915–35	1.27	21	1877–14	0.94	38
		1950–56	2.19	7	1936–49	0.63	14
					1957–69	0.60	13
					1970–88	0.92	19
Pacific Ocean slope	1921–1985	1921–31	1.08	11	1932–45	0.75	14
		1946–56	1.50	11	1957–85	0.91	29
Indian Ocean slope	1921–1985	1949–56	1.20	8	1921–48	0.92	28
		1973–85	1.15	13	1957–72	0.93	16
Endorheic basins	1921–1985	1921–31	1.23	11	1932–49	0.45	17
		1949–55	2.45	7	1956–85	0.82	30

Table 9.17. *Probability of water content transition from one grouping to another by ocean basins and principal river basins of Australia and Oceania*

Basin	Water content characteristics[a]	Probability of transition from one grouping to another, %			Average duration of water groups, years
		$\overrightarrow{1}$	$\overrightarrow{2}$	$\overrightarrow{3}$	
Murray–Darling River	1	61	12	27	2.6
	2	67	7	26	1.1
	3	35	19	46	1.8
Pacific Ocean slope	1	62	19	19	2.7
	2	8	33	59	1.5
	3	55	15	30	1.4
Indian Ocean slope	1	40	55	5	1.7
	2	29	57	14	2.3
	3	22	45	33	1.5
Endorheic basins	1	76	7	17	4.1
	2	57	0	43	1.0
	3	38	25	37	1.6

[a] 1, Low; 2, average; 3, high.

Table 9.18. *Distribution of ocean basin inflow by latitudinal zones ($km^3/year$)*

Latitude	Pacific Ocean	Indian Ocean
20–30° N	10.0	–
10–20° N	41.0	–
0–10° N	9.00	–
0–10° S	1152	370
10–20° S	192	117
20–30° S	59.4	13.0
30–40° S	144	36.0
40–50° S	195	38.4
Total	1800	574

portion of the population of the entire macro region of Australia and Oceania (UN, 1995a). This population growth is expected to be accompanied by intense economic development. These optimistic predictions for growth are based on the perception of a rapidly growing economy in the Pacific Rim region. Drysdale (1986) predicted that the Pacific region, with 43% of the world GNP in 1986, would increase its share to 50% by 2000.

Taking into account the growing role of environmental issues it can be expected that a large portion of the Australian GNP will be devoted to environmentally sustainable water resource management (Flemming and Daniell, 1994; Rijsberman, 2000). This will require restructuring of the water industry, development of new strategies and implementation of new practices for integrated water and land management, reorientation of society and the development of new and appropriate approaches and technologies for water supply, re-use of water and waste water treatment. Another important development is likely to be the emergence of a water market (Newman and Mouritz, 1992). The updating of irrigation systems will be a costly operation (Smith, 1983).

Since 1984, the economy of New Zealand has displayed new trends that can be attributed to the globalization of the world economy. New Zealand has opened its economy to the world, becoming an active player in various world markets. Foreign investments in New Zealand have grown, and the Government has encouraged rich immigrants from South East Asia. Trade and commerce with Australia continue to strengthen (Britton, 1991). These trends allow forecasts of the continuation of the high rate of growth of the

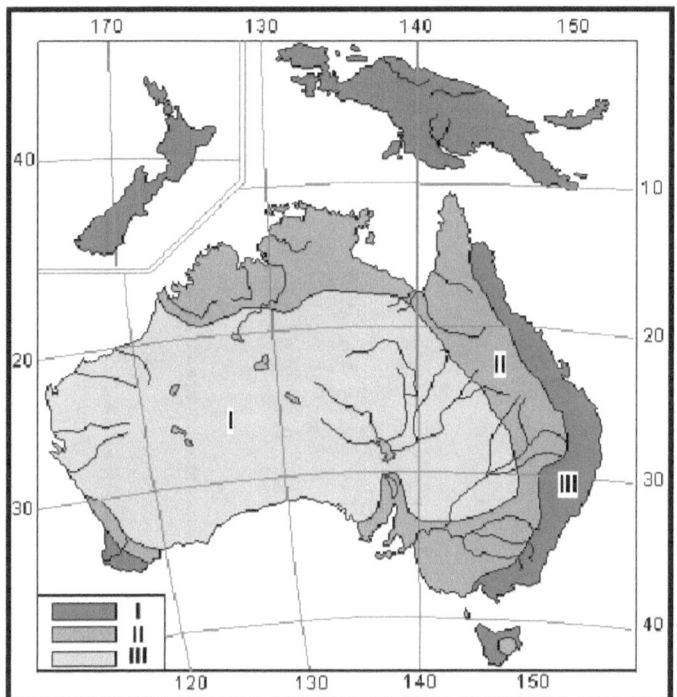

Fig. 9.9 Seasonal runoff distribution for the territory of Australia. I, Arid zone (temporary watercourses); II, semi-arid zone (streams with flow during one season only); III, humid zone (streams with permanent flow).

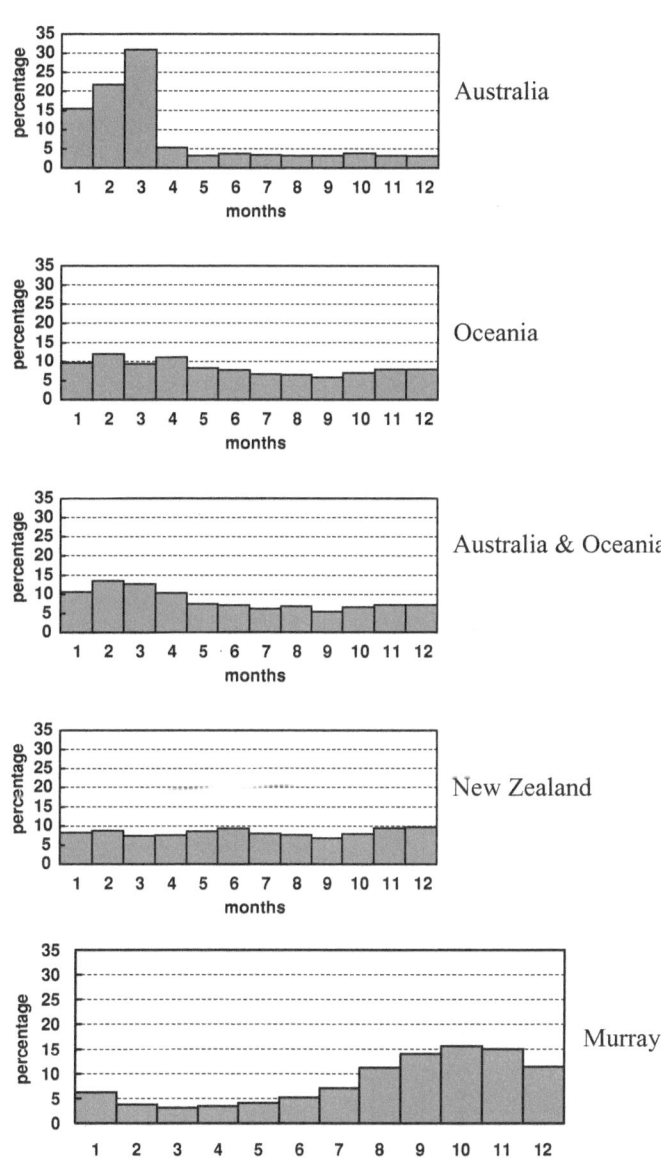

Fig. 9.10 Distribution by month of water resources of Australia and Oceania during a year with an average water content.

GNP from which it can be assumed the costs of water management can be readily met. The proposed doubling of the capacities of existing systems can be made without any foreseeable impact on the environment. Today the New Zealand economy is being restructured, and the attraction of investments is becoming more difficult. There is strong competition between the industrial, commercial and recreational needs in many areas. This conflict of interests might bring about inadequate and hence inefficient models of development (Government of New Zealand, 1988).

The approach in this Monograph is based on sustainable development (Strzepek and Bowling, 1995) and on this basis, the rate of growth in the GNP in both Australia and New Zealand is likely to be high during the coming decade (about 6%). If this figure is accepted, then by the year 2025 the Australian GNP can be expected to reach $US40 000 per capita, while in New Zealand it will exceed $US30 000. The per capita GNP of the Pacific Islands is likely to be much lower. Referring to the same *International Financial Statistics Yearbook* for 1950 to 1994 (IMF, 1994), in Papua–New Guinea it was $US1017 in 1990. It is assumed that between 1980 and 1999 some $US5–10 billion will have been invested in this subregion (Rubtsov, 1991).

Many Australian rivers have been regulated. There are also transfers from one basin to another and large volumes of water are used for irrigation. Unfortunately data on the water use and regulation of flow are not readily available, making it difficult to

distinguish the current trends in water use and causing considerable difficulties for forecasting. This is also true of New Zealand and of other countries in the region. Tourism brings in a large income but it is difficult to assess its effect on the growth of the GNP and its relationship with increased water use.

For the present study indirect methods of assessing water use were made to provide a basis for forecasting future levels of demand (Shiklomanov and Markova, 1987; Shiklomanov, 1988, 2000a, b; Penkova and Shiklomanov, 1998; Shiklomanov *et al.*, 2000). These methods are described in Section 3.3. They draw very extensively on the analysis of the available data and on estimates taken from the literature on abstractions and consumption for certain industries. The trends and changes in the use

Table 9.19. *Average monthly runoff distribution by zones of Australia and Oceania*

Territory	Station	Monthly runoff distribution, in % of mean annual value												Wet period		Dry period	
		1	2	3	4	5	6	7	8	9	10	11	12	%	Months	%	Months
Zones																	
Arid																	
Diamantina	Birasville	5.0	9.2	26.6	25.6	11.3	9.5	5.9	2.1	0.2	1.3	0.8	2.5	52.2	3–4	47.8	5–2
Attack Creek	Stewart Highway	21.5	49.5	22.6	0	1.4	0	0	0	0	0	0	5.0	98.6	12–3	1.4	4–11
Gascoyne	Fishy Pool	14.4	24.7	26.0	4.7	7.7	16.6	2.3	2.9	0.3	0	0.1	0.4	65.1	1–3	34.9	4–12
Humid																	
Barron	Myola	12.9	24.8	27.0	11.1	5.6	4.0	3.0	2.5	1.9	1.7	1.8	3.7	75.8	1–4	24.2	5–12
Clarence	Lilydale	12.2	15.5	16.4	9.3	7.6	11.9	8.3	4.2	2.4	3.8	3.0	5.3	44.1	1–3	55.9	4–12
Snowy	Jindabyne	3.8	2.4	3.0	4.0	5.9	7.0	7.5	9.4	14.6	22.0	13.4	7.0	50.0	9–11	50.0	12–10
Warren	Brockman Bridge	0.7	0.2	0.5	0.9	2.3	12.5	28.2	23.6	16.0	9.7	3.7	1.7	90.0	6–10	10.0	11–5
Semi-arid																	
Victoria	Coolibar Homestead	21.3	28.5	46.7	2.34	0.4	0	0	0	0	0	0.5	0.3	96.5	1–3	3.5	4–12
Fitzroy	Riversleigh	12.7	34.5	18.8	14.0	3.2	2.8	3.1	1.4	0.3	1.4	1.0	6.8	80.0	1–4	20.0	5–12
Torrens	George Weir	0.6	0.5	0.4	1.8	3.8	13.6	20.7	26.1	19.7	8.8	2.8	1.2	88.9	6–10	11.1	11–5
Islands																	
Tasmania																	
South R.	Launceston	2.2	2.5	3.8	5.2	7.2	12.2	15.9	16.9	12.4	11.4	6.1	4.2	68.8	6–10	31.2	11–5
New Guinea																	
Ramu R.	Vonki Dome	10.5	11.7	13.0	11.6	7.6	4.7	3.9	4.7	5.6	8.2	7.2	11.3	58.1	12–4	41.9	5–11
New Zealand (North Island)																	
Wanganui R.	Paetawa	4.9	5.2	5.0	5.4	9.7	12.7	12.6	10.2	8.2	8.0	9.5	8.6	35.5	6–8	64.5	9–5
New Zealand (South Island)																	
Waitaki R.	Mouth	11.6	12.4	9.9	9.6	7.1	5.9	5.3	5.0	5.4	7.8	9.4	10.6	63.5	11–4	36.5	5–10
New Caledonia																	
Rivière des Lacs	Goulet	11.1	18.8	9.5	17.8	8.2	7.6	5.0	5.6	3.8	2.1	5.4	5.1	57.2	1–4	42.8	5–12

Table 9.20. *Average monthly distribution of water resources by regions and ocean slopes of Australia and for the Murray River Basin*

Territory	Water resources, km³/year	Monthly runoff distribution, % of mean annual value											
		1	2	3	4	5	6	7	8	9	10	11	12
Australia and Oceania	2400	10.3	13.2	12.4	10.1	7.4	7.1	6.2	6.9	5.4	6.6	7.2	7.2
Australia	352	15.4	21.7	30.9	5.3	3.2	3.6	3.4	3.2	3.2	3.7	3.2	3.2
Oceania	2050	9.6	12.1	9.5	11.2	8.2	7.7	6.7	6.4	5.8	7.0	7.9	7.9
Pacific Ocean slope	114	11.5	19.2	20.9	9.6	6.2	6.5	5.0	3.9	3.7	4.9	3.7	4.9
Indian Ocean slope	180	17.9	23.2	37.2	2.5	1.3	1.8	2.3	2.8	2.8	3.0	2.9	2.3
Continent	304	15.4	21.7	30.9	5.3	3.2	3.6	3.4	3.2	3.2	3.7	3.2	3.2
Murray River Basin	24	6.3	3.8	3.1	3.4	4.1	5.2	7.1	11.2	14.0	15.5	14.9	11.4

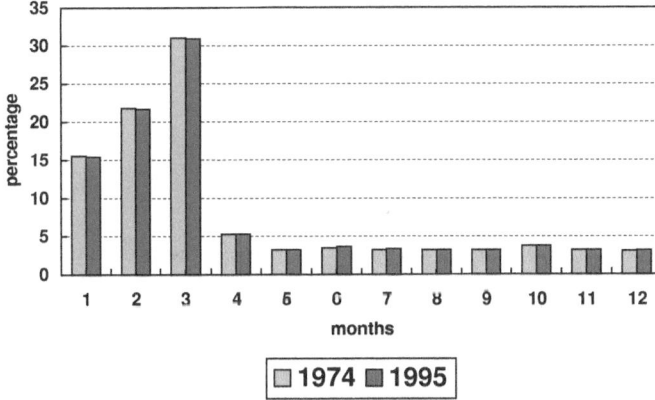

Fig. 9.11 Monthly distribution of annual water resources in Australia, as estimated by Korzun (1974b) and the present study (1995).

of water were studied in detail by the SHI (Shiklomanov and Markova, 1987; Shiklomanov, 1988). Further studies give estimates of water use for Australia and Oceania and forecasts for various future dates such as Frolova (1974), ESCAP (1978), Berezner (1981), Brown (1984), Dunk (1984) and Green (1984). The consumption for Oceania was also estimated from the literature: ECA (1978), ESCAP (1978) and Connell (1984), and use was also made of proxy data.

Data from these publications matched the SHI estimates to 1980 (Shiklomanov and Markova, 1987; Shiklomanov, 1988). Forecasts by SHI were given for 1990 but these have been updated with figures from international, Australian and New Zealand publications (Government of New Zealand, 1988; WRI, 1992, 1996, 2000; Andressian, 1993; UN, 1993; Margat, 1994; Strzepek and Bowling, 1995; ESCAP, 1998; Seckler *et al.*, 1998; Rijsberman, 2000). These data show the SHI forecasts to be slightly overestimated because demands for irrigation and by industry were overestimated. In fact industrial water use decreased by 3.8 km³/year from 1980 to 1990 due to the introduction of water-saving technologies, while water consumption kept growing.

Strzepek and Bowling (1995) showed that industrial water use in Australia and New Zealand should grow significantly, with abstractions increasing by factors of 2.6 and 2.7 by 2025. However, these coefficients seem to be considerably overestimated compared to present trends, taking environmental and resource issues into account, industrial restructuring with science-intensive industries taking precedence.

Considering the growth of telecommunications, the role of financial services, tourism and the like and taking into account these developments, the above coefficients have been decreased by 40% for this Monograph, while those for specific water use in the municipal sector have been decreased by 5%. Due to the lack of data New Zealand coefficients were used for the Pacific Island estimates. The growth of irrigation declined in both Australia and New Zealand, and this trend was taken as the basis for forecasts. An increase in agricultural water consumption has been assumed for other countries to be proportional to the average growth of the GNP per capita for the region (Strzepek and Bowling, 1995). It is difficult to obtain accurate estimates of the area irrigated in Australia because the hydrological regime and the soils are quite complicated and inadequately studied (Australian Academy of Science, 1964; Flemming and Daniell, 1994), but from available data this area seems to be between 4.2 and 4.8 million ha (Alekseyevsky, 1974). By 1995 about 42% had been developed, but further growth of irrigation is constrained by its environmental impact and the likely changes in river regimes (Brown, 1984; Warner, 1995). The irrigated area in New Zealand, as shown by slightly outdated data, is about 300 000 ha. If the current trends are sustainable, the growth of irrigation will slow in both countries. Forecasts for Australia for 2000, 2010 and 2025 indicate that 2.1, 2.25 and 2.4 million ha are likely to be irrigated, while in New Zealand the areas will be 326 000, 353 000 and 371 000 ha.

New abstractions for irrigation in Australia will be slightly higher than 7000 m³/ha, and in New Zealand they are likely to be

Table 9.21. *Dynamics of fresh water use in Australia and Oceania by type of economic activity (km³/year)*[a]

Water use	Assessment								Forecast		
	1900	1940	1950	1960	1970	1980	1990	1995	2000	2010	2025
Population, $\times 10^6$			11.8	16.0	20.3	24.0	27.8	29.6	31.4	34.8	38.7
Irrigated land, ha $\times 10^6$	0.05	0.43	0.66	1.20	1.58	1.67	2.18	2.31	2.43	2.60	2.77
Agriculture	0.46	3.50	5.20	9.40	12.5	13.0	14.7	15.5	16.3	17.4	18.5
	0.35	2.80	4.10	7.50	9.90	10.2	11.6	12.2	12.7	13.5	14.3
Industry	1.00	3.00	4.10	6.20	8.30	10.5	6.70	7.20	7.80	8.70	10.3
	0.20	0.45	0.50	0.64	0.69	0.78	0.46	0.62	0.77	1.44	2.18
Public Supply	0.14	0.33	0.75	1.10	1.50	2.80	3.10	3.30	3.50	3.90	4.60
	0.03	0.08	0.16	0.21	0.25	0.30	0.36	0.38	0.41	0.43	0.46
Reservoirs	0	0	0.34	0.70	1.00	3.10	4.00	4.50	5.00	5.60	6.20
Total	1.60	6.80	10.4	17.4	23.3	29.4	28.5	30.5	32.6	35.6	39.6
	0.60	3.30	5.10	9.00	11.8	14.4	16.4	17.7	18.9	21.0	23.1

[a] Nominator, withdrawals; denominator, consumption.

4300 m³/ha. These estimates for Australia have taken into account a 5–6% decrease due to better land management and a corresponding decrease in the amount of return waters. For New Zealand the area of irrigation is assumed to be constant for all future dates. Papua–New Guinea is the only country where abstractions may reach 0.189 km³/year by 2025.

9.6.2 Trends in the use of water

Estimates of use were made for this Monograph for Australia, New Zealand and other Pacific nations for various dates and then all the data were added together. The results are shown in Table 9.21 and Fig. 9.12 for each of the sectors. The trends in the consumption of water and abstractions are shown in Table 9.22 for various regions of Australia and Oceania. Taking into account current and possible future trends, abstractions for the Pacific region were expected to increase by 7% to 2000, and by 17% to 2010, in comparison with 1995. By 2025 there may be a 30% increase and the same rise is predicted for the consumption of water.

It is useful to compare different estimates of water use in Australia and Oceania with figures from SHI for 1974 (Korzun, 1974). It was indicated the 1975 total was likely to be 30 km³/year, rising to 60 km³/year by 2000. However, the rates of growth were then being significantly overestimated. Estimates from the USA made in 1980 (Barney, 1980) gave the 1977 water use as 29 km³/year with 25 km³/year for Australia (in a good agreement with the SHI data) and 4 km³/year for Oceania. UNEP (Prashchikin, 1986) estimated the annual water use in Australia for 1975 as 17.8 km³, and 1.2 km³ for New Zealand in 1980. Margat (1994) gave much lower values. He assumed that the total water use in the region was 14.2 km³/year in the late 1980s, with 8.5 km³/year being used by agriculture, 1.3 km³/year by industry (excluding thermal power) and 4.4 km³/year by the domestic sector. For 2025 these estimates were 18 km³/year, 11.4 km³/year, 0.7 km³/year and 5.9 km³/year, respectively, according to a high water demand scenario. Apparently the differences are due to the fact that the water used in energy generation and evaporation from reservoirs were not taken into account. Table 9.23 shows the estimates of current and future water use compared to the annual runoff. In Australia water use is 7.7% of the annual runoff, but it is only 0.17% in Oceania. By 2025 these figures are forecast to be 10.0% and 0.20%, respectively. Water use in Oceania, as a percentage of runoff, is the lowest in the world.

9.6.3 Water availability

Table 9.24 and Fig. 9.13 show the changes in the mean specific availability for the period 1950 to 2025. From these it is evident that Oceania has the world's highest water availability per unit area (except Guyana). Due to the low density of population, water availability per head is quite high in Australia. In 1995 it was similar to countries in Northern Europe (Sweden), in South American (Argentina, Ecuador) and in Central America (Honduras). By 2025 availability is expected to decrease by about 30% in Australia and Oceania. Dry years can be expected to exacerbate this decrease (Table 9.25). In a dry year the water availability per head is 2.2 times less in Australia, 2.0 times less in Oceania and 1.8 times less for Australia and Oceania together. Significant variations in availability can occur in Australia due to the variability of the climate and these in turn alter runoff, infiltration and other hydrological variables including evaporation from reservoirs and irrigated areas.

Table 9.22. *Dynamics of water withdrawal and water consumption by natural–economic regions of Australia and Oceania (km³/year)*[a]

Region	Assessment								Forecast		
	1900	1940	1950	1960	1970	1980	1990	1995	2000	2010	2025
Australia	1.5	6.2	9.4	16.0	21.4	27.0	25.5	27.2	28.9	31.7	35.2
	0.5	3.0	4.6	8.1	10.7	13.1	14.5	15.4	16.4	18.1	19.7
Oceania	0.1	0.6	0.9	1.4	1.9	2.4	3.0	3.4	3.7	3.9	4.4
	0.1	0.4	0.5	0.9	1.2	1.5	1.9	2.2	2.5	2.9	3.4
Australia and	1.6	6.8	10.3	17.4	23.3	29.4	28.5	30.5	32.6	35.6	39.6
Oceania	0.6	3.4	5.1	9.0	11.9	14.6	16.4	17.7	18.9	21.0	23.1

[a] Nominator, withdrawals; denominator, consumption.

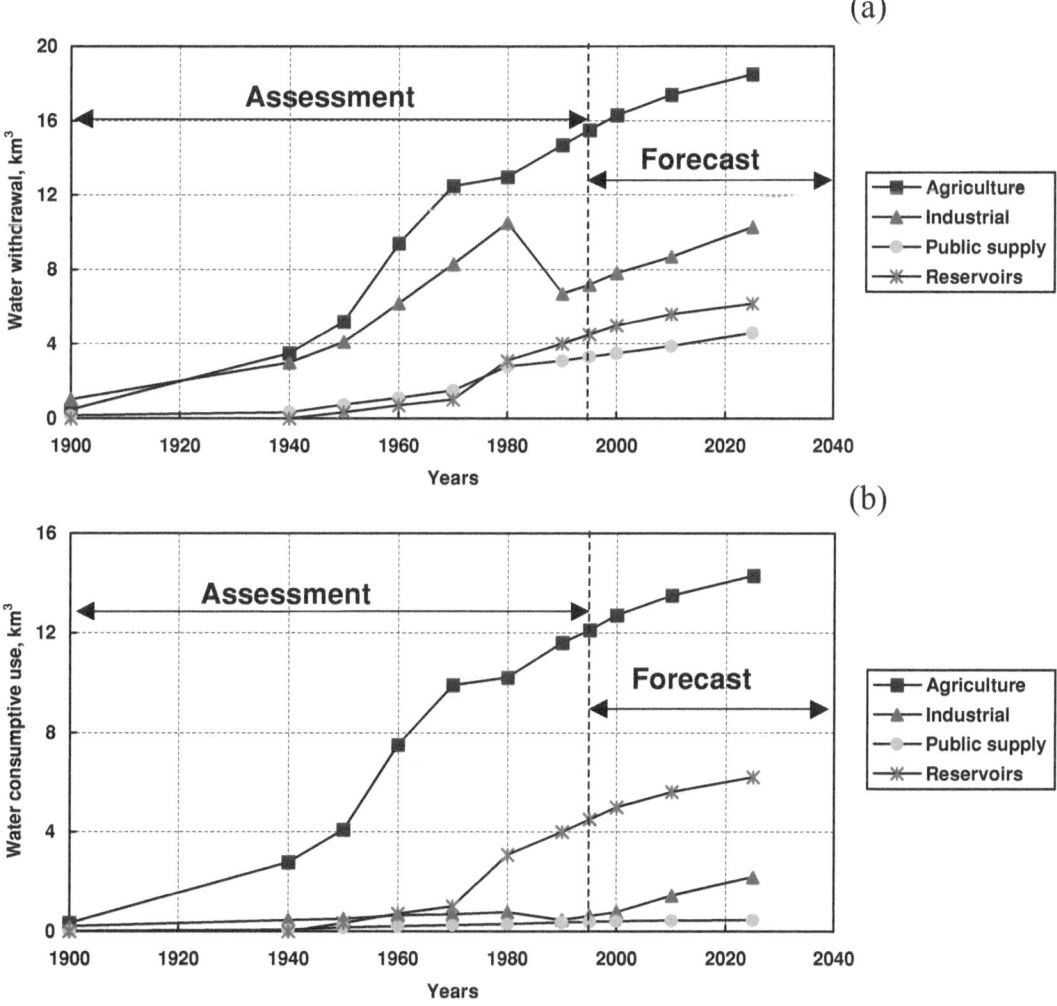

Fig. 9.12 Dynamics of water use in Australia. (a) Water withdrawal;
(b) water consumption.

Table 9.23. *Water use as a percentage of water resources by natural–economic regions of Australia and Oceania*

| Region | Local water resources, km³/year | Water use as a percentage of water resources | | | |
| | | 1995 | | 2025 | |
		Withdrawal	Consumption	Withdrawal	Consumption
Australia	352	7.70	4.40	10.0	5.60
Oceania	2050	0.17	0.11	0.20	0.17
Australia and Oceania	2400	1.30	0.70	1.60	1.00

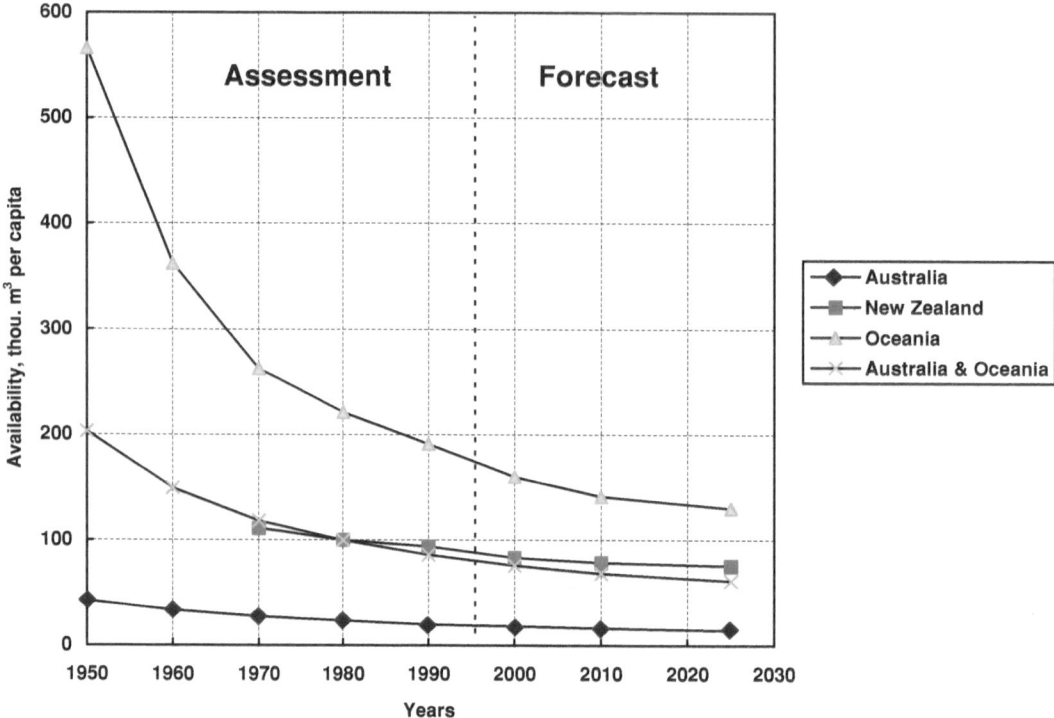

Fig. 9.13 Dynamics of water availability per head by countries and regions of Australia and Oceania.

9.7 MAJOR ECONOMIC PROBLEMS DUE TO STRESSES ON WATER RESOURCES

The Australian economy faces a range of problems because the continent's water resources are not evenly distributed in time and space, and experience greater variability than those of the rest of the world. To address these problems there has been considerable investment in water management. Various projects have been developed which aim to regulate flows in different parts of the continent and to use the water for irrigation and other purposes (Korzun, 1974b). One proposal is to turn Lake Eyre into a large permanent reservoir with a surface area of 9600 km² and to use this water body for irrigation and water supply.

The Snowy Mountains Hydroelectric Scheme, completed in 1974, is one of the best-known water management projects in the Southern Hemisphere. It transfers the waters of the Snowy River (draining the eastern slope of the Great Dividing Range) to the Murray Basin (Johnson and Miller, 1978; Berezner, 1982). The project was conceived in the late nineteenth century when the first proposal was made to transfer waters to irrigate the fertile Murray and Murrumbidgee Basins. The idea was not implemented however due to the dispute over the division of the flow of the Snowy River between the states of New South Wales, Victoria and the Australian Capital Territory (Alekseyevsky, 1974), but such an ambitious project could only have been funded by the Commonwealth Government. Its design and construction took 25 years starting in 1949 and its cost amounted to US$900 million. This

Table 9.24. *Dynamics of population and water availability by natural–economic regions and selected countries of Australia and Oceania*

Region	Area, km² × 10⁶	Year	Population, ×10⁶	Water availability, m³ × 10³/year per head
Australia and Tasmania	7.683	1950	8.2	42.5
		1960	10.3	33.4
		1970	12.5	27.3
		1980	14.7	23.1
		1990	17.1	19.7
		2000	18.6	18.1
		2010	20.3	16.4
		2025	23.0	14.4
New Zealand	0.265	1950	1.1	
		1960	2.4	
		1970	2.8	111
		1980	3.1	100
		1990	3.3	93.7
		2000	3.7	83.2
		2010	4.0	78.6
		2025	4.1	75.6
Oceania and New Zealand	1.267	1950	3.6	566
		1960	5.7	362
		1970	7.8	262
		1980	9.2	221
		1990	10.7	191
		2000	12.8	160
		2010	14.5	141
		2025	15.7	130
Australia and Oceania	8.950	1950	11.8	203
		1960	16.0	149
		1970	25.3	118
		1980	24.0	99.6
		1990	27.8	85.7
		2000	31.4	75.9
		2010	34.8	68.3
		2025	38.7	61.4

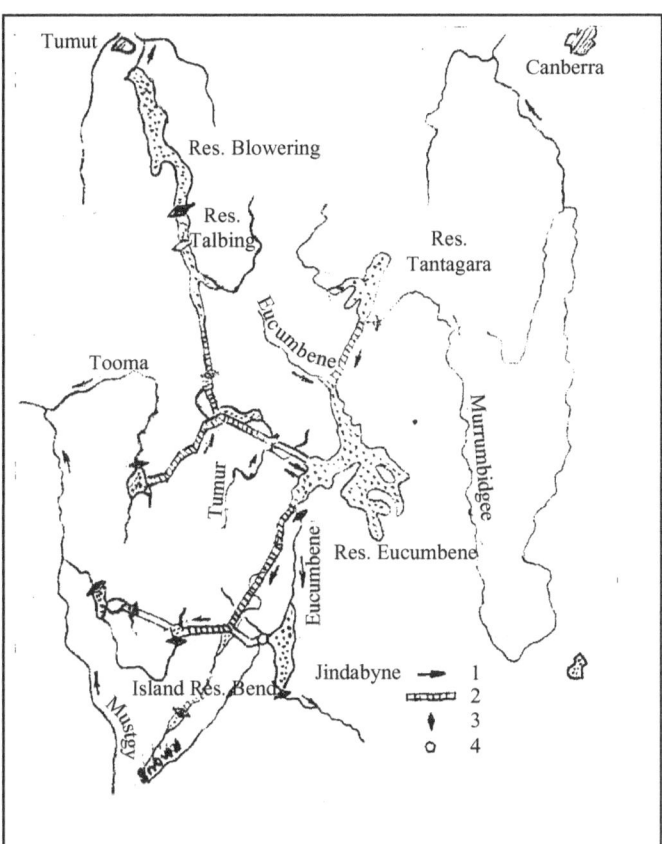

Fig. 9.14 The scheme of water transfer by the Snowy Mountains project. 1, Direction of water transfer; 2, tunnels; 3, dams; 4, pumping stations.

pipelines and some 140 km of tunnels (Fig. 9.14). The designed power output of the scheme is 5 billion kWh. The volume of water transferred to the Murray River Basin from the Snowy River waters is 2.4 km³/year, which allowed an extra 250 000 ha to be irrigated. The scheme consists of two subschemes: (1) Snowy–Tewmeet and (2) Snowy–Murray. The first subscheme transfers 1.39 km³/year to the Murrumbidgee and the second one transfers 0.99 km³/year into the upper reaches of the Murray.

The Eucumbene Reservoir is the largest water body but there are other reservoirs in the lower reaches of the rivers which are used for irrigation. The Blowering Reservoir on the Tumut regulates flows as does the enlarged Hume Reservoir. The scheme increased the flow to the Murrumbidgee and Murray by 100% and 60%, respectively. Compensation flows from the scheme during droughts increased by 5% in average conditions and by 34% in dry years. The scheme reduced the threat of droughts and allowed the development of areas with large flow variations and lacking water resources (Hall, 1992).

The recent developments in mining and oil production coincide with a growing awareness of the need to safeguard the environment and the government response with tighter

multi-purpose project supplies additional power for peak loads in the electricity grid of New South Wales and Victoria, and it also supplies water for irrigation to the basins of the River Murray and its tributary the Murrumbidgee.

The Snowy Mountains Hydroelectric Scheme consists of 16 large dams, seven hydropower plants, five reservoirs with a total capacity of 8.5 km³, two pumping stations, about 80 km of

Table 9.25. *Water availability throughout long-term dry periods and the driest years in Australia and Oceania*

| Territory | Population $\times 10^6$ | | Water resources during dry periods and years, km³/year | | | | Consumption, km³/year | | Water availability, m³ $\times 10^3$/year per head | | | |
| | | | | | | | | | Dry period | | Dry year | |
	1994	2025	Period	Water resources	Year	Water resources	1994	2025	1994	2025	1994	2025
Australia	17.9	23.0	1957–1970	267	1961	228	15.4	19.7	14.1	10.7	12.2	9.04
New Zealand	3.53	4.12	1930–1954	300	1950	246	1.25	1.71	84.7	72.3	69.4	59.2
Oceania	10.8	15.7	1921–1942	1910	1928	1515	2.2	3.4	177	122	141	96.9
Australia and Oceania	28.7	38.7	1921–1942	2260	1937	1850	17.6	23.1	78.1	58.0	63.8	47.3

regulations (Mabbut, 1986). These moves have affected some old industries in the more developed areas, and in other areas they have been added to by the opening of new national parks which now number 400 across Australia (Bambrick, 1994). In the large conurbations and their surrounding areas, environmental measures can be compared to those in other developed countries, and Australian environmental practice can be regarded as a model, particularly in a country which is a federation involving so many and various administrative bodies.

The agricultural development on the frontier of settlement that in many cases became possible due to irrigation today requires significant capital investment. Today, however, the country has a surplus of agricultural products, most of which are exported, and consequently the area under agriculture has not increased in size. Better management of the existing agricultural lands is the focus of attention, such as the use of better-quality seeds (Farian, 1987). The state Governments promote measures directed at restructuring irrigated agriculture, facilitating its deregulation and attracting foreign investments. One example (Robinson, 1995) is the production of sugar-cane in Queensland.

Recently environmental issues have become more important; for instance, the control of desertification has become a very important consideration. Land management problems arise that are related to the national economy, and they have an impact on the hydrology in the different regions. Desertification is a serious threat, one that has been known since the first European settlement. Local governments take measures to mitigate the so-called population stress on agricultural land and to encourage higher efficiency in agriculture such as using improved breeds of cattle. Measures are being taken which aim at comprehensive solutions to such problems as land reclamation, salinization, water pollution and irrigated land management. The basis for environmental legislation which is being developed is targeted at proper environmental management and conservation. The principles of social

and environmental monitoring are being developed (Mabbut, 1986).

New Zealand has recently adopted a number of environmental reforms (Menon, 1993). New Zealand timber exports have increased more than four times since 1969 (Wilson, 1995). Flood control is a serious problem in that country (Government of New Zealand, 1988). Soil erosion in New Zealand occurs at rates that are amongst the highest in the world (Alekseyevsky, 1974).

9.8 CONCLUSIONS

The studies carried out here have shown that the water resources of Australia are 352 km³/year, in Oceania they are 2050 km³/year and those of Australia and Oceania together are 2400 km³/year. The water availability for Australia is 45 800 m³/year per km², in Oceania 1 614 000 m³/year per km², and in Australia and Oceania together it is 268 000 m³/year per km². The potential water availability per head in Australia is 19 700 m³/year, while for Oceania, this value is the world's highest at 190 000 m³/year per head. Australian rivers are intensively used for industrial and agricultural purposes, although the amount of water consumed is not high (15.4 km³/year). In Oceania where water use is not so intensive, consumption is about 2.2 km³/year, while for 1995 for Australia and Oceania together some 17.7 km³/year of water is consumed.

The water resources of Australia and Oceania have been estimated previously and this Monograph has used these results where possible. The additional measurements of runoff made after 1970 have contributed greatly to more precise estimates. However these estimates and the estimates of water use are only approximate in a number of basins because of the poor hydrological network, also because of the unrepresentative nature of some hydrological stations and lack of measurements near the mouths of rivers.

Where time series contain gaps which have been filled by data determined from analogues or other methods described earlier, there are uncertainties involved. Similarly the forecasts of water resources, water use and water availability in Australia and Oceania are only approximate, and in future they may also be made even less reliable by climate change. Despite all these problems the estimates of water resources, water use and water availability in Australia and Oceania made here are the most precise available. Should new data appear then these estimates, of course, can be updated.

10 Global renewable water resources in space and time

10.1 REVIEW OF EXISTING ASSESSMENTS: INITIAL DATA

Since the end of the nineteenth century, estimates have been published of the magnitudes of the components of the global hydrological resource, including the total runoff and of the volume of the renewable fresh water resource. Baumgartner and Reichel (1972, 1973) collected many of these estimates, showing that the results differ considerably and by as much as a factor of three. During the last 30 years more of these estimates have been published, such as those of Nace (1967), L'vovich (1974b), SHI (Korzun, 1974b), Baumgartner and Reichel (1975), Berner and Berner (1987), Seckler *et al.* (1998), and they are regularly published by the World Resources Institute (WRI, 1992, 1996, 2000). The different sources indicate that the estimates of the total mean annual runoff for the Earth fall within the range of 37 400 to 44 500 km.

The most detailed and complete assessments of the world water balance and world water resources, have been presented by Russian and German scientists (Korzun, 1974b; Baumgartner and Reichel, 1975). These assessments have been widely used and considered to be the most reliable. When they are compared, however (Table 10.1), a difference is revealed of about 30–40% for the continents. This difference can be attributed mainly to the methods for determining the total runoff. In the Russian studies, the total runoff is determined by using the observational data from hydrological stations, in the German studies, runoff was estimated as the difference between precipitation and evaporation. Use of the latter method gives rise to large errors for small rivers and is not applicable for assessing water resources and their changes, particularly in arid and semi-arid areas.

It should be noted that many of the studies dealing with water resources on the global and lesser scales are not based on new data. They usually employ the data collected and used in the monograph on the world water balance (Korzun, 1974b). For example, Shiklomanov (1990) and Gleick (1993, 1998) employed these data. Then there is the problem that some studies use data from different sources and for different years and time periods, often without acknowledging this. For example, the World Resources Institute in Washington (WRI, 1992, 1996, 2000), has produced assessments compiled from different sources and referring to the years from 1975 to 1987, based mainly on data from the Institute of Geography, Russian Academy of Sciences and on the data collected by L'vovich from 1969 to 1972 (L'vovich, 1969, 1972). This problem is compounded because the World Resources Institute data have been used in many other studies of the world water resources and water use (Berner and Berner, 1987; Kulshreshtha, 1992; Postel, 1992; Seckler *et al.*, 1998; Rijsberman, 2000).

One example of these difficulties is the report *Sustaining Water: Population and the Future of Renewable Water Supplies* (Engelman and LeRoy, 1993), a document which has been disseminated widely by international organizations, and amongst scientists, decision-makers and politicians. This report presents the values for the annually renewable water resources in 149 countries and the availabilty of water per head for 1955, 1990 and 2025 with the forecast populations. This report appears to be accurate and reliable and its inferences seem to represent the best and most up-to-date knowledge on water resources and water availability for these countries. However, when the data cited are examined, it is obvious they were taken from a number of different sources, they are not new, and they contain no fresh and more precise information about the water resources of those countries. Furthermore the inferences do not reflect in full measure the present-day situation regarding the changes and trends in water availability. For most countries, the estimates of water resources and water availability require much consideration and correction. The main conclusions of this study are as follows.

1. To assess the water availability for a country in an international river basin, it is not correct to use the value of the total renewable resource including the runoff flowing in from outside. In this case the water resources for countries sharing the basin are explicitly overestimated because the runoff of one river is taken into account several times in different countries. The consequence is that the resource and availability estimates may need correction in a number of instances.

Table 10.1. *Total river runoff from continents, according to different authors*

Author	Year	Europe mm	Europe km³/year	Asia mm	Asia km³/year	Africa mm	Africa km³/year	North America mm	North America km³/year	South America mm	South America km³/year	Australia and Oceania mm	Australia and Oceania km³/year
L'vovich	1969	300	3180	286	12 440	139	4180	265	6440	445	7970	218	1950
L'vovich	1972	319	3380	293	12 750	139	4180	275	6680	583	10 440	226	2020
Korzun	1974	283	3000	324	14 100	153	4610	339	8240	661	11 830	280	2510
Baumgartner and Reichel	1975	282	2990	276	12 000	114	3430	242	5880	617	11 040	269	2410
World Resources Institute	1992	312	3310	324	14 100	126	3790	326	7920	588	10 520	263	2350
State Hydrology Institute (Shiklomanov)	1997	274	2900	311	13 510	134	4047	324	7870	672	12 030	268	2400

2. The estimates of water resources for countries and regions may be incomparable because they use data taken from different sources and for different years, and they may be determined by different methods. In some cases, these figures are for the total water resources, in others only for the areas where runoff originates. These are a number of the sources of the errors which together can increase the unreliability of the estimates.

3. To assess the actual water availability, in addition to the estimates of water resources and population, changes and trends in the use of water have to be taken into account.

In this Monograph, the method used was based on the utilization of observations collected from the stations in the global hydrological network, supplemented, where necessary, by meteorological data (monthly precipitation and temperature). This approach has been successfully applied before by Russian scientists to assess the global water balance and water resources, for example in the 1970s (Korzun, 1974b). The merit of this approach has, of course, increased, because time series are now at least 20 years longer and data from many regions of Africa, Asia and Latin America have become available when there were none before.

In the early 1990s there were about 64 000 river gauging stations in operation globally (Rodda, 1995). Records from about 40 000 of these stations have been used in this Monograph (Table 10.2). However the data derived from these stations is very diverse. Fragmentary data, unorganized data, data for individual years and even months have been included, as well as the data derived from maps with isolines of mean annual runoff. For most of these data, there was no measure of their reliability, no indication whether the gauges concerned complied with guidance provided by the World Meteorological Organization (WMO) or met the standards agreed by International Standards Organization (ISO). Consequently it was impossible to use of all these data to assess

Table 10.2. *Distribution of observation stations by continents*

Continent	Total number of hydrometric stations	Stations for assessment of water resources	Observation period, years
Europe	6 000	610	10–178
Asia	12 000	800	10–120
Africa	2 000	250	5–80
North America	12 000	300	10–130
South America	3 600	240	5–70
Australia and Oceania	3 000	200	5–80

world water resources and so a selection was made of hydrological stations as follows:

- those with the longest series of continuous observations
- those located on large and medium-size rivers, and if possible, evenly distributed
- those observations reflecting stations with the natural (or close to natural) regime.

The application of these criteria reduced the number of stations to about 2500 (Fig. 10.1, Table 10.2) and use was made of monthly and annual data from these stations. Unfortunately a considerable number the stations had to be selected which only had data for 5–10 years for the period from 1980 to 1988. These stations were in many cases located in developing countries of Africa, Asia and Latin America. Such differences in lengths of records caused problems when they were set against the aim of a single study period from 1921 to 1985. In contrast there were long records available for many countries of Europe and North America, including records for 1990 to 1994.

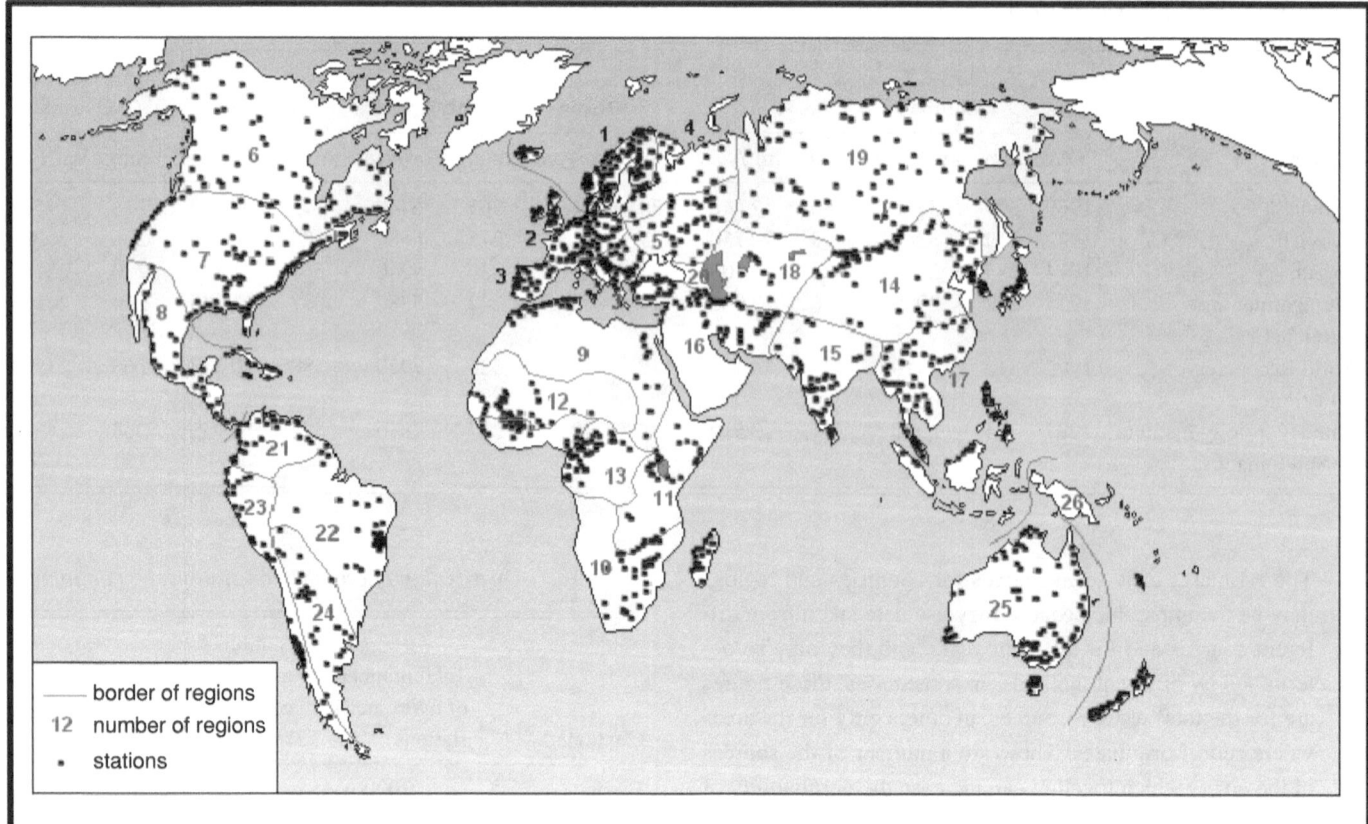

Fig. 10.1 The natural–economic regions of the world and the location of gauge stations.

Figure 10.1 shows the 26 regions of the world, each possessing relatively homogeneous physical conditions and a similar level, of economic development within its boundaries. In most cases the boundaries of the regions coincided with national borders, and thus the regions often included the entire territories of individual countries (from one to 15–17 countries). The exceptions were greatest in the largest countries namely Russia, China, the USA, as these could be subdivided. Some of these 26 regions are very large and approach 12–13 million km^2 (e.g. Siberia and the Far East of Russia, Canada and Alaska). While some are very small, such as the 190 000 km^2 of Transcaucasia, most regions extend over areas of 1 to 8 million km^2.

The adoption of a 65-year study period for all regions of the world was aimed at obtaining comparable mean values of water resources, and a representative sample of their extremes so that variability could be adequately established. However, many rivers lacked continuous data from 1921 to 1985 and this required an effort to fill the gaps in observations. The methods employed for filling these gaps have been described; these were methods which in some cases involved the use of meteorological data in addition to the hydrological data. There were also problems resulting from the lack of data from about 20% of the land area of the globe,

particularly the desert areas, the tropical regions and high latitudes. Then for many basins there were no data for the furthest downstream areas where discharge to the sea could be measured for the whole basin. The methods used to overcome these problems have also been discussed earlier. In addition, there were the problems in assessing the water resources of countries and regions, whose boundaries do not coincide with river basins, and the methods developed at SHI for overcoming these difficulties have also been described (Shiklomanov, 1997, 1998b, 2000a, b; Babkin, 1998; Shiklomanov *et al.*, 2000). Chapter 2 considers these methods and they are also referred to in sections 4.5.1, 5.5.1, 6.5.1, 7.5.1, 8.5.1 and 9.5.1 where the continents are considered. Of course runoff includes the water in the network of streams and rivers fed from rainfall and snowmelt, as well as from ground water seepage from upper aquifers. Some ground water also flows directly to the seas and oceans or evaporates and is not usually assessed in determining water resources from river discharges.

It is of considerable practical importance to estimate the volumes of renewable ground water not flowing into the river network as compared with the volume contained by the rivers. In certain regions ground water is far more important than surface water, such as areas of karst and similar highly permeable aquifers, particularly in arid and semi-arid areas. However the assessment of the contribution of ground water for the world as a whole is a

Table 10.3. *Renewable water resources and water availability by continents*

Continent	Area, $km^2 \times 10^6$	Population, $\times 10^6$	Water resources, km^3/year			Cefficient of variation, C_V	Potential water availability, $m^3 \times 10^3$/year	
			Average	Maximum	Minimum		Per 1 km^2	Per head
Europe	10.46	685	2 900	3 210	2 440	0.10	277	4.24
North America	24.3	453	7 870	8 820	6 660	0.10	324	17.4
Africa	30.1	708	4 047	5 082	3 073	0.10	134	5.72
Asia	43.5	3445	13 510	15 000	11 800	0.06	311	3.92
South America	17.9	315	12 030	14 350	10 330	0.07	672	38.3
Australia and Oceania	8.95	28.7	2 400	2 880	1 890	0.10	268	83.6
The World	135	5633	42 757	44 460	39 660	0.02	317	7.60

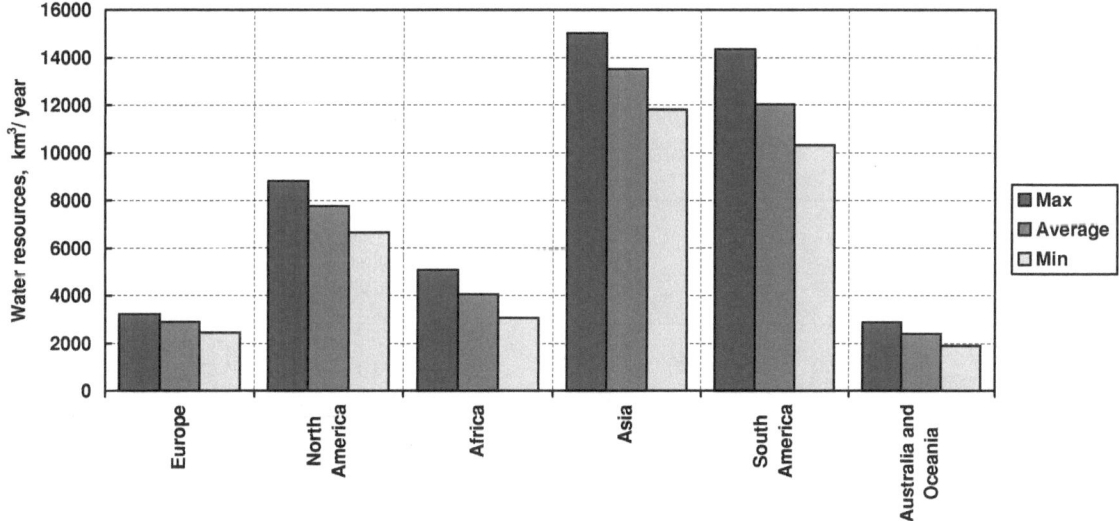

Fig. 10.2 Renewable water resources by continents.

very difficult problem, largely because of the lack of data and the knowledge derived from them. Nevertheless for certain areas and countries such assessments have been made, and these allow some generalizations to be undertaken at the global scale. One example is the study of Africa undertaken by FAO (1995a). From the FAO data, the total volume of renewable ground water not related to runoff is estimated to be 188 km^3/year for the continent as a whole, or 5% of the volume of runoff. However for certain countries, e.g. Egypt, Lebanon, Tunisia and Morocco, this percentage is larger and ground water is a very important component of renewable water resources.

10.2 ESTIMATING WATER RESOURCES

Estimates of water resources at the global and lesser scales are shown in Table 10.3 and Fig. 10.2. Using data for the period 1921 to 1985 the world's water resources are estimated to be 42 800 km^3/year (without the Antarctic). This figure is about 5000 km^3/ year greater than the Baumgartner and Reichel (1975) estimate, and 1800 km^3/year less than the previous SHI assessments (Korzun, 1974b). There are also differences from one continent to another (see Tables 10.1 and 10.3), and between the SHI assessments made in 1974 and 1997, of between 5% and 12%. These differences are probably largest for arid areas in Africa and Asia but they can also be large in areas lacking data such as Northern Canada. Another point is that runoff seems to have been underestimated for the last two decades, especially in Africa, Asia and Europe (Fig. 10.3).

Table 10.3 and Fig. 10.4 present estimates of the potential supply of water which is available from runoff to meet the demands of the total population of the globe. This figure was 7600 m^3/year per head for 1994, varying by the continents from 4000 for Asia to 38 000 for South America, and to 84 000 for Australia with Oceania. It is noteworthy that since the previous assessment in 1970 the world figure has decreased by a factor of 1.7, from

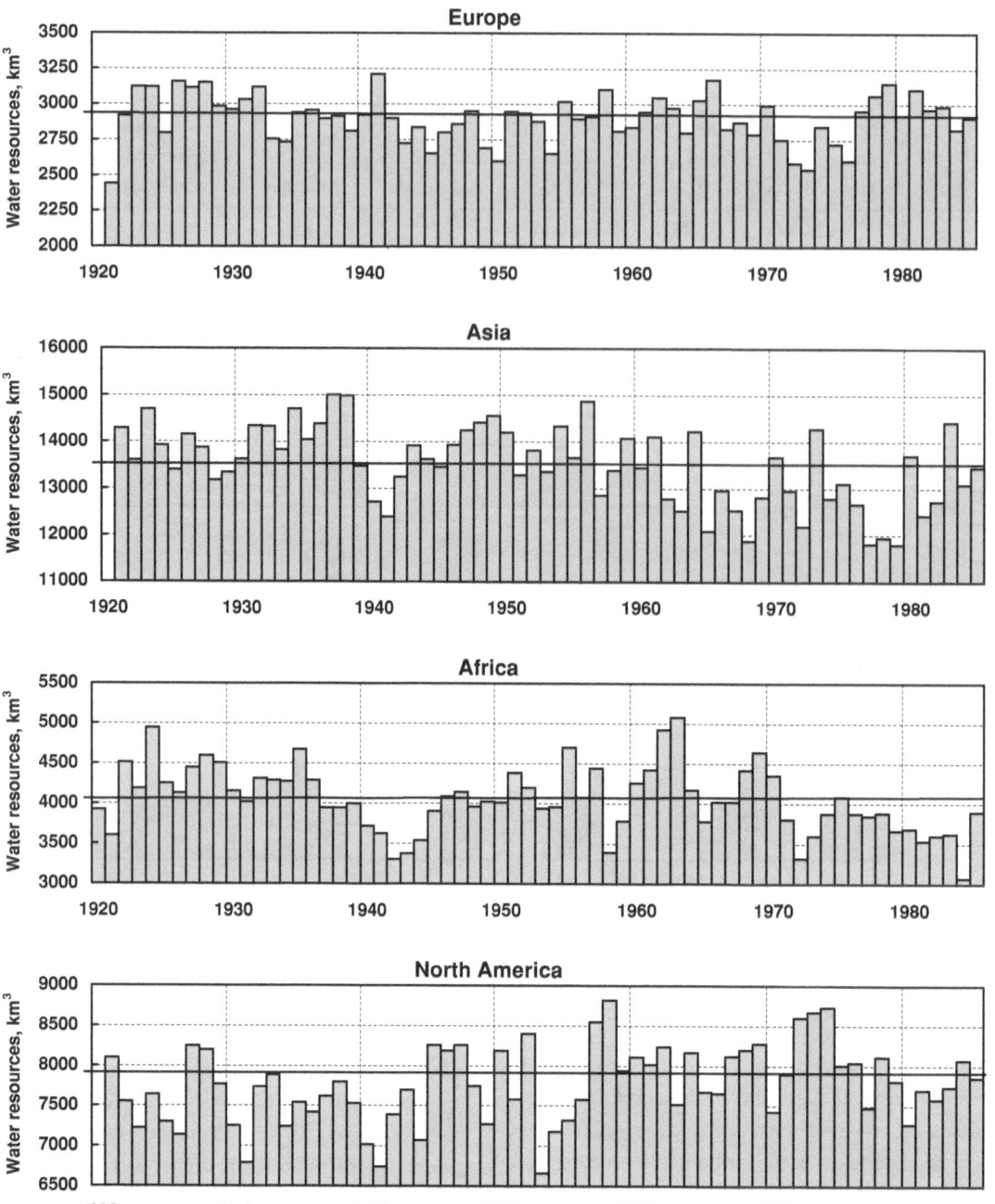

Fig. 10.3 Renewable water resources (km³/year) of the Earth and the continents.

12 900 to 7600 m³/year per head, due to the growth in population of almost 2 billion. The most marked decrease in availability has occurred in Africa (by 2.8 times), in Asia (by 2.0 times), and in South America (by 1.7 times), whereas the European decrease is only 16% (Korzun, 1974b). In addition to the growth in population, there has been a growth in water consumption per head in many areas, the growth in Africa, Asia and South America being especially significant.

The compilation of an archive of global runoff for the period 1921 to 1985 allowed these records to be analysed for cycles and trends and the results are presented in Fig. 10.3. This shows cycles of wet and dry years of differing magnitude and duration as well as the deviations from the average. Taking the globe as a whole, there are dry periods such as 1940–4, 1965–8 and 1977–9, when runoff was 1600–2900 km³ below average, as well as wet periods, for example 1926–7, 1949–52 and 1973–5, where there was a higher runoff. Apart from the cyclic variations in runoff, there is no trend in the record for the 65-year period. This is characteristic of all the continents; the increased runoff in South America during the last

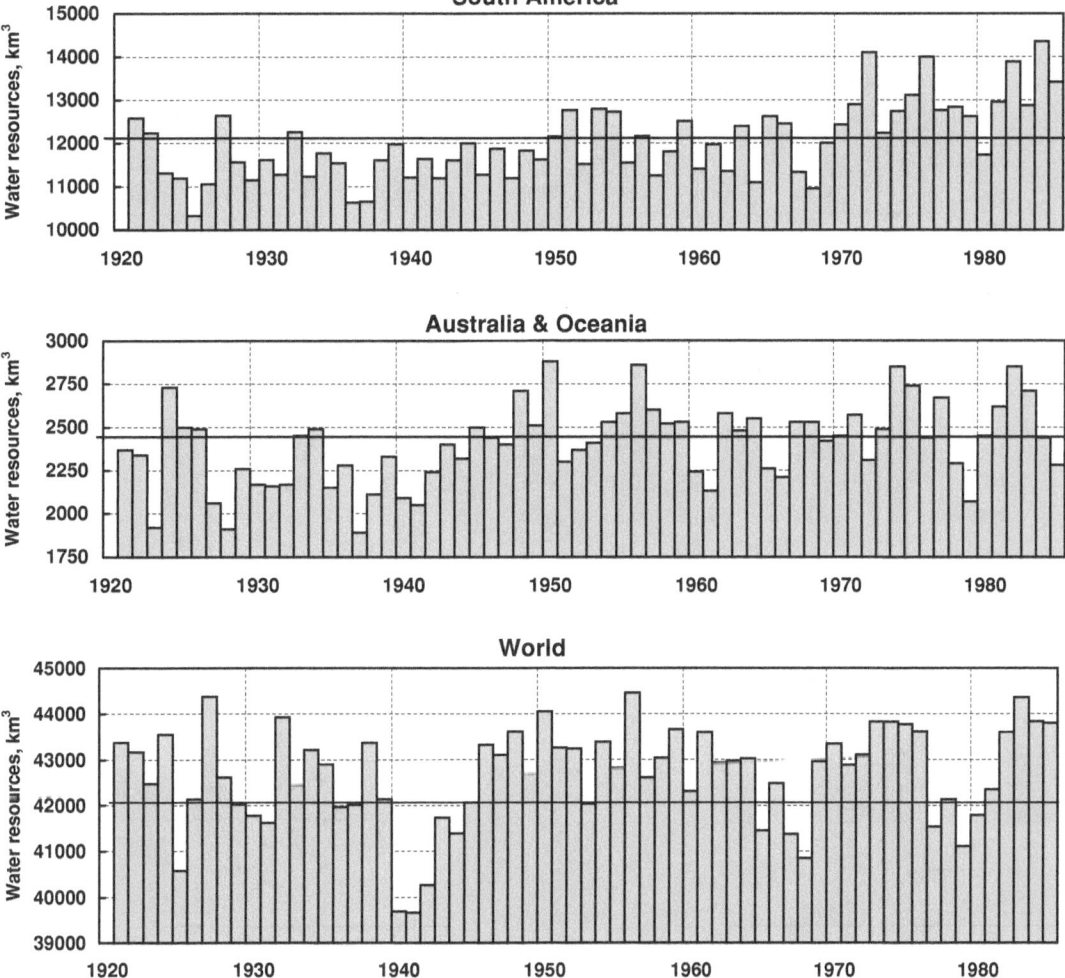

Fig. 10.3 (*cont.*).

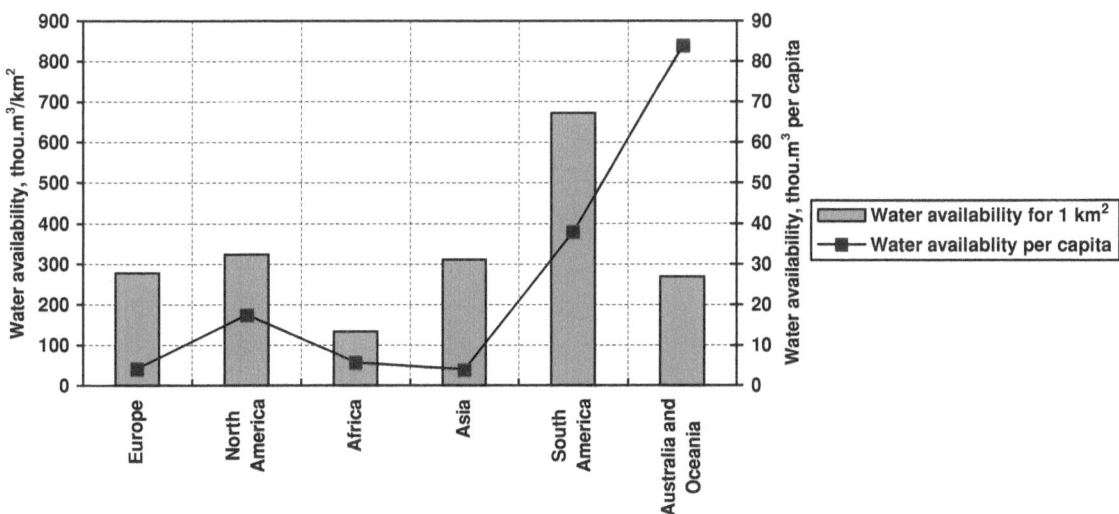

Fig. 10.4 Potential water availability by continents.

Table 10.4. *Streamflow distribution during a year for the continents by the data of SHI (Korzun, 1974b; Shiklomanov, 1997)*

Continent	Year of data	Mean annual water resources (local), km³/year	Months, as percentage of the mean annual value											
			1	2	3	4	5	6	7	8	9	10	11	12
Europe	1974		6.4	6.7	7.6	10.1	16.7	13.3	7.9	5.9	5.6	6.7	6.6	6.5
	1997	2 900	6.2	6.6	6.9	8.9	14.3	13.3	9.2	7.6	7.3	6.9	6.6	6.2
North America	1974		2.4	1.9	2.4	2.9	7.9	14.3	14.8	16.8	15.4	10.4	5.4	5.4
	1997	7 870	4.7	4.9	5.0	7.0	11.6	15.2	12.6	9.9	9.6	8.6	5.9	5.0
Africa	1974		14.9	11.5	14.6	12.6	12.3	8.1	5.1	4.1	4.1	4.3	4.5	3.9
	1997	4 047	8.4	7.5	7.0	7.1	7.5	6.6	6.1	6.1	8.0	10.6	12.7	12.4
Asia	1974		5.0	4.0	5.0	6.0	10.0	14.0	17.0	13.0	8.0	7.0	6.0	5.0
	1997	13 510	5.1	4.1	4.7	5.1	8.8	13.7	14.9	13.8	11.2	7.2	6.8	4.6
South America	1974		5.9	7.0	8.0	10.0	11.4	12.0	11.0	9.7	7.6	6.0	5.7	5.7
	1997	12 030	5.9	7.0	8.1	10	11.4	12.1	11.1	9.7	7.6	6.0	5.5	5.6
Australia and Oceania	1974		15.5	21.7	31.1	5.2	3.4	3.5	3.2	3.2	3.2	3.6	3.2	3.2
	1997	2 400	10.3	13.2	12.4	10.1	7.4	7.1	6.2	6.9	5.4	6.6	7.2	7.2
World	1974		5.9	5.6	6.8	7.3	10.3	12.4	12.1	11.4	9.5	7.5	5.7	5.5
	1997	42 757	5.9	6.1	6.5	7.6	10.2	12.5	11.7	10.4	9.0	7.4	6.8	5.9

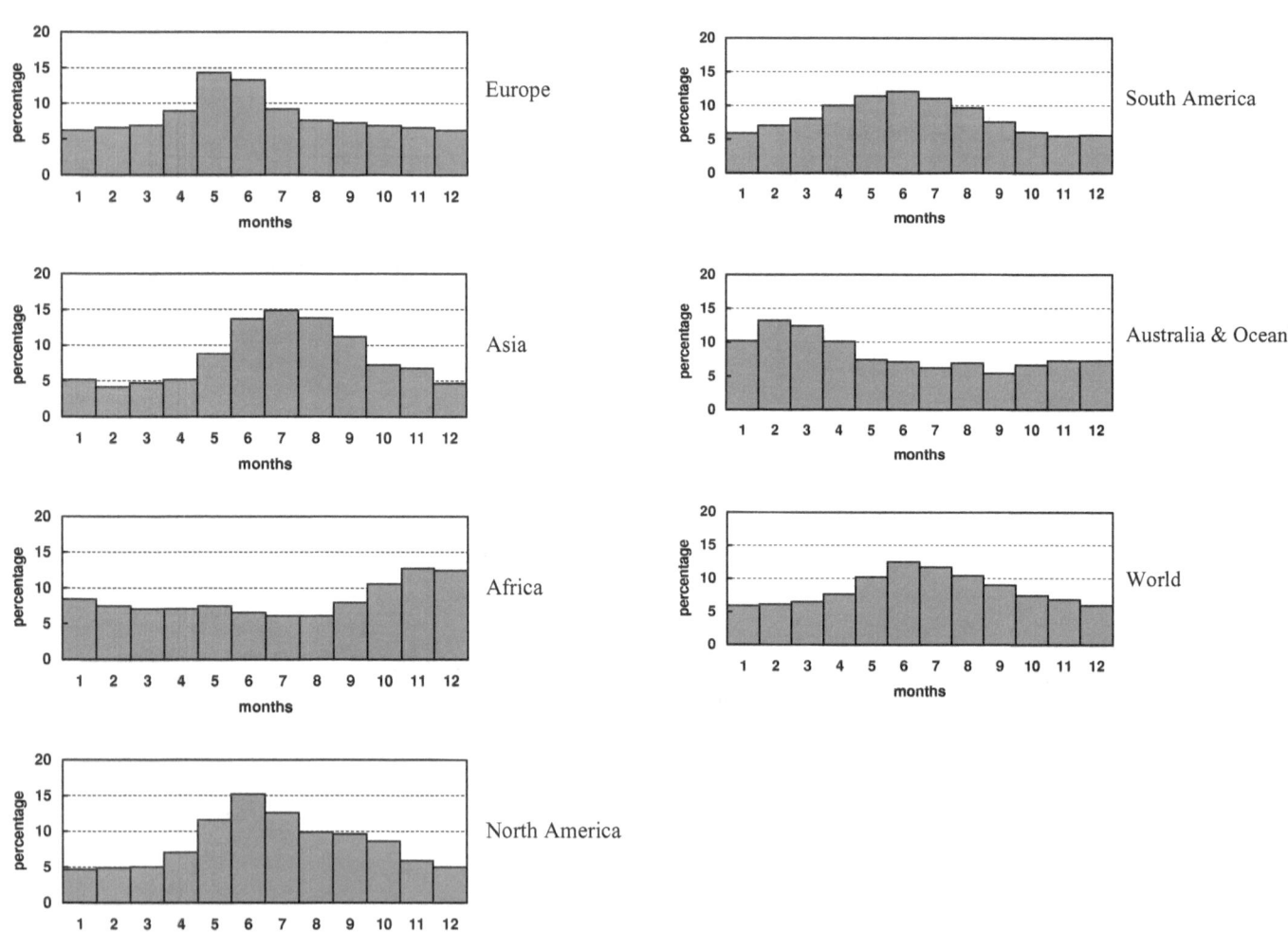

Fig. 10.5 Streamflow distribution by continents during an average year.

Table 10.5. *Renewable water resources and potential water availability of natural–economic regions of the world*

Region	Area, km² × 10⁶	Population (1994), ×10⁶	Water resources, km³/year				Coefficient of variation, C_v	Potential water availability, (m³ × 10³/year)	
			Inflow	Local				Per km²	Per head
				Average	Minimum	Maximum			
Europe	10.46	684.7		2 900	2 440	3 210	0.10	277	4.24
Northern	1.32	23.2		705	585	828	0.10	534	30.4
Central	1.86	293.0	6.0	617	353	836	0.21	332	2.12
Southern	1.79	188.0	109	546	377	838	0.18	305	3.19
Northern part of the European territory of the former Soviet Union	2.71	28.5	27	589	434	775	0.12	217	21.1
Southern part of the European territory of former Soviet Union	2.78	152.0	123	443	266	756	0.17	159	3.32
North America	24.3	453.0		7 870	6 660	8 820	0.1	324	17.4
Canada and Alaska	13.67	29.0	130.0	4 980	4 360	5 830	0.10	364	174
USA	7.84	261.0	70.0	1 800	950	2 480	0.17	230	7.03
Central America and Caribbean	2.74	163.0	2.5	1 090	530	2 000	0.20	398	6.69
Africa	30.1	708.0		4 047	3 073	5 082	0.1	134	5.72
North	8.78	157.0	140	41	19.0	96.0	0.34	4.67	0.71
Southern	5.11	83.5	86.0	399	270	549	0.14	78.1	5.29
Eastern	5.17	193.5	26.0	749	504	940	0.11	145	3.94
Western	6.96	211.3	30.0	1 088	581	1 948	0.28	156	5.22
Sahel	5.30	46.9	77.4	104	52.3	175	0.29	19.6	3.04
Central	4.08	62.8	80.0	1 770	1 453	2 263	0.09	434	28.8
Asia	43.5	3445		13 510	11 800	15 000	0.06	311	3.92
North China and Mongolia	8.29	482		1 029	590	1 735	0.23	124	2.13
Southern	4.49	1214	300	1 988	1 535	2 458	0.10	443	1.77
Western	6.82	232		490	227	931	0.35	71.8	2.11
South East	6.95	1404	120	6 646	5 342	7 607	0.09	956	4.77
Middle Asia	3.99	54	46.0	181	121	265	0.17	45.4	3.78
Siberia and Far East of Russia	12.76	42	218	3 107	2 628	3 500	0.06	243	76.6
Transcaucasia	0.19	16	12.1	68	51.5	88.8	0.12	358	4.63
South America	17.9	314.5		12 030	10 330	14 350	0.07	672	38.3
Northern	2.55	57.3		3 340	2 390	4 670	0.15	1310	58.3
Eastern	8.51	159.1	1900	6 220	5 200	7 640	0.08	731	45.1
Western	2.33	48.6		1 720	992	2 380	0.18	738	35.4
Central	4.46	49.4	720	750	531	1 310	0.17	168	22.5
Australia and Oceania	8.95	28.7		2 400	1 890	2 880	0.1	268	83.6
Australia	7.68	17.9		352	228	701	0.24	45.8	19.7
Oceania	1.27	10.8		2 050	1 515	2 570	0.10	1614	190
The world	135	5633		42 757	44 460	39 660	0.02	317	7.60

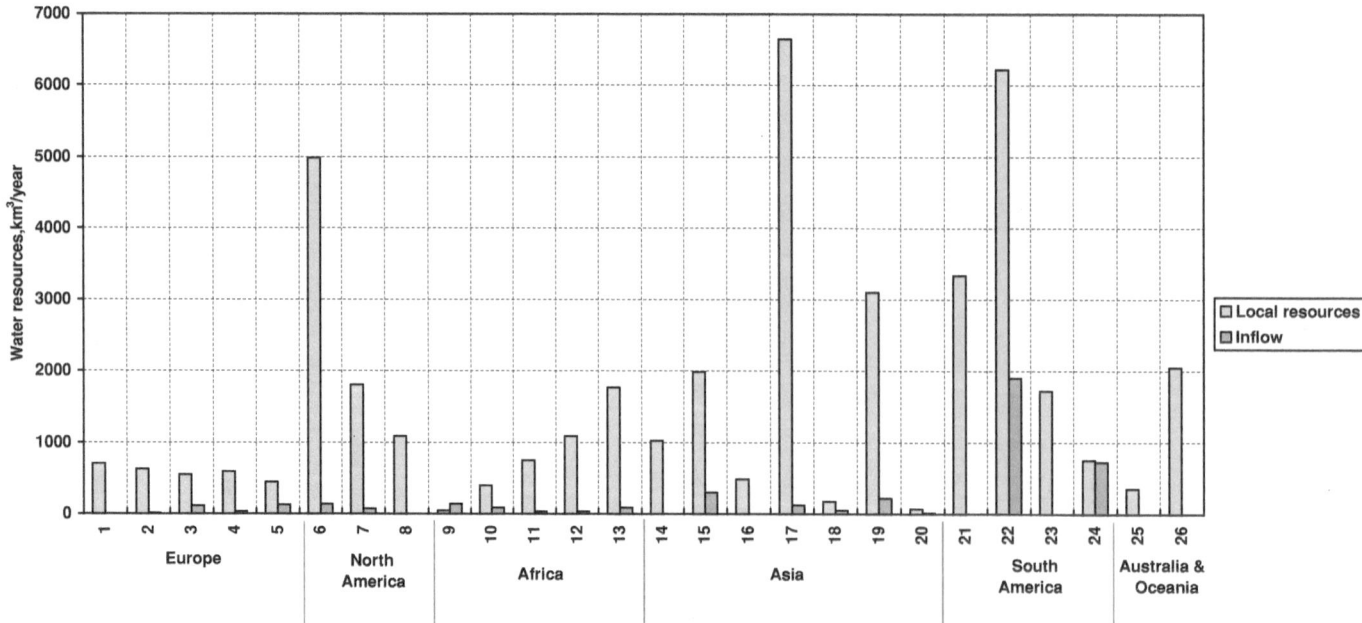

Fig. 10.6 Renewable water resources of physiographic and economic regions. Europe: 1, North; 2, Central; 3, South; 4, Northern part of European territory of the former Soviet Union (ETSU); 5, Southern part of ETSU. North America: 6, North; 7, Central; 8, South. Africa: 9, North; 10, Southern; 11, Eastern; 12, Western; 13, Central. Asia: 14, North China and Mongolia; 15, South Asia; 16, West Asia; 17, South East Asia; 18, Middle Asia; 19, Siberia and Far East of Russia; 20, Transcaucasia. South America: 21, North; 22, East; 23, West; 24, Central. Australia and Oceania: 25, Australia; 26, Oceania.

two decades has been compensated for by the decrease in African runoff over the same period (Fig. 10.3).

Lack of an overall trend raises doubts about the conclusion by the Intergovernmental Panel on Climate Change (IPCC) that one of the major causes of the rise in the level of the World Ocean observed since the beginning of the twentieth century is the rapid melting of mountain glaciers (IPCC, 1990, 1995, 2001a). The values of annual runoff per head and per km^2 are shown in Table 10.3 and Fig. 10.4 but these are not reliable indicators of the availability of water for the whole or part of a continent. In most regions runoff is very unevenly distributed throughout the year. For example some 60–70% of the runoff occurs during the spring and early summer in many regions. The consequence is that fresh water resources vary considerably throughout a year. Variations in runoff by months (as a percentage of the annual flow) are shown in Table 10.4 and Fig. 10.5. Comparison between these figures and those of Korzun (1974b) are reasonably close except for Africa where new data has given different results.

Much of the runoff in Europe occurs between April and July (46%), in Asia from June to September (54%), in Africa between

September and December (44%), North America from May to August (49%), South America from April to July (45%), and in Australia and Oceania between January and April (46%). Taking the globe's land surface as a whole, about 45% of the total runoff occurs between May and August. The irregular pattern of runoff leads to the need for reservoirs to provide a stable supply of water; some 37% of the total global runoff, on average (Korzun, 1974b), or approximately 16 000 km^3/year is in this category. Table 10.5 and Fig. 10.6 give average values of renewable water resources and specific water availability by regions for the period 1921 to 1985. For South Europe, the southern part of the European territory of the former Soviet Union (ETSU), South Africa, and Middle Asia, inflow is between reaches 20–25% of the water resources

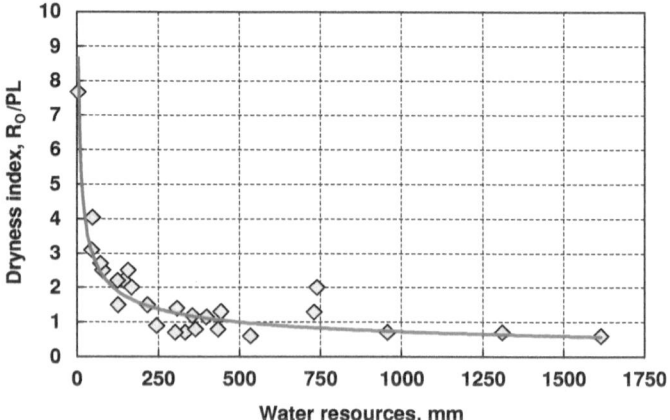

Fig. 10.7 The relationship between water resources (mm) and dryness index.

Table 10.6. *Renewable water resources and potential water availability of selected countries of the world*

Country	Area, km² × 10⁶	Population, ×10⁶	Inflow	Water resources, km³/year Local Average	Local Maximum	Local Minimum	Coefficient of variation, C_v	Potential water availability, m³ × 10³/year Per km²	Per capita
Australia	7.68	17.9	0	352	701	228	0.24	45.8	19.7
Albania	0.03	3.60	5.20	18.6	42.9	13.1	0.21	620	5.89
Algeria	2.38	27.3	0.4	13.9[a]				5.84	0.52
Argentina	2.78	34.2	623	270	610	150	0.27	97.1	17.0
Armenia	0.03	3.55	2.1	6.2	8.5	4.8	0.12	207	2.04
Azerbaijan	0.09	7.47	20.2	7.78	12.8	5.1	0.15	86.4	2.39
Belarus	0.21	10.3	21.7	34.4	59	20.4	0.22	164	4.39
Bolivia	1.10	7.2	155	361	487	279	0.13	328	60.9
Brazil	8.51	159	1900	6220	7640	5200	0.08	731	45.1
Burkina Faso	0.27	10.0	2.0	14.7				54.4	1.57
Canada	9.98	28.0	130	3290	3760	2910	0.10	330	120
Chad	1.28	6.18	28.3	15.8				12.3	4.85
Chile	0.76	14.0	0	354				466	25.3
China	9.60	1209	0	2700	3930	1970	0.12	281	2.23
Colombia	1.14	34.3	0	1200				1053	35.0
Cuba	0.11	11.1	0	34.5				314	3.11
Dominican Republic	0.05	7.2	0	20.0				400	2.78
Ecuador	0.28	11.2	0	265				946	23.7
Egypt	1.00	61.6	65.5	0.5				0.50	0.54
El Salvador	0.02	5.2	0	19.0				950	3.65
France	0.55	56.8	27.0	168	263	90.3	0.22	305	3.20
Gambia	0.01	1.08	4.70	3.20				320	5.14
Georgia	0.07	5.45	6.9	53.3	67.7	40.8	0.13	761	10.4
Guyana	0.22	0.8	0	270				1227	338
Guatemala	0.11	10.6	0	116				1055	10.9
Haiti	0.03	6.5	0	11.0				367	1.69
Honduras	0.11	5.49	0	95				864	17.3
India	3.27	919	581	1456	1794	1065	0.11	445	1.90
Italy	0.30	57.7	0	185				617	3.21
Jordan	0.1	5.20	0	0.96[a]				9.60	0.18
Jamaica	0.01	2.43	0	8.3				830	3.42
Kazakhstan	2.72	16.7	56	70.2	111	39.3	0.24	25.8	5.88
Kyrgyzstan	0.20	4.67	0	48.7	70.1	37.3	0.19	244	10.4
Lebanon	0.01	3.06	0	2.8[a]				280	0.92
Libya	1.76	5.2	0	5.29[a]				3.01	1.02
Madagascar	0.59	14.3	0	332				563	23.2
Mali	1.24	10.5	44.4	50.0				40.3	6.88
Mauritania	1.03	2.2	11.0	0.1				0.10	2.55
Mexico	1.97	94.8	2.5	346	645	229	0.18	176	3.66
Morocco	0.45	26.5	0	30.0[a]				66.7	1.13
Nicaragua	0.13	4.5	0	175				1346	38.9
Niger	1.27	8.85	30.4	3.0				2.36	2.06
Nigeria	0.92	108	43.6	274	437	148	0.26	298	2.74
New Zealand	0.27	3.50	0	313	405	246	0.11	1159	89.4
Pakistan	0.81	137	86	40	63.9	22.5	0.21	49.4	0.61
Panama	0.08	2.6	0	144				1800	55.4
Peru	1.28	23.3	144	1100				859	50.3

Table 10.6. (*cont.*)

| Country | Area, km$^2 \times 10^6$ | Population, $\times 10^6$ | Water resources, km^3/year | | | | Coefficient of variation, C_v | Potential water availability, m$^3 \times 10^3$/year | |
| | | | Inflow | Local | | | | Per km^2 | Per capita |
				Average	Maximum	Minimum			
Poland	0.31	39.2	6.4	49.5				160	1.34
Portugal	0.09	9.93	34.5	18.5	157	15.2	0.55	206	3.60
Russia	17.08	148	227	4059	4541	3533		238	28.2
Senegal	0.20	8.1	17.4	17.4				87.0	3.22
South Africa	1.22	40.6	5.2	47.4				38.9	1.23
Spain	0.51	39.6	0	108	253	27.2	0.47	212	2.73
Sudan	2.51	27.4	140	22.0				8.76	3.36
Surinam	0.16	0.4	0	230				1438	575
Sweden	0.45	8.34	12.2	164				364	20.4
Tajikistan	0.15	5.93	47.9	47.4	65.8	36.4	0.13	316	12.0
Thailand	0.51	58.2	0	199				390	3.42
Trinidad and Tobago	0.005	1.2	0	5.1				1020	4.25
Tunisia	0.16	8.73	0.42	3.52[a]				22.0	0.43
Turkmenistan	0.49	4.01	69.8	1.13	1.14	0.92	0.22	2.31	8.99
Uruguay	0.18	3.2	74.0	68.0				378	32.8
USA	9.36	262	146	2930	3680	1960	0.11	313	11.5
Uzbekistan	0.45	20.3	98.1	9.52	19.7	4.98	0.27	21.2	2.89
Zaire	2.34	42.6	313	987	1328	786	0.10	422	26.8

[a] Water resources include renewable ground water resources (FAO, 1995a).

originating in this region, which in North Africa and the Sahel the inflow is comparable with the local water resources, or several times greater.

Of course the magnitude of a regions renewable water resources is mainly determined by climate. Figure 10.7 shows a relationship between water resources in millimetres and a dryness index $R_0/(PL)$, where R_0 is the radiation balance of a wet surface, P is the precipitation, and L is the latent heat of evaporation. The value of the dryness index was determined for every region by using the most detailed global maps of radiation balance and precipitation (Budyko, 1956; Korzun, 1974b). The areas of greatest dryness correspond to those with the smallest water resources and vice versa. Of course, there are other factors to be taken into account such as topography, vegetation and soil. The year-to-year changes and trends in water resources are presented in Chapters 4 to 9. The variability from one year to the next can be quite significant, and these variations are often masked by the averaged data. This pertains especially to the arid and semi-arid regions, where the coefficients of variation (C_v) of annual discharges are 0.20–0.35 and the values of renewable water resources for particular years can be 1.5 to 2 times less than the means (Table 10.5). For wet regions the coefficients of variation are smaller ($C_v = 0.05$–0.15), and the difference between annual and long-term means of water resources are reduced to the range from 15% to 25%.

The use of water resources is determined by not only the year-to-year variability in the resource but also by the variation during the year. The distribution of the runoff within the year is shown in the different chapters. Most parts of the world experience runoff regimes which vary considerably through the year. For example, between 50% and 70% of the annual runoff may occur during the flood season, such as in the ETSU where 54% of the runoff takes place in the three flood months. This compares with 64–68% in Siberia and the Far East, Southern Asia, Western Africa and Australia. By way of contrast, many areas recorded as little as 4–10% of the total runoff in the three to four months of the dry season. For instance, for the three low-flow months in the north of the ETSU, Canada and Alaska, and Northern China and Mongolia, runoff is 8–9% of the annual, in Central America it is 6–7%, in Siberia and the Far East of Russia and Southern Asia it is only 4–5%.

For practical purposes it is important to make reliable estimates of renewable water resources but such estimates have only been made for this Monograph for 60 countries. Some of these are developing countries and others are developed countries. Some are large in area and population, while others are small. There are

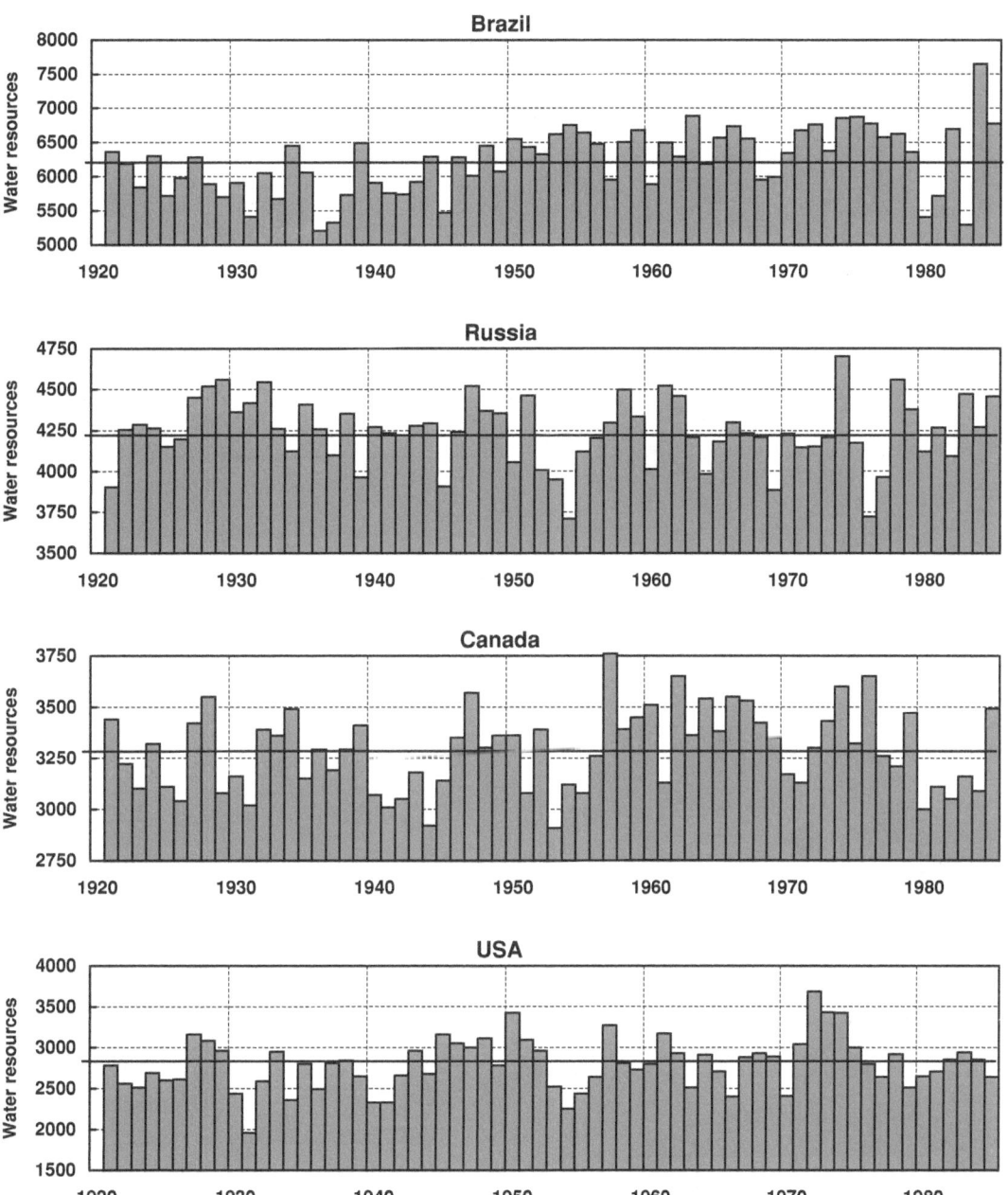

Fig. 10.8 Year-on-year variations in renewable water resources (km³/year) of selected countries of the world.

countries in the Northern and Southern Hemispheres, some with copious water resources, others with a shortage. Some 71% of the world's renewable water resources originate in them and they are home to 70% of the population of the world. Table 10.6 presents details of the resources of these countries including the maximum and minimum values for the study period. Values of the specific availability of water range from 100–5000 to 1 000 000–1 800 000 m³/year per km², and from 180–1 500 to 100 000–500 000 m³/year per head. Some of these countries depend to a large extent on the

flow of river water from upstream neighbours. These flows are often as great or greater than resources originating locally; for instance, in Argentina, Mauritania, Niger, Sudan and Uzbekistan, the inflow is 2 to 10 times greater than the local water resources. If such inflows are not taken into account water resources may be underestimated substantially.

Table 10.6 shows that six of the largest countries by area, namely Brazil, Russia, Canada, USA, China and India, have the largest water resources with more than 40% of the world's total annual runoff (Fig. 10.8). Alternating wet and dry periods of different duration characterize the records for all countries. For Brazil and

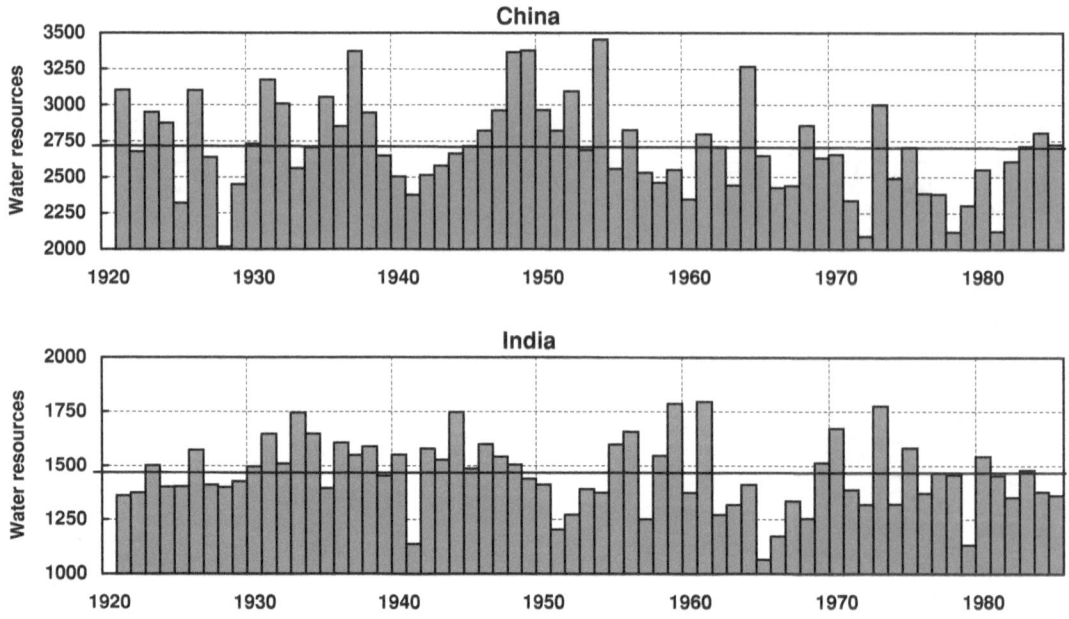

Fig. 10.8 (*cont.*).

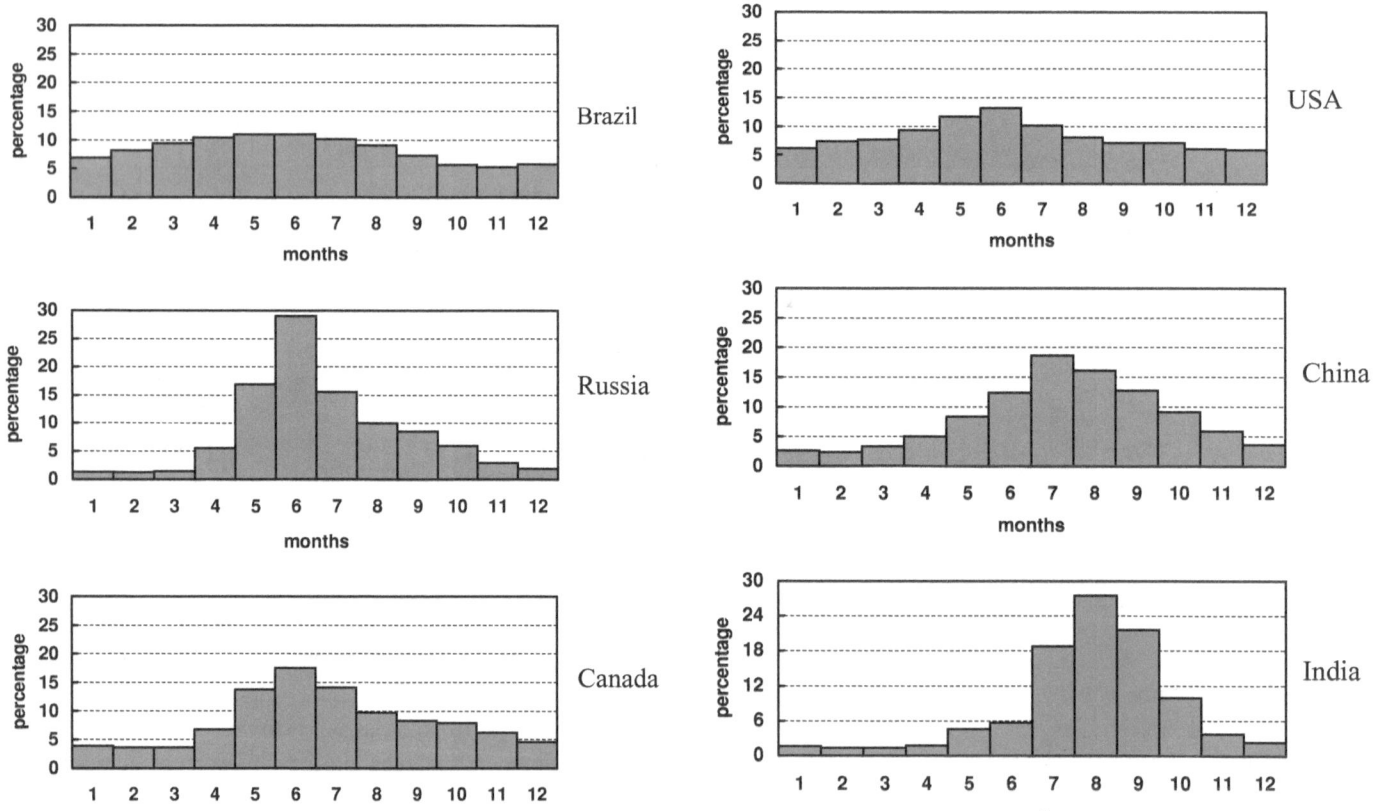

Fig. 10.9 Streamflow distribution during a year as percentage of annual runoff by selected countries.

Table 10.7. *Renewable water resources of selected rivers of the world*

River	Area, km$^2 \times 10^6$	Population, $\times 10^6$	Water resources, km^3/year			Coefficient of variation, C_v	Potential water availability, 1000 m^3/year	
			Average	Maximum	Minimum		Per km^2	Per head
Amazon	6.92	14.3	6920	8510	5790	0.08	1000	484
Ganges^a	1.75	439	1389	1690	1220	0.04	794	3.16
Congo	3.50	48.3	1300	1775	1050	0.10	371	26.9
Orinoco	1.00	22.4	1010	1380	710	0.15	1010	45.1
Yangtze	1.81	346	1003	1410	610	0.15	554	2.90
La Plata	3.10	98.4	811	1860	450	0.26	262	8.24
Yenisei	2.58	4.8	642	749	466	0.08	249	134
Lena	2.49	1.9	539	670	424	0.11	216	284
Mississippi	2.98	72.5	515	881	281	0.24	173	7.10
Mekong	0.79	100	505	610	376	0.16	639	5.05
Ob	2.99	22.5	404	586	270	0.16	135	18.0
Amur	1.86	4.5	355	538	225	0.21	191	78.9
Mackenzie	1.78	0.35	325	427	284	0.12	183	929
St Lawrence	1.03		320	405	242	0.10	312	
Niger	2.09	131	303	482	163	0.26	145	2.31
Volga	1.38	43.3	250	390	161	0.19	181	5.77
Columbia	0.67		237	331	144	0.18	355	
Danube	0.82	85.1	225	321	137	0.18	274	2.64
Indus	0.96	150	220	359	126	0.19	229	1.47
Yukon	0.85		196	335	122	0.26	231	
Aldan	0.73		168	212	117	0.16	230	
Nile	2.87	89.0	161	248	94.8	0.16	56.1	1.81
Pechora			136	124	115	0.12		
Kolyma	0.65		128	203	74.4	0.23	197	
Fraser	0.23		115	155	82	0.13	494	
North Dvina			105	92.4	81.8	0.17		
Khatanga	0.36		90.5	172	51.5	0.25	251	
Rhine			86.1	72.6	61.2	0.22		
Neva			75.7	66.5	58.6	0.16		
Amu-Darya	0.31	15.5	69.5	118	56.7	0.14	224	4.48
Huang He	0.75	82.0	66.1	97	22.1	0.38	88.1	0.81
Rhône			64.9	55.4	47.2	0.21		
Indigirka	0.36		54.8	89	39.4	0.21	152	
Dnieper	0.50	36.6	53.3	95	21.7	0.25	107	1.46
Po			50.3	42.5	36.3	0.26		
Ebro			38.9	23.5	12.6	0.49		
Syr-Darya	0.22	13.4	37.5	75	26.2	0.21	170	2.80
Yana	0.24		29.3	74	19.4	0.20	122	
Don	0.42	17.5	26.9	53	11.9	0.36	64.0	1.54
Murray	1.07	2.1	24	129	1.16	0.75	22.4	11.4
Colorado (middle reaches)	0.64		16	32	4	0.38	25.1	

^{a} Including the Brahmaputra and Meghna Rivers.

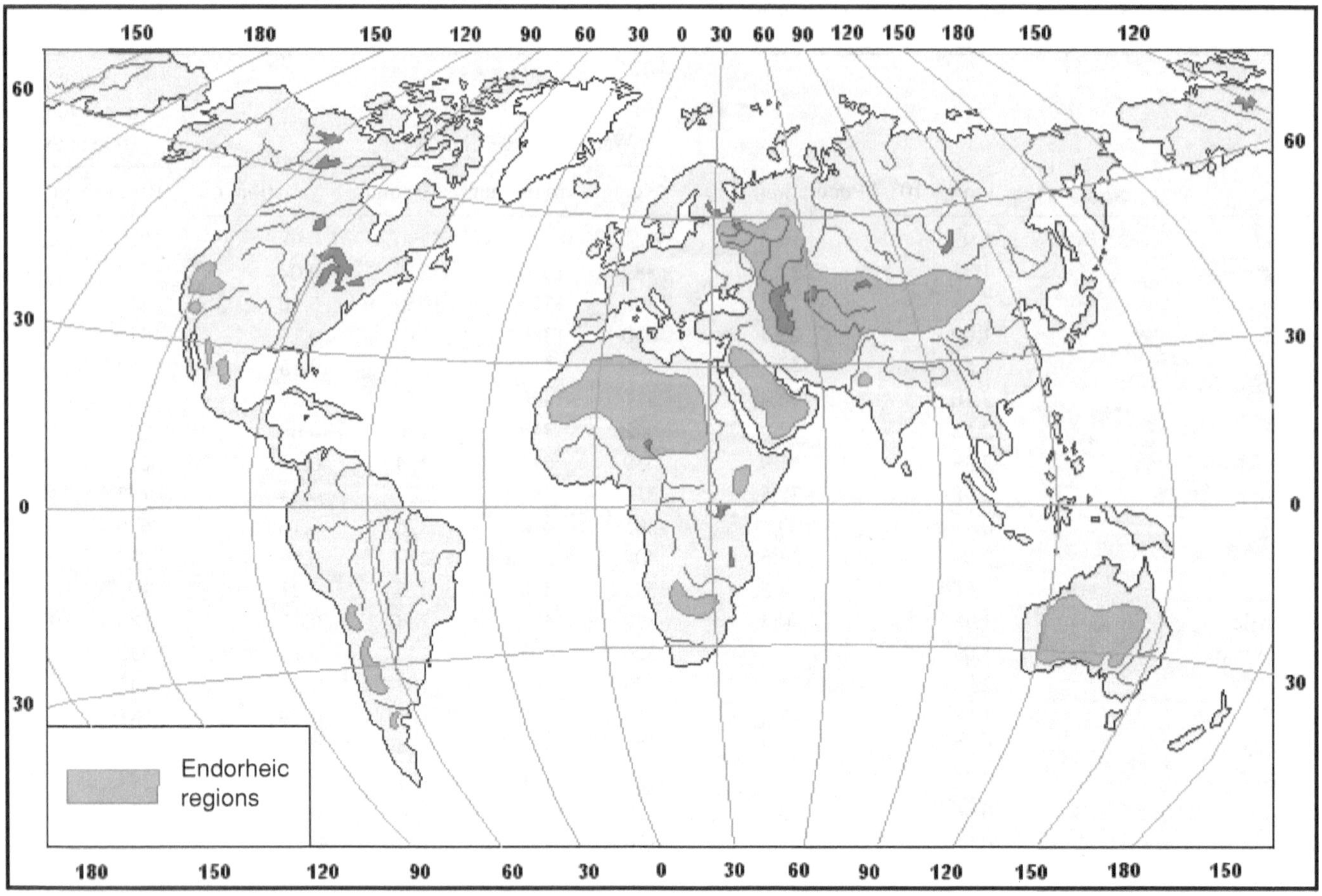

Fig. 10.10 The endorheic regions of the world.

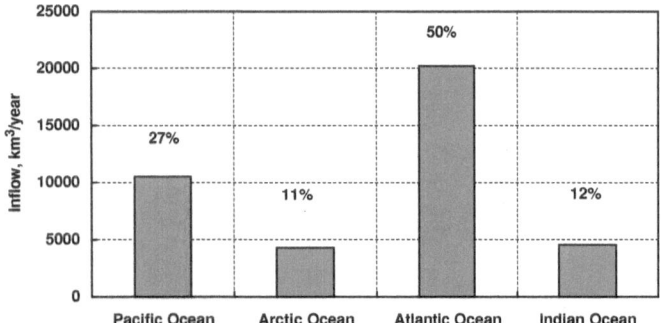

Fig. 10.11 River water inflow (km³/year) to the World Ocean.

Canada there is an insignificant upward trend, while there was a downward trend for India and China. Figure 10.9 presents the monthly distribution of runoff, showing that the variations are evened out over the course of the year between the wet and dry periods. For example, in Russia and Canada, 55–70% of the runoff occurs between May and August, while 47–65% of the runoff in India and China takes place between July and September. These values for the renewable water resources of selected countries are significantly different from assessments published elsewhere, for example, Korzun (1974b) and Gleick (1993, 1998). For a number of countries, the difference is more than 10%, and for others it is far greater.

10.3 FLOWS TO THE WORLD OCEAN

The basis for the assessment of the renewable water resources whether it is by country, region or continent is the runoff data collected from the hydrological network and processed by the hydrological service. Information about the flow of the principal rivers of the world is given in Table 10.7. The largest river of the world, the Amazon, provides 16% of the global total annual runoff, while the five largest river systems (Amazon, Ganges with the Brahmaputra, Congo, Yangtze and Orinoco) together account for 27% (Table 10.7). The data contained in Table 10.7 represent more than 60% of the world's water resources. Figures for the river basins, which are also shown, are for period 1921 to 1985; but longer periods of observation have been used for certain rivers. These data from the global hydrological network also appear in the appropriate chapters of this Monograph. Of course knowledge of the flow to the ocean is also important for oceanography.

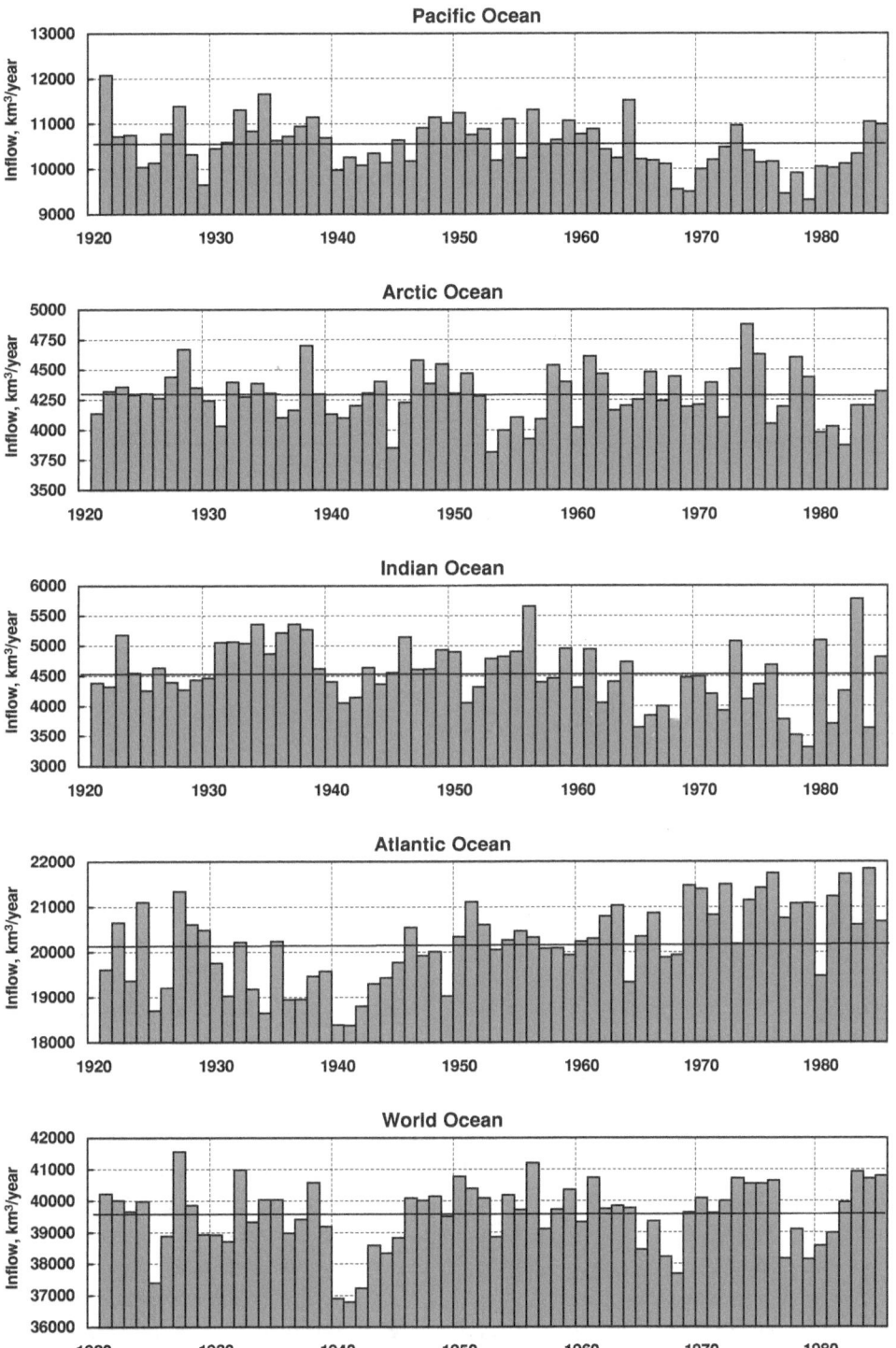

Fig. 10.12 Year-on-year variation in river water inflow (km³/year) to the World Ocean.

The inflow to the World Ocean is not precisely the total global water resource, for two reasons. First, there are a number of areas of inland drainage which are not connected to the World Ocean. They are shown in Table 10.8 and Fig. 10.10 and they cover about 30 million km² (20% of the total land area); however they produce only 2.3% (about 1000 km³/year) of the runoff. The largest of these areas are the basin of the Caspian Sea, most of Middle and Central Asia, the Arabian Peninsula, much of North Africa, and

Table 10.8. *Major characteristics of the endorheic regions of the world*

Continent	Area, km² × 10⁶	Water resources, km³/year
Europe	2.16	311
Asia	12.3	415
Africa	9.6	150
North America	0.88	15
South America	1.41	56
Australia and Oceania	3.92	9.4
World	30.27	956.4

Table 10.9. *Water inflow to the World Ocean*

Ocean	Volume, km³ × 10⁶	Inflow, km³/year			Coefficient of variation, C_v
		Average	Maximum	Minimum	
Atlantic	329.7	20 190	21 840	18 370	0.04
Pacific	710.4	10 530	12 080	9 310	0.05
Indian	282.6	4 530	5 780	3 310	0.11
Arctic	18.1	4 280	4 880	3 820	0.05
Total	1340.7	39 530	41 560	36 780	0.03

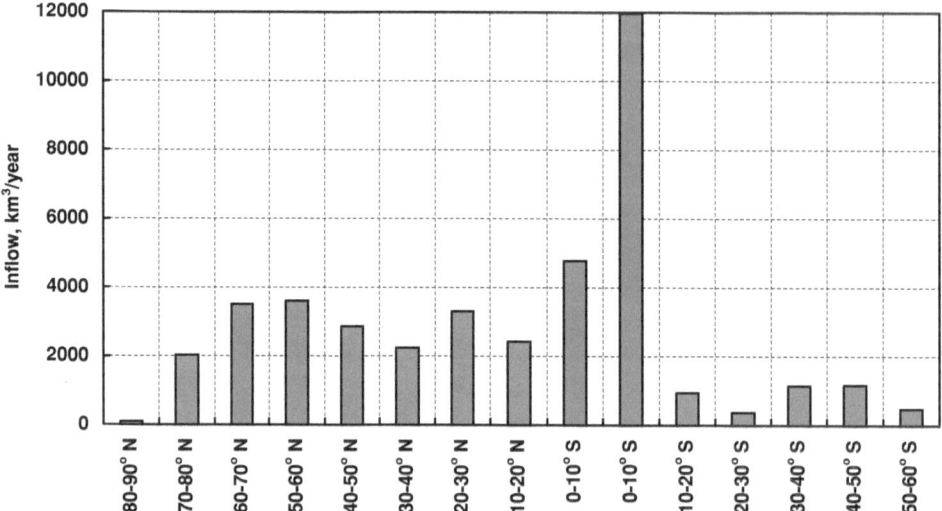

Fig. 10.13 River water inflow to the World Ocean by latitudinal zones.

central Australia. Second, there is a large amount of water lost by evaporation from the rivers that cross arid and semi-arid areas on their way to the sea, for example the Indus, the Niger, the Nile and the Colorado. About 1100 km³ of runoff per year is lost in this way, which includes 380 km³/year in Asia, 300 km³/year in Africa and 340 km³/year in North America. Thus, the total inflow to the World Ocean (Fig. 10.11, Table 10.9) is less than the world total water resource. Figure 10.12 shows the year-by-year changes in this figure (period 1921 to 1985). Approximately half the inflow reaches the Atlantic Ocean, a large volume through four of the six principal world rivers (the Amazon, Congo, Orinoco and Mississippi). The least (4300 km³) enters the Arctic Ocean; however, this fresh water is of great importance for the water balance of this ocean simply because it contains only 1.2% of the

water in the World Ocean but receives about 11% of global river inflow.

The analysis of the Fig. 10.12 data shows that the total inflow to the World Ocean is rather stable and does not contain any explicit trend for the period considered. At the same time, there is a downward trend in the flow to the Indian and Pacific Oceans, and an upward trend to the Atlantic Ocean. To model the dynamic processes in the oceans, it is important to take into account not only the total volume of inflow but also its distribution. This distribution is shown in Fig. 10.13 which demonstrates the uneven pattern of flow into the World Ocean. About 40% on average of the total runoff enters the ocean between 10° N and 10° S, but this figure alters from year to year and season to season.

11 World water use and water availability

11.1 ANALYSIS OF THE AVAILABLE ESTIMATES AND PREDICTIONS

Over the last 30 years there have been numerous attempts to assess the current global use of water by the different sectors and to make forecasts of the future demand. However, most countries lack measurements of water use, while the reliability of forecasts of future water needs is questionable. In view of the difficulties in collecting and analysing the available data world-wide, most studies tend to rely on indirect assessment techniques for evaluating water use. These techniques are usually based on the analysis of the salient factors which determine the use of water, such as the total population both urban and rural, the growth of industrial production, trends in the availability of water, the development of agriculture and the growth of irrigation in developed and developing countries. In this context it is noteworthy that reliable data on water use are available for only a few countries, sectors and for particular industries – a point that has already been mentioned in this Monograph. Comparison of the different assessments of water use is rather difficult. Some authors publish data on the world total water use; others give data for the continents and for the major uses. The time-scales also differ; some studies are of current use and of forecast use, others indicate only the present water use or give outlooks for the future. Most of the research does not differentiate between use and consumption of water and the losses of water due to evaporation from reservoirs is not taken into account. Then there is the fact that different initial data are employed together with different preconditions for the evaluation of past and future water use. The time periods studied are usually different as are the methods used. This complexity gives rise to a number of difficulties in making comparisons of the different studies.

Figures 11.1 and 11.2 present estimates of world water abstractions (withdrawal) and water consumption, collected from various studies carried out over the last 30 years. Doxiadis (1967) made one of the first attempts to estimate world water use and to forecast future demand. He gave estimates for 1960 and predictions for 1975, 2000, 2030 and 2060. The 1960 use figure was given as 1326 km^3/year, increasing to 2412 km^3/year by 1975 and to 6500 km^3/year for 2000. Assuming asymptotical growth he forecast the total would reach 23 000 km^3/year in 2090. He divided abstractions between those for irrigation, for industry and for domestic purposes (urban and rural) as 83.0%, 7.6%, 7.5% and 1.9% for 1975 and 70.8%, 14.6%, 13.1% and 1.5% for 2000. Doxiadis did not explain the methods he used nor other details, such as whether reservoir evaporation was included. However his figures for abstractions and use of water for irrigation agree fairly well with the more detailed estimates obtained later (Fig. 11.1).

Kalinin (1968) published forecasts for world water use for the year 2000, based on data and various parameters which were employed later by other researchers. They were:

- figures for the growth of population and forecasts of future growth
- forecasts of the changes in the ratio of the urban to the rural population
- data on the growth of industrial production and forecasts for its future changes
- trends in the area irrigated
- standard figures for the abstraction of water for the USSR and USA, the countries where the most detailed studies were made.

According to Kalinin the total world water abstraction would reach 9700 km^3/year by 2000 (without accounting for the water used for the dilution of discharges of waste water); the value includes the figure of 5500 km^3/year for the consumption of water. Irrigation would require 72% (7000 km^3/year) of the total abstracted and would make up 87% (4800 km^3/year) of the water consumed. It is obvious that Kalinin's estimates are too high, mainly due to very high rates being assumed for the expansion of irrigation – a tripling by 2000 to 630 million ha. His forecasts for industry were more successful, being closer to the real volumes. However, those forecasts fail to use information on the current patterns of water use by the different sectors and this casts doubts on the reliability of the forecasts.

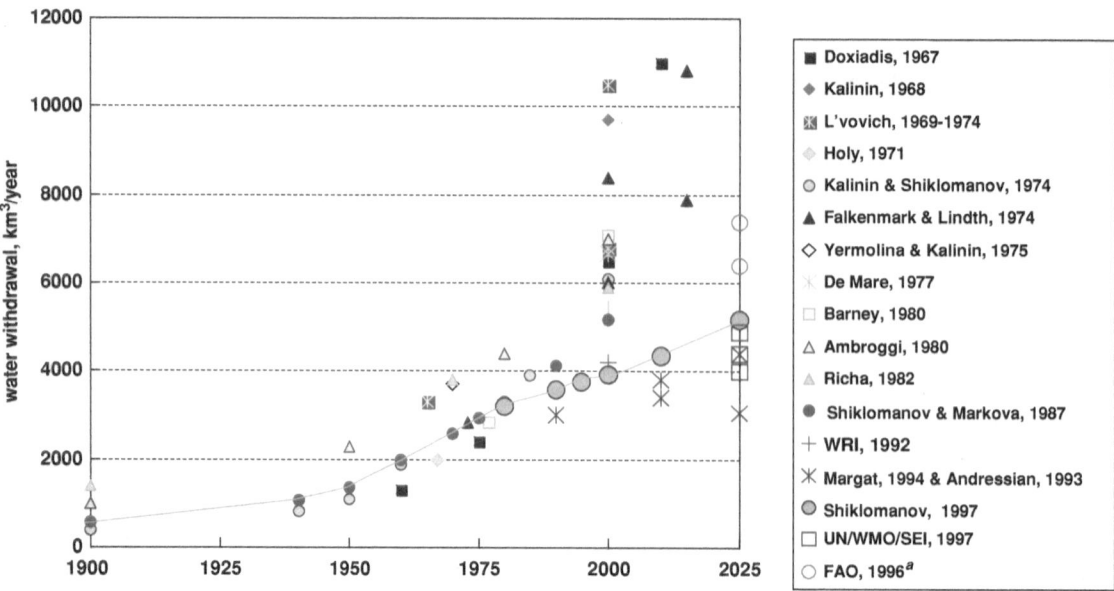

Fig. 11.1 Comparison of the values of world water withdrawal by the data of different authors. [a] Data given by Dr. J. B. Miller (WMO) by correspondence.

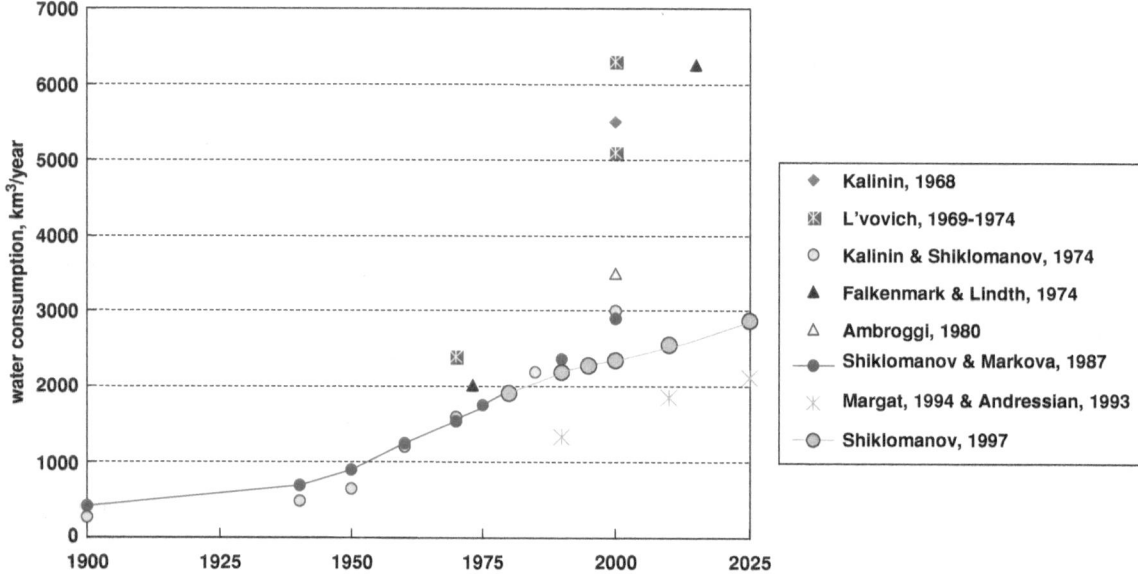

Fig. 11.2 Comparison of the values of world water consumption by the data of different authors.

L'vovich (1974) published estimates and forecasts of world water use that are well known and which have been used widely. These estimates were made employing indirect methods and they included the then-current situation and forecasts to the end of the twentieth century. The forecasts for all types of water use were given in two ways, one assuming that present practices continued and the other based on more rational principles. In the latter case it was assumed that waste waters would not be discharged to rivers, due to the implementation of closed-cycle systems of water supply to the industry and energy sectors, as well as the introduction of waterless technology ("dry" production). It was also assumed that irrigation and industries, which do not require water treated to drinking-water standards, would use treated waste waters. L'vovich calculated that the global total for abstraction amounted to 3300–3600 km^3/year, 2100–2400 km^3/year being consumed. L'vovich assumed that by 2000 abstractions would grow to 10 500–11 000 km^3/year provided the existing principles of water use persisted. If more rational principles were implemented by industry, for energy production and in the domestic sector, these figures would be 5100–6800 km^3/year for abstraction and the same for consumption. L'vovich (1974) assumed that the total area irrigated would be 500 million ha by 2000. Today L'vovich's estimates may appear rather arbitrary and lacking the confirmation of measurements of water used. The values he produced are overestimates but the forecasts are quite useful since they show that the concepts of the rational use of water resources and water pollution control are very important in determining the volumes used in the future.

In 1971 FAO published estimates and forecasts of world water use (Holy, 1971) which showed that for 1967 global abstraction totalled 1990 km^3/year. Abstractions were forecast to grow to 5450 km^3/year by 2000, with 51% being used for irrigation. These estimates contained for the first time actual data on water use for various purposes obtained from a number of countries. However, Holy did not identify evaporation from reservoirs and the water consumed in his figures, which nevertheless are in a fairly good agreement with other published figures.

The most complete study of world water use, including both abstraction and consumption estimates, sectoral usage and values for the separate continents was completed in the State Hydrological Institute (SHI) in 1974 (Korzun, 1974b). This study showed the trends in world water use from 1900 to 2000 (Kalinin and Shiklomanov, 1974) and an assessment for 1960 to 1970 based on actual water use data, water use per capita, industrial growth, changes in the irrigated area and population growth. Future abstractions were forecast for different continents and it was assumed that by 2000, 420 million ha would be irrigated. World total abstractions were 400 km^3/year in 1900, with the consumption being 270 km^3/year; in 1970 these figures were 2600 and 1600 km^3/year, and by 2000 they were expected to be 6000 and 3000 km^3/year, respectively (Fig. 11.1).

In 1974 Falkenmark and Lindth published their estimates and forecasts of world water use (Falkenmark and Lindth, 1974). They estimated abstractions and consumption by industry and irrigated agriculture together with domestic usage in 1973, 2000 and 2015. They forecast industrial water use both with treatment of waste water and recycling and without it. Their study showed that global abstractions in 1973 totalled 2850 km^3/year, and consumption was estimated as 2015 km^3/year. They assumed that in 2000 with re-use techniques, abstractions would increase to 6030 km^3/year, but without them this amount would increase to 8380 km^3/year. They also predicted that by 2015 abstractions with re-use would

be 7885 km^3/year, without it they would be 10 840 km^3/year; consumption of water was in both cases assumed to be 6265 km^3/year. Those results were obtained largely by indirect means and without water use data collected nationally. Neither did they include reservoir evaporation, nor figures for the separate continents. Nevertheless the estimates generally are in a good agreement with the SHI results of 1974, while their predictions for the years to come are obviously overestimated.

Yermolina and Kalinin (1975) carried out a detailed analysis of the abstractions in European countries and they also estimated world abstractions in 1970 to be 3700 km^3/year, increasing to 7000 km^3/year by 2000. De Mare (1977) predicted abstractions for the different countries for 2000. He divided the globe into 12 regions depending on the manner of abstractions and assessed future abstractions in each region, giving an absolute figure and a figure compared to the regional water resources. He predicted that world abstractions in 2000 would total 6680 km^3/year. His forecast was based on a limited number of parameters, and he only gave a world total rather than any sectoral figures. Batisse (1976) considered world water needs and world water resources and their optimum use, together with the analysis of available and future water resources and the water needs in the various regions. He estimated the global irrigated area to be 230, 310 and 420 million ha in 1970, 1985 and 2000, respectively, and abstractions as 3000, 4000 and 6000 km^3/year – figures which agree well with the SHI 1974 results.

In 1980 the US Environmental Protection Agency and the State Department issued a report named *The Global 2000 Report to the US President: Entering the Twenty-First Century* (Barney, 1980). The US Geological Survey calculated world abstractions for 1977 by continent and for certain countries using published data and data from Doxiadis (1967), Kalinin (1968), Holy (1971) and L'vovich (1974). The *Global 2000 Report* describes the methods employed in a very general way, e.g. abstractions for domestic purposes for the urban and rural population and the specific water use per capita; industrial abstractions based on industrial employment; abstractions for irrigation and the application of water per hectare. No analysis of data accuracy and reliability was attempted. Because no reference is made to it, it must be assumed that the authors were unaware of the SHI study of 1974.

Table 11.1 compares the estimates of abstractions for the continents and for major water users obtained by the US Geological Survey in 1980 (Barney, 1980) with those produced by the SHI in 1974 (Korzun, 1974a). The differences between global figures do not exceed 13–25%, while those for the continents are also similar, with the exception of Africa and South America. Considerable differences are to be found in the estimates for certain sectors, for certain countries and continents. Abstractions for industry for the former USSR were large in the US Geological Survey figures,

Table 11.1. *Water use by continents and selected countries, 1975–1977: comparison of SHI forecast (1974) and US Geological Survey assessment (1980)*[a]

Continent, country	Industrial water use, km^3/year	Irrigation		Municipal water use, km^3/year	Total water use, km^3/year
		Irrigated area, ha × 10^6	Water use, km^3/year		
Africa	6 / 15	10 / 6.4	120 / 61	6 / 12	170 / 88
Asia	80 / 99	187 / 147	1500 / 1400	50 / 98	1700 / 1600
Australia and Oceania	10 / 14	1.8 / 1.4	14 / 13	1.5 / 2	30 / 29
Europe	185 / 360	24 / 12	150 / 116	36 / 40	380 / 516
USSR	83 / 183	14 / 9.9	181 / 94	14 / 18	290 / 295
North America	340 / 308	27 / 22	230 / 205	46 / 38	640 / 551
USA	305 / 285	21 / 17	181 / 160	42 / 32	540 / 477
South America	12 / 11	8 / 3.7	60 / 35	7 / 11	90 / 57
World	633 / 807	257 / 193	2100 / 1830	150 / 201	3020 / 2840

[a]Numerator, SHI forecast; denominator, US Geological Survey assessment.

while abstractions for irrigation were half what they should have been. European industrial abstractions were also overestimated, while the water for irrigation was underestimated. Abstractions for irrigation for South America were underestimated by a factor of 2 (by comparison with the FAO data for 1977): irrigation covered 6.8 million ha rather than 3.7 million ha as stated in the *Global 2000 Report*. With respect to Africa, the area irrigated was underestimated by 30–40% and so was the volume of water abstracted.

The US report lacked data on the water that was consumed, evaporation losses from reservoirs and the historical increase in water use were not considered. During the period from 1975 to 2000 the report expected that abstractions would grow by 200–300%, mostly for irrigation. However, population growth alone in half of the countries of the world would double water use as compared to 1971, and even more, if the increase in the standard of living was taken into account. Comparison of the results in Table 11.1 allows the conclusion that provided there are data on water use in some countries and knowledge of the parameters that determine water use in others, it is possible to estimate total world water use and its distribution by sector. This technique, of course, does not permit the assessment of the use of water in particular countries or forecasts to be made.

Ambroggi (1980) analysed variations in water use in the twentieth century. He estimated total global abstractions in 1900, 1950 and 1980 as 1000, 2300 and 4400 km^3/year, respectively. He forecast that in 2000 world water use would amount to 7000 km^3/year, of which 3500 km^3/year would be consumed, 3000 km^3 would be polluted waste water while 500 km^3/year would remain unused. Ambroggi's estimates appear to be rather larger than the SHI results and those of the US Geological Survey. This was largely because Ambroggi's waste waters also include a certain amount of polluted natural waters.

Richa (1982) published a graph showing the growth of water use for the world from 1900 to 2000. Abstractions for 1900 and 1970 were shown as 1400 and 3800 km^3/year, and 6000 km^3/year for 2000. He forecasted that abstractions would increase to 22 000 km^3/year by 2200. These figures seem to be 1.5 to 3 times larger than those in Fig. 11.1 and they seem not to be based on data from countries, nor on estimates determined indirectly. Dukich (1982) presented a review of global water use by continents and by sectors which were in a very good agreement with the 1974 SHI results, probably because he based his study on them.

Shiklomanov and Markova (1987) produced the most complete and detailed evaluation of world water use that has yet been made.

For the first time use was made of the national assessments contained in more than 2000 publications from many countries, including those in Asia, Africa and South America. The authors were able to analyse the trends and changes in water use in greater detail than previously, not only by continent but also for the different regions. Estimates for water use for 1990 and 2000 were primarily based on the then widely popular long-term national development plans, which included irrigation, industry, growth of population and the rising standard of living. These plans are known to be extremely ambitious, particularly with respect to the extension of irrigation and this is regarded as a prime cause for the overestimation of future levels of world water use.

The 1987 SHI estimates of the total world water use for 1900, 1940, 1950, 1960, 1970 and 1980 were 580, 1060, 1360, 1990, 2580 and 3310 km^3/year, respectively, with consumption for the same years being 417, 700, 894, 1250, 1540 and 1950 km^3/year (Shiklomanov and Markova, 1987). The study predicted an increase in world abstractions to 4130 and 5190 km^3/year in 1990 and 2000 respectively, with consumption growing to 2360 and 2900 km^3/year. It is worth noting that these values are much lower than those given in the SHI study of 1974 (Figs. 11.1 and 11.2).

The developments in the value of assessments and forecasts over the last 30 to 40 years can be summarized as follows.

1. Assessments made in the 1960s and early 1970s were very rough, because of lack of data, while forecasts considerably overestimated future water use by comparison with current figures.
2. Assessments made in the late 1970s and early 1980s are increasingly reliable at the global and lesser scales but forecasts were still too large by comparison with today's values.
3. In the second half of the 1980s the assessments became more reliable, but were still handicapped by lack of data; forecasts of future volumes of water use continued to be overestimates, primarily because the development of irrigation slowed but also because of the increasing use of reuse of water by industry.

Lack of reliable data on water use continues to be a severe restraint to assessments and forecasts (Fig. 11.2). Unfortunately most of the publications that aim to summarize the global water situation, such as Gleick (1993, 1998), WRI (1992, 1996, 2000) and others, fail to acknowledge these restraints. Data on consumption are even fewer and less reliable than data on abstractions. The impact of these difficulties has wide ramifications for the pursuit of a rational and scientifically based water policy, globally, regionally, nationally and even for river basins.

Most forecasts of world water use made in the 1990s took into account the decrease in the growth of irrigation in most countries and more stable rates of industrial water use, and in some developed countries decreasing use. Some forecasts also accounted

for actual or planned measures to save water and to protect water resources from pollution. In the light of all these factors, WRI (1992, 1996, 2000) estimated the world volume of abstractions to be approximately 4200–4400 km^3/year by 2000.[1]

Andressian (1993) and Margat (1994) made forecasts of abstractions and consumption for 30 regions for 2010 and 2025 by comparison with their data from the 1980s. They used an extremely simple approach to estimate abstractions for domestic needs, irrigation and industry using factors such as population and the increase in the area irrigated. They did not allow for physical and socio-economic differences and they assumed that irrigation would use 75% of abstractions and industry 10%. The water used for generation of electricity was not included in the total for abstractions, while their forecasts assumed that by 2010 there would be a decrease in the rates of specific industrial abstractions by 40% and by 50% by 2025 in all developed countries (except the USA). The last of these assumptions is hardly realistic, and it contradicts the latest UNIDO data (Strzepec and Bowling, 1995) on trends in future industrial water use. Figure 11.1 shows these assessments, which are significantly lower than all earlier estimates, mainly due to the underestimation of abstractions for industry and because the abstractions for generation of electricity were not included in the total. They also omitted evaporation from reservoirs. The Andressian (1993) and Margat (1994) results differ from the more detailed SHI estimates given in Chapters 4 to 9 and those for certain regions. Nevertheless their results are of great interest and value for in the current debate on global water resources and use.

In 1997 Shiklomanov published (Shiklomanov, 1997) and then presented the material contained in Chapters 3 to 9 of this Monograph to the international symposium *Water: A Looming Crisis?*, held by UNESCO (Shiklomanov, 1998a). His results are contained in Fig. 11.1 and in Tables 11.2 and 11.3 and are in a very good agreement (before 1980) with the estimates made earlier (Shiklomanov and Markova, 1987). New data for 1990, 1995, 2000 and 2025 take into account all the latest trends in water use both in developed and developing countries and the latest long-term forecasts of world population growth. For 1995 abstractions have been calculated as 3750 km^3/year of which 2270 km^3/year represents consumed water. It is expected that by 2010 the abstractions will increase to 4300 km^3/year and consumption will be 2500 km^3/year, and that by 2025 these figures will reach 5100 and 2800 km^3/year, respectively. These values relate only to the mean climatic conditions and not to climate change and to the most probable scenario for world economic development. Allowing for the uncertainties in economic development, population

1 These data were given to the author of this chapter by Dr. J.B. Miller (World Meteorological Organization) in a personal communication.

Table 11.2. *Dynamics of world water use (km^3/year) by continents[a]*

Continent	Assessment								Forecast		
	1900	1940	1950	1960	1970	1980	1990	1995	2000	2010	2025
Europe	37.5	71	93.8	185	294	445	491	511	534	578	619
	17.6	29.8	38.4	53.9	81.8	158	183	187	191	202	217
North America	69.6	222	289	418	566	675	646	672	695	734	788
	29.2	83.8	108	146	193	232	229	241	250	267	286
Africa	41.0	49.0	56.0	86.0	116	168	199	215	230	270	331
	34.0	39.0	44.0	66.0	88.0	129	151	160	169	190	216
Asia	414	689	860	1222	1499	1784	2067	2157	2245	2483	3104
	322	528	654	932	1116	1324	1529	1565	1603	1721	1971
South America	15.2	27.7	59.4	68.5	85.2	111	152	166	180	213	257
	11.3	20.6	41.7	44.4	57.8	71.0	97.4	97.7	104	113	123
Australia and Oceania	1.6	6.8	10.3	17.4	23.3	29.4	28.5	30.5	32.6	35.6	39.6
	0.6	3.4	5.1	9.0	11.9	14.6	16.4	17.7	18.9	21	23.1
Total	579	1066	1369	1997	2584	3212	3584	3752	3917	4314	5139
	415	704	891	1251	1548	1928	2200	2268	2336	2514	2836

[a]Numerator, water withdrawal; denominator, water consumption.

Table 11.3. *Dynamics of world water use (km^3/year) by sector of economic activity[a]*

Sector	Assessment								Forecast		
	1900	1940	1950	1960	1970	1980	1990	1995	2000	2010	2025
Population × 10^6			2542	3029	3603	4410	5285	5735	6181	7113	7877
Irrigated land area, ha × 10^6	47.3	75.9	101	142	169	198	243	253	264	288	329
Agriculture	525	891	1125	1551	1834	2190	2412	2494	2570	2747	3113
	406	678	853	1177	1404	1699	1905	1945	1981	2109	2350
Industry	37.8	127	182	334	544	686	681	714	748	862	1105
	3.36	9.49	14.4	24.3	37.2	58.0	73.0	77.0	82.0	101	135
Municipal use	16	36.8	52.6	82.6	130	206	321	356	388	468	650
	4.17	9.04	13.9	20.1	31.0	41.8	53.5	58.3	64.1	70.1	82.9
Reservoirs	0.3	3.7	10.1	29.2	76.2	129	169	188	210	235	269
Total	579	1066	1369	1997	2584	3212	3584	3752	3917	4314	5139
	415	704	891	1251	1548	1928	2200	2268	2336	2514	2836

[a]Numerator, water withdrawal; denominator, water consumption.

growth and climate it can be assumed that by 2025 the total world abstractions would range from ±10–12% of the mean value, i.e. from 4600 to 5800 km^3/year.

During the course of the preparation of the report *Comprehensive Assessment of the Freshwater Resources of the World* for the UN Commission on Sustainable Development (UN/WMO/SEI, 1997), forecasts for 2025 of global world abstractions were made by the International Group of Experts under the Stockholm Environment Institute (SEI). These forecasts were based on the future "Conventional Development Scenario" created by the Group. At the same time the FAO also made a forecast of abstractions in 2025,[2] proceeding from the need for fresh water for producing food. The two forecasts for 2025 are shown in Fig. 11.1.

2 These data were given to the author of this chapter by Dr. J. B. Miller (World Meteorological Organization) in a personal communication.

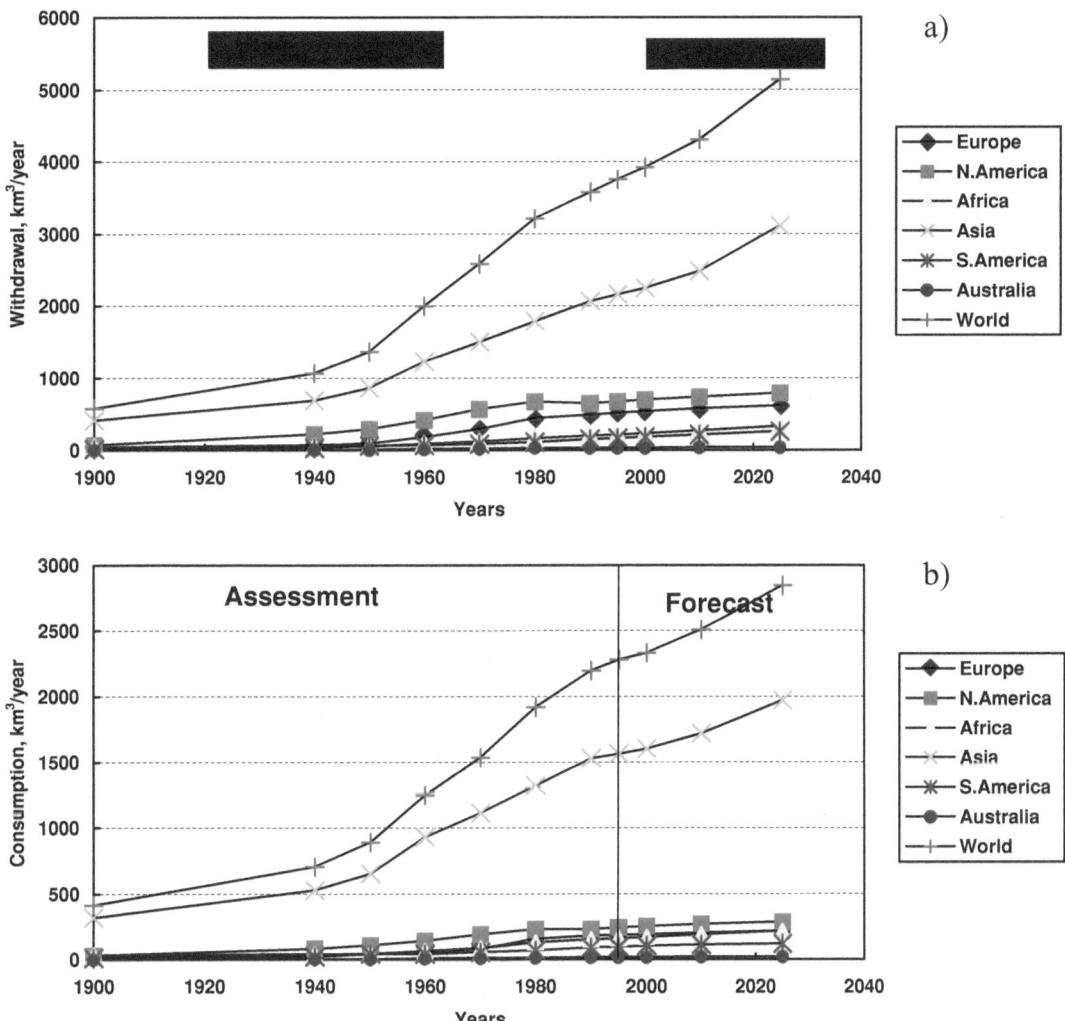

Fig. 11.3 The dynamics of (a) water withdrawal and (b) water consumption by continent.

These forecasts differ significantly: on average by from 4500 to 7000 km³/year. The earlier and independent forecasts made by SHI in 1997 fall between these extremes. The more detailed analysis of the SHI study on the changes and trends in world water use and water availability (Shiklomanov, 1997, 1998a, 1998b, 2000a, b; Penkova and Shiklomanov, 1998; Shiklomanov *et al.*, 2000), as well as their areal distribution, is discussed in the following sections.

11.2 TRENDS IN WORLD WATER USE

The trends and changes in world water use and use by continent for the twentieth century and to 2025 are shown in Table 11.2

and Fig. 11.3. In 1995 world abstractions were estimated to be almost 3750 km³/year, while consumption was 2270 km³/year, some 61% of abstractions. Abstractions are foreseen to increase by 10–12% per decade, so that by 2025 the figure could reach about 5140 km³/year, about 37% more that the 1995 total. Consumption would grow more slowly and would increase by 25%. Presently about 57% of abstractions and 69% of consumption take place in Asia, where irrigation is most widely practised. The fastest growth in the use of water is expected to occur in Africa and South America (50–60%) and the slowest in Europe and North America (about 20%). Table 11.3 and Figs. 11.4 and 11.5 show other details of these trends.

Presently 66% of world abstractions and 86% of consumption are for agriculture, but in future the importance of agriculture is likely to decrease, mostly due to growth in the industrial and domestic sectors. It is expected that by 2025 abstractions for agriculture will increase by 25%, for industry by 50% and for the domestic

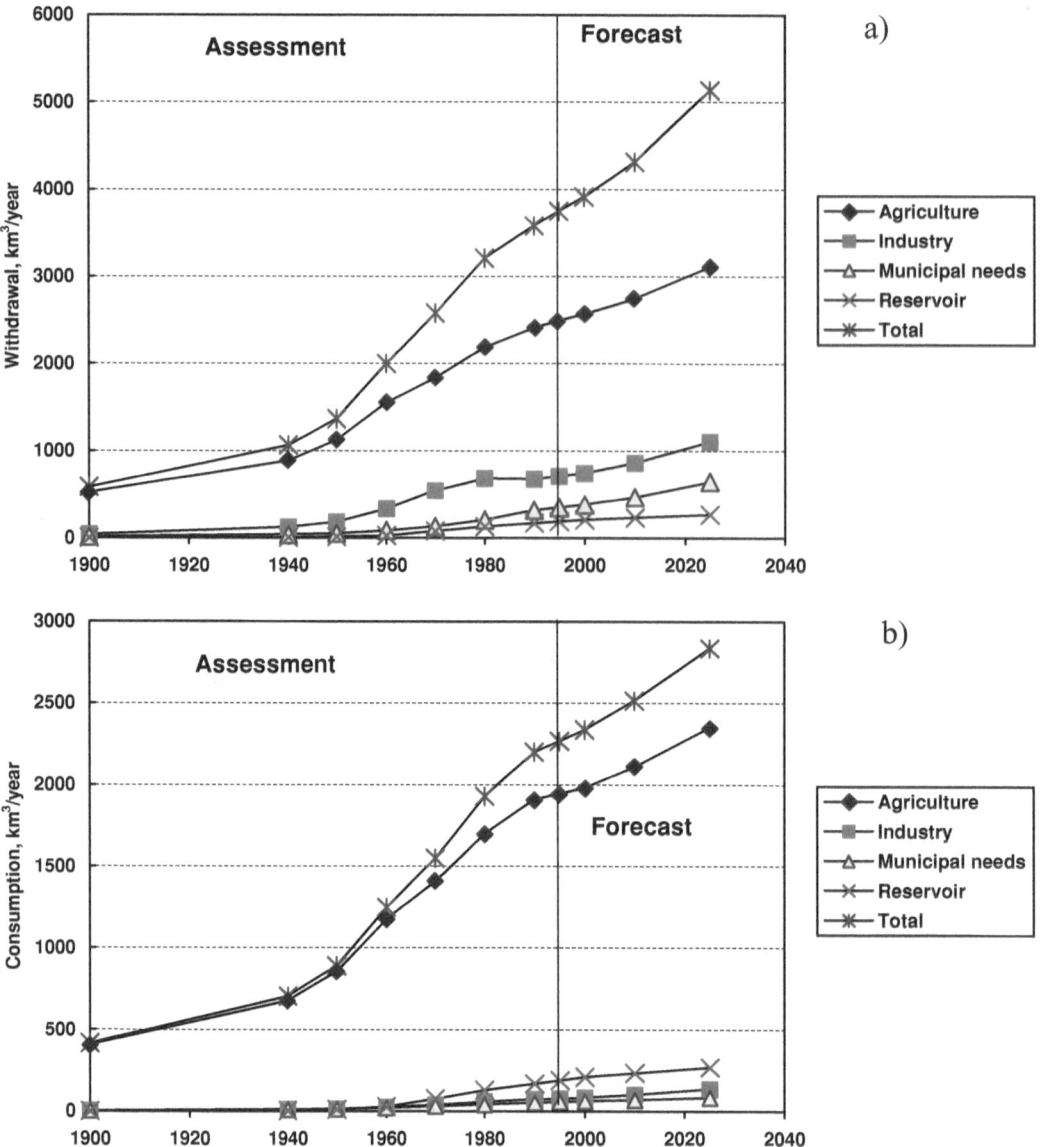

Fig. 11.4 Dynamics of (a) water withdrawal and (b) water consumption by type of economic activity.

sector by 80%. However, the additional evaporation from reservoirs constitutes an important component of world consumption; in fact it exceeds industrial and domestic consumption together. Currently irrigation covers 254 million ha across the world, and by 2010 it will probably increase to almost 290 million ha and to 330 million ha by 2025.

A number of forecasts are shown in Figs. 11.3 and 11.4 and when they are compared with those shown in Figs. 11.1 and 11.2, it is obvious that the latter are much higher. The major causes of such overestimation, as has already been discussed, are the following two factors. First, most forecasts made between the 1960s and the 1980s assumed a rapid increase in irrigation ahead of the growth

of population (Fig. 11.5). Second, the forecasts did not consider the decrease in abstractions by industry that occurred in the 1970s and 1980s in many developed countries. Unreliable data and poor methods also contributed to the errors in the forecasts.

The use of water by the different sectors and continents are shown in Tables 11.4 and 11.5 for 1950, 1995 and 2025. Europe and North America have similar patterns of water use, with industry as a major component. In 1995 European industry accounted for 45% of total abstractions, a figure expected to rise to 50%. In North America industry takes about 40% of total abstractions, falling to 39% by 2025 due to a significant rise in domestic water use because of population growth and urbanization, particularly in

Table 11.4. *The ratio of water withdrawal by sectors of economic activity[a] to total water withdrawal (in %) by continents*

Continent	1950				1995				2025			
	1	2	3	4	1	2	3	4	1	2	3	4
Europe	43.6	38.7	16.6	1.1	38.7	44.6	13.7	3.0	34.2	49.4	13.6	2.8
Asia	94.9	3.8	1.3	0.0	80.9	8.5	7.4	3.2	72.3	13.2	11.1	3.4
Africa	95.2	2.5	2.3	0.0	62.4	4.5	8.0	25.1	52.9	5.7	18.1	23.3
N. America	53.6	35.9	7.6	2.9	45.2	39.5	10.8	4.5	44.5	38.9	11.9	4.7
S. America	91.4	5.1	3.2	0.3	60.3	11.4	19.6	8.7	43.7	21.9	25.0	9.4
Australia and Oceania	50.0	39.5	7.2	3.3	50.8	23.6	10.8	14.8	46.7	26.0	11.6	15.7
World	82.2	13.3	3.8	0.7	66.5	19.0	9.5	5.0	60.7	21.5	12.6	5.2

[a] 1, agriculture; 2, industry; 3, domestic use; 4, reservoirs.

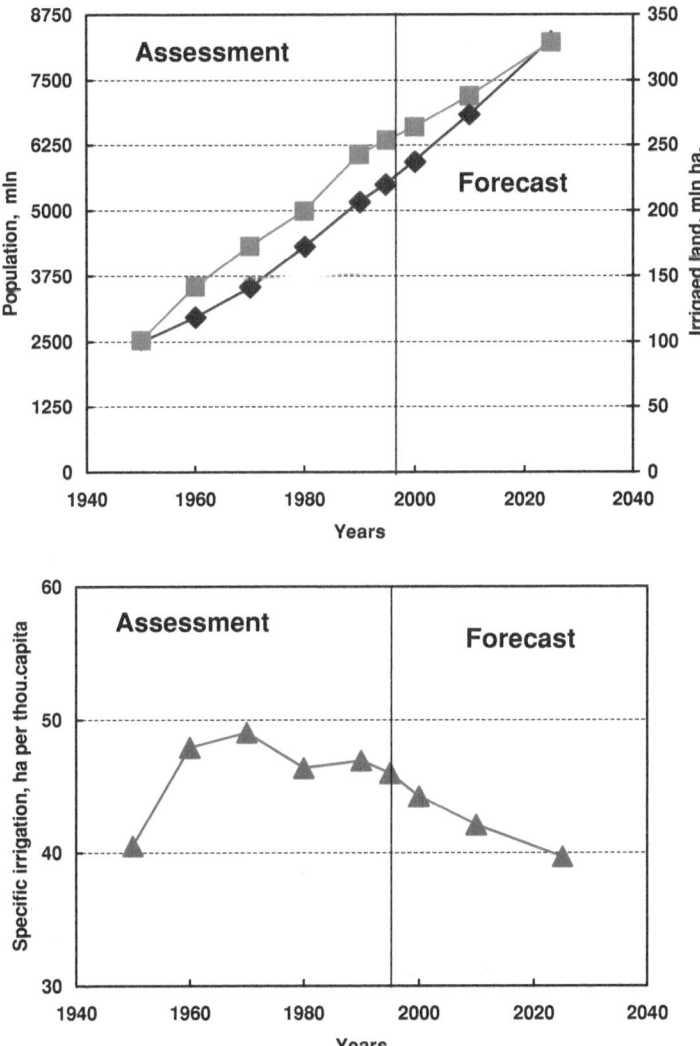

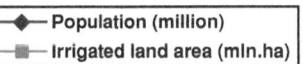

Fig. 11.5 The dynamics of population number and irrigated areas in the world.

Table 11.5. *The ratio of water consumption by sectors of economic activity[a] to total water consumption (in %) by continents*

Continent	1950				1995				2025			
	1	2	3	4	1	2	3	4	1	2	3	4
Europe	82.1	8.3	7.0	2.6	72.0	15.2	4.6	8.2	65.6	21.9	4.4	8.1
Asia	98.3	0.9	0.8	0.0	91.6	1.9	2.0	4.5	89.5	2.9	2.2	5.4
Africa	98.4	0.5	1.1	0.0	63.8	1.1	1.3	33.8	60.4	1.2	2.8	35.6
N. America	84.3	3.6	4.4	7.7	76.9	6.1	4.5	12.5	75.2	6.7	5.2	12.9
S. America	96.4	1.4	1.7	0.5	78.2	1.6	5.4	14.8	69.1	4.9	6.5	19.5
Australia and Oceania	80.4	9.8	3.1	6.7	69.0	3.5	2.1	25.4	61.8	9.4	2.0	26.8
World	95.7	1.6	1.6	1.1	85.7	3.4	2.6	8.3	82.8	4.8	2.9	9.5

[a] 1, agriculture; 2, industry; 3, domestic use; 4, reservoirs.

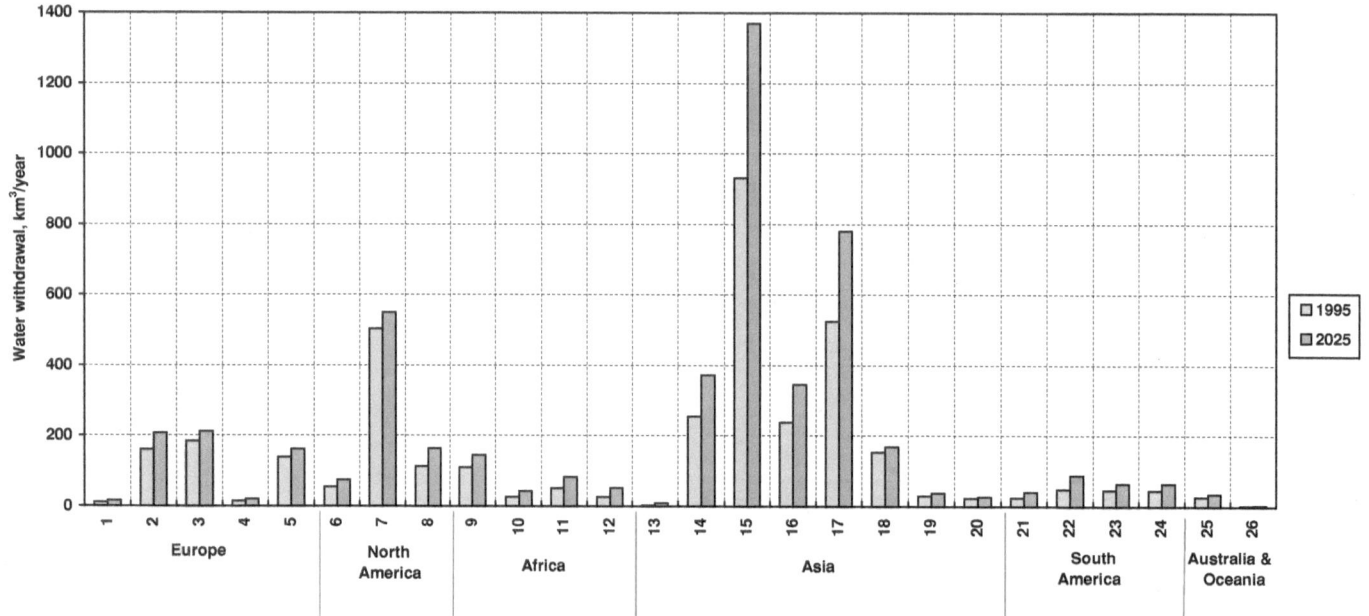

Fig. 11.6 Water withdrawal of natural–economic regions at 1995 and 2025. Europe: 1, North; 2, Central; 3, South; 4, northern part of the European territory of the former Soviet Union; 5, southern part of the European territory of the former Soviet Union. North America: 6, North; 7, Central; 8, South. Africa: 9, North; 10, Southern; 11, Eastern; 12, Western. 13, Central Asia; 14, North China and Mongolia; 15, South; 16, West; 17, South-East; 18, Middle; 19, Siberia and Far East of Russia; 20, Transcaucasia. South America: 21, North; 22, East; 23, West; 24, Central. 25, Australia; 26, Oceania.

Central American countries. With regard to consumption of water, both in Europe and North America agriculture is the major user, accounting for more than 70% of the total.

In Asia, Africa and South America in 1995 irrigation was responsible for 60–81% of abstractions and 64–92% of consumption. These estimates can be expected to change slightly by 2025,

but use of water by industry will probably increase by a factor of 2 or 3. This increase is likely to represent a 22% rise in abstractions in South America, a 13% rise in Asia and a 6% rise in Africa. Evaporation losses from reservoirs would account for about 25% of water consumption in Africa.

The variations in the patterns of water use are discussed in Chapters 4 to 9 and they are also summarized for 1995 and 2025 in Figs. 11.6 and 11.7. Naturally water use differs from region to region, but frequently within a region the location of the areas where demand for water is highest does not coincide with the resource-rich areas For instance, in Europe 94% of the water use occurs in the southern and central part of the continent; in North America the USA uses 75% of the water resources, in Australia

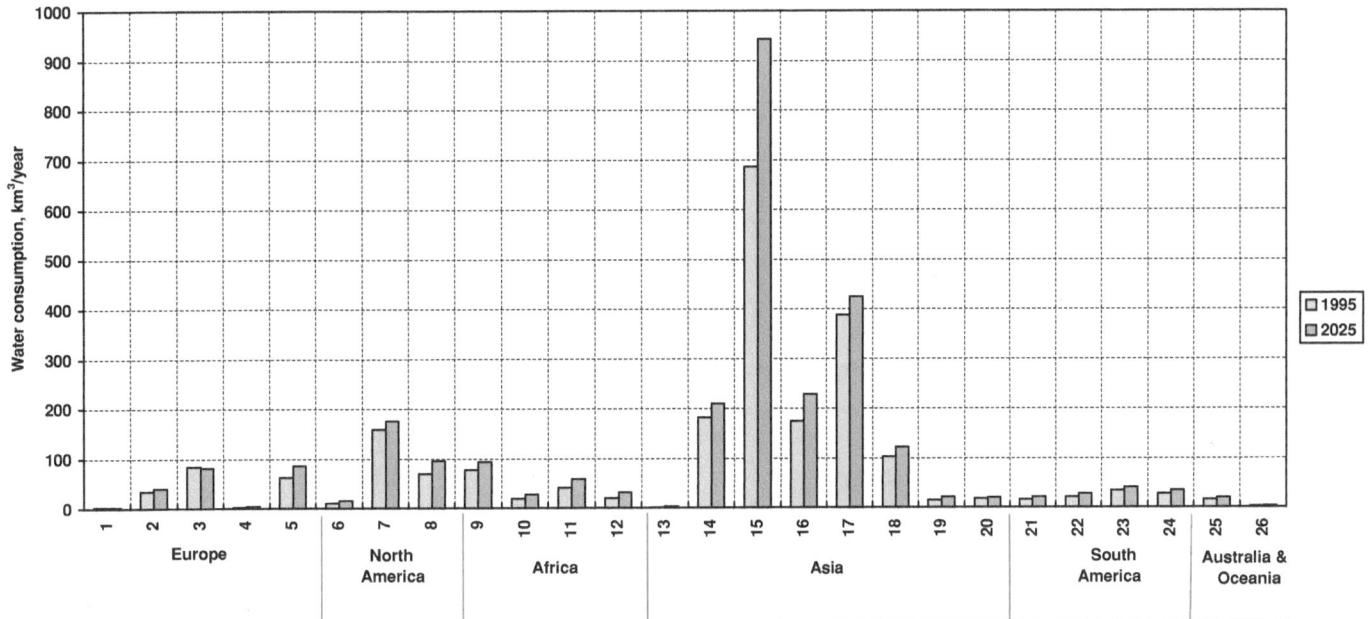

Fig. 11.7 Water consumption of natural–economic regions at 1995 and 2025. Europe: 1, North; 2, Central; 3, South; 4, northern part of the European territory of the former Soviet Union; 5, southern part of the European territory of the former Soviet Union. North America: 6, North; 7, Central; 8, South. Africa: 9, North; 10, Southern; 11, Eastern; 12, Western. 13, Central Asia; 14, North China and Mongolia; 15, South; 16, West; 17, South-East; 18, Middle; 19, Siberia and Far East of Russia; 20, Transcaucasia. South America: 21, North; 22, East; 23, West; 24, Central. 25, Australia; 26, Oceania.

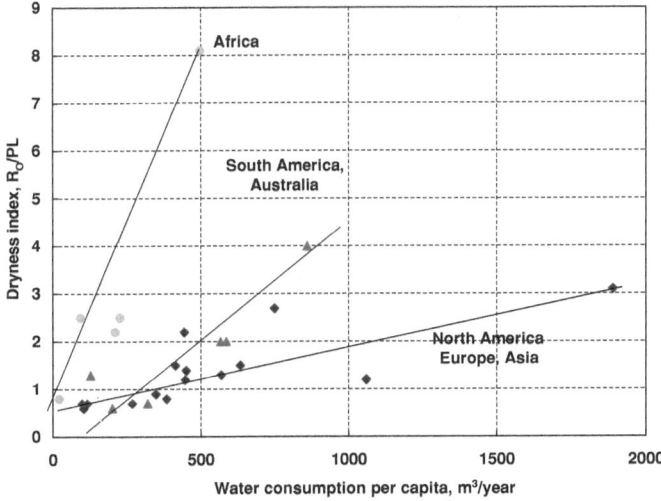

Fig. 11.8 The relationship between water consumption per capita and dryness index at 1995.

and Oceania 89% is used in Australia. In Asia, because of irrigation India, Pakistan, Bangladesh and the countries in South East Asia, including China, use most of the water. In Africa most of the water is used in the north, some 51% of total African abstractions. In South America water is used fairly uniformly across the continent.

By 2025 it is expected that in developed countries and in countries with limited water resources, water use will increase by between 15% and 35%, while in developing countries with abundant water resources, the use of water may grow by 200% to 300%. Of course, these values are subject to both the likely socio-economic and the likely climatic changes in the different regions as is shown in Fig. 11.8, which also shows the correlation between consumption per head and the dryness of the region. The highest use of water per head occurs in North America and Asia due to very high living standards in the former and the demands for irrigation in the latter. Africa is characterized by the lowest values of consumption, due to limited water resources, the low rate of their use and the low levels of socio-economic development in most of the countries.

It is valuable to compare estimates of abstractions with the estimates of renewable water resources (Fig. 11.9.). The data shown in Fig. 11.9 demonstrate how the use of water resources varies considerably from region to region from less than 1% to more than 100%. The regions (countries, basins) usually are divided into several relative categories by the rate of the their water resource use. In particular, according to UN/WMO/SEI (1997) there are four categories of the load on water resources, on the basis of coefficient K_w that show the ratio of the water withdrawal to the renewable water resources.

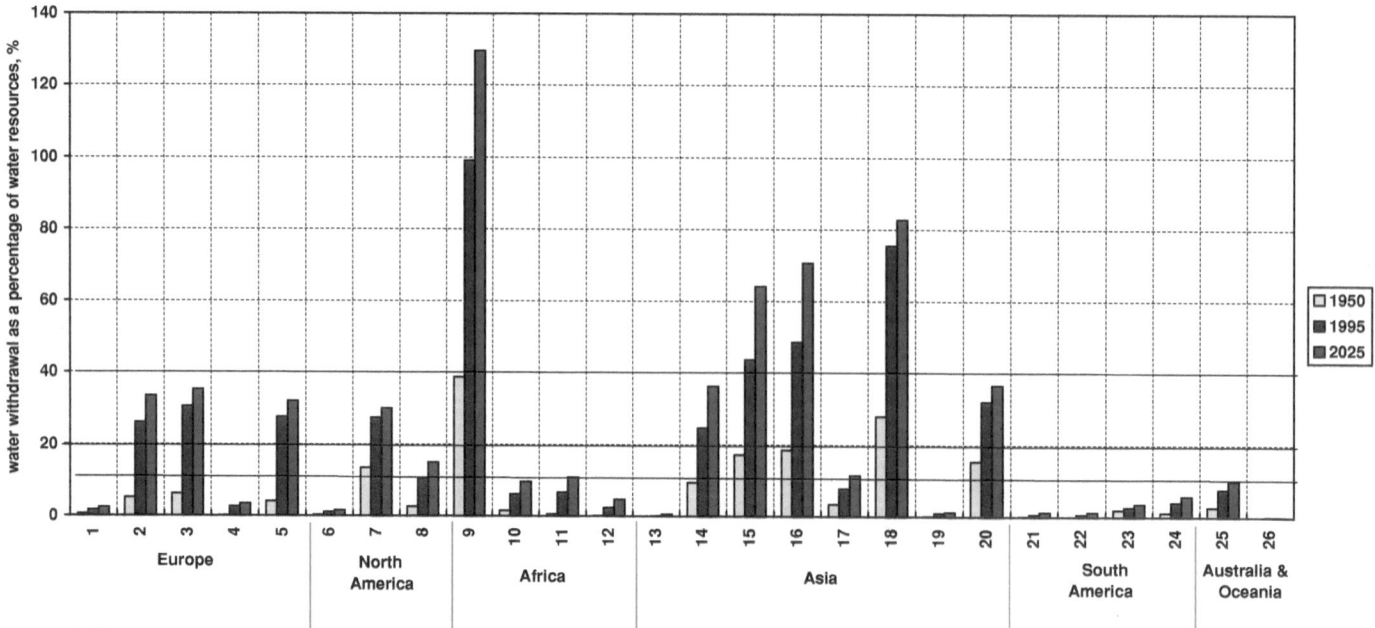

Fig. 11.9 Water withdrawal as a percentage of water resources for 1950, 1995 and 2025 by natural–economic regions. Europe: 1, North; 2, Central; 3, South; 4, northern part of the European territory of the former Soviet Union; 5, southern part of the European territory of the former Soviet Union. North America: 6, North; 7, Central; 8, South. Africa: 9, North; 10, Southern; 11, Eastern; 12, Western. 13, Central Asia; 14, North China and Mongolia; 15, South; 16, West; 17, South-East; 18, Middle; 19, Siberia and Far East of Russia; 20, Transcaucasia. South America: 21, North; 22, East; 23, West; 24, Central. 25, Australia; 26, Oceania.

First category: $K_w < 10\%$. Regions with the lowest load on the water resources. These regions, as a rule, are not subject to any significant load on their water resources.

Second category: $K_w = 10–20\%$. Regions with a moderate load on the water resources. The level of water availability can be regarded as a factor that puts certain constrains on the country's development, investments are required to make the water situation better.

Third category: $K_w = 20–40\%$. Regions with a heavy load on the water resources. To provide for a sustainable development water supply and demand should be adequately managed. Water problems should be specifically tackled, large investments are required, which for developing countries will cover a significant share of their GNP.

Fourth category: $K_w > 40\%$. These are the regions with a very high load on the water resources. Here we deal with a serious deficit of water; thus alternative water sources should be involved, which are very costly, water resources should be controlled and their use should be limited. Deficit of water resources becomes

a constrain on the economic growth of such a country adding an additional burden on the life of the people.

In 1950 the world had no problems with water availability on the basis of this classification. No part suffered a high stress on its water resources; category 3 (20–40% stress) existed in only two regions (North Africa and Middle Asia with Kazakhstan), while all other regions had a moderate or low stress on their water resources.

The situation has changed dramatically, however. At present many regions experience high or very high stress on their water resources and these regions are home to more than 70% of the world's population. By 2025 the situation will become even worse, particularly in developing countries on all continents. By that time more than 80% of the world population will experience high and very high stress on their water resources, with almost one-third facing conditions of catastrophically high stress ($K_w > 60\%$).

It needs to be recognized that on each continent (except South America) there are regions where the rate of water use is very high, while there are also regions where abstractions are very small by comparison with the available resources (Fig. 11.9). Contrast the present position in Southern and Central Europe with the North of the continent. In the former abstractions are estimated to be 24–30% of resources, in the latter they are less than 3%. In the north of North America abstractions are less than 1% of resources,

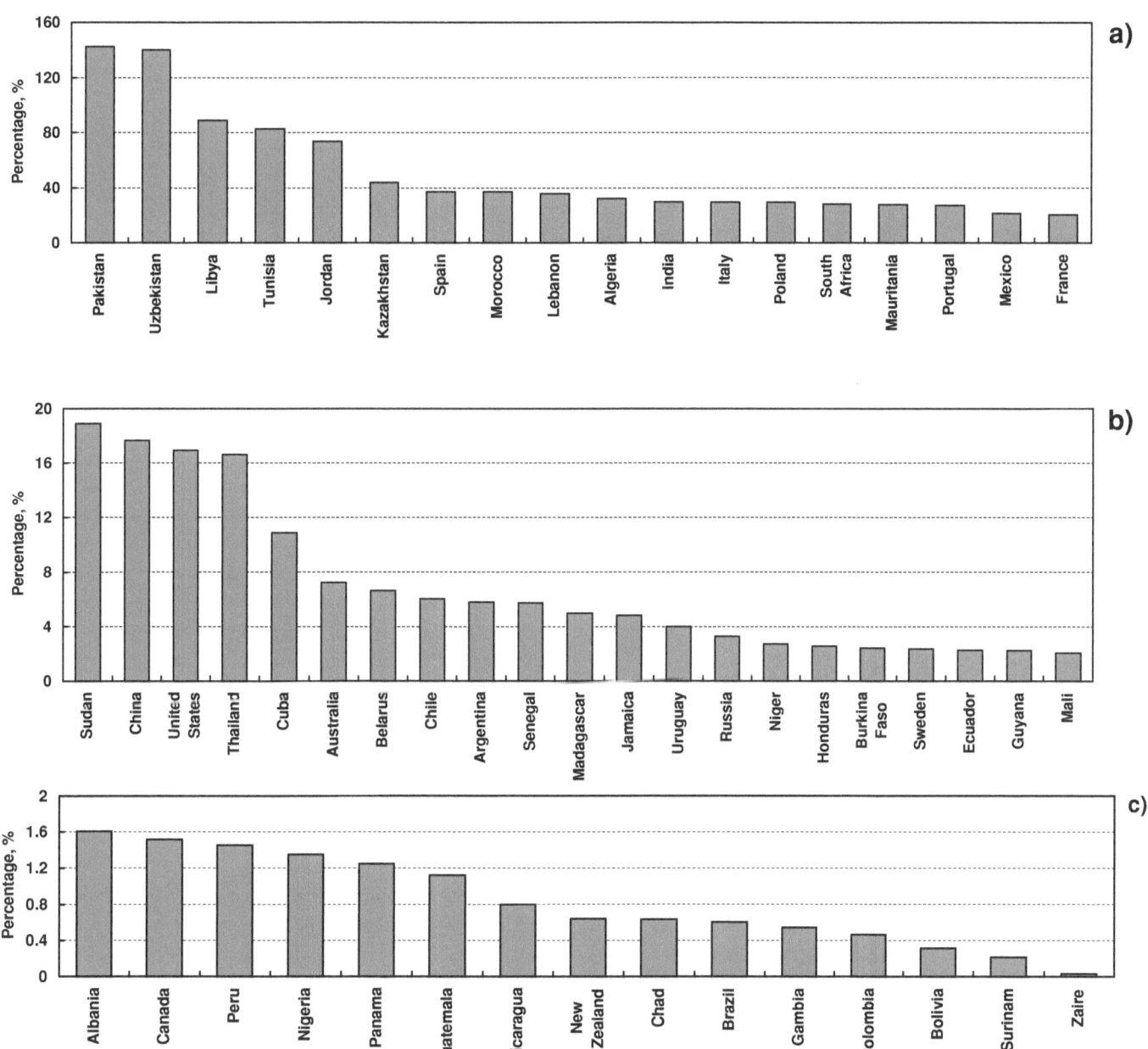

Fig. 11.10 Water abstraction as a percentage of water resources in individual countries of the world in 1995. (a) Water withdrawal >20%; (b) Water withdrawal 2–19.9%; (c) Water withdrawal <2%.

while for the conterminous USA they represent 27%. The situation in Africa and Asia is even worse. In North Africa today all the renewable resources are being used (99% of the resource is abstracted), while at the same time in other parts (particularly in Middle Africa) abstractions are very low. In Asia the contrast between, on the one hand Southern, Western and Middle Asia and Kazakhstan and on the other Siberia and the Far East is between up

to 84% and less than 1% of the resource being abstracted. Only in South America is this figure less than 4% for the whole continent.

By 2025, the uneven distribution of the available water resources and the use of water will become even more pronounced. Where water use as a percentage of water resources is fairly high today, in future it will increase and eventually will reach critical values. The contrast between northern parts of the globe and humid regions where water use is low and only a fraction of the resource will become more marked. Figure 11.10 shows the ratio of

Table 11.6. *Changes in precipitation and water resources of the continents due to water consumption*

Continent	Mean renewable water resources (R) km³/year	Water consumption (U_{ec}), km³/year		Sum of extra precipitation, (ΔP) km³/year		Runoff coefficient (L)	Volume of extra water resources (ΔR), km³/year		$\frac{\Delta R}{U_{ec}} \times 100$		$\frac{\Delta R}{R} \times 100$	
		1995	2025	1995	2025		1995	2025	1995	2025	1995	2025
Europe	2900	187	217	132	170	0.32	42	54	22.4	24.8	1.4	1.9
North America	7870	241	286	160	296	0.31	50	92	20.7	32.1	0.6	1.2
Africa	4047	160	216	208	248	0.12	25	30	15.6	13.8	0.6	0.7
Asia	13510	1565	1971	940	1270	0.39	367	495	23.4	25.1	2.7	3.7
South America	12030	97.7	123	0	0	0.41	0	0	0	0	0	0
Australia and Oceania	2400	17.7	23.1	0	0	0.34	0	0	0	0	0	0

abstractions to water resources for certain countries in 1995. Some countries have used their renewable resources completely and the greater part of inflows from upstream. Some of these countries also have to exploit their non-renewable ground water resources and resort to alternative water sources to meet their needs. However, Fig. 11.10(c) shows that there are some countries where the water resources are hardly touched ($K_w < 2\%$).

The intensity of human activities can lead not only to noticeable changes in the local climate over small areas but can also promote the addition of moisture to the atmosphere due to increased evaporation as compared with natural conditions. The volume of additional moisture put into the atmosphere can be assumed to roughly correspond to the consumption of water. This additional evaporation is expected to reach 2500–2800 km³/year early in the twenty-first century. It is a widespread effect caused mostly by the continued development of irrigation in arid and semi-arid regions, where under natural conditions the evaporation from dry surfaces would be insignificant.

This additional moisture in the atmosphere should cause increased precipitation to compensate for this increased evaporation. The extra precipitation could be quite significant in particular regions, but it is likely to occur over large areas. This phenomenon was first reported by Kalinin and Shiklomanov (1974) and by Drozdov *et al.* (1976). These estimates have recently been refined taking account of the new data available, such as data on water use, and the results obtained are shown in Table 11.6. Similar estimates for regions with intensive irrigation are presented in Table 11.7. As seen in Table 11.6, the total increase in precipitation in Europe, Asia, Africa and North America for 1995 to 2025 makes up between 60% and 130% of the corresponding values of consumption. For Australia and South America, there appears to be no rise in precipitation.

There are no difficulties in calculating approximately the extra mean annual runoff ΔR produced by the extra precipitation

ΔP using the values of the average runoff coefficients, for every continent, taken from Korzun (1974b). In 1995 the extra runoff amounted to a noticeable proportion of the volumes of consumption (from 16% to 23%). Further increases in the use of water are likely to cause an increase in the runoff. By 2025 it could be 14% to 32% of the total consumption (Table 11.6). Due to the large area of the continents, this extra precipitation may be expected to be redistributed widely whereas the extra evaporation occurs in particular places. However, the additional precipitation and runoff can be expected to occur in mountainous regions, unfortunately just those regions where hydrological networks are invariably the most sparse and where the errors of measurement are greatest. Table 11.7 gives estimates by region of the additional runoff, which could reach 30–40% of the total consumption by 2025. So the changing evaporation regime can be expected to result in some alterations in the water balance of the continents and their parts. Estimates of these changes are likely to be of great scientific interest and of practical importance for future water resources planning and management.

The values of ΔP and ΔR given in Tables 11.6 and 11.7 may be overestimated somewhat, due to the following assumptions:

- the extra evaporation resulting from the use of water, particularly for irrigation, has no effect on evaporation in those areas
- the volume of water used is equal to the extra evaporation.

The first assumption requires further consideration. Actually, in areas of inland drainage, the total evaporation is practically invariable, therefore an increased consumption results in a corresponding decrease in natural losses, and no additional moisture input to the atmosphere. In certain river basins evaporation can be decreased by the flood control measures which prevent flooding and reduce the area which may be inundated (Shiklomanov, 1979). While such alterations can be of great importance for assessing

Table 11.7. *Possible changes in precipitation and water resources by natural–economic regions of the world with developed irrigation under the influence of water consumption (by 2025)*

Region	Area, $km^2 \times 10^6$	Total water resources (R), km^3/year	Water consumption by 2025 U_{ec}, km^3/year	Additional precipitation (ΔP), km^3/year	Runoff coefficient (L)	Additional water resources (ΔR), km^3/year	$\dfrac{\Delta R}{U_{ec}} \times 100$	$\dfrac{\Delta R}{R} \times 100$
South Europe	1.79	655	117	88	0.33	29	25	4.4
Southern part of European territory of the former Soviet Union, including Transcaucasia	2.97	646	97	46	0.26	12	12	1.9
North China and Mongolia	8.29	1029	187	151	0.32	48	26	4.7
South Asia	6.95	2288	856	607	0.48	291	34	12.7
Western Asia	6.82	490	248	227	0.28	64	26	13.0
North Africa	8.78	181	97	75	0.08	6	6	3.3
West Africa	6.96	1118	32	50	0.24	12	38	1.1
East Africa	5.17	775	56	123	0.18	22	39	2.8
South Africa	5.11	485	31	31	0.11	3	11	0.6
Central America and the Caribbean	2.67	1092	77	89	0.34	30	40	2.7

changes in the water balance in individual river basins, they can be neglected in considering the possible future changes in runoff from the continents and their parts.

The second assumption can also cause errors, because both in industry and agriculture, some water is expended on additional evaporation and some, a much smaller volume, is included in the finished product. With domestic water use the larger part of the consumption (above 80%) is caused by evaporation during the transport of the water to the user.

The values given in Tables 11.6 and 11.7, as a whole, seem to be quite realistic, and quite close to the forecast values of water use. The values of extra precipitation and runoff given in Table 11.6 are quite significant as compared with the water used. It is of interest to consider, as a first approximation, the possible changes in the hydrological cycle for each of the continents due to the water used, such as those given by Drozdov *et al.* (1976). According to these data the increase in the mean precipitation that could be expected by 2000 would be insignificant as a whole, and consist of an increase of 1% for Africa, 2% for Europe and North America and 6% for Asia. No noticeable changes are to be expected in the general atmospheric circulation and in the rate of water vapour transport. The major characteristic of the transport rate, the transport coefficient, would show no change overall for the continents, but the extra evaporation could bring about some changes in the hydrological cycle locally. There is also the possibility of an increase in runoff of between 0.6% and 14% for the regions where irrigation is dominant and these figures are given in Table 11.7. Thus, the growth in the use of water would give rise to some changes in the relationships between individual elements of water balance of the different continents and their parts; however, it would be unlikely to result in any noticeable changes in global climate. While it is possible to estimate these changes to the water cycle, such changes are largely of scientific interest, although they might be acknowledged in large-scale water management. When compared to global climate change these effects are much smaller and likely to remain so.

11.3 WATER AVAILABILITY AND WATER STRESS

The world's water resources are distributed in a manner which does not match the distribution of the population. The per capita availability of water by region and country presents a different pattern of distribution. With the growth of the population and the increase in consumption of water the value of specific water availability will decrease. Changes of specific water availability from region to region and from country to country are shown in Fig. 11.11 and in Table 11.8 for the period from 1950

to 2025. In 1995 Canada, Alaska and Oceania had the highest specific water availabilty, namely 170 000–180 000 m³/year per head, while in the densely populated areas of Asia, Central and Southern Europe and Africa, present-day values range from 1200 to 5000 m³/year per head, against only 200–300 m³/year per head in North Africa and the Arabian Peninsula. These latter values fall into the categories of low and catastrophically low causing serious problems for domestic and agricultural use of water. To make the distribution of the specific water availability more apparent, it is shown for 1950, 1995 and 2025 in Fig. 11.12. The maps differentiate regions by hatching the following categories of specific water availability (in thousand cubic metres/year per head): ≤1, catastrophically low; 1.1–2.0, very low; 2.1–5.0, low; 5.1–10, average; 10.1–20, high; >20, very high.

In 1950 (Fig. 11.12a) the specific water availability over most of the globe was average or higher than average. Only North Africa had a very low value, while water availability was in the low category in Central and Southern Europe, Northern China and South Asia. No region of the world was in the catastrophically low category. By 1995 the situation had changed dramatically (Fig. 11.12b): water availability had decreased drastically. North Africa and the Arabian Peninsula were in the catastrophically low category, availability was very low in Northern China, Southern and Western Asia as well as in seven other regions of the world. Currently more than three-quarters of the world's population lives where specific water availability is lower than 5000 m³/year per head, while more than one-third of the population suffers very low and catastrophically low water availability.

The situation will become worse as the twenty-first century unfolds (Fig. 11.12c). By 2025 most of the world's population is likely to experience the conditions of very low or catastrophically low water availability, while about 30–35% of the population will have to live in conditions of catastrophically low water availability (less than 1000 m³/year per head). At the same time the specific water availability will be very high in Northern Europe, Canada and Alaska, as well as over most of South America, Central Africa, Siberia and the Far East, and Oceania. Such a scenario shows just how serious is the stress on the world's water resources and that action is needed now to avert a crisis. The pattern of the stresses on water resources across the world can be revealed by analysing trends and changes in the specific water availability. Such studies show that for the period 1950 to 2025 the decreases in water availability were and will be controlled by the rate of economic growth and the climate of a country or region (Fig 11.13). Three groups can be identified: (1) developed (industrial) countries; (2) developing countries with humid climates; and (3) developing countries with arid and semi-arid climates.

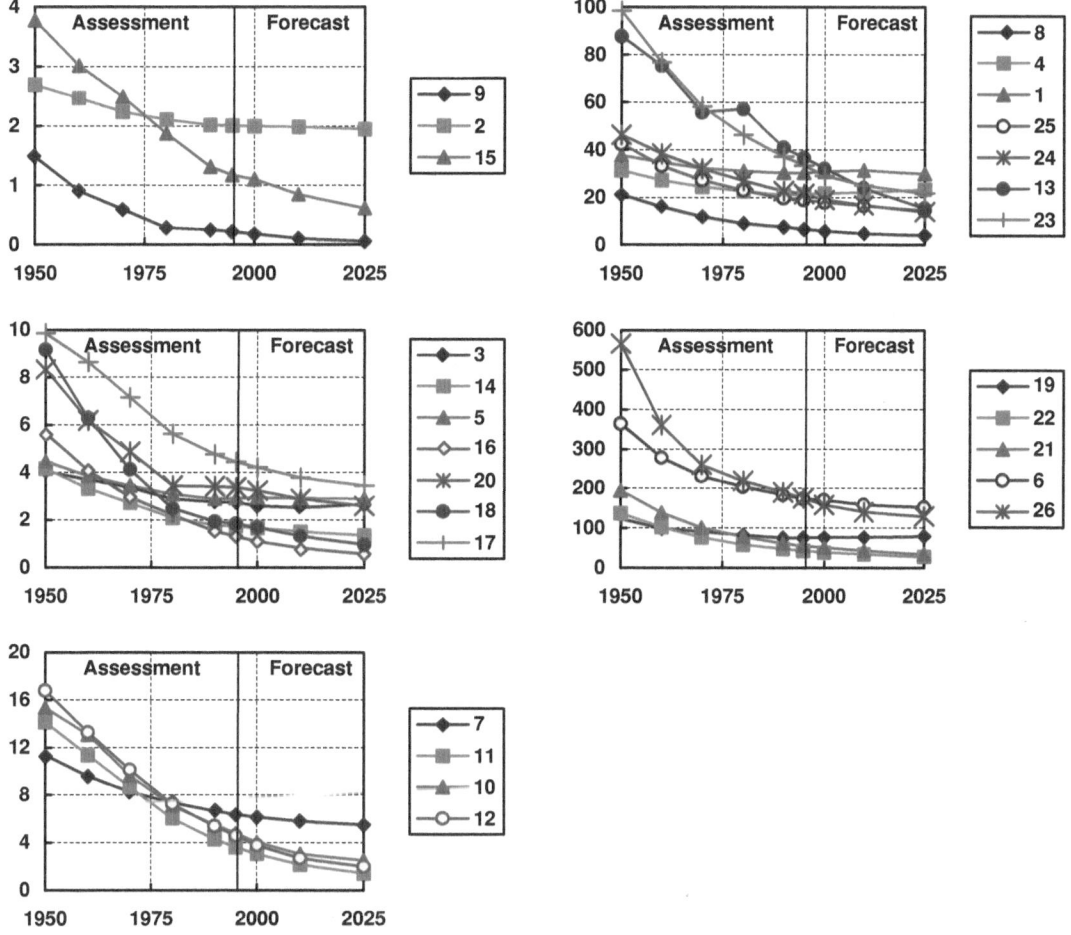

Fig. 11.11 Dynamics of specific water availability by natural–economic regions for the period 1950 to 2025. Europe: 1, North; 2, Central; 3, South; 4, northern part of the European territory of the former Soviet Union; 5, southern part of the European territory of the former Soviet Union. North America: 6, North; 7, Central; 8, South. Africa: 9, North; 10, Southern; 11, Eastern; 12, Western; 13, Central Asia; 14, North China and Mongolia; 15, South; 16, West; 17, South-East; 18, Middle; 19, Siberia and Far East of Russia; 20, Transcaucasia. South America: 21, North; 22, East; 23, West; 24, Central. 25, Australia; 26, Oceania.

Figure 11.13 shows that developed regions are likely to experience a reduction of about one-quarter in specific water availability which is largely independent the climate and the volume of the water resource, about 1.7-fold for the period in question, 70% occurring before 1995. For regions with developing countries the prevailing rates of decrease are much higher; for humid countries by 2025 availability is reduced to 4.5 times the 1950 figure and to 8 times for arid and semi-arid areas.

Thus the complications of natural variablity in the distribution of water resources over the globe and the resulting disparities in the patterns of water availability will become more pronounced as a result of economic development and population growth as time goes on. Similar disparities are observed when the specific water availability of selected countries is analysed (Fig 11.10). Slow rates of decrease in the specific water availability are characteristic of most developed countries, while developing countries will experience high rates of decrease due to the rapid growth of population and increasing consumption per capita. However there are some rich countries such as those of the Arabian Peninsula where conditions are rather different. Here the scant water resources are augmented by the exploitation of deep ground water as well as by desalination. Due to their wealth, these countries are able to meet their growing needs by employing costly alternatives, alternatives which are precluded from use in other arid and semi-arid countries because of cost.

Guidance to these alternatives and the specific water availability values for the seven countries of the Arabian Peninsula are

Table 11.8. *Dynamics of water availability in selected countries of the world*

Country	Population (1994), ×10^6	Water resources, km³/year		Water use (1990), km³/year		Water availability, m³ × 10³/year per head								
						Assessment							Forecast	
		Local	Inflow	Withdrawal	Consumption	1950	1960	1970	1980	1990	1995	2000	2010	2025
Albania	3.60	19	5.2	0.42	0.22				7.9	6.46	5.84	5.15	4.41	4.21
Algeria	27.3	16.4[a]	0.4[a]	4.50	2.63				0.78	0.55	0.48	0.42	0.36	0.27
Argentina	34.2	270	623	33.7	20.1			25.3	21.2	17.4	16.2	15.1	13.5	12.3
Australia	17.9	352	0	25.5	14.5	42.5	33.4	27.3	23.1	19.7	18.9	18.1	16.4	14.4
Bolivia	7.24	361	155	1.38	0.92					61.2	53.1	45.1	34.3	24.1
Brazil	159	6220	1900	43.0	18.0	138	103	77.5	59.2	48.2	43.3	40.1	34.7	29.3
Burkina Faso	10.0	14.7	2.0	0.38	0.26					1.73	1.59	1.45	1.07	0.60
Belarus	10.3	34.4	21.7	3.0	0.9	5.78	5.59	5.0	4.71	4.39	4.41	4.43	4.48	4.54
Canada	28.0	3290	130	51.6	10.3	240	184	154	137	123	119	113	107	103
Chad	6.18	15.8	28.3	0.19	0.13					5.36	4.71	4.06	3.16	2.21
Chile	14.0	354	0	21.4	15.4					25.7	24.3	22.9	20.7	17.1
China	1220	2700	0	477	365				2.38	2.02	1.92	1.83	1.68	1.54
Colombia	34.3	1200	0	5.55	3.62					37.0	34.3	31.5	27.3	22.1
Cuba	11.0	34.5	0	9.2	5.8	6.0	4.7	3.8	3.0	2.7	2.6	2.4	2.2	2.0
Ecuador	11.2	265	0	5.98	4.35					24.7	21.0	17.4	15	13.1
France	56.8	168	27	36.7	7.71	4.13	3.77	3.37	3.25	3.07	3.06	3.03	2.99	2.86
Gambia	1.08	3.2	4.7	0.03	0.02					6.36	6.25	6.14	4.86	2.93
Guyana	0.83	270	0	6.02	3.92					334	313	291	257	230
Guatemala	10.3	116	0	1.3	0.7	41.0	30.0	23.0	16.6	12.5	10.8	9.4	7.3	5.3
Honduras	5.49	95	0	2.6	1.8			40.2	27.3	19.5	16.9	14.3	11.1	8.58
India	935	1456	581	518	384			2.83	2.12	1.61	1.44	1.31	1.09	0.85
Italy	57.7	185	0	55.6	24.7				2.84	2.77	2.77	2.73	2.74	3.02
Jamaica	2.43	8.3	0	0.4	0.3			4.33	3.69	3.33	3.05	2.96	2.63	2.23
Jordan	5.20	0.964[a]	0	0.71	0.40				0.20	0.17	0.15	0.12	0.11	0.13
Kazakhstan	17.2	70	56	43	24	13.6	9.1	6.4	4.95	4.32	4.18	4.02	3.61	3.13
Lebanon	3.06	2.80[a]	0	1.00	0.66				0.85	0.78	0.63	0.49	0.36	0.28
Libya	5.22	5.29[a]	0	4.69	3.46				0.33	0.40	0.39	0.38	0.35	0.33
Madagascar	14.3	395	0	19.7	15.4					31.6	27.9	24.2	18.2	10.9
Mali	10.5	50.0	44.4	1.49	1.19					7.71	6.65	5.60	4.16	2.84
Mauritania	2.22	0.4[a]	11.0	1.63	1.22				3.65	2.60	2.00	1.69	1.13	0.79
Mexico	91.9	346	2.50	73.9	47	13.0	9.3	6.3	4.4	3.6	3.1	2.7	2.2	1.9

Country														
Morocco	26.5	33.0[a]	0	11.0	7.64				1.28	1.01	0.94	0.86	0.75	0.54
New Zealand	3.50	313	0	2	1.03	170	120	111	100	93.7	88.7	83.2	78.6	75.6
Nicaragua	4.27	175	0	1.4	0.7			96.0	64.0	47.0	38.7	33.0	26.0	19.0
Niger	8.85	3.0	30.4	0.50	0.35					2.31	2.07	1.82	1.31	0.78
Nigeria	108	274	43.6	4	2					2.71	2.26	1.81	1.29	1.04
Pakistan	141	40	186	242	180					0.09	0.07	0.05	0.03	0.01
Panama	2.58	144	0	1.8	1.1				73.0	59.0	54.2	49.0	43.0	37.0
Peru	23.3	1100	144	17	12.3	176	135	100		55.0	49.5	42.4	35.4	31.8
Poland	39.2	50	6.4	16.5	3.88				1.38	1.29	1.24	1.20	1.13	1.09
Portugal	9.93	19	34.5	10.2	3.92				3.22	3.15	3.14	2.84	2.71	2.91
Russia	148	4059	227	136	49.0	41.0	36.0	32.6	30.7	28.6	28.8	29.0	29.5	30.6
Senegal	8.10	17.4	17.4	1.50	0.97					2.68	2.34	2.00	1.52	0.97
South Africa	40.6	47.4	5.2	13.3	8.1			1.84	1.38	1.03	0.94	0.84	0.67	0.60
Spain	39.6	108	0	43.6	20.9			2.74	2.38	2.26	2.19	2.06	1.97	2.06
Sudan	27.4	22.0	140	17.4	12.4			5.02	4.21	3.16	2.78	2.4	1.88	1.3
Surinam	0.42	230	0	0.49	0.31					544	517	490	430	348
Sweden	8.34	164	12.2	4.1	0.55	24.1	22.7	21.1	20.4	19.8	20.2	20.7	20.2	19.8
Thailand	59.0	199	70	33.1	21.5				4.78	3.88	3.56	3.34	3.01	2.60
Tunisia	8.73	3.52[a]	0.42[a]	3.08	1.94				0.55	0.36	0.34	0.31	0.27	0.22
United States	261	2930	146	493	155	13.6	15.5	13.6	12.1	11.1	10.6	10.2	9.5	9.0
Uruguay	3.17	68.0	74.0	4.2	3.1					44.9	43.0	41	38	37.2
Uzbekistan	22.8	9.5	98	82.1	51.7	6.34	4.34	2.42	0.68	0.33	0.28	0.20	0.11	0.01
Zaire	42.6	987	313	0.36	0.08			30.6	30.6	27.3	24.0	17.9	11.4	

[a] Water resources include renewable ground water resources.

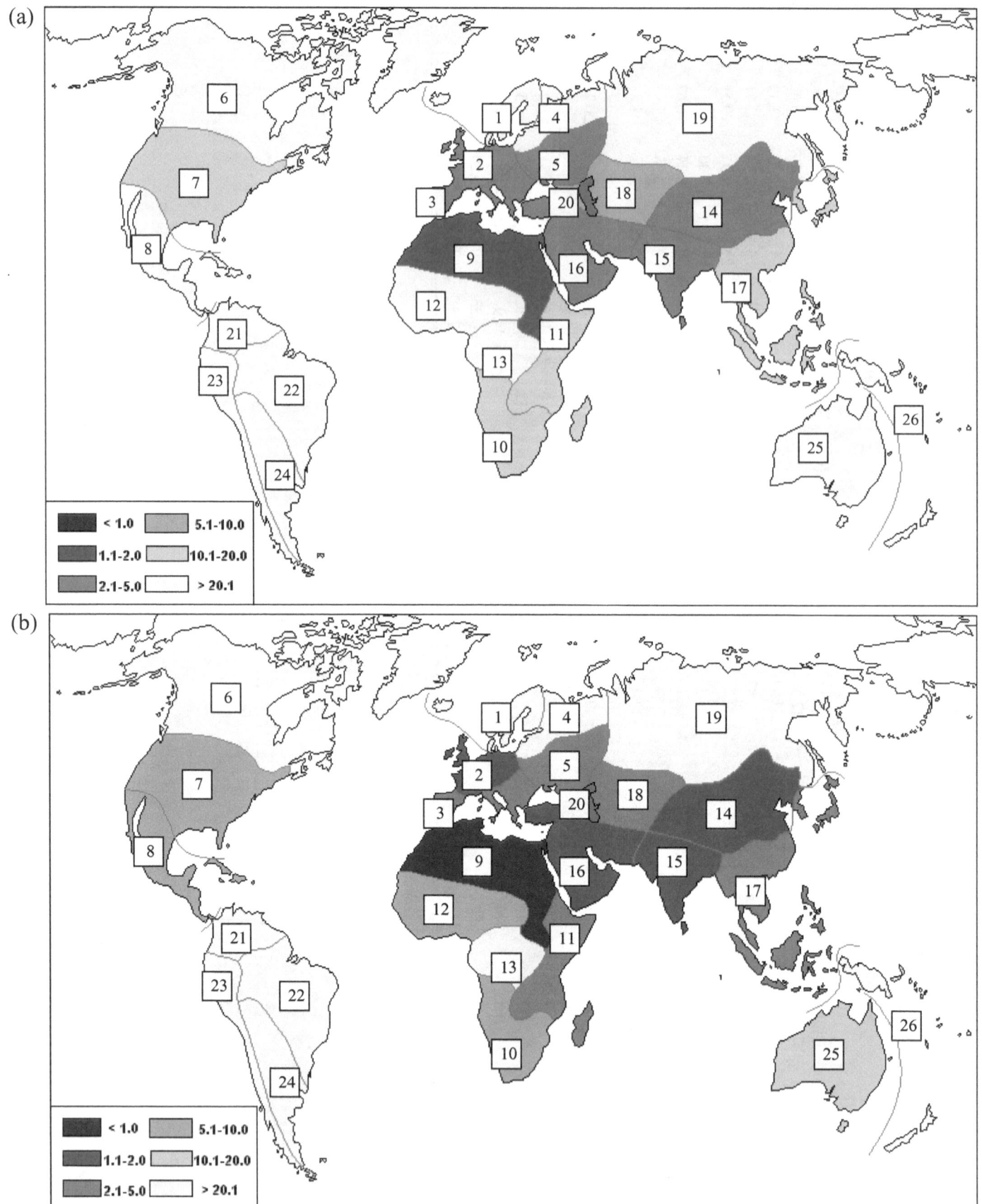

Fig. 11.12 Specific water availability (m^3 × 10^3/year per head) of world
regions (a) in 1950; (b) in 1995; (c) as projected for 2025.

(c)

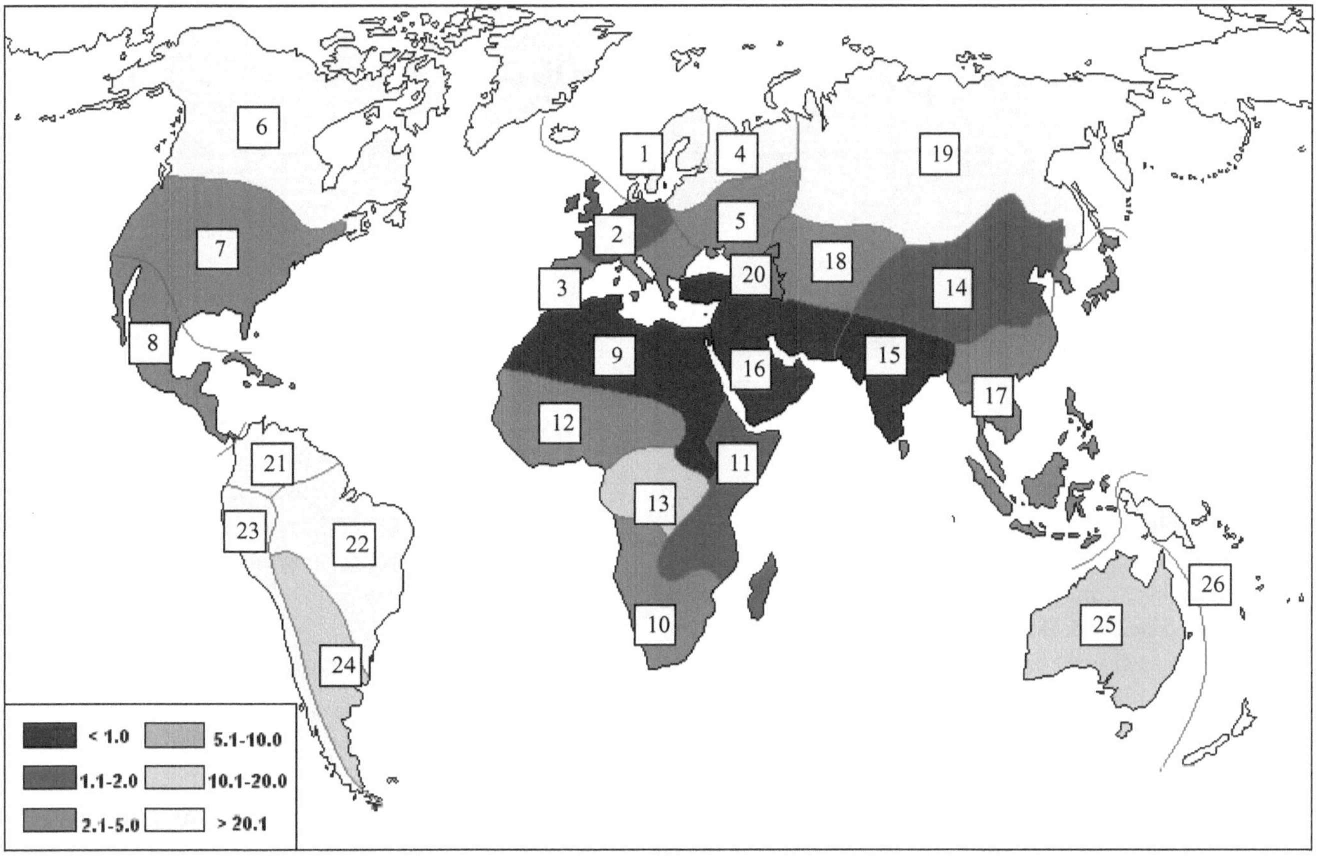

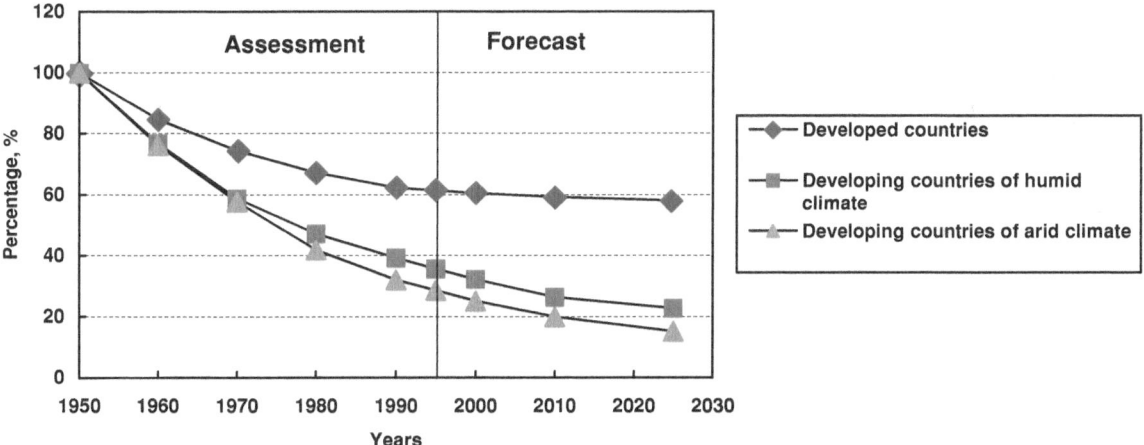

Fig. 11.12 (*cont.*).

Fig. 11.13 Dynamics of specific water availability (as a percentage of 1950 value) by natural-economic regions of the world, 1950–2025.

discussed in Chapter 5 and in Table 5.32. These data indicate increases in the available water resources in a number of countries (e.g. Bahrain, Qatar and Kuwait) which are well ahead of the growth of population and the increase in abstraction. Such measures require large financial investments in order to counteract the low values of specific water availability which are constraints on the future of these countries and to their sustainable development.

12 Climate change and water resources

12.1 INTRODUCTION

Water resources may be affected in a number of ways. The abstraction of water, the use of that water and its return, treated or untreated, to a river, lake or aquifer can alter the resource directly. Less directly, but as important, are the effects of land use change in the basins concerned. Deforestation, urbanization and agricultural practices such as the application of fertilizers and pesticides are particularly relevant in the context of changes to the quantity and quality of water resources. Such practices have modified water resources for several hundred years, but over the last 20 or 30 years, water resources have also been threatened by global climate change. The warmer world and rising sea levels that are forecast are likely to alter the hydrological cycle and water resources. However, there are also the more local effects on climate due to:

- development of urban areas
- construction of changes in the vegetation and land cover
- reservoirs and canals
- reduction in the frequency of flooding of river valleys
- irrigation and drainage of wetlands.

These measures can cause climate change principally by changing the Earth's albedo, changing the aerodynamic roughness, altering the evaporation and humidity, and modifying the soil and surface air temperatures. However, numerous studies have shown that these factors produce the most noticeable effect on the local climate and on the hydrological regimes of local water bodies. Changing the albedo and other local changes have an insignificant impact on the global climate, and they act in opposite directions, compensating for each other to a large extent. Alterations to the vegetation cover increase the albedo and lower the surface air temperature, while the development of irrigation and urbanization decrease the albedo and raise the global temperature. These local effects have a spatially restricted influence and have a small influence on future climate change and on changes to the water resources at the continental and regional scales. The greatest impacts on water resources at the global scale are expected to result from changes to the global climate and its influence on river flows and ground water. Changes in the global climate, of course, are due to the input to the atmosphere of greenhouse gases as a result of fossil-fuel burning and increases in water vapour because of evaporation from irrigated areas and from reservoirs (Budyko and Israel, 1987; Bolin et al., 1989; IPCC, 1990, 1992, 1995; Budyko et al., 1991).

The atmospheric gas is changing composition because of:

- the increasing concentration of CO_2 and CH_4
- the increasing concentration of rare gases such as freons, nitrogen compounds, etc.
- the increasing concentration of aerosols.

12.2 GLOBAL CLIMATE CHANGE: THE ESSENCE OF THE PROBLEM

Most meteorologists and climatologists are of the opinion that the increase in the concentration of CO_2 is the most important factor affecting the global climate. Carbon dioxide is almost transparent to short-wave solar radiation, but it attenuates considerably the long-wave radiation creating the so-called "greenhouse effect" in the atmosphere, promoting a rise in the temperature of the surface air layer. The continued increase in CO_2 in the Earth's atmosphere will lead to further warming. The effect is reinforced by the growing concentration of freons, nitrogen compounds and methane. Over the past millennia the Earth's climate has been comparatively sustainable; changes in the mean global temperature were not above a few tenths of a degree, and CO_2 concentration varied within 10% relative to the pre-instrumentation level (280 ppm) (Budyko et al., 1991).

Systematic observations of atmospheric CO_2 with reliable instrumentation were started during the International Geophysical Year in 1958. Then two stations were established: Mauna Loa (Hawaii) at 3400 m a.s.l., and at the South Pole. At present more than ten stations are making CO_2 concentration observations around the world. Using these direct measurements and by analysing air bubbles from ice cores, the concentration of CO_2

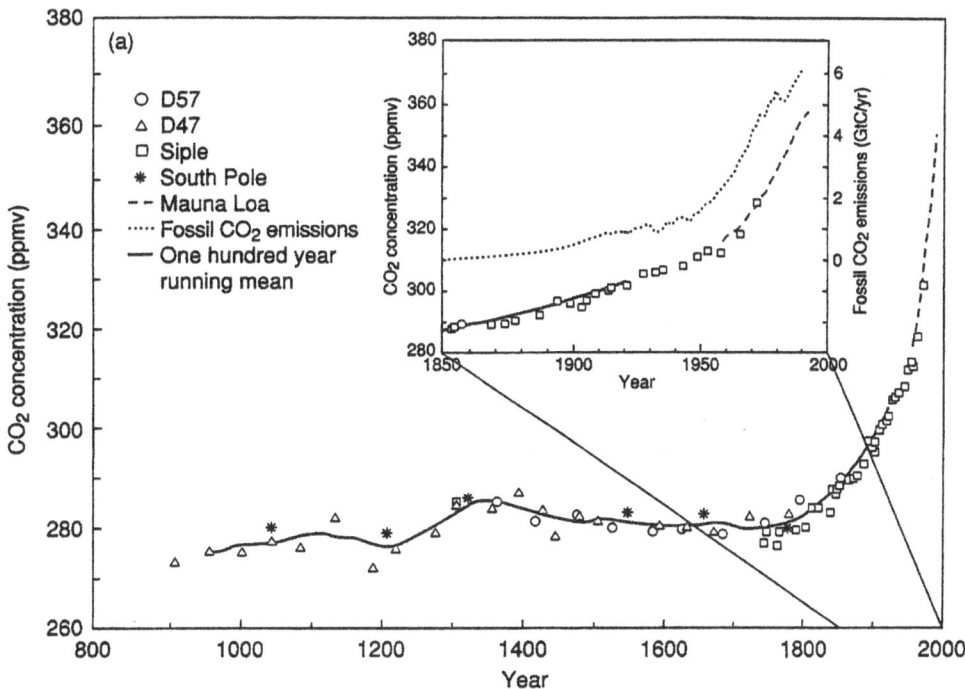

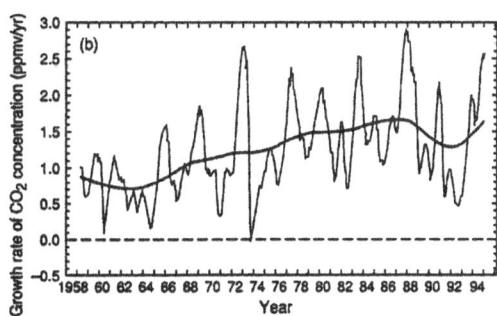

Fig. 12.1 (a) CO_2 concentrations over the past 1000 years from ice core records (D47, D57, Siple and South Pole; all in Antarctica) and (since 1958) from Mauna Loa, Hawaii. The smooth curve is based on a 100-year running mean. The rapid increase in CO_2 concentration since the onset of industrialization is evident and has followed closely the increase in CO_2 emissions from fossil fuels (see inset of period from 1850 onwards). (b) Growth rate of CO_2 concentration since 1958 in ppm per year at Mauna Loa. The smooth curve shows the same data but filtered to suppress variations on time-scales less than approximately 10 years.

has been found to have increased by 30%, from 280 ppm (at the end of the eighteenth century) to 370 ppm in 1999 (IPCC, 1995, 2001a). The rate of increase in CO_2 concentrations between 1980 and 1990 was approximately 1.5 ppm per year, while in 1994 it reached 2 ppm per year. In 1997 the concentration of atmospheric CO_2 was estimated to be 365 ppm. Figure 12.1 shows changes in CO_2 emissions to the atmosphere and its changing concentration. The major cause of the increasing concentration of CO_2 has

been the burning of fossil fuels (oil, gas and coal) and cement production. At present fossil fuel provides 80% of global energy requirements. The destruction of forests, particularly tropical forests, which leads to the decreasing use of CO_2 by plants during photosynthesis has reinforced the effects of fossil fuel burning.

In addition to CO_2, other gases released into the atmosphere can affect the chemical composition of the atmosphere and its thermal regime. Among these gases, of greatest importance are the so-called freons, widely used by the refrigeration industry and in manufacturing paint. The use of substances containing freons promotes their release to the atmosphere as gases, where they can have a long residence time (tens of years). The action of freons in the atmosphere is similar to that of CO_2; they are transparent to other wavelengths, but they intensively absorb infrared radiation. Even in small concentrations they can exert a pronounced effect on the surface air temperature, contributing to the greenhouse effect. The rate of increase in freon concentration in the atmosphere is much above that for CO_2 as it is for some of the other trace gases (e.g. methane). Budyko (1980) showed that the probability of climate change due to increases in trace gases is much less than that for CO_2, as their mass is insignificant, and their emissions can be comparatively easily controlled without disadvantaging the principal economic sectors. While these conclusions may be appropriate for the future, so far, there is an upward trend in their atmospheric concentrations. As mentioned in Bolin et al. (1989), their role in climate change is as important as that of CO_2.

There is also input to the atmosphere of tiny particles of various substances, or aerosols, which are increasing the natural aerosol concentration. According to estimates (IPCC, 1995) the mass of aerosol produced by human activities that has entered the atmosphere during the present epoch is approximately 200–400 million tonnes. This amounts to 10–20% of the total quantity of aerosols entering the atmosphere. The influence of this aerosol on climate is very complicated and is not completely understood; however, investigations show that on the whole, the increase in aerosols can cause cooling of the global climate. This effect is important where the aerosol-forced cooling (mainly due to emissions of sulphur compounds) might have already exerted effects on the observed rate of temperature change.

For at least 200 years observations of climate have been made by instruments at many locations across the globe. For the last 100 years, there are reliable records of climate, primarily air temperature, for many places in the Northern Hemisphere. What is the global air temperature response to the changing chemical composition of the atmosphere, in particular, to the increasing concentration of CO_2? To answer this question is very difficult, because the temperature variations are caused by a combination of natural and human factors. Recently attempts have been made to decipher the role of the individual factors, in particular the growth of CO_2 concentration. Independent studies carried out in different countries have come to a common conclusion that throughout the twentieth century the mean global air temperature has tended to rise, with a marked increase during the last 15 to 20 years. However, the inferences about the scale of this warming and its causes are ambiguous. For example, the international conference on the role of carbon dioxide and other gases causing the greenhouse effect in changing climate and related implications (Bolin et al., 1989) reached the conclusion that during the last 100 years the global mean air temperature has risen by 0.3–0.7 °C. Strictly speaking, it is not possible to associate this rise with the increase in the concentration of CO_2 and trace gases. However, these temperature rises agree with the results from global circulation models.

In the joint Soviet–American report *Prospects for Future Climate* (Budyko et al, 1991), the mean global air temperature was found to have risen by 0.4–0.5 °C from the late nineteenth century to the early 1990s (with an estimation error of about 0.1 °C). The major cause of this warming is the greenhouse effect, although it is possible that natural factors have increased or decreased the warming forced by the greenhouse effect during the last century by a few tenths of a degree Celsius. The 1995 report of the Intergovernmental Panel on Climate Change (IPCC, 1995) concluded that the mean air temperature has increased by 0.3–0.6 °C since the late nineteenth century. This conclusion was also reached by Jones (1988, 1990, 1994) using an extensive database of land-surface air temperatures. Over the last 30 to 40 years, temperatures have increased by 0.2–0.3 °C, and the 1980s and the 1990s included the warmest years since 1860 (Fig. 12.2). The possible influences on the warming of other factors of a human origin not related to the greenhouse effect are also discussed in the report. Urbanization is mentioned as one of these factors, leading to a warming over the land of no less than 0.05 °C, on average, for the twentieth century (Jones et al., 1990). A possible relation between climate change and desertification is also discussed. The warming was found to be much more noticeable in arid regions than for the land surface as a whole. However, bearing in mind that the arid regions comprise only 14% of the land area, the conclusion is that even if the additional warming in these regions is caused by desertification, it adds no more than a few hundredths of a degree to the total global warming. Undoubtedly, the empirical data cited is testimony to the critical importance of the growth of atmospheric CO_2 to global climate warming. Comparisons made between the theoretical and empirical estimates of temperature change for the last 100 years indicate the probability that the global warming occurring for this period is a consequence of increasing atmospheric CO_2.

Because of its significant temporal and spatial variations as well as the large errors of measurements, a much greater uncertainty arises in studying modern trends in precipitation. In spite of these problems recent studies of annual precipitation changes have yielded a number of important results (IPCC, 1995, 2001a; Carter et al., 2000). Principally, precipitation totals in the twentieth century were found to have increased in high latitudes and to have decreased in Northern Africa. According to Groisman et al. (1991) precipitation has increased by 10% in 100 years in the territory of the former Soviet Union. In Europe (not taking into account the European territory of the former Soviet Union) the changes during the past 100 years depend on the latitude. In Northern Europe (north of 55°) precipitation has increased since the 1960s (IPCC, 1995). In Central Europe changes in precipitation have been detected for the past 100 years (Brazdil, 1992), while in Southern Europe decreases have occurred (Palmieri et al., 1991; Dahlstrom, 1994).

There have also been increases in North America since the 1950s (Karl et al., 1993; Karl and Knight, 1998). In the USA, during the last 40 years, annual precipitation totals have risen by 5% (Karl et al., 1993; Groisman and Easterling, 1994) and in Canada by 10–15%. The greatest growth was observed in the eastern part of North America (Findlay et al., 1994; Lettenmaier

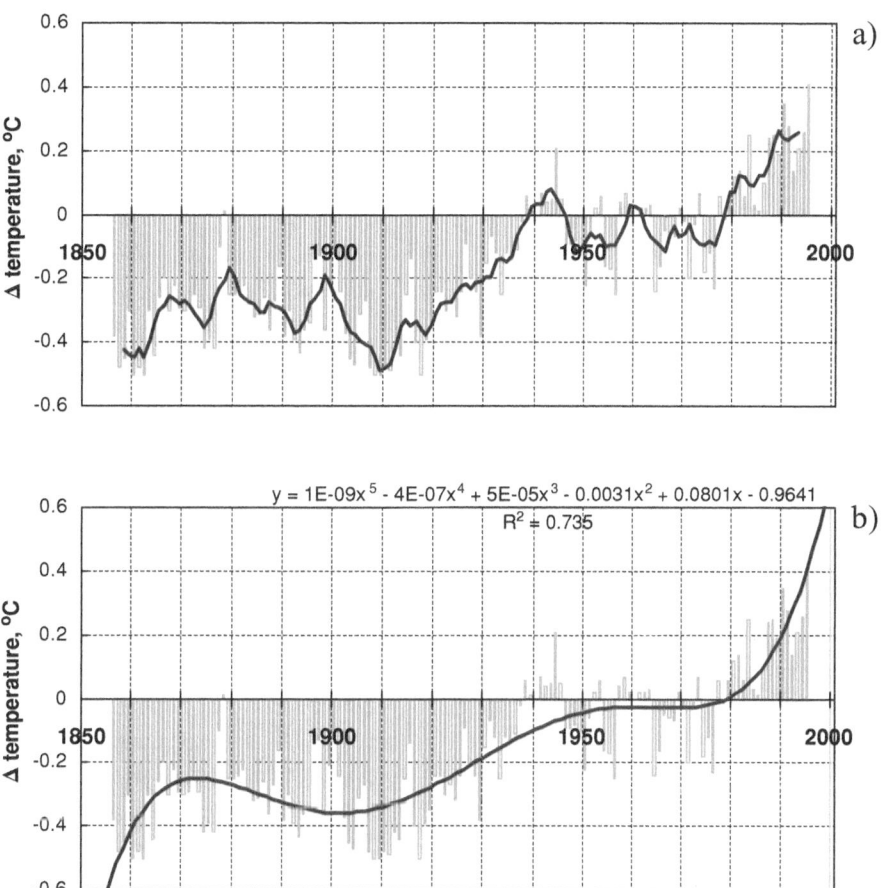

Fig. 12.2 Fluctuations of global surface air temperature from observation data. (a) Moving averaging by 5 years; (b) polynomial averaging in the power of 5.

et al., 1994). Overall precipitation has increased in North America and decreased in South America. Decreasing precipitation, especially since the middle of the twentieth century, has been recorded over large areas in the tropical and subtropical regions of North Africa, Eastern Asia, South East Asia and Indonesia (IPCC, 1995). Throughout the twentieth century, there has been an insignificant (1%) rise in precipitation over the globe's land surface. However, the land covers less than 30% of the surface of the globe, and measurements of precipitation at sea are virtually impossible to undertake. Knowledge of precipitation over the whole globe relies on a combination of gauge and satellite measurements. A relationship has been shown between variations in precipitation and temperature for the different latitudes. In the middle and high latitudes of the Northern Hemisphere, there is a simultaneous increase in precipitation and temperature, while in the tropics, subtropics and the middle latitudes of the Southern Hemisphere, the temperature rise is accompanied by a decrease in precipitation. This is a major feature of the temperature rise that started in the mid-1970s.

12.3 CLIMATE CHANGE SCENARIOS

The increasing concentrations of CO_2 and trace gases along with emissions of aerosols are changing the chemistry of the atmosphere and the global climate. The burning of fossil fuels and alterations of the vegetation and soil are major causative factors. The increase in plant productivity with the rise in CO_2 concentration may cause changes in the Earth's vegetation cover, but is likely to have only a small affect on the future global climate. Future CO_2 levels will continue to rise at a rate determined by the likely trends in the development of power generation and industry across the world as well as by the national measures put into force for restricting the release of CO_2 and other greenhouse gases. The growth in the number of road vehicles powered by petroleum products will also be a very significant factor. Forecasts indicate the rate of increase to be rather high during the coming decades, and throughout the twenty-first century, with a doubling of CO_2 concentrations expected by 2050 when a value of 700 ppm is likely to be reached (Fig. 12.3). However, there are doubts about the timing of the CO_2 doubling. The IPCC in its 1995 report (IPCC, 1995) gave somewhat low estimates of concentrations. One of the variants of these estimates is presented in Fig. 12.3. This forecast

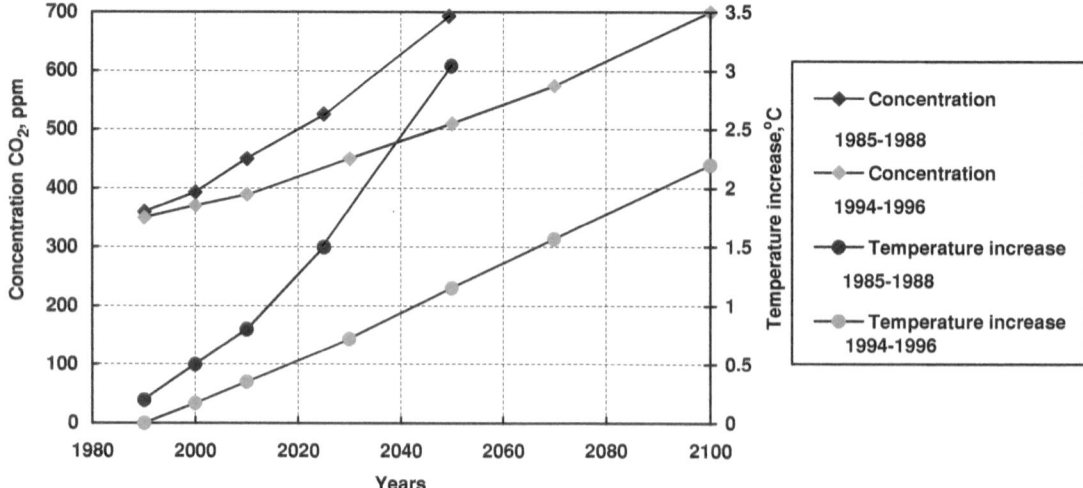

Fig. 12.3 An assessment of future atmospheric CO_2 concentration and global air temperature, based on observations obtained during 1985–8 and 1994–6.

shows a doubling of CO_2 concentration by the end of the twenty-first century. For this Monograph two separate approaches were developed for forecasting climate.

The first approach is theoretical and employs modelling of the general circulation of the atmosphere to forecast the future climate and to produce climate scenarios. The models are based on the laws of physics and take into account explicitly the dynamic, thermodynamic and hydrometeorological processes in the atmosphere and their interaction with the ocean and the land. These climatic models are being refined and improved and can accurately reproduce some of the variables of the global climate. However, they are based on general premises and take no account of geographical features. Therefore, considerable discrepancies may be observed in regional climate forecasts even if they are made within one scenario. And while the tropospheric aerosol resulting from human activities is distributed very unevenly in space, the models trying to take into account aerosol effects are based on a very simplified idea of these effects. In addition most models do not consider the factors characterizing changing land use, which are able to exert an influence on regional air temperature and precipitation, especially in the tropics and subtropics. Therefore the current state of development of climate models does not provide reliable forecasts of regional and seasonal changes in the climate.

The second approach, which is empirical, is based on analysing past climatic conditions. The main idea behind the use of palaeoclimatic data to forecast the future climate consists in searching for analogues in the past, when CO_2 concentrations were close to those forecast for the future, and using the climatic data of that period to describe the future climate. For instance, use is made of the Pliocene climatic optimum which occurred a few million years

ago as an analogue for the climatic conditions in 2050 AD. It was a moderately warm period with average air temperatures some 3–4 °C above the present, and with levels of CO_2 close to 700 to 1000 ppm. For this period, detailed climatic maps have been prepared for individual regions (Sinitsyn, 1967). SHI studies show that, for a 1 °C and 1.5 °C increase in the mean global air temperature, as compared to the current climate, the Holocene climatic optimum (5000 to 6000 years ago) and the last Mikulino interglacial (125 000 years ago), respectively, could serve as palaeoclimatic analogues.

There are a number of disadvantages in the empirical approach, however. Some variables in past warm epochs differ from the expected climate variables of future climates. For instance, at present, when rapid changes in the chemical composition of the atmosphere are occurring, the thermal inertia of the oceans can retard the warming and this makes the forecasts more complicated (Budyko *et al.*, 1991). There is also the point that the most effective use of this method is to study other climate-forming factors, which could have determined past climatic conditions. Consequently these assessments of future climatic conditions cannot really be considered as forecasts, but as possible scenarios of the future. So, for studies of the effects of climate change on water resources, it is most appropriate to use the results from the two methods to reveal the general trends of temperature and precipitation at the regional scale.

It is valuable to look at some of the results obtained during the last decade by general circulation models (GCMs) and from palaeoclimatic analogues. Bolin *et al.* (1989) conclude that a doubling of the CO_2 concentration is not expected earlier than the middle of the twenty-first century, and possibly only by the beginning of the twenty-second. In this case, the estimations of mean global warming based on GCMs gave values in the range from 1.5 to 4.5 °C. The warming is supposed to be more pronounced in the high latitudes where precipitation also increases. The joint

Soviet-American report (Budyko *et al.,* 1991) postulates a doubling of atmospheric CO_2 by the second half of the twenty-first century, as compared with the pre-industrial level. The results from a GCM based on this assumption, including a mixed ocean layer, show the mean global air temperature would rise by 2.8–5.2 °C. However, based on more recent model results, the calculated warming is only expected to reach 2.7 °C taking account of the cloudiness effects of water–ice transformations, and 1.9 °C with introduction of the changing optical properties of clouds.

The geographical distribution of changes in the surface air temperature for a doubled CO_2 calculated by the GCMs including an upper mixed ocean layer shows that the warming would occur everywhere in both winter and summer. In general, the warming would be a minimum in the tropics for both seasons and increase poleward. For precipitation there are both positive and negative changes, with most occurring between 30° S and 30° N. Poleward changes are mainly positive in both seasons. The same study presents estimates of the future climate obtained by analogues. A doubling of the CO_2 would result in a rise in mean global air temperature of about 3 °C, but the relationship has a logarithmic character. Warming would increase with the mean global temperature rising by about 0.75 °C for each 25 to 50 years depending on the rate of increase in greenhouse gases and on thermal inertia. During past warm epochs the temperature rise was largest in middle and high latitudes, with little or no rise in low latitudes and a small cooling in arid regions. During warm periods in the past an increasing temperature was accompanied by increasing precipitation. However, estimates of future changes in precipitation are far less certain than the corresponding estimates of temperature.

The results of the 1995 report of the IPCC (IPCC, 1995) are very pertinent to this Monograph. The forecasts of CO_2 concentrations that it gives were made by 18 different models of the carbon cycle. Depending on which scenario is accepted, as well as on model sensitivity, a doubled CO_2 can be expected at any time from the middle of the twenty-first century into the twenty-second century. To stabilize CO_2 concentrations below 750 ppm, emissions would have to be significantly below the present-day level. Stabilizing emissions at the 1980s level would be insufficient, as the increase in concentration will continue during the twenty-first century and well into the next. To stabilize at the 450 ppm level by 2200, emissions have to be reduced by a factor of 3 as compared to today.

Using different scenarios for CO_2 concentrations, the changes in the mean temperature were calculated for low, middle and high climate sensitivity to year 2100. Only the rise in CO_2 concentrations was considered (aerosol concentrations were accepted as constant at the 1990 level). Changes in the mean temperature were estimated to be from 1.0 to 4.5 °C for the different scenarios. This is less than in the 1990 IPCC report (Shiklomanov *et al.,* 1990), in part due to accounting for aerosols for the period up to 1990, in part due to revising the carbon cycle mechanism. Including the future effects of aerosols gives a 1.0–2.5 °C temperature rise by 2100. These forecasts indicate a smaller warming than the forecasts of the mid-1980s (Fig. 12.3). All of the models forecast that the largest warming is likely to occur in high latitudes in the Northern Hemisphere in autumn and winter. Precipitation would increase in high latitudes, and in most scenarios at mid-latitudes as well. In the tropics, the change in precipitation patterns varies from model to model. Nevertheless, many models anticipate an increase in precipitation in India and South East Asia. Most models forecast a decrease in precipitation in Southern Europe and little change in the dry subtropics. If the effects of aerosols are taken into account, there could be a minimal increase in mean global precipitation. Changes in precipitation patterns during the northern winter are mainly similar to those obtained for greenhouse gases alone, although less marked.

The following conclusions can be drawn:

1. Throughout the twenty-first century the global air temperature will be increasing due to growing concentrations of greenhouse gases in the atmosphere. However, to forecast these changes more precisely is impossible, because of the uncertainties surrounding these forecasts such as:
   - assessing the future levels of CO_2 releases by industry because of the political, technological and social factors involved
   - predicting the increase in CO_2, when using different emission scenarios because of insufficient knowledge of the exchange rates between the surface, intermediate and abyssal ocean depths
   - modelling the changing air temperature and other climate characteristics because of lack of knowledge of the feedbacks between cloudiness and albedo, temperature lapse rates and the humidity gradient with temperature.
2. All the GCMs show that the greatest warming is to be expected in high latitudes, especially in winter.
3. The different methods indicate an increase in the annual precipitation at high latitudes.

Scenarios of regional changes in climatic characteristics have been used to assess the expected changes in the hydrology and water resources of river basins (Greco *et al.,* 1994; Carter *et al.,* 2000). Most studies dealing with the hydrological implications of global warming are based on the scenarios obtained by the following GCMs:

- US Geophysical and Fluid Dynamics Laboratory (GFDL) model
- US Goddard Institute for Space Studies (GISS) model
- UK Meteorological Office (UKMO) model
- Canadian Climatic Centre (CCC) model
- German Max Planck Institute (MPI) model.

The results from the GCMs are presented as monthly values of the variables of climate on a regular grid covering the entire terrestrial globe. Several climate change scenarios have been developed based on the GCMs used:

- stationary, for a doubling of CO_2 (Manabe and Wetherald, 1987; Jenne, 1989; Mitchell *et al.*, 1989)
- non-stationary, for a 1% annual increase in CO_2 for different periods (Manabe *et al.*, 1991, 1992; Cubasch *et al.*, 1992; Murphy and Mitchell, 1995).

Every scenario is represented by two variants of the resulting data:

- basic, for a prescribed CO_2 increase
- control, for the current CO_2 concentration.

The difference between the assessments by these variants is the modelled change. Along with the scenarios based on GCMs, Russian scientists have used those based on palaeoclimate for 1 °C, 2 °C, and 3–4 °C warmings (Budyko and Israel, 1987; Budyko *et al.*, 1991; Anisimov and Poljakov, 1999).

12.4 THE LIKELY IMPACT OF CLIMATE CHANGE ON WATER RESOURCES

Assessing, forecasting and predicting future water resources have been among the most critical problems in scientific hydrology. They are directly associated with other important practical tasks such as population planning, developing and managing water supplies, and consequently with generalizing large-scale water management. Throughout the twentieth century various methods have been developed and practised for assessing water resources, for example those based on the theory of probability. Some of these methods rely on analysing runoff over the long term and are based on the hypothesis of stationarity of climate in the past and in the near future. Experience gained in designing and using water-management systems confirms that this hypothesis is valid (disregarding the human impact on runoff in river channels and watersheds). This result has also been obtained in studies of runoff variations using long-time observations from around the world made by Yevjevitch (1977), Schaake and Kaczmarek (1979), and for rivers in the former Soviet Union by Rozhdestvensky (1988). Chiew and McMahon (1995) analysed the variations in runoff in 142 rivers around the world with observations for more than 50 years in basins larger that $1000\,\mathrm{km}^2$. They found that statistically significant trends occurred in only a few regions and that they did not agree with each other. Similar studies made for this Monograph also show the absence of significant trends in long-term discharges and several other global studies have reported the same findings. Serious warnings of impending climate warming

due to the greenhouse effect came from the 1985 UNEP/WMO/ICSU Conference held in Villach, Austria (Bolin *et al.*, 1989).

ANALYSES OF LONG-TERM OBSERVATIONS OF RUNOFF AND CERTAIN METEOROLOGICAL VARIABLES

Studies of long-term observations are common. They analyse the hydrological characteristics of selected basins and regions and particularly the patterns of the departures from the long-term averages. Periods with significant trends and anomalies, either positive or negative, may be associated with climate warming. This type of study has been conducted in a number of basins in different parts of the world such as by Glantz and Ausubel (1988), Chunzhen (1989), Georgiyevsky *et al.* (1995, 1996), Lins and Slack (1999) and Shiklomanov *et al.* (2002). The results of these studies are of considerable scientific interest but how far they can be used to make forecasts remains unclear. During warm periods in the past precipitation increased and decreased. For instance, during the last 1000 years in Eastern Europe, there were two periods with a significant temperature rise: the 1930s and the 1980s to 1990s. However, the warming during the first period was accompanied by a noticeable decrease in precipitation, while precipitation increased during the second. So, there is still the question of what period can be considered as an analogue of possible future climatic conditions. Climatologists have recently developed an approach for assessing the future climatic conditions (Budyko and Groisman, 1989) based on records for the 1980s and 1990s. However, it is preferable to use the hydrological conditions observed during the current warming as an analogue, although the correctness of this analogy cannot be proved for the present.

THE USE OF REGRESSION RELATIONSHIPS BETWEEN RUNOFF CHARACTERISTICS AND METEOROLOGICAL FACTORS

Relationships between annual values of runoff, precipitation and temperature have been established for many basins around the world. These relationships have been used to assess possible changes in the hydrological regime due to climate warming (Stockton and Boggess, 1979; Speranskaya, 1988). This approach is very simple and requires comparatively little data. However, caution must be exercised with the inferences drawn from extrapolating into the future the results obtained from regression relationships based on historic data. These data correspond to a particular situation within the distribution of meteorological factors that are highly unlikely to be repeated in the future. Obviously, with the same annual values for temperature and precipitation, the annual runoff can vary considerably depending on the distribution of meteorological factors through the year. Use of the physical–statistical relationships that are widely applied to forecast runoff are more promising. For cold and temperate climates quite close relationships can be established between the spring flood discharge and the maximum snow storage and the soil temperature and moisture characteristics. However, there are difficulties over assessing these characteristics for the future climates.

USING THE WATER BALANCE METHOD OVER THE LONG-TERM PERIOD

The major problem in applying the water balance method is assessing the total evaporation from a basin, taking into account possible future changes in precipitation and temperature. This approach has been widely applied to assess future water resources with certain climatic scenarios for most parts of the globe (Glantz and Wigley, 1987; Vinnikov and Lemeshko, 1987; Griffiths, 1989; Vinnikov et al., 1990; Babkin et al., 1992; Budagovsky and Busarova, 1991; Busarova and Gusev, 1995; Arnell, 1999b). In many cases use was made of one or the other of the so-called "combination equations" to assess evaporation. Oldekop (1901) first put forward the idea that the mean annual evaporation from the surfaces of river basins is a function of maximum possible evaporation (potential evaporation) and precipitation and to suggest the nature of this relationship. Subsequently many other researchers have developed this approach (Budyko, 1956; Babkin, 1979; Budagovsky and Busarova, 1991). The main disadvantage of this approach for assessing future water resources is that it does not take into account possible changes in the distribution of precipitation and potential evaporation during the year. A possible way of taking these changes in precipitation into account has been proposed by Budagovsky and Busarova (1991). While this approach is promising, it results in large errors in arid regions.

APPLYING DETERMINISTIC HYDROLOGICAL MODELS

Some of the most detailed model studies have estimated the sensitivity of runoff to changing climatic variables (Gleick, 1986a, 1987, 2000; Flaschka et al., 1989; Kuchment et al., 1990; Bultot and Gellens, 1991; Saelthun, 1991; Georgiyevsky et al., 1996; Kaczmarek and Napiorowsky, 1996; Miller and Kim, 2000; Bergstrom et al., 2001). This approach examines the cause-and-effect relations in the "climate–water resources system", assessing the sensitivity of basins to changing climate characteristics, estimating possible changes in runoff under different natural conditions and planning future water management strategies in the light of regional climate forecasts. These mathematical models differ in their level of comprehensiveness in the descriptions of individual processes, the information needed, the time intervals dealt with, and in other ways. Their application proceeds in stages. To assess a basin's response to climate, the model is first calibrated on the data of that basin. Then for a particular climate the changes in the hydrological variables are estimated. By comparing the results obtained for the current and forecast climate conclusions can be

drawn about changes in runoff in the basin in question. These studies have been carried out predominantly in small and medium-size basins.

As an example, a water balance model has been developed at the State Hydrological Institute (SHI) and has been tested in various circumstances. It uses 10-day data for temperature and humidity as well as evaporation, soil moisture, snow accumulation and melt, ground water changes, surface and subsurface runoff.

The model has three layers: the active soil storage and the upper and lower ground water storages:

$$\frac{dW_S}{dt} = P - E - R_S - I_U + S \tag{12.1}$$

$$\frac{dW}{dt} = I_U - I_D - R_U \tag{12.2}$$

$$\frac{dW_D}{dt} = I_D - R_D \tag{12.3}$$

where W_S, W_U, and W_D are , respectively, the storage in the active soil and upper and lower ground water stores; R_S, R_U and R_D are the surface and ground water runoffs; E is the evaporation; P is the precipitation; S is the snowmelt rate; I_U and I_D are the infiltration to the ground water table.

The numerical solution of equations 12.1 to 12.3 with a 10-day time interval includes the estimation of E, I_U, I_D, S, R_U and R_D, while R_S is determined as a residual term in the water balance of the active soil layer. Evaporation and the active soil storage are estimated by the heat water balance method (Kharchenko, 1975). The estimates of evaporation are based on the relationship:

$$\frac{E}{E_0} = \beta \cdot \frac{W}{W_{lm}} \tag{12.4}$$

where E_0 is the potential evaporation, β is the parameter characterizing the vegetation, W is the average productive moisture content in the active soil layer for the design time interval, W_{lm} is the productive moisture storage of the active soil layer with the least water capacity. Evaporation from snow is estimated by Kouzmin's formula (Kouzmin, 1961) depending on the air humidity deficit, and the snowmelt rate by the temperature coefficient method.

For the study time interval, if the precipitation is greater than the evaporation, and the soil moisture is in excess of the maximum water-holding capacity, the moisture surplus (H) represents the future runoff. The distribution of discharge in the surface component (R_S) and recharge of the upper ground water reservoir (I_U) are carried out as follows:

$$R_S = H - I_U \tag{12.5}$$

$$I_U = \min(H, F_{max}) \tag{12.6}$$

where F_{max} is the maximum possible infiltration from the zone of aeration into the deeper layers depending on the type of soil, and at the beginning of the snowmelt period on the depth to which the upper soil layer is frozen. Then for the infiltration (I_U) to the water table, the runoff (R_U) and the infiltration to the lower aquifer (I_D), the runoff from both aquifers corresponds to a linear reservoir:

$$R_U = A_U \cdot W_U \qquad (12.7)$$

$$R_D = A_D \cdot W_D \qquad (12.8)$$

$$I_D = B \cdot W_U \qquad (12.9)$$

where R_U and R_D are the ground water inflows to the stream network, W_U and W_D are the water stored in the upper and lower aquifers; A_U, A_D and B are the coefficients of ground water discharge determined by the recession curves. The freezing depth is calculated by the formula:

$$L = a\sqrt{\sum(T-)} - bS \qquad (12.10)$$

where L is the freezing depth, $\sum(T-)$ and S are the sum of temperatures below zero and the snow storage at the beginning of snowmelt, respectively; a and b are the empirical coefficients. Infiltration into the frozen soil is calculated from the relationship:

$$F_{mf} = \begin{cases} F_{max} & L < L_{min} \\ F_{max}\dfrac{L - L_{min}}{L_{max} - L_{min}} & L_{max} > L > L_{min} \\ 0 & L > L_{max} \end{cases} \qquad (12.11)$$

where F_{mf} and F_{max} are the maximum possible, for the study time interval, infiltration into frozen and melting soil; L_{max} and L_{min} are the critical values of the freezing depth; and L is the real freezing depth. This model was used to assess possible changes in the hydrological regime and in water resources due to global warming, and applied to the river basins across the former Soviet Union (Shiklomanov and Georgiyevsky, 1992; Georgiyevsky *et al.*, 1995, 1996, 1997; Shiklomanov *et al.*, 2002).

THE APPLICATION OF GENERAL CIRCULATION MODELS

Three-dimensional models of the general circulation describing the interactions of the "atmosphere–land–ocean" system allow, in principle, the assessment of global climate variables for different concentrations of CO_2 and their regional characteristics, as well as changes in the hydrological cycle. Estimates of possible changes in runoff, soil moisture and evaporation have been obtained directly from these models for the USA and Canada, for example by Sanderson and Wong (1987) and by Singh (1987). However, the use of GCMs for this purpose seems to be rather limited. Most current models have a horizontal resolution of between $2.5°$ and $8°$ and are poor at incorporating the features of the land surface. Because of a large grid size, the physical features can be "lost", while the hydrological variables are prescribed only approximately. But these features are important controls of runoff

and knowledge of their spatial variability determines their parameterization. This problem of parameterization of subgrid processes, and the presentation on the subgrid scale depending on the values averaged over the spatial scale of the grid used in the GCM, has been considered by Kondratjyev (1992). There are a number of other unsolved problems surrounding the parameterization of the hydrological cycle, particularly:

- developing probabilistic methods for describing the spatial variations of "forcing parameters" (especially precipitation)
- improving parameterization of hydrothermal processes in snow cover (primarily this refers to the albedo of the snow)
- establishing relationships between hydraulic and thermal soil properties and the soil texture classes contained in global soil maps and determining analogous relations between the rooting depth, soil moisture and vegetation cover classes from season to season on global vegetation maps
- developing parameterization techniques for the state of vegetation both natural and agricultural (without and with irrigation) in terms of the dependence of the vegetation type and the density of cover, albedo and transpiration rate on soil moisture and roughness, rooting depth and seasonal vegetation patterns.

Such problems require investigation in test areas in basins located in different climatic zones coupled to remote sensing campaigns.

The prospects for building hydrological models coupled to GCMs have been considered by Kuchment and O'Connell (1993). They indicate that models on scales of 100 to 300 km (the resolution of contemporary GCMs) can be developed in a few years and used for improving GCMs and modelling the effects of climate change on runoff changes. Developing hydrological models for GCMs on scales of 30 to 100 km and less is a problem for the future. The use of general circulation models to assess the hydrological impacts of climate change at the regional scale is very promising. In a few years new results may be obtained that will allow reliable assessments of possible changes in water resources due to global warming.

12.5 CURRENT BASIN RESPONSES TO CLIMATE CHANGE

Since the late nineteenth century the global climate has warmed by about $0.5°C$: a distinct upward trend can be observed in the global temperature record. Evidence of a similar trend has been sought in other geophysical time series including records of the runoff. Unfortunately, these studies are not widespread, but they are extremely important as they can reveal tendencies and can provide time for developing and undertaking adaptation

Table 12.1. *Deviation of annual and seasonal runoff from the normal (%) on the Russian Plain for the period 1978–90*

River	Basin	Whole year	Winter	Summer, autumn	Spring
Pechora	Barents Sea	1	1	5	0
Onega	White Sea	2	14	4	−12
Sukhona		1	12	−3	6
Narva	Baltic Sea	19	25	−13	−2
Zap. Dvina		7	11	16	1
Neman		4	1	5	4
Volga	Caspian Sea	15	58	42	−12
Unzha		41	72	62	27
Kostroma		23	34	53	14
Vetluga		36	47	50	31
Vyatka		24	51	49	11
Kama		15	21	18	12
Chusovaya		6	19	35	−8
Belaya		5	35	19	−7
Oka		9	38	33	−6
Protva		28	68	72	6
Ugra		18	43	34	6
Dnieper	Black Sea	3	24	22	−10
Berezina		0	9	6	−10
Pripyat		13	24	25	2
Desna		16	38	53	−3
Don	Azov Sea	5	39	38	−11
Trudy		29	61	54	13
Olym		15	65	65	−9
Sosna		16	48	57	−3
Sev. Donets		5	34	27	−9
Tikhaya Sosna		14	14	62	−1

measures. The most detailed studies of contemporary changes in hydrology have been made in Russia and in neighbouring countries (Shiklomanov and Georgiyevsky, 1992; Georgiyevsky *et al.*, 1995, 1996, 1997; Shiklomanov *et al.*, 2002). Monthly runoff data were selected for stations on 80 rivers, located within the former Soviet Union. Each station had observations for 60 to 110 years and their basins were practically undisturbed. These are mainly medium-size basins from 5000 to 50 000 km² in area and representative of different geographical zones. The analysis of variations in runoff showed that, from the beginning of the second half of the 1970s to the present, over a considerable part of the area including the Eastern Europe and Western Siberia, extraordinary changes in the distribution of runoff have occurred. There has been an increase in flows during the summer, autumn and winter, previously the months of low flows. In the basins of Eastern Siberia and in mountainous rivers, there are no changes (Table 12.1). For the Volga, Don and Dnieper, the summer–autumn–winter flows were 20–40% above average over the last 15 to 20 years, while

for some regions (the northern part of the Volga and Oka basins) the increase was 50–70%. While the annual flows have increased, the spring flood discharges have decreased for most of the rivers (Table 12.1). Data for the past 100 years indicated that the principal control of annual runoff was the value of the spring discharge and this makes the hydrological conditions of the last 30 years somewhat unusual. Figure 12.4 shows the distribution of runoff in the Dnieper Basin. Similar patterns were found to exist for most basins in the East European Plain.

Additionally for the European territory of the former Soviet Union (ETSU), runoff data from 196 small rivers with basins of less than 3000 km² in area and with 40 to 60 years of records were investigated; these were basins whose hydrology was expected to be sensitive to climate change. The results of the analysis of these data fully confirmed the conclusions obtained for the medium-size basins. For the overwhelming majority of small rivers, discharges during the former period of low water (summer, autumn and winter) had increased by more than 20% from 1978 to 1990; the annual runoff had increased but the spring runoff had decreased in most small basins.

To investigate the causes of these changes data have been analysed from seven water-balance stations. These stations are located in the forest, forest–steppe and steppe zones of the ETSU and they recorded all the variables of the hydrological cycle from the late 1940s to the mid-1950s. The analysis showed that there had been considerable changes in precipitation and temperature (Yefimova *et al.*, 1994, 1996). For all water-balance stations, there was a significant increase in the annual precipitation (by 35–80 mm, or 5–15%), mostly during the summer and autumn, and a rise in temperature during the winter (1.0–1.2 °C). The rise in temperatures made thaws more frequent, the soil froze less often, and a considerable part of the meltwater infiltrated to increase the soil moisture content. Due to the increasing summer precipitation, the soil moisture content rose 10–30 mm during the summer and autumn. This caused infiltration and ground water recharge, and the levels of the water tables in the major aquifers were raised by 50–100 cm (Fig. 12.5). Increasing ground water resources led, correspondingly, to a greater baseflow component in the runoff. The increases in winter runoff were 15–25% in the forest–steppe zone and 50–60% in the forest zone. The summer–autumn runoff also increased by 50–60%, while the annual runoff was 5–25% greater.

It should be emphasized that the above data refer to the runoff in rivers draining major aquifers. As to temporary streams as well as the small rivers not draining major aquifers, in spite of the increased precipitation, their runoff was found to decrease in all seasons. This is explained by the fact that the water largely drains into aquifers, which then feed the larger rivers. The same conclusions have been reached in a number of other studies of runoff in Europe. For example, Schumann (1993) analysed long-term data

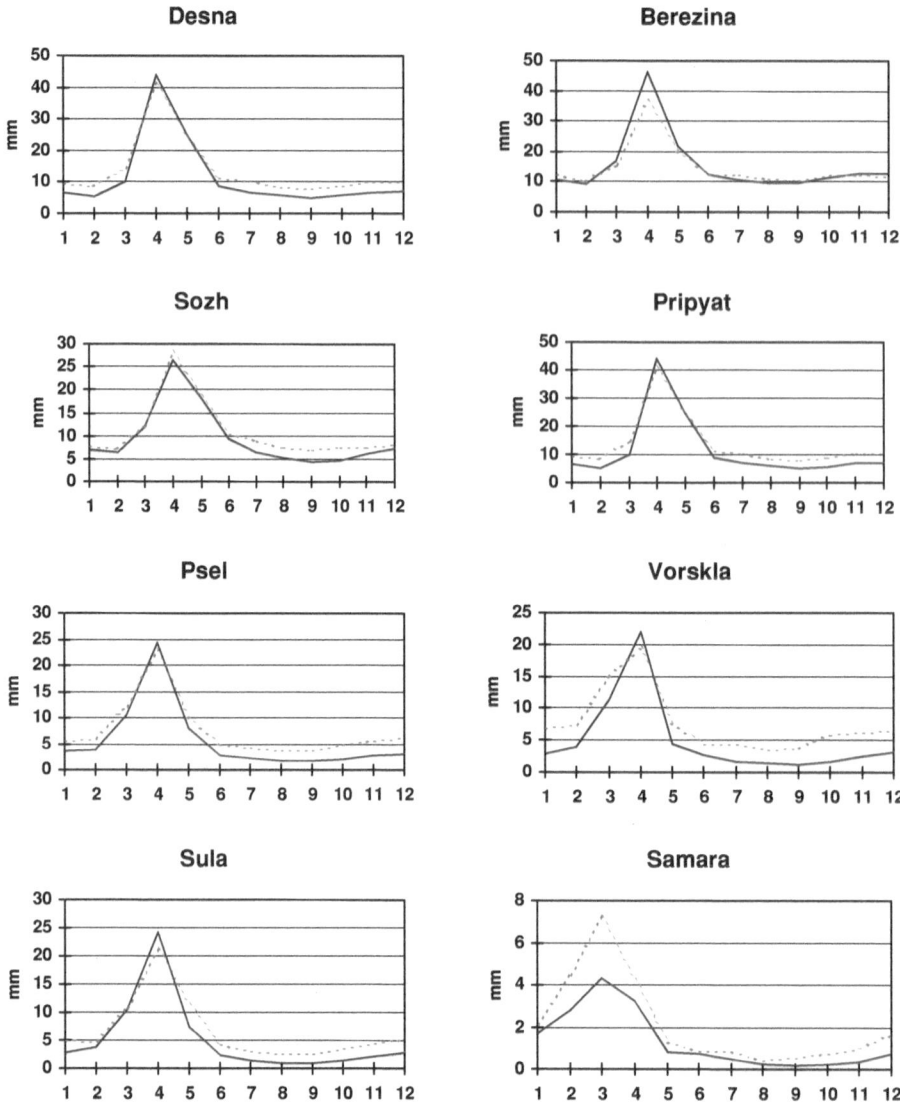

Fig. 12.4 Seasonal distribution of runoff in the Dnieper basin.
——— 1945–75; ...... 1976–92.

on runoff, including some stations with more than 150 years of records, for different regions of Germany. He showed that since the second half of the 1960s winter runoff in Alpine rivers has increased as a result of warming. Bergstrom and Carlsson (1993) and Tarend (1998) also found a significant increase in winter runoff for flows to the Baltic. In her study of variations in runoff over the most recent period Polutikov (1987) found there were no significant changes in the annual runoff, but there were changes on a seasonal basis. In the 10 basins investigated monthly runoff decreased by 50% in summer and increased 60% in winter, while the changes in annual runoff were within 5%.

The sensitivity of hydrological regimes to small changes in temperature (within 1 °C) has been studied for 49 river basins covering the different regions of Scandinavia (Krasovskaya, 1995, 1996).

Greatest sensitivity was shown to exist to relatively mild winters when runoff was at a minimum during the summer. The effects of temperature variations are most noticeable during warm years, when the winter runoff increases. Research conducted in North America has revealed similar results. For example a statistically significant increase has been discovered in autumn and winter runoff from 1948 to 1988 over almost the entire United States (Lins and Michaels, 1994; Lins and Slack, 1999). The largest increase was observed in the Rocky Mountains.

It can be concluded that there has been an increase in runoff in parts of the Northern Hemisphere caused, to a large extent, by rising temperatures especially during the winter. Precipitation has increased in most humid regions. Runoff was found to increase during the summer–autumn low water period in the East European Plain. Studies of the Yenisey (Shiklomanov, 1994), Russia's largest river basin with an area of 2.5 million km^2, show

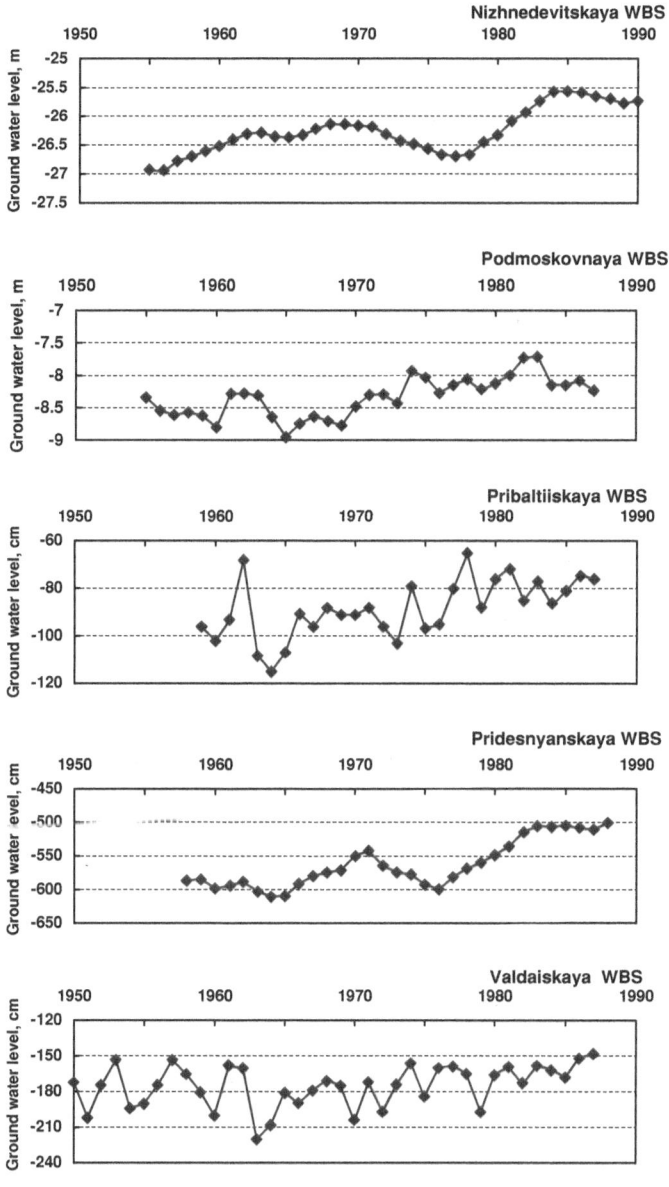

Fig. 12.5 Dynamics of the ground water level based on data from water-balance stations (WBS) in the European territory of the former Soviet Union.

that in the southern and middle parts of the basin, in spite of the 2–3 °C winter warming recorded during the last 15 years, there is no increase in winter runoff due to the very low winter mean temperature.

Marengo (1992, 1995) and Marengo *et al.* (1998) reviewed the results of research on changes in runoff in South America but found no general trends. However, there were some small-scale changes. The decrease in runoff that started in the 1970s in Colombia coincided with a similar trend on the Pacific Coast of Central America. Significant variations were recorded on rivers in northwest Peru. Flows in the Rio Negro increased slightly from

1973 due to increasing precipitation in the northwestern part of the Amazon basin. Budyko *et al.* (1994) also found increasing precipitation in certain regions of South America between 1981 and 1990, especially significant (100–300 mm) in the subequatorial zone, and rises in temperature of 0.5–1 °C. These changes resulted in a noticeable increase in the water resources of the continent (see Chapter 8). Thus, contemporary changes in river flows in South America are, probably, caused primarily by natural variations in regional precipitation (Waylen *et al.,* in press). According to Singh (1997) since the 1980s in the southern Caribbean there have been rising temperatures and decreasing precipitation, and this leads to changing water resources.

The decreasing trend in runoff in Africa is described in Chapter 6. Sircoulon (1990) found that African river flows to the Atlantic Ocean decreased by 17% between 1981 and 1990, as compared to the average for 1951 to 1990; while the runoff from arid Africa decreased by 27%. Attention has been given to the Sahel in particular, because the recent reduction in precipitation has created a critical situation in many areas. Sircoulon (1990) found that the 1970s and 1980s were the driest periods during the time with instrumental observations. The average reduction in runoff from 1970 to 1988 was 43%, while in certain years it reached 74%. Closely matching results have been obtained by Yoshino (1999).

As to the future hydrological regime of the Sahel, there is great uncertainty about what will happen. In recent years precipitation has increased and it is quite possible that the dry phase has been replaced by a wet one. The trends observed in the Sahel during the recent decades were experienced in some other arid and semi-arid regions. For example, there was a severe drought in the southern part of California from 1987 to 1992, when annual runoff was 23% below normal (IWR, 1994). Chunzhen (1989) showed that 1981 was the start of the warmest period for all the years of instrumental record in North China. During the 1980s the mean temperature was 0.5 °C above normal while precipitation was below normal, which resulted in a considerable decrease in water resources.

Decreasing river flows is a trend common to much of Australia. Thomas and Bates (1997) found that in southwest Australia precipitation has decreased significantly since 1975. On many rivers in the Darling Range runoff decreased by about 40% between 1975 and 1994, as compared with 1911 to 1974. Figure 12.6 shows the changes in runoff for one of the representative basins of the region. Since the late 1970s runoff has decreased dramatically due mainly to less winter precipitation, whereas summer precipitation has slightly increased. However, this has no effect on runoff, as it is almost completely lost by evaporation. Precipitation and runoff have tended to decrease in New Zealand since the mid-1970s as has been shown by Mosley's (1991) study of the Kaituna River, which used observations since 1906.

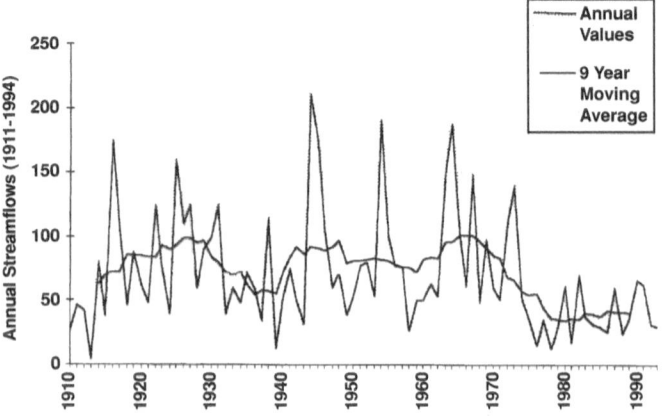

Fig. 12.6 Serpentine annual streamflows (1911–1994) in Australia. (From Thomas and Bates, 1997.)

12.6 THE INFLUENCE OF CLIMATE CHANGE ON WATER RESOURCES AND WATER REQUIREMENTS

Assessing the hydrological implications of global warming and determining the measures for adapting water and water-management systems to these implications are of especially great scientific and practical importance. This is not only because the hydrological regime is very sensitive to climate change, but also because hydrology and water resources directly determine development all over the world. They determine agriculture, the standard of living and the environmental condition. Assessments of future water resources can be obtained only by using the estimates of possible regional changes in climate (primarily, precipitation and temperature by seasons and months). However, such estimates are extremely unreliable even for the largest regions and river basins. To forecast changes in regional climate consequent on global warming, GCMs and palaeoclimatic reconstructions of climatic conditions in warm epochs in the past are the main methods and these were discussed above. During the last decade numerous studies have been made to assess the expected impact of climate change on basin hydrology, water resources and water requirements in the different regions and countries and some of their results and conclusions are given here.

12.6.1 Climate change and runoff

To assess regional implications, most of the studies have used the scenarios of global climate based on the GCMs listed earlier assuming a doubling of CO_2. Scenarios based on palaeoclimatic reconstructions (Budyko and Israel, 1987; Budyko *et al.*, 1991) have also been applied.

COLD AND TEMPERATE CLIMATES

Here, the greater portion of the annual runoff results from the snowmelt which takes place during the spring and early summer. This zone includes regions in northern North America, Northern and Eastern Europe, most of Russia, and Kazakhstan. Changes in runoff in 19 Finnish basins of between 600 and 33 000 km^2 in area were estimated with a doubled CO_2 model. This gave a 2–3 °C temperature rise and an increase in monthly precipitation of 20–30 mm. Mean annual runoff increased by 20–50%, and monthly runoff was more evenly distributed in most basins (Bergstrom, 1996). The maximum spring discharges were forecast to be decreased by up to 55%, but the flows in winter would increase by 100–300%. Especially significant changes are expected in the north and east of Finland.

Two GCM scenarios were used to assess the hydrological implications of global climate warming in Norway. Winter temperatures were assumed to increase by 3.0–3.5 °C, summer temperatures by 1.5–2.0 °C, and annual precipitation by 5–15%. Changes in runoff were calculated using a model developed at the Swedish Meteorological and Hydrological Institute (Bergstrom, 1996) for seven water-management regions in Norway for a 30-year period for the two climate scenarios. Runoff changes were estimated for: mountain territories, foothills, and lowlands. The results obtained showed a 10–15% increase in annual runoff in mountainous regions with a high annual precipitation and with an insignificant reduction in runoff in lowland forested basins. Changes in seasonal runoff were expected to be especially significant: the spring flood will decrease, the winter runoff will increase, summer runoff will decrease and the frequency of autumn and winter floods will rise (Saelthun *et al.*, 1990, 1998; Saelthun, 1991).

Different conclusions about the likely changes in annual runoff have been reached from studies carried out on the Varta River (basin area 54 530 km^2) as well as for parts of Poland and the country as a whole (Kaczmarek and Krasuski, 1991; Kaczmarek and Napiorowski, 1996) using climate scenarios GFDL R-15 and GISS with a doubled CO_2. The expected changes are shown in Fig. 12.7. The GFDL scenario gives a 4–6 °C rise in monthly temperature and a rise of less than 5% in annual precipitation together with a reduction of up to 20% in summer precipitation. The GISS scenario gives a 2–6 °C monthly temperature rise and a 20% rise in the annual precipitation. The former scenario is considered as "warm–dry", and the latter as "warm–wet" (Kaczmarek and Napiorowski, 1996). The effects of climate change on water resources were assessed from the water balance model CLIRU N3 devised by Kaczmarek (1993) and applied to 31 catchments in Poland. Figure 12.7c shows that for the GFDL scenario the annual runoff is expected to decrease by more than 20%, and the summer runoff by up to 35%; for the GISS scenario it is the opposite: annual runoff is expected to increase by more than 20%, including for the summer months. For the country as a whole with a

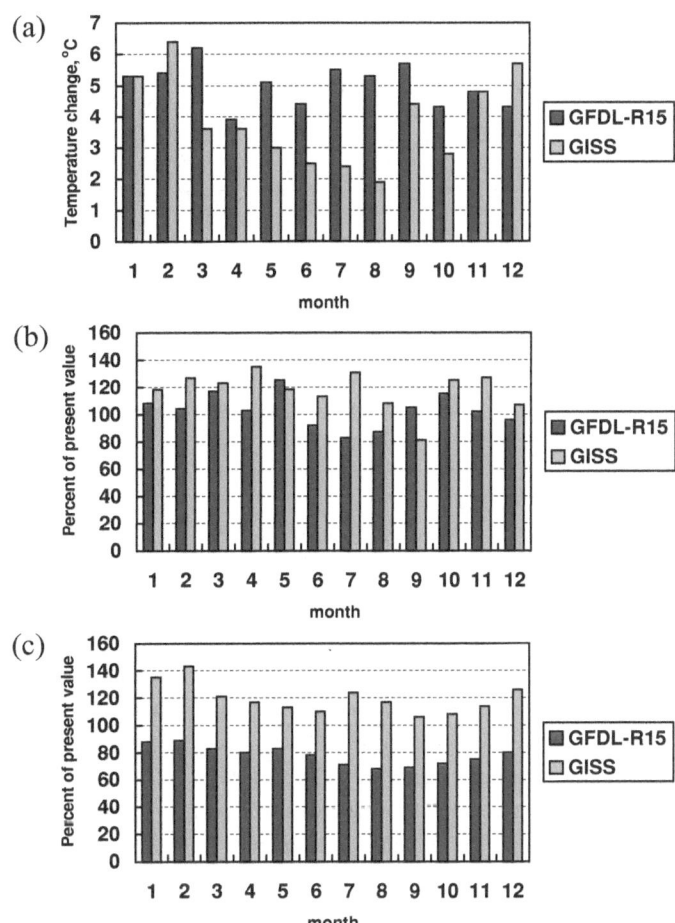

Fig. 12.7 Projected changes in (a) temperature, (b) precipitation and (c) runoff in Central Poland, with doubled concentration of CO_2. Two models were used: US Geophysical and Fluid Dynamics Laboratory R15 (GFDL-R15) and Goddard Institute for Space Studies (GISS).

doubled CO_2, the GFDL scenario shows that the water resources are expected to decrease by 13%, but using the GISS scenario they are expected to increase by 11%. Studies on the Varta (Kaczmarek and Krasuski, 1991) made with a doubled CO_2 scenario show the annual runoff to change slightly; however, the winter runoff is expected to increase by 21%, and the summer to decrease by 24%.

A similar result has been obtained for small and medium-size catchments in Belgium and Switzerland (Bultot and Gellens, 1991; Gellens, 1991a, b) based on model scenarios with a doubled CO_2 and a conceptual hydrological model with daily time-steps. Detailed studies for an experimental basin in Switzerland showed up the expected minor changes in annual runoff with increases in winter runoff of 10% and decreases in summer of 11% (Bultot and Gellens, 1991).

Beran (1986) showed the impact of a CO_2 doubling on runoff in the European Community countries using the UKMO model. He found a considerable increase in runoff in Britain, Denmark, Belgium and Holland, and a decrease in the central and southern

parts of Europe. Hulme and Jones (1988) estimated that in Britain the annual temperature was likely to rise by 3–5 °C and precipitation by 15–20%. Based on this scenario Arnell and Reynard (1989) determined the possible changes in annual and monthly runoff for four regions in Britain (southeast, southwest, northwest and Scotland) using relationships between runoff, precipitation and evaporation. The results obtained demonstrated that annual runoff can be expected to increase by more than 25% in northern and western regions and up to 10% in southern and eastern regions. Monthly runoff increases with increasing precipitation, especially during the winter and spring, and these results were confirmed by other studies Arnell (1998, 1999a).

A large number of studies have been carried out in the last decade in the USA and Canada on the possible effects of global warming effects on the different hydrological characteristics, and on water requirements and the economic, ecological and social consequences of using water resources and operating water-management systems under critical conditions. For most of Canada global warming is likely to result in increases in annual runoff and its redistribution during the year. This is especially significant for southern regions of the country, excluding the basin of Great Lakes, where the different scenarios show increases in temperature and decreases in precipitation, and, as a consequence, decreases in runoff which will lead not only to changes in the water balance and lake levels, but also to serious economic and ecological consequences (Sanderson and Wong, 1987; Chao and Wood, 1999; Chao et al., 1999).

Since the early 1980s many studies of the effects of climate change on water resources have been made for the USA. These studies have been widely reviewed particularly by Shiklomanov et al. (1990), Stakhiv et al. (1992) and Gleick (2000) as well as in the IPCC (IPCC, 2001b). The current GCMs give quite different estimates of the expected climate change, especially for monthly precipitation. According to Stakhiv and Lins (1989), of the four models which are most frequently applied, none is able to estimate annual precipitation with the accuracy required for hydrological analysis, and opposite results can be obtained for seasonal and monthly precipitation. The model results differ significantly in the warming forecast for the regions, especially for summer months. Kirshen and Fennessey (1992) analysed the possible effects of climate change on the water resources and water supplies for the Boston area based on four GCM scenarios (GISS, GFDL, UKMO and OSU) and the likely changes in water management, including the financial and economic consequences. The authors show that with the GISS and GFDL models the water supply is expected to reduce, but with the UKMO and OSU models the opposite is likely. Using these results the Massachusetts Water Resources Authority decided not to undertake any adaptation measures because of the high degree of uncertainty and the cost of the measures.

Table 12.2. *Changing annual river runoff in the Dnieper Basin under different climate change scenarios*

Basin	GFDL		UKMO		SHI 1 °C		SHI 2 °C		SHI 3–4 °C	
	mm	%	mm	%	mm	%	mm	%	mm	%
Upper Dnieper	4	2	136	64	−20	−9	9	4	88	42
Sozh	20	12	124	77	−20	−12	19	8	90	56
Desna	26	20	56	71	−12	−9	24	23	78	51
Soula	28	42	51	77	0	0	63	95	92	139
Psel	14	21	35	49	−8	−11	37	53	67	96
Vorskla	18	27	40	61	−9	−14	38	58	68	103
Samara	24	111	11	53	6	27	43	195	53	241
Berezina	36	19	152	81	−12	−6	30	16	101	54
Pripyat'	40	46	153	177	−1	−1	87	101	128	149
Teterev	0	0	38	53	−7	−10	51	73	89	127
Ros'	22	36	31	50	−2	−3	59	95	86	139

These differences between the GCM scenarios make it difficult to use them to assess the effects of climate change on water resources and this has resulted in an increase in temperature of 1–4 °C and changes of precipitation of between 10–20% becoming accepted as the likely outcome of global warming. These figures have been used to determine the sensitivity of water-management systems to climate change, and to provide a basis for adaptation measures and evaluating their cost, but these assumptions do not allow forecasts of the likely future effects of climate change. Nevertheless, by reviewing the results from a wide range of research, it is possible to reach the following conclusions about changes in water resources in the regions of the USA (Gleick, 1988b, 2000; Shiklomanov *et al.*, 1990; Shiklomanov and Lins, 1991):

- Northwest: a slight increase in annual runoff and in floods
- California: a considerable rise in winter runoff and a decrease in summer flows with an insignificant rise in the annual runoff
- Great Lakes: decreases in runoff
- Other regions: the changes in precipitation and water resources are not clear.

For Russia and neighbouring countries, detailed assessments of the expected changes in hydrology and water resources have been made for medium-size catchments in the Volga basin (15 catchments), Dnieper (12) and Yenisey (6). The climate scenarios based on palaeoclimatic reconstructions developed at SHI were employed in these studies, for the three most recent warm epochs corresponding to a global warming of 1 °C, 2 °C and 3–4 °C. In addition, three GCM scenarios were used (GFDL, UKMO and MPI) with the stationary variant of CO_2 doubling and non-stationary variant with a 10%, 50% and 80% CO_2 increase (Shiklomanov and Georgiyevsky, 1992, 1995; Shiklomanov, 1994;

Georgiyevsky *et al.*, 1995, 1996, 1997; Shiklomanov *et al.*, 2002). The analysis of the results from these simulations that are based on the water balance model are considered in Section 12.4. They allowed conclusions to be made for the intermediate variants with increases in CO_2 of 10%, 50% and 80%, but there was no agreement on the assessments of possible changes to water resources obtained by different GCMs. For instance, if CO_2 is increased by 80% in the Dnieper basin, for the scenarios based on the GFDL and MPI models, runoff is expected to decrease by 15–40%, whereas with the UKMO model the annual runoff in the basin is expected to rise by 30–50%. The major cause of these discrepancies is the considerable differences in the regional precipitation estimates obtained by the different models.

These differences in the model results and their corresponding but contrasting consequences for hydrology and water resources lead to the conclusion that with both the GCM scenarios and the palaeoclimatic scenarios of a 3–4 °C global warming, there is likely to be increasing precipitation and higher temperatures. However, the range of estimates for the changes in precipitation and temperature obtained by the different scenarios is large: these increases in precipitation are expected to be 10–20% in the Volga and Dnieper basins using palaeoreconstructions and the GFDL model, and to be 30–40% with the UKMO model. The expected rise in mean annual temperature is 7–9 °C for the UKMO scenario, and 3–5 °C with the GFDL scenario and palaeoreconstructions. For the Yenisey basin the palaeoclimatic method and the GFDL model both forecast an increase in annual precipitation of 15–25%, but the UKMO model gives an increase of 30–50%. The GFDL and UKMO models indicate a 6–7 °C rise in temperature but palaeoreconstructions give a 6 °C rise in the south and 11 °C in the north (Table 12.2).

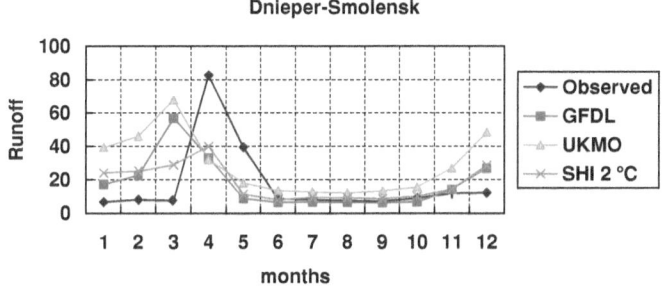

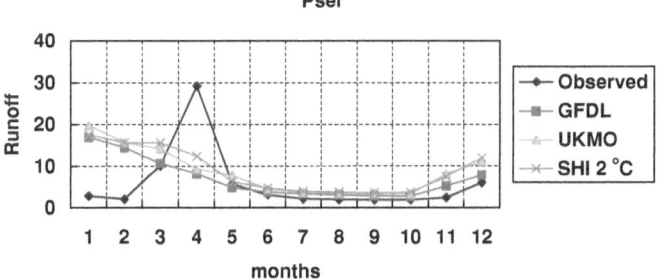

Fig. 12.8 Seasonal runoff distribution in Dnieper basin according different climate scenarios: GFDL, US Geophysical and Fluid Dynamics Laboratory; UKMO, UK Meteorological office; SHI, State Hydrological Institute.

The following conclusions can be reached for the different scenarios:

1. *GFDL $2 \times CO_2$*. As a result of the increases in precipitation the basin runoff will increase. Depending on the changes in precipitation and temperature, as well as on their seasonal distribution, the increase in runoff can be expected to be from 1–2% up to 42–46%. Runoff in the southernmost river, the Samara, located in the driest area, is likely to increase by 24 mm, i.e. 111%, as compared to its current value.

2. *UKMO $2 \times CO_2$*. The main feature of this scenario is the considerable increase in winter precipitation, which is expected for the entire basin.

3. *SHI 1 °C*. The annual runoff is expected to decrease by 1–14% (exclusive of the Samara). The major reason for this reduction is the expected rise in temperature.

4. *SHI 2 °C*. As a result of the greater precipitation, runoff is expected to increase by 16–23% in the northern forested part of the basin and by 50–100% in its forest–steppe and steppe areas.

5. *SHI 3–4 °C*. As precipitation is expected to increase by 30%, annual runoff is likely to rise by 40–150%.

All scenarios indicate that the seasonal distribution of runoff is very sensitive to both changes in precipitation and temperature and to what the time of year they occur. Figure 12.8 presents the distribution of runoff for the Dnieper at Smolensk and the

Psel River (a tributary to the Lower Dnieper) under current conditions and for the changes prompted by the different scenarios. At present there is a spring flood caused by snowmelt in the northern part of the area in April and May, and earlier in the south. Currently the major part of annual runoff occurs in the summer and autumn while the winter is the season of low flows. Under the modelled conditions, there will be two hydrological seasons: winter and summer. The winter is characterized by a large volume of runoff from snow and rain with a maximum in March. During the summer (May to October) low flows are likely with a minimum in July and August, when flow is maintained by ground water.

This hydrological regime will result from the rise in temperature. As a result of warming there is likely to be no snow cover, soil moisture will be close to or exceeding field capacity, and frequent floods will take place. During the summer more precipitation will be lost by evaporation, and river flows will be maintained by ground water. The runoff will be more evenly distributed than today, peak flows will be lower, and low flows higher. The results from the models for the Dnieper are, on the whole, typical of the forested part of the East European Plain, while those for the Psel River are representative of the forest–steppe. Analysis of the model results for changes in the patterns of the annual runoff in the Volga Basin (Table 12.3) and in its seasonal distribution led to the following conclusions:

- with the 1 °C SHI scenario no serious changes will occur in the annual runoff
- with the 2 °C SHI scenario annual runoff will increase over the entire basin; this increase will be 5–15% in the northern regions of the basin and 20–30% in the Lower Volga
- with the 3–4 °C SHI scenario annual runoff will increase by 20–35% in the northern part, and in the southern part by 50–60%
- with the GFDL $2 \times CO_2$ scenario, annual runoff will increase by 10–15% in the Upper Volga and Upper Oka Basins and by 80–100% in the Lower Volga Basin.

The UKMO $2 \times CO_2$ model predicts the most significant increase in water resources in the Volga Basin. The results for the seasonal distribution of runoff were very close to those obtained for the Dnieper Basin. Runoff is expected to increase significantly during the winter and decrease in the summer. The discharge of the Volga at Volgograd is not likely to change with a 1 °C warming because the small increase in precipitation will be compensated for by the rise in evaporation. With a 2 °C rise in temperature the runoff in the Volga is expected to increase by 23 km^3/year, and with a 3–4 °C rise by 70.0 km^3/year (27%). Using the GFDL model for a doubling of CO_2 the increase in the runoff is expected to be even more significant (86 km^3/year) than from the scenario obtained by palaeoclimatic analogues.

Table 12.3. *Changing annual river runoff in the Volga Basin under different climate change scenarios*

Basin	GFDL		SHI 1 °C		SHI 2 °C		SHI 3–4 °C	
	mm	%	mm	%	mm	%	mm	%
Kostroma	58	23	9	3	22	9	61	25
Vetluga	127	59	2	1	30	14	62	29
Upper Kama	92	36	20	3	31	12	65	25
Cheptsa	71	33	24	11	21	10	57	26
B. Kokshaga	30	20	−15	−10	14	10	50	34
Sok	90	82	−14	−13	9	9	39	36
Upper Volga	25	9	12	4	11	4	84	32
Upper Oka	26	15	−10	−5	12	7	56	33
Sura	95	84	−7	−6	29	26	59	52
B. Irgiz	52	106	−1	−1	15	31	39	80
Samara	63	89	1	1	15	22	44	62

Table 12.4. *Yenisey Runoff volume (km^3) of the Yenisey Basin with global climate scenario*

Climate scenario	Spring (months 4–6)	Summer–autumn (months 7–10)	Winter (months 11–3)	Whole year
Natural[a]	327	233	70	630
SHI 1 °C	350	243	94	687
SHI 2 °C	385	188	113	686
SHI 3–4 °C	325	171	228	724
GFDL 2 × CO_2	403	227	119	749
UKMO 2 × CO_2	553	238	126	917

[a] Average seasonal runoff distribution for the period before the beginning of active anthropogenic influence (before 1959).

The results for the Yenisey Basin are given in Table 12.4 and Fig. 12.9.

Southeastern areas (Lake Baikal basin)
All the scenarios show that annual runoff is likely to increase: by 20% using the 1 °C SHI scenario, and by 80% with the GFDL model. The volume of floods will increase most of all, while the flood season will last longer. The start of the flood season will be a month earlier in all cases except with the 1 °C temperature rise. Runoff in the autumn will also increase but only by a small amount.

Southern part (southern areas of the Krasnoyarsk and Irkutsk regions)
Palaeoclimatic scenarios show that the annual runoff will change very little but with the GFDL and UKMO scenarios it will increase by 5–7%. The SHI scenario with a 3–4 °C temperature rise and

the UKMO scenario indicated that the flood season would start a month earlier than at present.

Central part (including the Podkamennaya Tunguska basin)
All the above scenarios show an annual runoff increase of 8% and up to 50% using the results of the UKMO model, due to increases in the flood volume. The flood season is expected to take place a month earlier with the UKMO and the 3–4 °C SHI scenarios.

Northern part
Annual runoff is likely to increase by 15–20% with all scenarios and with the UKMO by 75%, mainly due to a greater volume of floods, although the maximum discharge decreases in the palaeoclimatic scenarios. Again the flood season is advanced by a month (excluding the 1 °C SHI) and two months earlier in the case of the 3–4 °C SHI.

The largest increase in runoff is likely to be observed in the central and northern parts of the basin, where about 70% of the flow of Yenisey originates. In this region with a more severe climate the increasing precipitation will have a greater significance as compared to the temperature rise. Runoff in the southern part will change little due to the smaller increase in precipitation and the dominance of the rise in temperature. The southeastern parts with the least precipitation and runoff are most vulnerable to a changing climate, especially in precipitation; however, less than 10% of the total runoff of the Yenisey originates there. Table 12.4 shows that depending on the climate change scenario accepted, the increase of the total annual runoff of the Yenisey River might amount to from 55 km^3/year (9%) to 287 km^3/year (46%).

The possible runoff increase for large rivers of the Northern Hemisphere, in particular those of Siberia, is also indicated by the results of approximate assessments by Miller and Russell (1992).

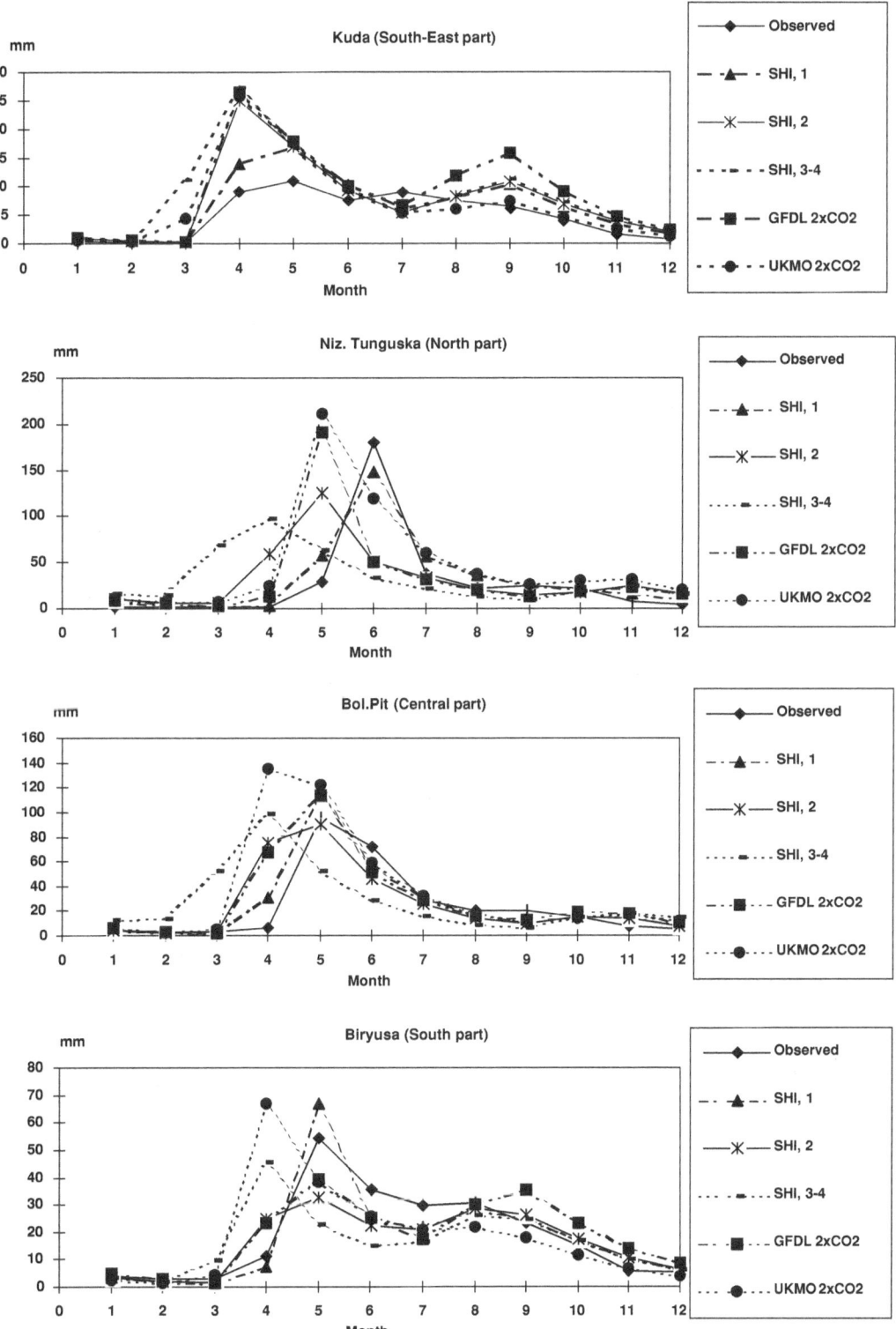

Fig. 12.9 Observed and computed hydrographs for different climate scenarios in various parts of the Yenisei Basin.

Table 12.5. *Impacts of climatic changes on mean annual runoff in semi-arid river basins in the USA*

Change in precipitation	Basin	Percent change in mean annual runoff with change in temperature			
		$T+1\,°C$	$T+2\,°C$	$T+3\,°C$	$T+4\,°C$
−10%	Peace River[a]	−50		−50	
	Great Basin[b]		−17 to −28		
	Sacramento River[c]		−18		−21
	Inflow to Lake Powell[d]		−24		−32
	White River[d]		−13		−17
	East River[d]		−19		−25
	Animas River[d]		−17		
0	Sacramento River		−3		−7
	Inflow to Lake Powell		−12		−21
	White River		−4		−8
	East River		−9		−16
	Animas River		−7		−14
+10%	Peace River	+50		+35	
	Sacramento River		+12		+7
	Inflow to Lake Powell		+1		−10
	White River		+7		+1
	East River		+1		−3
	Animas River		+3		−5

[a] All Peace River results from Nemec and Schaake (1982).
[b] All Great Basin rivers results from Flaschka *et al.* (1987).
[c] All Sacramento River results from Gleick (1986a, 1987).
[d] All Lake Powell inflow, White, East and Animas Rivers results from Nash and Gleick (1993).
Source: Frederick and Major (1997), adapted from Nash and Gleick (1993).

They used the GISS scenarios and calculated annual runoff for 33 large river basins of the world. By their calculations with $2 \times CO_2$ the Ob River discharge is expected to rise by 38%, Yenisey 42%, Lena 26%, Amur 12% and Kolyma 45%. Annual discharge of the large European rivers the Danube and Severnaya Dvina is also expected to grow, by 19% and 22%, respectively. The conclusions about expected increase in water resources of Siberia, the Far East and Northern Europe have been drawn by Shiklomanov and Babkin (1992). The results of the large number of studies of the effects of climate change on water resources shows that for the cold and temperate climates of the Northern Hemisphere water resources are expected to increase. In the basins where runoff is a product of snowmelt, winter runoff will increase and the volume of the spring flood will decrease so that the pattern of flow throughout the year will be more uniform. Temperature will be the driving factor rather than precipitation.

ARID AND SEMI-ARID REGIONS

In areas where water resources are currently under stress, it is especially important to have reliable data on the possible hy-drological effects of global warming. However, for these areas the results drawn from GCM scenarios and palaeoclimatic ana-logues vary considerably and may even be contradictory. Stud-ies of the hydrological effects of global warming in arid and semi-arid regions have been conducted for many parts of the world, probably in a most comprehensive manner for river basins in the United States. As an example, Table 12.5 shows assess-ments of the effects of climate change on the annual runoff of semi-arid basins in the United States taken from Frederick and Major (1997). These studies show that even relatively small changes in temperature and precipitation can lead to very signifi-cant changes in water resources in these dry regions. In this case runoff is more sensitive to changes in precipitation than those in air temperature.

Parallel studies have been presented by the IPCC (Shiklomanov *et al.,* 1990; Stakhiv *et al.,* 1992; Shiklomanov and Shiklomanov, 1999) as well as in studies made in Russia (Shiklomanov, 1989; Shiklomanov and Lins, 1991) and in a number of other countries. The use of a hydrological model (Chunzhen, 1989) shows that in the semi-arid regions of Northern China, with a 10% increase in precipitation and a 4% decrease in evaporation, runoff will rise by 27%, and with a 4% increase in evaporation by 18%. Similar changes in climate parameters for more arid regions will result in changes in runoff of 30–50% (Chang *et al.*, 1992). In general, these different studies show that catchments in arid and semi-arid regions are very sensitive to even quite small changes in climatic characteristics (Liu, 1998; Kang *et al.*, 1999). A temperature rise of 1–2 °C and decrease in precipitation by as much as 10% can result in reducing annual and seasonal runoff by 40–60%. Detailed studies of the hydrological effects of global warming have been made in Australia, in particular, for the Murray–Darling Basin (Close, 1988). In this study, quite optimistic conclusions have been drawn as to the future water resources and water supplies of the basin based on the assumption that precipitation will increase over much of Australia. These effects on runoff are also reflected in the ground waters of the Australian Continent (Chassemi *et al.*, 1991). Other studies in Australia (Australian Bureau of Meteorology, 1991) have focussed on the extremes of precipitation and runoff. These show that for a doubling of CO_2 the maximum precipitation tends to increase in summer and winter, while the minimum precipitation decreases. These trends seem to be typical of all of Australia, including areas where the annual precipitation is expected to decrease.

The severe drought of the 1970s and 1980s was a disaster for many countries in the Sahel. An analysis of the effects of climate change on water resources in this part of the world was made by Sircoulon (1990) and Sircoulon *et al.* (1999), who showed the river systems to be very sensitive to the climatic regime. However, the trends in the future water resources of the Sahel are uncertain because of the contradictory results from GCMs and from palaeoclimatic analogues. The latter indicate a sharp rise in precipitation in this region with an unchanged temperature regime leading to a considerable increase in water resources. However, from the three GCMs types with the greatest resolution (IPCC, 1990) for a doubling of CO_2, the temperature will rise by 1–2 °C, and winter precipitation will decrease by 5–10%, while summer precipitation will increase by 5%. The result will be that water resources will increase.

HUMID TROPICAL REGIONS

The effects of global warming have been assessed for basins in Venezuela (Postel, 1989), for the Rio de la Plata basins in Uruguay (Tucci and Damiani, 1991), for the Mekong (Mekong Secretariat, 1990), for basins and regions in Indonesia, Malaysia and Thailand (Parry *et al.*, 1991; Toth, 1993) as well as for some parts of Africa (Urbiztondo *et al.*, 1991), and for India and Sri Lanka (Leichenko, 1993; Nophadol and Hemantha, 1992). Three GCMs (GISS, GFDL and UKMO) were used in the studies of the Mekong (Mekong Secretariat, 1990). All showed a considerable increase in precipitation and runoff during the wet season, and a very small

reduction in precipitation during the dry season. However, the length of the dry season is forecast to increase. This will lead to an uneven distribution of runoff throughout the year and a greater need to manage water resources.

The hydrological implications of climate change for catchments in Indonesia, Malaysia and Thailand were investigated by Parry *et al.* (1991) and Toth (1993) using the GISS, GFDL and OSU scenarios with a doubling of CO_2. The results show an increase in precipitation in the wet season and less rain during the dry season. For three basins in Indonesia, Rozarri (1990) found that the monthly runoff would increase along with the erosion and the latter would cause soil productivity to decrease. Leichenko (1993) studied the two densely populated urban regions, Mumbai and Chennai, using the climatic scenarios from the GISS, GFDL and UKMO models with a doubling of CO_2 and assessing the hydrological implications of climate change by the water balance method. The results led to the conclusion about the very great uncertainty surrounding the future water resources and water availability for the two largest and most rapidly developing megalopolises in India. Nophadol and Hemantha (1992) compared the scenarios generated by the GISS, UKMO and GFDL models for Sri Lanka. The best results seemed to be obtained by using the scenarios from the GISS and UKMO models: with a doubling of CO_2 they gave increased daily and monthly precipitation during the wet season but with a very small change in annual precipitation. The current period of maximum precipitation (April to July) would shift to September to November, while the low water period would increase in length and the probability of drought would rise.

An analysis of the effects of climate change on the water resources of the Zambezi has been made by Urbiztondo *et al.* (1991) by using the GFDL, UKMO and GISS scenarios. For the GFDL and GISS scenarios with a doubling of CO_2 the discharge of the Zambezi is expected to decrease, while the UKMO indicates that it is likely to increase. Tucci and Damiani (1991) undertook a similar study using the scenarios from these three models for the Uruguay River. They found that all three models underestimate the present precipitation; however, the GISS model yielded the more acceptable results. With a doubling of CO_2 the GISS and UKMO scenarios showed that the discharge would decrease, by 11.7% and 6.4% respectively, while it would increase by 21.5% with the GFDL scenario. Studies of the effects of global warming on the changes in South American water resources were undertaken by Budyko *et al.* (1994) using scenarios based on palaeoclimatic reconstructions of past warm epochs and a simple water balance model. The results showed that for 1 °C and 2 °C increases in temperature the water resources are expected to increase by 8–10%; the effects are most pronounced for the Amazon basin, while in the Uruguay and Paraguay basins water resources are expected to decrease slightly.

For the humid tropics the changes in precipitation cause variations in the annual and seasonal discharges. For monsoon regions discharges are expected to increase during the rainy season and to decrease during the dry season while the probability of drought will rise (Shiklomanov and Shiklomanov, 1999).

12.6.2 Changes in climate, water requirements and water use

If the climate warms water resources and their seasonal distribution can be expected to change in many parts of the world. Changes are also likely to occur in the structure and character of water use, while competition between the different water users is also likely to intensify. With a large change in temperature and precipitation, plans for developing and managing irrigation are likely to require alterations, causing problems of the supply of water for this and other major users. Arid and semi-arid regions will be the most vulnerable areas and their limited resources will be placed under greater stress. Indeed as the major fresh-water user, irrigation is the most sensitive of these sectors to climate change and this area has been subject to a considerable amount of research, particularly in the United States. For example Peterson and Keller (1990) have studied the arid and semi-arid regions of the western USA. They concluded that for a doubling of CO_2 irrigation will be affected to the extent that the irrigated areas may decrease by 30%. Frederick (1991) compared water used for irrigation in Nebraska and Kansas in 1931 to 1940 with that for 1950 to 1980. The former was a dry and warm period when the temperature was $1\,^\circ$C higher and precipitation 100 mm less than in the 1950 to 1980 period. For the first period water use on the area of maize in Nebraska was 39% more than in the second period, and 14% greater in Kansas. From 1931 to 1940 the concentration of CO_2 increased from 350 to 450 ppm, and the water used for irrigation decreased by 7%. Similar results were obtained for this region by Easterling et al. (1991). Detailed studies on the temperate zone of the effects of changing climate, in particular the increase in CO_2 on the water used by crops, were carried out by McCahe and Wolock (1992). They used an irrigation model based on a modified Thornthwaite water-balance model and scenarios of changing precipitation and air temperature as well as stomatal resistance characterizing the increasing resistance to transport of water vapour with the rise in CO_2. The scenarios included temperature rises of 0, +2, +4 and +6$\,^\circ$C and precipitation increases of 0, 10% and 20% as well as increases in stomatal resistance. The results showed an increasing use of water associated with rising temperature and to a lesser extent with increasing precipitation. With a temperature rise of 2$\,^\circ$C and a precipitation increase of 20% water use increases. However water use by plants is more sensitive to changes in the stomatal resis-

tance than to temperature changes. When stomatal resistance increases to 20%, and the temperature rises by 4$\,^\circ$C, use of water will decrease.

Hatch et al. (1999) assessed irrigation water requirements in Georgia, USA, using a climate change scenario derived from HadCM2. This scenario produced a decrease in irrigation demand, ranging from just 1% by 2030 for soybean to as much as 20% by 2030 for maize. Along the Gulf Coast of USA, however, the same scenario implied an increase in irrigation demands (Ritschard et al., 1999).

Kaczmarek and Napiorowski (1996) showed that a temperature rise of 1$\,^\circ$C would lead to an increase in the water demands for irrigation by 12%; and if precipitation decreased by 10%, the demand for water would rise by 18%.

In the 1980s, an estimate was made of the effects of global warming on irrigation development in one of the basins of Lesotho (Institute of Hydrology, 1988; Nemec, 1989). The conclusion was that the demand for water for irrigation would increase by 65% or that the area under irrigation would halve.

The detailed studies of the effects of climate change have been conducted by SHI for the southern part of European Russia and the Ukraine, where irrigation is the major water user (Georgiyevsky et al., 1993, 1996). A model was used to calculate the irrigation regime (Shalygin, 1987) incorporating the physical processes occurring, the optimum conditions for crop growth, the application of water and the features of the soil. However, no account was taken of the influence of CO_2 on crop water use. The model was applied to parts of the Volga, Don, and Dnieper Basins for 1936 to 1990 with crops of cereals (winter), maize, and forage (alfalfa). Two approaches have been made to assess the sensitivity of the irrigation regime to climate change. In the first the irrigation regime for the dry period from 1936 to 1939 was compared with that for the wet period from 1978 to 1990. Table 12.6 shows the results for the Volga and Don. In the 1930s the water use increased, on average by 20–30%, but in contrast between 1978 and 1990 it fell by 20–30%. The controlling factor for water use appears to be the volume of precipitation during the growing season. The second approach consists of assessing the water requirements under the different climate change scenarios, both those based on palaeoclimatic reconstructions and on the GFDL and UKMO models with a doubling of CO_2. All these scenarios show an all-year rise in temperature and an increase in precipitation. For each of these scenarios the irrigation regimes have been determined for stations in the Volga, Don and Dnieper Basins with the following conclusions. The GFDL and UKMO scenarios and the palaeoreconstructions show that the water requirements for annual crops are expected to decrease by 10–30% due to the increase in precipitation during the summer and a shorter growing season. For perennial grasses, with the GFDL and UKMO scenarios, water needed for crops will increase by

Table 12.6. *Dynamics of irrigation norms for alfalfa in the European territory of the former Soviet Union*

Station	Average, mm	1936–9 mm	1936–9 %	1978–88 mm	1978–88 %
Astrakhan	699	808	16	653	−7
Bugul`ma	216	303	40	183	−15
Buzuluk	422	516	22	373	−12
Elat`ma	191	292	53	140	−27
Ershov	445	562	26	388	−13
Gotnya	222	297	34	167	−25
Kaluga	128	–	–	96	−25
Kazan`	221	339	8	145	−34
Kharabali	679	794	17	623	−8
Krasnodar	406	397	−2	368	−9
Krasnouphimsk	173	231	34	149	−14
Krasnyi Kut	543	614	13	536	−1
Millerovo	422	554	31	387	−12
Penza	259	360	39	252	−3
Pugachov	418	512	22	394	−6
Samara	351	493	40	326	−7
Saratov	460	–	–	483	5
Sengilei	311	405	30	271	−13
Seraphimovich	436	537	23	421	−3
Shakhty	427	525	23	380	−11
Tselina	402	523	30	310	−23
Valuiki	298	355	19	248	−7
Volgograd	597	633	6	556	−7
Volovo	172	309	80	124	−28
Voronezh	268	379	41	–	–
Yoshkar-Ola	169	272	61	129	−24
Zelenokum	444	586	32	402	−10

1–4% due to the longer summers and the increase in the number of harvests, but there was no change with the SHI scenarios.

Future use of water for irrigation will depend not only on changes in temperature and precipitation but also on the direct effects of increasing concentrations of CO_2 on photosynthesis. A rise in CO_2 promotes increasing stomatal resistance, which decreases transpiration per unit area of a leaf. Some experiments (Rosenberg *et al.*, 1990) show that a CO_2 doubling can decrease the transpiration rate by up to 50% on average. On the other hand, an increasing CO_2 concentration stimulates plant growth, resulting in an increase in the transpiring surface and thus more transpiration. One additional factor is the possible rise in leaf temperature as a result of the decreasing level of transpiration and a change in plant communities.

The direct effects of CO_2 on transpiration are quite complicated, and their impacts on water use will depend on the plant type and many other factors, such as climate, soil type and the depth to the water table. In a number of cases the effects of the rise in CO_2 and these factors can be very different, not only on the water used for irrigation but also on natural vegetation, and therefore, on the water balance and water resources. The likely effects of increasing CO_2 concentrations and its influence on plant transpiration were identified by Idso and Brasel (1984) and Aston (1987), respectively, for basins in Arizona and Australia. Their results were different from those of other researchers: doubling of CO_2 produced a 40–60% rise in the annual runoff in the Arizona rivers of the study, and a 60–80% rise for Australia. From these and similar studies, it can be concluded that the increasing of CO_2 is unlikely to have significant effects on runoff but it is likely to be important to the water used for irrigation. Frederick *et al.* (1993) developed the EPIC (Erosion Productivity Impact Calculator) model to determine the probable effects of an increasing CO_2 concentration on the runoff in the Missouri basin. The EPIC model includes a detailed hydrological processes model and components for soil erosion and crop production together with their interaction. The model was used to assess the effect of CO_2 concentrations on transpiration and growth and to calculate runoff from different parts of the catchment for the three climatic scenarios. The results show that because of increasing stomatal resistance, runoff will increase especially from the cropped areas. This effect will be compensated for in part by the temperature rise and the fall in precipitation, which are characteristic of the climatic scenarios used.

The above results are testimony to the problems of accounting for changes in CO_2: they cause additional uncertainties, in assessing the effects of global warming on water use by plants, on evaporation, and consequently, on runoff and on water resources. Of course climate change can produce effects on the demands for domestic and industrial water, however, to a lesser extent than for irrigation. Generalized results for the USA show that a 1 °C temperature rise increases domestic water use by between 0.02% and 3.8%, while a 1% fall in precipitation increases water use by 0.02–0.31%. This scatter of results is due to regional differences, the time of year, as well as on the ratio of the water used indoors to outdoors. Hughes *et al.* (1994) found that for urban areas in Utah a 2.2 °C rise in temperature increases public water needs by 2.8% in summer and by 8% during June. Herrington (1996) estimated that for Great Britain a warming of 1.1 °C will lead to a growth of 4% in domestic demand, but taking more showers can increase demand by 12%, and watering lawns and gardens by 19–35%. Water use in 2030 in the urban areas of Washington DC were studied by Boland (1997) for a stationary climate and for five climatic scenarios and the results are presented in Table 12.7. They show that climate change is less important to future water use than the growth of population and other socio-economic factors. These factors with

Table 12.7. *Annual water use forecasts for Washington DC Metropolitan Area (Conservation Policy)*

Climate scenario	1990, $m^3 \times 10^6$/day	2030		
		$m^3 \times 10^6$/day	Increase over 1990, %	Increase over scenario A, %
A: Stationary climate	2.35	4.71	+100.2	
B: GISS A	2.35	4.92	+109.3	+4.6
C: GISS B	2.35	4.35	+84.9	−7.6
D: GFDL	2.35	5.12	+117.9	+8.8
E: Max Planck Institute	2.35	5.16	+119.3	+9.6
F: UKMO	2.35	5.22	+122.2	+11.0

the progress of technology, economic development and social and political advancement must be taken into account, along with changes in temperature, precipitation and CO_2 concentrations for a comprehensive view of future water use in the different sectors.

12.7 CONCLUSIONS

Assumptions about the stationarity of runoff records in the long term, which have served as a basis for most hydrological methods of calculating major hydrological characteristics for civil engineering and water-management design, and for planning and long-term use of water resources, require a critical and thoroughgoing analysis.

This is related to the global climate warming occurring during the twentieth century and being strengthened over the last decades. Although there is no absolute confidence that the only cause of this warming is the current increase in atmospheric greenhouse gas concentrations of anthropogenic origin, one should recognize that most of climatologists support this point of view. The majority of scientists regard further climate warming as inevitable, although they disagree about its rate and magnitude, especially on a regional basis.

A changing climate will lead to changes in water resources and water use globally, regionally and locally. The distribution of these changes in space and time will determine the general strategy for the management of water resources, the supply of water and its use by different sectors. The assessment of future water resources and their use can be achieved from scenarios of the expected climate. Unfortunately, these forecasts of climate, even for a few decades ahead, are suspect on the scale most important for hydrology, namely the regional scale, especially for the variable most important in hydrology, namely precipitation.

As a result, there is currently doubt about assessing future water resources by the methods used in the past, which assume a stationary climate, and doubt about forecasts of future climate. This is on top of uncertainty surrounding the assessments of water resources that are due to perennial problems of degrading monitoring networks, unreliable data and other sources of error. Taking into account the current state of knowledge in this area, it is possible to propose directions of study to widen knowledge about the effects of climate change on water resources. First, there is the need for continued international co-operation in both research and operational hydrology such as the detailed hydrological monitoring of river basins, of the runoff regime including floods and droughts, evaporation from land and water surfaces, the water content in the upper soil horizons, in aquifers, in the snow cover and in lakes, etc. For this purpose data from national hydrological networks are very important as well as from experimental catchments and other research installations located to represent the different regions of the globe. Especially valuable are the long-term observations of all the variables in the hydrological cycle, their efficient storage in digital form, their accessibility and availability for use.

Detailed studies of hydrological processes will provide understanding of the changes in the storage in basins and their variations and help to establish the natural and human-induced causes of these changes. The success of these studies will allow the adaptation of water management systems to new conditions. A thorough analysis of hydrological variables in representative basins over a wide range of natural conditions, will demonstrate where hydrology and water resources are most sensitive to climate change and where the "CO_2 signal" can be most readily detected.

Assessing the possible range of changes in the hydrological variables that may result from the expected climate warming is an important line of research, together with the improvement of GCMs and palaeoclimatic analogues on the regional scale. These studies may provide ideas about the possible limits of changes to water resources with different concentrations of greenhouse gases and they should allow regions to be revealed, where water resources are most sensitive to climatic change. Additionally, there are some large river basins where results from GCMs and paleoclimatic reconstructions are available for assessing the likely changes to water resources from the climatic scenarios. Results from these basins can be used to demonstrate trends in hydrology and water resources, and the adaptation measures that need to be developed.

Models of the hydrological cycle can be used with GCMs to describe the processes of the hydrological cycle on the land surface and in the zone of active water exchange. This will permit GCMs to be improved so increasing the reliability of estimates of future water resources. This is a complex problem, however, not least because of the differences in the spatial and temporal scales of hydrological models and GCMs, and determining the transfer parameters of the effects of an increasing CO_2 concentration on transpiration and evaporation.

The present state of knowledge of the effects of climate change on water resources can be summarized as follows.

1. The scenarios of possible changes in regional climates are at present extremely uncertain and do not agree with each other, in particular in relation to changes in precipitation (the major factor determining water resources and water use. None of the existing climate scenarios provides a reliable basis for estimating changes in water resources at the regional and global scales.

2. Taking into account the rate of global warming and the large uncertainty surrounding climate scenarios with regard to assessing water resources, water use and water availability at the regional and global scales over the next 20 to 30 years, there is presently no basis for taking into account possible changes in climate.

3. The problem of providing reliable forecasts of global warming and its influence on water resources, water use and water availability is one of the most severe problems facing contemporary hydrology, taking into account the great sensitivity of the water system to even small changes in climate.

General conclusions: the outcome of the Monograph

To assess renewable water resources at the global scale, use has been made in this Monograph of the data from the global hydrological network (monthly and annual values). The stations in this network were located at 2500 sites across the continents; the data were for the period from 1921 to 1985. To update the series, to fill gaps in the observations and to calculate runoff from parts of the world not covered by observational data, various methods have been used, for example, in some instances, meteorological data on monthly precipitation and temperature. Where international frontiers do not coincide with watersheds, specially developed hydrological models have been used to determine water resources.

The estimated mean value of the renewable water resources of the world is 42 600 km^3/year, but this figure is very variable with time and space. The largest water resources in absolute terms are found to exist in Asia and South America (13 500 and 12 000 km^3/year), and the smallest in Europe and Australia with Oceania (2900 and 2400 km^3/year). For individual years the values of the world's water resources can vary within the range of ±15–25% of the mean. The year-to-year variations in world water resources do not contain a trend over the 65-year period from 1921 to 1985, and the same applies for the most part to the continents. The exceptions are the two last decades for Africa and South America. During this period, there has been an upward trend in the runoff from South America and a downward trend in runoff from Africa. For all regions runoff is very unevenly distributed during the year; some 60–70% occurs during the flood season, or period of high water.

The changes and trends in the water resources of 26 large regions and of 60 selected countries were also analysed. These included developing and developed countries, those whose economies are in transition, countries which are large in area and population and the reverse, together with some countries with copious water resources and others where today the demand for water outstrips the resource. These countries contain some 71% of the world's water resources and some 71% of the world's population. Brazil, Russia, Canada, USA, China and India are included in this list and they produce 40% of the world's total annual runoff. These data allow assessment to be made of the trends and changes

in the flow to the World Ocean. Approximately half the total inflow reaches the Atlantic Ocean, which is fed by four of the six largest rivers in the world. The smallest volume flows into the Arctic Ocean; however, this water is immensely important. The Arctic Ocean contains 1.2% of the water in the World Ocean, but it receives about 11% of the total runoff. The total flow to the World Ocean is fairly constant and exhibits no trend for the period in question. At the same time, there is a noticeable tendency towards decreasing flows to the Indian and Pacific Oceans and increasing flows to the Atlantic. Flows to the World Ocean are unevenly distributed in space with about 40% occurring between latitudes 10° N and 10° S.

Assessments of the use of water at the global scale have been made for the past and future, taking into account water use by the different sectors including the domestic, industrial (including power generation), and agricultural (irrigation) sectors, as well as water losses by evaporation from reservoirs. Such assessments were undertaken for 1940, 1950, 1960, 1970, 1980 and for the current period (1995), and forecasts were made for 2000, 2010 and 2025. Assessments of the use of water for countries were aggregated to the regions and then to continents. Where these data existed, preference was given to national data on actual or estimated water use. Without these data, assessments were made by proxy methods. Future water use was determined from trends and changes in the records of water use and from demographic and economic forecasts. Water availability (m^3/year per year) was estimated for the period 1950 to 2025, taking into account population and consumption. These estimates and forecasts assume a stationary climate.

In 1995 the total world abstraction of water was 3750 km^3/year, and consumption was 2280 km^3/year. Abstractions are expected to increase by approximately 10–12% every 10 years and to reach 5200 km^3/year (a 1.38-fold growth) by 2025. At present about 57% of the global water use and 70% of the global water consumption is in Asia, where the most irrigation takes place. However, the most rapid increase in abstractions is expected to take place in Africa and South America (a 1.5- to 1.6-fold growth) over the next 20 to 30 years, while Europe and North America will experience the slowest (a 1.2-fold growth). It is estimated that at present 67% of the total use and 86% of the total consumption is for agriculture. It is expected that agricultural water use will decrease in future, due to the growth in other sectors. In 1995, 254 million ha were irrigated worldwide but by 2010 this total is expected to increase to 290 million ha and by 2025 to 330 million ha.

The changes and trends in water use and the extent to which the available water resources are exploited are presented at the global and lesser scales. Even at present there are many countries whose water resources are completely utilized in terms of local water resources and in terms of inflows from the adjacent countries. About 75% of the world's population lives in countries of this category, and in regions where more than 20% of the water resources are used.

Since water resources and population are unevenly distributed around the world, the patterns of specific availability in space and time between 1950 and 2025 are also very uneven. Specific availabilty was largest in 1995 in some regions of Canada, Alaska and Oceania, where it reached 170 000–180 000 m^3/year per head. In the densely populated regions of Asia, Central and Southern Europe and Africa, the water availability in 1995 ranged from 1200 to 5000 m^3/year per head, while in North Africa and in the Arabian Peninsula, it was as little as 200–300 m^3/year per head compared to the values of 2000 m^3/year per head and 1000 m^3/year per head for the very low and catastrophically low categories of water stress. In these areas serious problems arise in all sectors. In 1995, 76% of the globe's population experienced a specific water availability of less than 5000 m^3/head, and 35% were in the very low or catastrophically low classes. As the twenty-first century unfolds the situation will worsen. By 2025 the greater part of the population of the globe will live under the conditions of very low or catastrophically low water availability, and approximately 30–35% of the world's population will have a supply of water classed as catastrophically low.

For many developed countries the decrease in specific water availability has been and is expected to be small over the period 1950 to 2025. For developing countries the specific water availability will fall dramatically and is expected to decrease 4.5 times for humid areas and 8.5 times for arid and semi-arid countries. The continuing growth of population will enhance the disparities and differences in the global pattern of water availability. However, when new data on the world's water resources appear and forecasts of the future situation are presented, questions will arise about their reliability and uncertainty. Many factors contribute to these attributes which differ from country to country, but particularly the state of the hydrological network: the number of hydrological stations, their spatial distribution, the duration and continuity of the observations, the quality of the measurements and the nature of the data management. WMO has analysed the state of hydrological networks globally: more than half the discharge

stations are located in Europe and North America and they have the longest observational series. Most of these stations (70%) are equipped with recorders and data-loggers which can capture detailed records. Consequently the estimates of water resources for these countries are the most reliable and match well with previous estimates.

Estimates of water resources for Africa, southern and southeast Asia, and for areas around the Arctic Ocean are far less reliable because they have very poor hydrological networks. More reliable estimates need improved hydrological networks: more and better stations, quality control of observing practices and of data, and computer-based processing and archiving. It is also evident that the quality of hydrological data has not been uniform for the period from 1925 to 1995. It gradually improved from the start of this period, suffered in many areas during the Second World War, but improved after the war to reach a peak in the 1970s. During the 1980s it declined especially in Africa, and this decline continued into the 1990s as the Soviet Union became a number of separate nations.

In many developing countries, the hydrological network is poorly developed, and the number of stations is declining. The time delay between the actual measurements, the processing and the publication of data, and the transmission of these data to national, regional and international centres is increasing. For a number of countries there are institutional obstacles in the operational exchange of hydrological information. It is far from simple, because of the network problems and these obstacles, to estimate the world's water resources. There is a marked contrast with meteorological data and allied information which is collected, exchanged and used for forecasting in a matter of minutes and hours, then generalized for the world in only a few months. To change this situation for hydrology and water resources needs a continuing effort by national bodies and international organizations, particularly funding institutions, to radically improve the state of the global hydrological network, and the collecting, processing and exchange of hydrological information. The situation is worse with respect to the problems of collecting and processing data on water use. For most developing countries these data are not available and published data are based on estimates made from population figures, the area irrigated and other indirect means.

There are other difficulties in estimating world water resources, such as the neglect of ground water in many areas. It has been estimated that globally about 600–700 km^3 of ground water is used annually, particularly for irrigation and domestic purposes, especially in countries with no runoff (e.g. the countries of Arabian Peninsula). For these countries estimates of water use and availability have to take into account ground water, which is not always the case. However the general pattern of global water availability will not change much with the inclusion of ground water. First, abstractions of ground water are about 15% of total abstractions and approximately the same percentage for the continents and the different regions. Second, at least half the ground-water that is used is hydraulically connected to surface runoff and abstractions of ground water simply reduce the runoff.

Consequently it seems more realistic to conclude that estimates of ground water can be decreased by half, and that ignoring ground water in assessments of water availability does not produce improper results. What is very evident is that estimates of resources and of specific water availability determined globally, regionally and nationally must consider their variability from year to year and within the year. Specific water availability (Figs. 11.12–11.14) can decrease by 1.2 to 2.0 times depending on the particular pattern of climate affecting the region and country. Average values may provide too optimistic a view of the current water situation. The same optimism may accrue from water resources estimates that neglect the quality of the resource. Pollution of water resources is an increasing threat in many parts of the world, particularly in industrially developed and densely populated regions of the globe without effective treatment of waste water. Industry and agriculture are major sources of pollution and domestic wastes are equally so. In 1995 the waste water volume was approximately 325 km^3/year in Europe, 430 km^3/year in North America, 590 km^3/year in Asia and 55 km^3/year in Africa. Much of the waste water was discharged untreated into rivers, lakes and aquifers, which then cannot be employed as sources of supply. Every cubic metre of waste water discharged into a water body renders unfit some 8 to 10 cubic metres of previously unpolluted water.

Estimates of world water resources for the period 2010 to 2025 require special attention. In part they are based on the trends observed in the past and, in part, on long-term demographic forecasts and forecasts of global economic development. The reliability of these forecasts determines the estimates of future abstractions and water availability. However, there are many sources of errors that can affect these estimates and there are factors in other areas, such as climate change, which add further complications. As well as forecasts of the future population, the likely growth of industry and thermal power production have to be taken into account in considering water availability up to 2025. Taking these factors into account, by 2025 world water abstractions are expected to be within ±10–15% of the cited value of 5100 km^3/year, i.e. approximately from 4600 to 5800 km^3/year. These figures match those from the international group of experts, including representatives of UNESCO, WMO and other UN agencies working with the Stockholm Environment Institute, who have given figures of 4500 to 7000 km^3/year.

More reliable and more detailed data on water resources and water use are vital to the improvement of these estimates. Acquisitions of such data needs the continuing attention of the scientific community, in partnership with bodies which can liberate funds for the installation of better data management systems and the upgrading of existing ones. Without this attention, future estimates of world water resources and the use of water will be as unreliable as those made today. However, future circumstances, when the population is much greater, will not allow for the luxury of the errors and uncertainties that exist today.

References

Abdel-Khalik H. A., Abdel-Gawad S. T. and Bons K. (1998). National environmental measures and their impact on drainage water quality in Egypt. In: *Water: A Looming Crisis?* (ed. H. Zebidi), *Proc. Int. Conf. on World Water Resources at the Beginning of 21st Century (UNESCO, Paris, 3–6 June 1998)*. IHP-Y Technical Documents in Hydrology, 18, 235–240.

Abdul Razzak M. J. (1995). Water supplies versus demand in the countries of the Arabian Peninsula. *J. Water Res. Plann. Mgmt*, May/June 1995, 227–234.

Abramova I. O. (ed.) (1993). *Tunisian Republic: Handbook*. Moscow: Nauka Publ. (in Russian)

Abu-Zeid M. and Abdel-Dayem S. (1991). Variation and trends in agricultural drainage water reuse in Egypt. *Water International*, 6(4), 247–253.

Abu Zeid M. and Biswas A. K. (1990). Impacts of agriculture on water quality. *Water International*, 15 (3), 160–167.

Afanasjyev A. N. (1967). *Variations in Hydrological Regimes in the USSR Territory*. Moscow: Nauka' Publ. (in Russian)

Alayev E. B. and Kolosov V. A. (eds.) (1989). *Territorial Economy Structure of Developed Capitalist Countries during the Scientific and Technological Revolution: Shifts and Tendencies*. Moscow: Nauka' Publ. (in Russian).

Alekseyevsky Ye. Ye. (ed.) (1974). *Irrigation and Drainage in the Countries of the World*. Moscow: Kolos' Publ. (in Russian)

Alexandratos N. I. (ed.) (1988). *World Agriculture: Towards 2000 – An FAO Study*. Chichester, UK: John Wiley & Sons.

Alexandratos N. I. (ed.) (1995). *World Agriculture: Towards 2010 – An FAO Study*. Chichester, UK: John Wiley & Sons.

Ali Ayub M. and Kuffner U. (1994). Water management in the Maghreb. *Finance and Development*, 31(2), 28–29.

Alpatjyev A. M. (1969). *Water Cycles in Nature and their Transformations*. Leningrad: Hydrometeoizdat. (in Russian)

Alpatjyev A. M. (1974). Water cycles in nature and their transformations. *Trans. SHI*, 221, 259–266. (in Russian).

Alpatjyev A. M. (1983). *Development, Transformation and Environment protection*. Leningrad: Nauka' Publ. (in Russian)

Alpatov V. M. (Ed.) (1992). *Japan: Handbook*. Moscow: Respublika' Publ. (in Russian)

Altunin V. S., Kupriyanova Ye. I. and Tursunov A. A. (1991). Inner resources for stabilization of the Aral Sea and reconstruction of ecological equilibrium in its basin. *News USSR Acad. Sci.*, Ser. Geogr., 4, 118–124. (in Russian)

Aluminium Company of Canada Ltd. (1982). *Alcan smelters and chemicals: Kemano, British Columbia Attitudes about Quality of Life and Reactions to Change in a Remote Community*. Toronto: Aluminium Company of Canada.

Alvarez L. W. (1987). Mass extinctions caused by large bolide impacts. *Phys. Today*, July, 24–33.

Ambroggi R. P. (1980) Water. *Scientific American*, 243 (5), 90–96, 100–101.

Amiotte Suchet P. (1995). Cycle du carbone, érosion chimique des continents et transfers vers les océans. *Mémoires Sciences Geologiques*, 97, 1–156.

Anaya M. (1967). Mexico and its water resources policy. *Proc. Int. Symp. "Water for Peace" (Washington, 23–30 May 1967)*, vol. 8, pp. 682–691.

Andressian V. P. (1993). *Forecasting Water Requirements for the Beginning of the 21st Century: A World-Scale Study*. Paris: UNESCO.

Anisimov O. A. and Poljakov V. Y. (1999). Predicting changes of the air temperature in the first quarter of the 21st century. *Meteorol. Hydrol.*, 2, 25–31. (in Russian)

Anonymous (1983). République du Niger. *Europe Outre-Mer*, 59(643–644), 171–179.

Anonymous (1984a). *Population of World Countries (1984)*. Moscow: Finance and Statistics Publ. (in Russian)

Anonymous (1984b). *Water Control in Viet Nam*. Hanoi: Foreign Language Publ. House.

Anonymous (1988). Problems of water supply in the countries of the Near and Middle East. *Bull. Foreign Commercial Information*, 93, 2. (in Russian)

Anonymous (1989). *Features of Water Management Economy in Developing Countries*. Moscow: MGMI Publ. (in Russian)

Anonymous (1990). Water management in the Nile delta in Egypt. *SCAN*, 1, 19–52.

Anonymous (1992a). *The Far East and Australasia*. London: Europa Publ.

Anonymous (1992b). The water industry of Sweden. *Effluent and Water Treatment*, 23 (12), 502–505.

Anonymous (1993). Land reclamation and water management in the CIS countries. *Land Reclam. Water Mgmt*, 1, 2–15. (in Russian)

Anonymous (1995a). *Africa Today: Facts, Events, Opinions*, No. 7. Moscow: Nauka' Publ. (in Russian)

Anonymous (1995b). Potential and planning priorities in the Lower Mekong basin. *Water Res. J.*, C/187, 26–30.

Antonov V. I. (2000). The main thing in overcoming the Aral crisis. *Land Reclam. Water Mgmt*, 1, 36–39. (in Russian)

Aoki R. (1987). Integrated water development project of Gravatai River basin, Brasil. *Water Sci. Technol.*, 19(9), 59–68.

Arar A. (1987). The role of use of sewage effluent for irrigation development in the Near East region. In: *Water for the Future: Water Resources Development in Perspective* (eds. W. O. Wunderlich and J. E. Prins), *Proc. Int. Symp. on Water for the Future (Rome, 6–11 April 1987)*, pp. 251–254. Rotterdam, The Netherlands: A. A. Balkema.

Arduino G. (1990). Hydrologia de la Plata. *Interciencia*, 15(5), 373–377.

Arkhipova M. A. (ed.) (1986). *Africa in Figures: A Statistical Handbook*. Moscow: Nauka Publ. (in Russian)

Arnell N. W. (1998). Climate change and water resources in Britain. *Climatic Change*, 39, 83–110.

Arnell N. W. (1999a). The effect of climate change on hydrological regimes in Europe: A continental perspective. *Global Environ. Change*, 9, 5–23.

Arnell N. W. (1999b). Climate change and global water resources. *Global Environ. Change*, 9, 31–49.

Arnell N. W. and Reynard N. (1989). Estimating the impacts of climatic change on river flow: Some examples for Britain. In: *Proc. Conf. on Climate and Water (Helsinki, 11–15 September 1989)*, vol. 1, pp. 426–436. Helsinki: Valtion Painatuskeskus.

Arthington A. H., Brizga S. O. and Kennard M. J. (1998). *Comparative Evaluation of Environmental Flow Assessment Techniques: Best Practice Framework*. Occasional Paper No. 26/98. Canberra, Australia: Land and Water Resources Research and Development Corporation.

Artyushkov E. V. (1970). Differentiation with respect to the Earth matter density and the related phenomena. *News Acad. Sci. USSR*, Ser. Physics of the Earth, 5, 25–37. (in Russian)

Aston A. R. (1987). The effect of doubling atmospheric CO_2 on stream flow: a simulation. *J. Hydrol.*, 67, 273–280.

Australian Academy of Science (1964). *Water Resources Use and Management, Proc. Symp. (Canberra, 9–13 September 1963)*. Melbourne, Australia: University of Melbourne Press.

Australian Bureau of Meteorology (1991). The impact of climate change on hydrology and water resources. In: *Climate Change Impacts Workshop (Melbourne, June 1991)*. 76.

Australian Bureau of Statistics (1971). *Official Yearbook of Australia*, vol. 57. Canberra, Australia: Australian Bureau of Statistics.

Australian Bureau of Statistics (1977). *Official Yearbook of Australia*, vol. 62. Canberra, Australia: Australian Bureau of Statistics. 783.

Australian Government Publishing Service (1978). *Variability of Runoff in Australia*. Canberra, Australia: Australian Government Publishing Service.

Australian Water Resources Council (1965). *Review of Australia's Water Resource: Streamflow and Underground Resources*. Canberra, Australia: Australian Water Resources Counci, Department of National Development.

Australian Water Resources Council (1971). *Stream Gauging Information. Australia, December 1969*. Canberra, Australia: Australian Water Resources Council, Department of National Development.

Australian Water Resources Council (1974). *Hydrological Series*. Canberra: Department of the Environment and Conservation.

Australian Water Resources Council (1976). *Review of Australia's Water Resources 1975*. Canberra, Australia: Australian Water Resources Council, Department of National Development.

Australian Water Resources Council (1987). *Review of Australia's Water Resources and Water Use by 1985: Water Resources Data Set*, vol. 1, *Canberra, November 1987*, vol. 2, *Canberra, November 1987*. Canberra, Australia: Australian Water Resources Council.

Avakyan A. B. and Sidoruk Ye. B. (1993). The problems of using water resources of the Rio de la Plata basin. *Wat. Resources*, 20(2), 251–261. (in Russian)

Avakyan A. B., Saltankin V. P. and Sharapov V. A. (1987). *Water Reservoirs*. Moscow: Mysl' Publ. (in Russian)

Babkin V. I. (1979) Interrelation between water balance elements in river basins. *Trans. SHI*, 260, 26–38. (in Russian)

Babkin V. I. (1998). Methodological principles for assessing renewable water resources of the terrestrial globe. In: *Water: A Looming Crisis?* (ed. H. Zebidi), *Proc. Int. Conf on World Water Resources at the Beginning of the 21st Century (UNESCO, Paris, 3–6 June 1998)*. IHP-Y Technical Documents in Hydrology, 18, 15–23.

Babkin V. I. and Serkov N. K. (1974). Modelling of hydrological characteristics by using Markov chains. *Meteorol. Hydrol.*, 7, 55–59. (in Russian)

Babkin V. I. and Sokolov A. A. (1992). Water problems in the CIS at the threshold of the 21st century. *Trans. SHI*, 360, 11–17. (in Russian)

Babkin V. I. and Voskresensky K. P. (1977). Methodological principles for the computation of water resources and water balance of the USSR territory. *Trans. SHI*, 241, 11–28. (in Russian)

Babkin V. I., Budyko M. I. and Sokolov A. A. (1990). Water resources of the USSR and their use at the present time and in the future. In: *Water Resources and Water Balance* (eds. A. A. Sokolov and V. I. Babkin), *Proc. 5th All-Union Hydrological Congress*, vol. 1, pp. 98–120. Leningrad: Hydrometeoizdat. (in Russian)

Babkin V. I., Grigorkina T. E. and Postnikov A. N. (1992). The influence of possible anthropogenic climate change on the annual river runoff in the European part of CIS. In: *Proc. 16th Conf. of Danube Countries (Kellheim, Germany, April 1992)*, pp. 479–484. Kellheim, Germany: UNESCO/WMO.

Babkin V. I., Gusev O. A. and Novikova N. A. (1974). Methodology for averaging and interpolation of hydrometeorological characteristics. *Trans. SHI*, 217, 175–186. (in Russian)

Babkin V. I., Gusev O. A. and Roumyantsev V. A. (1972). Modelling of characteristics of runoff and river catchments in the Don basin by using the factor analysis. *Trans. SHI*, 200, 3–25. (in Russian)

Babkin V. I., Kolokolov A. G., Grigorkina T. Ye. and Polad-Zade L. I. (1991). Current natural water avialbility in the USSR. *Trans. SHI*, 352, 3–21. (in Russian)

Badruddin M. (1987). Policy framework and institutions for water resources planning, development, and management in Pakistan. In: *Water Resources Policy for Asia* (eds. M. Ali, G. E. Radosevich and A. Ali Khan), *Proc. Reg. Symp. on Water Res. Policy in Agro-Socio-Economic Development (Dhaka, Bangladesh, 4–8 August 1985)*, pp. 105–117. Rotterdam, The Netherlands: A. A. Balkema.

Badry M., M, Mehdi M. S. and Khawar J. M. (1980). *Water Resources in Iraq*. Baghdad: Irrigation and Agricultural Development International Expert Consultancy.

Balek J. (1977). *Hydrology and Water Resources in Tropical Africa*. New York: Oxford University Press.

Balek J. (1983). *Hydrology and Water Resources in Tropical Regions*. New York: Oxford University Press.

Bambrick S. (ed.) (1994). *The Cambridge Encyclopaedia of Australia*. Cambridge, UK: Cambridge University Press.

Barney G. O. (ed.) (1980). *The Global 2000 Report to the President of US: Entering the Twenty-First Century*, vols. 1–3. Washington, DC: US Government Printing Office.

Batisse M. (1976). Balancing needs and resources in use of water. *Pontif. Acad. Sci.*, Ser. varia, 40, 2–23.

Baumgartner A. and Reichel E. (1972). Preliminary results of new investigation of the world's water balance. In: *World Water Balance, Proc. Symp. on the World's Water Balance (Reading, July 1970)*, vol. 3, IAHS Publ., 93, 580–592.

Baumgartner A. and Reichel E. (1973). Eine neue Bilanz des globalen Wasserkreislaufs. *Umschau in Wissenschaft und Technik*, H.2a, 631–632.

Baumgartner A. and Reichel E. (1975). *The World Water Balance*. Munich: R. Oldenburg Verlag.

Baxter R. M. (1980). Environmental effects of dams and impoundments in Canada: Experience and prospects *Can. Bull. Fish. Aquat. Sci.*, 205, 34.

Beard D. P. (1994). Remarks of D. P. Beard, Commissioner US Bureau of Reclamation, before the International Commission on Large Dams (Durban, South Africa, 9 November 1994).

Bekker A. P. (1983). Water use in South Africa, and estimated future needs. *The Covol. Engin. in South Africa*, 24(12), 653, 655–659, 666.

Belousov V. V., Lyustih E. N. and Shantser E. V. (1972). Hydrosphere. In: *Great Soviet Encyclopedia* (ed. A. M. Prokhorov), vol. 9, *Earth*, pp. 480–482. Moscow: Soviet Encyclopedia. (in Russian)

Beran M. (1986). The water resources impact of future climate change and variability. In: *Effect of Changes in Stratospheric Ozone and Global Climate* (ed. J. Titus), vol. 1, *Overview*, pp. 299–330. Washington, DC: US Environmental Protection Agency/UNEP.

Beran M. A. (ed.) (1986). *The Water Resources Impact of Future Climate Change and Variability (Geneva, 10–14 November 1986)*. Geneva, Switzerland: WMO.

Berezner A. S. (1981). Australian Water Resources. *Hydraul. Land Reclam.*, 5, 33–90. (in Russian)

Berezner A. S. (1982). The Hydropower system with interbasin diversion of runoff "Snow Capped Mountains" (Australia). *Hydrotech. Construct.*, 2, 53–56. (in Russian)

Berg L. S. (1908). The Aral Sea: experience of physical geography. *News of the Turkmen Branch of the USSR Geogr. Society*, 5 (9), 1–508. (in Russian)

Bergstrom S. (1996). *Development and Application of a Conceptual Runoff Model for Scandinavian Catchments*. Norkopping, Sweden: SMHI.

Bergstrom S. and Carlsson B. (1993). Hydrology of the Baltic Basin. *Swedish Met. Hydr. Inst. Repts Hydrology, 7, 28*.

Bergstrom S. and Carlsson B. (1994). River runoff to the Baltic Sea: 1950–1990. *Ambio*, 23 (4–5), 280–287.

Bergstrom S., Carlsson B., Gardelin M., Lindstrom G., Petterson A. and Rummukainen M. (2001). Climate change impacts on runoff in Sweden: Assessment by global climate models, dynamic downscaling and hydrological model. *Climate Research*, 16, 101–112.

Berner E. K. and Berner R. A. (1987). *The Global Water Cycle: Biochemistry and Environment*. Englewood Cliffs, NJ: Prentice-Hall.

Bertoldi G. L. (1992). Subsidence and consolidation in alluvial aquifer systems. In: *Changing Practices in Groundwater Management* (ed. J. J. DeVries), *Pros and Cons of Regulation, Proc. 18th Biennial Conference on Groundwater (Sacramento, 16–17 September 1991)*, University of California Report 77, 62–75.

Bertranou A. V., Llop A. and Vazquez A. K. J. (1983). Water use and reuse management in arid zones: the case of Mendoza, Argentina. *Water International*, 8(1–2), 1–12.

Bessonov S. A. (ed.) (1994). *Africa: Regional Aspects of Global Problems*. Moscow: Nauka' Publ. 152. (in Russian)

Bhath M. A. and Kijne I. W. (1991). Irrigation allocation problems at tertiary level in Pakistan. *Water Res. J.*, C/169, 94–54.

Biswas A. K. (1991). Discussion. *Water International*, l6(3), 185–189.

Blyutgen I. (1972). *Geography of Climates*, vol. 1. Moscow: Progress Publ. (in Russian)

Boer P. (2000). Huge water supply scheme for Perth. *Civil Eng. Australia*, 72(10), 58–98.

Bogacheva O. (1995). Features of economic rise in the 1990s in the USA. *Int. Econ. Int. Relations*. 1, 118–124 (in Russian)

Boland J. J. (1997). Assessing urban water use and the role of water conservation measures under climate uncertainty. In: *Climate Change and Water Resources Planning Criteria* (eds. K. D. Frederick, D. C. Major and E. Z. Stakhiv), vol. 37, pp. 157–176. Dordrecht, The Netherlands: Kluwer Academic Publishers.

Bolin B., Dees B. R., Yager J. and Warrik R. (eds.) (1989). *Greenhouse Effect, Climate Change, and Ecosystems*. Leningrad: Hydrometeoizdat. (in Russian) (Translated into English, Chichester, UK: John Wiley & Sons, 1990)

Bolotin B. M. and Sheinis V. L. (1983). *Economy of Developing Countries in Figures, 1950–1985*. Moscow: Economy Publ. (in Russian)

Borisenkov Ye. P. and Borisov L. Ye. (1987). El-Niño and the southern oscillation and an anomalous character of winters in Eurasia. *Trans. SHI*, 513, 3–16. (in Russian)

Borodavchenko I. I. and Mikhura V. I. (1981). Comprehensive use and protection of the Danube water. CMEA *Inf. Bull. on Water Management*, 2 (28), 38–41. (in Russian)

Borozdina R. I. (1978). Agriculture development and the food problem. *Latin America*, 3, 22–36. (in Russian)

Bortnik V. N. and Chistyayeva S. P. (eds.) (1990). *Hydrometeorology and Hydrochemistry of the USSR Seas*, vol. 7, *The Aral Sea*. Leningrad: Hydrometeoizdat. (in Russian)

Bortnik V. N., Kuksa V. I. and Tsytsarin A. G. (1991). Current state and possible future of the Aral Sea. *News USSR Acad. Sci.*, Ser. Geogr., 4, 62–68. (in Russian)

Botzak M. (1988) Water losses in irrigation systems. *Hidrotechnica*, 33 (11), 394–398. (in Romanian)

Brazdil R. (1992). Fluctuation of atmospheric precipitation in Europe. *Geojournal*, 27, 275–291.

Breinzer L. A. (1991). Hydrological consequences of tropical forest landscape transformations. *Nature and Resources*, 27(3–4), 42–53.

Brenning M. H. and Platon S. W. (1983). Land use planning in Denmark. *Nordic Hydrol.*, 14 (5), 267–276.

Briese D. (1981). Probleme der Wasserwirtschaft in Sudwestafrika (Namibia). *Wasser u. Boden*, 33(11), 520–522.

Britton S. (1991). Recent trends in the internationalism of the New Zealand economy. *Austral. Geogr. Studies*, 29(1), 3–25.

Bromley Yu. V. (ed.) (1980). *North America*. Moscow: Mysl' Publ. (in Russian)

Bromley Yu. V. (ed.) (1982). *Africa: An Overview – Northern Africa*. Moscow: Mysl' Publ. (in Russian)

Bromley Yu. V. (ed.) (1983). *South America*. Moscow: Mysl' Publ. (in Russian)

Brown I. A. H. (1984). The assessment magnitude and utilization of Australian water resources. *Inst. Eng. Austral. Nat. Cont. Publ.*, 84(1), 291–298.

Brown I. A. H. and Bergman I. (1987). *Resource Assessment: Australian Hydrology 1975–1986*. Paris: UNESCO.

Brown L. R. (1990). *The Changing World Food Prospects: The Nineties and Beyond*. Washington, DC: Worldwatch Institute.

Brown L. R. (2001). *Eco-Economy: Building an Economy for the Earth*. New York: W. W. Norton & Co.

Bruce J. P. (1974). Lakes: The vital water resource in Canada. *J. Fish. Res. Board Canada*, 31(5), 505–512.

Budagovsky A. I. and Busarova O. B. (1991). Principles of an assessment method for changing soil water resources and river runoff by different scenarios of climate change. *Water Resources*, 2, 5–16. (in Russian)

Budyko M. I. (1956). *Energy Balance of the Earth's Surface*. Leningrad: Hydrometeoizdat. (in Russian)

Budyko M. I. (1971). *Climate and Life*. Leningrad: Hydrometeoizdat. (in Russian)

Budyko M. I. (1980). *Climate in the Past and Future*. Leningrad: Hydrometeoizdat. (in Russian)

Budyko M. I. (1980). *The Earth's Climate: Past and Future*. Leningrad: Hydrometeoizdat. (in Russian) (Translated into English, New York: Academic Press, 1980)

Budyko M. I. and Groisman P. Ya. (1989). The 1980s warming. *Meteorol. Hydrol.*, 3, 5–10. (in Russian)

Budyko M. I. and Israel Yu. A. (eds.) (1987). *Anthropogenic Climate Changes*. Leningrad: Hydrometeoizdat. (in Russian) (Translated into English, Tucson, AZ: University of Arizona Press, 1991)

Budyko M. I., Borzenkova I. I., Menzhulin G. V. and Shiklomanov I. A. (1994). Cambios antropogénicos del clima en America del Sur. *Ser. Acad. Nac. Agronom. Veterin.*, 19, 224.

Budyko M. I., Israel Yu. A., MacCracken M. C. and Hecht A. D. (eds.) (1991). *Prospects for Future Climate*. A Special US/USSR Report on Climate and Climate Change. Leningrad: Hydrometeoizdat. (in Russian) (Translated into English, Chelsea, MI: Lewis Publishers, 1993)

Bultot F. and Gellens D. (1991). *Repercussions of a CO$_2$ Doubling on the Water Balance: A Case Study in Switzerland*. Berne: Swiss National Hydrological and Geological Survey.

Busarova O. V. and Gusev V. M. (1995). Using the results of climate change modelling to assess changes in total evaporation in the territory of Europe. *Meteorol. Hydrol.*, 10, 29–34. (in Russian)

Busby M. W. (1963). *Yearly Variations in Runoff for the Conterminous United States, 1931–60*. US Geological Survey Water Supply Paper No. 1669-S. Washington, DC: US Government Printing Office.

Ca'ceres L., Gruttner V. E. and Confreas D. R. (1992). Water recycling in arid regions: The Chilean case. *Ambio*, 21(2), 138–144.

Calvo Jullo C. (1990). Water resources development in Costa Rica 1970–2000. *Hydrol. Sci. J.*, 35(2), 185–196.

Campbell K. O. (1964). An assessment of the case for irrigation in Australia. In: *Proc. Symp. on Water Resources Use and Management (Canberra, 9–13 September 1963)*, pp. 450–457. Melbourne, Australia: University of Melbourne Press.

Canadian Fish and Wildlife Service (1979). *Salmon Studies Associated with the Potential Kemano II Hydroelectric Development*, vol. 1, *Summary*. Vancouver, BC.

Carter T. R., Hulme M., Crossley J. F., Malyshev S., New M. G., Schlesinger M. E. and Tuomenvira H. (2000). *Climate Change in the 21st Century: Interim Characterizations Based on the New IPCC Emissions Scenarios*. Helsinki: The Finnish Environment.

Central Water Commission (1994). *Water and Related Statistics*. New Delhi: Statistics Directorate.

CEPAL (1998). Caida del crecimiento economico en la region. *Notas de la CEPAL, Comision Económica para America Latina y el Caribe (Mexico City, 1 November 1998)*. Mexico: CEPAL.

Ceplecha V. J. (1971). The distribution of the main components of the water balance in Australia. *Austral. Geographer*, 11(5), 455–462.

Chan Than Suan (1978). *Land Hydrology (Hydrophysics and Hydraulics)*, vol. 7, *Spatial distribution of the Runoff over the Territory of Viet Nam*. Leningrad State Hydrometeorology Institute Publ. No. 20. Leningrad: Hydrometeoizdat. (in Russian)

Chang L. H., Hunsaker C. T. and Draves J. D. (1992). Recent research on effects of climate change on water resources. *Water Res. Bull. (AWRA)*, 28(2), 273–286.

Chang W. (1987). Large lakes of China. *J. Great Lakes Res.*, 13 (3), 235–249.

Chao P. T. and Wood A. W. (1999). *Water Management Implications of Global Warming*, vol. 7, *The Great Lakes–St. Lawrence River Basin*. Alexandria, VA: US Army Corps of Engineers.

Chao P. T., Hobbs B. F. and Venkatesh B. N. (1999). How climate uncertainty should be included in Great Lakes management: Modelling Workshops Results. *J. Am. Water Resources Assoc.*, 35, 1485–1497.

Chassemi F., Jacobsen G. and Jakeman A. S. (1991). Major Australian aquifers: Potential climate change impacts. *Water International*, 16(1), 38–44.

Chaturvedi M. C. (1991). *Water Management: Eastern Water Study*. New Delhi: Centre for Policy Research.

Chaturvedi M. C. (1993). Transboundary river basin management and sustainable development in developing countries: – case study of the Ganges–Brahmaputra–Meghna basin. In: *Transboundary River Basin Management and Sustainable Development* (eds. J.-C. Van Dam and J. Wessel), *Proc. Lustrum Symp. (Delft, The Netherlands, 18–22 May 1992)*, vol. 1, pp. 237–263. Paris: UNESCO.

Chaturvedi M. C. (2000). Water for food and rural development: developing countries. *Water International*, 25(1), 40–53.

Chebotarev A. I. (1978). *Hydrological Glossary*. Leningrad: Hydrometeoizdat. (in Russian)

Chen J. and Wu G. (1987). Water resources development in China. In: *Water Resources Policy for Asia* (eds. M. Ali, G. E. Radosevich and A. Ali Khan),

Proc. Reg. Symp. on Water Res. Policy in Agro-Socio-Economic Development (Dhaka, Bangladesh, 4–8 August 1985), pp. 51–60. Rotterdam, The Netherlands: A. A. Balkema.

Chernogayeva G. M. (1969). Water resources of Europe. *News Acad. Sci. USSR*, Ser. Geogr., 5, 58–67. (in Russian)

Chiew F. H. S. and McMahon T. A. (1995). Trends and changes in historical annual streamflow volumes and peak discharge of rivers in the world. In: *Proc. Int. Congr. on Modelling and Simulation (Newcastle, Australia, November 1995)*, pp. 140–150.

Chomchai P. (1993). Intitutional aspects of planning the development of a river basin: a Mekong case study. In: *Transboundary River Basin Management and Sustainable Development* (eds. J.-C. Van Dam and J. Wessel), *Proc. Lustrum Symp. (Delft, The Netherlands, 18–22 May 1992)*, vol. 2, pp. 135–147. Paris: UNESCO.

Chou Vichitkh (1994). Economics of rice production in Southeast Asia. *Int. Agricult. J.*, 3, 20–23. (in Russian)

Chudodeyev A. (1993). A Magnificent Eight. *New Time*, 47, 22–25. (in Russian)

Chunzhen L. (1989). The study of climate change and water resources in North China. Beijing: Ministry of Water Resources. Unpublished manuscript.

Chyba C. F. (1987). The cometary contribution to the oceans of primitive Earth. *Nature*, 330, 632–635.

Clark A. K. and Anid M. (1994). Canal irrigation and development opportunities for the Indus right bank in Sindh and Baluchistan. *Water Res. J.*, C/183, 84–94.

Clark R. H. (1987). Interbasin water transfers and the Canadian water resources. In: *Proc. 8th Can. Hydrotechn. Conf. (Montreal, 19–22 May 1987)*, pp. 307–328.

Close A. F. (1988). Potential impact of the greenhouse effect on the water resources of the River Murray. In: *Greenhouse: Planning for Climate Change* (ed. G. Pearman), pp. 312–323. Canberra, Australia: CSIRO.

CMEA (1977). *Prediction of Water Management Development in CMEA Member States until 1990*. Council of Leaders of Water Management Bodies of CMEA Member States. Moscow: CMEA. (in Russian)

CMEA (1981). *Norms of Water Use and Water Draining*. Meeting of Leaders of Water Management CMEA Organizations. Moscow: CMEA. (in Russian)

CNC/IHD (1978). *Hydrological Atlas of Canada*. Ottawa, Ontario: Canadian National Committee for the International Hydrologic Decade, Supply and Services.

Collier S. (ed.) (1992). *The Cambridge Encyclopaedia of Latin America and the Caribbean*, 2nd. edn. Cambridge, UK: Cambridge University Press.

Condie K. (1989). Origin of the Earth's crust. *Palaeogeog., Palaeoclimatol., Palaeoecol.*, Global and planetary change section, 75, 58–81.

Conley A. M. and Hansmann J. G. G. (1985). A scarce resource allocation strategy and decision support system for water management, in the Republic of South Africa. In: *Scientific Basis for Water Resources Management* (ed. M. Diskin), *Proc. Jerusalem Symp. (Jerusalem, September 1985)*. IAHS Publ. 153, 65–78.

Connell J. and Pritchard B. (1990). Tax havens and global capitalism: Vanuatu and the Australian connection. *Austral. Geogr. Studies*, 29(1), 38–50.

Connell L. (1984). Islands under pressure: Population growth and urbanization in the South Pacific. *Ambio*, 12(5–6), 306–308, 310–312.

Conway D. and Hulme M. (1993). Recent fluctuations in precipitation and runoff over the Nile subbasins and their impact on main Nile Discharge. *Climate Change*, 25, 127–151.

Crabb P. (1988). Managing the Murray–Darling Basin. *Austral. Geographer*, 19(1), 64–88.

Creton D. (1991). Geopolitics and the Euphrates water resources. *Geography*, 76 (331), Part 2, 157–159.

CSIRO (1982). *Murray–Darling Basin Project Development Study*, Stage 1, *Working Papers*. Canberra, Australia: CSIRO Division of Water and Land Research.

Cubasch U., Hasselmann K., Hock H., Maier-Reimer E., Mikolajewicz U., Santer B. and Sausen R. (1992). Time-dependent greenhouse warming computations with a coupled ocean–atmosphere model. *Climate Dynamics*, 8, 55–69.

Czaya E. (1981). *Strome der Erde*. Leipzig: Edition Leipzig.

Dahlstrom B. (1994). Short-term fluctuations of temperature and precipitation in Western Europe. In: *Climate Variations in Europe* (ed. R. Heino), pp. 30–38. Helsinki: Academy of Finland.

Daniel I. R. K. (1990). An analysis of streamflow pattern in Guyana. *Water International*, 5(3), 134–143.

Danilevsky A. (1987). Development of the Rio de la Plata system. *J. Water Res. Plann. Mgmt* 113(6), 761–778.

David E. L. (1990). *Manufacturing and Mining Water Use in The United States, 1954–1983: National Water Summary 1987*. US Geological Survey Water Supply Paper 2350. Washington, DC: US Government Printing Office.

David G. (1979). *Ordinal Statistics*. Moscow: Nauka' Publ. (in Russian)

Davis G. H. (1985). *Water and Energy Demands and Effects*. Studies and Reports in Hydrology No. 42. Paris: UNESCO.

Davydov A. D. (ed.) (1985). *Irrigation in the Countries of the Near and Middle East*. Tashkent: FAN Publ. (in Russian)

Day J. C. and Quinn F. (1987). Dams and diversions: Learning from Canadian experience. In: *Impacts and Research Needs for Canada* (eds. W. Nicholaichuk and F. Quinn), *Proc. Symp. on Interbasin Transfer of Water (Saskatoon, Canada, 9–10 November 1987)*, pp. 43–58.

De Mare L. (1977). An assessment of the world water situation by 2000. *Water International*, 2(9), 31–46.

Dengo M. B. (1982). Development in hydrometry in Central America and Brazil. In: *Advances in Hydrometry* (ed. J. A. Cole), *Proc. 1st IAHS Symp. (Exeter, July, 1982)*, pp. 335–344.

Dengo M. B. (1992). Water balance of the Amazon and its probable changes. *WMO Bulletin*, 41 (1), 85–95. (in Russian)

Denisov V. M. and Denisov Yu. M. (1961). The theoretical scheme to calculate the hydrograph of melt water of mountainous rivers. *News Acad. Sci. USSR*, Ser. Tech. Sci., 5, 49–60. (in Russian)

Department of National Development (1968). *Surface Water Resources: Atlas of Australian Resources*, 2nd Series, *Geographic Section* Canberra, Australia: Department of National Development.

Derpgolts V. F. (1971). *Water in the Universe*. Leningrad: Nedra' Publ. (in Russian)

Dingelshtedt H. (1893). *Experience of Studying Irrigation of the Turkestan Area*. St. Petersburg: Russian Geography Society Publ. (in Russian)

Direccion Nacional de Hidrografía (1994). *Aprovechamiento de los Recursos Hidrocos Superficiales: Inventario Nacional 1993–1994*. Montevideo: Division Recursos Hidricos.

Dmitriyevsky Yu. D. (1967). *Inland Waters of Africa and their Use*. Leningrad: Hydrometeoizdat. (in Russian)

Dmitriyevsky Yu. D. (1990). Current tendencies in developing water resources of Africa. *Wat. Resources*, 4, 172–181. (in Russian)

Dobroumov B. M. and Ustyuzhanin B. S. (1980). *Transformation of Water Resources and River Regimes in the Centre of ETSU*. Leningrad: Hydrometeoizdat. (in Russian)

Dobrynin V. F. (1943). *Physical Geography of Western Europe*. Moscow: Uchpedgiz' Publ. (in Russian)

Dourojeanni A. and Nelson M. (1982). Developments in river basin management. *Water Sci. Technol.*, 19(9), 201–210.

Doxiadis S. A. (1967). Water for the human environment. In: *Proc. Int. Conf. "Water for Peace" (Washington, 23–31 May 1967)*, vol. 1, pp. 33–60. Washington DC: US Government Printing Office.

Dreier N. N. (1978). *Water Balance of North America*. Moscow: Mysl'/Nauka' Publ. (in Russian)

Drinkwater K. F. (1987). The effect of freshwater discharge on the marine environment. In: *Impacts and Research Needs for Canada* (eds. W. Nicholaichuk and F. Quinn), *Proc. Symp. on Interbasin Transfer of Water (Saskatoon, 9–10 November 1987)*, pp. 415–430.

Drozdov O. A., Grigorieva A. S. and Sorochan A. S. (1981). *Water Cycle in Nature*. Leningrad: Znaniye' Publ. (in Russian)

Drozdov O. A., Sorochan O. G. and Shiklomanov I. A. (1976). Preliminary estimation of changes in global water cycle due to economic activities. *Water Resources*, 6, 45–55. (in Russian)

Drysdale P. (1986). The Pacific Basin and its economic vitality. In: *The Pacific Basin: New Challenges for the United States* (ed: James W. Morley), pp. 11–22. New York: Academy of Political Sciences.

Dubeau D. (1991). Etude sur les effets environnementaux cumulatifs du Plan des Installations. *Rev. Energ. (France)*, 42(428), 147–152.

Dubovikov F. G. (1922). Influence of water interception in the upper reaches of Syr Darya and in its lower reaches. *Cotton Plant Business*, 9–10, 67–84. (in Russian)

Dubreuil P. (1985). Problemas hidricos en Brasil. *La Houille Blanche*, 2, 111–122.

Dukhovny V. A. and Sokolov V. I. (1994). Interstate common information system for the use, management and influence of water resources of the Aral Sea Basin. In: *Water: Ecology and Technology, Proc. Materials Int. Congress (Moscow, 6–9 September 1994)*, vol. 1, pp. 127–135. Moscow: Ministry of Agriculture and Foodstuffs. (in Russian)

Dukhovny V. A., Avakyan I. S., Prikhod'ko V. T. and Ruziyev M. T. (2000). Basin of the Aral Sea and irrigation in the Central Asia. *Land Reclam. Water Mgmt*, 3, 12–15. (in Russian)

Dukich D. (1982). Trends in some water losses and protection from the beginning to the end of the 20th century. *Glavn. Ser. Geogr. Sci*, 62(1-c), 13–22C.

Dunin-Barkovsky L. V. (1967). Development of irrigation and the fate of the Aral Sea. In: *Problems of Nature Transformation in the Middle Asia* (ed. I. P. Gerasimov), pp. 75–84. Moscow: USSR Academy of Sciences Publ. (in Russian)

Dunk W. P. (1984). Water resources and conservation in Australia. *Bhagirath*, 21(3), 71–77.

Duthie H. C. (1979). Limnology of sub-arctic Canadian lakes and some effects of impoundment. *Arct. Alp. Res.*, 2(2), 145–158.

Duvanin A. I. (1981). On the choice of strategy for using water resources. *News Moscow State Univ.*, Ser. 5, Geography, 2, 30–33. (in Russian)

DWR (1993). *California Water Plan Update: November 1993*. Draft. DWR Draft Bulletin No. 160–93. Sacramento, CA: Department of Water Resources.

DWR (1994). *California Water Plan Update*. DWR Final Bulletin No. 160–93. Sacramento, CA: Department of Water Resources.

Eagleson P. S. (1991). Hydrologic science: a distinct geoscience. *Rev. Geophys.*, 29 (25), 237–248.

Earmme S. Y. (1979). A water use projection model for the North Saskatchewan River basin, Alberta, 1980–85: An input–output approach. PhD dissertation. Edmonton, Canada: University of Alberta.

Easterling W. E., McKennly M., Rosenberg N. J. and Leman K. (1991). *A Farm Level Simulation of the Effects of Climate Change on Crop Productivity in the MINK Region*, Washington, DC: US Department of Energy.

ECA (1978). Regional Report of Economic Commission for Africa for the UN Water Conference "Water Development and Management" (Mar-del-Plata, Argentina, March 1977). In: *Water Development, Supply and Management* (ed. Asit K. Biswas), vol. 1, Part 2, pp. 519–637. Oxford, UK: Pergamon Press.

ECLAC (1978a). Report of the Regional Preparatory Meeting of the Countries of Latin America and the Caribbean for the UN Water Conference "Water Development and Management" (Mar-del-Plata, Argentina, March 1977). In: *Water Development, Supply and Management* (ed. Asit K. Biswas), vol. 1, Part 2, pp. 419–465. Oxford, UK: Pergamon Press.

ECLAC (1978b). The Water Resources of Latin America: Regional Report for the UN Water Conference "Water Development and Management" (Mar-del-Plata, Argentina, March 1977). In: *Water Development, Supply and Management* (ed. Asit K. Biswas), Vol. 1, Part 2, pp. 721–790. Oxford, UK: Pergamon Press.

ECLAC (1984). Water Resources in Latin America. *Water International*, 9(1), 26–36.

ECLAC (1995). *The State of Water Management in Latin America and the Caribbean*. Regional Report for the 3rd Meeting of the Committee on Natural Resources (24 November 1995). Santiago: ECLAC, Devision of Environment and Natural Resources, Natural Resources Unit.

Eden M. I. (1974). Irrigation system and development of peasant agriculture in Venezuela. *Tijdschr. Econ. Soc. Geogr.*, 65(1), 48–54.

EEA (1999). *Sustainable Water Use in Europe*, Part I, *Sectoral Use of Water*. Copenhagen: European Environment Agency.

El Hares H. and Aswed M. (1979). Trends in Libyan desalination aid water reuse policy. *Desalination*, 30(1–3), 163–173.

Eley F. J. and Lawford R. G. (1987). Potential climatic impacts of water transfers. In: *Impacts and Research Needs for Canada* (eds. W. Nicholaichuk and F. Quinn), *Proc. Symp. on Interbasin Transfer of Water (Saskatoon, 9–10 November 1987)*, pp. 319–333.

Elpiner L. I. (1992). Medical–ecological problems in the Aral Region. In: *Proc. Int. Conf. on The Aral Sea Crisis: Environmental Issues in Central Asia (9–13 July 1990)*. Bloomington, IN: mimeo.

Encyclopedia Britannica (1980). 15th edn. Chicago: Encyclopedia Britannica, Inc. Energy, Mines and Resources Canada (1974). *The National Atlas of Canada*. Ottawa, Canada: McMillan.

Engelman R. and LeRoy P. (1993). *Sustaining Water: Population and the Future of Renewable Water Supplies*. Washington, DC: Population and Environment Program, Population Action International.

Environment Agency (1987). *Quality of the Environment in Japan*. Tokyo: Environment Agency.

Eramov R. A. (1973). *Physical Geography of Foreign Europe*. Moscow: Mysl' Publ. (in Russian)

ESCAP (1978). Report of the ESCAP Regional Preparatory Meeting for the UN Water Conference "Water Development and Management" (Mar-del-Plata, Argentina, March 1977). In: *Water Development, Supply and Management* (ed. Asit K. Biswas), vol. 1, Part 2, pp. 371–418. Oxford, UK: Pergamon Press.

ESCAP (1987). *Water Resources Development in Asia and the Pacific*. New York: ESCAP.

ESCAP (1989a). *Water Use Statistics in the Long-Term Planning of Water Resources Development*. Water Resources Series No. 64. New York: ESCAP.

ESCAP (1989b). *Development and Conservation of Groundwater Resources and Water-Related Natural Disasters and their Mitigation in Selected Least Development Countries and Developing Island Countries in the ESCAP region*. Water Resources Series No. 66. New York: ESCAP.

ESCAP (1991). *Assessment of Water Resources and Water Demand by User Sectors in Thailand*. New York: UN.

ESCAP (1992). Floods in Pakistan. *Water Res. J.*, C/172, 76–82.

ESCAP (1994). Asian and Pacific Experience. *Water Res. J.*, C/180, 75–96.

ESCAP (1998). *Sources and Nature of Water Quality in Asia and the Pacific*. New York: UN.

ESCAP/UN (1989). *Guidelines for the Preparation of National Master Water Plans*. New York: UN.

ESCWA (1978). Report of the Economic Commission for Western Asia Regional Preparatory Meeting for the UN Water Conference "Water Development and Management" (Mar-del-Plata, Argentina, March 1977). In: *Water Development, Supply and Management* (ed. Asit K. Biswas), vol. 1, Part 2, pp. 639–658. Oxford, UK: Pergamon Press.

Fairbridge R. W. (ed.) (1966). *The Encyclopedia of Oceanography*. New York: Reinhold. (Translated into Russian, Leningrad: Hydrometeoizdat, 1974.)

Falkenmark M. (ed.) (1977). *Water in Sweden*. National Report to the United Nations Conference "Water Development and Management" (Mar-del-Plata, Argentina, March 1977). Stockholm: Ministry of Agriculture.

Falkenmark M. and Lindth G. (1974). How can we cope with the water resources situation by the year 2015? *Ambio*, 3(3–4), 114–121.

FAO (1982). *Agroclimatological Data for Africa*. Rome: FAO.

FAO (1993). *China Desk Study: Preliminary Results*. Rome: Global Water Information System, FAO.

FAO (1995a). *Production Yearbooks 1965–1994*, vols. 10–47. Rome: FAO.

FAO (1995b). *Water Resources of African Countries: A Review*. Rome: FAO.

FAO (1995b). *Irrigation in Africa in Figures*. Water Report No. 7. Rome: FAO.

FAO (1997). *Water Resources of the Near East Region: A Review*. Rome: FAO.

FAO (2000). *Aquastat Information System*. http://www.FAO.org/waicent/faoinfo/agricult/aglw/aquastat

Farian A. (1987). Institutional arrangement for the planning and management of water supply in Nigeria. In: *Water for the Future: Water Resources Development in Perspective* (eds. W. O. Wunderlig and J. E. Prins), *Proc. Int. Symp. on Water for the Future*, pp. 317–333. Rotterdam, The Netherlands: A. A. Balkema.

Fiering M. B. (1983). *Methodology for Water Resources Assessment*. US Geological Survey Project No. 14-08-0001-204159. Cambridge, MA: Harvard University Press.

Findlay B. F., Gullett D. W., Malone L., Reycraft J., Skinner W. R., Vincent L. and Whitewood R. (1994). Canadian national and regional standardized annual precipitation departures. In: *Trends '93: A Compendium of Data on Global Change* (eds. T. A. Boden, D. P. Kaiser, R. J. Sepanski and F. W. Stoss), Oak Ridge, TN: Oak Ridge National Laboratory.

Finlayson B. L. and McMahon T. A. (1991). Runoff variability in Australia. *Nat. Cong. Publ. Inst. Eng. Austral.*, 91(2), 504–571.

Finn D. (1983). Land use and abuse in the East African Region. *Ambio*, 12(6), 296–301.

Fitzgibbon J. (1987). Summary of the Workshop proceedings. In: *Impacts and Research Needs for Canada* (eds. W. Nicholaichuk and F. Quinn), *Proc.*

Symp. on Interbasin Transfer of Water (Saskatoon, 9–10 November 1987), pp. 497–502.

Flashka I. M., Stockton C. W. and Boggess W. R. (1989). Climatic variation and surface water resources in the Great Basin region. *Water Res. Bull.*, 23, 45–57.

Flemming N. S. and Daniell T. M. (1994). Sustainable water resources management: an Australian perspective. *Water Res. J.*, 183, 16–23.

Folland O. K., Palmer T. N. and Parker P. E. (1986). Sahel rainfall and world-wide sea temperatures, 1901–85. *Nature*, 320, 602–606.

Forcasiewicz J. and Margat J. (1980). *Tableau mondial de donnés nationales d'économie de l'eau, ressources et utilisation*. Orléans, France: Departement Hydrogéologique.

Foxworthy L. and Moody D. W. (1986). *National Perspective on Surface-Water Resources: National Water Summary 1985*. US Geological Survey Water Supply Paper No. 2300. Washington, DC: US Government Printing Office.

Frazer D. (1986). Water crisis threatenes to dry up China's future. *New Straits Times*, 8 May.

Fread D. F. (1976). *A Dynamic Model of Stage Discharge Relations Affected by Changing Discharge*. Washington, DC: National Oceanic and Atmospheric Administration.

Frederick K. D. (1986). *Scarce Water and Institutional Change: Resources for the Future*. Washington, DC: US Government Printing Office.

Frederick K. D. (1991). *Processes for Identifying Regional Influences of and Responses to Increasing Atmospheric CO_2 and Climate Change*. The MINK Project Report No. 4 – *Water Resources*. Washington, DC: US Department of Energy.

Frederick K. D. and Major D. C. (1997). Climate change and water resources. In: *Climate Change and Water Resources: Planning Criteria* (eds. K. D. Frederick, D. C. Major and E. Z. Stakhiv), pp. 7–23. Dordrecht, The Netherlands: Kluwer Academic Publishers.

Frederick K. D., McKennly M. S., Rosenberg N. J. and Balser D. K. (1993). Estimating the effects of climate change and carbon dioxide of water supplies in the Missouri River Basin. In: *Resources for the Future*, Discussion Paper No. ENR 93–18. Washington, DC: US Department of Energy.

FRIEND (1993). *Flow Regimes from International Experimental and Network Data (FRIEND)* (eds. G. Rees, L. Roald and J. Dixon), vol 2, *Hydrological Data*. Wallingford, UK: Institute of Hydrology.

Frolov Yu. S. (1971). New fundamental data on the morphometry of the World Ocean. *News Moscow State Univ.*, Ser. 5, Geography, 1, 85–90. (in Russian)

Frolova L. G. (1974). Water resources of Australia. In: *Australia and Oceania: History, Geography, Culture* (ed. K. V. Malakhovsky), pp. 123–142 Moscow: Nauka' Publ. (in Russian)

Frostell B. and Ramier I. L. (1994). *Water Supply and Sewage Treatment in Chile*. Stockholm: Swedish Environmental Research Institute.

Fyurin P. (1966). *Water Problems on the Earth*. Leningrad: Hydrometeoizdat. (in Russian)

Ganshin G. A. and Remyga V. N. (eds.) (1993). *Modern China: Reform and Development*. Moscow: Institute Far East RAS Publ. (in Russian)

Gaphurov B. G. (ed.) (1977). *Modern Asia: A Handbook*. Moscow: Nauka' Publ. (in Russian)

Garcia N. O. and Vargas W. M. (1996). The spatial variability of runoff and precipitation in the Rio de la Plata basin. *Hydrol. Sci. J.*, 41(3), 279–299.

Garcia-Quintero A. (1950). Hydrology of Mexico. *Proc. Am. Soc. Civil Eng.*, 76 (Supplement 38), 17.

Garmonov I. V., Konoplyantsev A. A. and Lushnikova N. P. (1974). Water storage in the upper part of the earth's crust. In: *The World Water Balance and Water Resources of the Earth* (ed. V. I. Korzun), pp. 51–54. Leningrad: Hydrometeoizdat. (in Russian)

Gavrilov A. M. (1950). India (ed. A. P. Domanitsky). *Trans. SHI*, 8, 1–255. (in Russian)

Gavrilov L. V. (ed.) (1989). *Africa on the Way to the 21st Century*. Moscow: Nauka' Publ. (in Russian)

Gavrilova T. S. and Tarasov K. S. (1985). *Latin America: Ecology and Politics*. Moscow: Institute for Latin American Studies, *RAS* Publ. (in Russian)

Gellens D. (1991a). Impact of CO_2-induced climate change on river flow variability in three rivers in Belgium. 12.

Gellens D. (1991b). Sensitivity of the stream flow and of the sizing of flood and low flow control reservoirs to the 2x CO_2 climate change and related hypotheses: Study of two catchments in Belgium. 17.

Geller S. Y. (1969). Some aspects of the problem of the Aral Sea. In: *Problems of the Aral Sea* (ed. I. P. Gerasimov), pp. 5–25. Moscow: Nauka' Publ. (in Russian)

Georgiyevsky V. Yu. and Vladimirova T. I. (1991). The resources of surface water of Amu Darya and its change. In: *Monitoring of Natural Environment in the Aral Sea Basin* (eds. Yu. A. Israel and Yu. A. Anokhin), pp. 52–58. St. Petersburg: Hydrometeoizdat. (in Russian)

Georgievsky V. Yu., Ezhov A. V. and Shalygin A. L. (1997). Assessment of river runoff change under the effect of man's activity and global climate warming. In: *Runoff Computations for Water Projects* (ed. A. V. Rozhdestvensky), *Proc. of the St Petersburg Symposium (30 October – 3 November 1995)*, Part 2, pp. 75–81. Paris: UNESCO.

Georgiyevsky V. Yu., Ezhov A. V., Shalygin A. L., Shiklomanov I. A. and Shiklomanov A. I. (1996). Evaluation of possible climate change impact on hydrological regime and water resources of rivers of the former USSR. *Meteorol. Hydrol.*, 11, 89–99. (in Russian)

Georgiyevsky V. Yu., Shalygin A. L. and Doganovskaya T. M. (1993). Modern and future dynamics of crop requirements in irrigation due to global climate change. *Meteorol. Hydrol.*, 12, 81–86. (in Russian)

Georgiyevsky V. Yu., Zhuravin S. A. and Ezhov A. V. (1995). Assessment of trends in hydrometeorological situation on the Great Russian Plain under the effect of climate variations. In: *Proc. American Geophysical Union* (ed. H. J. Morel), *15th Annual Hydrology Days (Fort Collins, 3–7 April 1995)*, pp. 47–58. Atherton, CO: Hydrology Day Publ.

Gerasimov I. P. (ed.) (1986). *Arid Land Development and Desertification Control: Comprehensive Approach*. Moscow: UNEP/SCST. (in Russian)

Glantz M. (1994). Creeping environmental phenomena in the Aral Sea Basin. *Paper presented at Conf. "Creeping Environmental Problems and Societal Responses to Them" (Boulder, 7–10 February 1994)*.

Glantz M. (ed.) (1999). *Creeping Environmental Problems and Sustainable Development in the Aral Sea Basin*. New York: Cambridge University Press.

Glantz M. H. and Ausubel J. H. (1988). Impact assessment by analogy: Comparing the impacts of the Ogallala aquifer depletion and CO_2-induced climate change. In: *Societal Responses to Regional Climatic Change: Forecasting by Analogy*, pp. 428–441. Boulder, CO: Westview Press.

Glantz M. H. and Wigley T. M. (1987). Climatic variations and their effects on water resources. In: *Resources and World Development: Report of Dahlem Workshop (Berlin, 27 April – 2 May 1986)*, Part B, *Water and Land*, pp. 625–641.

Gleick P. H. (1986). Methods for evaluating the regional hydrologic impacts of global climatic changes. *J. Hydrol.*, 88, 99–116.

Gleick P. H. (1986). Regional water availability and global climatic change: The hydrologic consequences of increases in atmospheric carbon dioxide and other trace gases. PhD dissertation Berkeley, CA: University of California.

Gleick P. H. (1987). Regional hydrologic consequences of increases in atmospheric CO_2 and other trace gases. *Climatic Change*, 10, 137–161.

Gleick P. H. (1988). Climate change and California: Past, present and future vulnerabilities. In: *Social Responses to Regional Climatic Change: Forecasting by Analogy*, pp. 307–327. Boulder, CO: Westview Press.

Gleick P. H. (1988). The effects of future climatic changes on international water resources: The Colorado River, the United States and Mexico. *Policy Sci.*, 21, 23–39.

Gleick P. H. (1998). *The World's Water 1998–1999: The Biannual Report on Fresh Water Resources*. Washington, DC: Island Press.

Gleick P. H. (2000). *Potential Consequences of Climate Variability and Change for the Water Resources of the United States*. Report of the Water Sector Assessment Team of the National Assessment of the Potential Concequences of Climate Variability and Change. Oakland, CA: Pacific Institute for Studies on Development, Economics, and Security.

Gleick P. H. (2000). Water Futures: A Review of Global Water Resources. In: *World Water Scenarios: Analysis* (ed. F. R. Rijsberman), pp. 27–45. London: Earthscan Publ.

Gleick P. H. (ed.) (1993). *Water in Crisis: A Guide to the World's Fresh Water Resources*. New York: Oxford University Press.

Gleick P. H., Loh P., Gomes S. V. and Morrison J. (1995). *California Water 2020: A Sustainable Vision*. Oakland, CA: Pacific Institute for Studies on Development, Environment, and Security.

Glushkov V. G. (1929). Tenth anniversary of State Hydrological Institute. *Trans. SHI*, 25, 3–21. (in Russian)

Goldsmith E. and Hildyard N. (1985). *The Social and Environmental Effects of Large Dams*. San Francisco: Sierra Club Books.

Golubev S. M. and Leontyev O. A. (1988). Water management of India. *Land Reclam. and Water Mgmt*, 10, 59–61. (in Russian)

Golubev V. S. and Zmeikova I. V. (1991). Interannual changes in the conditions of evaporation in the Near-Aral region. In: *Monitoring of Natural Environment in the Aral Sea Basin* (eds. Yu. A. Israel and Yu. A. Anokhin), pp. 80–86. St. Petersburg: Hydrometeoizdat. (in Russian)

Gomes-Amaral J. C. and Day J. C. (1987). The Kemano Diversion: A hindsight assessment. In: *Impacts and Research Needs for Canada* (eds. W. Nicholaichuk and F. Quinn), *Proc. Symp. on Interbasin Transfer of Water (Saskatoon, 9–10 November 1987)*, pp. 137–152.

Gonzalez P. (in press). Desertification and a shift of forest species in the West African Sahel. *Climate Research*.

Gordeev V. V. and Tsirkunov V. (1998). River discharge of dissolved and suspended substances into sea basins from the territory of the former USSR (FSU). In: *From Dniepr to Baikal: Water Quality in the Former Soviet Union* (eds. V. Kimstach, M. Meybeck and E. Baroudy). London: E. & F. N. Spon.

Gordunio G., Mestre E. and Tapia F. (1984). Large-scale runoff diversion within the framework of the general plan of using water resources of Mexico. In: *Zonal Redistribution of Water Resources* (ed. M.G. Khublaryan), pp. 52–65. Moscow: Stroiizdat' Publ. (in Russian)

Gorshkov V. G., Kotlyakov I. M. and Losev K. S. (1994). Economical growth, environmental state, wealth and poverty. *News Acad. Sci. USSR*, Ser. Geogr., 1, 7–13. (in Russian)

Gottschalk L. (1985). Hydrological regionalization of Sweden. *J. Hydrol. Sci.*, 30, 65–84.

Government of India (1981). *Sixth 5-Year Plan 1981–1985*. New Delhi: Government of India.

Government of New Zealand (1988). *New Zealand Official Yearbook (1986/1987)*. Wellington, New Zealand: Central Office of Statistics.

Grabs W. T. and De Couet P. J. (1996). *Freshwater Fluxes from the Continents into the World Ocean based on Data from the Global Runoff Database*. Bundesanstalt fur Gewasserkunde (Federal Institute of Hydrology) Report No. 10. Koblenz, Germany: GRDC/WMO.

Grammatikati A. A. (1974). Development of irrigation in the East African countries. *Trans. VNIIGiM*, 1, 102–107. (in Russian)

Grave N. A. (1968). The frozen crust of the Earth. *Nature*, 1, 46–53. (in Russian)

Greco S., Moss R. H., Viner D. and Jenne R. (1994). *Climate Scenarios and Socio-economic Projections for IPCC Working Group II Assessment*. Working document for WG II lead authors, printed by Consortium of International Earth Science Information Network, May 1994.

Green K. A. (1984). A perspective on Australia's water resources to the year 2000. *Inst. Eng. Austral. Nat. Publ.*, 283–290.

Griffiths G. A. (1989). *Water resources: New Zealand Report on Impacts of Climate Change*. Canterbury, New Zealand: North Canterbury Catchment Board.

Grigorkina T. Ye. (1973). Hydrography and water resources of Australia. *News USSR Geogr. Soc.*, 105(4), 339–343. (in Russian)

Grigorkina T. Ye. (1974). Australia and Oceania. In: *World Water Balance and Water Resources of the Earth* (ed. V. I. Korzun), pp. 383–422. Leningrad: Hydrometeoizdat. (in Russian)

Grigorkina T. Ye. (1975). Runoff regime and New Zealand water balance. *News USSR Geogr. Soc.*, 107(2), 150–154. (in Russian)

Grigorkina T. Ye. (1976a). Dead heart of Australia. *News USSR Geogr. Soc.*, 108(5), 442–446. (in Russian)

Grigorkina T. Ye. (1976b). Long-term variations and seasonal distribution of the Australia and Oceania runoff. In: *Proc. 4th All-Union Hydrol. Congr.*, vol. 2, *Water Resources and Water Balance*, pp. 397–405. Leningrad: Hydrometeoizdat. (in Russian)

Grigorkina T. Ye. (1977). Australian groundwaters and their use. *News USSR Geogr. Soc.*, 109(16), 96–100. (in Russian)

Grigorkina T. Ye. (1980). Hydrography, water resources and water balance of Oceania. In: *Countries of the Southern Seas: History, Economics, Ethnog-*

raphy, Geography (ed. K. V. Malakhovsky), pp. 240–253. Moscow: Nauka' Publ. (in Russian)

Grigoryev A. A. (ed.) (1961). *Short Geographical Encyclopedia*, vol. 2. Moscow: Soviet Encyclopedia Publ. (in Russian)

Grima A. P. (1972). *Residential Demand for Water: Alternative Choices for Management*. Toronto, Canada: University of Toronto Press.

Groisman P. Ya. and Easterling D. R. (1994). Variability and trends of precipitation and snowfall over the United States and Canada. *J. Climate*, 7, 184–205.

Groisman P. Ya, Koknaeva V. V., Belokrylova T. A. and Karl T. R. (1991). Overcoming biases of precipitation measurement: A history of the USSR experience. *Bull. Am. Meterol. Soc.*, 72, 1725–1733.

Gromyko A. A. (ed.) (1987). *Africa: Encyclopedic Handbook*, vols. 1 and 2. Moscow: Soviet Encyclopedia Publ.

GUGK (1964). *Physical and Geographic Atlas of Asia*. Moscow: GUGK Publ.

Gusakov B. L. and Petrova N. A. (1987). *Facing the Great Lakes*. Leningrad: Hydrometeoizdat. (in Russian)

Hackl P. and Mulamoottil G. (1987). Ecological Impacts of the GRAND Canal Scheme on James Bay. In: *Impacts and Research Needs for Canada* (eds. W. Nicholaichuk and F. Quinn), *Proc. Symp. on Interbasin Transfer of Water (Saskatoon, 9–10 November 1987)*, pp. 407–414.

Hall A. (1992). The Snowy Mountains hydro electric scheme and its role in the development of Australian water resources. *WMO Bulletin*, 41(1), 59–69. (in Russian)

Hanchey J. R., Schilling K. E. and Stakhiv E. Z. (1987). Water resources planning under uncertainty. In: *Proc. 1st North American Conf. Preparing for Climate Change*, pp. 394–405. Fort Bevoir, VA: Government Institutes Inc.

Harden P. O. and Sundborg A. (1992). *The Lower Mekong Basin Suspended Sediment Transport and Sedimentation Problems*. Bangkok: Mekong Secretariat.

Hardison C. H. (1972). Potential United States water-supply development. *Proc. Am. Soc. Civil Eng., J. of Irrigation and Drainage Division*, Paper 9214, 479–492.

Hare F. K. (1984). *The Impact of Human Activities on Water in Canada: Inquiry on Federal Water Policy*. Research Paper No. 2. Ottawa, Canada: Environment Canada, Inland Waters Directorate.

Harpaz Y. (1994). Artificially enhanced precipitation to reduce risks provoked by overexploitation of groundwater. In: *Future Groundwater Resources at Risk* (eds. J. Soveri and T. Suokko), *Proc. Int. Conf. (Helsinki, June 1994)*. IAHS Publ. 222, 31–35.

Hatch U., Jagtar S., Jones J. and Lamb M. (1999). Potential effects of climate change on agricultural water use in the Southeast US. *J. Amer. Water Resources Assoc.*, 35, 1551–1561.

He Shangde (1989). Utilization and development of water resources of the Yellow River. In: *Water Use Statistics in the Long-Term Planning of Water Resources Development*. Water Resources Series No. 64, pp. 69–77. New York: ESCAP.

Hecky R. E. (1987). An environmental overview of the Churchill River Diversion. In: *Impacts and Research Needs for Canada* (eds. W. Nicholaichuk and F. Quinn), *Proc. Symp. on Interbasin Transfer of Water (Saskatoon, 9–10 November 1987)*, pp. 103–104.

HELCOM (1986). *Water Balance of the Baltic Sea*. Baltic Sea Environment Proceedings No. 16. Helsinki: Government Printing Office.

Herrington P. (1996). *Climate Change and the Demand for Water*. London: Department of the Environment.

Hess P. J. (1986). *Groundwater Use in Canada, 1981*. NHRI Paper No. 28/IWD Technical Bulletin No. 140. Ottawa, Canada: Environment Canada, National Hydrology Research Institute/Inland Waters Directorate.

Higgins G. M., Dielman P. J. and Abernethy C. L. (1988). Trends in irrigation development and their implications for hydrologists and water resources engineers. *Water Res. J.*, 158, 14–21.

Hoang Tien (1978). Water economy of Viet Nam. *Hydraul. Land Reclam.*, 4, 114–120. (in Russian)

Holda I., Osrodka L. and Wojtvlak M. (1989). Estimation of changes of runoff from urbanized and industrialized catchments. In: *Friends in Hydrology* (eds. L. Roald, K. Nordseth and K. A. Hassel), *Proc. Int. Conf. on Flow Regimes from International Experimental and Network Data (Bolksjo, Norway, 1–6 April 1989)*, pp. 403–408. IAHS Publ. 197. Wallingford, UK: IAHS Press.

Holing K. S. (ed.) (1981). *Ecological Systems: Adaptive Evaluation and Management*. Moscow: Mir' Publ. (in Russian)

Holland H. D. (1989). *The Chemical Evolution of the Atmosphere and Oceans.* Moscow: Mir' Publ. (in Russian)

Holy M. (1971). *Water and the Environment.* Rome: FAO.

Horie T., Baker J. T., Nacagava H. and Matsui T. (2000). Crop ecosystem responses to climatic change on rice yield in Japan. In: *Climate Change and Global Crop Productivity* (eds. K. R. Reddy and H. F. Hodges), pp. 81–106. Wallingford, UK: CAB International.

Hoyd Y. W. (1974). Importance of water resources in development of the Chilean copper industry. *Trans. Inst. Mining Metall.,* A.83, A63–A66.

Hoyle F. (1978). *The Cosmogony of the Solar System.* Cardiff, UK: University College of Cardiff Press.

Hubert P. and Carbonnel J. P. (1987). Approche statistique de l'aridification de l'Afrique de l'Ouest. *J. Hydrol.,* 95(1/2), 165–183.

Hughes T., Wang Y. M. and Hansen R. (1994). *Impacts of Projected Climate Change on Urban Water Use: An Application Using the Wasatch Front Water Demand and Supply Model.* Provo, UT: US Bureau of Reclamation.

Hulme M. and Jones P. D. (1988). *Climatic Change Scenarios for the UK.* Norwich, UK: University of East Anglia, Climatic Research Unit.

Hulme M., Doherty R. M., Ngara T., New M. G. and Lister D. (in press). African climate change: 1900–2100. *Climate Research.*

Hydrosphere (1960). *Short Geographical Encyclopedia* (ed. A. A. Grigoryev), vol. 1. Moscow: Soviet Encyclopedia. (in Russian)

Hydrosphere (1993a). *Encyclopedia Britannica.* Macropedia, 20, 715–731.

Hydrosphere (1993b). *Encyclopedia Britannica.* Micropedia, 6, 195.

ICID (1981). *Irrigation and Drainage in the World: A Global Review* (ed. K. K. Framji), vol. 1. New Delhi: ICID.

ICID (1982). *Irrigation and Drainage in the World: A Global Review* (ed. K. K. Framji), vol. 2. New Delhi: ICID.

ICOLD (1987). *Water Power and Dam Construction: Handbook.* Sutton, UK: ICOLD.

ICOLD (1992). *Water Power and Dam Construction: Handbook.* Sutton, UK: ICOLD.

IDB (1999). *Facing Up to Inequality in Latin America: Economic and Social Progress in Latin America.* Washington, DC: Inter-American Development Bank.

Idso S. and Brasel A. (1984). Rising atmospheric carbon dioxide concentrations may increase streamflow. *Nature,* 312, 51–53.

IIED/WRI (1987). *World Resources 1987: An Assessment of the Resource Base that Supports the Global Economy.* New York: Basic Books.

ILEC/UNEP (1987–9). *Data Book of World Lake Environments: A Survey of the State of World Lakes.* Otsu, Japan: International Lake Environments Committee/United Nations Environment Programme.

ILEC/UNEP (1991). *Data Book of World Lake Environments: A Survey of the State of World Lakes.* Otsu, Japan: International Lake Environments Committee/United Nations Environment Programme.

ILEC/UNEP (1993). *Data Book of World Lake Environments: A Survey of the State of World Lakes.* Otsu, Japan: International Lake Environment Committee/United Nations Environment Programme.

IMF (1994) *International Financial Statistics Yearbook 1994.* Washington, DC: International Monetary Fund.

Institute of Hydrology (1988). *Effects of Climatic Change on Water Resources for Irrigation: An Example for Lesotho.* Report prepared for FAO. Wallingford, UK: Institute of Hydrology.

Institute of Hydrology (1995). *Annual Report 1993/1994.* Wallingford, UK: Institute of Hydrology.

IPCC (1990). *Climate Change: The IPCC Scientific Assessment* (eds. J. T. Houghton, G. J. Jenkins and J. J. Ephraims). Cambridge, UK: Cambridge University Press.

IPCC (1992). *Climate Change: The Supplementary Report to the IPCC Scientific Assessment* (eds. J. T., Houghton, B. A., Callander and S. K. Varney). Cambridge, UK: Cambridge University Press.

IPCC (1995). *Climate Change 1995: The Science of Climate Change* (eds. J. T. Houghton, L. G. Meira Filho, B. A. Callander, N. Harris, A. Kattenberg and K. Maskell). Contribution of Working Group I to the 2nd Assessment Report of the Intergovernmental Panel on Climate Change. Cambridge, UK: Cambridge University Press.

IPCC (2001a). *Climate Change 2001: Impacts, Adaptation, and Vulnerability* (eds. J. J. McCarthy, O. F. Canziani, N. A. Leary, D. J. Dokken and K. S. White). Contribution of Working Group II to the 3rd Assessment Report of the Intergovernmental Panel on Climate Change. Cambridge, UK: Cambridge University Press.

IPCC (2001b). *Climate Change 2001: The Scientific Basis* (eds. J. T. Houghton, Y. Ding, D. J. Griggs, M. Nouger, P. J. Van der Linden, X. Dai, K. Maskell and C. A. Johnson), Contribution of Working Group I to the 3rd Assessment Report of the Intergovernmental Panel on Climate Change. Cambridge, UK: Cambridge University Press.

Isachenko A. G. and Shlyapnikov A. A. (1989). *Landscapes.* Moscow: Mysl' Publ. (in Russian)

Isayev V. A. (ed.) (1990). *The State of Kuwait: A Handbook.* Moscow: Nauka' Publ. (in Russian)

Israel Y. A., Yanshin A. L. and Polad-Zade P. A. (1988). Current state and proposals for a cardinal improvement of the ecological and sanitary–epidemiological situation in the Aral Sea regions and the lower reaches of the Amu Darya and Syr Darya Rivers. *Meteorol. Hydrol.,* 9, 5–22. (in Russian)

Ivanov K. Ye. and Penkova N. V. (1987). Analysis of some studies on forest effects on the mean annual river runoff by using an objective water-balance criterion. In: *Collection of Studies on Hydrology,* vol. 18, pp. 3–17. Leningrad: Hydrometeoizdat. (in Russian)

Ivanov S. A. and Luzhetsky A. N. (1973). Natural conditions and water management problems of Algeria. *Hydraul. Land Reclm.,* 5, 104–106. (in Russian)

Ivanova L. V. (1992). The hydrological aspects of the problem of the Aral Sea. *Water Resources,* 2, 39–49. (in Russian)

IWPDC (1991). *Handbook of International Water Power and Dam Construction.* Sutton, UK: Reed.

IWR (1994). Executive summary of lessons learned from the California drought (1987–1992). *IWR Report 94-NDS-6,* October 1994, 36.

Izmailova A. V. (1998). Water availability and water scarcity in the countries of Central America and Caribbean region. In: *Water: A Looming Crisis?* (ed. H. Zebidi), *Proc. Int. Conf. on World Water Resources at the Beginning of 21st Century (UNESCO, Paris, 3–6 June 1998),* IHP-V Technical Documents in Hydrology, 18, 254–255.

Izmailova A. V. (1999). Water resources, water use and water availability of North America. PhD dissertation, State Hydrological Institute, St Petersburg.

Izmailova A. V. and Moiseenkov A. I. (1998). Water resources and water availability in countries of Central America and Caribbean at present and in the future with taking into account possible climate change. In: *Proc. 2nd Int. Conf. on Climate and Water (Espoo, Finland, 17–20 August 1998),* vol. 3, pp. 1632–1642.

Jacobs K. (1992). Rationale for groundwater management in Arizona: A critique. In: *Changing Practices in Groundwater Management* (ed. J. J. DeVries), *Pros and Cons of Regulation, Proc. 18th Biennial Conf. on Groundwater* (Sacramento, 16–17 September 1991), University of California Report 77, 113–118.

Jaensch R. (1995). Pakistanisches Wasserkraftwerk kommit voran. *Nachrichten Anssenhand,* 58, 104.

Jalal K. F. (1987). Regional water resources situation: quantitative aspects. In: *Water Resources Policy for Asia* (eds. M. Ali, G. E. Radosevich and A. Ali Khan), *Proc. Reg. Symp. on Water Res. Policy in Agro-Socio-Economic Development (Dhaka, Bangladesh, 4–8 August 1985),* pp.13–36. Rotterdam, The Netherlands: A. A. Balkema.

Jang Jicheng (1986). Hydrologic applications of remote sensing in China. In: *Hydrologic Applications of Space Technology* (ed. A. I. Johnson), *Proc. Int. Workshop (Cocoa Beach, FL, August 1985).* IAHS Publ. 160, 269–273.

Jenne R. L. (1989). *Data from Climate Models: The CO_2 Warming.* Reston, VA: US Geological Survey.

Jiasheng P. and Jiasheng Z. (1994). Hydropower development in China. *Water Res. J.,* C/180, 75–97.

Jing Zhang, Wei Wen Huang and Mao Chong Shi (1990). Huang Ho (Yellow River) and its estuary: sediment origin, transport and deposition. *J. Hydrol.,* 120, 203–223.

Johnson K. E. and Miller F. C. (1978). Mass transfer of water into inland Australia for irrigation and power by the Snowy Mountains hydro electric scheme. In: *Proc. 8th Int. Congr. on Irrigation and Drainage,* vol. 7, pp. 115–138. Athens:ICID.

Jones B. L. (1966). *Effects of Agricultural Conservation Services on the Hydrology of Korea.* New York: Oxford University Press.

Jones P. (1975). Future water resources for England and Wales. *Geography,* 60 (4), 298–300.

Jones P. D. (1988). Hemispheric surface air temperature variations: Recent trends and update to 1987. *J. Climate*, 1, 654–660.

Jones P. D. (1990). Antarctic temperatures over the present century: a study of the early expedition record. *J. Climate*, 3, 1193–1203.

Jones P. D. (1994). Hemispheric surface air temperature variations: A reanalysis and an update to 1993. *J. Climate*, 7, 1794–1802.

Jones P. D., Groisman P. Ya., Coughlan N., Plummer N., Wang W.-C. and Karl T. R. (1990). Assessment of urbanization effects in time series of surface air temperature over land. *Nature*, 347, 169–172.

Jorgulesku F. (1979). Comprehensive use and protection of water resources in the SRR. C *MEA Inf. Bull. on Water Management*, 1, 27–30. (in Russian)

Kaczmarek Z. (1980). Water resources of Poland and principles of their rational use. *J. Polish Acad. Sci.*, 23 (4), 33–44.

Kaczmarek Z. (1993). Water balance model for climate impact analysis. *Acta Geophys. Polon.*, 41(4), 423–427.

Kaczmarek Z. and Krasuski D. (1991). *Sensitivity of Water Balance to Climate Change and Variability*. Luxemburg: International Institute for Applied Systems Analysis.

Kaczmarek Z. and Napiorkowski J. (1996). Water resources adaptation strategy in an uncertain environment. In: *Adapting to Climate Change: An International Perspective*, (eds. J. B. Smith, N. Bhatti, G. V. Menzhulin, R. Benioff, M. Campos, B. Jallow, F. Rijsberman, M. I. Budyko and R. K. Dixon), pp. 211–224. New York: Springer-Verlag.

Kadyrov A. A. and Shalomayev M. I. (1984). Water management of the Socialist Republic of Viet Nam. *Land Reclam. Water Mgmt*, 5, 30–31. (in Russian)

Kadzayev M. B. (1984). Agricultural exploration of lands in the People's Democratic Republic Yemen. *Hydraul. Land Reclam.*, 12, 75–77. (in Russian)

Kaisl Ch. (1972). *The Analysis of Time Series of Hydrological Data*. Leningrad: Hydrometeoizdat. (in Russian)

Kalinin G. P. (1968). *Problems of the Global Hydrology*. Leningrad: Hydrometeoizdat. (in Russian)

Kalinin G. P. (1972). Some questions of the theory of managing the land water regime. In: *The Problems of Studying and Using Water Resources* (eds. S. N. Kritsky and M. F. Menkel), pp. 7–49. Moscow: Nauka' Publ. (in Russian)

Kalinin G. P. (1974). Some problems of water exchange in the nature. In: *Water Exchange in Nature and its Role in Forming Fresh Water Resources* (ed. G. P. Kalinin), pp. 27–40. Moscow: Stroyizdat' Publ. (in Russian)

Kalinin G. P. and Shiklomanov I. A. (1974). Water resources development. In: *World Water Balance and Water Resources of the Earth* (ed. V. I. Korzun), pp. 575–604. Leningrad: Hydrometeoizdat. (in Russian)

Kalinin G. P. and Shiklomanov I. A. (1978). Exploitation of the earth's water resources. In: *World Water Balance and Water Resources of the Earth* (ed. V. I. Korzun), pp. 592–621. Paris: UNESCO.

Kammerer J. C. (1990). *Large Rivers of the World*. US Geological Survey Water Supply Paper No. 2350. Washington, DC: US Government Printing Office.

Kang E., Cheng G., Lan Y. and Jin H. (1999). A model for simulating the response of runoff from the mountainous watersheds of northwest China to climate change. *Science in China*, 42.

Karagodin N. A. and Elyanov A. Yu. (eds.) (1992). *Experience of Economic Reforms in Developing Countries*. Moscow: Nauka' Publ. (in Russian)

Karasik G. Ya. (1970). *Water Balance of Africa*. Moscow: VINITI Publ. (in Russian)

Karasik G. Ya. (1974). *Water Balance of South America*. Moscow: Soviet Radio' Publ. (in Russian)

Karavayev A. P. (1987). *Capitalism in Brazil: Past and Present*. Moscow: Nauka' Publ. (in Russian)

Karev V. and Shtyka V. (1973). Economical aspects of water resources utilization and conservation in France. *Hydraul. Eng. Land Reclam.*, 8, 100–101. (in Russian)

Karl T. R. and Knight R. W. (1998). Secular trends of precipitation amount, frequency and intensity in the United States. *Bull. Am. Meteorol. Soc.*, 79, 231–241.

Karl T. R., Groisman P. Ya., Knight R. W. and Heim R. R. Jr. (1993). Recent variations of snow cover and snowfall in North America and their relation to precipitation and temperature variations. *J. Climate*, 6, 1327–1344.

Kashef A. A. (1981a). The Nile: one river and nine countries. *J. Hydrol.*, 53, 53–71.

Kashef A. A. (1981b). Technical and ecological impacts of the High Aswan Dam. *J. Hydrol.*, 53, 73–84.

Katz A. L. (1964). A two-year cyclicity in the equatorial stratosphere and general circulation of the atmosphere. *Meteorol. Hydrol.*, 16, 3–10. (in Russian)

Kayastha S. L. (1980). An appraisal of water resources of India and need for national water policy. In: *Proc. 24th Int. Geogr. Congress (Tokyo, Japan, 23 April –10 September 1980)*, Main Session Abstracts, vol. 4, pp. 112.

Keller R. (1952). *Gewasser und Wasserhaushalt das Festlandses*. Leipzig: Academia Verlag.

Kendall M. and Student A. (1976). *Statistical Inferences and Relations*. Moscow: Nauka' Publ. (in Russian)

Kennedy D. N. (1992). California's water resources: Looking to the future. In: *Changing Practices in Groundwater Management* (ed. J. J. DeVries), *Pros and Cons of Regulation, Proc. 18th Biennial Conf. on Groundwater* (Sacramento, 16–17 September 1991), University of California Report 77, 9–14.

Khadam M. A., Shammas N. Kh. and Al-Feraiheedi Y. (1991). Water losses from municipal utilities and their impacts. *Water International*, 16 (4), 254–261.

Kharchenko S. I. (1975). *Hydrology of Irrigated Lands*. Leningrad: Hydrometeoizdat. (in Russian)

Khaydarova V. A., Kucherova V. A. and Penkova N. V. (1998). On assessment of changes in water balance components under global warming and possible consequences for CIS agricultural areas. In: *Proc. 2nd Int. Conference on Climate and Water (Espoo, Finland, August 1998)*, vol. 3, pp. 1643–1652. Helsinki: Helsinki University of Technology.

Khomenko A. N. and Yemelyanova V. P. (1991). Characteristics of the pollution on the Amu Darya and Syr Darya Rivers. In: *Monitoring of Natural Environment in the Aral Sea Basin* (eds. Yu. A. Israel and Yu. A. Anokhin), pp. 109–115. St Petersburg: Hydrometeoizdat. (in Russian)

Kierans T. W. (1980). Thinking big in North America the GRAND-canal conception. *Futurist*, 14(6), 29–32.

Kierans T. W. (1987). Recycled water from the North, the alternative to interbasin diversions. In: *Impacts and Research Needs for Canada* (eds. W. Nicholaichuk and F. Quinn), *Proc. Symp. on Interbasin Transfer of Water (Saskatoon, 9–10 November 1987)*, pp. 59–70.

Kirgizov G. (1982a). Water economy of Hungarian People's Republic. *Hydraul. Eng. Land Reclam.*, 9, 73–74. (in Russian)

Kirgizov G. (1982b). Water resources of Finland and their use. *Hydraul. Eng. Land Reclam.*, 10, 79–80. (in Russian)

Kirshen P. H. and Fennessey N. M. (1992). *Potential Impacts of Climate Change upon the Water Supply of the Boston Metropolitain Area*. Draft Report to US Environmental Protection Agency. Washington, DC: US Government Printing Office.

Kitchen H. M. (1975). *A Statistical Estimation of a Demand Function for Residential Water*. Ottawa, Canada: Environment Canada, Inland Waters Directorate, Water Planning and Management Branch.

Klemes V. (1985). *Sensitivity of Water Resource Systems to Climate Variations*. World Climate Applications Programme No. WSP-98. Geneva: WMO.

Klige R. K. (1980). *The Ocean Level in the Geological Past*. Moscow: Nauka' Publ. (in Russian)

Klige R. K. (1982). Tendencies in changes of surface water hydrosphere. *Water Resources*, 3, 92–105. (in Russian)

Klige R. K. (1992). The change of water exchange in palaeo- and historical time period. *Wat. Resources*, 4, 5–6. (in Russian)

Klige R. K., Danilov I. D. and Konishchev V. N. (1998). *Hystory of the Hydrosphere*. Moscow: Scientific World Publ. (in Russian)

Kliot N. (1994). *Water Resources and Conflict in the Middle East*. London: Routledge.

Klochkowsky L. L. (ed.) (1991). *Latin America in UNESCO Studies: Prospects of Development of Science, Education and Culture to 2000*. Moscow: UNESCO/Institute for Latin American Studies Russian Academy of Science Publ. (in Russian)

Komar Yu. I. (ed.) (1990). *The Town in the Modern Afro-Asian World*. Moscow: INION RAS Publ. (in Russian)

Kondratjyev K. Ya. (1992). *Global Climate*. St. Petersburg: Nauka', Publ. (in Russian)

Konstantinov A. R. (1968). *Evaporation in Nature*. Leningrad: Hydrometeoizdat. (in Russian)

Kopanev I. D. and Shvert Ts. A. (1991). *The Applied Aspects of Using Climatic and Hydrological Information for Siberia and the Far East*. Leningrad: Hydrometeoizdat. (in Russian)

Korkunov I. N. (ed.) (1992). *Modern Chinese Villages: Main Tendencies of Social–Economical development*. Moscow Institute of Far East, Russian Academy of Science Publ. (in Russian)

Korzun V. I. (ed.) (1974a). *Atlas of the World Water Balance*. Leningrad: Hydrometeoizdat. (in Russian)

Korzun V. I. (ed.) (1974b). *World Water Balance and Water Resources of the Earth*. Leningrad: Hydrometeoizdat. (in Russian)

Kosicheva L. A. and Odesser S. V. (1979). Water resources of the countries of Iberian Peninsula and specific features of their use. *Water Resources*, 4, 176–185. (in Russian)

Kotlyakov V. M. (1979). The problems of the present glaciology. *Nature*, 9, 27–90. (in Russian)

Kotlyakov V. M. (1984). Snow cover. In: *Glaciological Dictionary* (ed. V. M. Kotlyakov), pp. 416–418. Leningrad: Hydrometeoizdat. (in Russian)

Kotlyakov V. M. (ed.) (1991). The main adeas of the concept on preservation and restoration of the Aral Sea, and on normalization of the ecological, sanitary–hygienic, medico-biological and socio-economic situation in the Aral Region. *News Acad. Sci. USSR*, Ser. Geogr., 4, 8–21. (in Russian)

Kotlyakov V. M. (ed.) (1997). *Atlas of Snow and Ice Resources of the Earth*. Moscow: Inst. Geogr. RAS Publ. (in Russian)

Kotwicki V. (1991). Water in the Universe. *Hydrol. Sci. J.*, 36 (1), 49–66.

Kouchment L. S. (1980). *Models of River Runoff Formation Processes*. Leningrad: Hydrometeoizdat. (in Russian)

Kouchment L. S. and O'Connell P. E. (1993). Constructing the models of the land hydrological cycle at the global scale: The analysis of modern condition and prospects. *Water Resources*, 2, 149–159. (in Russian)

Kouchment L. S., Motovilov Yu. G. and Nazarov N. A. (1990). *The Sensitivity of Hydrological Systems*. Moscow: Nauka' Publ. (in Russian)

Koupriyanov V. V. (1977). *Hydrological Aspects of Urbanization*. Leningrad: Hydrometeoizdat. (in Russian)

Kouzin P. S. and Babkin V. I. (1979). *Geographical Features of River Hydrology*. Leningrad: Hydrometeoizdat. (in Russian)

Kouzmin P. P. (1961). *The Snow Cover Melt Process*. Leningrad: Hydrometeoizdat. (in Russian)

Krasovskaya I. (1995). Quantification of stability of river flow regimes. *Hydrol. Sci. J.*, 40(5), 587–597.

Krasovskaya I. (1996). Sensitivity of the stability of river flow regimes to small fluctuations in temperature. *Hydrol. Sci. J.*, 41(4), 251–264.

Krestovsky O. I. (1986). *The Influence of Deforestation and Afforestation on River Water Content*. Leningrad: Hydrometeoizdat. (in Russian)

Kuenen P. H. (1950). *Marine Geology*. New York: John Wiley & Sons.

Kuksa V. I. (1994). *The Southern Seas (the Aral Sea, the Caspian Sea, the Sea of Azov and the Black Sea) under Conditions of Anthropogenic Stress*. St Petersburg: Hydrometeoizdat. (in Russian)

Kulp J. L. (1951). The origin of the hydrosphere. *Bull. Geol. Soc. Amer.*, 62, 326–330.

Kulshreshtha S. N. (1992). *World Water Resources and Regional Vulnerability: Impact of future changes*. Luxemburg: International Institute for Applied Systems Analysis.

Kump L. R. (1989). Chemical stability of the atmosphere and ocean. *Palaeogeog., Palaeoclimatol., Palaeoecol.*, Global and planetary change section, 75, 123–136.

Kundzewicz Z. W. and Parry M. L. (2001). Europe. In *Climate Change 2001: Impacts, Adaptation and Vulnerability* (eds. J. J. McCarthy, O. F. Canziani, N. A. Leary, D. J. Doten and K. S. White), pp. 641–692. Cambridge, UK: Cambridge University Press.

Kurbatov V. P. (1993). Water management construction in China. *Land Reclam. Water Mgmt*, 2, 48–50. (in Russian)

Kutsobin P. N. (ed.) (1990). *India 1988 Yearbook*. Moscow: Nauka' Publ. (in Russian)

Kuznetsov A. A. (1989). *Encyclopedia of New China*. Moscow: Progress Publ. (in Russian)

Kwadijk J. and van Deursen W. (1994). Scenarios for the discharge of the Rhine: estimates and uncertainties. *Clim. Change*, 27, 14–23.

Langbein W. B. (1949). *Annual runoff in the United States*. US Geological Survey Circular No. 5. Washington, DC: US Department of the Interior.

Langbein W. B. (1982). *Dams, Reservoirs and Withdrawals for Water Supply: Historic Trends*. US Geological Survey Open-File Report No. 82–256. Washington, DC: US Government Printing Office.

Laslo D. (1984). Experience of water management prediction in Hungary. *Hydrotechn. Construct.*, 4, 50–54. (in Russian)

Lebedev A. N. and Kopanev I. D. (eds.) (1975). *Climates of Foreign Asia*. Leningrad: Hydrometeoizdat. (in Russian)

Leichenko R. M. (1993). Climate change and water resource availability: An impact assessement for Bombay and Madras, India. *Water International*, 18(3), 147–156.

Lettenmaier D., Wood E. F. and Wallis J. R. (1994). Hydroclimatological trends in the Continental United States (1948–88). *J. Climate*, 7, 586–607.

Levchenko G. P., Sumarokova V. V. and Tsytsenko K. V. (1990). Study of water consumption and return water from the irrigated lands of arid areas of the USSR. In: *Hydrological Basis for Water Management* (eds. I. A. Shiklomanov and V. Yu. Georgiewsky), *Proc. 5th All-Union Hydrological Congress*, vol. 4, pp. 511–518. Leningrad: Hydrometeoizdat. (in Russian)

Levin A. P. (1973). *Water Factors in the Location of Industrial Production*. Moscow: Stroiizdat' Publ. (in Russian)

Levitanus A. Yu. (1986a). Current problems of water economical development of India. *News Acad. Sci. USSR*, Ser. Geogr., 4, 53–60. (in Russian)

Levitanus A. Yu. (1986b). *River Basins of India: Problems of Economic Exploration*. Moscow: Inst. of Geography USSR Acad. Sci. Publ. (in Russian)

Library of Congress (1991). *Egypt: A Country Study*. Washington, DC: US Government Printing Office.

Lindsley R. K. (1985). Models "Precipitation–runoff". In: *The System Approach to Water Resources Management* (ed. A. K. Biswas), pp. 25–29. Moscow: Nauka' Publ. (in Russian)

Lins H. F. (1985a). Interannual streamflow variability in the United States based on Principal components. *Water Resources Res.*, 21, 691–701.

Lins H. F. (1985b). Streamflow variability in the United States, 1931–78. *J. Climate Appl. Meteorol.*, 24, 463–471.

Lins H. F. and Michaels P. J. (1994). Increasing streamflow in the United States. *EOS*, 75, 281–286.

Lins H. F. and Slack J. R. (1999). Streamflow trends in the United States. *Geophys. Res. Letters*, 26, 227–230.

Liu C. (1998). The potential impact of climate change on hydrology and water resources in China. In: *Proc. 2nd Int. Conf. on Climate and Water (Espoo, Finland, August 1998)*, vol. 1, pp. 1420–1445. Helsinki: Helsinki University of Technology.

Loh P. (1994). *(De)Constructing the California Water Plan: Science, Politics, and Sustainability*. Oakland, CA: Pacific Institute for Studies in Development, Environment, and Security.

Losev K. S. (1989). *Water*. Leningrad: Hydrometeoizdat. 270 (in Russian)

Lukichev G. A. (1990). *The Countries that Have Become Free: Use of Resources for the Purpose of Development*. Moscow: DN Univ. Publ. (in Russian)

L'vovich M. I. (1945). *Elements of the Water Regime of World's Rivers*. Moscow: Hydrometeoizdat. (in Russian)

L'vovich M. I. (1969). *Water Resources of the Future*. Moscow: Prosveshchenie' Publ. (in Russian)

L'vovich M. I. (1972). Water balance of the Earth' s continents and a balance assessment of the world resources of freshwater. *News Acad. Sci. USSR*, Ser. Geogr., 5, 5–20. (in Russian)

L'vovich M. I. (1974). *World Water Resources and Their Future*. Moscow: Mysl' Publ. (in Russian)

L'vovich M. I. (1986). *Water and Life*. Moscow: Mysl' Publ. (in Russian)

Mabbut G. (1986). Desertification in Australia. In: *Arid Land Development and Desertification Control: Comprehensive Approach*, pp. 105–118. Moscow: UN Environment Programme. (in Russian)

MacKichan K. A. (1951). *Estimated Water Use in the United States, 1950*. US Geological Survey Circular No. 115. Washington, DC: US Department of the Interior.

MacKichan K. A. (1957). *Estimated Water Use in the United States, 1955*. US Geological Survey Circular No. 398. Washington, DC: US Department of the Interior.

MacKichan K. A. and Kammerer J. C. (1961). *Estimated Use of Water in the United States, 1960*. US Geological Survey Circular No. 456. Washington, DC: US Department of the Interior.

Maclaine D. J. (ed.) (1981). *Tasmanian Yearbook*, vol. 15. Hobart, Tasmania: Australian Bureau of Statistics, Tasmanian Office.

MacLean D. J. and Beckstead G. R. E. (1987). Long-term effects of an interbasin diversion on the Milk River. In: *Impacts and Research Needs for Canada* (eds. W. Nicholaichuk and F. Quinn), *Proc. Symp. on Interbasin Transfer of Water (Saskatoon, 9–10 November 1987)*, pp. 295–318.

Maddock T. S. and Hines W. G. (1995). Meeting future public water supply needs: A Southwest perspective. *Water Res. Bull.*, April 1995, 317–329.

Maganza C. H. D. (1996). Climate change: some likely multiple impacts in southern Africa. In: *Climate Change and World Food Security* (ed. T. E. Downing), pp. 449–483. Dordrecht, The Netherlands: Springer-Verlag.

Maganza C. H. D. (2000). Climate change impacts and human settlements in Africa: prospects for adaptation. *Environmental Monitoring*, 61, 193–205.

Maiga S.B. (1984). Mali: Le Plan de redressment de l'office-du-Niger. *Europe Outre-Mer*, 60(648–649), 26–27.

Main Air Force Headquarters (1992). *Atlas of Hydrometeorological Data*, vol. 2. St. Petersburg: MAFH. (in Russian)

Makarov A. V. (1976). Development of water economy and land exploration in the Syrian Arab Republic. *Hydraul. Land Reclam.*, 2, 109–114. (in Russian)

Makarov A. V. and Marchenko A. A. (1982). Land Reclamation in the KDPR. *Hydraul. Land Reclam.*, 4, 78–84. (in Russian)

Malakhovsky K. V. (1988). *Australia: Is the Country in Transition?* Moscow: Nauka' Publ. (in Russian)

Malik L. K. (1990). *Geographical Forecasts of the Consequences of Hydropower Construction in Siberia and the Far East*. Moscow: Institute of Geography USSR Acad. Sci. Publ. (in Russian)

Mamilton P. and Maizels J. (1989). Flows to forecast future. *Geogr. Magazine*, 61, 28–37.

Manabe S. and Wetherald R. T. (1987). Large-scale changes of soil wetness induced increase by an increase in atmospheric carbon dioxide. *J. Atmos. Sci.*, 44, 1211–1235.

Manabe S., Spelman M. J. and Stouffer R. J. (1992). Transient responses of a coupled ocean–atmospheric model to gradual changes of atmospheric CO_2, Part 2: Seasonal response. *J. Climate*, 5, 105–126.

Manabe S., Stouffer R. J., Spelman M. J. and Brian K. (1991). Transient responses of a coupled ocean–atmospheric model to gradual changes of atmospheric CO_2, Part 1: Annual mean response. *J. Climate*, 4, 785–818.

Mann L. J. (1985). *Groundwater-Level Changes in Five Areas of the United States: National Water Summary 1984*. US Geological Survey Water-Supply Paper No. 2275. Washington, DC: US Government Printing Office.

Marcinek Y. (1987). Die grosten Strome der VR China. *Geograph. Berichte*, 122, 1–11.

Marengo J. A. (1992). Interannual variability in surface climate in the Amazon basin. *Int. J. Climatol.*, 12, 853–863.

Marengo J. A. (1995). Variations and change in South American streamflow. *Climate Change*, 31, 99–117.

Marengo J. A., Tomasella J. and Uvo C. R. (1998). Trends in streamflow and rainfall in tropical South America: Amazonia, eastern Brasil and southwestern Peru. *J. Geophys. Res.*, 103, 1775–1783.

Margat J. (1990). *Prospects for Resources and Needs in the Mediterranean Countries*. Contribution to the "Plan Bleu". Paris: European Commission.

Margat J. (1992). *L'Eau dans le Baissin Méditerranéen*. Fascicule du Plan Bleu No. 6. Paris: Economica.

Margat J. (1994). *Water Use in the World: Present and Future*. Contribution au Projet M-1-3 du Programme Hydrologique International. Paris: PHI-IV/UNESCO.

Margat J. (1995). Prospective des pénuries d'eau au Maghreb. *Coll. Int. Eau: Gestion de la Rareté*, Rabat, Morocco, vol. 10.

Margat J. and Vallée D. (2000). *Water for the 21st Century: Vision to Action – Mediterranean Vision on Water, Population and Environment*. Monaco: JS Communication.

Markov K. K. (1960). *Palaeogeography*. Moscow: Moscow University Publ. (in Russian)

Markova A. N. (ed.) (1995). *History of Global Economics: Economical Reforms of 1920–1990*. Moscow: VINITI Publ. (in Russian)

Maslennikova I. N. (1987). *Irrigation Landscapes of Southwest Asia*. Moscow: Moscow University Publ. (in Russian)

Matalas N. C. and Fiering M. B. (1977). Water-resource systems planning. In: *Climate, Climatic Change and Water Supply* (ed. J. Wallis), pp. 99–110. Washington DC: National Research Council.

Matalas N. C., Landwehr J. M. and Wolman M. G. (1982). Prediction in water management. In: *Scientific Basis of Water-Resource Management*, pp. 118–127. Washington DC: National Academy Press.

Matsuyama X., Oki T. and Masuda K. (1993). The water budget in the Amazon River basin during the FGGE period. In: *Hydrology of Warm Humid Regions* (ed. J. S. Gladwell), *Proc. Yokohama Symp. (Yokohama, July 1993)*, IAHS Publ. 216, 35–42.

McCahe G. L. and Wolock D. M. (1992). Sensitivity of irrigation demand in a humid–temperate region to hypothetical climate change. *Water Res. Bull.*, 28(3), 535–543.

McCruady W. (1988). Left bank outfall drain in Pakistan. *Water Res. J.*, C/156, 68–72.

McDonald W. S., Cocks K. D., Wood N. H., Ive J. R. and Yapp G. A. (1993). The future population of Australia's coastal lands. *Austral Geogr. Studies*, 31(2), 177-188.

McNeil R. and Tate D. M. (1991). *Guidelines for Municipal Water Pricing*. Ottawa, Canada: Environment Canada, Inland Waters Directorate, Water Planning and Management Branch.

McQuenn C. (1992). Marketing in Pakistan. *Overseas Bus. Repts*, 3, 1–53.

Meade R. H. (1994). *Suspended Sediments of the Modern Amazon and Orinoco Rivers*. Contribution 65 of the CAMREX Project – INQUA. Amsterdam: Elsevier.

Meade R. H., Rayol H., Daconceicao S. C. I., Nativi A. E. (1991). Backwater effects in the Amazon River basin of Brazil. *Environ. Geol. Water Sci.*, 18(2), 105–114.

Meangus I. (1993). Anglian Water International spreads its wings. *Water Bull.*, 573, 21–22.

Medvedev A. P. (1989). *Agropotential of Maghreb Countries*. Moscow: Institute of Geography RAS Publ. (in Russian)

Meinzer F. C. and Zhu J. (1998). Nitrogen stress reduces the efficiency of the C_4–CO_2 concentrating system, and therefore quantum yield, in *Saccharum* (sugarcane) species. *J. Exp. Bot.*, 49, 1227–1234.

Meko D. M. and Stockton C. W. (1984). Secular variations in streamflow in the Western United States. *J. Climate Appl. Meteorol.*, 23, 889–897.

Mekong Secretariat (1990). *A Study on Impact of Climate Change on Water Resources in the Lower Mekong Basin*. Report prepared for US Environment Protection Agency. Boulder, CO: University of Colorado.

Mengxiong C. and Zunhuang C. (1994). Risks for development of groundwater resources in urban areas of China. In: *Future Groundwater Resources at risk* (eds. J. Soveri and T. Suokko), *Proc. Int. Conf. (Helsinki, June 1994)*. pp. 471–480. IAHS Publ. 222. Wallingford, UK: IAHS Press.

Menon P. A. (1993). *Keeping New Zealand Green: Recent Environmental Reforms*. Dunedin, New Zealand: Otago University Press.

Meybeck M. (1988). How to establish and use world river budget material. In: *Physical and Chemical Weathering in Geochemical Cycles* (eds. A. Lerman and M. Meybeck) pp. 247–272. Dordrecht, The Netherlands: D. Reidel. (reprinted by permission)

Meybeck M. and Ragu A. (1995) *River Discharges to the Oceans: An Assessment of Suspended Solids Major Ions and Nutrients*. Nairobi: UN Environment Programme.

Mikhura V. I. (1982). Water usage in EEC Member Countries of the UNO. *Hydrotech. Construct.*, 4, 53–57. (in Russian)

Mikulski Z. (1988). Baltic Sea as a hydrological system and its water balance. In: *Water Resources and Water Budget* (eds. A. A. Sokolov and V. I. Babkin), *Proc. 5th All-Union Hydrological Congress*, vol. 2, pp. 638–649. Leningrad: Hydrometeoizdat. (in Russian)

Milko R. J. (1987). The GRAND Canal: Potential ecological impacts to the North and research needs. In: *Impacts and Research Needs for Canada* (eds. W. Nicholaichuk and F. Quinn), *Proc. Symp. on Interbasin Transfer of Water (Saskatoon, 9–10 November 1987)*, pp. 85–99.

Miller J. R. and Russell G. L. (1992). The impact of global warming on river runoff. *J. Geophys. Res.*, 97 (D3), 2757–2765.

Miller N. L. and Kim J. (2000). Climate change sensitivity analysis for two California watersheds. *J. Am. Water Res. Assoc.*, 36, 657–661.

Milliman J. D. (1990). Fluvial sedimentation in coastal seas: Flux and fate. *Nature*, 26(4), 12–22.

Milliman J. D. and Meade R. H. (1983). World-wide delivery of river sediment to the oceans. *J. Geol.*, 91(1), 1–21.

Milliman J. D., Rutkowsky C. and Meybeck M. (1995). *River Discharges to the Sea: A Global River Index (GLORI)*. Texel, The Netherlands: Netherlands Institute for Sea Research.

Milovsky V. V. (1990). Development of irrigation and water economy of Afghanistan. *Land Reclam. Water Mgmt*, 8, 58–61. (in Russian)

Ministry for the Environment (1999). *Making Every Drop Count*. Wellington, New Zealand: Ministry for the Environment.

Ministry of Water Resources of China (1987). *Assessment of Water Resources of China*. Beijing: MWRC. (in Chinese)

Ministry of Works (1970). *Hydrological Statistics: Monthly, Annual and Long-Term Flows by 31.12.69.* Wellington, New Zealand: Investigations Section, Power Design Office.

Mirzayev S. Sh. and Rachinsky A. A. (1991). The Aral Sea is our mutual concern. *News Acad. Sci. USSR*, Ser. Geogr., 4, 113–117. (in Russian)

Mitchell B. and McBean E. (1985). *Water Resources Research in Canada: Issues and Opportunities.* Research Paper No. 16. Ottawa, Canada: Inquiry on Federal Water Policy, Ontario Environment Canada Inland Water Directorate.

Mitchell J. F. B., Senior C. A. and Ingram W. J. (1989). CO_2 and climate: missing feedback? *Nature*, 341, 132–134.

Monin A. S. (1977). *The History of the Earth.* Leningrad: Nauka' Publ. (in Russian)

Monin A. S. and Shishkov Yu. A. (1979). *The History of Climate.* Leningrad: Hydrometeoizdat. (in Russian)

Moore M. R., Crosswhite W. M. and Hostetler J. E. (1990). *Agricultural Water Use in the United States, 1950–1985: National Water Summary 1987.* US Geological Survey Water-Supply Paper No. 2350. Washington, DC: US Government Printing Office.

Morozova M. Yu. (1983). Agronatural resources of Pakistan and problems of their exploration. *Geog. Nat. Res.*, 4, 108–115. (in Russian)

Mosley M. P. (1991). Climate change: Impacts on water resources. In: *Climate Change: The New Zealand Response*, pp. 70–81.

Mosley M. P. (1996). Water use problems in the Southwest Pacific. *WMO Bulletin*, 45(4), 421–429. (in Russian)

Muranova A. (1993). Resources of economical development of the countries of Asia, Oceania and Australia. *Asia and Africa Today*, 6–7, 66–69, 64–67. (in Russian)

Muranova A. (1994). Gross domestic product and branch structure of its production in the countries of Asia, Oceania and Australia. *Asia and Africa Today*, 1, 36–39. (in Russian)

Murphy J. M. and Mitchell J. F. B. (1995). Transient response of the Hadley Centre coupled ocean–atmosphere model to increasing carbon dioxide, Part 2: Spatial and temporal structure of the response. *J. Climate*, 8, 57–80.

Murray C. R. (1968). *Estimated Use of Water in the United States, 1965.* US Geological Survey Circular No. 556. Washington, DC: US Department of the Interior.

Murray C. R. and Reeves E. B. (1972). *Estimated use of water in the United States, 1970.* US Geological Survey Circular No. 676. Washington, DC: US Department of the Interior.

Murray C. R. and Reeves E. B. (1972). *Estimated Use of Water in US in 1970.* US Geological Survey Circular 67b. Washington, DC: Government Printing Office.

Murray C. R. and Reeves E. B. (1977). *Estimated Use of Water in the United States, 1975.* US Geological Survey Circular No. 765. Washington, DC: US Department of the Interior.

Murray–Darling Basin Commission (1999). *The Salinity Audit of the Murray–Darling Basin: A 100-Year Perspective.* Canberra, Australia: Murray–Darling Basin Commission.

Nace R. (1967). *Are We Running out of Water?* US Geological Survey Circular No. 536. Washington, DC: US Government Printing Office.

Nash L. L. and Gleick P. H. (1993). *The Colorado River Basin and Climatic Change: The Sensitivity of Streamflow and Water Supply to Variations in Temperature and Precipitation.* Washington, DC: US Environment Protection Agency.

Nemaltsev A. S. (1969). *Mean Multiyear Runoff and its Distribution over the Area of the Globe.* Moscow: Moscow University Publ. (in Russian)

Nemec J. (1989). Impact of climate variability and change of water resources management in agriculture. In: *Proc. Conf. on Climate and Water (Helsinki, 11–15 September 1989)*, vol. 1, pp. 15–23. Helsinki: Valtion Painatuskeskus.

Nemec J. and Schaake J. (1982). Sensitivity of water resource systems to climate variation. *J. Hydrol. Sci.*, 27, 327–343.

Neronov V. M. (ed.) (1990). *Sahara Desert.* Moscow: Nauka' Publ. (in Russian)

Newman P. and Mouritz M. (1992). The urban/greenfield and rural planning context. In: *Managing Stormwater: The Untapped Resource* (ed. I. Bergman), *Workshop Proc. Envir. Tech. Comm. Research Report*, 3, pp. 20–42. Melbourne.

New Zealand Journal of Agriculture (1980). Special Topic Edition: Horticulture. *New Zealand Journal of Agriculture*, 141(4), 5–61.

Nguyen Din Tien (1980). *The Forming and Regime of Mekong Basin River Runoff.* Moscow: Moscow University Publ. (in Russian)

Nguyen Van Ky (1990). *Dynamics of Mouth Areas of Rivers of the Socialist Republic of Viet Nam.* Moscow: Moscow University Publ. (in Russian)

Nikitin M. (1984). Exploration of the Amazon basin. *International Life*, 4, 111–114. (in Russian)

Nikolayev N. K. (1971). Water resources of Peru and the possibilities of their use. *Hydraul. Construct.*, 10, 44–49. (in Russian)

Nikolayeva G. M. and Chernogayeva G. M. (1977). *Water Balance of Asia.* Moscow: Soviet Radio Publ. (in Russian)

Nikolayeva G. M. and Chernogayeva G. M. (1979). Water resources of Asia. In: *Resources of Economic Development of the Countries of Asia and Africa* (eds. N. A. Dlin and V. T. Zaychikov), pp. 21–23. Moscow: Nauka' Publ.

Nimmo W. H. R. (1949). The world's water supply and Australia's portion of it. *J. Inst. Eng., Austral.*, 21, 29–34.

Nizskaya L. O. (1989). *The Niger Republic.* Moscow: Nauka' Publ. (in Russian)

Noor Rahman Rahmani (1989). *Water Resources Development in Afghanistan.* New York: ESCAP.

Nophadol L. and Hemantha E. J. (1992). Impact assessment of global warming on rainfall–runoff characteristics in a tropical region (Sri Lanka). In: *Managing Water Resources During Global Change* (ed. R. Herrmann), *Proc. AWRA Symposium*, pp. 547–556.

O'Donnell T. (1986). Deterministic catchment modelling. In: *River Flow Modelling and Forecasting* (eds. D. A. Kraijenhaff and J. R. Moll), pp. 11–37. Dordrecht, The Netherlands: D. Reidel.

Ojo J. (1987). Hydroclimatic consequence of climatic events in West Africa: the lessons of the 1969–1984 Sahelian droughts. In: *Water for the Future* (eds. J. C. Rodda and N. C. Matalas), IAHS Publ. 164, 229–238.

Oldekop E. M. (1901). *On Evaporation from Surface of River Basins.* Report prepared by students at the Meteorological Observatory at the Yurjyev University, No. 4. Yurjyev: Yurjyev University Publ. (in Russian)

Ongweni G. S., Kithila S. M., Denga F. O. and Abwao P. O. (1993). Environmental and hydrological implications of the development of multipurpose reservoirs in some catchment of Kenya. In: *Sediment Problems, Strategies for Monitoring, Prediction and Control* (eds. R. F. Hadley and T. Mizuyama), *Proc. Int. Symp. Yokohama, July 1993.* IAHS Publ. 217, 207–215.

Oniango Ogembo B. (1980). The water balance and use of water resources of Kenya. *News Acad. Sci. USSR*, Ser. Geogr., 1, 90–105. (in Russian)

Onuphriyev Yu. G. (1971). *Agriculture and Agrarian Relations in the Countries of Latin America.* Moscow: Nauka' Publ. (in Russian)

Oranas G. (1978). The hydroelectric resources of Colombia. *Water Power and Dam Constr.*, 30, 41–45.

Oyebande L. (1991). Problems and prospects of large river basin management in West Africa in 1990s. *Mitteilunglsbl. des Hydrograph. Dienstes in Österreich*, 64/66, 29–30.

Palmer A. R. and van Rooyen A. F. (1998). Detecting vegetation change in the southern Kalahari using Landsat[TM] data. *J. Arid Environments*, 39, 143–153.

Palmieri S., Siani A. M. and Agostina A. D. (1991). Climate fluctuations and trends in Italy within the last 100 years. *Ann. Geophys.*, 9, 769–776.

Panov V. G. (ed.) (1990). *Yearbook of Great Soviet Encyclopedia: 1990.* Moscow: Soviet Encyclopedia Publ. (in Russian)

Panovsky G. A. and Brier G. V. (1972). *Statistical Methods in Meteorology.* Leningrad: Hydrometeoizdat. (in Russian)

Parnikel' B. B. (ed.) (1995). *The Towns Giant of Sunantra.* Moscow: "Sunantar" Soc. Publ. (in Russian)

Parry M. L., Blantran de Rozari M., Chong A. L. and Panich S. (eds.) (1991). *The Potential Socio-Economic Effects of Climate Change in Southeast Asia.* Nairobi, Kenya: UNEP.

Pavlov A. N. (1977). *Geological Water Cycle on the Earth.* Leningrad: Nedra' Publ. (in Russian)

Pearce F. (1996). *Wetlands and Water Resources.* Conservation of Mediterranean Wetlands (MedWet) (series eds. J. Skinner and A. J. Crivelli) No. 5. Arles, France: Tour du Valet.

Pearse P. H., Bertrand F. and MacLaren J. W. (1985). *Currents of Change.* Final Report. Ottawa, Canada: Inquiry on Federal Water Policy.

Pebayle R. (1981). L'irrigation dans le Nord-Este du Brasil. *Notes et études*, Doc. 4609–4610, 84–109.

Peet S. E. and Day J. C. (1980). The Long Lake diversion: an environmental evaluation. *Can. Water Res. J.*, 5(3), 34–48.

Pelenda Appukhamilage Piyasiri Karunatilaka (1990). *Genesis of Freshwater Resources of Sri Lanka, their Spatial and Temporal Typical Features.* Moscow: Moscow University Publ. (in Russian)

Penkova N. V. (2000a). Regional aspects of water – economic development in a transdisciplinary dimension: unique units and general methodology. In: *Transdisciplinarity: Joint Problem-Solving among Science, Technology and Society* (eds. R. Haberli, R. W. Scholz, A. Bill and M. Wolti), *Proc. Int. Conf. Transdisciplinarity 2000 (Zurich, 27–29 February and 1 March 2000)*, Workbook I: Dialogue Sessions and Idea Market, pp. 461–466. Zurich, Switzerland: Swiss Federal Inst. of Technology.

Penkova N. V. (2000b). Structural dynamics of integral socio-economic systems and their effectiveness as related to water use and availability in different natural regions. In: *Water Security for the 21st Century: Innovative Approaches, Proc. of the 10th Stockholm Water Symposium (14–17 August 2000)*, pp. 47–50. Stockholm: Stockholm International Water Institute.

Penkova N. V. and Shiklomanov I. A. (1998). Methodological principles for assessment and prediction of water use and water availability in the world. In: *Water: A Looming Crisis?* (ed. H. Zebidi), *Proc. Int. Conf on World Water Resources at the Beginning of the 21st Century (UNESCO, Paris, 3–6 June 1998)*. IHP-Y Technical Documents in Hydrology, 18, 25–36.

Penn A. F. (1987). River diversions in Northern Quebec: Learning from the La Grande Complex. In: *Impacts and Research Needs for Canada* (eds. W. Nicholaichuk and F. Quinn), *Proc. Symp. on Interbasin Transfer of Water (Saskatoon, 9–10 November 1987)*, pp. 153–168.

Peterson D. and Keller A. (1990). Irrigation. In: *Climate Change and US Water Resources* (ed. P. Waggoner), pp. 269–307. New York: John Wiley & Sons.

Peterson D. H., Cayan D. R., Dileo-Stevens J. and Ross T. G. (1987). Some effects of climate variability on hydrology in western North America. In: *The Influence of Climate Change and Climate Variability on the Hydrologic Regime and Water Resources* (eds. S. I. Solomon., M. Beran and W. Hogg), IAHS Publ. 168, 45–62.

Plan Nacional Hidraulico de Mexico (1981). Mexico, DF.

Playle R. C., Williamson D. A. and Duncan D. A. (1987). Water chemistry changes following diversion, impoundment and hydroelecric development in Northern Manitoba. In: *Impacts and Research Needs for Canada* (eds. W. Nicholaichuk and F. Quinn), *Proc. Symp. on Interbasin Transfer of Water (Saskatoon, 9–10 November 1987)*, pp. 337–352.

Plekhach V. (1976). Potential development of water management in the Csechoslovakia until 1990. C*MEA Inf. Bull. on Water Management*, 17, 14–19. (in Russian)

Plotnikov N. I. (1976). *Underground Water: Our Wealth.* Moscow: Nedra' Publ. (in Russian)

Polad-Zade P. A. (1994). Water without boundaries. In: *Materials Int. Congress "Water: Ecology and Technology"*, vol. 1, pp. 35–46. Moscow: Ministry of Agriculture and Foodstuffs. (in Russian)

Polutikov J. P. (1987). Some possible impacts of greenhouse gas induced climatic change on Water resources in England and Wales. In: *The Influence of Climate Change and Climatic Variability on the Hydrologic Regime and Water Resources, Proc. Vancouver Symp. (Vancover, August 1987)*, IAHS Publ. 168, 585–596.

Polyak I. I. (1975). *Numerical Methods for Analysis of Observations.* Leningrad: Hydrometeoizdat. (in Russian)

Polyak I. I. (1975a). Estimating linear trend of time meteorological series. *Trans. Main Geophys. Observ.*, 364, 51–55. (in Russian)

Pope D. L. (1992). Groundwater management in Kansas. In: *Changing Practices in Groundwater Management* (ed. J. J. DeVries), Pros and Cons of Regulation, *Proc. 18th Biennial Conference on Groundwater (Sacramento, 16–17 September 1991)*, University of California Report 77, 103–113.

Popov I. V. (1958). *The Nile River.* Leningrad: Hydrometeoizdat. (in Russian)

Postel S. (1989). *Water for Agriculture: Facing the Limits.* Worldwatch Paper 93, Washington, DC: Worldwatch Institute.

Postel S. (1992). *Last Oasis: Facing Water Scarcity.* New York: W. W. Norton & Co.

PPWB (1982). *Prairie Provinces Water Demand Study: Historical and Current Water Uses in the Saskatchewan–Nelson Basin.* Regina, Canada: Prairie Provinces Water Board.

Prashchikin A. V. (1986). Water resources and Australian runoff changes. In: *Peculiarities and Regularities of Land Water Formation: Water Exchange Processes* (eds. R. K. Klige and I. S. Zekzer), pp. 190–205. Moscow: Institute for Water Problems RAS. (in Russian)

Prebish P. (1993). *Periphery Capitalism: Is There an Alternative?* Moscow: Nauka' Publ. (in Russian)

Pretro G. A. and Fedorov M. P. (1993). Utilization of the potential water resources of the world at hydro-power developments. *Hydrotech. Construct.*, 8, 1–8. (in Russian)

Prime Minister's Science, Engineering and Innovation Council (1999). *Moving Forward in Natural Resource Management: The Contribution the Science, Engineering and Innovation Can Make.* Report prepared for the Prime Minister's Science, Engineering and Innovation Council. Canberra, Australia: Australian Government Publishing Science.

Probst J. L. and Tardy Y. (1987). Long-range streamflow and world continental runoff fluctuations since the beginning of this century. *J. Hydrology*, 94, 289–310.

Prokhorov A. M. (ed.) (1987). *Soviet Encyclopedic Dictionary*, 4th edn. Moscow: Sov. Encyclopedia' Publ. (in Russian)

Prowse T. D. and Ommaney C. S. L. (1990). *Northern Hydrology: Canadian Perspective.* NHRI Report No. 1. Sakatoon, Canada: Environment Canada.

Pulyarkin Ya. G. and Lipetz V. M. (eds.) (1991). *Territorial Structure of the Economy of Developing Countries.* Moscow: Nauka' Publ.

Pyatigorsky A. A. (1990). Water economy of Ethiopia. *Land Reclam. Water Mgmt*, 9, 60–62. (in Russian)

Quinn F. J. (1981). Water transfers – Canadian style. *Can. Water Resources J.*, 6(1), 64–76.

Ragab R. (2001). Climate change and water resources management in the southern Mediterranian and Middle East countries. In: *Proc. Int. Conf. on Integrated Water Management (Nicosia, 11–13 May 2001)*. (ed. I. Papadopoulos), pp. 9–40. Nicosia, Cyprus: Agricultural Research Institute of Cyprus.

Rahman A. (1989). Near surface hydrological processes under changing land use in humid equatorial conditions. In: *Friends in Hydrology* (eds. L. Roald, K. Nordseth and K. A. Hassel), *Proc. Int. Conf. on Flow Regimes from International Experimental and Network Data (Bolksjo, Norway, 1–6 April 1989)*, pp. 331–340. IAHS Publ. 187. Wallingford, UK: IAHS Press.

Rakhmanov V. V. (1973) River runoff and agricultural technology. *Trans. SHI*, 114, 200. (in Russian)

Raskin P., Hansen E., Zhu Z. and Slavisky D. (1992). Simulation of water supply and demand in the Aral Sea region. *Water International*, 17 (2), 55–67.

Rastyannikov V. G. and Shirokov G. K. (eds.) (1995). *Capitalism in the East in the Second Half of the 20th Century.* Moscow: Russian Academy of Science Publ. (in Russian)

Ratkovitch D. Ya. (1976). *Long-Term Variations of River Runoff.* Leningrad: Hydrometeoizdat. (in Russian)

Ratkovitch D. A. (1993). *Hydrological Grounds of Water Supply.* Moscow: Institute for Water Problems RAS Publ. (in Russian)

Raumer F. (1977). Project Poza Honda und Zielvorstellung: Verwicklichung eines wasserwirtschaftlichen Mehrzweckprojektes in Ecuador. *Ber. Versuch. Wasserbau. Techn. Univ. München. O. Miller Inst.*, 33, 55–87.

Razumikhin N. V. (1976). Basic features of the evolution of Mesocenozoic era palaeoclimates and the problems of palaeohydrology. In: *Problems of Palaeohydrology* (eds. G. P. Kalinin and R. K. Klige), pp. 267–274. Moscow: Nauka' Publ. (in Russian)

Reboucas A. C. (1998). Outlines on water crisis in Latin America. In: *Water: A Looming Crisis?* (ed. H. Zebidi), *Proc. Int. Conf. on World Water Resources at the Beginning of 21st Century (UNESCO, Paris, 3–6 June 1998)*, IHP-Y Technical Documents in Hydrology 18, 385–390.

Remenda V. and Davis E. (1987). Interbasin transfer: impacts on water quality. In: *Impacts and Research Needs for Canada* (eds. W. Nicholaichuk and F. Quinn), *Proc. Symp. on Interbasin Transfer of Water (Saskatoon, 9–10 November 1987)*, pp. 353–366.

Renzetti S. (1987). *The Economic Aspects of Industrial Water Use.* Ottawa, Canada: Environment Canada, Inland Waters Directorate.

Revelle R. (1955). On the history of the ocean. *J. Marine Res.*, 14, 446–461.

Revelle R. and Waggoner P. E. (1983). *Effects of a Carbon-Dioxide Induced Climatic Change on Water Supplies in the Western United States in Changing Climate.* Washington, DC: National Academy Press.

Ricca V., Simmons P. W, McGuinness G. L. and Taiganides E. P. (1970). Influence of land use on runoff from agricultural watersheds. *Trans. ASAE*, 13 (2), 187–190.

Richa I. (1982). Contribution to the analysis of the hydrological cycle and of the water consumption cycle. *Study CSAVL Akademia* (Prague), 12–111.

Richey J. E., Mertes L. A. K. and Dunne T. (1989a). Sources and routing of the Amazon river flood wave. *Global Biochem. Cycles*, 3(3), 191–204.

Richey J. E., Nobre C. and Deser C. (1989b). Amazon river discharge and climate variability: 1903 to 1985. *Science*, 246(4926), 101–103.

Riebsame W. E. and Smith D. J. (1986). *Sensitivity and Adjustment of Water Resource Systems to Climate Impacts: General Principles and a California Case Study*. Draft Manuscript CO 80309. Department of Geography, University of Colorada. Boulder, CO:.

Riebsame W. E., Strzepek K. M., Wescoat J. L. Jr., Perrit R., Graile G. L., Jacobs J., Leichenko R., Maganza C., Phien H., Urbiztondo B. J., Restepo P., Rose W. R., Saleh M., Ti L. H., Tucci C. and Yates D. (1995). Complex river basins. In: *As Climate Changes, International Impacts and Implications* (eds. K. M. Strzepek and J. B. Smith), pp. 57–91. Cambridge, UK: Cambridge University Press.

Rijsberman F. R. (ed.) (2000). *World Water Scenarios: Analysis*. London: Earthscan Publ.

Rijsberman F. R. (ed.) (2001). 2nd World Water Forum: Session Reports. *Water Policy*, 3 (Supplement), 215.

Ritschard R. L, Cruise J. F. and Hatch L. U. (1999). Spatial and temporal analysis of agricultural water requirements in the Gulf Coast of the United States. *J. Am. Water Res. Assoc.*, 35, 1585–1596.

Robbroec T. (1979). Inter-basin water transfers in South Africa. *Civ. Eng. S. Africa*, 21(2), 29–35.

Robinson G. M. (1995). Deregulation and restructuring of the Australian cane sugar Industry. *Austral. Geogr. Studies*, 33(2), 212–227.

Robinson M. (1989). Small catchment studies of man's impact on flood flows: agricultural drainage and plantation forestry. In: *Friends in Hydrology* (eds. L. Roald, K. Nordseth and K. A. Hassel), *Proc. Int. Conf. on Flow Regimes from International Experimental and Network Data (Bolksjo, Norway, 1–6 April 1989)*, pp. 299–308. IAHS Publ. 187. Wallingford, UK: IAHS Press.

Rodda J. C. (1995). Guessing or assessing the world's water resources? *J. Chart. Inst. Water Environ. Mgmt*, 9, 360–368.

Romanova E. P. (1968). Water resources of Italy and their use. *News Moscow State Univ.*, Ser. Geogr., 2, 21–30. (in Russian)

Romanova E. P., Alekseyev B. A. and Medvedev A. V. (1994). Agronatural potential of landscapes of the Earth. *Geog. Nat. Res.*, 3, 5–14. (in Russian)

Rosenberg D. M. and Barton D. (1986). The Mackenzie River System. In: *The Ecology of River Systems* (eds. B. R. Davis and K. F. Walker), pp. 425–433. Dordrecht, The Netherlands: W. Junk Publ.

Rosenberg N. J., Kimball B. A., Martin P. and Cooper C. F. (1990). From climate and CO_2 enrichment to evapotranspiration. In: *Climate Change and US Water Resources* (ed. P. Waggoner), pp. 151–175. New York: John Wiley & Sons.

Ross L. (1983). Changes in water policy in the People's Republic of China. *Water Res. Bull.*, 19(I), 69–72.

ROSTAS (1994). *Water Resources Assessment in the Arab Region*. Delft, The Netherlands: UNESCO-ROSTAS.

ROSTAS (1995). *Rainfall Water Management in the Arab Region*. State of Art Report. Cairo: ROSTAS.

Roy D. and Messier D. (1987). Repercussions du transfer des eaux des rivières Eastmain–Opinaca et Caniapiscau dans La Grande Rivière (Quebec). In: *Impacts and Research Needs for Canada* (eds. W. Nicholaichuk and F. Quinn), *Proc. Symp. on Interbasin Transfer of Water (Saskatoon, 9–10 November 1987)*, pp. 169–183.

Rozanov A. S. (1967). On the problem of arid area water resource use: Study Case – Australian groundwaters *News Acad. Sci. USSR*, Ser. Geogr., 1, 133-139. (in Russian)

Rozarri M. B. (1990). *Socioeconomic Impacts of Climate Change*. Indonesian Report Submitted to United Nations Environment Programme.

Rozhdestvensky A. V. (ed.) (1988). *Spatial–Temporal Variations in USSR River Runoff*. Leningrad: Hydrometeoizdat. (in Russian)

Rozdestvensky A. V. and Chebotaryev A. I. (1974). *The Statistical Methods in Hydrology*. Leningrad: Hydrometeoizdat Publ. (in Russia)

Rubey W. W. (1951). Geologic history of seawater. *Bull. Geol. Soc. Amer.*, 62, 111–114.

Rubtsov B. B. (1987). *New Zealand*. Moscow: Nauka' Publ. (in Russian)

Rubtsov B. B. (1991). *Oceania*. Moscow: Nauka' Publ. (in Russian)

Rush R. Y. (1988). *Hoover Dam and the Central Arizona Project – A Milestone Year: National Water Summary 1986*. US Geological Survey Water-Supply Paper No. 2325.

Russell D. F. and Woodcock C. P. N. (1993). What will water rates be like in the 1990s? *Water Resources J.*, 177, 45–51.

Russel G. L. and Miller J. R. (1990). Global river runoff calculated from a global atmospheric general circulation model. *J. Hydrol.*, 117 (1–4), 241–254.

Russel G. L. and Richard L. S. (1990). *The Global Historical Climatology Network: Long-Term Monthly Temperature, Precipitation, Sea Level Pressure and Station Pressure (NDP-41)*. Washington, DC: Department of Energy, CDIAC, US.

Russel N. E. (1984). The South African problem: youthful perspectives. *Civ. Eng. S. Africa*, 26(2), 63–64.

Russian Academy of Science (1989). *Latin America in Figures: A Handbook*. Moscow: Institute for Latin American Studies of RAS Publ. 202 (in Russian)

Russian Academy of Science (1994). *Japan 1991/2: Yearbook*. Moscow: RAS Publ. (in Russian)

Russian Academy of Science (1995). *China 1995: Analysis and Forecast of the Economic Situation*. Moscow: Institute of Far East, RAS Publ. (in Russian)

Ryabchikov A. M. (ed.) (1976). *Natural Resources of Foreign Territories of Europe and Asia*. Moscow: Mysl' Publ. (in Russian)

Ryabchikov A. M. (ed.) (1988). *Physical Geography of the Continents and the Oceans*. Moscow: High School Publ. (in Russian)

Ryan R. (Ed.) (1972). *Encyclopaedia of Papua and New Guinea*, vols. 1–3. Melbourne, Australia: University of Melbourne Press.

Saelthun N. R. (1991). Climatic change impact on the hydrological regimes of Norway. In: *Proc. Conf. on Nordic Hydrology and the Greenhouse Effect (Reykjavik, 3–5 April 1991)*.

Saelthun N. R., Aittoniemi P., Bergstrom S., Einarsson K., Johannesson T., Lindstrom G., Ohlsson P.-O., Thomsen T., Vehrilainen B. and Aamodt K. O. (1998). Climate change impacts on runoff and hydropower in the Nordic countries. *TemaNord*, 552, 170.

Saelthun N. R., Rogen I. *et al.* (1990). *Climate Change Impact on Norwegian Water Resources*. Oslo: Norwegian Water Resources and Energy Administration.

Sagoyan L. Yu. (1993). *Republic of Chad Handbook*. Moscow: Nauka' Publ. (in Russian).

Salih A. M. A. (1978). Irrigation and water resources in Sudan. *Proc. Int. Conf. "Water Resources Development and Management" (Bangkok, 1978)*, pp. 1025–1040.

Samaha A. H. M. (1979). The Egyptian master water plan. *Water Supply Mgmt*, 3(4), 251–266.

Samarakoon A. and Gifford R. M. (1996). Elevated CO_2 effects on water use and growth of maize in wet and drying soil. *Austral. J. Plant Physiol.*, 23, 53–62.

Sanderson M. and Wong L. (1987). Climatic change and Great Lakes water levels. In: *The Influence of Climate Change and Climatic Variability on the Hydrologic Regime and Water Resources, Proc. Vancouver Symp. (Vancover, August 1987)*, IAHS Publ. 168, 477–487.

Sarker N. N. and Sarker R. I. (1979). Agricultural mechanization strategies in Bangladesh. *Agr. Mech. Asia*, 10(2), 22–28.

Sarukhanyan E. I. and Smirnov P. P. (1971). *Long-Term Volga Runoff Fluctuations*. Leningrad: Hydrometeoizdat. (in Russian)

Schaake J. K. and Kaczmarek Z. (1979). Climate variability and designing and exploiting water management systems. In: *Proc. 1st World Climate Conference (Geneva, February 1979)*, pp. 208–230. Geneva, Switzerland: WMO.

Schliephake K. and Deparage F. (1977). Die landwirtschaftliche Bewasserung in Nordafrika – Stand und Ziele. *Wasser u. Boden*, 29(9), 262–266.

Schmidt M. (1985). Wassernutsungen in der Region Darfur (Sudan). *Wasser u. Boden*, 37(8), 376–380.

Schopf T. J. M. (1980). *Palaeooceanography*. Cambridge, MA: Harvard University Press.

Schumann A. H. (1993). Changes in hydrological time series: A challenge for water management in Germany. In: *Hydrology of Warm Humid Regions* (ed. J. S. Gladwell), *Proc. Yokohama Symp. (Yokohama, July 1993)*, IAHS Publ. 216, 95–102.

Scogerboe G. V. (1983). Agricultural Water Management and the Environment. In: *Long-Distance Water Transfer: A Chinese Case Study and International Experiences* (eds. Asit K. Biswas, Zuo Dakang, James E. Nickum and Lin Changming), pp. 35–64. Dublin: Tycooly.

Seagel G. S. (1987). Pacific to Arctic transfer of water and biota: The McGregor Diversion Project in British Columbia, Canada. In: *Impacts and Research Needs for Canada* (eds. W. Nicholaichuk and F. Quinn), *Proc. Symp. on Interbasin Transfer of Water (Saskatoon, 9–10 November 1987)*, pp. 431–438.

Seckler D., Amarsinghe U., Molden D., De Silva R. and Barker R. (1998). *World Water Demand and Supply, 1990 to 2025: Scenarios and Issues.*

IWMI Research Report No. 19. Colombo: International Water Management Institute.

Sedunov Yu. S. (ed.) (1991). Atmosphere: A Handbook. Leningrad: Hydrometeoizdat. (in Russian)

Sehmi N. (1996). Mysterious Nile. WMO Bulletin, 45(3), 339–343.

Semyonov V. A. (1986). Climatic changes in USSR river runoff. Trans. VNIIGMI-MCD, 133, 59–84. (in Russian)

Senghor L. S. (1977). Senegal 1977. Africa – India Infrastructure, 137, 42–68.

Sergeant I. P. (1964). Preliminary assessment of Australian surface water resources. In: Proc. Symp. on Water Resources Use and Management (Canberra, 9–13 September 1963), pp. 136–146. Melbourne, Australia: University of Melbourne Press.

Sewell W. R. D. and Roueiche L. (1974). The potential impact of peak load pricing on urban water demands. Victoria, British Columbia, a case study. In: Priorities in Water Management (ed. F. M. Leversedge), pp. 141–161. Victoria, BC, Canada: University of Victoria.

Shahin M. (1989). Review and assessment of water resources in the Arab region. Water International, 14(4), 206–219.

Shahin M. (1996). Hydrology and Scarcity of Water Resources in the Arab region. IHE Monograph No. 1. Rotterdam, The Netherlands: A. A. Balkema.

Shahin M. M. A. (1985). Discussion of the paper entitled "Ethiopian Interests in the Division of the Nile River Waters". Water International, 11(1), 16–22.

Shakhbazyan G. S. (1979). Use of water resources of Iran for developing national economy. In: Resources of Economic Development of the Countries of Asia and Africa, pp. 114–124. Moscow: Nauka' Publ. (in Russian).

Shalash S. (1980). The effect of the High Aswan Dam on the hydrological regime of the River Nile. In: The Influence of Man on the Hydrological Regime with Special Reference to Representative and Experimental Basins, Proc. Int. Symp. (Helsinki, 23–26 June 1980). IAHS Publ. 130, 251–256.

Shalash S. (1986). Effect of long-term capacity reservoir on a large river with special reference to the High Aswan Dam of Egypt. In: Proc. Int. Symp. on the Impact of Large Water Projects on the Environment, 24.

Shalygin A. L. (1987). Improved mathematical model for optimum crop irrigation regime. Trans, SHI, 326, 36–41. (in Russian)

Shengquan G., Guohui Y. and Yuhen W. (1993). Distributional features and fluxes of dissolved nitrogen, phosphorus and silicon in Hanzhou Bay. Marine Chem., 43, 65–80.

Sheremetyev I. K. (ed.) (1994). Latin America: Structural Reconstruction of Economy. Moscow: Institute for Latin American Studies RAS Publ. 240 (in Russian)

SHI (1967). Water Resources and Water Balance of the Soviet Union Territory. Leningrad: Hydrometeoizdat. (in Russian)

SHI (1984). Handbook for Determining the Calculated Hydrological Characteristics. Leningrad: Hydrometeoizdat. (in Russian)

SHI (1987). Water Resources of the USSR and their Use. Leningrad: Hydrometeoizdat. (in Russian)

Shiklomanov A. I. (1994). The influence of anthropogenic changes in global climate on the Yenisey river runoff. Meteorol. Hydrol, 2, 84–93. (in Russian)

Shiklomanov I. A. (1976). Hydrological Aspects of the Caspian Sea Problem. Leningrad: Hydrometeoizdat.

Shiklomanov I. A. (1979). Anthropogenic Changes in the Water Content of Rivers. Leningrad: Hydrometeoizdat. (in Russian)

Shiklomanov I. A. (1988). Studying Land and Water Resources: Results, Problems, and Outlook. Leningrad: Hydrometeoizdat. (in Russian)

Shiklomanov I. A. (1989a). Man's impact on River Runoff. Leningrad: Hydrometeoizdat. (in Russian)

Shiklomanov I. A. (1989b). Hydrological Studies in the USSR. Hydrometeoizdat. (in Russian)

Shiklomanov I. A. (1990). Global water resources. Nature and Resources, 26(3), 34–43. (in Russian)

Shiklomanov I. A. (ed.) (1997). Assessment of Water Resources and Water Availability in the World. Background Report for the Comprehensive Assessment of the Freshwater Resources of the World. Paris: WMO/SEI.

Shiklomanov I. A. (1998a). World Water Resources: A New Appraisal and Assessment for the 21st Century. Paris: UNESCO.

Shiklomanov I. A. (1998b). Global renewable water resources. In: Water: A Looming Crisis? (ed. H. Zebidi), Proc. Int. Conf. on World Water Resources at the Beginning of 21st Century (UNESCO, Paris, 3–6 June 1998), IHP-V Technical Documents in Hydrology, 18, 3–14.

Shiklomanov I. A. (2000a). World water resources and water use: present assessment and outlook for 2025. In: World Water Scenarios: Analysis (ed.: F. R. Rijsberman), pp. 160–203. London: Earthscan Publ.

Shiklomanov I. A. (2000b). Appraisal and assessment of world water resources. Water International, 25 (1), 11–32.

Shiklomanov I. A. and Babkin V. I. (1992). Climate changes and water resources. Water International, 25 (1), 11–32.

Shiklomanov I. A. and Georgievsky V. Yu. (1992). Problems of the effect of anthropogenic climate changes on hydrological parameters and water management. Meteorol. Hydrol, 8, 38–43. (in Russian)

Shiklomanov I. A. and Georgievsky V. Yu. (1995). The influence of anthropogenic factors on river runoff of the former USSR. In: Geographical Orientations in Hydrology (eds. N. I. Koronkevich and G. M. Chernogaeva), pp. 96–107. Moscow: RAS Publ. (in Russian)

Shiklomanov I. A. and Lins H. (1991). Climate variation effect on hydrology and water management. Meteorol. Hydrol, 4, 51–65. (in Russian)

Shiklomanov I. A. and Markova O. L. (1987). Specific Water Availability and River Runoff Transfers in the World. Leningrad: Hydrometeoizdat. (in Russian)

Shiklomanov A. I. and Shiklomanov I. A. (1999). Assessment of the impacts of climate variability and change on the hydrology of Asia and Australia. In: Impacts of Climate Change and Climate Variability on Hydrological Regimes (ed. J. C. van Dam), pp. 85–107. Cambridge, UK: Cambridge University Press.

Shiklomanov I. A. and Penkova N. V. (2000). World Water Resources at the Beginning of the 21st Century: Role of Socio-Economic and Climatic Factors. In: Problems of Hydroclimatology and Environment at the Beginning of the 21st Century (eds. Ye. A. Genikhovich, V. P. Meleshko and B. Ye. Shneyerov), Proc. Int. Theoret. Conf. (St Petersburg, June 1999), pp. 112–140. St Petersburg: Hydrometeoizdat. (in Russian)

Shiklomanov I. A., Lins H. and Stakhiv E. (1990). Hydrology and water resources. In: The IPCC Impact Assessment (eds. W. Tegart, G. Sheldon and D. Griffiths), pp. 4.1–4.42. Canberra, Australia: Australian Government Publishing Service.

Shiklomanov I. A., Lammers R. B., Peterson B. J. and Vorosmarty C. (2002). The dynamics of river water inflow to the Arctic Ocean. In: The Freshwater Budget of the Arctic Ocean, The Netherlands: Kluwer Academic Publishers.

Shoemaker E. M. (1984). Large body impacts through geological time. In: Patterns of Change in Earth's Evolution, pp. 15–40. Berlin: Springer-Verlag.

Showers V. (1989). World Facts and Figures, 3rd edn. New York: John Wiley & Sons.

Shepa B. G. (1973). Water Economy of Iran. Review Information No. 2. Moscow: Ministry for Water Management. (in Russian)

Shultz V. L. (1965). Rivers of the Middle Asia. Leningrad: Hydrometeoizdat. (in Russian)

Shuttleworth W. J. (1991). The role of hydrology in global science. In Hydrological Interactions between Atmosphere, Soil and Vegetation (eds. G. Kienitz, M. M. Th. Van Genichten and D. Rosbjerg), Proc. Symp. IUGG (Vienna, August 1991), IAHS Publ. 204, 361–375.

Shuttleworth W. J., Gash J. H. C., Roberts J. M., Nobre C. A., Molion L. C. B. and Ribeiro G. (1991). Post-deforestation Amazonia Climate: Anglo-Brazilian research to improve prediction. J. Hydrol, 129(1–4), 71–85.

Shvartsev S. L. (1995). Water resources management system in France. Water Resources, 22 (4), 466–469. (in Russian)

Simoniya N. A. (ed.) (1976). Modern Thailand: A Handbook. Moscow: Nauka' Publ. (in Russian)

Simoniya N. A. (ed.) (1983). Indonesia: A Handbook. Moscow: Nauka' Publ. (in Russian)

Simoniya N. A. (ed.) (1990). The Town in the Formation and Development of the Countries of the East. Moscow: Nauka' Publ. (in Russian)

Sineishchikova I. G. (1989). The problems of developing agriculture of Latin America, in proceedings of Latin Americans. In: Issues on Economics and Political Geography of Foreign Countries, vol. 10, pp. 167–177. Moscow: Nauka Publ. (in Russian)

Singh B. (1987). The impacts of CO$_2$-induced climate change on hydroelectric generation potential in the James Bay territory of Quebec. In: The influence of Climate Change and Climate Variability on the Hydrologic Regime and Water Resources, Proc. Vancouver Symp. (Vancouver, August 1987), IAHS Publ. 168, 403–418.

Singh B. (1997). Climate-related global changes in the Southern Caribbean: Trinidad and Tobago. In: Proc. 15th Conf. on Global and Planetary Change (3–4 October), pp. 93–111. Amsterdam: Elsevier.

Singh G. (1980). Geography of India. Moscow: Progress Publ. (in Russian)

Singh V. P., Penkova N. V., Khaydarova V. A., Zalataev V. S. and Novikova N. M. (1998). Identification of hydrologic parameters of the ecosystems of drought-prone regions. Arid Ecosystems, 4, 58–73. (in Russian)

Sinitsyn V. M. (1967). Introduction to Palaeoclimatology. Leningrad: Nedra' Publ. (in Russian)

Sioli H. (ed.) (1984). The Amazon. Amsterdam: Elsevier.

Sircoulon J. (1990). Impact possible des changements climatiques à venir sur les ressources en eau des régions arides et sémiarides. WMO/TD, 380, 88.

Sircoulon J., Lehel T. and Arnell N. W. (1999). Assessment of the impacts of climate variability and change on the hydrology of Africa. In: Impacts of Climate Change and Climate Variability on Hydrological Regimes (ed. J. C. van Dam), pp. 67–84. Cambridge, UK: Cambridge University Press.

Smerdon E. T. (1982). Water: its role in from now to the year 2000. Nat. Res. J., 22 (4), 907–914.

Smith D. I. (1998). Water in Australia: Resources and Management. Melbourne, Australia: Oxford University Press.

Smith R. C. G. (1983). Technical efficiency of irrigation agriculture: Water technology, reuse and efficiency. In: Water 2000, Consultants' Report No. 10, pp. 61–81. Canberra, Australia: Australian Government Publishing Service.

Smith S. E. and Al-Rawahy H. M. (1990). The Blue Nile: Potential for conflict and alternatives for meeting future demands. Water International, 15(4), 217–222.

Snowy Mountains Hydroelectric Authority (1971). Streamflow Records of the Snowy Mountains Region, Australia, 1966–1970. Cooma, NSW, Australia: Snowy Mountains Hydroelectric Authority.

Sofer M. and Skirymonskaya B. (1994). Water under ground. Science and Life, 8, 1–19. (in Russian)

Sokolov A. (1986). Water: Problems at the Threshold of the 21st Century. (in Russian)

Sokolov A. A. (1960). Hydrology of China, vol. 1. St. Petersburg: SHI. (in Russian)

Sokolovsky D. L. (1972). River Runoff. Leningrad: Hydrometeoizdat. (in Russian)

Soons J. M. and Selby M. J. (eds.) (1992). Landforms of New Zealand. Auckland, New Zealand: Longman Paul.

Sorokhtin O. G. (1974). Global Evolution of the Earth. Moscow: Nauka' Publ. (in Russian)

South African Information Service (1993). Meet South Africa. Pretoria: SAIS.

Soyuzvodproekt (1988). Handbook of Reservoirs of the USSR, Part 1, Reservoirs with a Capacity of 10 million m³ and more (ed. B. V. Orlov). Moscow: Soyuzvodproekt Publ. (in Russian)

Soyuzvodproekt/SHI (1984). Recommendations for Determining the Runoff Volumes of Irrigation and Return Waters from Irrigated Lands. Moscow: Soyuzvodproekt Publ. (in Russian)

Spengler O. A. (1964). Frances Hydrological Sketch. Leningrad: Hydrometeoizdat. (in Russian)

Speranskaya N. A. (1988). The features of changing annual river runoff of the Soviet Union with changing global thermal regime. Trans. SHI, 330, 120–125. (in Russian)

Stakhiv E. Z. and Lins H. (1989). Impacts of climate change on US water resources with reference to the Great Lakes Basin, USA. Presented at IPCC Workshop (Geneva). 15.

Stakhiv E., Lins H. and Shiklomanov I. A. (1992). Hydrology and water resources. In: The Supplementary Report to the IPCC Impact Assessments (eds. W. Tegard and G. Sheldon), pp. 71–83. Canberra, Australia: Australian Government Publishing Service.

Stamenov S. (1977). Bulgarian water resources, their utilization at present and in the long run. CMEA Inf. Bull. on Water Management, 1, 28–33. (in Russian)

Stancik A. and Jovanovich S. (1988). Hydrology of the River Danube. Moscow: UNESCO/Priroda Publ. (in Russian)

Stanners D. and Bourdeau P. (1995). Europe's Environment: The Dobris Assessment. Copenhagen: European Environment Agency.

Staudacher T. and Allegre C. J. (1982). Terrestrial xenology. Earth Planet Sci. Lett., 60, 389–406.

Stepanov V. N. (1983). Oceansphere. Moscow: Mysl' Publ. (in Russian)

Sterling T. (1980). Der Amazonas. Amsterdam: Elsevier.

Stockton C. and Boggess W. R. (1979). Geohydrological implications of Climate Change on Water Resource Development. Fort Belvoir, VA: US Army Coastal Engineering Research Center.

Strakhov N. M. (1963). Types of Lithogenesis and their Evolution in the Earth's History. Moscow: Geotekhizdat' Publ. (in Russian)

Strzepek K. and Bowling P. (1995). UNIDO Global Industrial Water Use Assessment. Draft. New York: UNIDO.

Subramaniam V. (1987). Environmental geochemistry of Indian river basins: A Review. Geol. Soc. India, 29, 205–220.

Sumarokova V. V. and Degtyarev G. M. (1985). Water use of agricultural crops and return water in the Amu Darya Basin. Meteorol. Hydrol., 11, 93–102. (in Russian)

Sumarokova V. V. and Tsysenko K. V. (1978). Decrease of river runoff in the Aral Sea basin. Soviet Hydrology, Selected Papers, 17(4), 323–328. (in Russian)

Sumarokova V. V. and Tsysenko K. V. (1991). Reconstruction of water management in Middle Asia is necessary! In: Man and Natural Elements (ed. A. I. Ugryumov), pp. 88–90. (in Russian)

Sumarokova V. V., Bakhina L. P. and Krovensova V. Ye. (1991). Landscape parameterization of the low reaches of Amu Darya based on interpretation of aerospace information. In: Monitoring of the Natural Environment in the Aral Sea Basin (eds. Yu. A. Israel and Yu. A. Anokhin), pp. 200–208. St Petersburg: Hydrometeoizdat. (in Russian)

Suzuki T. (1973). Improvement in exploration of rivers of Japan. In: Energetics of the World (ed. P. S. Neporozhniy), p. 305. Moscow: Energy Publ.

Szestay K. (1982). River basin development and water management. Water Qual. Bull., 17, 155–162. (in Russian)

Szestay K. (1982). River basin development and water management. Water Qual. Bull., 7, 155–162.

Takeuchi K., Jayawardena A. W. and Takahasi Y. (1995). Catalogue of Rivers for Southeast Asia and the Pacific, vol. 1. Hong Kong: UNESCO-IHP Regional Steering Committee for Southeast Asia and the Pacific.

Talling J. F. and Lemoalle J. (1998). Ecological Dynamics of Tropical Inland Waters. Cambridge, UK: Cambridge University Press.

Tarasov K. S. (ed.) (1986). Latin America: Economic Development of New Territories. Moscow: Institute for Latin American Studies of RAS Publ. (in Russian)

Tarasov K. S. (ed.) (1992). Forest Resources and their Use. Moscow: Institute for Latin American Studies of RAS Publ. 152 (in Russian)

Tarasov K. S. (ed.) (1993). Latin America: Economy and Ecology of Nature Use. Moscow: Institute for Latin American Studies of RAS Publ. 200 (in Russian)

Tarend D. D. (1998). Changing flow regimes in the Baltic States. In: Proc. 2nd Int. Conf. on Climate and Water (Espoo, Finland, August 1998), vol. 1, pp. 109–117. Helsinki: Helsinki University of Technology.

Tate D. M. and Scharf D. N. (1992). Water Use in Canadian Industry, 1986. Ottawa, Canada: Environment Canada, Ecosystem Sciences and Evaluation Directorate, Economic and Conservation Branch.

Tate D. M. (1977). Manufacturing Water Use Survey: A Summary of Results. Ottawa, Canada: Inland Waters Directorate, Fisheries and Environment.

Tate D. M. (1983). Water Use in the Canadian Manufacturing Industry, 1976. Water Planning and Management Branch. Canada, Ottawa: Environment Canada, Inland Waters Directorate, Water Planning and Management Branch.

Tate D. M. (1984). Industrial water use and structural change in Canada and its regions: 1966–1976. PhD dissertation. Ottawa, Canada: University of Ottawa.

Tate D. M. (1986). Structural change implications for industrial water use. Water Resources Res., 22(11), 1526–1530.

Tate D. M. (1989). *Municipal Water Rates in Canada, 1986: Current Practices and Prices.* Ottawa, Canada: Environment Canada, Inland Waters Directorate. Water Planning and Management Branch.

Tate D. M. (1990). *Water Demand Management in Canada: A State-of-the-Art Review.* Ottawa, Canada: Environment Canada, Inland Waters Directorate. Water Planning and Management Branch.

Tate D. M. and Lacelle D. M. (1978). Municipal water use in Canada. *Can.* Water Planning and Management Branch.

Tate D. M. and Lacelle D. M. (1987). Municipal Water Use in Canada, 1983. *Water Res. J.,* 3(2), 61–78.

Tate D. M. and Lacelle D. M. (1992). *Municipal Water Rates in Canada.* Planning and Management Branch.

Tate D. M. and Scharf D. N. (1985). *Water Use in Canadian Industry, 1981.* Ottawa, Canada: Environment Canada, Ecosystem Sciences and Evaluation Directorate, Economic and Conservation Branch.

Tate D. M., Renzetti S. and Shaw H. A. (1992). *Economic Instruments for Water Management: The Case for Industrial Water Pricing.* Ottawa, Canada: Environment Canada, Ecosystem Sciences and Evaluation Directorate, Economic and Planning and Management Branch.

Taylor R. T. and Smith I. (1997). *The State of New Zealand's Environment 1997.* Wellington, New Zealand: Ministry of the Environment.

Tempelmann G. (1994). Office du Niger, Mali: Netherlands intervention in an irrigated rice scheme in Western Africa. *Ned. Geogr. Stud.,* 186, 243–249.

Tenant D. L. (1976). Instream flow regimes for fish, wildlife, recreation and related environmental resources. *Fisheries,* 1(4), 6–10.

Terehov N. M. (ed.) (1981). *Small Atlas of the World.* Moscow: GUGK USSR (in Russian)

Thomas I. F. and Bates B. C. (1997). Responses to the variability and increasing uncertainty of climate in Australia. In: *Risk, Reliability, Uncertainty and Robustness of Water Resource Systems. Proc. 3rd IHP/IAHS Kovacs Colloquium (Paris, 1996).* Paris: UNESCO.

Thomas I. F. J. (1959). *Nelson River Drainage Basin in Canada, 1953–56. Industrial Water Resources of Canada.* Water Survey No. 10. Ottawa, Canada: Department of Mines and Technical Surveys.

Thompson D. (1987). Three converging and complementary techniques. *Environmental Impact Assessment, Strategic Planning and Uncertainty Management,* 1987, 217–235.

Timashev I. Ye. (1983). Quantitative assessment of water resources in the USA. *News Moscow State University,* 1, 67–74. (in Russian)

Timofeyev P. P., Kholodov V. P. and Zverev V. P. (1988). Hydrosphere and evolution of the Earth. *News Acad. Sci. USSR, Ser. Geol.,* 6, 3–19. (in Russian)

Tolokonnikova A. A. (1972). *Geography of Foreign Countries,* vol. 1. *Land and Water Resources of Foreign Asia.* Moscow: Nauka? Publ. (in Russian)

Toth P. L. (1993). *Policy Responses to Climate Change in Southeast Asia.* Luxembourg: International Institute for Applied Systems Analysis.

Tryostnikov V. F. (ed.) (1988). *Geographical Encyclopaedic Dictionary: Notions and Terms.* Moscow: Soviet Encyclopaedia Publ.

Tsinzerling V. V. (1927). *Irrigation on the Amu Darya.* Moscow: Water Management Administration of Middle Asia Publ. (in Russian)

Tsytsarin A. G. and Bortnik V. N. (1991). Current problems of the Aral Sea and perspectives of their solution. In: *Monitoring of the Natural Environment in the Aral Sea Basin* (eds. Yu. A. Israel and Yu. A. Anokhin), pp. 7–28. St Petersburg: Hydrometeoizdat. (in Russian)

Tsytsenko K. V. and Sumarokova V. V. (1999). Change of the rivers' flow in the Aral Sea basin (in connection with the problem of quantitative assessment and consideration of environmental after-effects). In: *Creeping Environmental Problems and Sustainable Development in the Aral Sea Basin* (ed. M. Glantz), pp. 191–203. New York: Cambridge University Press.

Tsytsenko K. V. and Vonsovskaya O. G. (1985). Current and prospective assessment of water consumption and return water in the Syr Darya Basin. *Trans. SHI,* 278, 3–17. (in Russian)

Tucci C., Silveira A., Sanchez J., and Albuquerque F. (1995). Flow regionalization in the upper Paraguay basin, Brazil. *Hydrol. Sci. J.,* 40(4), 485–497.

Tucci G. E. M. and Damiani A. (1991). International studies on climate change impacts: Uruguay River basin. In: *Workshop on Analysis of Potential Climate Change in the Uruguay River Basin.* Federal University of Rio Grande do Sul. Porto Alegre, Brazil: Institute of Hydraulic Research.

Tuddenhorzh D. and Myagmarzhav B. (eds.) (1985). *Atlas of Climate and Surface Water Resources of the Mongolian People's Republic.* Moscow: GUGK USSR Publ.

Turnock D. (1986). The Danube–Black Sea canal and its impact on Southern Romania. *Geojournal,* 12 (1), 65–73.

Tuffuor K. A. (1987). Towards effective future water resources policies for socio-economic development in Africa. In: *Water for the Future* (eds. J. C. Rodda and N. C. Matalas). IAHS Publ. 164, 289–298.

UN (1985). *Statistical Yearbook.* New York: UN.

UN (1985). *The Water Resources of Latin America and the Caribbean: Planning, Hazards and Pollution.* Report on progress in the application of the Mar del Plata Action Plan. Santiago, Chile: ECLAC.

UN (1990a). *The Water Resources of Latin America and the Caribbean: Planning, Hazards and Pollution (1990).* Santiago, Chile: ECLAC.

UN (1990b). *World Urbanization Prospect.* New York: UN.

UN (1991). Global consultation on safe water and sanitation for the 1990s. *Water Res. J.,* C/168, 1–7.

UN (1992). *Water Resources Database in the ESCWA region.* New York: UN.

UN (1993). *United Nations Statistical Yearbook 1992.* New York: UN.

UN (1995a). *World Population Prospects: The 1994 Revision.* New York: UN.

UN (1995b). *A Review on the Adverse Effects of Water Resources Development and Use on Groundwater and Aquifers.* New York: UN.

UN/DPCSD (1995a). *A Review of the Adverse Effects of Water Resources Development and Use on Groundwater and Aquifers.* New York: UN.

UN/DPCSD (1995b). *Implications of Agenda 21 for Integrated Water Management in the ESCWA Region.* Amman: UN (DPCSD).

UN/WMO/SEI (1997). *Comprehensive Assessment of the Freshwater Resources of the World.* Report prepared for the 5th Session of the UN Commission on Sustainable Development. Stockholm: UN/World Meteorological Organization/Stockholm Environment Institute.

UNDP (1997). *Aridity Zones and Dryland Populations: An Assessment of Population Levels in the World's Drylands.* New York: UNDP Office to Combat Desertification and Drought.

UNDP (1998). *Human Development Report 1998.* New York: Oxford University Press.

UNEP (1988a). *Problems of Desertification,* No. 4. Moscow:UNEP. (in Russian)

UNEP (1988b). *Problems of Desertification,* No. 5. Moscow:UNEP. (in Russian)

UNEP (1988c). *Problems of Desertification,* No. 6. Moscow: UNEP (in Russian)

UNEP (1994). *Global Freshwater Assessment: Past and Ongoing Freshwater Assessment Activities – Latin American and Brazilian Perspectives.* Geneva: UN Environment Programme, Regional Office for Latin America.

UNEP (1997). *World Atlas of Desertification,* 2nd edn. London: UNEP/Edward Arnold

UNEP/WHO (1995). *GEMS/WATER Report on Past and Current Water Quality Issues: Status and Trends.* Geneva: WHO

UNESCO (1970). *Climatic Atlas of Europe.* Budapest: UNESCO.

UNESCO (1993). *Discharges of Selected Rivers of the World,* vol. 2, Part 2, *Monthly and Annual Discharges Recorded at Various Selected Stations: 20-year Catalogue (1965–1984).* St Petersburg: Hydrometeoizdat.

UNESCO (1995a). *Groundwater Protection in the Arab Region.* Paris: UNESCO.

UNESCO (1995b). *Discharge of Selected Rivers of Africa.* Paris: UNESCO.

UNESCO (1996a). *Global River Discharge Database (RivDIS v1.0),* III: Europe. Paris: UNESCO/IHP.

UNESCO (1996b). *Global River Discharge Database (RivDIS v1.0),* vol. II, Asia. Paris: UNESCO.

UNESCO (1996). *Global River Discharge Database,* vol. 5, South America. Paris: UNESCO.

Urazov Ye. I. (ed.) (1991). *Turkey: New Tendencies of Economical Development in the 1980s.* Moscow: Nauka Publ. (in Russian)

Urbizondo B. I., Rose W. R. and Restrepo P. (1991). *Potential Climate Change Impacts in the Middle Zambezi River Basin.* Report prepared for US Environment Protection Agency, Boulder, CO: University of Colorado.

US Geological Survey (1986). *National Water Summary 1985: Hydrologic Events and Surface-Water Resources.* US Geological Survey Water-Supply Paper No. 2300. Washington, DC: US Government Printing Office.

US Geological Survey (1990). *National Water Summary 1987: Hydrologic Events and Water Supply and Use*. US Geological Survey Water-Supply Paper No. 2350, Washington, DC: US Government Printing Office.

US Geological Survey (1991). *National Water Summary 1991: Hydrologic Events and Issues*. US Geological Survey Water-Supply Paper No. 2375.

US Water Resources Council (1968) *The Nation's Water Resources: Summary Report*. US Government Printing Office.

US Water Resources Council (1978) *The Nation's Water Resources: 1975–2000*, vol. 1, Summary. Washington, DC: US Government Printing Office.

US Water Resources Council (1980) *The Second National Assessment*. US Government Printing Office.

USAID (2000). *FEWS Current Vulnerability Assessment*. Washington, DC: US Government Printing Office.

USEPA (1984) *Potential Climatic Impacts of Increasing Atmospheric CO_2 with Emphasis on Water Availability and Hydrology in the United States*. Report prepared for EPA by Goddard Institute for Space Studies. US Environmental Protection Agency. Washington, DC. Available online at http://www.fews.org/va/vanome.html

Usmanov U. U. and Ignatikov I. A. (1990). State and perspectives of using groundwater of Afghanistan. In: *Republic of Afghanistan: Experience and Tendency of Development* (ed. M. A. Babakhodjayev), pp. 74–103. Tashkent: FAN Publ. (in Russian)

Vaithiyanathan P., Ramanathan A. and Subramanian V. (1992). Sediment transport in the Cauvery river basin. *J. Hydrol*, 139, 197–210.

Van Alphen I. G. (1979). A note on the irrigation of red soils in Upper Tana Catchment, Kenya. *Annual Report. Int. Inst. Land Reclam. and Improv.* (1979), 39–42.

Van der Leeden F. (1975). *Water Resources in the World: Selected Statistics*. New York: Water Information Center.

Van der Leeden F., Troise F. L. and Todd, D. K. (eds.) (1990). *The Water Encyclopedia*, 2nd edn. Leipzig: Lewis Publ.

Van Kooten G. C., Kulshreshtha S. N. and Holmes C. (1987). The economics of interbasin water transfers: Selected methodological issues. In: *Impacts and Research Needs for Canada* (eds. W. Nicholaichuk and F. Quinn), *Proc. Symp. on Interbasin Transfer of Water* (Saskatoon, 9–10 November 1987).

Vanchura J. (1979). Water use and protection in CSSR. *CMEA Inf. Bull. on Water Management*, 1, 31–33. (in Russian)

Vasilevsky E. K. (1994). Evolution of structure of the American economy. In: *The USA: Economy, Politics, Ideology* (ed. V. S. Mikheev), pp. 8–9, 17–42. (in Russian)

Vasilyev Yu. S. and Pietro G. A. (1989). Use of the Danube river and Gabchikovo–Nagymaros water power development complex. *Hydrotech. Construct*, 3, 33–34. (in Russian)

Vasquez A., Bianco I. and Sanchez L. (1989). The lower Orinoco floodplain system. In: *Proc. Conserv. et Dev.: Gestion Integrée Zones Humides, 5ème Cons., Int. Zones Humides (Paris, 19–23 September 1989)* pp. 61–62. Paris: UNESCO.

Valkar V. T. (1989). Water resources of India and their development. *Water Resour. Ser. (ESCAP)*, 64, 78–94.

Venema H. D., Schiller E. J., Adamowski K. and Thizy J. M. (1997). A water resources planning response to the climate change in the Senegal River Basin. *J. Envir. Mgmt*, 49(1), 125–155.

Vernadsky V. I. (1967). *The History of the Earth's Crust Minerals: Natural Water History*, Moscow: Academy of Sciences of the USSR Publ. (in Russian)

Vinikov K. Ya. and Lemeshko N. A. (1987). Soil moisture content and runoff for the USSR territory with global warming. *Meteorol. Hydrol*, 12, 96–103. (in Russian)

Vinikov K. Ya., Lemeshko N. A. and Speranskaya N. A. (1990). Soil water content and runoff of the extratropical part of the Northern Hemisphere with global warming. *Meteorol. Hydrol*, 3, 5–10. (in Russian)

Vinogradov A. P. (1959). *Chemical Evolution of the Earth*. Moscow: Academy of Sciences of the USSR Publ. (in Russian)

Vinogradov A. P. (1973). The cycle of matters. In: *Great Soviet Encyclopedia*, vol. 13, pp. 488–491. Moscow: Soviet Encyclopedia. (in Russian)

Vinogradov Yu. B. (1988). *Mathematical Modelling of Runoff Formation Processes*. Leningrad: Hydrometeoizdat. (in Russian)

Vinte J. (1979). 30th anniversary of CMEA and development of water management in HPR. *Bull. Inf. on Water Management*, 1, 3–6. (in Russian)

Vitukhina G. O and Onuchko V. G. (1989). *The Mali Republic*. Moscow: Nauka Publ. (in Russian)

Vodogretsky V. Ye. (1979). *The Influence of Agro-Forest Ameliorations on Annual Runoff*. Leningrad: Hydrometeoizdat. (in Russian)

Voltsun I. B., Sumarokova V. V. and Tsytsenko K. V. (1988a). On the change of structure of river runoff losses in the irrigated zone of the basins of Amu-Darya and Syr-Darya. *Wat. Resources*, 3, 117–123. (in Russian)

Voltsun I. B., Sumarokova V. V. and Tsytsenko K. V. (1988b). Water resources of the Aral Sea Basin: state and perspectives of use. In: *Water Resources and Water Budget* (eds. A. A. Sokolov and V. I. Babkin), *Proc. 5th All-Union Hydrological Congress*, vol. 2, pp. 197–204. Leningrad: Hydrometeoizdat. (in Russian)

Vol'skaya B. A. and Yagya V. S. (eds.) (1991). *Countries and Peoples of the East*, vol. 27, Africa: Geography, History, Culture, Economy. Moscow: Nauka Publ. (in Russian)

Vol'sky V. V. (ed.) (1990). *Latin America: Reference Book*. Moscow: Nauka Publ. (in Russian)

Voronchuk M. M. (1970). Mathematical analysis of running averaging methods and difference curves. *Trans. UKRNIGMI*, 88, 105–114. (in Russian)

Voropayev G. V. (1992). Is the problem of reconstruction of the Aral Sea acute today? *Wat. Resources*, 2, 5–11. (in Russian)

Voropayev G. V. and Gerasimov I. P. (1982). The problem of redistribution of water resources in the middle region: Forecasting of the change of natural conditions. *News Acad. Sci. USSR, Ser. Geogr.*, 6, 24–27. (in Russian)

Voropayev G. V. and Vendrov S. L. (eds.) (1979). *Water Reservoirs of the World*. Moscow: Nauka Publ. (in Russian)

Voropayev G. V., Blagovyerov B. G. and Ismaiylov G. Kh. (1989). *Economic and Geographic Aspects of the Formation of Territorial Units in Water Management of the Country*. Moscow: Nauka Publ. (in Russian)

Voropayev G. V., Ratkovich D. Ya., Budagovsky A. I. and Ivanova L. V. (1992). The problems of the Aral Sea Basin. *Wat. Resources*, 2, 5–49. (in Russian)

Vorosmarty C. J. (1989). Continental scale models of water balance and fluvial transport: An application to South America. *Global Biogeochem. Cycles*, 3, 241–265.

Vorosmarty C. J. and Moore B. (1992). Osziott parameteru vismerieg es folyami transzportmodeller a globalis eghajlatvaltozasok vizsgalatahos. *Vizugyi Kozlemenyek*, 3, 297–318.

Vuglinsky V. S. (1991). *Water Resources and Water Balance of Large Reservoirs of the USSR*. Leningrad: Gidrometeoizdat. (in Russian)

Waggoner P. E. (ed.) (1990). *Climate Change and US Water Resources*. New York: John Wiley & Sons.

Wallace J. (1996). Hydrological processes. *Inst. Hydrol. Ann. Rep.*, 1994–5, 25–34.

Warner R. F. (1995). Guest Editorial: Human impact on Australian rivers. *Austral. Geogr. Studies*, 33(1), 3–115.

Warren C. R. (1963). Further light on river floods reaching Lake Eyre. *Trans. Roy. Geogr. Soc. Austral*, S.A. Branch, 64.

Water International (1991). Special Topic issue on "International drinking water supply and sanitation decade". *Water International*, 16 (3), 215.

Water Resources Council (1968) *The Nation's Water Resources: The First National Assessment of the Water Resources Council*. Washington, DC: Government Printing Office.

Waylen P. R. and Caviedes C. N. (1990). Annual and seasonal fluctuations of precipitation and streamflow in the Aconcagua river basin, Chile. *J. Hydrol*, 120(1–4), 79–102.

Waylen P., Compagnucci R. H. and Caffera M. (in press). Inter-annual and inter-decadal variability in stream flow from the Argentine Andes. *Physical Geography*.

WHO/UNEP (1990) *Global Freshwater Quality: A First Assessment*. Geneva: WHO.

WHO (1991). *Report on Water Quality: Progress in the Implementation of the Mar del Plata Action Plan*. Geneva: WHO.

Wilson G. A. (1995). Wood chipping of indigenous forest on private lands in New Zealand 1969–1993. *Austral. Geogr. Studies*, 32(2), 256–273.

WMO (1986). *The Global Climate System: Autumn 1984 – Spring 1986*. Geneva: WMO.

WMO/UNESCO (1991). *Report on Water Resources Assessment: Progress in the Implementation of the Mar del Plata Plan and a Strategy for the 1990s*. Paris: UNESCO.

Wolman M. G. (ed.) (1990). *The Geology of North America*, vol. 0–1, *Surface Water Hydrology*. Baltimore, MD: Johns Hopkins University, Department of Geography and Environmental Engineering.

Woo Ming-Koo and Waylen P. R. (1987). Effects of flow diversion on the runoff regimes of Northern Ontario rivers. In: *Impacts and Research Needs for Canada* (eds. W. Nicholaichuk and F. Quinn), *Proc. Symp. on Interbasin Transfer of Water (Saskatoon, 9–10 November 1987)*, pp. 261–278.

World Bank (1986). *World Development Report 1986*. New York: Oxford University Press.

World Bank (1993). *World Development Report 1992: Development and Environment*. New York: Oxford University Press.

World Bank (1994). *Water Resources Management*. Washington, DC: World Bank.

World Bank (1995). *World Development Report 1994: Infrastructure for Development*. New York: Oxford University Press.

World Bank (1996). *World Bank Environmental Projects, July 1988 – June 1996*. Washington, DC: World Bank.

World Dams Today (1970). Tokyo: Japan Dams Association.

WRI (1990) *World Resources 1990–91: A Guide to the Global Environment – Toward Sustainable Development*. New York: Oxford University Press.

WRI (1992) *World Resources 1992–93: A Guide to the Global Environment – Toward Sustainable Development*. New York: Oxford University Press.

WRI (1996) *World Resources 1996–1997: A Guide to the Global Environment – Toward Sustainable Development*. New York: Oxford University Press.

WRI (2000), *World Resources 2000–2001: A Guide to the Global Environment – Toward Sustainable Development*. New York: Oxford University Press.

WRI/IIED (1986) *World Resources 1986: An Assessment of the Resource Base that Supports the Global Economy*. New York: Basic Books.

Wunderlich W. O. and Prins J. E. (eds.) (1987). *Water for the Future: Water Resources Development in Perspective, Proc. Int. Symp. on Water Resources Development in Perspective, Proc. ... the Future (Rome, 6–11 April 1987)*. Rotterdam, The Netherlands: A. A. Balkema.

Yakovlev A. G. (ed.) (1995). *China in the World and Regional Politics*. Moscow: Institute of Far East RAS Publ. (in Russian)

Yaskovyak A. (1979). Water management development in the PPR. CMEA *Inf. Bull. on Water Management*, 1, 23–26. (in Russian)

Yefimova N. A., Strokina L. A., Baikova I. M. and Malkova I. V. (1994), Variations in air temperature and cloudiness during 1967–1990 in the territory of the former USSR. *Meteorol. Hydrol.*, 6, 66–69. (in Russian)

Yefimova N. A., Strokina L. A., Baikova I. M. and Malkova I. V. (1996), Changes in major climate elements in the USSR territory for 1967–1990. *Meteorol. Hydrol.*, 4, 34–41. (in Russian)

Yermolina N. A. and Kalinin G. P. (1975). Water use and its influence on land water. In: *Global Water Exchange* (eds. G. P. Kalinin and R. K. Klige), pp. 24–40. Moscow: Nauka' Publ. (in Russian)

Yermolina N. A. and Klige R. K. (1979). The use of water resources of foreign Europe. *Water Resources*, 4, 162–175. (in Russian)

Yevjevitch V. (1977). *Fluctuations of Wet and Dry Years: An Analysis of Variance Spectrum*. Denver, CO: Colorado State University.

Yevseyeva L. S., Zhuk V. A. and Nguyen Lan Chau (1988). *Hydrological Regime of the Red River in the Territory of Viet Nam*. Moscow: VINITI Publ. (in Russian)

Yoshino F. (1999). Studies on the characteristics of variation and spatial correlation of the long-term annual runoff in the world rivers. *J. Japan Soc. Hydrol. Water Resources*, 12, 67–72.

Young J. R. and Bennett A. M. (1992). Groundwater management in Arkanas. In: *Changing Practices in Groundwater Management* (ed. J. J. DeVries), *The Pros and Cons of Regulation, Proc. 18th Biennial Conference on Groundwater (Sacramento, 16–17 September 1991)*, University of California Report 77, 119–123.

Yunusov G. R. (1974). Dynamics of the river runoff of the Aral Sea and Lake Balkhash in connection with the development of irrigation. *Trans. SHI*, 221, 128–159. (in Russian)

Yuri G. K. (1959). The initial atmospheres of planets. *Nature*, 4, 25–29. (in Russian)

Zair-Bek I. A. (1988). *Ecological-Geographical Problems of Using Inland Waters of Africa*. Leningrad: Hertsen State Pedagogic Inst. Publ. (in Russian)

Zaretskaya I. P. (1998). State-of-the-art and expected water availability and water use in the Danube Basin. In: *Water: A Looming Crisis?* (ed. H. Zebidi), *Proc. Int. Conf. on World Water Resources at the Beginning of the 21st Century (UNESCO, Paris, 3–6 June 1998)*. IHP-V Technical Documents in Hydrology, 18, 159–160.

Zaroubayev N. V. (1976). *The Composite Use and Protection of Water Resources*. Leningrad: Stroizdat' Publ. (in Russian)

Zenkov L. (1982). Urgent problems of water use in PRB. *Problems Geogr.* 3, 17–20. (in Russian)

Zevelev I.A. (1985). *Southeast Asia: Urbanization and Problems of Social Development*. Moscow: Nauka' Publ. (in Russian)

Zezhen Z. and Shangshi D. (1987). The development of irrigation in China. *Water International*, 12(2), 46–52.

Zhang J., Huang W. W., Liu S. M., G. and Wang J. H. (1992). Transport of particulate heavy metals towards the China Sea: a preliminary report. *Marine Chem.*, 40, 161–178.

Zhang Qishun and Zhang Xiao (1995). Water issues and sustainable social development in China. *Water Res. J*, C/187, 20–26.

Zonn I. S. and Nosenko P. P. (1981). Modern level and prospects for developing land reclamation in the countries of the world. *Hydraul. Land Reclam.*, 1, 82–86. (in Russian)

Zubenok L. I. (1970). A specified water balance scheme of the continents. *Trans. Main Geophys. Observatory*, 263, 79–82. (in Russian)